# 14th Latin American Workshop on Plasma Physics

# (LAWPP 2011)



**Journal of Physics: Conference Series Volume 370**

## Mar del Plata, Argentina
## 20-25 November 2011

**ISBN: 978-1-62276-804-2**
**ISSN: 1742-6588**

**Printed from e-media with permission by:**

Curran Associates, Inc.
57 Morehouse Lane
Red Hook, NY 12571

**Some format issues inherent in the e-media version may also appear in this print version.**



Printed by Curran Associates, Inc. (2013)

For permission requests, please contact the Institute of Physics
at the address below.

Institute of Physics
Dirac House, Temple Back
Bristol BS1 6BE UK

Phone: 44 1 17 929 7481
Fax: 44 1 17 920 0979

techtracking@iop.org

**Additional copies of this publication are available from:**

Curran Associates, Inc.
57 Morehouse Lane
Red Hook, NY 12571 USA
Phone: 845-758-0400
Fax: 845-758-2634
Email: curran@proceedings.com
Web: www.proceedings.com

# TABLE OF CONTENTS

# Preface

These proceedings present the written contributions from participants of the Latin American Workshop on Plasma Physics (LAWPP), which was held in Mar del Plata, Argentina, on 20–25 November 2011. This was the 14th session of the series of LAWPP biennial meetings, which started in 1982.

The five-day scientific program of LAWPP 2011 consisted of 32 talks and various poster sessions, with the participation of 135 researchers from Argentina, Brazil, Canada, Chile, Colombia, Mexico, Puerto Rico, USA, Venezuela, as well as others from Europe and Asia. In addition, a School on Plasma Physics and a Workshop on Industrial Applications of Plasma Technology (AITP) were organized together with the main meeting. The five-day School held in the week previous to the meeting was intended for young scientists starting their research in Plasma Physics. On the other hand, the objective of the AITP Workshop was to enhance regional academic and industrial cooperation in the field of plasma assisted surface technology.

Topics addressed at LAWPP 2011 included space plasmas, dusty plasmas, nuclear fusion, non-thermal plasmas, basic plasma processes, plasma simulation and industrial plasma applications. This variety of subjects is reflected in these proceedings, which the editors hope will result in enjoyable and fruitful reading for those interested in Plasma Physics.

It is a pleasure to thank the Institutions that sponsored the meeting, as well as all the participants and collaborators for making this meeting possible.

*The Editors*
**Luis Bilbao, Fernando Minotti and Hector Kelly**

**Participants of the 14th Latin American Workshop on Plasma Physics, 20–25 November 2011, Mar del Plata, Argentina**

Journal of Physics: Conference Series **370** (2012) 011001 doi:10.1088/1742-6596/370/1/011001

## International Scientific Committee

Carlos Alejaldre, Spain
María Virginia Alves, Brazil
Ibere Caldas, Brazil
Luis Felipe Delgado-Aparicio, Peru
Mayo Villagrán, Mexico
Kohnosuke Sato, Japan
Héctor Kelly, Argentina
Edberto Leal-Quirós, Puerto Rico
George Morales, USA
Julio Puerta, Venezuela
Leopoldo Soto, Chile
Michael Tendler, Sweden
Carlos Varandas, Portugal
Henry Riascos, Colombia
Ivan Vargas-Blanco, Costa Rica

## Local Organizing Committee

Luis Bilbao (Chairman)
Fernando Minotti (Vice-Chairman)
Luis Bernal, UNMDP
Alejandro Clausse, PLADEMA-CNEA
Graciela Gnavi, INFIP, CONICET-UBA
Fausto Gratton, INFIP, CONICET-UBA
Diana Grondona, INFIP, CONICET-UBA
Héctor Kelly, INFIP, CONICET-UBA
Adriana Márquez, INFIP, CONICET-UBA
María Milanese, UNCPBA
César Moreno, INFIP, CONICET-UBA

## Sponsors

Instituto de Física del Plasma (INFIP)
Consejo Nacional de Investigaciones Científicas y Técnicas (CONICET)
Comisión Nacional de Energía Atómica (CNEA)
Agencia Nacional de Promoción Científica y Tecnológica (ANPCyT)
Centro Latino-Americano de Física (CLAF)
Universidad Nacional de Mar del Plata (UNMP)
Universidad Nacional del Centro de la Provincia de Buenos Aires (UNICEN)
Academia Nacional de Ciencias de Buenos Aires (ANCBA)

Journal of Physics: Conference Series **370** (2012) 011001 doi:10.1088/1742-6596/370/1/011001

# LAWPP 2011

## Latin American Workshop on Plasma Physics

20/25 November 2011 - Mar del Plata - Argentina

## Proceedings

### 14th Latin-American Workshop on Plasma Physics

**20 - 25 November 2011, Mar del Plata, Argentina**

www.lawpp2011.org.ar

CONICET INFIP CONICET UBA CLAF CNEA UNICEN AGENCIA Universidad Nacional de Mar del Plata

# Peer review statement

All papers published in this volume of *Journal of Physics: Conference Series* have been peer reviewed through processes administered by the proceedings Editors. Reviews were conducted by expert referees to the professional and scientific standards expected of a proceedings journal published by IOP Publishing.

14th Latin American Workshop on Plasma Physics (LAWPP 2011) IOP Publishing
Journal of Physics: Conference Series **370** (2012) 012001 doi:10.1088/1742-6596/370/1/012001

# Transport barriers in plasmas

**I L Caldas[1], J D Szezech Jr[1], T Kroetz[1], F A Marcus[2], M Roberto[3], R L Viana[4] and S R Lopes[4]**

1 Institute of Physics, University of São Paulo, São Paulo, Brazil
2 PIIM, Université de Provence, F-13397 Marseille cedex 20, France
3 Aeronautic Institute of Technology, São José dos Campos, Brazil
4 Physics Department, Federal University of Paraná, Curitiba, Brazil

E-mail: ibere@if.usp.br

**Abstract.** We discuss the creation of transport barriers in magnetically confined plasmas with non monotonic equilibrium radial profiles. These barriers reduce the transport in the shearless region (i.e., where the twist condition does not hold). For the chaotic motion of particles in an equilibrium electric field with a nonmonotonic radial profile, perturbed by electrostatic waves, we show that a nontwist transport barrier can be created in the plasma by modifying the electric field radial profile. We also show non twist barriers in chaotic magnetic field line transport in the plasma near to the tokamak wall with resonant modes due to electric currents in external coils.

## 1. Introduction

Recently there is a growing interest in non monotonic systems [1, 2, 3], for which the twist condition is not fulfilled for all points in the domain of interest [4]. The non twist condition have been found in many dynamical systems of physical interest, like fluids and plasmas [5, 6, 7, 8, 9, 10].

Nontwist systems present internal transport barriers that play a major role in the study of diffusion in fluids and plasmas [11]. In particular, internal transport barriers are important to separate regions with qualitatively different dynamical behavior. Even after the internal barriers are broken the remaining dimerized islands created by nontwist systems may present high stickness configurations that decreases the transport in the nontwist region [1].

In this work, initially, we introduce the barriers in the standard nontwist map and then we analyze two kinds of problems in plasmas with non-monotonic field profiles: the first is the chaotic particle drift motion caused by electrostatic drift waves; the second problem is the chaotic magnetic field line transport in plasmas with external resonant perturbations.

## 2. Nontwist standard map

Initially, we present the nontwist transport barrier in the most used symplectic nontwist map, namely, the standard nontwist map (SNTM) introduced in Ref. [6]. The winding number of this map is not monotonic with a twistless surface surrounded by two chain of dimerized islands. This map is given by

$$x_{n+1} = x_n + a(1 - y_{n+1}^2), \qquad (1)$$

$$y_{n+1} = y_n + b\sin(2\pi x_n) \qquad (\mathrm{mod}\ 1), \qquad (2)$$

where $x \in [-1/2, +1/2)$, $y \in \mathbb{R}$, $a \in (0,1)$, and $b > 0$.

The function $a(1 - y_{n+1}^2)$ is the winding number of the unperturbed phase-space trajectories lying on nested tori, its derivative being the so-called shear function. If the shear function is monotonically increasing or decreasing for all values of interest, the shear does not change sign and the twist condition is satisfied. The lines where winding number changes sign define the shearless curves in phase space.

Recently there is a growing interest in nontwist maps, some of whose properties are quite different from those known for twist maps, such as the KAM and Aubry-Mather theory [2]. For example, certain periodic islands can interact without overlapping as the perturbation is increased, rather experiencing either a separatrix reconnection process or mutual annihilation after they collide [2,3]. One of the conspicuous features of nontwist maps is the robustness of shearless curves against increasing perturbations, which leads to transport barriers not present in twist maps [1]. In the latter there is a well-known transition from local (separatrix) to global chaos, whereas in nontwist maps separatrix reconnection may or may not lead to global transport depending on the map parameters [1]. The transport barriers of nontwist maps are examples of structures which affect in a global way the transport over a phase space region.

The mathematical properties of SNTM have been extensively investigated over the past two decades [1, 2, 3, 11]. In the following we outline some of these properties, referring to the literature for a more complete coverage of them. In the unperturbed case ($b = 0$) the twist condition is violated at the point $y = 0$, what defines a shearless curve. The quadratic term in (1) leads to two invariant curves with the same winding number at both sides of the shearless curve.

As the perturbation becomes nonzero ($b \neq 0$) two periodic island chains appear at the two invariant curve locations, and the former shearless curve becomes a shearless invariant tori separating these two island chains [1, 2]. There are also chaotic layers attached to both island chains, as expected from the presence of homoclinic crossings therein. These chaotic layers are not connected, though, as far as there are invariant curves near the shearless invariant tori acting as dikes, preventing global transport [1, 2, 11].

In figure 1, where a Poincaré section of the SNTM is depicted for $a = 0.52$ and $b = 0.32$, we observe two island chains with two islands each. The local maxima of the perturbed winding number profile define a shearless invariant curve, whose existence can be inferred between the two island chains. The island chains bordering the shearless invariant curve are transport barriers, since chaotic trajectories above and below do not mix at all [11].

If the parameters are further modified another noteworthy feature of nontwist maps can emerge, depending on the parameter space region. In one scenario (generic reconnection) the island chains with the same winding number approach each other and their unstable and stable invariant manifolds suffer reconnection. In the region between the chains, there appear new invariant tori called meandering curves (which are not KAM tori, though, since the latter must be a graph over $x$, while meanders are not). The periodic orbits remaining eventually coalesce and disappear, leaving only meanders and the shearless torus.

Further growth of the $a$-parameter causes the breakup of the transport barrier and the consequent mixing of the chaotic orbits formerly segregated on both sides of the shearless invariant torus.

## 3. Transport barrier at tokamak plasma edge

In tokamaks the plasma confinement can be improved when the radial electric field profile is modified by a bias electrode [12, 13].

We use a Hamiltonian model to show that the reported transport reduction is caused by displacement of a transport barrier from the scrape-off layer to the plasma edge. For that we consider a large aspect ratio tokamak with $\mathbf{B} = B_0\mathbf{e_z}$ , and the particle guiding center drift in

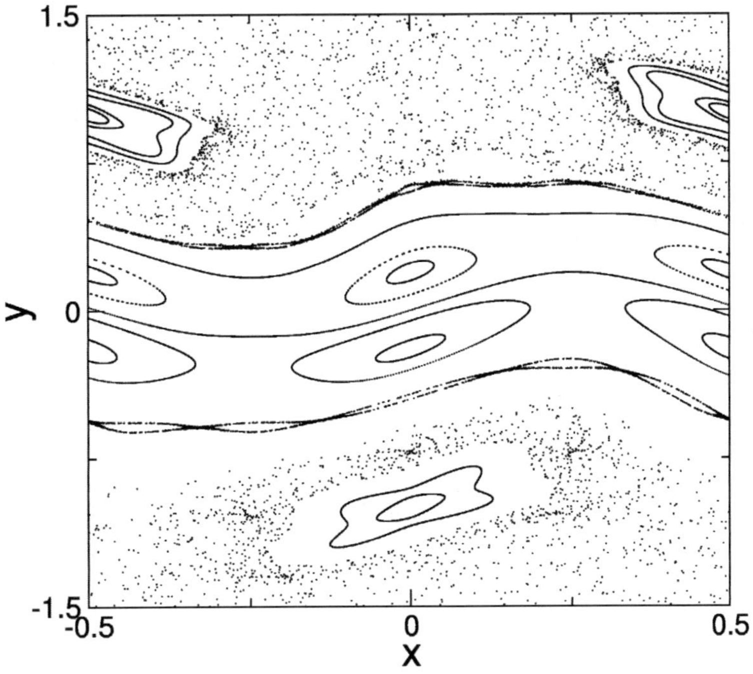


**Figure 1.** Nontwist map. Phase space with transport barrier for $a = 0.52$ and $b = 0.32$.

poloidal direction given by

$$\mathbf{v}_E = \frac{\mathbf{E}\times\mathbf{B}}{B^2} = \frac{E}{B}\mathbf{e_z}, \tag{3}$$

where $x$ and $y$ stand for the radial and poloidal coordinates respectively. We consider that the observed particle transport at the plasma edge is manly caused by $\mathbf{E}\times\mathbf{B}$ drift [14, 15].

Due to the particle density gradient in the radial direction, a perturbed electric field appears changing the electric field configuration to $\mathbf{E} = E(x)\mathbf{e_x} + \tilde{E}\mathbf{e_y}$ , and consequently giving rise to the electrostatic waves, which are responsible to the radial drifts of the particles. We assume an electrostatic potential of the form [16]:

$$\phi(x, y, t) = \phi_0(x) + \sum_i A_i \sin(k_{x_i} x)\cos(k_{y_i} y - \omega t), \tag{4}$$

which is composed of a background equilibrium electrostatic potential, $\phi_0(x)$, with the superposition of a collection of waves propagating in the poloidal direction. By the (3) with $\mathbf{E} = -\nabla\phi(x, y, t)$, we describe the system by the Hamiltonian formalism defining the Hamiltonian by $H(x, y, t) = \frac{\phi}{B_0}$ . In this model we use only two waves since this is the minimum to ensure the chaos in the phase space,

$$H(x, y, t) = \phi_0(x) - u_1 x + A_1 \sin(k_{x_1} x)\cos(k_{y_1} y) + A_2 \sin(k_{x_2} x)\cos(k_{y_2}(y - ut)), \tag{5}$$

with $u_1 = \frac{\omega_1}{k_{y_1}}$ being the phase velocity of the first wave and $u$ is the phase velocity difference between the two waves. For equilibrium electric field profiles that are not monotonic, the Hamiltonian system violates the twist condition and it possesses robust invariant tori or barriers that inhibit transport [1]. From the Hamiltonian formalism, we introduce in (6) the trapping profile U(x) [17]. This function is used to describe the regions with high chaos activity in the phase space where we associate it with high radial particle transport

$$U(x) = \frac{1}{A_1 k_{x1}} \left[ \frac{d\phi_0}{dx} - u_1 B_0 \right]. \tag{6}$$

The resonant condition occurs at $U(x) = 0$. Around this value the particle transport is effective, which implies from (6) that the wave phase velocity matches the background drift velocity, $v_E \approx u_1$.

In figure 2 we depict the phase space for a bias discharge. The $U(x) \approx 0$ regions are marked with horizontal sequences of dots, and following those dots to the phase space, we see the chain islands near to the limiter ($x = 1$). Another important issue to notice is the trapping profile maximum point that states the nontwist barrier and it is located inside the plasma ($x = 1$). The dominant wave parameters are the following: $A_1 = 0.084$, $\omega_1 = 7.5$, $k_{y1} = 5.4$, $k_{x1} = 28.3$. The $U(x) = 0$ regions are marked with horizontal sequences of dots, and following those dots to the phase space, we see the chain islands near to the limiter ($x = 1$). Another important issue to notice is the trapping profile maximum point that states the nontwist barrier and it is located inside the plasma ($x = 1$).

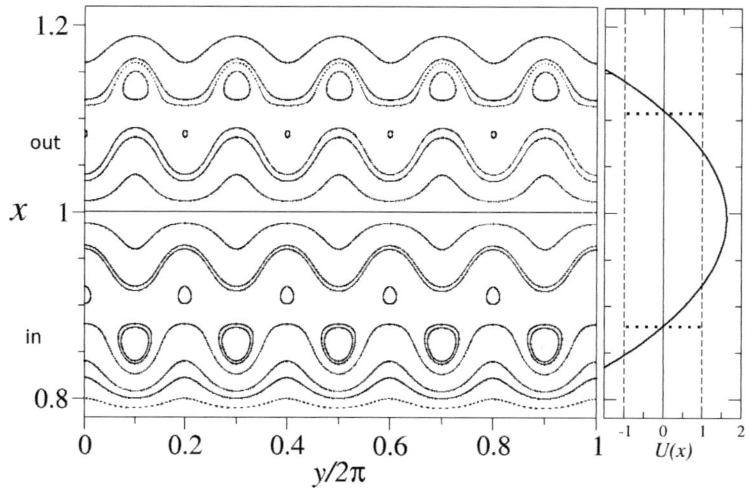


**Figure 2.** Phase spaces with the trapping profile $U(x)$ aside. The nontwist barrier can be seen where the trapping profile is maximum.

## 4. Magnetic field line transport

Chaotic field lines at the plasma edge play a key role on plasma-wall interaction in tokamaks [18, 19]. Since charged plasma particles follow magnetic field lines to leading order, one of the undesirable effects of chaotic field lines is the concentration of heat and particle loadings on the tokamak wall that deteriorates the overall plasma confinement quality [20]. Here, we show how the magnetic field line escape to the tokamak wall is affected by the action of a resonant helical perturbation on a tokamak plasma in MHD equilibrium.

In MHD equilibrium the field lines lie on magnetic surfaces, with topology of nested tori. The magnetic surfaces are characterized by a surface quantity which is an approximated analytical solution of the Grad-Shafranov equation in terms of the toroidal non-orthogonal coordinates $r_t$ and $\theta_t$, such that $\Psi \approx \Psi(r_t)$ [21]. The intersections of the flux surfaces $\Psi = constant$ with a toroidal plane are not concentric circles but rather present a Shafranov shift toward the exterior equatorial region [21]. For the considered coordinate system, $\Psi(r_t)$ is obtained for a monotonical toroidal current density profile, commonly observed in tokamaks discharges, given by $J_\phi(r_t) = \frac{I_p R_0}{\pi a^2} \frac{(\gamma+2)(\gamma+1)}{\beta+\gamma+2} \left(1 + \beta \frac{r_t^2}{a^2}\right) \left(1 - \frac{r_t^2}{a^2}\right)^\gamma$, where $I_p$ and $a$ are the total plasma current and plasma radius, respectively, $R_0$ is the major tokamak radius, and $\gamma$ is a positive constant.

We introduce a Hamiltonian formalism, with action and angle variables, to describe the considered integrable equilibrium. However, for a small helical magnetic perturbation the Hamiltonian system becomes almost-integrable [22]. In this work we use an ergodic magnetic limiter (EML) to generate a small localized perturbation, where the explicit dependence on $\phi$ reflects the break of symmetry [22, 23]. The EML creates resonances in the plasma and it can be designed to excite resonances closer or farther from the tokamak wall, depending on the mode numbers chosen for the limiter winding. To show the perturbed field lines distribution in space, we present a Poincaré map that represents the field line intersections with the toroidal plane $\phi = 0$.

For the present numerical simulations we choose an equilibrium characterized by the following parameters: $a/R_0 = 0.26$ and $b/R_0 = 0.35$, where $b$ is the minor tokamak radius. We also choose $q(a) = 5$ and $q(0) = 1$, corresponding to the safety factors at the plasma edge and magnetic axis, respectively, for which $\gamma = 0.78$ and $\beta = 3.0$. Figure 3 shows the safety factor profile for the considered MHD equilibrium.

Next, we present an example of a transport barrier created by the resonant perturbation considered in this work. We consider that the poloidal and toroidal mode numbers are, respectively, $m_0 = 4$ and $n_0 = 1$. We also normalize lengths to the minor radius ($b = 1$). The results are presented in terms of $r$ and $\theta$ variables, more adequate to the comparisons with experimental results. Figure 4 shows a Poincaré map for the limiter currents $I_h/I_p = 2.73$. The island chains due to these resonances although partially destroyed, can be recognized. In the external part of the plasma there are mixed regions with islands and chaotic lines, while more inside the plasma a transport barrier (in red in figure 4) can be identified.

Recent experiments have shown that the radial structure of the electron temperature and density at different times of the discharge reveals a correlation between the field line connection lengths (the number of toroidal turns it takes for a field line to reach the tokamak wall) and the heat flux [20]. There has been observed that most of the heat content is brought from the plasma core wall by the field lines with relatively large connection lengths. The regions with short and long connection lengths are also called laminar and ergodic, respectively [20].

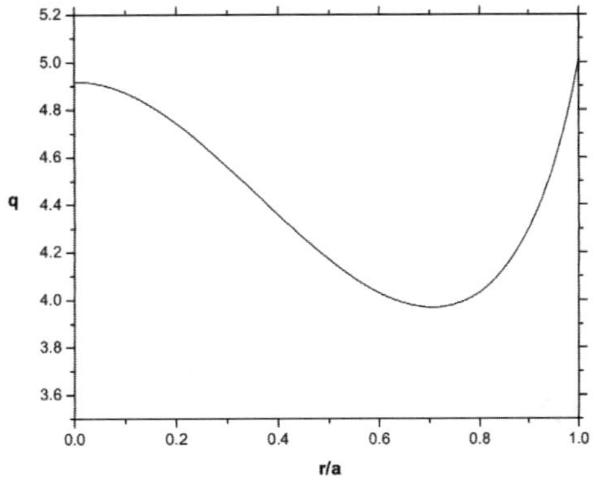


**Figure 3.** Safety factor profile for the equilibrium considered in this work.

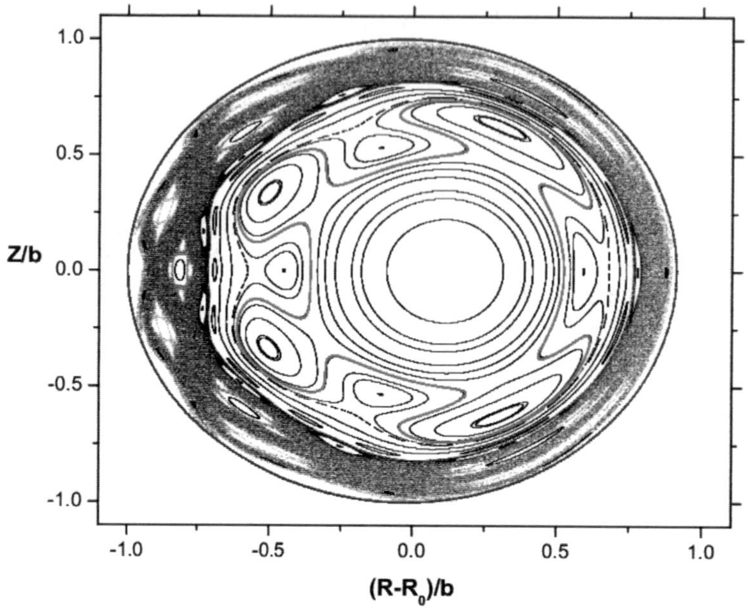


**Figure 4.** Poincaré section of magnetic field lines inside a tokamak plasma in MHD equilibrium perturbed by an EML.

## 5. Conclusions

Several applications of the theory of chaos have been reported in the last years to interpret different phenomena observed in tokamak plasmas [24]. Here, we apply the Hamiltonian theory of nontwist dynamical systems to predict the existence of nontwist transport barriers in plasma physics. In [11] we presented examples of such barriers described by plasma models described by symplectic maps. In this article, initially, we give an example of such barriers present in the non Twist Standard Map, a well known paradigm that can be used to investigate qualitatively the onset of transport barriers and their break-up in several systems. Next, we introduce two dynamical systems describing transport in tokamak plasmas with non-monotonic profiles [25]. One system describes the chaotic particle drift motion caused by electrostatic waves; the other system describes the chaotic magnetic field line distribution in plasmas with external resonant perturbations. For these two systems we show how non-monotonic field profiles give rise to transport barriers.

## References

[1] Szezech Jr J D, Caldas I L, Lopes S R, Viana R L, and Morrison P J 2009 *Chaos: An Interdisciplinary Journal of Nonlinear Science* **19** 043108
[2] Wurm A, Apte A and Morrison PJ 2004 *Braz. J. Phys.* **34** 1700
[3] Wurm A, Apte A, Fuchss K and Morrison PJ 2005 *Chaos: An Interdisciplinary Journal of Nonlinear Science* **15** 023108
[4] Morrison P J 2000 *Phys. Plasmas* **7** 2279
[5] Kerner W and Tasso H 1982 *Phys. Rev. Lett.* **49** 654
[6] del-Castillo-Negrete D and Morrison P J 1993 *Phys. Fluids* A **5** 948
[7] Horton W, Park H B, Kwon J M, Strozzi D, Morrison P J and Choi D I 1998 *Phys. Plasmas* **5** 39103917
[8] Oda G A and Caldas I L 1995 *Chaos Soliton Fractals* **5** 15
[9] del-Castillo-Negrete D 2000 *Phys. Plasmas* **7** 17021711
[10] Kroetz T, Roberto M, da Silva E C, Caldas I L and Viana R L 2008 *Phys. Plasmas* **15** 092310
[11] Caldas I L, Viana R L, Szezech Jr J D, Portela J S E, Fonseca J, Roberto M, Martins C G L and da Silva E J 2012 *Comm. Non. Science Num. Sim.* **17** 2021

[12] Nascimento IC *et al* 2005 *Nuclear Fusion* **45** 796
[13] Van Oost G et al. 2003 *Plasma Physics Control. Fusion* **45** 621
[14] Ritz C P *et al* 1989 *Physical Review Letters* **62** 1844
[15] Ritz C P, Lin HT, Rhodes L, and Wootton AJ 1990 *Physical Review Letters*, **65** 2543
[16] Horton W 1985 *Plasma Physics and Controlled Fusion* **27** 937
[17] Marcus FA, Caldas I L, Guimaraes-Filho Z O, Morrison P J, Horton W, Kuznetsov Y K and Nascimento I C 2008 *Physics of Plasmas* **15** 112304
[18] Jakubowski M W *et al* 2006 *Phys. Rev. Lett.* **96**, 0350041
[19] Evans T E *et al* 2004 *Phys. Rev. Lett.* **92** 235003
[20] Jakubowski M W, Abdullaev S S, Finken K H, Lehnen M and TEXTOR Team 2005 *J. Nucl. Mat.* **176** 337
[21] Kucinski M Y and Caldas I L 1987 *Z. Naturforsch A: Phys. Sci* **42** 1124
[22] Caldas I L, Viana R L, Araujo M S T, Vannucci A, da Silva E C, Ullmann K and Heller M V A P 2002 *Braz. J. Phys.* **32** 980
[23] Ghendrih Ph *et al* 2002 *Nucl. Fusion* **42**, 1221
[24] Viana R L , da Silva E C, Kroetz T, Caldas I L, Roberto M and Sanjuan M A F *Phil. Trans. Royal Society* (to be published)
[25] Kroetz T, Marcus F A, Roberto M, Caldas I L and Viana R L 2009 *Computer Physics Communications* **180** 642

# Laboratory astrophysics experiments studying hydrodynamic and magnetically-driven plasma jets

**F Suzuki-Vidal[1], S V Lebedev[1], M Krishnan[2], M Bocchi[1], J Skidmore[1], G Swadling[1], A J Harvey-Thompson[1,5], G Burdiak[1], P de Grouchy[1], L Pickworth[1], L Suttle[1], S N Bland[1], J P Chittenden[1], G N Hall[1], E Khoory[1], K Wilson-Elliot[2], R E Madden[1], A Ciardi[3] and A Frank[4]**

[1] Department of Physics, Imperial College London, Prince Consort Road, London SW7 2BW, UK
[2] Alameda Applied Sciences Corporation, San Leandro, CA 94577, USA
[3] LERMA, Université Pierre et Marie Curie, Observatoire de Paris and École Normale Supérieure, UMR 8112 CNRS, France
[4] Department of Physics and Astronomy, University of Rochester, Rochester, NY 14627-0171, USA
[5]Present address: Sandia National Laboratories, Albuquerque, NM, USA

E-mail: `f.suzuki@imperial.ac.uk`

**Abstract.** Laboratory astrophysics is a novel approach to study different types of astrophysical phenomena by the means of carefully scaled laboratory experiments. Particularly, the formation of highly supersonic, radiatively cooled plasma jets for the study of protostellar jets is an active area of research at present. At Imperial College London, different experimental configurations allow producing plasma flows which are scalable to protostellar jets. The plasma is produced by introducing a ~1.4 MA, 250 ns current pulse from the MAGPIE generator into a load. By varying the geometry of the load it is possible to study different regions of interest in the jet. For instance, the effect of magnetic fields in the launching and collimation of the jet, and the propagation of the jet far away from the launching region as it interacts with the ambient medium. Two main experiments can address such regions of interest: radial wire arrays and radial foils. By using a radial wire array it is possible to produce a jet driven by a predominant toroidal magnetic field on the axis of a magnetic "bubble", which expands with velocities up to ~300 km/s. In a radial foil the wires are replaced by a continuous disk allowing to produce a hydrodynamic jet, i.e. a jet in which magnetic fields are not dynamically significant. With this particular configuration it is possible to introduce a neutral gas above the foil in order to study jet-ambient interactions. Experimental results from different diagnostics will be presented together with 3-D MHD simulations using the GORGON code.

## 1. Introduction

Laboratory plasma astrophysics deals with the experimental modeling of astrophysical processes, including shock compression of materials under ~Mbar pressures to study the equation of state of planetary interiors, laser-driven blast waves in atomic cluster media to study radiative cooling instabilities, and the formation/propagation of jets in regimes of interest to flows observed from newly forming stars. For a recent review of these and other applications see e.g. [1].

We are concerned with jets launched from young stars, known as Herbig-Haro objects, which play an essential role during star formation [2]. These are high Mach number flows whose dynamics are governed by magnetic fields and exhibit complex shock features from their interaction with the interstellar medium. In general these jets can be well approximated as ideal, magnetized, compressible plasmas, where viscosity, resistivity and thermal conduction are negligible (i.e. the dimensionless Reynolds, magnetic Reynolds, and Peclet numbers are $\gg$1 respectively). The plasma flow can thus be described by ideal magnetohydrodynamics (MHD), with the same equations describing both the astrophysical jet and its laboratory counterpart [3, 4].

The investigation of jet formation mechanisms in young stars requires dynamically significant magnetic fields to be included. For jet propagation studies and jet-ambient interaction the jet can be considered as a hydrodynamic flow, as it expected that far away from the launching source magnetic fields play a less important role in the dynamics of the jet.

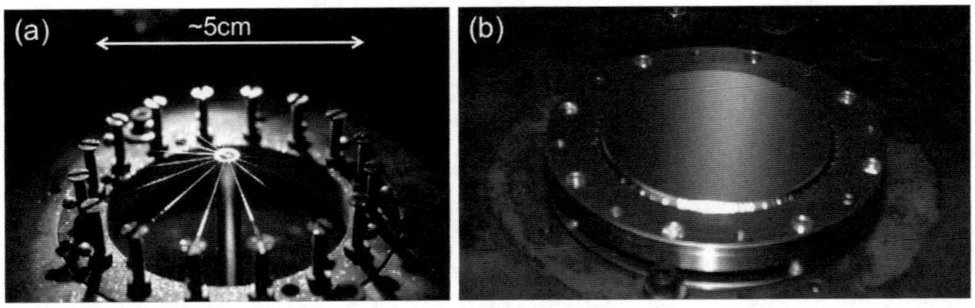


**Figure 1.** (a) A radial wire array, (b) a radial foil.

## 2. Experimental setup

For studying magnetically-driven or hydrodynamic jets, the experimental arrangements are very similar, and are presented in figure 1 respectively. Their dynamics are presented schematically in figure 2. Two concentric electrodes drive a 1.4 MA, 250 ns current pulse from the MAGPIE generator [5] into a load, which depending on the experiment, will be made of discrete metallic wires (radial wire array) or a continuous metallic disk (radial foil). The current flows along the central electrode (with a typical diameter of 3–6 mm) and along the wires/foil, producing a toroidal magnetic field with a magnitude decreasing inversely proportional to the radius from the axis. As the current increases, the wires/foil are heated and converted into plasma, with the Lorentz force ablating plasma in a direction normal to their surface producing a background, low-density plasma. The ablation rate is maximum at the edge of the central electrode, where the magnetic field is the strongest. The absence of ablation above this electrode leads to ablated plasma from larger radii to converge on the axis due to radial pressure gradients, forming a plasma jet on the axis. The formation of this hydrodynamic jet is independent of the load, and continues until the end of the current pulse in radial foils (see figures 2(a)-(b)) [6, 7].

After the hydrodynamic jet is formed, radial wire arrays behave differently to radial foils. Radial wire arrays have less available mass for ablation in comparison to radial foils, and thus wires can reach full ablation near the central electrode, where the Lorentz force is the strongest. This leads to the formation of a radial gap, driving plasma above the initial position of the wires and forming a "magnetic cavity" (see figures 2(c)-(d)). Current can now flow along the plasma on the axis of the cavity and return along its walls, producing a toroidal magnetic field inside the cavity that confines and collimates a plasma jet. This marks the formation of a magnetically-driven jet [8, 9, 10].

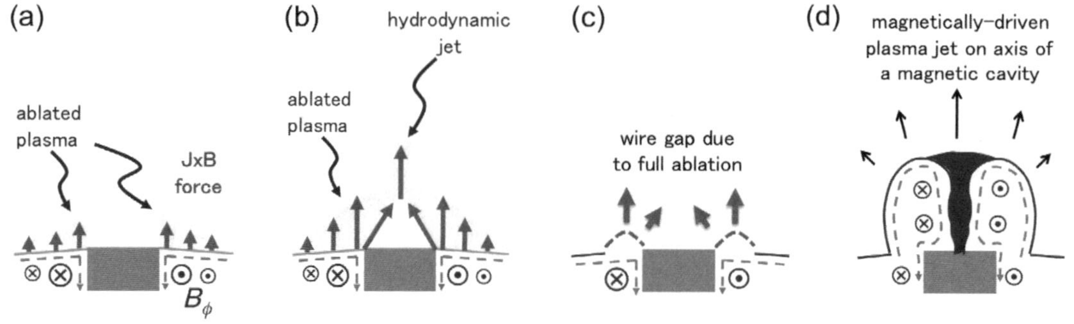


**Figure 2.** Schematic evolution for radial wire arrays (a)−(d), radial foils (a)−(b).

## 3. Magnetically-driven jets from radial wire arrays

The dynamical evolution of a radial wire array consisting of 16 tungsten (13 $\mu$m diameter each) is presented from time-resolved (3 ns gate) XUV self-emission images of the plasma, presented in figure 3. The results were obtained from 2 nominally identical experiments. The early-time images show the formation of the magnetic cavity, which expands axially and radially. Typical expansion velocities are ∼100's km/s, with the maximum axial velocity reaching ∼300 km/s . The formation of a jet on the axis of the cavity is observed at ∼200 ns. As the jet is compressed by the toroidal magnetic field, is affected by current-driven instabilities. These instabilities lead the jet to detach from the base of the cavity at 240 ns, though it continues to propagate axially. The series of images from 358−386 ns used a larger field of view of the XUV cameras, allowing to follow the late-time evolution of the jet after detaching. It is seen that the jet becomes a series of highly-emitting clumps, which remain collimated as they propagate, indicating the possibility of trapped magnetic field in these plasma clumps. The presence of magnetic field in these clumps could also be an indication of high magnetic Reynolds numbers present in the plasma.

## 4. Hydrodynamic jet-ambient interaction in radial foils

The interaction of the jet with an ambient medium was achieved by injecting cold, neutral argon gas above the initial position of an aluminium foil (15 $\mu$m thick) using a solenoid-valve with a supersonic gas nozzle [11]. The tip of the nozzle was positioned 55 mm above the foil and 15 mm from the axis in order to prevent direct impact of the jet onto the nozzle. The valve was opened typically ∼350 $\mu$s before the start of the current pulse, with a duration of ∼200 $\mu$s. The nozzle released argon with a Mach number of ∼9 and a measured gas density at the position of the foil of $\rho \sim 3-5\times10^{-3}$ kg/m$^3$. Numerical simulations of the experiments were performed using the GORGON code [9], an explicit, parallel code designed to solve the resistive MHD equations on a three-dimensional Cartesian grid. The resolution of the computational grid was 200 $\mu$m.

The main features of a jet from a radial foil in *vacuum* and with the addition of argon are shown in figures 4(b)-(c) respectively. Both images were obtained at ∼430 ns using optical laser interferometry ($\lambda$=532 nm, ∼0.3 ns pulse duration) and represent 2-D maps of line-electron density, i.e. electron density integrated along the length of the plasma region. Without the presence of argon (figure 4(b)), the jet is a well defined, highly-collimated column with a typical diameter of ∼4 mm, and an aspect ratio (jet length to radius) >20. Abel inversion of these data results in typical electron densities in the jet of $\sim10^{19}$ cm$^{-3}$, surrounded by lower density plasma with a typical electron density of $<10^{18}$ cm$^{-3}$. Results from optical Thomson scattering show the jet is supersonic (Mach∼3) and radiatively cooled, i.e. its cooling parameter <1.

The main features of the jet interacting with an argon ambient are presented in figure 4(c). A

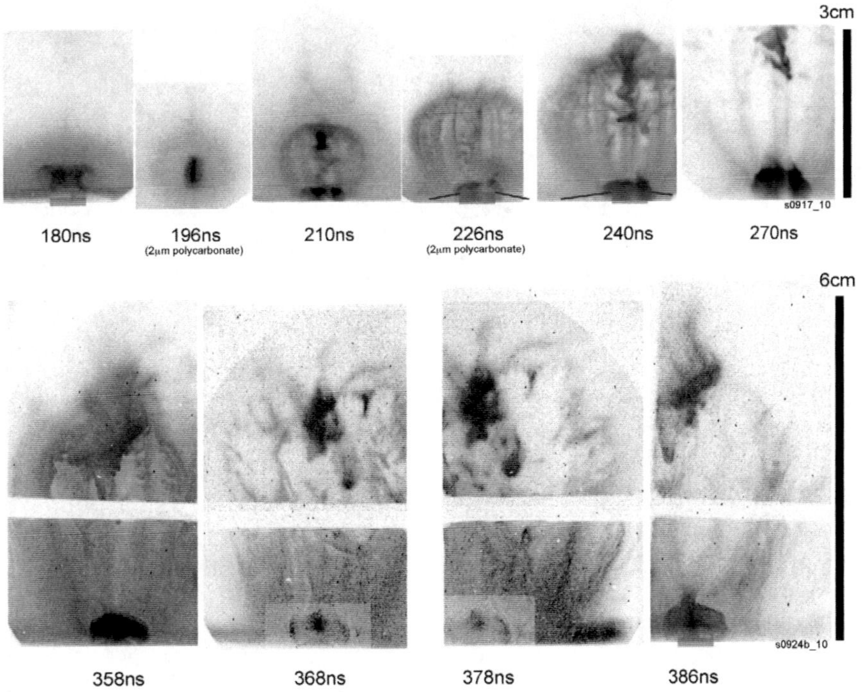


**Figure 3.** XUV self-emission images showing the dynamics of the formation of a magnetically-driven jet from a radial wire array.

jet with similar characteristics to the case in vacuum is observed, however the presence of argon leads to the formation of two prominent shock features. The first shock feature extends from the tip of the jet towards the edges of the foil in an "inverted cone" shape. The second shock feature is observed at this time ahead of the tip of the jet, forming a quasi-hemispherical shape and more resemblant to a bow-shock. The dynamics of the formation of these two shock features can be observed in detail in figure 5(a) from time-resolved images recording the self-emission of the plasma in the XUV (h$\nu$>30 eV). These results allow measuring the axial tip velocities of both shock features, resulting in approximately constant axial velocities of $V_{jet}\sim$70 km/s and $V_{bow}\sim$90 km/s, i.e. the bow-shock moves faster by $\sim$30%.

Numerical simulations of the XUV emission are presented in figure 5(b), and show a very good agreement with the experimental data. The simulations indicate that the plasma flow from the jet and from the foil converge at the tip of the jet due to the presence of a radial component of the J$\times$B force, i.e. some fraction of the current flows along the jet and along the inverted-conical shock. The convergence of current at the tip of the jet produces a compressing effect, which increases the density, temperature and pressure in this region, and explains the strong emission observed at the tip of the jet on the XUV images. The formation of the bow-shock ahead of the tip of the jet is due to the localized increase of pressure in this region, which creates strong axial pressure gradients, causing an acceleration of the plasma above it and producing a "nozzle-like" configuration. This allows plasma to propagate ahead of the tip of the jet at a higher velocity and interact with undisturbed argon.

Journal of Physics: Conference Series **370** (2012) 012002 doi:10.1088/1742-6596/370/1/012002

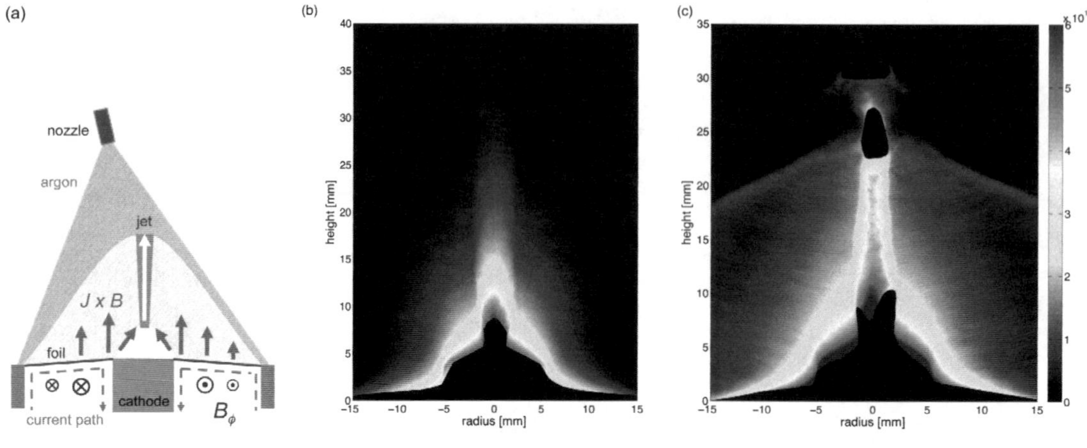


**Figure 4.** (a) Schematic setup for jet-ambient interaction using a radial foil. (b)−(c) Line-electron density results (in units of $cm^{-2}$, at 430 ns) showing the jet in vacuum and in argon respectively. The height is measured from the initial position of the cathode.

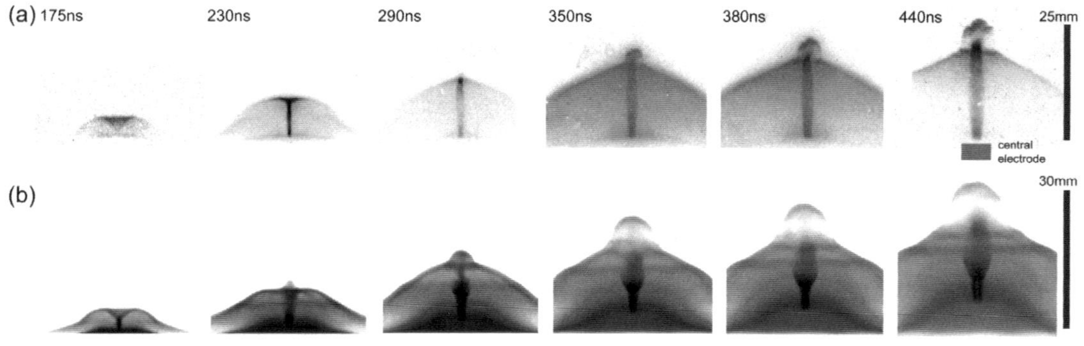


**Figure 5.** Time-resolved, XUV self-emission images of a jet from a radial foil interacting with argon from (a) experiments, (b) numerical simulations.

## 5. Future work

The experiments presented here open new possibilities for the study of supersonic, radiatively-cooled plasma jets relevant to the physics of jets from young stars. Experiments using radial wire arrays allow producing a jet on the axis of a magnetic cavity, where a toroidal magnetic field is responsible for confining and driving the jet. This mechanism is similar to astrophysical scenarios where magnetic fields are responsible for the collimation of jets in young stars. The use of radial foils allows investigating the physics of the interaction of hydrodynamic jets, i.e. jets without a dynamically significant magnetic field, with the interstellar medium. The formation and interaction of the bow-shock ahead of the jet with the undisturbed argon ambient is very promising for the investigation of the effects of radiative cooling on the properties of forming working surfaces, which is highly relevant to astrophysical jets.

## Acknowledgments
This work was supported by the EPSRC Grant No. EP/G001324/1, by the NNSA under DOE Cooperative Agreements No. DE-F03-02NA00057 and No. DE-SC-0001063, by DOE SBIR Grant DE-FG02-08ER85030, and by a Marie Curie European Reintegration grant.

## References
[1] Remington B A, Drake R P and Ryutov D D 2006 *Rev. Mod. Phys.* **78** 755-807
[2] Reipurth B and Bally J 2001 *Ann. Rev. Astron. Astrophys.* **39** 403-55
[3] Ryutov D, Drake R P, Kane J, Liang E, Remington B A and Wood-Vasey W M 1999 *ApJ* **518** 821-2
[4] Ryutov D D, Drake R P and Remington B A 2000 *ApJS* **127** 465-8
[5] Mitchell I H, Bayley J M, Chittenden J P, Worley J F, Dangor A E, Haines M G and Choi P 1996 *Rev. Scient. Inst.* **67** 1533-41
[6] Suzuki-Vidal F, Lebedev S V, Bland S N, Hall G N, Swadling G, Harvey-Thompson A J, Chittenden J P, Marocchino A, Ciardi A, Frank A, Blackman E G and Bott S C 2010 *Phys. Plasmas* **17** 112708
[7] Gourdain P, Blesener I C, Greenly J B, Hammer D A, Knapp P F, Kusse B R and Schrafel P C 2010 *Phys. Plasmas* **17** 012706
[8] Lebedev S V, Ciardi A, Ampleford D J, Bland S N, Bott S C, Chittenden J P, Hall G N, Rapley J, Jennings C A, Frank A, Blackman E G and Lery T 2005 *Month. Not. R. Astron. Soc.* **361** 97–108 (*Preprint* arXiv:astro-ph/0505027)
[9] Ciardi A, Lebedev S V, Frank A, Blackman E G, Chittenden J P, Jennings C J, Ampleford D J, Bland S N, Bott S C, Rapley J, Hall G N, Suzuki-Vidal F A, Marocchino A, Lery T and Stehle C 2007 *Phys. Plasmas* **14** 056501
[10] Suzuki-Vidal F, Lebedev S V, Bland S N, Hall G N, Harvey-Thompson A J, Chittenden J P, Marocchino A, Bott S C, Palmer J B A and Ciardi A 2010 *IEEE Trans. Plasma Sci.* **38** 581-8
[11] Krishnan M, Wright J and Ma T 2009 *AIP Conference Proceedings* **1086** 264–269 URL http://link.aip.org/link/?APC/1086/264/1

Journal of Physics: Conference Series **370** (2012) 012003 doi:10.1088/1742-6596/370/1/012003

# Velocity shear instability and plasma billows at the Earth's magnetic boundary

**F T Gratton[1,2], G Gnavi[1-3], C J Farrugia[4], L Bilbao[1-3] and R Torbert[4]**

[1]Instituto de Física del Plasma, Consejo Nacional de Investigaciones Científicas y Técnicas y Universidad de Buenos Aires, Buenos Aires, Argentina
[2]Departamento de Física, Facultad de Ciencias Fisicomatemáticas e Ingeniería, Pontificia Universidad Católica Argentina, Buenos Aires, Argentina
[3]Departamento de Física, Facultad de Ciencias Exactas y Naturales, Universidad de Buenos Aires, Buenos Aires, Argentina
[4]Space Science Center, University of New Hampshire, Durham, NH, 03824, USA

E-mail: fgratton@arnet.com.ar

**Abstract.** The Kelvin-Helmoltz instability (KH) with formation of vortices appears in a wide variety of terrestrial, interplanetary, and astrophysical contexts. We study a series of iterated rolled-up coherent plasma structures (15) that flow in the equatorial Earth's boundary layer (BL), observed on October 24, 2001. The data were recorded during a 1.5 hour-long Wind crossing of the BL at the dawn magnetospheric flank, tailward of the terminator ($X \approx -13\ R_E$). The interplanetary magnetic field (IMF) was radially directed, almost antiparallel to the magnetosheath (MS) flow. This configuration is expected to be adverse to the KH instability because of the collinearity of field and flow, and the high compressibility of the MS. We analyze the BL stability with compressible MHD theory using continuous profiles for the physical quantities. Upstream, at near Earth sites, we input parameters derived from an exact MHD solution for collinear flows. Further downtail at Wind position we input measured parameters. The BL is found KH unstable in spite of unfavorable features of the external flow. On the experimental side, the passage of vortices is inferred from the presence of low density - hot plasma being accelerated to speeds higher than that of the contiguous MS. It is further supported by the peculiar correlation of relative motions (in the bulk velocity frame): cold-dense plasma drifts sunward, while hot-tenuous plasma moves tailward. This event differs from many other studies that reported BL vortices under strongly northward IMF orientations. This is a case of KH vortices observed under an almost radial IMF, with implicit significance for the more common Parker's spiral fields, and the problem of plasma entry in the magnetosphere.

## 1. Introduction

### *1.1. Foreword*

The first part of the (plenary conference) paper is an introduction to the Kelvin-Helmholtz (KH) instability in terrestrial plasmas. The importance of KH phenomena relies on the enhancement of momentum, energy, and eventually the mass transport they cause. While the presence of this instability is pervasive in terrestrial, interplanetary, and cosmic scenarios, here we limit the study to the Earth´s magnetosphere. The physics that we learn via *in situ* observations can also provide lessons for remote scenarios of the universe that cannot be explored directly.

The "*the flow-driven*" instability and its effects occur throughout the solar system. We call to mind the presence of KH in our atmosphere, and at the bottom of the oceans. The phenomenon has been observed on planets and at the Sun itself. The first part describes several aspects of KH in the magnetosphere of Earth based on our past investigations which offer promising prospects for further studies. A summary review of magnetohydrodynamic (MHD) properties of the instability is included, to support the second part that contains a new research on a peculiar KH event, observed on October 24, 2001 by the spacecraft Wind.

The topics treated in the paper are the following. 1) Notes on manifestations of velocity gradient instabilities in the solar system. 2) Basic notions on the KH instability, and elements of a MHD treatment. 3) Aspects of Kelvin-Helmholtz research in the space environment of Earth with potential for deeper studies. 4) An unusual event of magnetopause instability with almost collinear magnetic field and flow. 5) The experimental evidence of KH vortices in the boundary layer during the Wind traversal of October 24, 2001.

### *1.2. The ubiquitous Kelvin-Helmholtz instability*

Vortex formation, breaking of sea waves, and billows in atmosphere clouds are part of our common experience. There are frequent observations of Kelvin-Helmholtz vortices in our atmosphere. Their importance for meteorology has been appreciated since the second part of the XIX century. The KH mechanism is present in the billowy wake of wind blowing over a mountain ridge (see photos in Tritton's textbook,' and also NASA [1]). Kelvin-Helmholtz clouds resemble the breaking of waves in the ocean. Evaporation and condensation patterns help the visualization of the capped tops and cloudless troughs of the billows. In the following we use the words vortex and billow as synonimous (To our knowledge, the first use of billow in a MP context was made in [67]).

The vortices generated by the KH effect appear in lee waves downstream of an island of the southern Pacific Ocean. Observed by Landsat (1999) [2] from above, the Alexander Selkirk island is a striking example. Rising steeply, the island's highest point is 1600 m above sea level. As wind-driven clouds encounter this obstacle, they flow around it to form large, spinning eddies. KH due to velocity gradients appears also inside hurricanes. Kelvin–Helmholtz billows were photographed (from an aircraft at an altitude of 4500 m) in the eye-wall of Hurricane Erin (at 18:15 UTC, 10 Sep 2001 [3]).

A classic example of a KH phenomenon is the formation, growth, and break of water waves in wind - sea interaction. Kelvin-Helmholtz vortices are also found deep under water, in oceanic floors. Non-linear waves breaking on the slope of submarine mountains of the Atlantic ocean, 1100 km south of the Azores were recently reported (2011 [4]). A line of temperature probes at a depth of 500 m recorded the passage of the structures, with wavelengths ≈ 75 m, and periods ≈ 50 s.

Proceeding to the planets, KH structures in the Saturn atmosphere were observed by the Cassini probe from 5.9 million km, on October 9, 2004 (NASA [5]). The KH vortices are due to the interaction of two belts of Saturn's atmosphere. This phenomenon is estimated to be frequent in the gaseous giant planets because of the alternated jets and the temperature differences of atmospheric strips and belts. Another case of an atmospheric vortex is Jupiter's giant red spot, which was observed to last for at least 300 years.

The Cassini satellite also observed a plasma vortex in Saturn's dayside magnetosphere [6]. The probe encountered the vortex on 13 December 2004 as it was travelling toward the planet. The spacecraft crossed the magnetopause three times, before being immersed in the low-latitude boundary layer (BL). During the transition between the boundary layer and the magnetosphere proper, the spacecraft observed a deflected boundary layer plasma, and a twisted magnetic field topology. The data are interpreted as an encounter with a vortex on the inner edge of the boundary layer, an interface susceptible to the growth of the KH instability due to its low magnetic shear. We return to this point in section 4.

In space plasmas magnetic effects add to the complexity, and the variety of KH phenomena. The Solar Dynamics Observatory (SDO), presently the most recent and advanced spacecraft devoted to solar studies, provides detailed views at high resolution of plasma filaments that rise from Sun's spots, of plasma arcades sustained by magnetic fields, and solar eruptions. On the subject of the Kelvin–Helmholtz instability at the Sun itself see Foullon et al., 2011 [7]. During the readjustments of flow and field of the high solar atmosphere, following a strong commotion produced by a coronal mass ejection (CME) KH phenomena were also recorded by SDO, 2011 [7].

The solar wind is a continuous expansion of the coronal plasma that fills interplanetary space. It reaches supersonic speeds already close to the Sun. The planets constitute occasional obstacles that generate local shock waves in the solar wind flow. Magnetized planets, like the Earth, have a bow shock, after which the plasma flow is subsonic for some stretch, and have also a magnetosphere. The magnetosphere is dominated by the magnetic field of the planet. A current sheet separates the shocked plasma in the magnetosheath from the magnetosphere. At the interface between the magnetosheath and the magnetosphere – the magnetopause – the KH instability may occur.

A general comprehensive study, theory and observations, of the physics of the Sun´s atmosphere and the interplanetary plasma is available in the monograph edited by Russel and Kivelson (1995) [8]. During the outward flow of solar wind there are several occasions where the KH instability may occur. For instance, recently the Messenger mission published magnetic records that are interpreted as the formation of KH vortices at Mercury [9].

Moreover, the solar wind has a bimodal structure. Fast solar wind streams, from 700 – 800 km/s and more, originate in coronal holes. Slow solar winds 300 – 400 km/s come mainly from Sun's equatorial belt. At the interface between plasmas moving with different speeds (for example, [10]) the KH instability can grow.

### *1.3. Terrestrial magnetosphere.*

The possibility of significant perturbations of the Earth's magnetopause caused by the KH instabiliy has a long history, and a large literature. It was first suggested by Dungey (1954) [11], who was also the first to hypothesize the reconnection of magnetic field lines in the frontside under IMF south, and beyond the polar cusps during periods of IMF north (Dungey (1961) [12]). The transition between magnetosheath and magnetosphere where velocity and density gradients appear is the MP boundary layer (BL).

During several decades the possibility of energy and momentum transfer from the solar wind to the magnetosphere by KH effects has been considered in a number of theoretical and observational papers. We recall some significant examples, Axford and Hines (1961) [13], Southwood (1968) [14], Sonnerup (1980) [15], Hones et al., (1981) [16], Ogilvie y Fitzenreiter (1989) [17], among many others, and draw the reader's attention to the references in reviews of the subject. For instance, Belmont y Chanteur (1989) [18], Miura (1995) [19], Fitzenreiter y Ogilvie (1995) [20], Kivelson y Chen (1995) [21], Farrugia et al. (2001) [22], *inter alia*.

The evidence for the presence of the instability is either direct, from data analysis of *in situ* records of spacecraft crossing the MP, or indirect, by inference from measurements of ground magnetometer chains at high geographical latitudes. Examples of the former are the studies of multiple boundary layer (BL) crossings by a spacecraft, consistent with large amplitude surface waves, likely of KH origin, moving downtail, Lepping and Burlaga [23], Seon et al. [24]. Regarding the latter method, it is now generally accepted (see [8]) that the KH instability causes pulsed oscillations of the geomagnetic field in the Pc5 range (Samson and Rostoker [25]) with periods in the range 150-600 s (e.g., [26])) Seon et al. [24], for instance, quote periods 120-240 s attributed to observed KH waves.

The knowledge of the KH instability and its evolution is complemented by numerical simulations that provide data analysis with important interpretive concepts and schematics. MHD computational studies of the KH in the MP context started with Miura (1984) [27]. The plausibility of MHD for space physics was argued, and explained by E. N. Parker in the 1960s. Since then MHD has been a powerful help to study the solar wind and its interaction with the terrestrial magnetosphere. Steady

90progress can be appraised in a set of reviews of the topic (for instance:[18], [19], [21]), and several recent research papers, Fairfield and Otto (2000) [28], Nykyri and Otto (2001; 2003) [29], Smets et al. (2002) [30], Hasegawa et al. (2004; 2009) [31], Gratton et al. (2006, 2009, 2011) [32], Bilbao 2009 [32], Nakamura et al., (2006; 2008) [33]), Hwang et al., (2011) [34], amongst many.

### *1.4. The cold dense plasma sheet*

The transport process through the MP that produces the so called cold dense plasma sheet (CDPS), located around the central equatorial plane of the magnetospheric tail is at the focus of a debate since the discovery of the phenomenon (Fujimoto and Teresawa (1994) [35]). The CDPS is a plasma at a lower temperature, and higher density than the ordinary hot and tenuous magnetospheric plasma. Space scientists investigate possible mechanisms by which plasma enters the magnetosphere under different conditions of geomagnetic configuration, solar wind, and solar cycle. Because of the frequent observation *in situ* of the CDPS during periods of northward IMF the subject engages the attention of the magnetospheric community.

During periods of northward IMF the KH instability is an important mechanism of momentum and energy transfer from the solar wind to the magnetosphere. Recent studies (Otto and Fairfield (2000) [36]) maintain that the non-linear stage of the KH instability, accompanied by large vortices that roll-up plasma of the magnetosheath and mingle it with magnetospheric plasma in the MP flanks, generate inside the big eddies favorable conditions to plasma entry during north IMF phases (see also [28], [29], [30], [31], [32], [33]). The transfer processes inside the large eddies include diffusion and/or, under certain conditions, convection and stretching of magnetic field lines.The latter progress until thin current sheets are formed, which may form localized, small-scale, magnetic reconnections, with the possibility of mass entry. We report observations of current sheets formed in a row of vortices of the magnetospheric BL in sections 4 – 5.

Other investigators support the hypothesis that reconnection of magnetic field lines beyond the polar cups is the main mechanism of entrance of mass during periods of IMF north. A simultaneous reconnection in the vicinity of both, northern and southern, cusps generate flux tubes full of magnetosheath plasma that penetrating in the magnetosphere would sink gradually inside, while they are being dragged towards the tail [37], [38], [39].

Scientists are still divided on this question while the evidence available is not yet decisive. Many papers discuss the mechanisms of plasma entry in relation to the CDPS (for example, [35], [38] [39] [40], [41], [42], [43], [44], [45]).

## 2. KH properties and elements of MHD theory

### *2.1. Basics of the KH instability.*

The process was studied by Kelvin during the second part of the XIX century in parallel with (but independently of) Helmholtz. The instability of flows with velocity gradients carries the name of the two pioneers. Lord Kelvin is the famous Scottish physicist associated with the second law of thermodynamics, the mathematical physics of electricity and magnetism, and the first transoceanic telegraphic cable. He was also active in the study of the conservation of circulation in fluids, the instability of strata moving with different velocities, and the formation of sea waves (Kelvin, 1871). Helmholtz, the eminent German physicist who made important contributions to acoustics, thermodynamics, magnetism, and physical physiology, also studied the laws of vorticity in fluid mechanics. He investigated the instability of layers in relative motion, with applications to the case of clouds (Helmholtz, 1868) (The references to the original papers can be found in Chandrasekhar [46]).

A simple idea of the instability mechanism is the case of wind blowing over water with the growth of sea waves. The positive feedback is due to the Bernoulli's effect in subsonic flows: $p + \frac{1}{2} \rho u^2 = const.$; this is sketched in figure 1. The slowing down at the wave bottom becomes a place of higher pressure, while the increased speed at the wave top is a site of lower pressure than average.

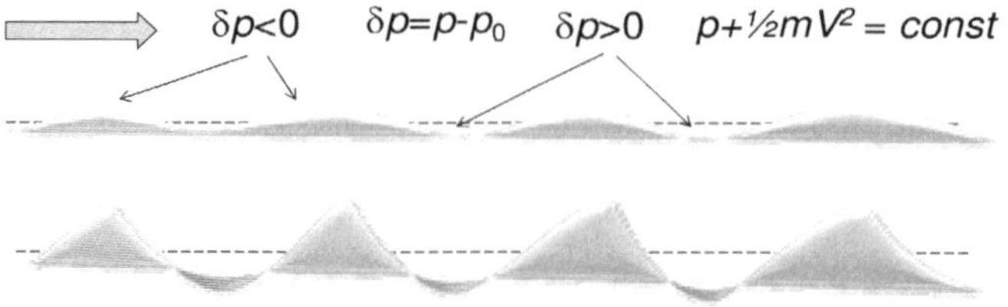


**Figure 1**. Wind-sea interaction. The wave growth arises from the positive feedback due to Bernouilli's effect for incompressible flows. In a wave hollow the speed slows down, and the pressure is larger than in the unperturbed state. By contrast, at a wave crest the speed increases, and the pressure is smaller than the equilibrium state.

The KH excitation can be examined from different points of view. In fluid dynamics considerable insight is gained from the study of the vorticity mechanism [46]. The energy flow approach is also illuminating, specially in plasmas because it can be extended to branches of the plasma wave spectrum other than the low frequency MHD oscillations, which are our main concern here. The presence of free energy in a velocity gradient flow can also be a source of growth for high frequency plasma waves. (for example, [47])

The energy method is described in figure 2. It represents a perturbed interface of counter-streaming flows, eventually observed from a moving frame, so that the velocity of the upper layer is parallel and opposite to that of the bottom. The physical quantities on either side of the interface are denoted by sub-indices 1, 2. The motion is along $\pm x$, and the equilibrium surface is the plane $y=0$. When perturbed by a wave, a sector one wavelength along $x$ and unit length in $z$, that extends along $y$ on one side, say $y>0$, has a smaller energy content than in the absence of undulation.

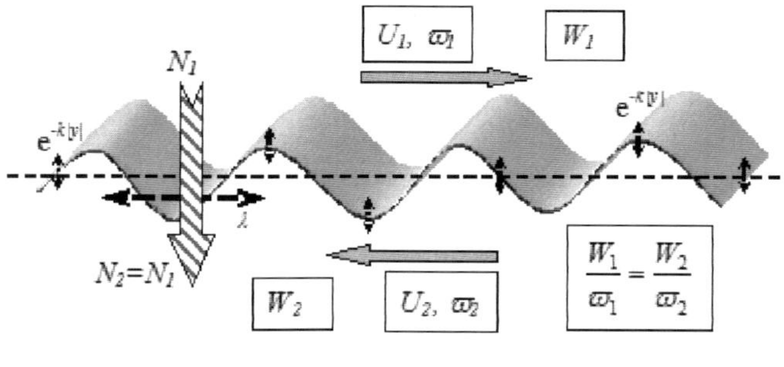


**Figure 2**. Energetics of the KH instability. Interface between two opposite moving layers with speeds $U_1$, $U_2$. The perturbation with wavenumber $k=2\pi/\lambda$ is localized near the interface, and decreases as $\exp(-k|y|)$ on both sides; $\varpi_1$, $\varpi_2$: Doppler shift frequencies, $W_1$, $W_2$, perturbation energy densities. Energy is transferred from one side to the other (zebra arrow) coupling negative with positive energy densities. The amplitude grows with no external energy sources. The process conserves the number of energy quanta, $N_1$, $N_2$ (see text).

Conversely, the twin sector that extends on the opposite side, $y<0$, has more energy than in equilibrium, and by the same amount. Accordingly, we can say that the energy of the perturbed state on the $y>0$ side is negative, while that of the other side $y<0$ is positive. The instability couples positive with negative energies across the surface. Energy flows from one side to the other, and the amplitude of the perturbation grows on both sides of the interface, with no need of external agents.

It can be shown that the ratio of Doppler-shifted frequencies on either sides, $\varpi_1/\varpi_2=(\omega-kV_1)/(\omega-kV_2)$ is negative in the unstable situation, and is equal to the ratio of the respective energy densities, $W_1/W_2$. This can be interpreted as the conservation of the number of energy quanta $N_1=W_1/\hbar\varpi_1=W_2/\hbar\varpi_2 = N_2$, in this process (for example, [48]).

Compressibility allows a coupling of KH perturbations with acoustic waves. This reduces the efficiency of the KH mechanism. To begin with, compressive KH modes have a penetration distance (normal to the interface) much larger than $\lambda/2\pi$, the range of influence of incompressible perturbations. When the magnetoacoustic coupling becomes significant, the same amount of energy is spread in a wider plasma volume, and the growth rate of the mode is reduced. Eventually, at sufficiently large supersonic speeds, the energy is radiated away from the surface by magnetoacoustic waves, and the mode cannot grow any further.

Figure 3 illustrates the effect: on the left, below a critical velocity $V$, a coupling of negative and positive perturbations leads to the instability; on the right, at a higher (supersonic) speed, the emission of a pair of acoustic waves (with positive and negative energy densities) takes place. The argument explains qualitatively the stabilizing effect of compressibility. The energy flux does not feed the growth of the surface perturbation, but goes away with the waves instead (e.g. [47], [49]).

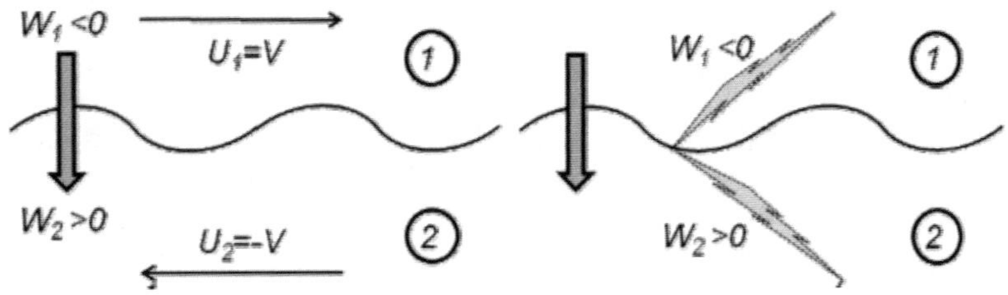


**Figure 3**. Schematic to illustrate the stabilizing effect of compressibility. Left panel: the energy flux feeds the growth of the interface perturbation for $V$ less than a critical value. Right panel: at a larger speed the energy flux is radiated away by a pair of positive-negative energy acoustic waves.

Precise conditions for the KH instability in compressible MHD require an elaborate description. ([50], [51]). Briefly put, the BL is likely to be unstable when the magnetic shear angle through the transition is small, and/or when the Alfvénic Mach number is large, $M_A = \Delta U/V_A >>1$ ($\Delta U$: velocity jump through the layer; $V_A$: Alfvén speed); we return to compressible KH in subsection 2.4.

*2.2. MHD stability theory*

In a region of flow with velocity shear the KH instability is driven by the projection of the velocity on the $\boldsymbol{k}$ vector, the wave-number of a perturbation mode. It is opposed by tensions associated with the magnetic field projection on the same $\boldsymbol{k}$-mode. The velocity and magnetic field components in the $\boldsymbol{k}$ direction are denoted with $V_k$, $B_k$, respectively,

$$V_k = \mathbf{V} \cdot \hat{\mathbf{h}}, \quad B_k = \mathbf{B} \cdot \hat{\mathbf{h}} \tag{1}$$

where $\hat{\boldsymbol{h}} = \boldsymbol{k} / |\boldsymbol{k}|$ is the unit vector along $\boldsymbol{k}$. The physical quantities on either side of the velocity gradient region are indicated with sub-indices 1, 2. In a MP context 1 is for the magnetosheath, and 2 for the magnetosphere.

In the simplest model the interface has no thickness, and the stability is computed in the incompressible MHD approximation. It is a planar tangential discontinuity (TD) with a difference of parallel velocities across it. Incompressible theory predicts stability of any given $\boldsymbol{k}$-mode when the following (necessary and sufficient) condition is satisfied,

$$\left\langle \left(V_k - \langle V_k \rangle\right)^2 \right\rangle \le \left\langle V_{Ak}^2 \right\rangle, \tag{2}$$

(Gratton et al, 2004 [52]). The bracket < > indicates a weighted average of any physical quantity, *f*, defined as

$$\langle f \rangle = \left(\rho_1 f_1 + \rho_2 f_2\right)/\left(\rho_1 + \rho_2\right) \tag{3}$$

where $\rho$ is the mass density. In condition (2) $V_{Ak}^2 = B_k^2/4\pi\rho$ is the square of the Alfvén speed computed with the magnetic field projection $B_k$, which embodies the effect of magnetic tensions. When the average of the square of the projected Alfvén speed is larger than the mean square of the projected velocity deviations of the average value, the mode is stable.

Stability condition (2) can be set in a more widely known [46], [53] equivalent form:

$$\rho_R \left(\Delta \mathbf{V} \cdot \hat{\mathbf{h}}\right)^2 \le \left[\left(\mathbf{B}_1 \cdot \hat{\mathbf{h}}\right)^2 + \left(\mathbf{B}_2 \cdot \hat{\mathbf{h}}\right)^2\right] \Big/ 4\pi \tag{4}$$

where $\rho_R$, defined by $1/\rho_R = (1/\rho_1 + 1/\rho_2)$, is a kind of "*reduced*" mass density, and $\Delta \boldsymbol{V} \equiv \boldsymbol{V}_1 - \boldsymbol{V}_2$ is the velocity difference across the surface. The TD model, also called as "thin" model in Farrugia et al., (1998) [54] is fully stable when the stability condition holds for all the directions of $\boldsymbol{k}$. The thin model stability does not depend on the wavelength.

Interestingly, condition (2) can be extended to a "thick" BL model [53], with a stratified planar geometry where the physical quantities are spread over $(x, z)$ planes, and are represented by continuous functions of the coordinate $y$ normal to the strata. The functions connect the physical values on both sides of a velocity gradient interface with finite width $D$. The form of the stability condition (2) for a thick model remains unchanged if the weighted average (3) is replaced by a new average with a weight $\mu_k(y)$:

$$\langle f \rangle_k = \int_{-\infty}^{+\infty} f(y) \mu_k(y) dy, \tag{5}$$

where the integration is over the whole range $-\infty < y < +\infty$ that extends across the BL model [52]. However, now the weight function $\mu_k(y)$ is mode-dependent. Here $\zeta_k = \zeta_k(y)$ is the $\boldsymbol{k}$-mode amplitude, and it is the solution of a boundary value problem for the stability equation of the thick model, equation 6. Stability of thick models depends not only on the direction of $\boldsymbol{k}$ but also on the wavelength.

The equation of the amplitude $\zeta$ (we now omit the sub-index for simplicity) in the incompressible approximation, is [53]

$$\frac{d}{dY}\left[H \frac{d\zeta}{dY}\right] - k^2 H \zeta = 0, \tag{6}$$

where

$$H(y) = \rho\left[\left(c - V_K(y)\right)^2 - V_{AK}{}^2(y)\right], \tag{7}$$

and $c$ is the phase velocity of the modes, defined at equation (11), subsection 2.3 (where the compressible treatment is given). The term $\rho(y) V_{Ak}^2(y)$ represents the magnetic tension $B_k^2/4\pi$. In fact, the roots of $H = 0$ are transverse Alfvén waves that propagate with respect to the plasma flow

$$c - V_K = \pm V_{AK}(y), \tag{8}$$

The function $V_{Ak}^2(y)$ is a stabilizing agent, and cannot be switched off except when the magnetic field is unidirectional. A case in which modes with $B_k = 0$ for all $y$ are possible (called *flute modes* in the literature). However, in general, the direction of the magnetic field is different on either side of the interface, i.e., a rotation of the magnetic field direction occurs. Magnetic shear prevents the existence of flute modes, and is always a stabilizing factor.

Condition (2) takes the same form for both thin and thick models, although with a different averaging procedure, equations (3) and (5) respectively. It is, therefore, possible for a TD with given quantities on both sides of the interface to satisfy the stability condition, while a companion thick model connecting the same values with continuous functions may predict the presence of unstable modes.

This may seem odd since, as we approach the TD limit by decreasing the width of a thick model, the velocity gradient increases and it looks like the interface should become prone to the instability. When the thin model is unstable, the associated continuous model is certainly unstable, albeit with a reduced growth rate. But the converse is not true. A continuous model can be unstable at moderate wavelengths, comparable with the interface width, but long wavelengths may be stable. In the latter case the TD approximation is found to be stable [53].

The reason for the different behavior resides in a degree of flexibility allowed to $\zeta$, the solutions of equation (6), which may assume shapes that reduce the contribution to the averages of condition (2) of $y$-ranges with large magnetic tensions, and enhance instead the effect of $y$-intervals with a dominant velocity gradient. This is in contrast to the rigidity of density as averaging weight. Of course, the increased freedom associated with $\zeta$ is possible only in continuous models, and is absent in the thin approximation case.

*2.3. Stability equation and characteristic value problem.*

In ideal (non resistive) compressible MHD the amplitude of the Fourier modes of KH perturbations is governed by the second order differential equation,

$$\frac{d}{dy}\left[H\left(1-\frac{1}{M}\right)\frac{d\zeta}{dy}\right]-k^2 H\zeta=0, \tag{9}$$

(derived by Gratton et al., 1988 [55]). The modes are of the form

$$\Xi=\zeta(y)\exp(-i\omega t+ik_x x+ik_z z), \tag{10}$$

where $\Xi$ is the $y$ component of the Lagrangian displacement of a plasma element from the equilibrium position, and $\zeta$ is the corresponding amplitude. The wave-vector of the mode is represented by $\boldsymbol{k}=(k_x,0,k_z)$; the absolute value $k=|\boldsymbol{k}|$ and the angle $\varphi$ between $\boldsymbol{k}$ and $\boldsymbol{V}$, also define the wave-vector properties.The (complex) angular frequency of the modes is denoted by $\omega=\omega_r+i\gamma$. The real part $\omega_r$ is the frequency of the perturbation, and the imaginary part $\gamma$ is the temporal growth rate.

A complex phase velocity $c\equiv\omega/k$ is defined so that the coefficient $M(y)$ of the differential equation for $\zeta$ can be written as follows

$$M(y)=1-\frac{\left(V_A{}^2+c_s{}^2\right)}{(c-V_K)^2}+\frac{V_{AK}{}^2 c_s{}^2}{(c-V_K)^4}, \tag{11}$$

while the function $H(y)$ is the same as defined in subsection 2.2, equation (7). In equation (11) $V_A=B/\sqrt{4\pi\rho}$ is the Alfvén speed, $c_s$ is the speed of sound, $V_k$, $B_k$ are projections already defined (equation (1)), $V_{Ak}=B_k/\sqrt{4\pi\rho}$ is the projected Alfvén speed (subsection 2.1), and all these quantities are functions of $y$. We use hyperbolic functions with a scale length $d$, for example $V_x=\frac{1}{2}U_1\,[1+\tanh(y/d)]$ for the velocity field (where $U_1$ is the magnetosheath velocity). The $x$-axis is chosen along the flow direction. Similar expressions connect the other scalar and vector fields across the BL. A detailed description of the BL models for MP stability studies is given in Gnavi et al., 2009 [56].

The stability analysis belongs to the temporal class, in which the (real) wave-number $\boldsymbol{k}$ is given (a Fourier component of the initial perturbation), and the response of the system determines the unknown (complex) value of $c$. This is a characteristic value problem for equation (9) subject to the boundary conditions, $\zeta \to 0$ for $y \to \pm\infty$.

When $c_s \to \infty$, the coefficient $M \to \infty$, and equation (9) reduces to the incompressible approximation, equation (6). In this case the asymptotic behavior of $\zeta$ is fixed on both sides as $\exp(\pm ky)$, $y \to \pm\infty$. If the transition layer is very thin with respect to the wavelength, that is when $kd<<1$ (where $d = ¼D$ is the scale length of the hyperbolic functions, associated to the width $D$ of the BL), a dispersion relation can be obtained integrating equation (6) across the discontinuity (assumed to be at $y=0$). Using decreasing exponentials solutions on either sides, we obtain the dispersion relation of the thin model

$$H_1 + H_2 = 0 , \tag{12}$$

where $H_1$ and $H_2$ are the values taken by $H$ on each side. Equation (12) then readily leads to the stability conditions (2)-(4).

### 2.4. *Influence of compressibility*

Condition (2) for thin or thick models is restricted to the incompressible approximation. Compressibility can stabilizes KH modes as noted in subsection 2.1. When $\rho$ is constant across the TD, and $\Delta \boldsymbol{V} \cdot \boldsymbol{B} = 0$ (or in the absence of a magnetic field; $\Delta \boldsymbol{V}$ is the velocity difference across the interface), the stability condition with $\boldsymbol{k}$ parallel to the flow is $M_s > 2$ (derived by Landau, 1944 [57]) where $M_s$ is the sonic Mach number defined by $M_s=|\Delta \boldsymbol{V}|/c_S$.

But this result is valid only for $\boldsymbol{k}$ aligned with the flow. When $\Delta V_k=\Delta \boldsymbol{V}\cdot\hat{\boldsymbol{h}}$ is less than the speed of sound, and this occurs when $\boldsymbol{k}$ turns away from $\Delta \boldsymbol{V}$, the corresponding modes are still unstable, although with reduced growth rates because the instability is not driven by the full impact of the velocity difference. In addition, when a magnetic field is also present as in MHD, the instability that remains in modes with $\boldsymbol{k}$ at an angle with $\Delta \boldsymbol{V}$ may still be suppressed. This happens when a rotation of $\boldsymbol{k}$ away from $\Delta \boldsymbol{V}$ brings $\boldsymbol{k}$ closer to the $\boldsymbol{B}$ direction, so that adverse magnetic tensions ($B_k^2/4\pi$) are switched-on, or augmented. Moreover, the critical value $M_s > 2$ for parallel $\boldsymbol{k}$ stability holds only for equal densities across the interface. But the critical condition is very sensitive to changes of the density ratio $\rho_2/\rho_1$, as we shall find in section 4.

For large $|y|$ ($|y|/d >> 1$) all quantities attain their constant values on both sides of the boundary layer, and indicating with $M_1$, $M_2$ the constant values of the coefficient $M$ on each side, the solutions of equation (9) tend asymptotically to

$$\zeta \to A_1 \exp\left( -ky\sqrt{\frac{M_1}{M_1 - 1}} \right), \qquad y > 0$$

$$\zeta \to A_2 \exp\left( +ky\sqrt{\frac{M_2}{M_2 - 1}} \right), \qquad y < 0, \tag{13}$$

The KH modes are perturbations localized at the interface, and tend to zero on both sides of the MP, that is, $\zeta \to 0$ when $y \to \pm\infty$. Therefore, the branch of the square root must be defined as

$$\kappa \equiv \Re\left( k\sqrt{\frac{M_a}{M_a - 1}} \right) \geq 0, \qquad a = 1,2. \tag{14}$$

In the case of unstable modes, $\Im(c)>0$ the $M_a$ are complex valued, and the square root has also an imaginary part. Therefore, when $|y|/d >> 1$ while the mode amplitude tends to zero, it also oscillates spatially along $y$. Ordinarily $\kappa \sim O(k)$, so that the penetration depth of the perturbation on either side is usually of order $1/k=\lambda/2\pi$. Except when $M_a\approx 0$, when the influence of the perturbation extends far on both sides of the transition.

The roots of $M$=0, the coefficient that distinguish compressible (9) from incompressible (6) behavior, are the phase velocities of MHD fast and slow modes (with respect to the projected plasma velocity),

$$(c-V_K)^2 = \frac{1}{2}\left(V_A{}^2 + c_s{}^2\right) \pm \frac{1}{2}\sqrt{\left(V_A{}^2 + c_s{}^2\right)^2 - 4V_{AK}{}^2 c_s{}^2} \tag{15}$$

Thus, when $M$ takes small values, $M$ <<1, a coupling of the KH mode with magnetoacoustic waves is expected The perturbation extends over a larger volume than in the incompressible case, and eventually energy is lost by emission of acoustic waves (subsection 2.1).

The extension from the incompressible thin case to the (more complicated) compressible thin model was studied by González and Gratton, 1994 [50]. The compressible dispersion relation is given by

$$H_1\sqrt{\frac{M_1}{M_1-1}} + H_2\sqrt{\frac{M_2}{M_2-1}} = 0 \tag{16}$$

where the square roots must have positive real parts, condition (14). Due to this constraint the dispersion relation is non algebraic. When we express this as a polynomial, spurious roots are introduced. Numerical work is required, and for this reason the compressible thin model does not lead to a simple exact formula, as the incompressible thin case. An instability condition like (2) or (4) is not available. The role of magnetic tensions in counteracting the velocity difference drive is subject to the intervention of magnetoacoustic effects.

The stability conditions for supersonic magnetospheric flanks are intricate because of density and temperature asymmetries between magnetosheath and magnetosphere. Theory indicates that the stabilizing action of compressibility is a strong function of the density ratio $r_d = \rho_2/\rho_1$ across the transition. Moreover, the KH stability condition of a BL modeled by continuous profiles for $\boldsymbol{B}$, $\boldsymbol{V}$ and $\rho$ can be very different from the stability predicted by the TD approximation in common use [53]. In addition, the efficiency of the stabilizing effects depends on the local values of $M_s$ and $M_A$, and the angle between $\boldsymbol{B}$ and $\boldsymbol{V}$. A stability treatment tailored to each specific configuration under study is necessary. These points rise in the study of the special event examined in sections 4 and 5.

## 3. Topics of magnetopause research

### *3.1. Instability bands on the dayside MP*

A basic tenet of the front-side model of Farrugia et al. [22], [54], for a northward IMF with small $|B_y|/B_z|$, is that KH activity resides in two bands whose width increases with decreasing $B_y$ (field components are in the GSM system; see for example [8]). The inputs for the stability study are based on a theoretical MHD model of the dayside MP with a paraboloid shape that interacts with the solar wind. The external flow is matched with a compressed geomagnetic dipole field inside the diurnal magnetosphere.

Views of the velocity field with black arrows, and local magnetic shear angle with a color scale over the dayside MP surface, are shown in figure 4 for two IMF clock angles (angle $\theta$ of the IMF with the geomagnetic north; positive when $B_y$>0), 20° left panel, and 5° right panel. In the figure the surface is a frontside view (from $X$>0) of the MP model with a solar wind stagnation point flow at the center. The coordinates $X$, $Y$, $Z$ are in the GSM system with center at the Earth; $X$ points to the Sun, $Z$ to the geomagnetic north, and $Y$ completes the orthogonal triad, with $Y$>0 pointing to dusk.

An important feature of this model is that the MHD stagnation point flow is compelled to move normal to the local magnetic field, so that the KH instability is excited mainly with modes $\boldsymbol{k} \perp \boldsymbol{B}$. The KH-active strips are bands of low magnetic shear (i.e.,with nearly-parallel magnetic fields on either sides of the MP). In the northern hemisphere they extend eastward (and vice-versa in the southern). As $\theta$ decreases, the bands widen until, for due northward IMF, all the low latitude MP away from noon is KH-active, and the MP near noon is KH active to higher latitudes (compare the two figures). Clearly, north-south and dawn-dusk asymmetries in dayside KH activity are implied by this model. Their existence poses a challenge to any interpretation of mass transfer by mechanisms other than KH. Moreover, the unstable bands are sources of KH waves that are convected toward the flanks, and while moving tailward they may develop, under favorable conditions, into full size vortices.

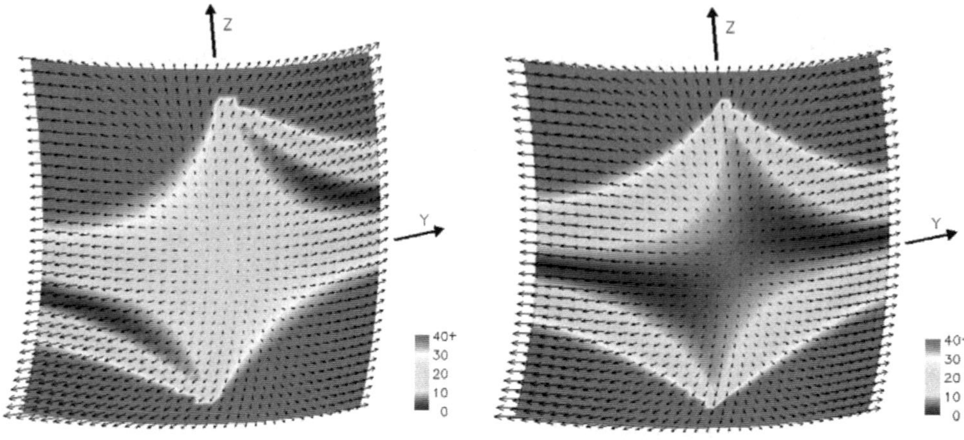


**Figure 4.** An illustration of the dayside instability bands according to Farrugia et al. [54]. The surface is a front side view (from $X>0$) of a MP model, with the solar wind stagnation point flow at the center; $X$, $Y$, $Z$ are GSM coordinates. Small black arrows represent the velocity field, and the color scale of different areas indicates the local magnetic shear angle. The left panel is for the IMF clock angle of θ=20°, while the right panel is for θ=5°.

Indirect experimental support for the unstable bands theory was found via the agreement of changes of the Pc5 power spectra with variations of the clock angle [58].

KH perturbations of the MP can couple with resonances in the front side magnetospheric cavity. Surface perturbations are transmitted into the magnetosphere, and resonate at magnetic shells, known as $L$-shells, where resonance conditions for transverse Alfvén waves are satisfied [59]. The wave that couples the surface waves with resonating magnetic lines is a magnetosonic mode, since it must transport energy across the field. The excitation of the resonant magnetic lines was confirmed, both by spacecraft observation and by ground magnetometers, for ULF (*ultra low frequency*) magnetospheric waves. They correspond to Pc5 geomagnetic perturbations (2.5 to 10 minutes periods).

The CME which passed Earth on April 11, 1997 afforded an excellent opportunity of studying KH activity at the dayside magnetopause as a function of the clock angle $\theta$ of the IMF. Figure 5 shows data taken by Wind on April 11, 1997, the $z$ component of the field (GSM system) and $\theta$ from 8 to 20 UT. The magnetic field is strongly north in the interval 10 to 14 UT, which is favorable for KH excitation.

From, approximately, 10 to 12 UT the average clock angle is 18° and after that, in the following lapse 12-14 UT, the average clock angle decreases to 6°. We can appreciate in figure 4 the important increase of low magnetic shear areas, from the first to the second period.

The Pc5 power spectral density of the high latitude chain of ground magnetometers, IMAGE, computed for the two mentioned time intervals, is represented in figure 6. The plots clearly display an increase of power, and a shift to lower frequencies, when the second time period is compared with the first. The dayside band theory gives an interpretation of these facts: for θ = 18° only short wavelengths that fit into small band areas can be unstable, while for θ = 6° with larger areas available longer wavelenghts for unstable modes are possible. The KH frequency is proportional to the inverse of the wavelength, so that more room for the instability at lower frequencies is available.

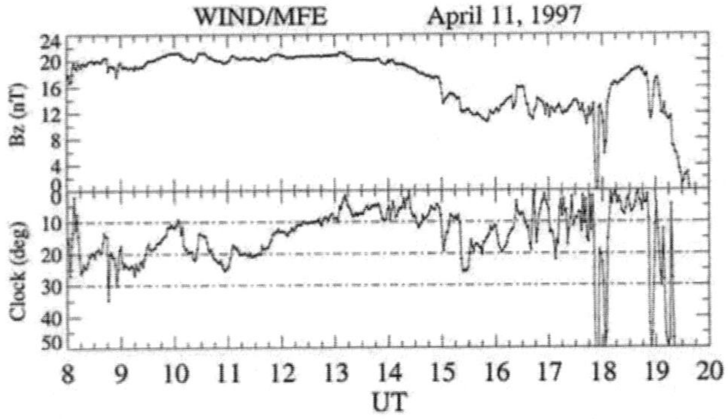


**Figure 5**. Data from Wind on April 11, 1997: the $z$ component of the magnetic field (GSM system) and the clock angle of the IMF, $\theta$, from 8 to 20 UT (see text).

In our research of the KH excitation on the dayside MP the presence of a plasma depletion layer (PDL) was included in the MP model [53], [54], [60]. This is a layer with a strong magnetic field (due to convection pile-up) and low density that appears in the subsolar regiόn of the magnetopause, usually under northward IMF conditions. The *PDL* should be considered in studies of dayside MP stability as the outer boundary condition. The *PDL* also affects the critical $M_A$ value predicted by Fuselier et al., 2000 [45] that regulates the competition between KH and tearing mechanisms (on which we return in subsection 3.3), as well as the density ratio across the MP, and it is therefore a significant element to bear in mind in stability studies for this region.

In modeling the MP transition the use of continuous field profiles is necessary. As we have shown [52], [53], the oft-quoted criterion for KH instability of the MP, which assumes it to be a TD, can be misleading when predicting stability. For a subsonic dayside physical regime, we found that there are unstable modes with $\lambda \sim 2.5\ D$ (where $D$ is the thickness of the transition) with significant growth rates, which are stable under the same boundary values according to the TD approximation. Hence, the stability analysis with a TD model can miss the KH excitation altogether.

With Cluster and Themis crossings of the dayside MP under steady northward IMF conditions, and on both sides of noon, further progress can be made on this matter. More advanced global models could be a valuable help. Another point is to find the ionospheric signatures of the KH-active bands at the MP. This requires the mapping of the KH unstable bands of the dayside MP to the ionosphere.

*3.2. Stability of dayside pristine MPs: the effect of transient periods of compression and expansion.*
The difference in ground response to Kelvin-Helmholtz instability of the dayside MP for two different IMF $\theta$ (the clock angle) orientations described in subsection 3.1 is reduced by the presence of a low latitude boundary layer (LLBL) [54]. The dependence of KH activity on $\theta$ should be most clearly observed in pristine magnetopauses, i.e., those that occur without an adjoining magnetospheric boundary layer. Pristine MP are about a 10 % of a statistical study of data crossings of the frontside magnetopause [61]. They show very sharp density gradients, with a thickness $\delta$ for density transition, much smaller than the thickness $\Delta$ of the current layer, $\delta \ll \Delta$. Ratios 1:5 and less has been observed.

We comment on calculations that take into account two length scales, $\Delta$ and $\delta$, in the equilibrium fields. Together with a velocity gradient, the effect of a transient acceleration $g$ of the MP is included, which is the driver of the Rayleigh-Taylor (RT) instability (Gratton et al., 1996 [62]). The acceleration of the dayside surface happens during lapses of fast expansion, or compression of the magnetosphere, following changes of solar wind properties. At small wavelengths, the results with $\delta/\Delta \ll 1$ differ substantially from those for $\delta/\Delta = 1$, and suggest new tests of correlations between spacecraft observations, and ground magnetometer data.

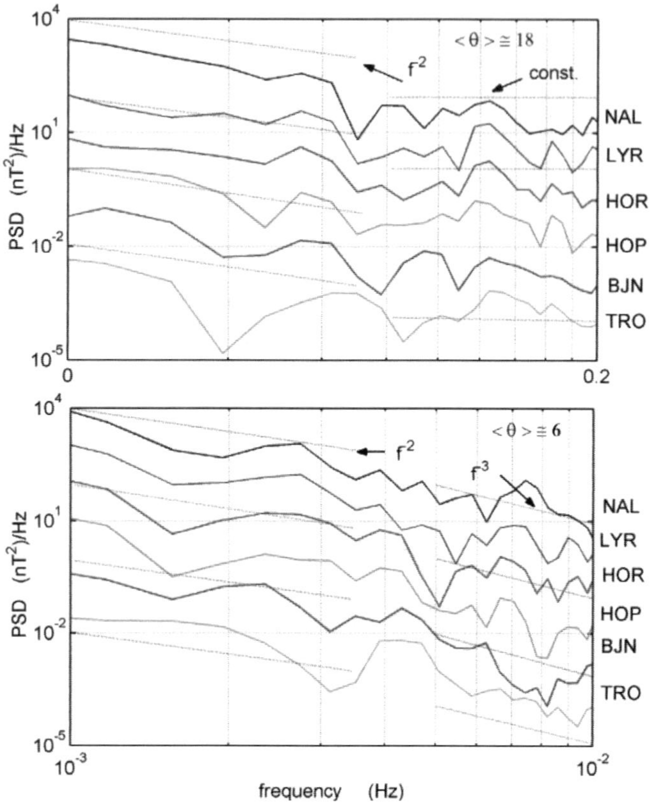


**Figure 6.** Power spectral density in the Pc5 range of a chain of high latitude ground magnetometers for two time intervals, from 10 to 12 UT, upper panel, and from 12 to14 UT, lower panel. The plots show an increase of power, and a shift to lower frequencies, from the second to the first time period.

MHD perturbations of a simple MP model without low latitude boundary layer, display the basic trends effortlessly. This is a model where changes of the physical quantities take place over two very different length scales as in figure 7. The schematic show a velocity gradient transition with a small scale density drop, represented by a discontinuity. The profile for $B_\kappa$ is similar to $V_\kappa$.

The Kelvin - Helmholtz instability operates over relatively long stretches of time, in sites where the action of the velocity gradient prevails over stabilizing magnetic forces. Conversely, the Rayleigh-Taylor instability is a transient process which appears during episodes of magnetopause acceleration, as a result of rapid changes of solar wind dynamic pressure. Since the magnetosheath expands or contracts together with the magnetopause, in a neighborhood of this boundary nothing changes locally, except for the acceleration of the site in toto.

For ordinary fluids with stratified density, G. I. Taylor showed that regarding stability the effect of a global acceleration was mathematically equivalent to the effect of gravitation studied earlier by Lord Rayleigh (references cited in [46]). Physically, this is an example of Einstein's equivalence principle.

In the simplified theory the density profile is represented by a discontinuity. WKB approximations, with $\lambda \ll \Delta$, are used in the velocity gradient regions at either side of the discontinuity, and matched across the density jump. It is easy to derive the dispersion relation,

$$\omega = kV_k + \frac{1}{2}\left(\frac{\rho_1 - \rho_2}{\rho_1 + \rho_2}\right) DV_k \pm \sqrt{\frac{1}{4}\left(\frac{\rho_1 - \rho_2}{\rho_1 + \rho_2}\right)^2 (DV_k)^2 + \left[\frac{2k^2 B_k{}^2}{4\pi(\rho_1 + \rho_2)} - gk\frac{\rho_1 - \rho_2}{\rho_1 + \rho_2}\right]} \quad (17)$$

where the growth rate is given by $\gamma = Im(\omega) > 0$. Here $V_k$ and $B_k$ stand for the projections of $\boldsymbol{u}$ and $\boldsymbol{B}$, on the $\boldsymbol{k}$ vector of a Fourier mode. $DV_k$ denotes the $V_k$ gradient across the magnetopause. The quantities $B_k$, $V_k$, $DV_k$, assumed to be continuous, are computed at the density jump; density appears with both side values, $\rho_1$ for the magnetosheath, and $\rho_2$ for the magnetosphere $\rho_1 > \rho_2$; $g > 0$ corresponds to a sunward acceleration (MP expansion).

The dispersion relation equation (17) is computed in the incompressible limit for subsonic speeds (but compressibility corrections to the main instability branch can ordinarily be neglected on the dayside MP). The whole derivation is discussed in Gratton et al., 2003 [63].

If $\gamma = 0$ the modes are stable, and when $g = 0$ the formula shows the typical absence of the KH instability for small $\lambda$. Interestingly, when $g > 0$ the Rayleigh-Taylor effect is stabilized, or at least reduced, by the presence of the velocity gradient $DV_k$, even for flute modes where $B_k = 0$. On the other hand, in the phase of acceleration towards the Earth, $g < 0$, the modes with $\lambda \ll \Delta$ are stable, flute or not. Equation (17) can be contrasted with the long wavelength, $\lambda \gg \Delta$, subsonic dispersion relation

$$\omega = k\frac{\rho_1 V_{\kappa 1} + \rho_2 V_{\kappa 2}}{\rho_1 + \rho_2} \pm \sqrt{\left[\frac{B_{\kappa 1}^2 + B_{\kappa 2}^2}{4\pi(\rho_1 + \rho_2)} - \frac{\rho_1 \rho_2}{(\rho_1 + \rho_2)^2}(V_{\kappa 1} - V_{\kappa 2})^2\right]k^2 - gk\frac{\rho_1 - \rho_2}{\rho_1 + \rho_2}} \quad (18)$$

which follows directly from the incompressible stability equation of the thin model (including $g$),

$$H_1 + H_2 + kg(\rho_1 - \rho_2) = 0 \quad (19)$$

The roles of the velocity gradient in the former case, and the velocity difference in the latter, are inverted. Therefore, the instability due to the joint factors has a transition from large $\lambda \gg \Delta$, where the KH and RT (with $g > 0$) effects add together to amplify the perturbation, to small $\lambda$ where the velocity gradient works in opposition to the RT mechanism, and may even produce stability in a definite range of small $\lambda \approx \Delta$ (although for very small $\lambda \ll \Delta$ the RT becomes dominant again).

From the scaling law of the equations it follows that the influence of the RT action as an activating mechanism of the KH instability is more intense during a slow solar wind period than in a fast one. Within lapses of fast solar wind streams the importance of transient accelerations on KH excitation is expected to be reduced [63].

### *3.3. KH and tearing modes at the MP*

That velocity shears can hinder the onset of tearing modes [64] was noted in plasma physics with the discovery in the 1990s of the H-mode Tokamak operation. It was found that a differential rotational flow in the device could suppress reconnection of magnetic surfaces and improve plasma confinement. A number of theoretical studies have been published since, having to do not only with Tokamak physics but also with more generic shear flows of plasmas dealing with KH and tearing effects ([65] and references therein). For space plasmas early influential studies were those of La Belle-Hamer et al., 1995 [41], and Chen et al., 1997 [44].

About the same time this question became a hot topic in space physics because of the great interest aroused by encounters with cold, dense plasma sheets under northward IMF conditions. As mentioned in section 1.3, CDPS compositional studies showed to be of magnetosheath origin [66] (Fuselier et al., 1999). Reconnection poleward of the cusps is a possible candidate. But under northward IMF conditions the KH instability can be operative as well. Many subsequent papers on the issue have struggled with these entry mechanisms ([39], [41], [43], [44], [45], [66], [68], [69], [70], [71]). In the absence of crucial evidence, the question is whether other processes such as diffusive entry [42] and the KH instability [28], [31], [35], may contribute to the formation of the CDPS.

A basic conclusion of the theoretical work for flows with constant density, is that the presence of velocity shear indeed interferes with the tearing process when the Alfven Mach number, $M_A > 1$. The tearing modes are stable in super-Alfvenic flows, where the KH instability predominates. An example of this assertion is illustrated by figure 8 which shows the competition between KH and tearing modes in a plasma model with magnetic shear and a velocity gradient flow (Bender, Ph.D. Thesis, 1997 [72], and Bender et al. 1998 [73]). The theory is MHD equations with resistivity and viscosity included.

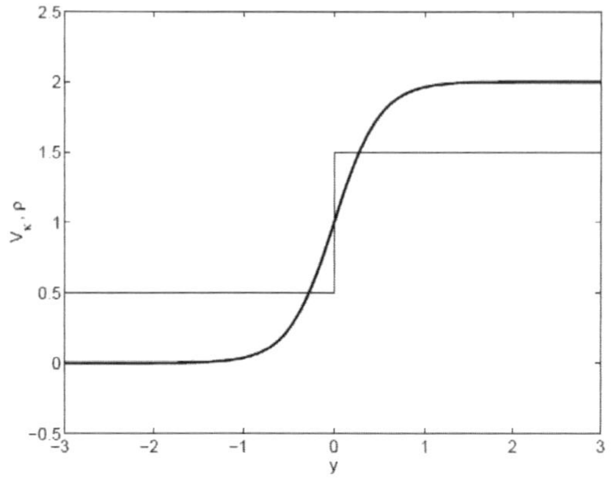


**Figure 7.** A velocity gradient interface with a small scale density drop, represented by a discontinuity. Field profiles for a WKB analysis: velocity $V_K$ and density ρ versus *y* (*y>0* points toward the MS). The trend for $B_K$ is similar to $V_K$.

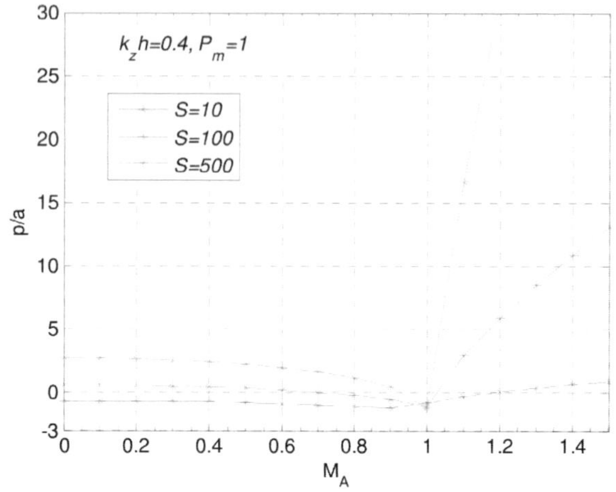


**Figure 8.** Competition between KH and tearing modes. The plot shows a normalized growth rate *p*/*a* versus $M_A$, for a fixed normalized wavenumber $k_z h$=0.4 and magnetic Prandtl number $P_m$=1, for three values of the Lundquist number *S* (see text). For $M_A$ <1 there is tearing instability and no KH excitation, vice versa when $M_A$ >1 tearing is quenched and there is only KH instability.

Here the perturbation modes have the functional dependence $A(y)exp(pt+ik_z z)$, where $A(y)$ stands for the modes amplitude. Figure 8 shows the normalized growth rate *p*/*a* versus $M_A$, for a given normalized wavenumber $k_z h$=0.4 and a fixed magnetic Prandtl number $P_m$=1.

Across a boundary layer model, stratified in *y* and layered on planes (*x*, *z*) which is at the same time a current sheath and vorticity layer, the main flow $V_z=V_z(y)$ is counter-streaming, and the magnetic field $B_z=B_z(y)$ changes sign at *y*=0. The resistive and viscous structure is maintained in steady state against diffusive effects by an additional stagnation point flow $V_x=ax$, $V_y=-ay$, where *a* is the strain rate of this secondary flow ($e_{xx}=\partial V_x/\partial x=a=-\partial V_y/\partial y=-e_{yy}$). The boundary layer scale length along *y* is represented by *h*, and is used also to normalize $k_z$. The magnetic Prandtl number $P_m=\nu/\nu_m$ is the ratio of viscous and resistive kinematic diffusivities.

The colored traces in figure 8 are for several values of the Lundquist number, $S=V_A/(a/\nu_m)^{1/2}$. When $M_A$ <1, only the tearing modes are excited, with a growth rate that diminishes as the resistive effect decreases. By contrast, when $M_A$ >1 the tearing modes are stabilized by the flow, and only KH modes that overcome magnetic tensions are amplified. It was noted, however, that the precise value of $M_A$ at which tearing is inhibited depends on the density ratio across the layer.

In his thesis work Bender [72] found also that the critical $M_A$ value that suppress the tearing effect depends in a significant way on the magnetic Prandtl $P_m=\nu/\nu_m$. [73]. The classic theory of plasma transport predicts a strong dependence of $P_m$ with temperature [74]. However, in a non-collisional plasma as the interplanetary medium it is expected that $P_m$ is determined mainly by anomalous effects related to the micro-fluctuation level of wave fields. On waves in the space environment of the Earth see, for example, [58], [75], [76], and the literature quoted therein.

On the other hand a small-scale tearing process may enhance the efficiency of diffusive transport inside large KH eddies formed in the BL. This is a possible cooperation of tearing with KH (in the non-linear stage) in the sense that rolled-up vortices may stretch and bring into contact oppositely-directed field lines which then reconnect. In the MP context this effect was emphasized by e.g. Otto and Fairfield (2000) [28], and others. The concept was known in MHD much earlier, by way of 2D simulations of vortices that wound weak magnetic fields into coil shapes [77] (Weiss 1966). The magnetic field lines yield passively to the strength of the flow. It was found that the whirling motion intensify the dissipation rate of the magnetic field, via a sequence of tearing and reconnection events.

To conclude, despite important research progress the knowledge of the actual influence of velocity shear on tearing, and vice versa, in the magnetopause context is still incomplete. Some elements of the problem have not yet received proper attention, but can make substantial contributions in assessing the merits of the mass entry mechanisms.

### *3.4. Gaps of KH amplification at the flanks.*

The MP flanks have been repeatedly visited by spacecraft under different magnetosheath conditions, and at different latitudes and downtail distances. The many configurations encountered may now be catalogued as a function of these quantities, and their stability examined systematically. The stability of flows with velocity shear is strongly affected by compressibility when the flow becomes supersonic, as it does on the flanks.

The main question is to find when is a flank region is KH-unstable. What are the onset conditions for supersonic flows? How does the broadening of the flank BL, produced by the KH instability, depend on (i) the local orientation of $\boldsymbol{B}$ and $V$, (ii) the physical conditions at the boundary, and (iii) its distance from the subsolar point?

KH amplification may have gaps, i.e., regions of local stability, on the flanks (Gnavi et al. 2009 [56]). Consider the solar wind with a northward IMF in the magnetosheath at a near Earth flank position, where the flow is again moving at supersonic speeds. We investigate the theoretical conditions required for the Kelvin-Helmholtz modes of the magnetopause to be stable. The stability is studied with an equilibrium model of the low latitude boundary layer (LLBL) described by hyperbolic tangent profiles for the main scalar, and vector fields, ρ, $T$, $\boldsymbol{V}$, $\boldsymbol{B}$, of a MHD treatment. The BL thickness $D$=4$d$ (as in subsection 2.2) is a free parameter of the calculation that can be adjusted at the end, according to observation inputs or judicious conjectures.

KH stability (or negligible growth rate) is possible when the sonic Mach number $M_s=U_l/c_s$ ($U_l$, $c_s$, flow speed and sound speed of the adjacent MS, respectively) is larger than but close to one, $M_s\sim1.3$, while the Alfvénic Mach number $M_A=U_l/V_A$ (where $V_A$ is the MS Alfvén velocity) remains a bit smaller than one, $M_A\leq1.2$. Local stability of the LLBL is found when, combined with the quoted Mach number values, there is a significant magnetic shear angle between the IMF and the geomagnetic field. Solar wind with physical conditions favorable to the stability of near Earth flanks, are cold and not too dense plasmas, with rather strong magnetic fields. Such that $M_A$ is rather smaller, while $M$ is somewhat larger, than the average values of ordinary solar winds. These conditions are often realized during the passage of the tail of CMEs by Earth.

We computed several stability cases of the equatorial flanks. The boundary layer parameters are inputs from spacecraft crossings, complemented by judicious extrapolations of the physical conditions expected upstream, and downstream of the orbits. We reach the conclusion that when the above-mentioned features materialize a gap between KH amplifying regions at the near equatorial flank can be expected. The KH instability of the unstable strips of the front side [54] convected to the flanks may traverse a neutral amplification region before reaching the MP tail. The distant flanks are again unstable because of the enfeeblement of the geomagnetic field toward the magnetosphere tail.

On the other hand, when large-scale vortices are observed in the flank BL the question is how far upstream they are generated. Under the conditions discussed it may happen that the origin is not distant a few amplification lengths upstream, but a far away area in the unstable bands of the dayside.

An example is shown in figure 9. In this case the geomagnetic field is somewhat stronger than the outer field. The flow is supersonic $M = U_l/c_s=1.4$, and slightly super Alfvénic $M_A= U_l/V_A=1.2$, while the magnetic shear angle across the LLBL is 45 °. After optimization over $k$, and the angle φ of $\boldsymbol{k}$ with the flow, the (normalized) maximum growth rate (for φ=15°) is already very small. An e-folding time of growth, assuming $D = 6000$ km and $U_1 = 300$ km/s, can be estimated as $\tau$ = ½ hour. Moving a degree upstream, where $M$ =1.2, $M_A$ =1, which corresponds to a $U_1$ decrease by a 20 % (keeping the other fields at the same values) no growth is found, the LLBL is locally stable. A stability gap, between the dayside unstable bands and the unstable distant MP tail, is present.

Note that for very long wavelengths (the range near $kd$=0) the growth rate becomes zero at a small $kd$ value. This means that a thin model of the LLBL (a tangential discontinuity, *TD*) fails to perceive the existence of an instability for $M$=1.2, $M_A$=1.4. What we called a "*pitfall*" of the *TD* model [53] occurs also in supersonic flows. It is worth keeping this in mind at downstream sites, where $\gamma$ becomes significant while a TD approximation may still forecast a stable MP. Finally, further downstream near the tail, the instability chances improve because $M_A$, $M$, increase together with $U_1$ while the magnetic tensions in the BL become weaker with the decline of the geomagnetic field.

The parameters of the event of January 11, 1997 studied in [67] provide another example. This is again a possible scenario with a non-amplifying gap in the near flank. The magnetosheath values are: $U_1$=300km/s, $B_1$=30 nT, $\theta_1$=100°, $n_1$=10 p/cm$^3$, $T_1$=0.130 keV. On the magnetosphere side: $B_2$=20 nT, $\theta_2$=30°, $n_2$=1 p/cm$^3$, $T_2$=1.9 keV. The Mach numbers in this case are $M_s$=1.90, $M_A$=1.67. The maximum growth rate occurs for $kd$=0.65, $\lambda\approx$4.8 $R_E$, assuming a width ≈1 $R_E$, and the e-folding time is $\tau_e\approx$3.5 min. The ***k***-***V*** angle for the maximum growth rate is $\varphi$=- 4.33°, and modes with a deviation of Δφ ≈±15° from the optimal ***k*** orientation are stable [56].

Figure 10 represent the normalized growth rate versus changes of $M_s$. The ratios of the physical quantities, and the angles of the fields are maintained constant, except the two Mach numbers that vary together keeping a constant ratio $M_s/M_A$=1.14.

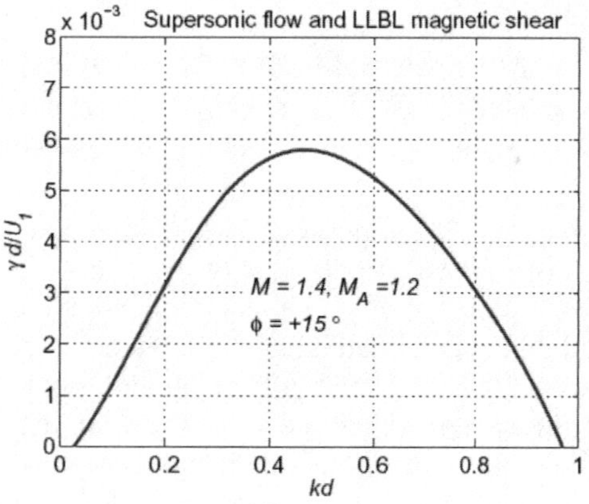


**Figure 9.** Normalized growth rate $g$=$\gamma d/U_1$ as function of non-dimensional wave number, *kd*. The model is specified with hyperbolic functions. The magnetic shear angle across the LLBL is 45 °, and the ratio $B_2/B_1$ is a bit larger than 1.The ***k*** direction is for the maximum growth rate (ϕ=15° with the *x*-axis). The Mach numbers are $M_s$ = $U_1/c_s$ =1.4, and $M_A$= $U_1/V_A$ =1.2. Note that for small *kd* values (≤0.03) the modes are stable (see text). The instability is weak and not far upstream, where $U_1$ is 20% smaller, there is no growth

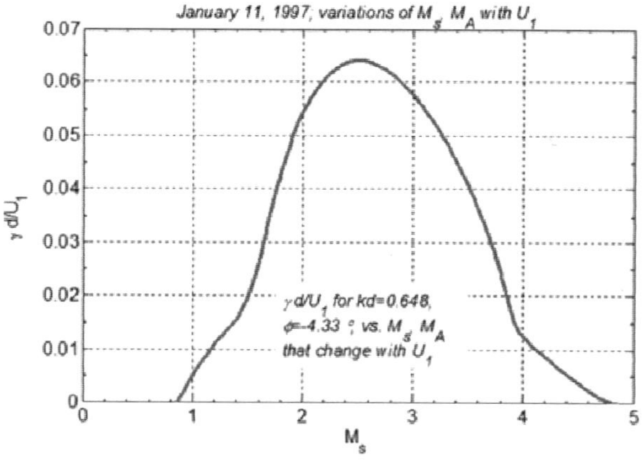


**Figure 10.** Scenario for the January 11, 1997, event [67]. Another case that shows possible stable gaps in the flanks. Normalized growth rate versus changes of $M_s$. All the other quantities are maintained constant while both Mach numbers vary jointly (see text).

This can be a consequence of changes of the magnetosheath velocity $U_1$. The change may be associated with variations of the solar wind speed, or with a choice of different positions along the flank. The growth rate has a maximum for $M_s \approx 2.5$, but the instability declines both for $M_s < 2.5$, and $M_s > 2.5$ (upstream or downstream, respectively, if we interpret the variations as changes of site). For $M_s > 2.5$ the stability is due to the influence of compressibility. For $M_s < 2.5$ the effect of magnetic tensions prevails. The results emphasize the importance of using data from actual spacecraft crossings to initiate stability studies with realistic features. It also bears on the size and width of the dayside KH-active bands, mentioned in subsection 3.1, which continuously feed KH perturbations that travel tailward with the flow. Examples of the interplay of these elements can be found in [51], [56], [67], [78], [79], [80]. Several years of data from flank crossings (including MP skimming orbits) mainly from Cluster and Geotail, and recently Themis are now available to address these issues. Cluster and Themis can provide simultaneous multi-point observations at key entry locations, which may allow boundary conditions to be set from measurements made near the locale whose stability is being examined. A noteworthy attempt was done by Paschmann et al. (2005) [81], who characterized the MP and BL in the dawnside near-tail, based on 24 crossings in a single Cluster skimming orbit. A systematic survey of flank crossings, classifying them by their characteristics and input parameters for BL models, along the lines mentioned, especially during CMEs passage, is desirable so that stability criteria for each representative class can be properly investigated.

## 4. The event of October 24, 2001 and stability analysis

### *4.1. ACE and Wind observations*

We report a research on vortices observed in the BL on October 24, 2001, and the stability of that layer. The magnetosheath field was approximately parallel to the flow direction (albeit pointing sunward), as studied in detail by Farrugia et al., 2010 [82]. The interest of this event resides in the peculiar field orientation, different from the northward IMF of most investigations on the KH instability, and the recognition that it can provide knowledge concerning also the much more frequent case of Parker's spiral field.

A plasma flow with a parallel magnetic field is not a favorable configuration for the Kelvin-Helmholtz instability, because magnetic tensions in full action operate to stabilize the configuration. In particular, this happens at the stagnation point region of the dayside MS. Near the subsolar point the solar wind slows down to $M_A<1$, and is dominated by magnetic tensions. From there, the MS plasma speeds up again along the flanks, where $M_A$ increases and eventually exceeds unity. Hence the KH mechanism could start t but the issue is not straightforward, because the motion at the flanks is supersonic, and compressibility cooperates with magnetic tensions to favor stability (as noted in section 2). The instability may occur perhaps too far tailward to explain the origin of the vortices observed in the BL during a Wind´s crossing. This is the problem we adress here.

The magnetosheath is shaped by the solar wind characteristics, and the bow-shock transition. The event we consider occurs during the last phase of the passage of a multiple ICME group by Earth. A description of the interplanetary physics during the days 21-25 October, 2001 is given in [82], [83]. The distinctive trait of the late hours of October 24, 2001, is a quiet and steady plasma, with an almost radial speed and magnetic field.

Figure 11 displays the solar wind data registered from 18 to 21 UT by ACE during the event. This period comprises the interval of interest, from 19 to 20:30 UT, when vortical perturbations were observed in a Wind crossing of the BL.

We focus our attention on this time interval. ACE's time scale has been adjusted by the time delay for the interplanetary plasma to reach the region of Wind's orbit. The first row of Figure 11 presents: the magnetic field intensity $B$ (black), almost constant at 6 nT, and the three components in GSM coordinates $B_x$, (red) $B_y$ (green) and $B_z$ (blue). In the interval 19 to 20:30 UT $B_x$ is also quite steady, less than $B$ only by a small extent, and positive. The other components are small, and the angle of the field with respect to the flow oscillates slightly between 164°-159° (shear angle, seventh row). The field is approximately antiparallel to the flow, which is almost radial (antisunward) with a moderate speed, $V \approx 370$ km/s, indicated in the fourth row. The second and third rows are for particle density, $N_p \approx 4$ p/cm$^3$, and temperature $T \approx 4 \times 10^3$ °K, respectively. The plasma is very cold and field-dominated, with a plasma beta $\beta \approx 0.02$ (fifth row). These physical conditions are frequent in CMEs. The dynamic pressure is shown in the sixth row, $P_d \approx 1$ nP.

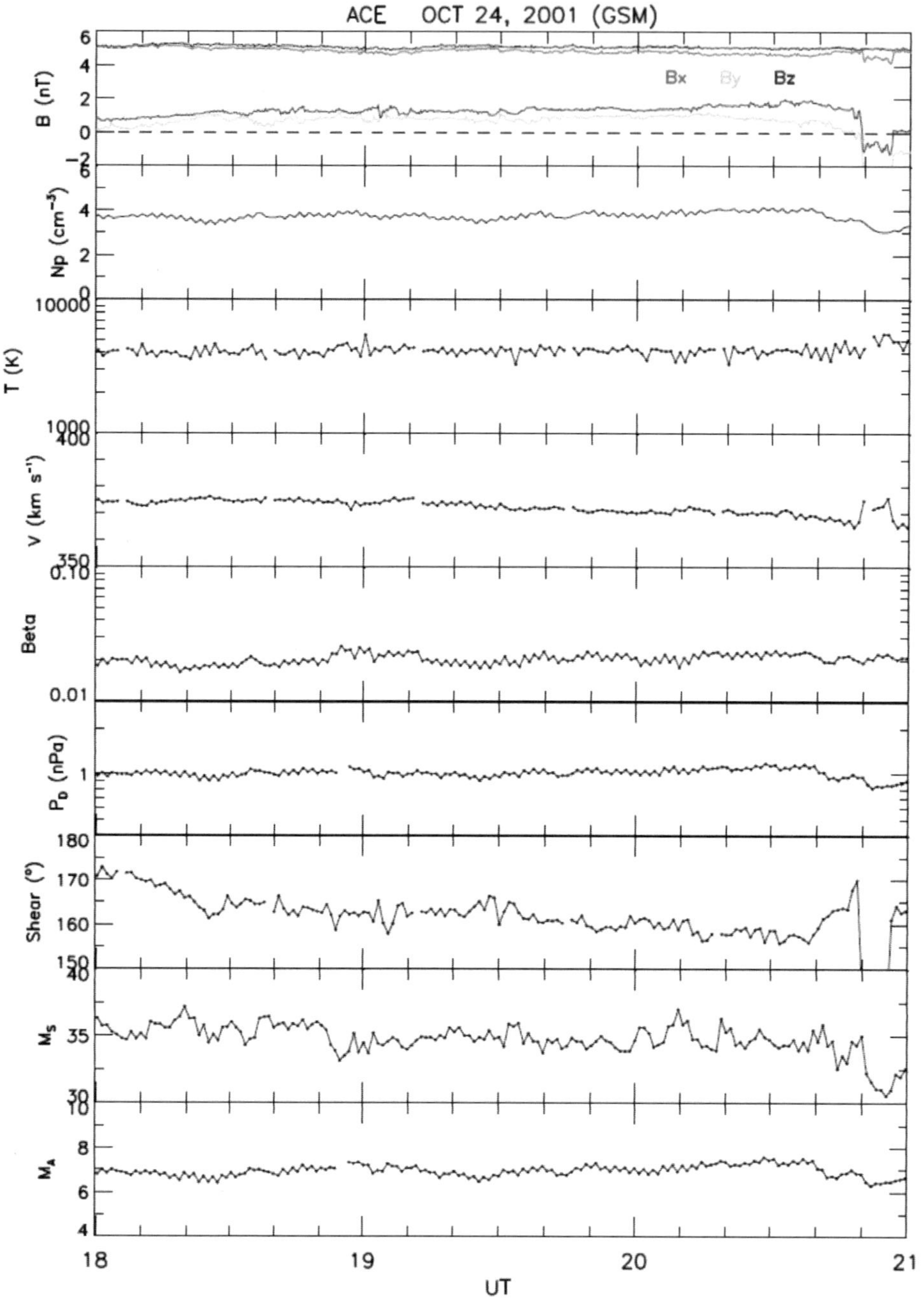


**Figure 11.** ACE interplanetary data from 18 – 21 UT, October 24, 2001, covering the time interval 19 – 20:30 of Wind traversal of the boundary layer. ACE time is shifted by the delay of the solar wind to reach Wind's position (for details see text).

To conclude: the quiet solar wind recorded by ACE has a moderate speed, but given the prevailing magnetic field strength, density, and very low plasma temperature, it flows steadily at an unusually high sonic Mach number, $M_s \sim 34$, and a rather large Alfvenic Mach number, $M_A \sim 6.5$ (last two rows). The Mach numbers $M_s$, $M_A$ of the solar wind are connected to the corresponding magnetosheath Mach numbers by a MHD theory of collinear flows (subsection 4.3).

We come now to Wind's observations. The orbit (see a plot in [82]) is easily described as follows. At 19 UT Wind enters the BL from the magnetospheric side at (-14.1, -22.8, 6.1) $R_E$ (GSM coordinates) and at (-13.7, -25.0, 5.3) $R_E$, at 20:30 UT, exits into the magnetosheath. Wind was moving mainly dawnward at a downtail distance X ~ -14 Re, an orbit's segment roughly normal to the local boundary layer. It is an orbit ideal to observe the dynamics during a complete traverse, and to evaluate the width of the BL.

Crossing the equatorial dawnside plasma sheet/LLBL, Wind observed the quasi-periodic field and flow disturbances shown in figure 12. From top to bottom, the proton density, p/cm$^3$ (with a logarithmic scale), the plasma speed (absolute value) km/s, the proton temperature °K (logarithmic scale), and the magnetic field (absolute value) nT. Then follow the velocity components (km/s), and the magnetic field components (nT), respectively, in GSM coordinates X,Y, Z.

The pulsed character of $N$ and $T$, with the quasi-periodic, alternating hot-tenuous and cold-dense plasma, is already hinting at a motion of organized structures over the spacecraft. A series of 15 perturbations of this type are recorded. The impression of a passage of billows with two classes of organized plasma is accentuated by the well correlated, and periodical, spiky pattern of $B_x$, $B_y$, and also (with somewhat less definition) of $V_x$, $V_y$. Note that while the magnetic record is uninterrupted, the plasma measurements show a few gaps due to instrument problems. The following step is to find a plausible explanation of the BL observations

### *4.2. Stability of collinear magnetic field and flow*

The MP from dayside onward is subject to the impact of waves and fluctuations coming from the magnetosheath. The waves transfer momentum across MP, from the solar wind into the boundary layer. The result is equivalent to a viscous drag of the adjacent magnetospheric plasma (the so called "pseudo-viscosity"). Plasma close to the boundary is dragged along by the solar wind. Therefore, a velocity shear flow is established inside the boundary layer.

On the equatorial dayside up to the terminator, the flow near the inner edge of the boundary layer is practically normal to the geomagnetic field, and it is subsonic (because of the reduced speed in the LLBL, and the higher temperature of the magnetosphere). The physical conditions at the inner edge side of the BL are very different from those of the outer limit, contiguous to the field-aligned flow. On the magnetospheric side of the BL, although the velocity is reduced with respect to MS, the development of the KH is easier when we consider flute modes. By this we intend modes with $\boldsymbol{k}$ normal to the (geomagnetic) field, and therefore parallel to the flow. Thus, we expect that in the October 24, 2001 event the inner part of the boundary layer is the first to become unstable.

The vector and scalar fields chosen for the discussion of the KH instability on the front side of the magnetosphere, where the BL perturbations observed further down-tail start, are of necessity based on ACE observations supplemented by theory, in view of the absence of spacecrafts monitoring the dayside magnetosheath. In the unperturbed model the magnetic field $\boldsymbol{B}$ is assumed collinear with the flow $\boldsymbol{V}$, anti-parallel to it as revealed (approximately) by ACE's observations. As in section 2, the analysis uses a local $x$-axis set in the direction of $\boldsymbol{V}$ (so that only the $V_x$ component is different from zero). The $y$-axis is normal to the MP pointing outward. Completing the triad, the $z$-axis is oriented in the same sense as the geomagnetic north (the $x$-axis direction is opposite to that of the GSM X coordinate). The magnetic field in the adjacent magnetosphere is assumed to be perpendicular to the flow direction, in the sense of the geomagnetic field. Thus, the magnetic shear angle of the model is 90° (clockwise, looking from $+y$).

Presumably, this angle is not exactly 90° and varies with the distance to the subsolar point. We expect that at equatorial latitudes the deviations from 90° are not substantial. Anyway, this choice is not particularly biased in favor of the instability, because we intend to switch-off, or at least reduce, the magnetic tensions on the magnetospheric side of the BL by considering $\boldsymbol{k}$-modes normal to the local geomagnetic field. This $\boldsymbol{k}$ orientation maximizes the $V_k$ contribution to the instability, but exposes the mode to the full stabilizing capacity of the magnetosheath field ($B_k$ component).

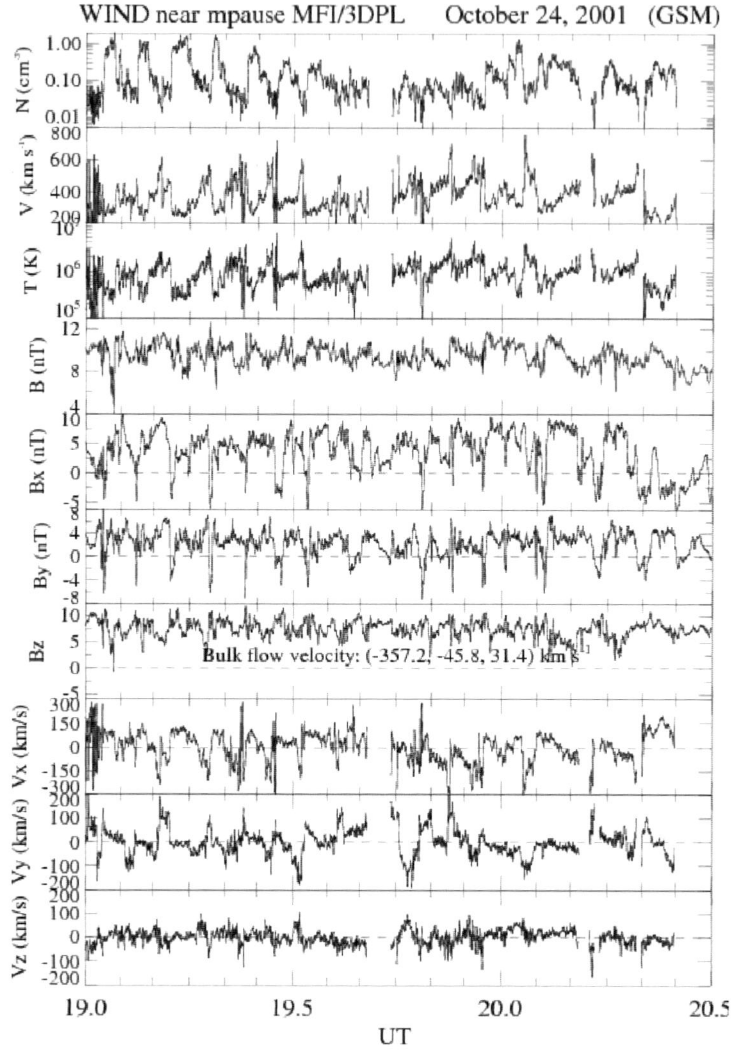


**Figure 12.** General view of Wind's records during the BL crossing of October 24, 2001, from 19.0 to 20.5 UT, decimal hours. From top to bottom, proton density p/cm$^3$ (logarithmic scale), plasma speed (absolute value) km/s, proton temperature °K (logarithmic scale), magnetic field intensity, nT. Then follow the velocity components (km/s), and the magnetic field components (nT), respectively, in GSM coordinates.

At the near-Earth MP we assume a typical particle density ratio $n_2/n_1$=0.1, and magnetic field intensity ratios $B_2/B_1$=1.4-1, with a stronger field in the magnetosphere. The highest ratio is an upper bound enforced by the need to preserve pressure balance across the MP. All quantities are normalized with their MS values, and both Mach numbers, $M_s$, $M_A$ are computed with parameters of that region. Figure 13 illustrates an example of basic functions used in the stability calculation. In the first panel are represented $n$, $V_x$, $B$=|$\boldsymbol{B}$|, with blue, green, and red lines respectively, and in the third panel, $B^2$, $\chi$

with blue and green lines, respectively. Here $\chi$ denotes the magnetic shear angle (the angle of $\boldsymbol{B}$ with the $x$-axis, since field and flow are collinear). Together with $n$ and $B^2$, $\chi$ determines the temperature change across the BL via a pressure balance equation. The second (central) panel of figure 13 shows the projections $V_k$, $B_k$. They are important functions that decide the outcome of the stability study, as we have seen in section 2.

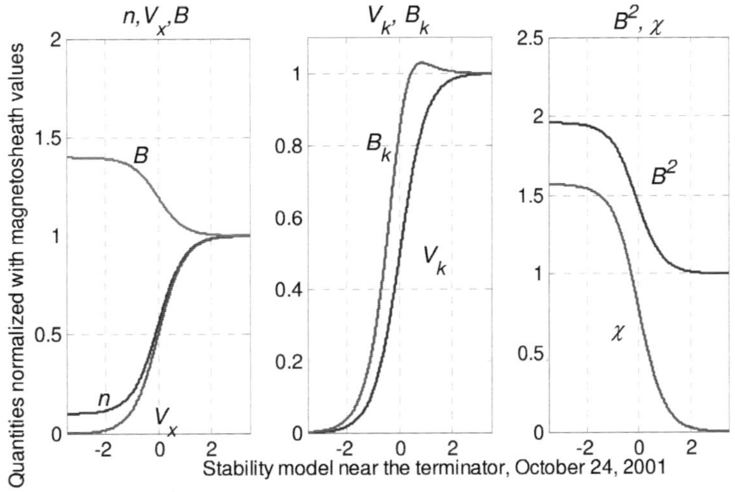


**Figure 13**. An example of the set of functions that appear in the equations of the compressible MHD stability analysis for a BL model. Right panel: $B$, $n$, and $V_x$ as functions of $y/d$, red, blue and green lines, respectively. Center panel: $B_k$, $V_k$ vs. $y/d$, green and blue lines, respectively. Left panel, $B^2$, $\chi$, vs. y/d, blue and green lines, respectively. See text for the meaning and role of these functions.

*4.3. Stability near the terminator.*

For near-Earth scenarios the physical quantities for the MS, and the Mach numbers, are based on the Spreiter and Rizzi 1974 [84] theory, which gives an approximate representation of the transition from solar wind before the bow-shock to the magnetosheath, for collinear MHD flows. In the absence of direct experimental information we use the Spreiter-Rizzi solution with solar wind data from ACE as inputs. Close to Earth at the MP terminator, the theory predicts approximately $M_s$=7.7, $M_A$ =4.9.

We begin with an interesting result of the thin approximation shown in figure 14, computed from equation (16) with these Mach numbers, and other quantities as in subsection 4.2. Keeping other parameters fixed the density ratio is changed. The instability appears only in a restricted range of density ratios, $\rho_2/\rho_1 \sim 0.03 - 0.43$. The equal density case is seen to be stable in agreement with Landau's theory. The growth rate $\gamma$ has a maximum near $\rho_2/\rho_1 \sim 0.1$; $\gamma$ is normalized with $kU_1$ since a thin model has no width. For a rule of thumb comparison with a thick model at the maximum growth rate, one may take roughly $k \sim 1/D$ ($D$: width of the BL).

The fact that a supersonic (and super Alfvénic) TD model may be stable both for small, and large $\rho_2/\rho_1$ values, is in line with results of the case of Wind's data input. As discussed at the end of this section, with a density ratio ~ 0.01 (but somewhat different parameters) the region of Wind's crossing is also found stable in the long wavelength limit.

The analysis for an hyperbolic functions model at the terminator is computed with the field and flow properties of subsection 4.2 and the previous Mach numbers. We assume a magnetic field ratio $r_B \equiv B_2/B_1 = 1$, which is sufficiently representative ($\gamma$ is not very sensitive to $r_B$ changes in the range 1-1.4). The magnetic tensions, whose influence decrease with the square of $M_A$, are reduced in their importance with respect to subsolar regions with small velocities near the stagnation point.

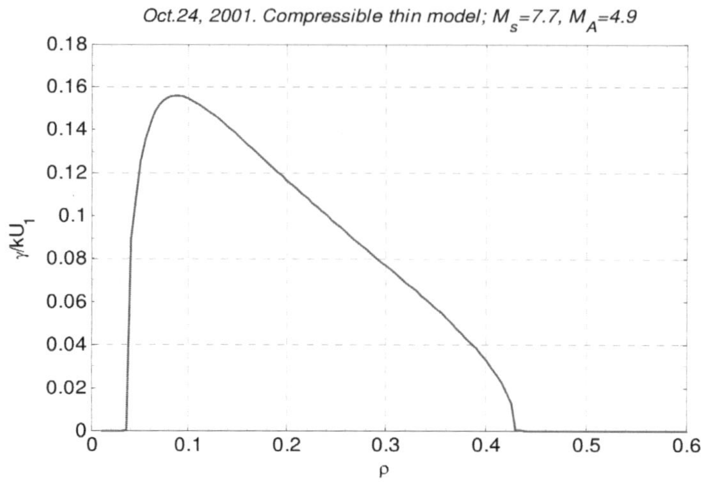


**Figure 14.** Growth rate γ (normalized with $kU_1$) versus variations of density ratio $\rho=\rho_2/\rho_1$. Thin model results with inputs from Spreiter-Rizzi theory applied to the magnetosheath near the MP terminator.

The result is given in two plots. Figure 15(a) presents the (normalized) growth rate, $\gamma d/U_1$ as a function of *kd* (modeled with hyperbolic functions with scale length *d*), while figure 15(b) shows $\Im(c)=\gamma/kU_1$, the imaginary part of the characteristic value *c*. An interesting feature appears in this range of parameters. The imaginary part of the KH dispersion relation, $\Im$(c), is a linear function of *kd* in most of the interval (for $kd\geq0.15$). The growth rate is therefore a quadratic function *k*. For $kd$= 0.245 the (normalized) growth rate attains a maximum, $\gamma d/U_1$=0.0195. From this value we may estimate the e-folding time $\tau=1/\gamma\sim137$ s for sites near the terminator. We use $D=0.5R_E$, assuming that the BL has not been yet broadened by the instability, and $U_1$=300 km/s. Hence, in a boundary layer not yet expanded by KH vortices, the growth of the perturbation proceeds at a fair pace.

The *kd* value for maximum γ is smaller than that of the cases of subsection 3.4, and leads to $\lambda\sim6.41D$ for the KH waves of fastest growth. Given the long spatial size of KH perturbations over the MP, where wavelengths about tenfold times $R_E$ have been reported [23], it is opportune to note that a stretching effect acts on the wavelengths. This is due to the speed increment from the dayside to the flanks. The effect is already present in the linear phase of the instability. For instance, starting near the subsolar point with a ***k*** vector pointing east, the wave front is convected downtail faster than the slow moving rear of the perturbation [85]. This generates an effective increment of the wavelength over the large extensions of the magnetosphere flanks.

### *4.4. Stability at the position of Wind's orbit*

In the following case the model is prepared with data recorded by Wind along the orbit across the BL. The scenario is built from average magnetospheric measurements, before but near the entrance to the LLBL, including some initial part of the passage through it. On the other hand, we use averages of data recorded later, when Wind is in the magnetosheath.

The LLBL is already perturbed by the passage of the vortical structures, formed at some place near Earth. It could be argued that a local stability test is insecure because one inputs quantities that are already modified by the instability. Obviously, the remark does not apply to stable MP regions. When the site is ostensibly perturbed, (a) one should use only asymptotic values on either side, and leave the width of the transition as a free parameter of the study, and (b) with averages of perturbed quantities across the BL as input, one can still estimate whether the background posses amplifying power. The latter point means that in the case of intermittence, the KH activity may resume spontaneously.

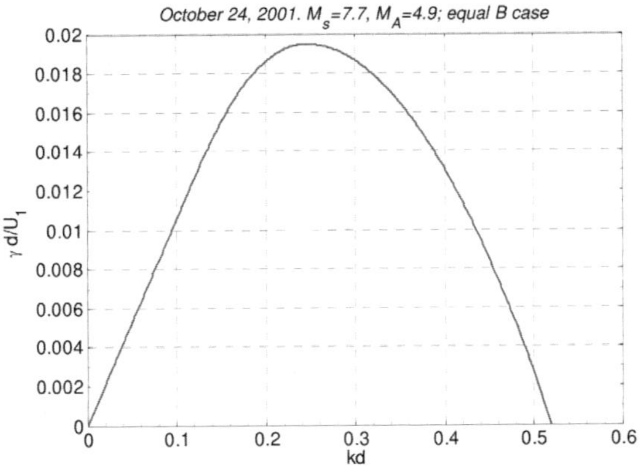


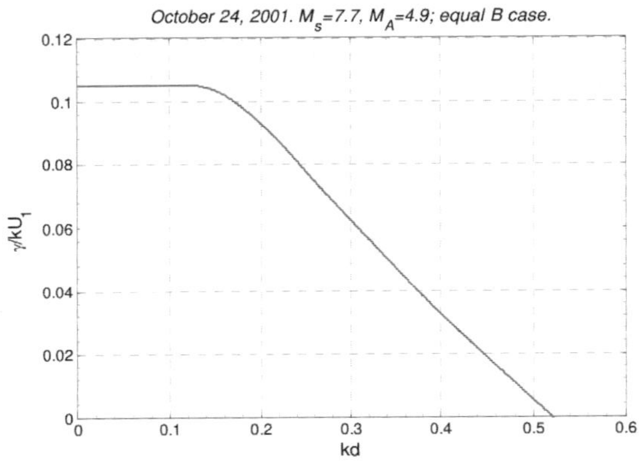


**Figure 15.** (a) (left) Normalized growth rate as a function of *kd* for the hyperbolic functions BL model at the terminator (see text). (b) (rigth) Imaginary part of the phase velocity *c* versus *kd*, for the same case of figure 15(a). $\Im(c)$ is a (decreasing) linear function of *k*, for *kd*>0.15.

The averages include long stretches of time on either sides of the LLBL, because the instability is found for long wavelenghts, of several $R_E$. At Wind's position, with a magnetosheath average velocity of 300 km/s, the Mach numbers are $M_s$=5.6 and $M_A$=6.8. The $c_s$ that enters in the sonic Mach number is the plasma ion sound speed $c_s=(\Gamma k_B T_e/m_i)^{1/2}$, defined with the electron temperature and the ion mass. We assume $T_e=T_i$, (because $T_e$ is not available in the data) and the specific heaths ratio is $\Gamma=5/3$. Unlike the terminator example, the magnetic shear is not 90° but rather 71° with respect to the sunward direction (derived from the magnetospheric side ***B*** average). Observations confirm that ***B*** is still approximately collinear with the magnetosheath flow. To sum up, the input parameters for the stability analysis are as follows. Magnetosheath: $V_{1x}$=300 km/s, $n_1$=5 p/cm³, $T_1=2.5\times10^5$ °K, $B_1$=4.5 nT, $\chi_1=0°$, $M_s$ =5.6 , $M_A$ =6.8. Magnetosphere: $V_{2x}$=0 km/s, $n_2=5\times10^{-2}$ p/cm³, $T_2=2.0\times10^6$ °K, $B_2$=9.5 nT, $\chi_2=129°$ (the values comply approximately with pressure balance).

Note that the average magnetosheath speed adjacent to the BL is smaller than the ACE records ~370 km/s, and smaller than the bulk speed inside the BL (again close to 370 km/s) where vortical processes are going on. In the magnetosheath the gas presssure and the magnetic pressure are of the same order, but on the magnetosphere side the pressure is almost purely magnetic, the gas pressure being very small.

Figure 16 illustrates the results for this case. The plot at 16(a) shows the normalized growth rate $\gamma d/U_1$, as a function of *kd*. This is a case of stability at long wavelengths, i.e., $kd \ll 1$. For $\rho_2/\rho_1=10^{-2}$ , the growth rate is zero when $0 < kd < 0.07$. It has a maximum $\gamma d/U_1=0.033$ at $kd=0.43$ and goes to zero at $kd\approx0.07$ (long λ) and $kd=0.84$ (short λ). Figure 16(b) displays the behaviour of the imaginary part $\Im(c)$ as a function of *kd*.

Figure 17 presents the results of a thin model with the present local parameters. The growth rate (normalized "ad hoc" with *kd*~0.3) and $\Im(c)$ are plotted as functions of the change of both Mach numbers keeping fixed their ratio $M_s/M_A$ =1.21 (for instance, as the consequence of changes of $U_1$). The thin model is stable for $M_s$ =5.6, in agreement with the continuous model, and we note that the long wavelength approximation becomes unstable only when $M_s=1.21M_A\geq7.7$.

This analysis intends to show that a BL model, endowed with quantities represented by mean values of magnetospheric, and magnetosheath data, is unstable. The result is an indication that at Wind's orbit position, the boundary retains the property to augment fluctuations. If the passage of vortices is temporarily suspended, the background has the capability to rise perturbations, and to roll-over the layer again.

To conclude, the theory shows that the long series of large amplitude perturbations observed by Wind is compatible with a KH origin. The instability is possible in spite of the adverse configuration of field and flow. The outcome clears the way for the interpretation of data in terms of KH vortices.

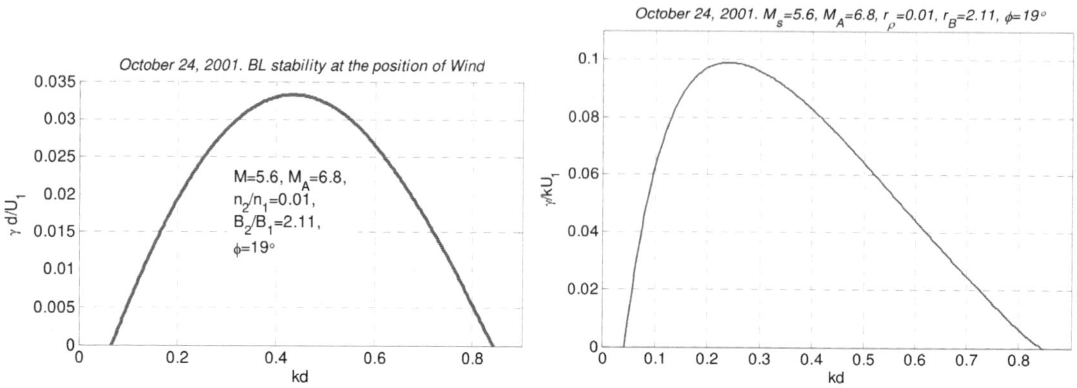


**Figure 16.** (a) (left) Normalized growth rate. (b) (right) Imaginary part of the phase velocity (normalized). Data inputs from Wind´s crossing.

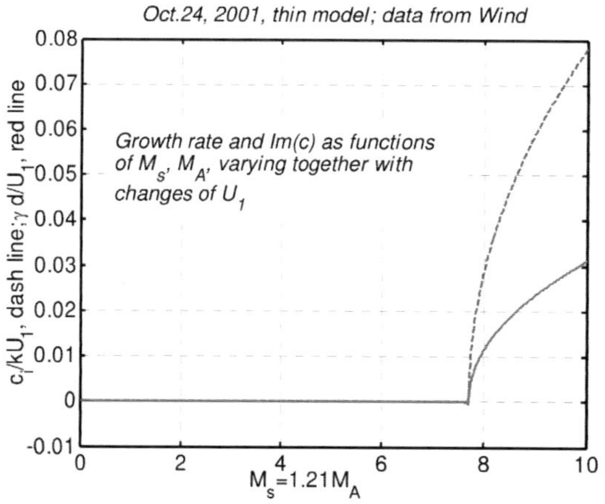


**Figure 17.** Thin model results with data from Wind. BL stability with Wind's input as a function of changes of Mach numbers (see text). The TD model becomes unstable only when $M_s$=1.21 $M_A \geq 7.7$.

## 5. Experimental evidence of vortices

### *5.1. Data record details and interpretive key.*

The data plotted in figure 18, with physical details of the third structure observed by Wind since the entrance in the BL, is representative of many other members of the series. The time interval (19.20 – 19.31 UT, approximately) includes a characteristic episode of cold-dense and hot-tenuous plasma alternation. The velocity components $V_x$, $V_y$, are averaged over the interval and the deviations $dV_x$=<$V_x$-<$V_x$>>, $dV_y$=<$V_y$-<$V_y$>> are plotted (km/s) with dashed blue, and dashed red lines respectively (the vertical scale is in km/s). The magnetic field components $B_x$, $B_y$ are the solid green, and solid magenta lines, respectively (the magnetic field scale is enlarged 20 times, 20 units = 1 nT ). In the same figure a dotted black line represents the number of particle density (scale enlarged 100 times, 100 units= 1 p/cm$^3$). The latter is added to visualize the phases of dense and tenuous plasma.

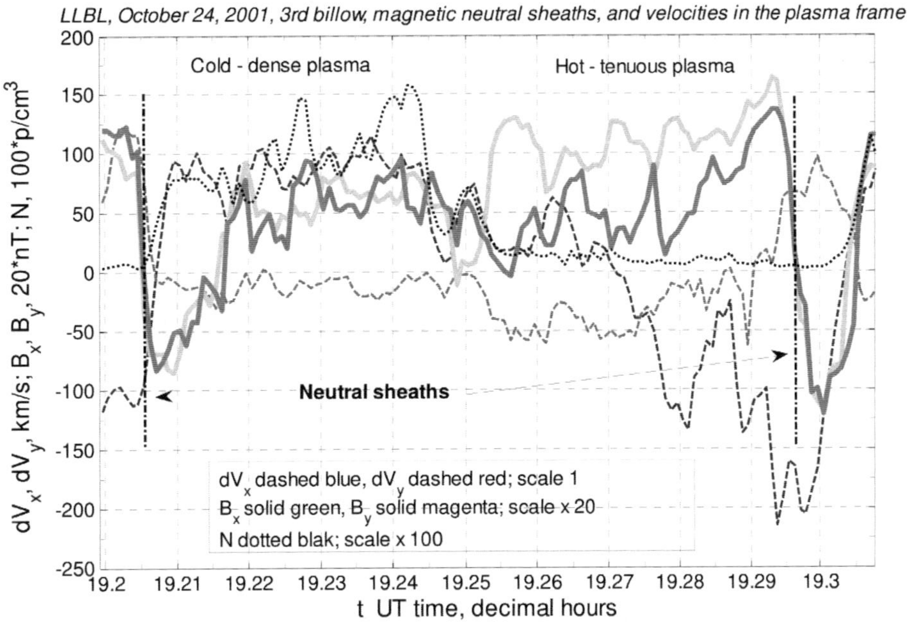


**Figure 18.** Details of the physical state of the third vortex observed by Wind. The presence of current layers is emphasized with black dashed lines. These are neutral sheaths that separate field components with opposite signs.

The dense plasma lags behind, in the average velocity frame it appears to move sunward. Conversely, the tenuous plasma moves in advance of the average, and proceeds fast tailward. The deviation $dV_x$ attains important positive (+150 km/s), and negative (-200 km/s) values. Note that the average $<V_x> \approx$ -360 km/s is already moving faster than the adjacent magnetosheath flow average (-300 km/s), close to the solar wind speed before the bow-shock. The plasma structures are subject to an acceleration process operating in the BL.

The other important point of figure 18 is the presence of current sheaths, thin neutral layers across which the magnetic field components change sign (indicated by dashed vertical lines). The current sheaths appear in the transitions from hot-tenuous to cold-dense plasmas, they are absent in the reverse passage. The $B_x$-$B_y$ components are fairly well correlated, $B_x$ is mostly positive, in agreement with the collinear field of the magnetosheath, $B_y$ generally points duskward (to the magnetosphere). Exceptions occur at hot-tenuous to cold-dense passages, where the spiky changes of magnetic sign take place. The current sheaths are signatures of a strain process of the collinear magnetic field lines, which are frozen-in the plasma and are being convected by the structures.

Figure 19 helps to visualize better the change of directions of the *x-y* magnetic and velocity components for the same third billow observed by Wind. We show arrow plots for the magnetic field $B_x$-$B_y$ in the upper panel, and for the velocity deviations $dV_x$-$dV_y$ ,in the lower part. The rotational pattern of the *x-y* velocity components at the transition from hot-tenuous to cold-dense plasma is apparent. It occurs at the positions where the *x-y* magnetic field components change sign, and the current sheaths are formed. The dashed lines in the bottom panel mark the separation of plasma in two classes by density and temperature.

### *5.2. Consequences of scatter plot analysis*

Clearly, rolled-up vortices are a central concern in magnetopause physics. We have an MP crossing by a single spacecraft. What then, is it possible vortex detection by a single spacecraft? A peculiar signature (Takagi et al., 2006 [86]) allows rolled-up vortices to still be identified. Their simulations show that the tailward speed of a fraction of low-density magnetospheric plasma exceeds that of the magnetosheath flow. This feature appears only after a vortex is rolled up. We apply this rule to the special field and flow configuration observed on October 24, 2001.

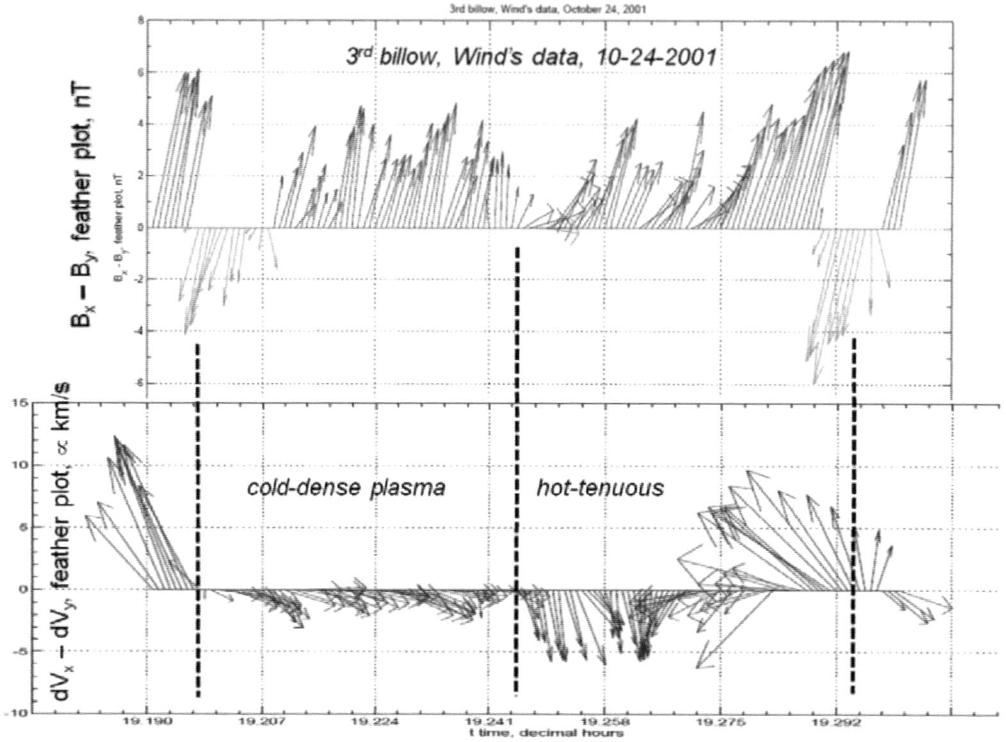


**Figure 19.** Arrow plots for the magnetic field $B_x$-$B_y$ and the velocity $dV_x$-$dV_y$ components, corresponding to the third vortex observed by Wind. The rotational pattern of the *x-y* velocity components at the transition from hot-tenuous and cold-dense plasma is apparent. It occurs at the same position where the *x-y* magnetic field components change sign, and the current sheaths are formed.

The vortex hypothesis accepted, qualitative pictures like figure 20 can be used as interpretive data tools. Figure 20 shows a vortex turn-over example, adapted from a MHD simulation. The cold-dense matter (in violet colors) of the upper layers (moving to the right) rolls back with a dense tongue, which embraces a hot-tenuous protuberance (in blue colors) that emerges from the bottom. The black dashed line indicates the (apparent) spacecraft motion relative to the vortex. The alternate cold-dense and hot-tenuous series makes sense immediately as the passage of a row of vortices that stir magnetosheath with magnetosphere plasmas. Another consequence is that the collinear field convected by the plasma motion tend to be coiled-up in the vortices.

Statistically more pregnant is the analysis of the whole set of LLBL data, represented in figure 21. We present a scatter-plot of $dV_x$ (km/s) versus $n$ p/cm$^3$ with a logarithmic scale ($dV_x$ is the deviation of $V_x$ from the mean value). To improve the visibility of the temperature trend, both the marker's size and the color are associated with $T$. The size is $\propto T$, large size for hot, and small size for cold matter. The color is $\propto \ln(T)$ (logarithmic scale is used because of the great difference of values), red for high, and blue for low temperatures.

Clearly, a significant fraction of the low-density, high-temperature plasma is flowing (tailward) at speeds well in excess of the bulk average of the BL, and much faster than the contiguous magnetosheath speed. In agreement with the rule of [86] we have a sequence of rolled-up vortices passing by the spacecraft. We can also note that in the average frame, an important population of cold-dense plasma is moving sunward at considerable speeds. It is evident that a fraction of cold-dense magnetosheath plasma is turning back, and a majority of magnetospheric plasma is being accelerated toward the tail at high speeds. Even without the support of a simulation, the data alone give compelling evidence of a rotational motion superposed to a bulk translation.

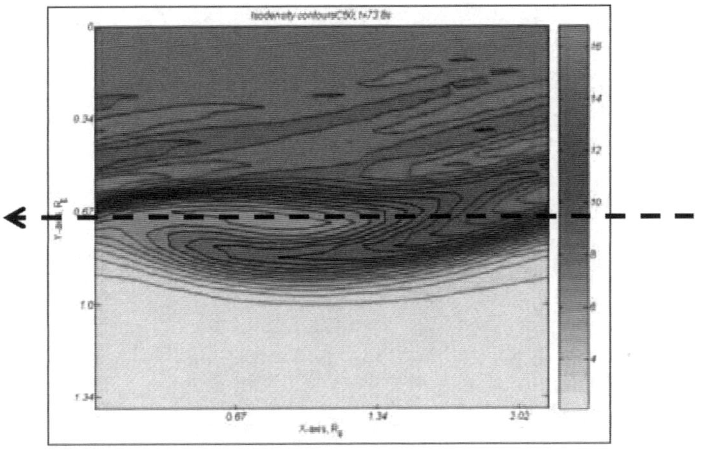


**Figure 20.** A vortex turn-over-over example (from a MHD simulation). Cold-dense matter (violet colors) from the upper layers rolls back with a dense tongue that embraces a hot-tenuous protuberance (blue colors) originated in the bottom layers. The black dashed line indicates the (apparent) spacecraft motion relative to the vortex. The cartoon helps as interpretive data tool.

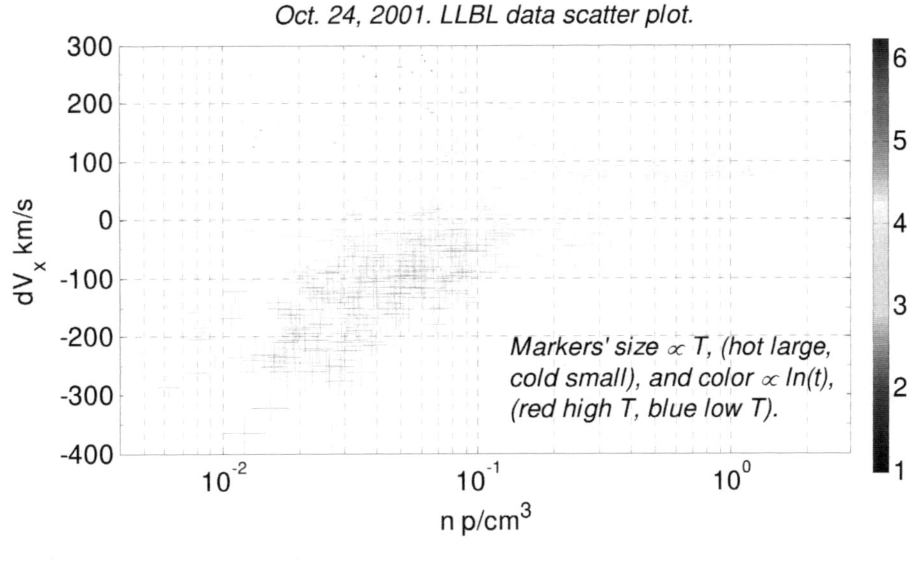


**Figure 21.** Scatter plot of $dV_x$ km/s versus n p/cm3 for the whole LLBL data set. $dV_x$ is the deviation of $V_x$ from the mean value. Both, marker's size and color are associated with T, to improve the visibility of the temperature trend. The size is $\propto T$, large for hot, and small for cold matter. The color is $\propto \ln(T)$ (logarithmic scale because of the great difference of values), red for high, and blue for low $T$.

The statistics is complemented by the scatter-plot of figure 22. This is a correlation of $dV_y$ versus $dV_x$ also with the whole BL data set. To add information the marker's size is proportional to the number of particle density (p/cm$^3$) and the color (blue to violet scale) is again proportional to *ln(T)*. At first sight for every $dV_x$ bin there is approximately as much plasma moving to dawn (toward the magnetosheath) as moving to dusk (to the magnetosphere). At a closer look we may notice a slight bias. Hot-tenuous matter tends to move out, toward the magnetosheath, conversely cold-dense matter tends to move in, toward the magnetosheath. This trend is lesser in extent, more significant is that figure 21 suggests that Wind encounters the vortices at a stage of at least one roll-over, because the two classes of plasma move both, outward and inward, in almost equal proportion.

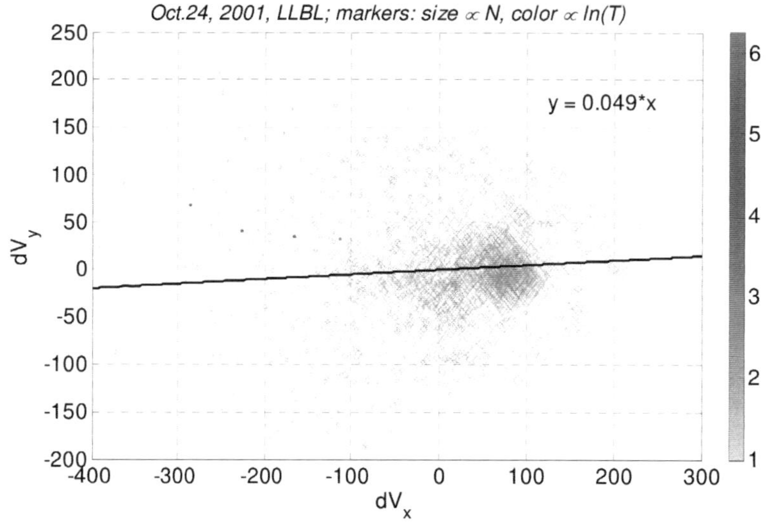


**Figure 22**. Scatter plot $dV_y$ versus $dV_x$ (in km/s) for the whole LLBL data set. The markers size is $\propto N$ (here denotes particle density) and the color is$\propto$ ln($T$) (like in figure 21). The spread of positive – negative $dV_y$ values is equiprobable at a first approximation, which can be interpreted that the vortices have done at least one turn-over. But a small bias exists, hot plasma has a slight tendency to move dawnward (toward the MS), while cold plasma shows a small preference to move toward dusk (inside the magnetosphere).

A more detailed statistical study is possible by grouping the data of the different vortices. In the last figure 23, the scatter-plots corresponds to the first five vortices observed by Wind, starting at 19 UT. Five different markers (+, ×, *o*, etc..) and colors (see the inset in the plot) identify the vortices by order of appearance in the observations. The figure format is a multivariate group representation in the fashion of a matrix, with 2×3 particular scatter-plots, in which the identity of every vortex can be distinguished. Some correlations are well defined in all five vortices, like those of $dV_x$-$T$, and $dB_y$-$dB_x$. The correlation $dV_x$ - $N$ is also fairly good, with the exception of the first vortex that shows more spread. In some other plots there is more disparity of behavior between vortices, like in the $dB_y$ - $N$, and $dV_x$ - $dB_x$ correlations.

The scatter plot analysis gives decisive evidence: the observations correspond to a series of vortices. Unlike many other investigations on KH at the MP we have quoted, the case investigated is for an IMF directed mainly along +*X*. We suggest that the inner edge of the LLBL, where the magnetic shear is small and ***k*** perpendicular to the geomagnetic field is a right choice, become KH unstable.

Using compressible MHD theory, we have shown in section 4 that the inner edge is indeed KH unstable. Section 5 consolidated the KH vortex thesis from the experimental side. The event studied gives important clues for the evaluation of the KH stability and vortex formation of the more common Parker's spiral fields and flows.

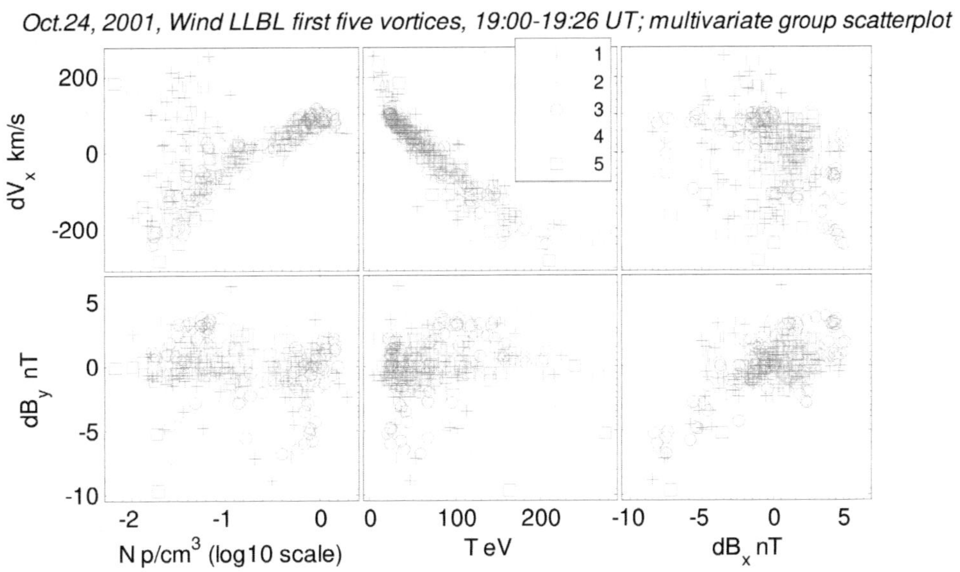


**Figure 23.** A multivariate scatter plot for the first five vortices observed by Wind. The markers indicate the vortices by order of appearance (see the inset). The scatter plots of pairs of quantities are shown in a matrix format. Several correlations are evident in all five vortices, like those in the scatter plots $dV_x - N$, $dV_x - T$, $dB_y - dB_x$. In other plots a different behavior between vortices is noted, as in $dB_y - N$, and $dV_x$ - $dB_x$.

## 6. Conclusions

In the first part of this paper, the pervasive action of the KH mechanism in the solar system with a series of recent examples is described. The importance of the phenomenon for the physics of the Earth magnetosphere is emphasized, and basic concepts with equations of the MHD theory of the KH instability are outlined.

Possible objections on the limits of MHD as a plasma model lose relevance in comparison with the explanatory power of data interpretation in space plasma dynamics. The use of more sophisticated plasma theories in the context of data analysis is often an unjustified effort. It must be bear in mind that the validation of theoretical, or computational models, in space physics is inevitably indirect and limited (in contrast with the circumstances of terrestrial fluid dynamics). It is a consequence of the coherence of interpretations in a variety of observations (often qualitative rather than quantitative) and the compatibility with other theoretical models.

We then describe a set of research topics of magnetopause stability that show promise for further investigations. The instability band theory of dayside MP would benefit from further experimental work with Cluster and Themis dayside crossings under steady northward IMF conditions, and should be revisited with the help of present day global simulations. The peculiar features of pristine MP stability can be tested with data from events of sudden expansion of the magnetosphere under a drop of solar wind dynamic pressure, especially during slow solar wind periods. Magnetopause accelerations can help the start of the KH instability on the front side. The subject of the competition of KH and tearing, and its consequences, is a large area of research that in spite of important advances recorded in the literature, is still open in the directions we point out. The KH instability of the MP flanks, with the supersonic dynamics aspects on which we expatiate, and a significant case to which sections 4 – 5 are devoted, is also a rich field for further experimental and theoretical work.

In the second and last part of the paper we report a study of rolled-up vortices generated by the KH instability in the near-tail dawn flank when the IMF is pointing nearly in a radial direction (sunward). Theory models of MP at the terminator, and further downtail at Wind orbit, indicate that on October

24, 2001 the BL is Kelvin-Helmholtz unstable. This happens in spite of (*i*) the large magnetic shear across the boundary and (*ii*) the large supersonic speed (both stabilizing agents). This is possible because the Alfvén Mach number at both locales is large enough (i.e., weak magnetic tension forces), and the instability starts inside the boundary layer.

During the rise and development of the billows of this event, the collinear field is engaged as participant. Then one must face one of the two horns of a dilemma. Either the radial field component is convected and rolled-up in the vortex, or the frozen-in condition ($\boldsymbol{E}+\boldsymbol{V}\times\boldsymbol{B} = 0$) breaks-down, and does not apply (or only partially). In both cases there are important consequences. In the first case there is coiling of field lines in the vortex with formation of internal current sheaths, which have been observed by Wind (as emphasized in section 5) that are prone to tearing and mass diffusion. In the second case one accepts from the start that the boundary is not a perfect shield and is porous to external mass entry.

This event differs from several other studies that reported BL vortices under strongly northward IMF orientations. This is a case of KH vortices observed under an almost radial IMF, with implicit significance for the more common Parker's spiral fields, and the problem of plasma entry in the magnetosphere.

**Acknowledgements**
This research was supported by CONICET grant 11220090100608 PIP 2010 - 2012 ("*La Magnetopausa en Campo Magnético Interplanetario Norte: Análisis de Datos, Teoría y Simulación de Grandes Vórtices*"). The work at University of New Hampshire was supported by NASA grant NNX08AD11G.

**References**

[1] Tritton D J 1988 *Physical Fluid Dynamics* (Oxford: Clarendon Press) NASA/JPL/Malin Space Science Systems site
[2] Landsat 7 WRS Path 6 Row 83, center: -33.18, -79.99.; Karman vortices, image taken 9/15/1999
[3] *Monthly Weather Review* 2006 **134** 1036
[4] *Geophysical Research Letters* 2011 Feb. 6
[5] Source http://photojournal.jpl.nasa.gov/catalog/PIA06502
[6] Masters A et al. 2010 *J. Geophys. Res.* **115** A07225 (doi:10.1029/2010JA015351)
[7] Foullon C et al. 2011 *Astrophys. J. Lett.* **729** L8; SDO/AIA, Ofman L and Thompson B J 2011 *arXiv*:1101.4249v2 [astro-ph.SR] Jan 25. NASA, SDO
[8] ed C T Russel and M G Kivelson 1995 *Introduction to Space Physics* (New York: Cambridge University Press)
[9] Sundberg T et al. 2011 *Planet. Space Sci.* doi:10.1016 /j.pss. 2011.05.008.
[10] Bochsler P et al. 2010 *Ann. Geophys.* **28** 491-7
[11] Dungey J W 1954 Pennsylvania State University, Ionosphere Research Laboratory, *Scientific Report 69*
[12] Dungey J W 1961 *Phys. Res. Lett.* **6** 47
[13] Axford W I and Hines C O 1961 *Can. J. Phys.* **39** 1433
[14] Southwood D J 1968 *Planet. Space Sci.* **16** 587
[15] Sonnerup B U O 1980 *J.Geophys. Res.* 85 2017
[16] Hones et al. 1981 *J.Geophys. Res.* **86** 814-20
[17] Ogilvie K W and R J Fitzenreiter 1989 *J. Geophys. Res.* **94** 15113
[18] Belmont G and Chanteur G 1989 *Physica Scriptae* **124** 124
[19] Miura A 1995 in *Physics of the Magnetopause*, Geophys. Monogr. Ser. **90**, 285 ed Song et al. (Washington DC: AGU)
[20] Fitzenreiter and Ogilvie 1995 in *Physics of the Magnetopause*, Geophys. Monogr. Ser. **90**, 277 ed Song et al. (Washington DC: AGU)

[21] Kivelson M G and Chen S H 1995 in *Physics of the Magnetopause*, Geophys. Monogr. Ser. **90** 257 eds Song et al. (Washington DC: AGU)
[22] Farrugia C J, Gratton F T and Torbert R B 2001 *Space Sci. Rev.* **95** 443
[23] Lepping R P and L F Burlaga 1979 *J. Geophys. Res.* **84** 7099
[24] Seon J et al 1995 *J. Geophys. Res.* **100** 907
[25] Samson J C and Rostoker G 1972 *J. Geophys. Res.* **77** 6133
[26] Kivelson M G and D J Southwood 1986 *J. Geophys. Res.* **91** 4345
[27] Miura A 1984 *J. Geophys. Res.* **89** 801
[28] Fairfield D H et al. 2000 *J. Geophys. Res.* **105** 21159
[29] Nykyri K and Otto A 2001 *Geophys. Res. Lett.* **28** 3565; Otto A and Nykyri K 2003 *Geophys. Monogr.* **133** 53 ed P T Newell and T Onsager (Washington DC: AGU)
[30] Smets R, Delcourt D, Chanteur G and Moore T E 2002 *Ann. Geophysicae* **20** 757
[31] Hasegawa H, Fujimoto M, Phan T et al. 2004 *Nature* **430** 755; Hasegawa H, Retinò A, Vaivads A et al 2009 *J. Geophys. Res.* **114** A12207 doi: 10.1029/2009JA014042
[32] Gratton F T et al. 2006 *Am. Inst. of Physics* **C P 875** Plasma and Fusion Science pp 300-3 ed. Herrera Vázquez J J E Washington DC; Gratton F T et al. 2009 *J. of Physics CS* **166** 012023 doi:10.1088/1742-6596/166/1/012023; Bilbao L 2009 *J. of Physics CS* **166** 012020 doi:10.1088/1742-6596/166/1/012020; Gratton F T et al. 2011 *J. of Physics CS* **296** 012006 doi: 10.1088/1742-6596/296/1/012006
[33] Nakamura T K M et al. 2006 *Geophys. Res. Lett.* **33** L14106 doi:10.1029/2006GL026318; Nakamura T K M et al. 2008 *J. Geophys. Res.* **113** A09204 doi:10.1029/2007JA012803
[34] Hwang K-J et al. 2011 *J. Geophys. Res.* **116** A08210 doi:10.1029/2011JA016596
[35] Fujimoto M and Terasawa T 1994 *J. Geophys. Res.* **99** 8601
[36] Otto A and Fairfield D H 2000 *J. Geophys. Res.* **105** 21175
[37] Sandholt P E, Farrugia C J, Cowley S W H *et al.* 1999 *Geophys. Res. Lett.* **26** 2833
[38] Onsager T G et al. 2001 *J. Geophys. Res.* **106** 25467
[39] Øieroset M et al. 2005 *Geophys. Res. Lett.* **32** L12S07 (doi:10.1029/2004GL021523)
[40] Gosling J T et al. 1991 *J. Geophys. Res.* **96** 14097
[41] La Belle-Hamer A L et al. 1995 *J. Geophys. Res.* **100** 11875
[42] Terasawa T *et al.* 1997 *Geophys. Res. Lett.* **24** 935
[43] Fujimoto M et al. 1998 *J. Geophys. Res.* **103** 2297; Fujimoto M et al 1998 *J. Geophys. Res.* **103** 4391
[44] Chen Q, Otto A and Lee L C 1997 *J.Geophys. Res.* **102** 151-61
[45] Fuselier S A, Petrinec S M and Trattner K J 2000 *Geophys. Res. Lett.* **27** 473
[46] Chandrasekhar S 1981 *Hydrodynamic and Hydromagnetic Stability* (Dover: Mineola); Batchelor G K 1970 *An Introduction to Fluid Dynamics* (Cambridge University Press)
[47] Gratton J and Gratton F T 1971 *Plasma Physics* **13** 567-85
[48] Kadomtsev B B 1983 *Plasma Physics* (Moscow: MIR); Pierce J R 1974 *Almost All About Waves* (Boston: MIT Press)
[49] Duhau S, Gratton F T and Gratton J 1971 *Phys. Fluids* **14** 2067-8
[50] González A G and Gratton J 1994 *J. Plasma Phys.* **51** 43; González A G and Gratton J 1994 *J. Plasma Phys.* **52** 233
[51] González A G et al. 2002 *Braz. J. Phys.* **32** 945
[52] Gratton F T et al. 2004 *Braz. J. Phys.* **34** 1804
[53] Gratton F T et al. 2004 *J. Geophys. Res.* **109** A04211 (doi:10.1029/2003JA010146)
[54] Farrugia C J et al. 1998 *J. Geophys. Res.* **103** 6703; Farrugia C J et al 1998 *Polar Cap Boundary Phenomena* ed J Moen et al (Kluwer Acad. Publ.)
[55] Gratton F T, Gratton J and González A G 1988 *Plasma Phys. Controlled Fusion* **30** 435
[56] Gnavi G et al. 2009 *J. of Physics C. S.* **166** 012022 (doi:10.1088/1742-6596/166/1/012022)
[57] Landau L D 1944 *Dokl. Akad. Nauk. SSSR* **44** 151
[58] Farrugia C J et al. 2003 *Adv. Space Res.* **31** 1105

[59] Southwood D J 1974 *Planet.Space Sci.* **22** 483; Kivelson M G and Southwood D J 1986 *J.Geophys.Res.* **91** 4345-51
[60] Contin J E, Gratton F T and Farrugia C J 2003 *J Geophys. Res.* **108** 1227
[61] Eastman T E, Fuselier S A and Gosling J T 1996 *J. Geophys. Res.* **101** 49
[62] Gratton F T, Farrugia C J and Cowley S W 1996 *J. Geophys. Res.* **101** 4929
[63] Gratton F T et al. 2003 *Planetary and Space Science* **51** 769-83
[64] Furth H P, J Killeen and M N Rosenbluth 1963 *Phys. of Fluids* **6** 495
[65] Ofman L 1991 *Phys. Fluids* **B3** 1364; Ofman L 1992 *Phys. Fluids* **B4** 2751; Einaudi G 1999 *Plasma Phys.Control. Fusion* **41** A293
[66] Fuselier S A, Elphic R C and Gosling J T 1999 *J. Geophys. Res.* **104** 4515
[67] Farrugia C J et al 2000 *J. Geophys. Res.* **105** 7639
[68] Gosling et al. 1991 *J. Geophys. Res.* **96** 14097
[69] Phan T D, Lin R P, Fuselier S A and Fujimoto M 2000 *J. Geophys. Res.* **105** 5497
[70] Øieroset et al. 2002 *Earth's Low-Latitude Boundary Layer* **133** ed P T Newell and T G Onsager (Washington DC: AGU) p 253; Øieroset M et al, 2008 *Geophys. Res. Lett.* **35** L17811 (doi:10.1029/2008GL033661)
[71] Li W et al. 2005 *Geophys. Res. Lett.* **32** L12S08 (doi:10.1029/2004GL021524)
[72] Bender L *Procesos disipativos rápidos en magnetohidrodinámica* 1997 Physics PhD Thesis, Facultad de Ciencias Exactas y Naturales, Universidad de Buenos Aires, Biblioteca FCEyN: Tesis n 2957
[73] Bender L et al. 1998 *Anales AFA* **8** ed R. Gratton et al. (Tandil: ISSN 0327-358X)
[74] Gratton F T and Krasnopolsky R 1994 *Proc. The Solar Wind-Magnetosphere System* ed H. K. Biernat et al. (Vienna: Austrian Academy of Sciences ISBN 3-7001-2127-X) pp 343-57
[75] Gnavi G, Gratton F T and Farrugia C J 2000 *J. Geophy. Res.* **105** A9 20973-87; Gnavi G, Farrugia C J and Gratton F T 2000 *Proc. ICPP 2000* vol 3 (Quebec) p 1060; Gnavi G, Farrugia C J and Gratton F T 2004 *Braz. J. Phys.* **34** 1797-803
[76] Farrugia C J et al. 2004 *J. Geophys. Res.* **109** 1-16; Farrugia C J et al. 2005 *Annales Geophysicae* **23** 1317-33
[77] Weiss N O 1966 *Proc. R. Soc. Lond.* A **293** 310-28 (doi:10.1098/rspa.1966.0173)
[78] Farrugia C J et al. 2008 *J. Geophys. Res.* **113** A03208 (doi:10.1029/2007JA012800)
[79] Farrugia C J and Gratton F T 2009 *J. Atmos. Solar-Terr. Phys* (doi:10.1016/j.jastp.2009.10.008)
[80] Foullon C et al 2008 *J. Geophys. Res.* **113** A11203 (doi:10.1029/2008JA013175); Foullon C et al 2009 *J. Geophys. Res.* **114** A10201 (doi:10.1029/2009JA014444); Foullon C et al 2010 *J. Geophys. Res.* **115** A09203 (doi:10.1029/2009JA015189)
[81] Paschman G S et al. 2005 *Ann. Geophys.* **23** 1481
[82] Farrugia C J et al. 2010 *J. Geophys. Res.* **115** A08227 (doi:10.1029/2009JA015128)
[83] Farrugia C J et al. 2007 *Ann. Geophys.* **25** 1-15
[84] Spreiter J R and Rizzi A W 1974 *Acta Astron.* **1** 15–35
[85] Chen S H et al. 1993 *J. Geophys. Res.* **98** 5727
[86] Takagi K C et al. 2006 *J. Geophys. Res.* **111** A08202 (doi:10.1029/2006JA011631); Hasegawa H et al. 2006 *J. Geophys. Res.* **111** A09203 (doi:10.1029/2006JA011728)

# Erratum

An earlier version of this paper was published in error, which did not include some updates to the references. The updated references can be found below.

[4] Van Haren H and Gostiaux L 2010 *Geophys. Res. Letters* **37** L03605

[32] Gratton F T et al 2006 *AIP Conference Proceedings* **875** 300-3; Gratton F T et al. 2009 *J. of Physics CS* **166** 012023 doi:10.1088/1742-6596/166/1/012023; Bilbao L 2009 *J. of Physics CS* **166** 012020 doi:10.1088/1742-6596/166/1/012020; Gratton F T et al. 2011 *J. of Physics CS* **296** 012006 doi: 10.1088/1742-6596/296/1/012006

[75] Gnavi G, Gratton F T and Farrugia C J 2000 *J. Geophys. Res.* **105** 20973-87; Gnavi G, Farrugia C J and Gratton F T 2000 *Proc. ICPP* 2000 vol 3 (Quebec) p 1060; Gnavi G, Farrugia C J and Gratton F T 2004 *Braz. J. Phys.* **34** 1797-803

Journal of Physics: Conference Series **370** (2012) 012004 doi:10.1088/1742-6596/370/1/012004

# Spacecraft charging in flowing plasmas; numerical simulations

**W J Miloch[1], V V Yaroshenko[2], S V Vladimirov[2,3,4], H L Pécseli[1] and J Trulsen[5]**

[1] Department of Physics, University of Oslo, Box 1048 Blindern, N-0316 Oslo, Norway
[2] Max-Planck-Institut für Extraterrestrische Physik, 85741 Garching, Germany
[3] School of Physics, The University of Sydney, Sydney, New South Wales 2006, Australia
[4] Joint Institute for High Temperatures, Russian Academy of Sciences, Moscow, Russia.
[5] Institute of Theoretical Astrophysics, University of Oslo, Box 1029 Blindern, N-0315 Oslo, Norway

E-mail: `w.j.miloch@fys.uio.no`

**Abstract.** The density and potential variations at and in the vicinity of a spacecraft in flowing plasmas are studied by numerical simulations. The spacecraft charging, wake formation, and their role for the diagnostics of the ambient plasma are addressed. It is demonstrated that the wake features, such as ion focusing, can be altered by photoemission due to directed sunlight. For multiple ion species, the wakes for ions with different ion masses can be separated, with light ions contributing to the ion focus, and heavy ions giving a geometrical shape of the wake. As a specific case, the Cassini spacecraft approaching Saturn is considered. The simulations are carried out with *DiP2D* and *DiP3D* codes, the two- and three-dimensional particle-in-cell codes.

## 1. Introduction

Understanding of spacecraft-plasma interaction is of importance for all space missions and experiments [1-3]. A spacecraft is charged by plasma and also other currents, such as photoemission. With respect to the plasma potential, any isolated object is at floating potential, at which the net current to the object surface is zero. If the spacecraft is much smaller that the Debye length $\lambda_D$, which is a characteristic length in plasma, its potential can often be approximated by the orbit-motion-limited (OML) theory [1]. However, in intricate plasma environments, such as plasmas with photoelectrons, or multispecies plasmas, the analytic formulas can give incorrect results [4].

Charged objects disturb the surrounding plasma. In the vicinity of the object's surface a plasma sheath forms, while relative motion of the object and the plasma gives rise to a plasma wake downstream from the object [5-8]. The sheath and wakefield can influence the plasma diagnostics with instruments onboard the spacecraft. Thus, for the analysis of such data, it is necessary to account for the spacecraft potential and its variations in different regions in space. A study of this problem with analytical models is difficult and usually employs linearized models. This has encouraged the use of numerical models, such as particle-in-cell (PIC) simulations [9], which allow us to study the plasma-object interactions self-consistently on a detailed kinetic level, based on first principles.

Journal of Physics: Conference Series **370** (2012) 012004 doi:10.1088/1742-6596/370/1/012004

Numerical simulations reveal that in the presence of a plasma flow, the ion focus can form downstream from the object [5, 10]. Ion focusing is due to electrostatic lensing of ions in the wake region by a negatively charged object. It depends on the material and shape of the object and is more pronounced for large electron to ion temperature ratios [11]. The resulting potential distribution gives rise to a local maximum in the wake, that is located further downstream than the maximum in ion density, consistent with the Poisson equation. Ion focusing has important implications in, for instance, dusty plasmas. It can explain the formation of dust chains from grains suspended in the sheath and presheath of the discharge, where an ion flow is present [12].

We employ *DiP2D* and *DiP3D* numerical codes [13, 14], to study the spacecraft-plasma interaction in different plasma environments. The problems addressed here include wake formation in subsonic and supersonic flows, also in the presence of photoemission due to directed sunlight, as well as multiple ion species. As a specific case we consider a set of parameters corresponding to the Cassini spacecraft approaching Saturn.

## 2. Numerical codes

The *DiP2D* and *DiP3D* codes are two- and three-dimensional PIC codes, which have been described in detail in previous works [13, 14], and here we present briefly the main concepts of the model. In *DiP* codes, two and multicomponent plasmas can be simulated, with the plasma particles (electrons and ions) followed in self-consistent electric fields. For computational efficiency, the codes use a grid for solving the field equations. At each time step, the plasma charge density is built on the grid by weighting individual particles to the nearest grid points. After solving the field equations, the fields are interpolated from the grid points to plasma particles. The particle trajectories are then advanced with the *leap frog* method characterized by a staggered time mesh [9]:

$$\begin{aligned} x_k(t+\Delta t) &= x_k(t) + v_k(t+\Delta t/2)\Delta t \\ v_k(t+\Delta t/2) &= v_k(t-\Delta t/2) + f_k(t)\Delta t/m_k \end{aligned}$$

where $k$ refers to a plasma particle, $f_k$ is a force projected from the nearest grid points on the $k$-th particle of mass $m_k$, and $\Delta t$ is the computational time step.

The plasma particles are initially uniformly introduced with Maxwellian velocity distributions in the simulation box of several electron Debye lenghts, $\lambda_{\mathrm{De}}$ in each direction, and they are also injected into the simulation box at each time step according to prescribed particle fluxes. To simulate an open plasma system, the particles can leave freely through the boundaries of the simulation box. The codes use Dirichlet boundary conditions for the potential, and the plasma potential is $\Phi_{pl} = 0$ V.

A spacecraft, or other finite-sized object, is placed inside a simulation box, and constitutes internal boundary conditions for the plasma particles. The object is initially uncharged, and it is charged self-consistently during the simulation. It can be either made of perfectly insulating or conducting material. For a conducting object, each plasma particle that hits the surface is lost, and it is redistributed on the grain surface to cancel internal electric fields. For an insulating object, the plasma particle that hits the surface remains there for all later times contributing to the local electric fields.

In addition to the plasma currents, the object can be charged by photoemission current due to directed photon flux [15]. The angle of irradiance, energy, and intensity of the monoenergetic photons can be varied. When a photon hits the surface of the object, a photoelectron of energy $E = E_{h\nu} - W$ is produced at distance $l = sv\Delta t$ from the dust grain surface, where $s$ is a uniformly distributed random number $s \in (0, 1]$, and $v$ is the photoelectron speed. For conducting objects, we use the work function $W < 4.5$ eV, which is typical for many metallic materials [16]. Photoelectron velocity vectors are uniformly distributed over the hemicircle and directed away from the dust grain surface, that is in accordance with Lambert's law.

## 3. Results and discussion

With *DiP* codes we have addressed spacecraft charging under various collisionless plasma conditions. Usually, a spacecraft is in relative motion with respect to plasma, and we focus our analysis on flowing plasmas. For generality, we assume a spherical spacecraft made of conducting material. Thus, the results can also be applied to other plasma-surface interaction problems, such as the plasma probe performance, or dust grain charging [17]. While extended parts, solar panels, or booms can modify the plasma dynamics in the closest vicinity to the spacecraft surface, the spherical approximation generally holds further away from the surface [13].

When a conducting, spherical spacecraft is charged by plasma currents only, its potential can be well approximated with the OML model with stationary electrons and flowing ions, where the current densities to the surface are given by [1]:

$$\begin{aligned} J_i &= n_i q_i v_d \left[ \left( 1 + \frac{1}{2\xi^2} - \frac{q_i \Phi_{fl}}{kT_i \xi^2} \right) \operatorname{erf}(\xi) + \frac{1}{\sqrt{\pi}\xi} \exp\left(-\xi^2\right) \right], \\ J_e &= n_e e \sqrt{\frac{kT_e}{2\pi m_e}} \exp\left[ -\frac{e\Phi}{kT_e} \right], \end{aligned} \tag{1}$$

where $\xi = v_d/\sqrt{2kT_i/m_i}$, erf is the error function, $v_d$ ion drift speed, $T_{e/i}$ electron/ion temperature, $e < 0$ electron charge, $q_i$ ion charge, $n_{e/i}$ electron/ion density, and $\Phi_{fl}$ spacecraft potential.

The floating potential from the simulations of a spherical, conducting object, with radius $r$ being small in terms of the Debye length ($r \approx 0.1\lambda_{\mathrm{De}}$), is shown in figure 1. The floating potential calculated with the OML model is also plotted for comparison. The agreement between the results from the simulations and theoretical calculations is very good for supersonic velocities and $v_d \leq 0.3C_s$, where $C_s$ is the speed of sound. For $v_d \in (0.6, 0.9)$ the agreement is worse with deviations up to 10%. Simulations for $v_d \in (0.3, 0.5)$ did not converge and thus they are not included in figure 1. Note that for insulating objects immersed in flowing plasmas, an electric dipole moment develops [13].

A negatively charged object will act as an electrostatic lens for the plasma particles, and will focus ions into the wake [11, 12]. The ion focus refers to a locally enhanced ion density in the wake of the object, an example is shown in figure 2. The maximum in the ion density is at a distance of few electron Debye lengths $\lambda_{\mathrm{De}}$ downstream from the object, being located further away for increasing velocities. Ion focusing is more pronounced for large electron to

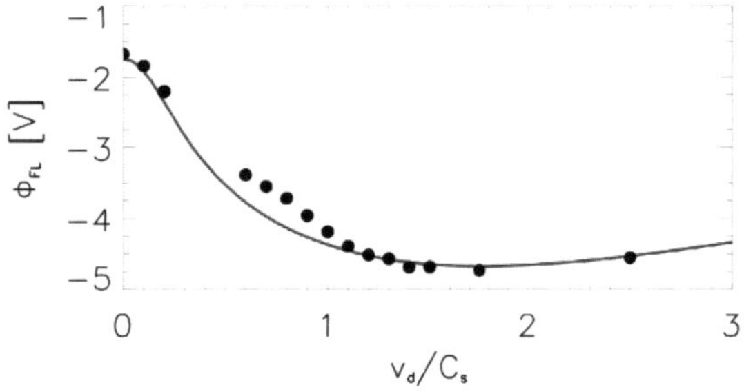


**Figure 1.** Floating potential $\Phi_{fl}$ on a conducting spherical object as a function of $v_d$, $n = 10^{13}\mathrm{m}^{-3}$, $T_e/T_i = 100$. The data from simulations with *DiP3D* code are marked with •, line corresponds to the OML model calculations.

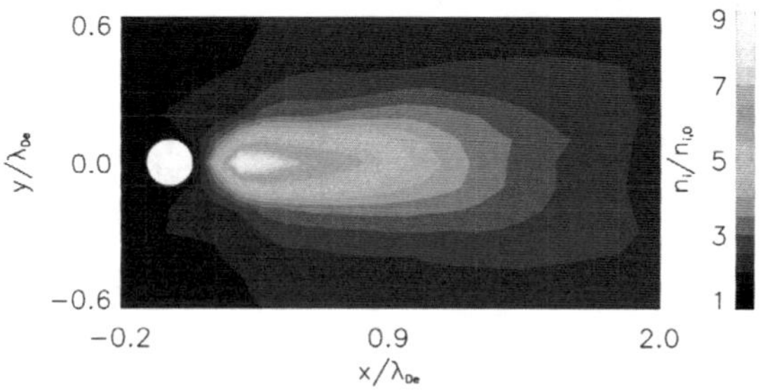


**Figure 2.** Ion density $n_i$ in the wake of a negatively charged conducting object. The flow of $v_d = 1C_s$ is in the positive $x$ direction, $n_{i,0} = 10^{13}\mathrm{m}^{-3}, T_e/T_i = 100$. Data is from *DiP3D* code.

ion temperature ratios [11], and thus it can be important for the plasma measurements by a spacecraft in planetary environments [18]. Ion focusing can be influenced by the shape of the object, and elongated objects can lead to asymmetric focusing [19].

The plasma flow gives rise to a wake in the potential distribution. In particular, ion focusing can lead to a local potential enhancement. The potential distribution in the wake is shown in figure 3 for subsonic (a) and supersonic (b) flow velocities. For the supersonic flow, the potential distortion in the wake is limited by the Mach cone. Similarly to the linear response calculations of the wake pattern behind the object [20, 21], the simulations with *DiP* codes show a possibility of forming several potential maxima in the wake. The wake pattern is more pronounced for cold ions, as for the warm ions the focusing is weak and the Landau damping reduces the wake oscillations.

A radiative environment can significantly modify the spacecraft potential [1]. Positively charged objects and photoelectrons will substantially alter the wake pattern [15]. An example of such a wake behind a conducting object is shown in figure 4 [15], where photoemission is due to unidirectional, monoenergetic photons. In contrast to the negatively charged object, there is no ion focusing. Instead, ions are slowed down by the positively charged object, which gives rise to an enhanced ion density in front of the spacecraft, and density depletion in the wake. Photoelectrons contribute to the electron population, and the electrons are no longer Boltzmann distributed in the vicinity of the spacecraft [15]. Charging by photoemission can have important implications for the insulating parts of the spacecraft. Here the angle of incidence with respect to the plasma flow becomes important, and for some angles of incidence, a weak double layer can be formed [22], while spinning objects can lead to strong modulations in the signal received by the spacecraft instruments [23].

The plasma shielding and wakefield behind a charged spacecraft, and associated ion dynamics or photoemission currents, can substantially modify the measurements with the instruments onboard spacecraft, such as electrostatic probes on booms. The results of such measurements can correspond to a locally disturbed plasma, and not reflect the plasma conditions further away from the spacecraft. Both sheath and wake effects need to be accounted for when analyzing the

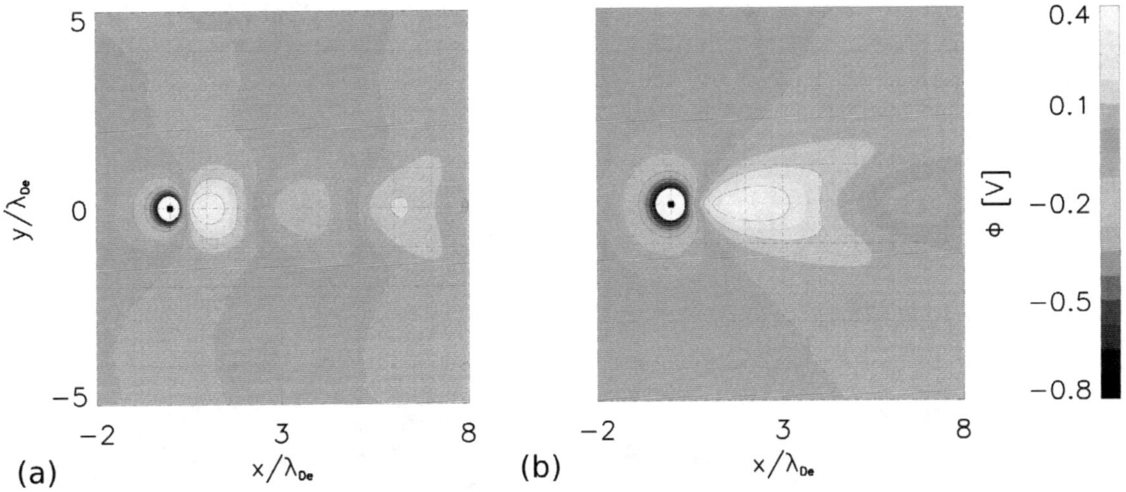


**Figure 3.** Potential distribution in the vicinity of a negatively charged object in the plasma flowing with $v_d = 0.8C_s$ (a), and $v_d = 0.8C_s$ (b). The potentials with values $\Phi < -0.8$ V are colored white. The flow is in the positive $x$ direction. $n = 10^{13}\mathrm{m}^{-3}, T_e/T_i = 100$, data is from *DiP3D* code.

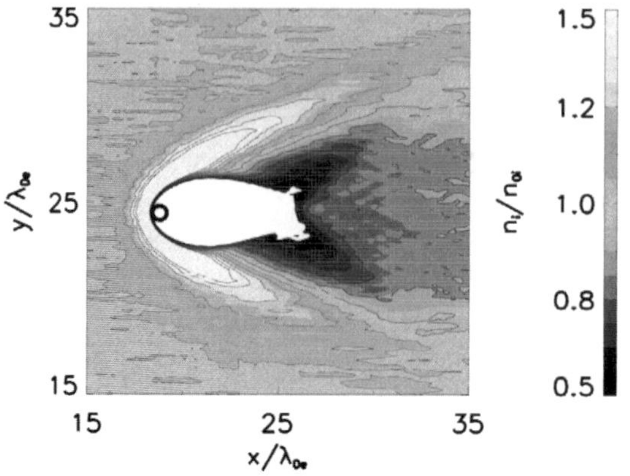


**Figure 4.** Ion density distribution around the object exposed to a supersonic plasma flow and photon flux. Densities with $n_i/n_{i,0} < 0.5$ are colored white. Data is from simulations with *DiP2D* code [15].

data from satellites or rockets. In particular, the ionospheric sounding rockets are the fast spinning objects that can move at sub- or supersonic velocities with respect to the ambient plasma. However, in the studies of ionospheric plasma, one needs to account for multispecies plasma, as well as collisions with neutral background.

To demonstrate the role of multispecies plasma for spacecraft-plasma interaction, we have considered the Cassini spacecraft approaching Saturn [24] with the plasma parameters given in table 1 for $d/R_s = \{3.95, 6.25, 10.2\}$, where $d$ is the distance from the center of Saturn, and $R_s$ is Saturn's radius. The considered plasma has several components, and we account here for electrons, hydrogen ions, as well as water vapor ions. The water vapor in the Saturn environment is due to icy Enceladus moon [25]. The plasma is co-rotating with Saturn, thus there is a plasma flow with respect to the Cassini spacecraft. Photoemission due to directed sunlight is also included, and the photon angle $\alpha$ changes with $d$.

The study of the Cassini wake reveals that for $d/R_s = 3.95$ and $d/R_s = 6.25$, heavy water vapor ions give rise to geometric wake, while much lighter protons form the ion focus region, see

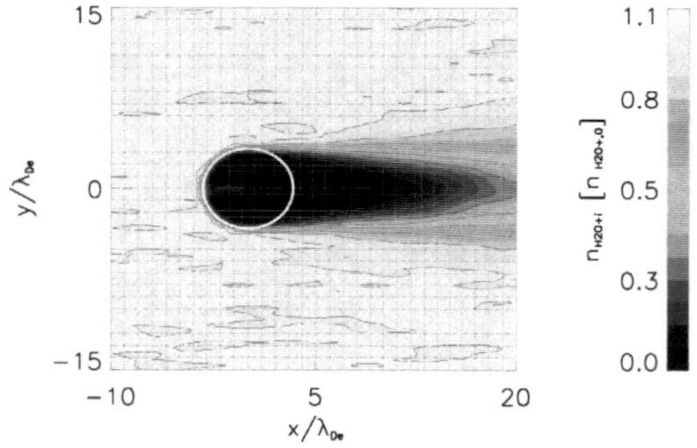


**Figure 5.** The water ion density around Cassini at $r/R_s = 3.95$. Results obtained with *DiP3D* code [24].

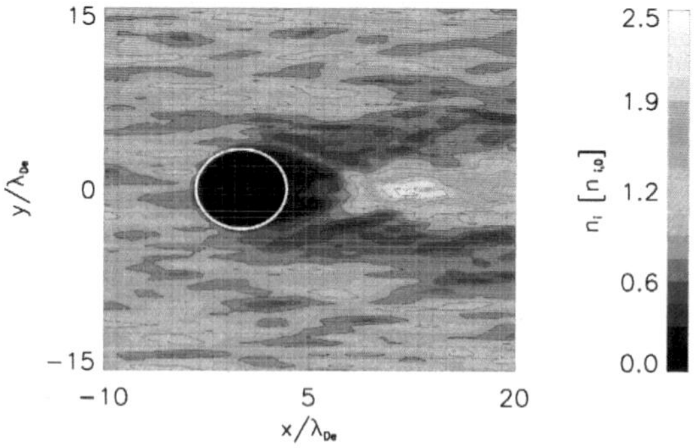


**Figure 6.** The proton density around Cassini at $r/R_s = 3.95$. Results obtained with *DiP3D* code [24].

**Table 1.** The plasma parameters for the Cassini spacecraft approaching Saturn.

| $d/R_s$ | $n_e$ ($\mathrm{cm}^{-3}$) | $n_{H^+}/n_{H_2O^+}$ | $T_e$ (eV) | $T_{H^+}$ (eV) | $T_{H_2O^+}$ (eV) | $\alpha$ (deg) |
|---|---|---|---|---|---|---|
| 3.95 | 50 | 0.05 | 1 | 0.2 | 3.3 | 62 |
| 6.25 | 30 | 0.5 | 6 | 0.7 | 8.5 | 52 |
| 10.2 | 0.1 | 2 | 35 | 2.3 | 22.0 | 35 |

figures 5 and 6. Thus, there is a complex ion wake in the vicinity of the Cassini, extending up to few $\lambda_{\mathrm{De}}$ (note that $\lambda_{\mathrm{De}}$ changes with $d$). An overall wake in the potential is characterized by a potential reduction extending up to several $\lambda_{\mathrm{De}}$, see figure 7. The floating potential of Cassini in this multispecies plasma is $\Phi_{fl} \approx -2.5$ V for $d/R_s = 3.95$, and $\Phi_{fl} \approx -13$ V for $d/R_s = 6.25$. This large difference over a relatively short distance is due to changes in plasma conditions. For the largest considered distance $d/R_s = 10.2$, $\Phi_{fl} \approx 7$ V, meaning that the photoemission current gives an important contribution to the charging in this dilute plasma environment, and hence the wake structure is substantially modified, as shown in figure 8.

## 4. Conclusions

By numerical PIC simulations it has been shown that relative motion of a spacecraft and plasma leads to the formation of a wake in the density and potential distributions. The details of the wake properties depend on the flow velocity, plasma constituents, and floating potential of the spacecraft. In case of the multispecies plasma, different plasma species can give rise to separate wakes, with heavy ions forming a geometrical wake, and light ions being focused. Photoemission currents can lead to a positively charged spacecraft, thus modifying significantly the plasma properties close to the spacecraft. The floating potential of the spacecraft as well as its wakefield need to be accounted for when analyzing the data from instruments onboard the spacecraft during space missions, in particular for spacecrafts entering planetary magnetospheres, when large differences in the potential and plasma parameters can occur at relatively small distances.

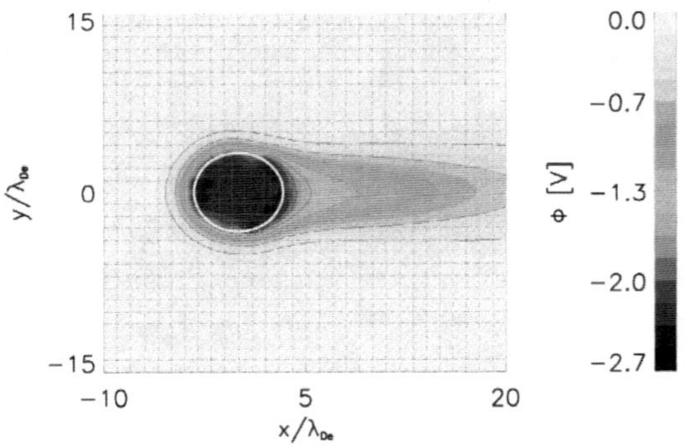


**Figure 7.** The wake in potential of Cassini at $r/R_s = 3.95$. Results obtained with *DiP3D* code [24].

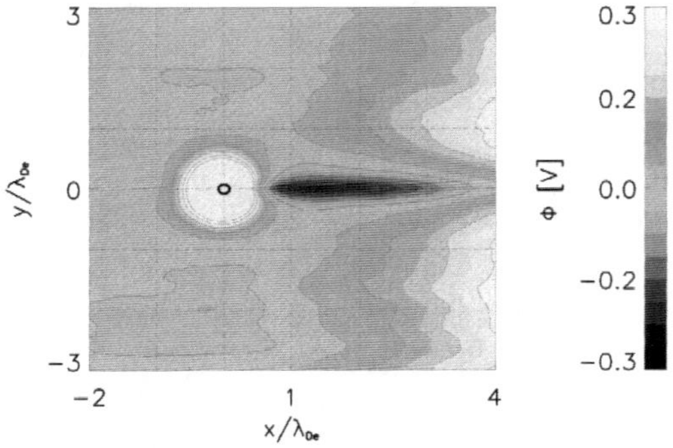


**Figure 8.** The wake in potential of Cassini at $r/R_s = 10.2$. Values above $\Phi = 0.3$ V are colored yellow. Results obtained with *DiP3D* code [24].

## Acknowledgments
This work was partially supported by the Russian-German Partner Group for Complex Plasmas, the Australian Research Council, and the Norwegian Research Council, NFR.

## References
[1] Whipple E C 1981 *Rep. Prog. Phys.* **44** 1197
[2] Svenes K R and Trøim J 1994, *Planet. Space Sci.* **42** 81
[3] Roussel J F and Berthelier J J 2004 *J. Geophys. Res.* **109** A01104
[4] Allen J E 1992 *Phys. Scr.* **45** 497
[5] Melandsø F and Goree J 1995, *Phys. Rev.* E **52** 5312
[6] Vladimirov S V and Ishihara O 1996 *Phys. Plasmas* **3** 444
[7] Guio P and Pécseli H L 2004 *Geophys. Res. Lett.* **31** L03806
[8] Kimura S and Nakagawa T 2008 *Earth Panets Space* **60** 591
[9] Birdsall C K and Langdon A B 1985 *Plasma Physics Via Computer Simulation* (New York: John Wiley)
[10] Maiorov S A, Vladimirov S V and Cramer N F 2001 *Phys. Rev.* E **63** 017401
[11] Miloch W J, Trulsen J and Pécseli H L 2008 *Phys. Rev.* E **77** 056408
[12] Melzer A, Schweigert V A, Schweigert I V, Homann A, Peters S and Piel A 1996 *Phys. Rev.* E **54**, R46
[13] Miloch W J, Pécseli H L and Trulsen J 2007 *Nonlin. Processes Geophys.* **14** 575
[14] Miloch W J, Kroll M, Block D 2010 *Phys. Plasmas* **17** 103703
[15] Miloch W J, Vladimirov S V, Pécseli H L and Trulsen J 2009 *New J. Phys.* **11** 043005
[16] Rosenberg M, Mendis A and Sheehan D P 1996 *IEEE Trans. Plasma Sci.* **24** 1422
[17] Fortov V E, Ivlev A V, Khrapak S A, Khrapak A G and Morfill G E 2005 *Phys. Rep.* **421** 1
[18] Rusell C T (ed.) 2007 *The Mars Plasma Environment* (Dordrecht: Springer)
[19] Miloch W J, Vladimirov S V, Pécseli H L and Trulsen J 2008 *Phys. Rev.* E **78** 036411
[20] Vladimirov S V and Ishihara O 1996 *Phys. Plasmas* **3** 444
[21] Lampe M, Joyce G and Ganguli G 2005 *IEEE Trans. Plasma Sci.* **33** 57
[22] Miloch W J, Rekaa V L, Pécseli H L and Trulsen J 2010 *J. Plasma Phys.* **76** 429
[23] Miloch W J and Vladimirov S V 2009 *Geophys. Res. Lett* **36** L18110
[24] Yaroshenko V V, Miloch W J, Vladimirov S, Thomas H M, and Morfill G E 2011 *J. Geophys. Res.* **116** A12218
[25] Sittler E C Jr. et al. 2008 *Planet. Space Sci.* **56** 3

# Alfvén waves and current relaxation: attenuation at high frequencies and large resistivity

**F E M Silveira**

Centro de Ciências Naturais e Humanas, Universidade Federal do ABC,
Rua Santa Adélia, 166, Bairro Bangu, 09210-170, Santo André, São Paulo, Brazil

E-mail: francisco.silveira@ufabc.edu.br

**Abstract.** The dispersion relations of Alfvén waves propagating in a resistive plasma are explored by assuming a finite relaxation time for the current density. It is shown that the proposed approach is consistent with the hydromagnetic approximation. An extension for the equation governing the space and time evolution of Alfvén waves is provided. New results are found at high values of the wave frequency $\omega$: for a small resistivity, the wavelength increases as the cube of the equilibrium magnetic field but decreases with the cube of $\omega$; for a large resistivity, the wave attenuation does not depend on $\omega$, saturating to a finite value which is fully determined by the relaxation time of the current density. A transition frequency, $\omega_t$, between two sharply distinct regimes of the perturbation is identified: for $\omega < \omega_t$, the disturbance propagates in the resistive plasma as an attenuated oscillation; for $\omega > \omega_t$, the wave ceases very rapidly to oscillate (in space), its amplitude saturating to a finite value. The results presented here may be relevant for investigations of some transient phenomena in plasma physics such as the reconnection of magnetic field lines.

## 1. Introduction

A simple kind of hydromagnetic wave, first explored by Alfvén [1], develops in an initially static, homogeneous and isotropic infinite plasma, subjected to a constant and uniform magnetic field, when a small disturbance of the fluid flow is produced in a direction perpendicular to the field. Such a perturbation gives rise to an infinitesimal magnetic field in the same direction of the flow, which, however, propagates parallel to the equilibrium field. This phenomenon is commonly referred to as an Alfvén wave.

Within the classical approach to the problem, the current density $\vec{J}$, induced in a plasma of finite conductivity $\sigma$, is related to the electric $\vec{E}$ and magnetic $\vec{B}$ fields, and to the fluid velocity $\vec{v}$, through the standard Ohm's law,

$$\vec{J} = \sigma \left( \vec{E} + \vec{v} \times \vec{B} \right) . \tag{1}$$

Such an expression describes satisfactorily the dissipative processes which occur in the conductive plasma, provided the characteristic frequency of the perturbation is low enough. Particularly, the inertia of the charged species can be neglected.

At higher frequencies, however, inertial effects should become important. In the simple case of a plasma with two species (electron and ion), the standard Ohm's law may be extended to (actually, there are three other terms in (2) to follow, proportional to the electron and ion

Journal of Physics: Conference Series **370** (2012) 012005 doi:10.1088/1742-6596/370/1/012005

pressure gradients, and to the Lorentz force density - the so called Hall effect -, but they do not play any relevant role within our model) [2]

$$\frac{m_e m_i}{Z\left(n_e m_e + n_i m_i\right) e^2} \frac{\partial \vec{J}}{\partial t} = \vec{E} + \vec{v} \times \vec{B} - \eta \vec{J}, \tag{2}$$

where $\eta = \sigma^{-1}$ is the plasma resistivity, $e$ is the electron charge, $Z$ is the charge number, $m_e$ ($m_i$) and $n_e$ ($n_i$) are the electron (ion) mass and number density, respectively.

To see the main consequence of the introduction of the left hand side of (2) into the problem, we consider the simple case of a plasma with a singly ionized ion, $Z = 1$. Then, if the electromagnetic field is suddenly removed from the presence of the fluid, the neutrality condition, $n_e = Z n_i$ , together with the approximation $m_e \ll m_i$ imply

$$\vec{J} = \vec{J_0} e^{-t/\tau}. \tag{3}$$

This means that the current density relaxes in the plasma, from some initial value, $\vec{J_0}$ say, in a time scale of the order of

$$\tau = \frac{m_e}{n_e e^2 \eta}. \tag{4}$$

In this case, the so called ideal plasma approximation (a perfectly conductive fluid) should be properly achieved only by fully neglecting the electron mass, that is, by taking the limits $\eta \to 0$ ($\sigma \to \infty$) and $m_e \to 0$ simultaneously. As a result, the relaxation time $\tau$ in (4) would be always finite.

The considerations above suggest strongly that, at sufficiently high frequencies, for which inertial effects of charged species can not be neglected, the dissipative processes that occur in conductive plasmas should be described appropriately by the following extension of (1):

$$\left(1 + \tau \frac{\partial}{\partial t}\right) \vec{J} = \sigma \left(\vec{E} + \vec{v} \times \vec{B}\right). \tag{5}$$

where $\tau$ is the relaxation time of the current density. A previous work had explored the attenuation and damping of electromagnetic fields in rigid conductive media, with basis on a generalization similar to that given by (5) [3]. Subsequently, the skin effect and dissipative processes in rigid conductors were investigated [4]. Recently, the metallic absorptivity at normal incidence was examined, thereby leading to an extension of Hagen-Rubens relation up to the near infrared [5]. More recently, the resistive heating and magnetic diffusion in good conductors have been rediscussed [6].

This work explores the dispersion for Alfvén waves, propagating in a plasma of finite resistivity, with basis on the extended Ohm's law (5). First, the proposed approach is shown to be consistent with the hydromagnetic approximation. Then, an extension for the equation governing the space and time evolution of Alfvén waves is provided. As a consequence, new results for the wavelength and wave attenuation are found at high values of the wave frequency. Much interestingly, a transition frequency between two sharply distinct regimes of the perturbation is identified.

## 2. Hydromagnetic approximation

So far, the importance of inertial effects at high frequencies has been stressed. However, the following question arises naturally: how high should be the frequency such that the hydromagnetic approximation could be still valid? To address this issue, we start by considering the Ampère-Maxwell law,

$$\frac{1}{\mu} \nabla \times \vec{B} = \vec{J} + \epsilon \frac{\partial \vec{E}}{\partial t}, \tag{6}$$

where $\epsilon$ and $\mu$ are the electric permittivity and magnetic permeability, respectively. By combining (5) and (6), we get

$$\left(1+\tau\frac{\partial}{\partial t}\right)\frac{1}{\mu}\nabla\times\vec{B}-\sigma\vec{v}\times\vec{B}=\sigma\vec{E}+\left(1+\tau\frac{\partial}{\partial t}\right)\epsilon\frac{\partial\vec{E}}{\partial t}. \tag{7}$$

For our purposes, it proves convenient to assume that the electric and magnetic fields vary harmonically in time, as $\sim e^{-\imath\omega t}$ say. In this case, (7) leads to

$$\frac{1}{\mu}\nabla\times\vec{B}-\frac{\sigma}{1-\imath\omega\tau}\vec{v}\times\vec{B}=\left(\frac{\sigma}{1-\imath\omega\tau}-\imath\epsilon\omega\right)\vec{E}. \tag{8}$$

Now, if we ask what condition should be satisfied by the frequency $\omega$ such that the displacement current $\epsilon\partial_t\vec{E}$ could be neglected in (6), the obvious answer is

$$\frac{\sigma}{\mid 1-\imath\omega\tau\mid}\gg\epsilon\omega. \tag{9}$$

As one may easily check, condition (9) leads to

$$\omega\ll\frac{1}{\tau\sqrt{2}}\left[\sqrt{1+\left(\frac{2\sigma\tau}{\epsilon}\right)^2}-1\right]^{1/2}. \tag{10}$$

Condition (10) determines the domain of validity of the hydromagnetic approximation by defining an upper limit for the characteristic frequency of the electromagnetic field. In the limit $\tau\to 0$, inertial effects can be neglected and condition (10) recovers the standard upper limit for the frequency [7],

$$\omega\ll\frac{\sigma}{\epsilon}, \tag{11}$$

which does not depend on the relaxation time.

## 3. Perturbative mechanism

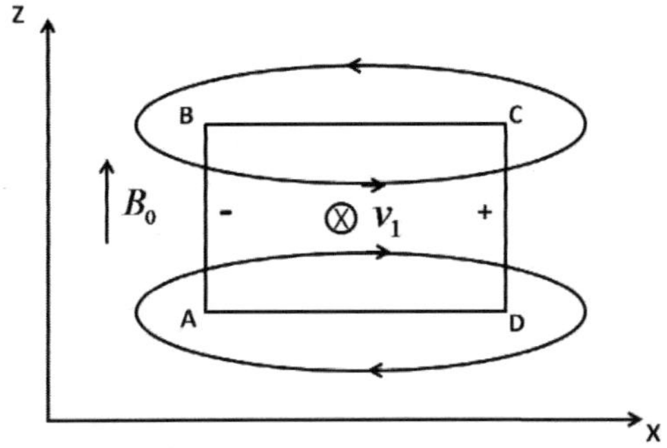


**Figure 1.** An initially static, homogeneous and isotropic infinite plasma, subjected to a constant and uniform magnetic field $B_0\hat{z}$. The small rectangular plasma section ABCD, extending along the $y$-direction, suddenly acquires an infinitesimal velocity $v_1\hat{y}$. Then, due to the forces $\sim v_1B_0\hat{x}$, the segment AB becomes negatively charged and CD, positively charged. Since we are dealing with a conductive medium, the plasma external to ABCD completes the electric circuit. Typical (solenoidal) current lines are shown.

A simple mechanism for the production of Alfvén waves can be described by following standard lines [8]. We start by considering an initially static, homogeneous and isotropic infinite plasma, subjected to a constant and uniform magnetic field $B_0\hat{z}$ (see figure 1). Then, if the small rectangular plasma section ABCD, extending along the $y$-direction, acquires an infinitesimal velocity $v_1\hat{y}$, the charged species experience forces $\sim v_1 B_0 \hat{x}$, which tend to separate electrons and ions. As a result, the segment AB becomes negatively charged and CD, positively charged. However, since we are dealing with a conductive medium, the plasma external to ABCD completes the electric circuit. Typical (solenoidal) current lines are shown in figure 1. Now, the induced current density $J_1\hat{x}$ interacts with the equilibrium magnetic field $B_0\hat{z}$, such that the force density $-J_1 B_0 \hat{y}$ decelerates ABCD while the plasma external to it is accelerated by $J_1 B_0 \hat{y}$. Eventually, ABCD transfers its motion to the neighboring external plasma. However, the original mechanism still operates and the whole process continues to develop, thereby propagating the disturbance along the $z$-direction. Therefore, if $E_1\hat{x}$ and $B_1\hat{y}$ are the induced electric and magnetic fields, respectively, satisfying Faraday's law - see the second of (12) in the next section - one may explore the most simple wave motion, characterized by the perturbative quantities $v_1(z)\,\hat{y}$, $J_1(z)\,\hat{x}$, $E_1(z)\,\hat{x}$, and $B_1(z)\,\hat{y}$, provided the remaining components, all vanish. This is the Alfvén wave.

## 4. Alfvén waves

Given the perturbative mechanism, as shown in the preceding section, we pass now to describe the propagation of Alfvén waves in a plasma with finite conductivity. The set of basic equations can be recast by the extended Ohm's law (5), together with the Ampère (calculations are carried out within the hydromagnetic approximation, see Sec. 2) and Faraday laws,

$$\nabla \vec{B} = \mu_0 \vec{J}, \qquad \frac{\partial \vec{B}}{\partial t} = -\nabla \times \vec{E}, \tag{12}$$

respectively, where $\mu_0$ is the vacuum magnetic permeability (we restrict ourselves to non magnetic plasmas), and the continuity and Euler (an inviscid fluid is assumed) equations,

$$\frac{\partial \rho}{\partial t} = -\nabla \cdot (\rho \vec{v}), \qquad \frac{\partial \vec{v}}{\partial t} + (\vec{v} \cdot \nabla)\,\vec{v} = -\frac{\nabla p}{\rho} + \frac{\vec{J} \times \vec{B}}{\rho}, \tag{13}$$

respectively, where $\rho$ is the mass density and $p$ is the fluid pressure.

According to (5), (12), and (13), we see that the first order perturbative quantities $v_1$, $J_1$, $E_1$, and $B_1$ satisfy the relations

$$\left(1 + \tau \frac{\partial}{\partial t}\right) J_1 = \sigma\,(E_1 + v_1 B_0), \tag{14}$$

$$-\frac{\partial B_1}{\partial z} = \mu_0 J_1, \qquad \frac{\partial B_1}{\partial t} = -\frac{\partial E_1}{\partial z}, \tag{15}$$

$$\frac{\partial v_1}{\partial t} = -\frac{J_1 B_0}{\rho_0}, \tag{16}$$

where we have assumed that the fluid is incompressible (the mass density is constant and uniform, $\rho_0$ say, so that the velocity field is solenoidal, $\nabla \cdot \vec{v} = 0$). Particularly, the pressure gradient is determined by $\partial_z p = J_1 B_1$, which is a second order perturbative term, such that it may be excluded of the foregoing discussion.

By combining (14), (15), and (16), we get

$$v_A^2 \frac{\partial^2 B_1}{\partial z^2} = \frac{\partial^2 B_1}{\partial t^2} - \left(1 + \tau \frac{\partial}{\partial t}\right) \frac{1}{\mu_0 \sigma} \frac{\partial^3 B_1}{\partial t \partial z^2}, \tag{17}$$

where we have introduced the Alfvén speed $v_A = B_0\,(\mu_0 \rho_0)^{-1/2}$. In the limit $\tau \to 0$, inertial effects can be neglected and (17) recovers the standard equation governing the propagation of Alfvén waves in a plasma with finite conductivity [8].

Journal of Physics: Conference Series **370** (2012) 012005 doi:10.1088/1742-6596/370/1/012005

## 5. Dispersion relations

If we assume that the perturbative magnetic field varies harmonically in space and time, as $\sim \exp(\imath kz - \imath\omega t)$ say, (17) leads to the general dispersion relation

$$v_{\mathrm{A}}^2 k^2 = \omega^2 \left[ 1 + \imath (1 - \imath\omega\tau) \frac{k^2}{\mu_0 \sigma \omega} \right]. \tag{18}$$

### 5.1. *Small resistivity*

When the resistive correction term $\sim \sigma^{-1}$ in (18) is small, the wave number $k$ satisfies

$$v_{\mathrm{A}} k = \omega \left[ \left( 1 + \frac{\omega^2 \tau}{2\mu_0 \sigma v_{\mathrm{A}}^2} \right) + \imath \frac{\omega}{2\mu_0 \sigma v_{\mathrm{A}}^2} \right]. \tag{19}$$

*5.1.1. Low frequencies* At low values of the wave frequency $\omega$, the relaxation term $\sim \tau$ in (19) just slightly corrects the wavelength (proportional to the inverse of the real part of the wave number) that increases as $\sim B_0$ but decreases with $\sim \omega$. In the limit $\tau \to 0$, the inertial correction vanishes and (19) recovers the standard wave number describing the propagation through a slightly resistive plasma [7],

$$k = \frac{\omega}{v_{\mathrm{A}}} + \imath \frac{\omega^2}{2\mu_0 \sigma v_{\mathrm{A}}^3}, \tag{20}$$

for which the wave attenuation (the imaginary part of the wave number) increases as $\sim \omega^2$ but decreases with $\sim B_0^3$.

*5.1.2. High frequencies* At high frequencies, the unit can be neglected in (19) and the wave number is given by

$$k = \frac{\omega^3 \tau}{2\mu_0 \sigma v_{\mathrm{A}}^3} + \imath \frac{\omega^2}{2\mu_0 \sigma v_{\mathrm{A}}^3}. \tag{21}$$

As it appears, the wave attenuation is the same as that of (20) at low frequencies. However, the wavelength increases as $\sim B_0^3$ but decreases with $\sim \omega^3$ now. This is the fundamental, new limiting result of (18) for a slightly resistive plasma, at high frequencies, due to the introduction of the finite relaxation time $\tau$ for the current density in (5).

### 5.2. *Large resistivity*

In the opposite extreme for which the resistive term dominates, the wave number is given by the vanishing of the factor between squared brackets on the right hand side of (18),

$$(1 - \imath\omega\tau) k^2 = \imath \mu_0 \sigma \omega. \tag{22}$$

Note that the wave number does not depend on $B_0$ now.

*5.2.1. Low frequencies* At low frequencies, the wave number in (22) is given by

$$k = \sqrt{\frac{\mu_0 \sigma \omega}{2}} \left( 1 - \frac{\omega\tau}{2} \right) + \imath \sqrt{\frac{\mu_0 \sigma \omega}{2}} \left( 1 + \frac{\omega\tau}{2} \right). \tag{23}$$

This shows that the imaginary part of the wave number is slightly larger (by the small relaxation term $\sim \omega\tau$) than the real one. Therefore, both parts increase approximately as $\omega^{1/2}$. In the limit $\tau \to 0$, the inertial correction vanishes and the wave number acquires equal real and imaginary parts. In this case, both parts increase exactly with $\omega^{1/2}$ and (23) recovers the standard wave number describing the propagation through a strongly resistive plasma [7].

*5.2.2. High frequencies* At high frequencies, the wave number in (22) is given by

$$k = \frac{1}{2\omega\tau}\sqrt{\frac{\mu_0\sigma}{\tau}} + \imath\sqrt{\frac{\mu_0\sigma}{\tau}}. \tag{24}$$

As it appears, the wave attenuation does not depend on $\omega$ any longer, saturating to a finite value now, which is fully determined by $\tau$. Additionally, the real part of the wave number becomes much smaller (by the large factor $\sim \omega\tau$) than the imaginary one. Particularly, since the wavelength increases with no limit as $\sim \omega$, for all practical purposes, we can consider that the perturbation does not oscillate (in space), being purely attenuated as $\sim \tau^{-1/2}$. This is the fundamental, new limiting result of (18) for a strongly resistive plasma, at high frequencies, due to the introduction of the finite relaxation time $\tau$ for the current density in (5).

## 6. Transition from low to high frequencies

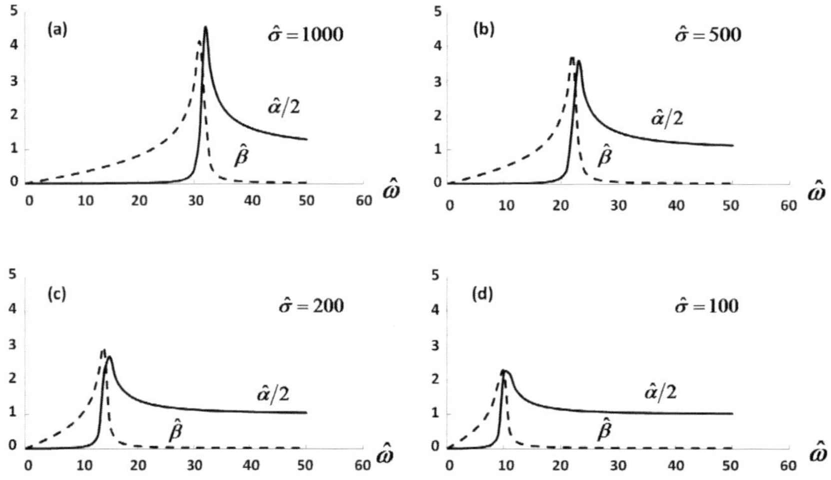


**Figure 2.** The normalized propagation constant $\hat{\beta}$ (dashed lines) and attenuation coefficient $\hat{\alpha}/2$ (full lines) as functions of the normalized frequency $\hat{\omega}$, for selected values of the normalized conductivity $\hat{\sigma}$: (a) $\hat{\sigma} = 1000$; (b) $\hat{\sigma} = 500$; (c) $\hat{\sigma} = 200$; (d) $\hat{\sigma} = 100$. Note the tendency for saturation of $\hat{\alpha}/2$ when both $\hat{\omega}$ and $\hat{\sigma}^{-1}$ attain large values, as predicted by (24).

As it has been shown by (24) in the preceding section, the attenuation for an Alfvén wave propagating through a strongly resistive plasma tends to a finite asymptotic value at high frequencies, due to inertial effects of charged species. Such a result enables us to rewrite the general dispersion relation (18) in the form

$$\left[\left(\hat{\sigma} - \hat{\omega}^2\right) - \imath\hat{\omega}\right]\hat{k}^2 = \hat{\omega}^2, \tag{25}$$

where we have introduced the normalized quantities

$$\hat{\omega} = \omega\tau, \qquad \hat{\sigma} = \sigma\tau\mu_0 v_{\mathrm{A}}^2, \qquad \hat{k} = k\sqrt{\frac{\tau}{\mu_0\sigma}}. \tag{26}$$

Note that all normalization constants in (26) are fully determined by the relaxation time $\tau$ for the current density. This means that such a procedure is possible only within the framework of the extended Ohm's law (5).

By introducing the propagation constant $\beta$ and attenuation coefficient $\alpha/2$ in the usual way, that is, as the real and imaginary parts, respectively, of the wave number $k$, we can write $\hat{k} = \hat{\beta} + \imath\hat{\alpha}/2$ for the corresponding normalized quantities. In this case, (25) leads to (see figure 2)

$$\left\{\hat{\beta};\ \frac{\hat{\alpha}}{2}\right\} = \frac{\hat{\omega}}{\sqrt{2}}\left[\frac{\sqrt{(\hat{\sigma}-\hat{\omega}^2)^2+\hat{\omega}^2} \pm (\hat{\sigma}-\hat{\omega}^2)}{(\hat{\sigma}-\hat{\omega}^2)^2+\hat{\omega}^2}\right]^{1/2}, \qquad (27)$$

where the $\pm$ signs stand for $\hat{\beta}$ and $\hat{\alpha}/2$, respectively.

Figure 2 shows $\hat{\beta}$ and $\hat{\alpha}/2$ as functions of the normalized frequency $\hat{\omega}$, for some selected values of the normalized conductivity $\hat{\sigma}$. Note the tendency for saturation of $\hat{\alpha}/2$ when both $\hat{\omega}$ and $\hat{\sigma}^{-1}$ attain large values, as predicted by (24).

Now, a much interesting result of the presently developed model is also shown by figure 2: for each fixed $\hat{\sigma}$, $\hat{\beta}$ and $\hat{\alpha}/2$ both intercept at the same finite $\hat{\omega}$. From (27), we see that the condition $\hat{\beta} = \hat{\alpha}/2$ is satisfied for $\hat{\omega} = \hat{\sigma}^{1/2}$. By plugging back the physical quantities, as given by (26), this means that the propagation constant and attenuation coefficient coalesce at

$$\beta = \frac{\alpha}{2} = \left(\frac{\mu_0\sigma}{\tau}\right)^{3/4}\sqrt{\frac{v_A\tau}{2}} \qquad (28)$$

when the wave frequency attains the value

$$\omega_t = v_A\sqrt{\frac{\mu_0\sigma}{\tau}}. \qquad (29)$$

The quantity $\omega_t$, given by (29), can be physically interpreted as the transition frequency between two sharply distinct regimes of the perturbation: as long as $\omega < \omega_t$, the disturbance propagates in the resistive plasma as an attenuated oscillation; as soon as $\omega > \omega_t$, the wave ceases very rapidly to oscillate (in space), its amplitude saturating to a finite value. In the limit $\tau \to 0$, inertial effects can be neglected, all transition quantities, as given by (28) and (29), diverge, and the perturbation exhibits a unique regime, the former one, for which the condition $\omega < \infty$ is trivially satisfied for the whole frequency range that is admissible within the hydromagnetic approximation (see section 2).

## 7. Conclusion

The dispersion relations of Alfvén waves propagating in a resistive plasma have been explored by assuming a finite relaxation time for the current density. The proposed approach has been shown to be consistent with the hydromagnetic approximation - see condition (10). An extension for the equation governing the space and time evolution of Alfvén waves has been provided - see (17). New results have been found at high values of the wave frequency $\omega$:

- for a small resistivity, the wavelength increases as the cube of the equilibrium magnetic field but decreases with the cube of the wave frequency - see (21);
- for a large resistivity, the wave attenuation does not depend on the wave frequency, saturating to a finite value which is fully determined by the relaxation time for the current density - see (24).

A transition frequency, $\omega_t$, between two sharply distinct regimes of the perturbation has been identified - see (29) -:

- for $\omega < \omega_t$, the disturbance propagates in the resistive plasma as an attenuated oscillation;
- for $\omega > \omega_t$, the wave ceases very rapidly to oscillate (in space), its amplitude saturating to a finite value.

It should be interesting to apply the results presented here for investigations of some transient phenomena in plasma physics such as the reconnection of magnetic field lines. Actually, a lower limit for the thickness of a reconnection layer can be determined by the inverse of the square root of the Lundquist number, a generally very large quantity, which is proportional to the plasma conductivity $\sigma$ [9]. Additionally, within one of the most basic approaches to this phenomenon, the Sweet-Parker model, the magnetic field lines may be shown to slip through the plasma to reconnect with a speed that is also proportional to the inverse of the square root of $\sigma$ [10].

Now, the time Fourier transform of the extended Ohm's law (5) is readily seen to be given by

$$\vec{J} = \frac{\sigma}{1 - \imath\omega\tau}\left(\vec{E} + \vec{v} \times \vec{B}\right). \tag{30}$$

This shows that the complex coefficient of proportionality between the electromagnetic field and the current density can be physically interpreted as an effective conductivity depending on the wave frequency. Therefore, we argue that relevant corrections to space and time scales describing the reconnection of magnetic field lines could be expected by applying the presently developed model at sufficiently high frequencies.

## Acknowledgments

The author is grateful to G. J. Morales for calling his attention to the verification of the consistency of inertial corrections to dispersion relations of Alfvén waves and the hydromagnetic approximation. This work is partially supported by FAPESP, Brazilian government agency, under grant number 2011/51221-1.

## References

[1] Alfvén H 1942 *Ark Mat. Astr. Fysik* **29B** No. 2
[2] Spitzer L 2006 *Physics of Fully Ionized Gases* 2nd ed (New York: Dover)
[3] Silveira F E M and Lima J A S 2009 *J. Phys. A: Math. Theor.* **42** 095402
[4] Silveira F E M and Lima J A S 2010 *J. Electromagn. Waves Appl.* **24** 151
[5] Silveira F E M and Kurcbart S M 2010 *Europhys. Lett.* **90** 44004
[6] Silveira F E M and Kurcbart S M 2012 *Int. J. Eng. Sci.* **52** 22
[7] Jackson J D 1975 *Classical Electrodynamics* 2nd ed (New York: Wiley)
[8] Reitz J R, Milford F J and Christy R W 1993 *Foundations of Electromagnetic Theory* 4th ed (Reading: Addison-Wesley)
[9] Priest E and Forbes T 2000 *Magnetic Reconnection* (Cambridge: Cambridge University Press)
[10] Sweet P A 1958 *Electromagnetic Phenomena in Cosmical Physics* (IAU Symposium) ed B Lehnert vol 6 (London: Cambridge University Press) p 123; Parker E N 1957 *J. Geophys. Res.* **62** 509

Journal of Physics: Conference Series **370** (2012) 012006 doi:10.1088/1742-6596/370/1/012006

# Determination of transport and reaction swarm coefficients from the analysis of complex transient pulses from the pulsed Townsend experiment

**A Bekstein[1], J de Urquijo[1], O Ducasse[2], J C Rodríguez-Luna[1] and A M Juárez[1]**

[1]Instituto de Ciencias Físicas, Universidad Nacional Autónoma de México, P.O. Box 48-3, 62251, Cuernavaca, Mor., Mexico
[2]Université de Toulouse, Laplace UMR CNRS 5213, UPS, 118 Route de Narbonne, 31062 Toulouse, cedex 9, France

E-mail: jdu@fis.unam.mx

**Abstract.** We present in this paper the interpretation and analysis of transient pulses from a pulsed Townsend experiment by solving the continuity equations of the charged carriers (electrons and ions) involved in the avalanche. The set of second order partial differential equations is solved by SIMAV, a simulator designed specifically for the pulsed Townsend avalanche. Complex situations involving processes such as electron detachment, ion-molecule reactions, Penning ionization and secondary electron emission from ion impact at the cathode, virtually impossible to solve analytically, are discussed here to illustrate the capability of the simulator to help explain the various reaction processes involved in the avalanche, and also to derive some of the transport and reaction coefficients.

## 1. Introduction

Electron and ion interactions with gases at low energies have been studied for several decades. In essence, two kinds of experiments have been performed, namely single collision, beam experiments, aimed at measuring cross sections, and multicollisional swarm experiments from which swarm coefficients are normally obtained. Both cross sections and swarm coefficients are related to each other [1], and in many cases measured swarm coefficients are used to either derive or validate cross section sets.

Nowadays there is an increasing need for either electron and ion swarm transport and reaction coefficients or the respective cross sections in view of the ever growing low temperature plasma applications and basic research on atmospheric physics, for instance. The swarm technique has been very helpful in providing a wealth of data for many gases and mixtures, and in particular the pulsed Townsend technique (PTT), since it permits the observation of the temporal development of the discharge and the derivation of transport and reaction data from it. Nevertheless, the swarm techniques have so far provided transport coefficients such as electron and ion drift velocities, diffusion coefficients and/or reaction coefficients such as electron impact ionization and attachment. Only in very few cases electron detachment coefficients and reaction rates have been derived. The reason for this stems from the fact that modelling the Townsend avalanche is difficult since it involves the solution of continuity equations for the charge carrier species involved in the process, hence this has only been possible for relatively simple cases such as electron/ion drift, ionization and attachment.

Journal of Physics: Conference Series **370** (2012) 012006 doi:10.1088/1742-6596/370/1/012006

Even simple cases such as electron detachment have been successfully solved for the electron component of the discharge, the ion part remaining unsolved thoroughly because of its analytical complexity. Thus, the need to solve the set of continuity equations numerically, taking into account the special boundary and initial conditions is called for.

A pulsed Townsend avalanche simulator, SIMAV, has been developed to solve the equations of continuity for up to six different charged carrier species (electrons and ions) moving and reacting in a gas under the action of a homogeneous electric field in a parallel plate arrangement [2-5]. SIMAV is capable of solving a set of second-order partial differential equations accounting for the drift, diffusion and reaction of all the charge carriers involved in a particular pre-breakdown condition. The solver is based on a finite difference approximation that calculates the temporal evolution of the density of each species and from these, the individual and total currents flowing in the discharge gap, the latter being accessible to measurement [3,4]. SIMAV has been successfully used and examples of this are presented in this paper. Its usefulness has been twofold since it has helped explain complex reaction schemes such as electron detachment and ion molecule reactions in $SF_6$, Penning ionization in $C_2F_4$-Ar mixtures, charge transfer in Xe-Ne mixtures, negative ion clustering in $O_2$, or even predict complex avalanche processes due to secondary electron emission by ion impact at the cathode.

## 2. The pulsed Townsend experiment

This section is dedicated to provide the essentials of the technique by means of illustrative examples that may help the reader appreciate the importance of resorting to the numerical simulation of the processes for the complex situations referred to above.

### *2.1. Experimental setup*

The pulsed Townsend apparatus used for this research has been described fully in detail elsewhere [2]. Briefly, the experiment relies on the measurement of the displacement current due to the electron and ion swarms (positive and/or negative) drifting and reacting under the action of a highly uniform, homogeneous electric field produced by a parallel-plate capacitor. The initial photoelectrons are released from the cathode by the action of a UV flash (337 nm, 1ns duration). These photoelectrons and their progeny drift toward their corresponding electrodes. Thus, the time-dependent total current flowing through the gap is amplified by a transimpedance amplifier and registered on a digital oscilloscope connected to a computer for data acquisition and further analysis. A sketch of the layout of the experiment is given in figure 1.

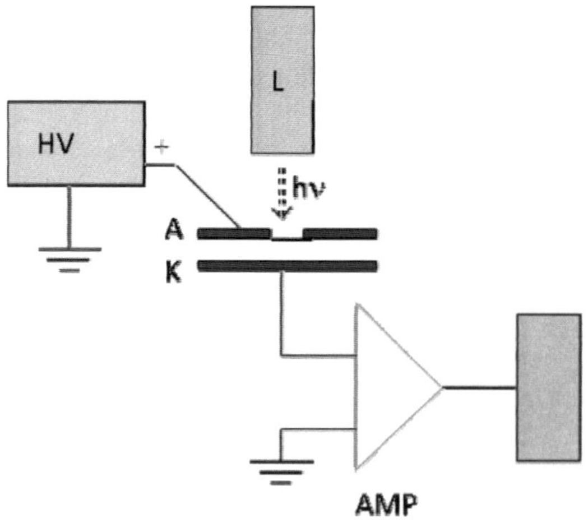


**Figure 1.** Schematic layout of the pulsed Townsend experiment. HV: high voltage power supply; L: pulsed laser; A: anode; K: cathode; AMP: transimpedance amplifier.

*2.2. Simple analysis*

The analytical expressions useful to derive the electron drift velocity $v_e$, the effective ionization coefficient $(\alpha-\eta)$ ($\alpha$ and $\eta$ are the ionization and attachment coefficients, respectively), and possibly the negative and positive ion drift velocities $v_p$ and $v_n$ are obtained readily from a simple solution of the continuity equations for the above species [3-5] provided that the ions are non-reactive. Even the longitudinal diffusion coefficient $D_L$ can be obtained from a curve fitting procedure to the measured electron component of the avalanche [6]. Because of their much higher mobility than that of the ions, the electron signal can be completely differentiated from the ionic one, and therefore the analysis of the former component proceeds readily by means of a curve-fitting procedure to the expression

$$I_e(t) = I_0 \exp\left[(\alpha-\eta)\ v_e t\right] \text{ with } I_0 = \frac{n_o q_o}{T_e} \tag{1}$$

where $I_0$ is the initial current produced by the release of photoelectrons from the cathode, $n_o$ is photoelectron number, $q_o$ is elementary charge, and $T_e = d/v_e$ is the electron transit time through the discharge gap of width $d$. When longitudinal diffusion is considered, a more complex expression [4,6] to analyze the measured electron component of the avalanche current is

$$I_e(t) = \frac{I_0}{2}\exp(\alpha_{eff} v_e t)\left\{\left(1 - erf\left(\frac{(v_e+\alpha_{eff}D_L)t-\mathrm{d}}{\sqrt{4D_L t}}\right)\right) + \exp\left(\frac{v_e+\alpha_{eff}D_L}{D_L}\mathrm{d}\right)\left(erf\left(\frac{(v_e+\alpha_{eff}D_L)t+\mathrm{d}}{\sqrt{4D_L t}}\right)-1\right)\right\} \tag{2}$$

where

$$erf(x) = \frac{2}{\sqrt{\pi}}\int_0^x \exp(-u^2)du \tag{3}$$

is the error function of argument $x$. Examples of the calculated waveforms are given in figure 2 for the cases $(\alpha-\eta) > 0$, $(\alpha-\eta) = 0$, and $(\alpha-\eta) < 0$. Also illustrated in figure 2 is the assessment of the electron transit time $T_e$ from which the electron drift velocity is obtained. For further details of the analysis the reader is referred to earlier work [3-5].

The equations for the positive and negative ionic components are, respectively [3,7]

$$I_p(t) = \frac{q_0}{T_p} n_0 \frac{\alpha}{\alpha_{eff}}\left[\exp(\alpha_{eff} d) - \exp(\alpha_{eff} v_p t)\right] \tag{4}$$

$$I_n(t) = \frac{q_0}{T_n} n_0 \frac{\eta}{\alpha_{eff}}\left[\exp(\alpha_{eff}(d - v_n t)) - 1\right] \tag{5}$$

where $T_p = d/v_p$ and $T_n = d/v_n$ refer respectively to the positive and negative ions transit time, i.e. $v_p$ and $v_n$ refer respectively to the drift velocity of positive and negative ions.

A sketch of the whole avalanche containing the fast electron component and the slower ionic ones is shown in figure 3.

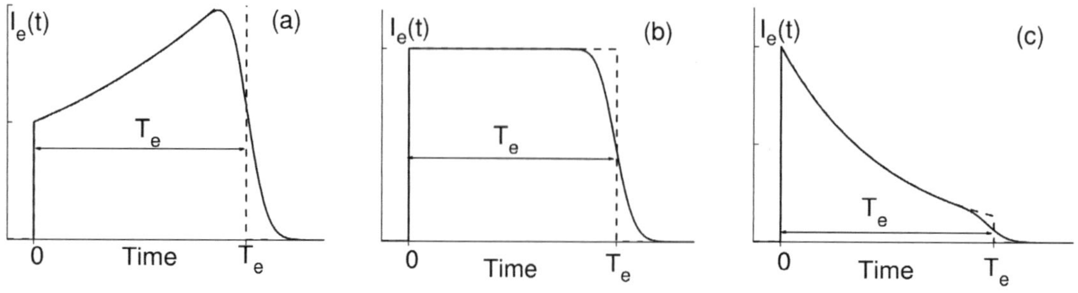


**Figure 2.** Calculated electron current $I_e(t)$ (including diffusion) for the cases (a) $(\alpha-\eta)>0$, (b) $(\alpha-\eta)=0$, and (c) $(\alpha-\eta)<0$. The assessment of the electron transit time $T_e$ is indicated.

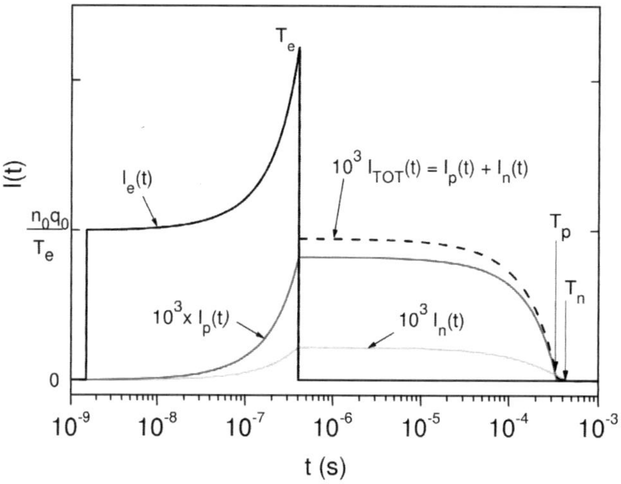


**Figure 3.** A sketch of the whole avalanche containing the fast electron component and the slower ionic one is shown in figure 3, calculated with equations (1), (4) and (5). Note that the time scale is logarithmic and that the ionic currents have been enlarged by three orders of magnitude.

## 3. The avalanche simulator (SIMAV)

In this section we briefly discuss the essential features of the simulator that solves a set of partial differential equations, first and second order, using a first order finite difference approximation. The one-dimensional (1D) simulation code calculates the density of all charged particle species and the current in the external circuit that can be used directly for comparison with the transient waveforms obtained from the pulsed Townsend experiment. For a thorough description the interested reader is referred to [2].

The general form of the continuity equations of the charged carrier species of density $n$ is [2,8]

$$\frac{\partial n_i(z,t)}{\partial t}+\frac{d\, j_i(z,t)}{dz}=\sum_{k=1}^{N_s} a_{i,k}\, j_k(z,t) \qquad (6)$$

where

$$j_i(z,t) = v_i n_i(z,t) - D_i \frac{dn_i(z,t)}{dz} \tag{7}$$

is the current density. The terms $a_{i,k}$ refer to the spatial coefficients of species $i$ due to charge carrier $k$, $z$ is the drift coordinate, $v_i$ and $D_i$ are the drift velocity and longitudinal diffusion coefficients of species $i$.

The total current which corresponds to the measurable one is given by

$$I_{Tot}(t) = \sum_i I_i(t) = \sum_i \frac{q_i}{T_i} \int n_i(z,t) dz \tag{8}$$

where $T_i = d/v_i$ is the transit time and $v_i$ the drift velocity.

The processes thus far included in the simulator are electron/ion drift and diffusion, electron impact ionization and attachment, electron detachment from negative ions, ionic conversion, Penning ionization, and secondary electron emission by positive ion impact at the cathode.

## 4. Case studies

In this section we shall discuss the capabilities of the simulator to solve complex reaction schemes dealing with electron detachment, ionic conversion, Penning ionization and secondary emission.

### *4.1. Ion clustering in $O_2$*

The measurement of an ionic avalanche in $O_2$ such as that shown in figure 4 (wriggled, gray curve), bearing a peculiar trailing edge, falling vertically, led us to suspect the presence of at least a second ionic species. Further inspection of the pulse suggested that this would be a third species since a straightforward measurement of the transit time (broken curve ending at about 660 μs) rendered a negative ion drift velocity compatible with that of $O_4^-$ [9]. Thus we were led to assume that the third ionic species was $O_6^-$. The full reaction scheme leading to the formation of this species is electron attachment (three body) leading to the formation of $O_2^-$

$$e^- + 2\,O_2 \rightarrow O_2^- + O_2, \tag{9}$$

followed by ionic conversion leading to the formation of $O_4^-$ from $O_2^-$ and $O_6^-$ from $O_4^-$,

$$O_2^- + 2\,O_2 \rightarrow O_4^- + O_2 \tag{10}$$

$$O_4^- + 2\,O_2 \rightarrow O_6^- + O_2 \tag{11}$$

The number of parameters needed to calculate the avalanche pulses is large, namely four drift velocities (electrons and three ion species), the attachment coefficient for reaction (9) and two more reaction rates for processes (10) and (11). Fortunately, all drift velocities were known, as well as the electron attachment coefficient and the reaction rate for process (10), thereby leading us to fit the measured curve with only one free parameter. The absence of $O_2^-$ in the ionic component of figure 4 is due to the very fast conversion process leading to the formation of $O_4^-$ during the first 200 ns of discharge development, as it is shown in figure 5.

A plot of the reduced mobilities of $O_4^-$ and $O_6^-$ in $O_2$ is given in figure 6, together with the calculated mobility curves resulting from using the cross section sets of figure 7, where the momentum transfer cross section $Q_m$ for $O_4^-$ and $O_6^-$ in $O_2$ was derived by using the measured mobilities to estimate its value over the low energy collision range.

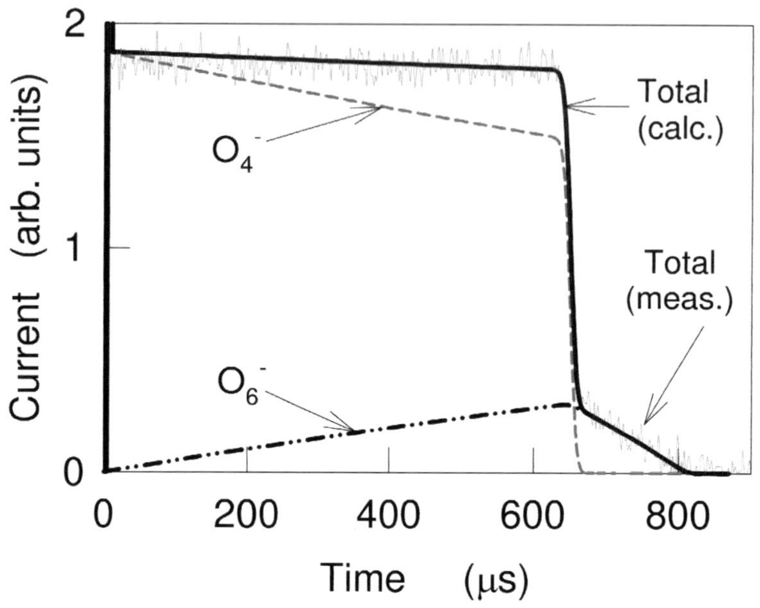


**Figure 4.** A measured ionic avalanche pulse in $O_2$ (gray, wriggled curve) with its total current, calculated curve (solid line) and the contribution of the individual components of $O_4^-$ and $O_6^-$ (see text). The conditions of the experiment were $p$ = 600 Torr and $E/N$ = 8 Td (1 Townsen*d* = $10^{-17}$ V cm$^2$) [9].

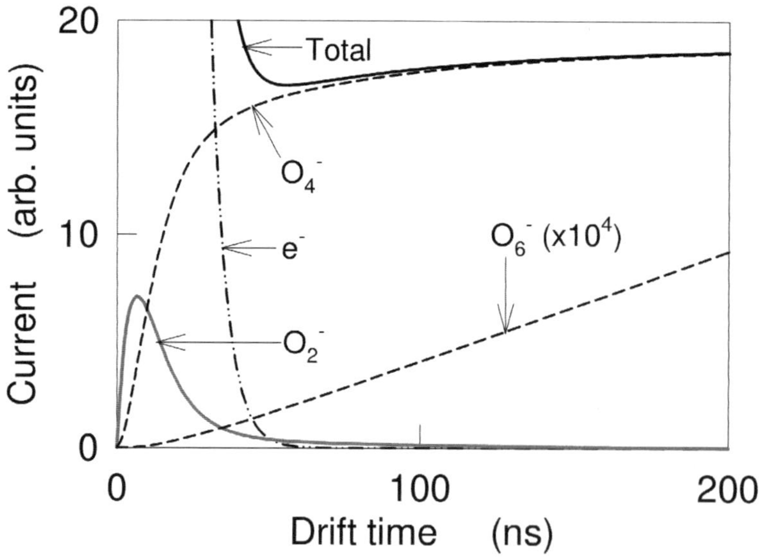


**Figure 5.** The temporal development of charge carriers in the $O_2$ avalanche during the first 200 ns. Note that at this time practically the only ion in the discharge gap is $O_4^-$ since the current due to $O_6^-$ is over four orders of magnitude smaller. ($p$ = 600 Torr and $E/N$ = 8 Td) [9].

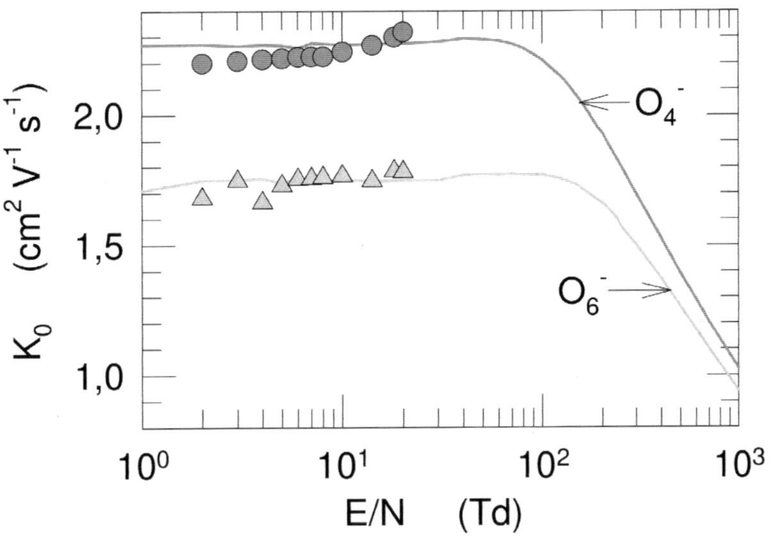


**Figure 6.** The reduced mobilities of $O_4^-$ and $O_6^-$ in $O_2$ as a function of E/N [9].

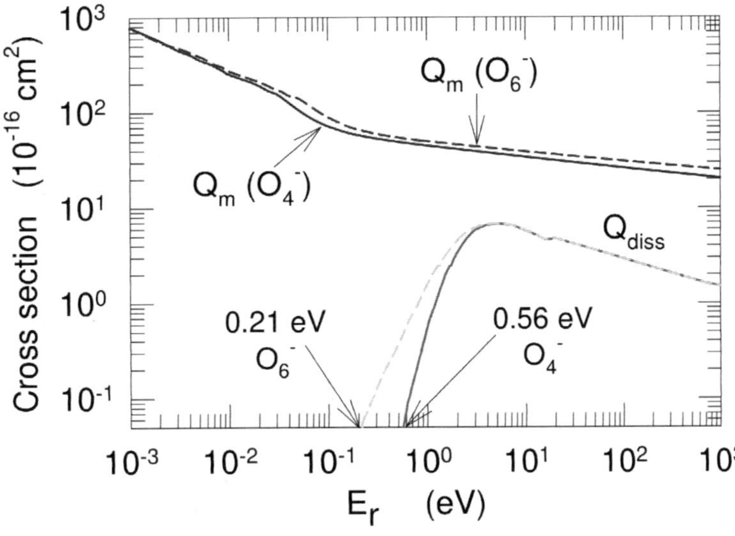


**Figure 7.** The collision cross section sets for $O_4^-$ and $O_6^-$ in $O_2$ as a function of the relative energy. $Q_m$ and $Q_{diss}$ are the momentum transfer and dissociation cross sections, respectively [9].

*4.2. Charge transfer in Xe-Ne mixtures*

Our second case deals with the formation of $Xe^+$ by charge transfer from $Ne^+$ in the Xe-Ne mixture. This study was due to prove that this process had minimal influence on the measured, total current so that the measured transit time, out of which the drift velocity and mobility were derived, was that of

Journal of Physics: Conference Series **370** (2012) 012006 doi:10.1088/1742-6596/370/1/012006

$Xe^+$, which was previously formed by electron impact. The conditions of the experiment were such that no more than 1% $Ne^+$ was present in the discharge [8].

Both $Ne^+$ and $Xe^+$ are initially formed by electron impact by the processes

$$e^- + Xe \rightarrow Xe^+ + 2\,e^- \tag{12}$$

$$e^- + Ne \rightarrow Ne^+ + 2\,e^- \tag{13}$$

with ionization coefficients that are well known. The charge transfer reaction

$$Ne^+ + Xe \rightarrow Xe^+ + Ne \tag{14}$$

was also considered. The values of the ionization coefficients and drift velocities involved in the calculation are given in [8].

Figure 8 shows the smallness of the $Ne^+$ component makes it practically negligible so that the drift velocity measured from the end of the total current is that of $Xe^+$. Once more, the usefulness of SIMAV was established by letting us model the reaction scheme represented by equations (12)-(14), thereby leading us to state that the ion in question was, undoubtedly, $Xe^+$. To illustrate this further, the density-reduced mobilities μN are shown plotted in figure 9 as a function of *E/N*. Note that the measured values (points) fit very well with an independent calculation using cross section data [8].

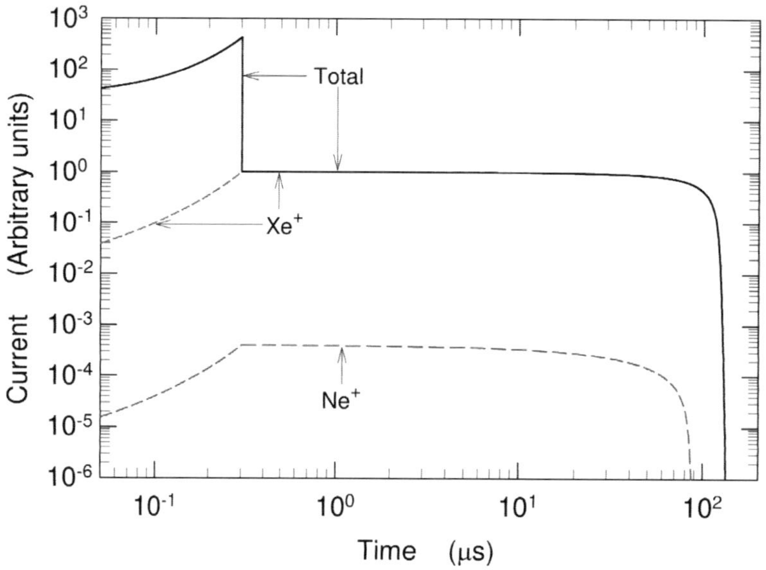


**Figure 8.** Calculated currents of $Xe^+$ and $Ne^+$ in the Xe-Ne mixture and the total current. Note that the scale is logarithmic. The conditions of the simulation are a 20% Xe-80% Ne mixture, $p$ = 5 Torr and *E/N* = 55 Td.

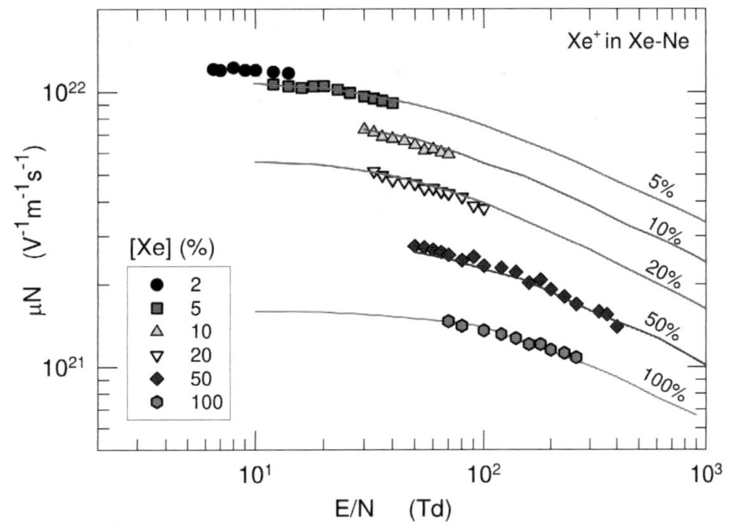


**Figure 9.** The density-reduced mobility, μN, of $Xe^+$ in Xe-Ne as a function of E/N. The points are the measured values while the solid lines are calculated values [8].

### *4.3. Penning ionization in $C_2F_4$-Ar mixtures*

Perhaps one of the least studied phenomena in Townsend discharges is that of Penning ionization in spite of its importance when molecular gases with a relatively low ionization threshold (IT) are mixed with rare gases (or some other) the level of metastable states of which exceed the IT of the molecular gases. Then positive ions from the molecular gas are formed, thereby constituting a second source of ionization in the realm of the discharge. It has already been shown that this extra source of ionization can be influential over certain E/N regions. This is hardly ever taken into account when modelling discharges mainly because of the lack of data such as metastable and Penning ionization rates.

Our case study deals with the ionization of $C_2F_4$ both by electron impact and Penning effect in the $C_2F_4$-Ar mixture. Measured electron transients indicated a marked pressure dependence of the electron drift velocity, a fact that is hard to understand. Thus we resorted to the following discharge scheme to consisting of direct ionization and attachment

$$e^- + Ar \rightarrow Ar^+ + 2\,e^- \tag{15}$$

$$e^- + C_2F_4 \rightarrow F^+ + 2\,e^- + C_2F_3 \tag{16}$$

$$e^- + C_2F_4 \rightarrow F^- + C_2F_3 \tag{17}$$

followed by the formation of metastable Ar*

$$e^- + Ar \rightarrow Ar^* + e^- \tag{18}$$

and, finally, Penning ionization

$$Ar^* + C_2F_4 \rightarrow F^+ + e^- + Ar + C_2F_3 \tag{19}$$

Figure 10 displays the measured and calculated electron currents in the 1%$C_2F_4$-Ar mixture at several pressures in the range 6-50 Torr, where the effects of Penning ionization are clearly distinguished by an enhancement of the apparent electron transit time in the measured curves (solid), and the simulated curves (dotted) that fit the measurements very closely. A single electron drift velocity was used. The enhancement of the apparent electron transit time is due to the presence of Ar* metastable that in turn produce a further positive ion cloud along the discharge gap. A film showing the evolution of the densities is available by writing to the authors.

*4.4. Electron detachment and ion conversion in $SF_6$*

Our next case deals with a study of the relative importance of electron detachment and negative ion conversion in the $SF_6$ avalanche [10]. The reaction scheme of $SF_6$ is very complex since it involves both positive and several negative ion species and electron detachment from $F^-$. Figure 11 shows the reaction scheme used in the simulations to be discussed further on.

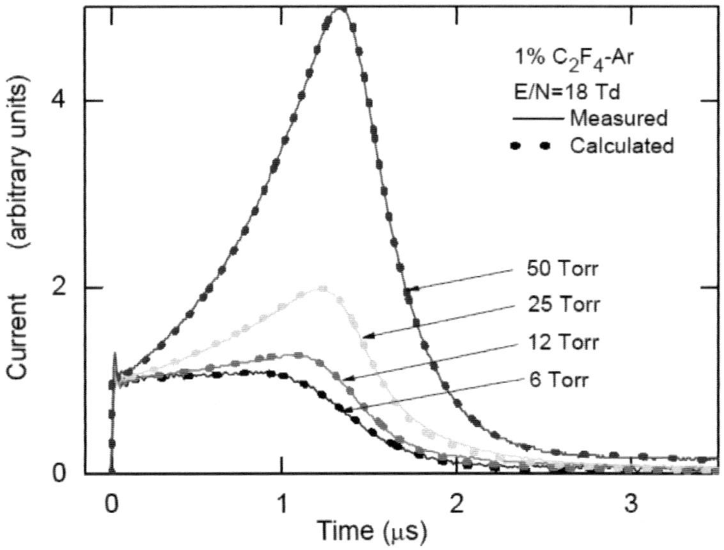


**Figure 10.** Measured electron transients (solid curves) in the 1% $C_2F_4$-Ar mixture and their fitting (doted curves) after using SIMAV with the discharge scheme of reactions (15) to (18). Conditions of the measurements are indicated on the inset of this figure.

One can see that there are 10 reactions and 6 charged species, thereby making it again impossible to model such scheme analytically. The calculated ionic avalanches are shown in figure 12. Looking at figure 12 one can see how the agreement between measurement and calculation improves as one includes more processes in the simulation. Note, in particular, the influence of electron detachment from $F^-$ on the positive ion current and that of $SF_6^-$ on the formation of the cluster species $SF_6$-$SF_6^-$ that accounts very well for the latter portions of the measured curve. All swarm data necessary for the simulations were already available in the literature with the exception of $\beta$, which was the only free parameter. Also, all parameters used for the simulations were the same with the exception of the electron detachment coefficient for which some pressure dependence was observed. A film showing the evolution of the densities is available by writing to the authors.

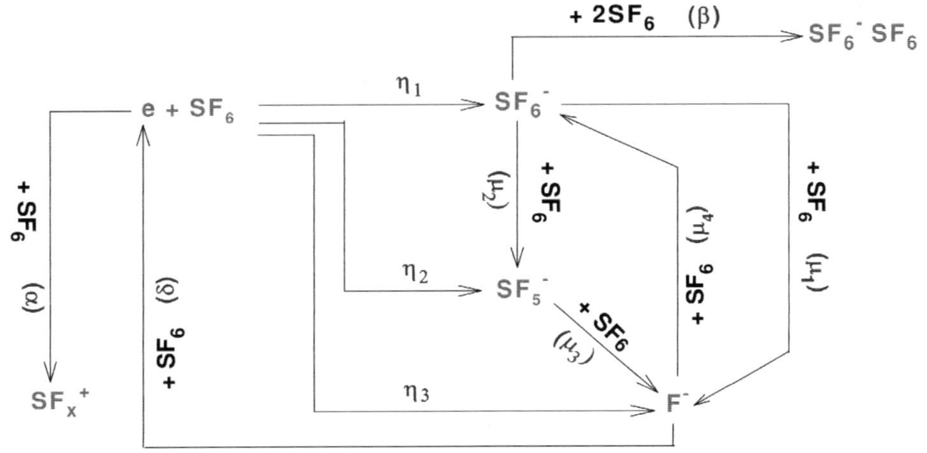


**Figure 11.** The discharge scheme used to simulate ionic avalanches in $SF_6$. $\alpha$ is the ionization coefficient, $\eta_1$ to $\eta_3$ are attachment coefficients, $\mu_1$ to $\mu_4$ are ion conversion coefficients (two-body), $\beta$ is a three-body ion conversion coefficient, and $\delta$ is the detachment coefficient.

*4.5. Secondary ionization by positive ion impact at the cathode*

Secondary electron emission processes in gases either by ion impact at the cathode or from photoionization are very important in controlling, among other processes, gas breakdown. The techniques most commonly used to study these phenomena are steady state experiments. Important efforts were made in the decade of 1960 [3] to study transient phenomena due to both secondary ionization processes. Unfortunately, for some reason these studies were not continued and, perhaps one of the reasons was, once more, the analytical difficulty in modelling these complex processes. Figure 13 shows an excellent oscillogram of a series of ionic avalanches produced in Ar by ion impact at a Ni cathode.

We have started simulating the process of secondary electron emission by ion impact at the cathode in $N_2$ gas. The pulses shown in figure 14 correspond to simulations with secondary emission coefficients of $\gamma=3\times10^{-7}$ and $\gamma=5\times10^{-7}$, showing a strikingly different behavior. While the avalanche for $\gamma=3\times10^{-7}$ shows a decaying series of positive ion successors, that for $\gamma=5\times10^{-7}$ indicates an upward behavior suggesting conditions close to breakdown. Also displayed are the contributions of electrons and positive ions. A film showing the evolution of the densities is available by writing to the authors.

## 5. Discussion

By means of five examples, four of them contrasting measurement and simulation, we have shown the extraordinary ability and usefulness of the avalanche simulator. Undoubtedly, SIMAV has let us gain more insight into collision processes hitherto not attempted because of the formidable analytical difficulties to solve the continuity equations of charge carriers subject to complex reaction schemes in which the equations became coupled, let alone the very particular initial and boundary conditions imposed by the experiment. Thus far SIMAV can only deal with situations in which space charge effects are either negligible or absent, such as those prevailing in the experiments herein described. Further attempts to simulate processes close to breakdown would demand full consideration of the additional contribution of local fields.

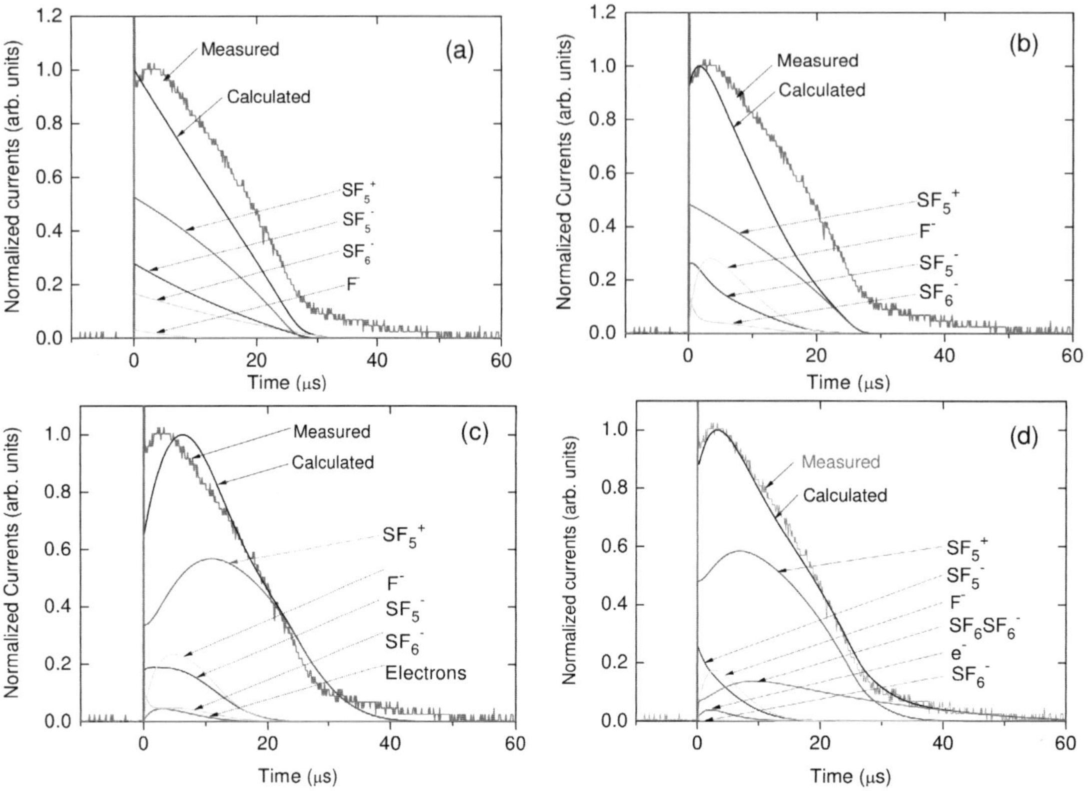


**Figure 12.** Measured (wriggled, red curves) and calculated (solid curves) currents in the $SF_6$ ionic avalanche at $p$ = 15 Torr, $E/N$ = 360 Td, $d$ = 2 cm, showing the evolution in the calculation as the discharge scheme becomes more complex [10]. (a) Only ionization and attachment are considered; (b) ionization, attachment and ion conversion; (c) ionization, attachment, ion conversion and electron detachment from $F^-$, and (d) the full scheme of figure 11 with ionization, attachment, ion conversion and electron detachment from $F^-$ and the three-body clustering reaction leading to $SF_6$-$SF_6^-$.

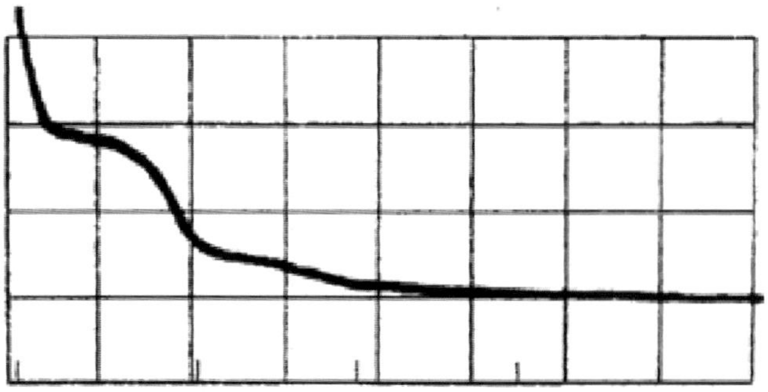

**Figure 13.** An ionic avalanche and its successors in Ar produced by positive ion impact at a Ni cathode with $p$ = 1.25 Torr, $E/N$ = 483 Td, $d$ = 1 cm, and a time base of 5 μs/div. The small vertical bars at the lower axis indicate the positions of the ion transit times of secondary avalanches as well as the start of the whole discharge. Taken from [3].

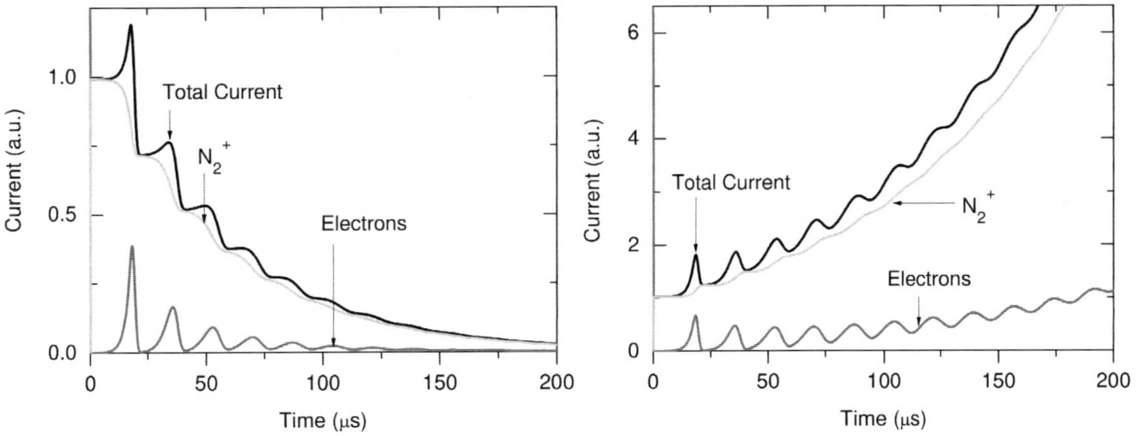


**Figure 14.** Ionic avalanches with successors in $N_2$ with secondary ionization coefficients of $\gamma=3\times10^{-7}$ (left) and $\gamma=5\times10^{-7}$ (right). Conditions of the simulations are $p$ = 780 Torr, $E/N$ = 130 Td, and $d$ = 1cm.

**Acknowledgements**
This work has been partially supported by PAPIIT-UNAM IN 111611. A. Bekstein acknowledges a postdoctoral grant from UNAM. Thanks are due to A. Bustos for technical assistance.

**References**
[1] Christophorou L G 1984 in *Electron-Molecule Interactions and their Applications* (Orlando:Academic Press)
[2] de Urquijo J, Juárez A M, Rodríguez-Luna J C and Ramos-Salas J 2007 *IEEE Trans. Plasma Sci.* **35** 1204
[3] Raether H 1964 *Electron Avalanches and Breakdown in Gases* (London: Butterworths)
[4] Hernández-Ávila J L, Basurto E and de Urquijo J 2002 *J. Phys. D: Appl. Phys.* **35** 2264
[5] de Urquijo J, Alvarez I and Cisneros C 1999 *Phys. Rev. E* **60** 4990
[6] Brambring J 1964 *Z. Physik* **179** 532
[7] de Urquijo J, Arriaga C, Alvarez I and Cisneros C 1999 *J. Phys. D: Appl. Phys.* **32** 41
[8] de Urquijo J, Basurto E and Bekstein A 2011 *J. Physics D: Appl. Phys.* **44** 325202
[9] de Urquijo J, Bekstein A, Ducasse O, Ruíz-Vargas G, Yousfi M and Benhenni M 2009 *Eur. Phys. J. D* **55** 637
[10] Bekstein A, Rodríguez-Luna J C, Ducasse O and de Urquijo J 2011 *Proc. XXX ICPIG* Belfast, UK

Journal of Physics: Conference Series **370** (2012) 012007 doi:10.1088/1742-6596/370/1/012007

# Measurements of ions and neutral particles in a plume generated by laser ablation of copper

**J B de Matos[1], C A B da Silveira[2] and N A S Rodrigues[2]**

[1]Technological Institute for Aeronautics-ITA, P.O. Box 6050, 12.228-900, Brazil
[2]Institute for Advanced Studies-IEAv, P.O. Box 6044, 12.231-970, Brazil

E-mail: julianamatos_ita@yahoo.com.br

**Abstract.** This paper presents the results for the measurement of neutral atoms and ion density in a Cu plume generated by laser ablation using electrostatic probes and a piezoelectric sensor. A tripled frequency Nd YAG laser (355 nm, 2 kHz, 23 ns, 10 kW) has been used to produce a plasma plume of copper in vacuum atmosphere ($10^{-4}$ Torr). It will be shown the effect of laser fluence on the relative densities of neutrals and ions departed from target. It was also observed that the ion density increases with laser fluence while neutral atoms remain roughly constant in the range of our experimental conditions. It was measured a drift velocity around 3-7 km/s and translational temperature in the range of 2.5-4.4x$10^4$(K).

## 1. Introduction

The ejection of particles from a solid or liquid surface by laser pulses is well known as laser ablation. In this process short laser pulses with high power densities lead to the formation of a plume made of neutral atoms, ions, clusters and droplets of liquid material [1]. Laser ablation is widely present in different fields of science and industry and a number of techniques have been employed over decades to analyze the plasma plume composition and dynamics, such as electrostatic probes [2], optical emission [3], spectroscopy [4], and spectrometry [5] among others. In this paper, we present results of study of relative population of neutral and ionized atoms of copper samples irradiated by Nd:YAG laser using pressure transducer device and electrostatic probes.

## 2. Experimental setup

A plasma plume was produced by laser ablation of copper samples inside a vacuum chamber kept at $10^{-4}$Torr. Nd:YAG laser pulses were focused onto the metallic target at 45° as sketched in figure 1. In order to measure total particles of plume, a pressure transducer, PVDF, was used at 150 mm from the target surface. The polyvinylidene fluoride (PVDF) film is a ferroelectric polymer which has pyro- and piezoelectric properties and gives information about plume drift velocity and temperature. Two parallel electrostatic plates were placed at 140 mm far from the target surface. For those distances no internal mechanism of plume acceleration was detected, it means that it remained in thermal equilibrium and expands according to a Maxwellian Velocity Distribution. A power supply (up to 1kV) was used to bias the electrode plate with negative voltages to extract ions of the plume.

Two experiments were performed, one using only PVDF to measure drift velocity and translational temperature of all particles of the plume, with values corresponding to 3-7 km/s and 2.5-4.4x$10^4$ K, respectively. In order to investigate the ratio of neutrals particles to ions in the plume, another experiment was accomplished with PVDF and simultaneously a pair of electrodes biased with different polarization voltages. The PVDF electrical signal was recorded in an oscilloscope for each polarization voltage as function of the incident laser fluence (20 up to 40 J/cm$^2$). In order to prevent degradation on the copper surface by prolonged radiation, the target was constantly in movement. Each measurement set was performed only after shot surface cleaning. Copper target was attached in controlled XY table to insure suitable velocity and precision to produce a plume with same characteristics for all laser ablation experiments. Figure 2 shows two typical PVDF curves concerning the effect of potential applied between the electrostatic plates.

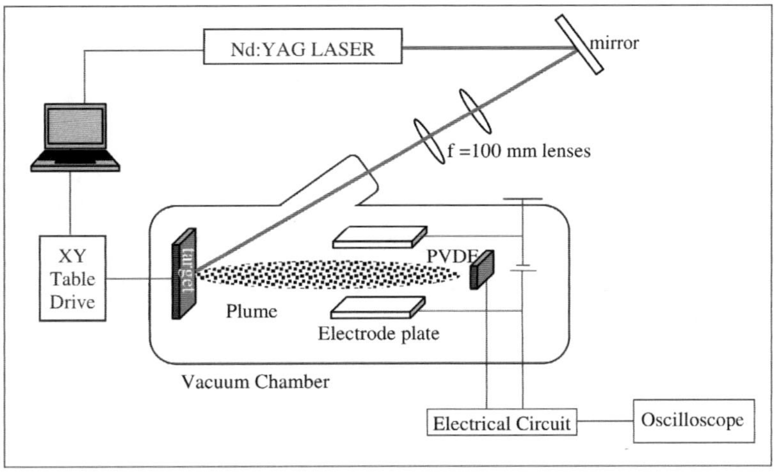


**Figure 1.** Experimental setup.

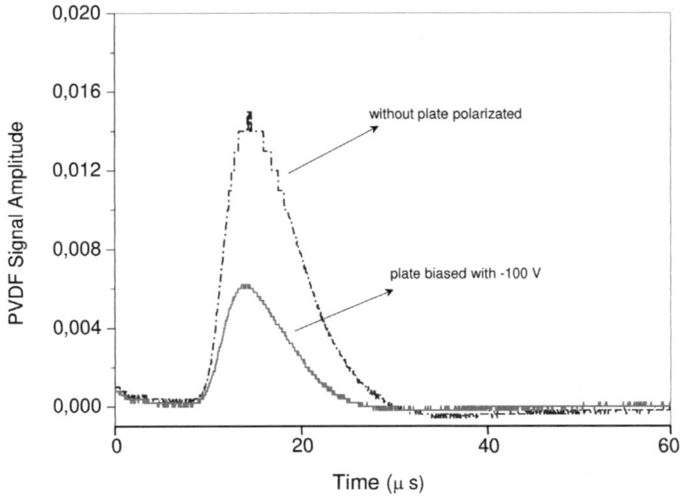


**Figure 2.** PVDF signal with polarized plate (solid line curve) and without potential between plates (dashed line curve).

Nd:YAG laser was used due to some special characteristics, in particular because it emits in low energy and high repetition rate, as shown in table 1. The laser beam illuminated a 25 μm diameter spot size by a set of two 100 mm focal length lens. It is well-known that not only laser characteristics but also chemical and physical properties of target and surrounding atmosphere play a very important role in laser ablation experiments. Cu samples were systematically polished and washed with acetone prior to the ablation experiments in order to remove any surface roughness, impurity or contamination.

**Table 1.** Nd:YAG laser parameters for laser ablation experiments

| Characteristics | Value | Unit |
|---|---|---|
| Wavelength | 355 | nm |
| Repetition Rate | 2 | kHz |
| Pulse Width | 23 | ns |
| Pulse Energy | 124-240 | μJ |
| Peak power | 5-10 | kW |
| Intensity | 0.8-1.7 | $GW/cm^2$ |
| Fluence | 20-40 | $J/cm^2$ |

## 3. Results and discussion

It was shown that the PVDF electrical signal is proportional to the linear moment transferred by particles of the plume to the sensor surface. The time-of-flight (TOF), when triggered by laser pulses, is given by [6]

$$S(t) \propto \frac{1}{t^5} \cdot \exp\left[ -\frac{m}{2K_B \cdot T} \cdot \left( \frac{l}{t} - v_0 \right)^2 \right] \tag{1}$$

where $l$ is the distance between target and sensor, the particles arrive to the sensor surface at the instant $t$ with velocity $v = l/t$, $m$ being the mass and $K_B$ Boltzmann constant. We assume the plume as a bunch of free particles far from Knudsen layer (150mm), and it expands according to Maxwellian Velocity Distribution with a drift velocity $v_0$ and translational temperature $T$. By adjusting the equation (1) to the PVDF electrical signal it is possible to infer the drift velocity and translational temperature, as it is shown in figure 3.

Some summarized results of $v_0$ and $T$ are presented in table 2 as function of laser fluence.

**Table 2.** Drift velocity ($v_0$) and translational temperature ($T$) measured with PVDF sensor as function of fluence in laser ablation experiments of copper target.

| Fluence ($J/cm^2$) | $v_0$ (km/s) | $T$ ($10^4$K) |
|---|---|---|
| 20 | 3 | 3.2 |
| 27 | 5.3 | 2.5 |
| 40 | 7.05 | 4.4 |

The relative abundance in ion density and neutral atoms present in laser ablation depend strongly on laser fluence. PVDF sensor was employed together with electrostatic plates to show how the laser energy density may affect directly on the ratio of ions to neutral atoms. Results present in figure 4 were obtained taking into account the maximum PVDF electrical signal and its dependence with plate polarization for different fluences. Each data represents the maximum peak of PVDF signal and all of them correspond to an average of 64 curves measured three times. As expected, it was observed that the amplitude of PVDF electrical signal tends to decrease as function of polarization in the electrode plates. Despite fluctuation observed in first data for 40 J/cm$^2$, it is evident a decreasing and it tends to a value of 90% of total particles as plate polarization increases up to 650V. It is possible to infer that the others 10% of the plume are made of neutral particles which are not affected by plates potential, as one can notice in the plateau for all curves. The same trend is observed for fluences between 20 J/cm$^2$ and 27 J/cm$^2$, and the reduction is in the range of 70% and 80% of total particles of the plume up to 30 V and 250 V applied, respectively. To our experimental conditions we conclude that ion density increases with fluence while neutral atoms remain roughly the same.

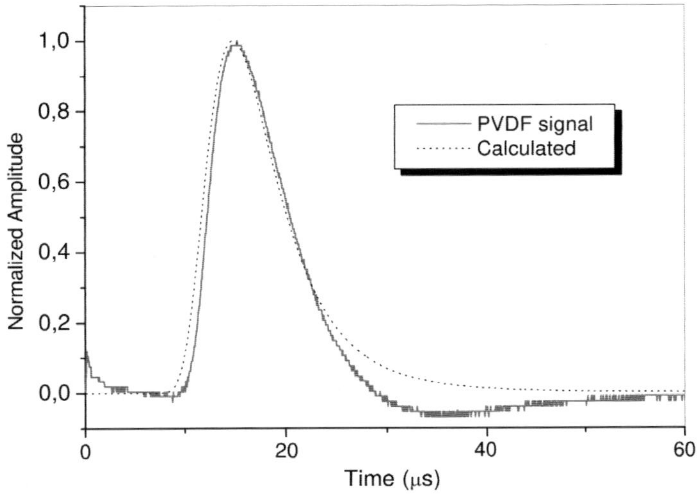


**Figure 3**. The theoretical curve (dashed line) was adjusted to PVDF signal (solid line) to obtain $v_0$ (m/s) and $T$(K).

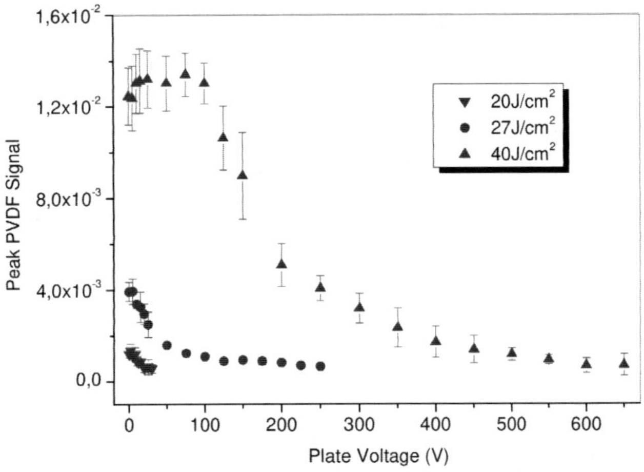


**Figure 4**. Effect of laser fluence in laser ablation experiments by using PVDF as a pressure sensor and electrostatic plates.

## References

[1] Amoruso S, Bruzzese R, Spinelli N and Velotta R 1999 *J. Phys. B: At. Mol. Opt. Phys.* **32** R131
[2] Donnelly T, Lunney J G, Amoruso S, Bruzzese R, Wang X and Ni X 2010 *J. Appl. Phys. A* **100** 569
[3] Guillermin M, Liebig C, Garrelie F, Stoian R, Loir A-S and Audouard E 2009 *Appl. Surf. Sci.* **255** 5163
[4] Knight A, Scherbarth N, Cremers D and Ferris M 2000 *Appl. Spectrosc.* **54** 331
[5] Russo R, Mao X, Gonzalez J and Mao S 2002 *J. Anal. At. Spectrom.* **17** 1072
[6] Gião M, Rodrigues R, Riva R and Schwab C 2004 *Rev Sci. Instrum.* **75** 5213

# Plasma species dynamics in a laser produced carbon plasma expanding in low pressure neutral gas background

**H M Ruiz, F Guzmán, M Favre, H Bhuyan, H Chuaqui and E Wyndham**

Departamento de Física, Pontificia Universidad Católica de Chile, Av. Vicuña Mackenna 4860, Santiago, Chile

E-mail: hmruiz@uc.cl

**Abstract**. We present time and space resolved spectroscopic observations of a laser produced carbon plasma, in an argon background. An Nd:YAG laser pulse, 370 mJ, 3.5 ns, at 1.06 μm, with a fluence of 6.8 J/cm$^2$, is used to produce a plasma from a solid graphite target, at a base pressure of 0.5 mTorr, and with 80 mTorr Argon background. The spectral emission in the visible is recorded with 15 ns time resolution. 20 ns time resolution plasma imagining, filtered at characteristic carbon species emission wavelengths, is used to study the dynamics of the expanding plasma. Two different fronts with ionic or molecular compositions are seen to detach from de laser target plasma.

## 1. Introduction

Several investigations have been reported regarding fundamental processes of carbon plasma dynamics and plasma chemistry in laser produced carbon plasmas [1,2]. Carbon and diamond like (DLC) thin film deposition using graphite targets has been investigated, using different parameter regimes [3]. In this context we have investigated carbon film deposition using a graphite target laser plasma in an argon background. Preliminary analysis of the resulting films indicates a correlation between films properties, as inferred from AFM and other standard materials science diagnostics, with the pressure of the argon background. In order to establish a relationship between the carbon plasma and the resulting films properties, we have studied the time and space evolution of the laser produced carbon plasma. Here, we present time and space resolved spectroscopic observations of the laser plasma.

## 2. Experimental set-up

The experiments were performed in a vacuum chamber at a base pressure of 0.5 mTorr. An Nd:YAG laser pulse, 370 mJ, 3.5 ns, at 1.06 9m, operating at 10 Hz was applied to a rotating graphite target, at approximately 45$^\circ$ to the normal. The focal spot was 0.0535 cm$^2$, thus resulting in a fluence of de 6.87 J/cm$^2$. Time resolved spectral observations of the carbon plasma plume were performed, at base pressure and at 80 mTorr argon background. The visible spectra were obtained with a Spectra Pro 275 (1200 g/mm) spectrometer, with a gated avalanche array with a time window of 15 ns. Light emission from the carbon plasma plume was collected with a f-f fiber optic arrangement, focusing at 2.5, and 4.5 mm, from the target surface. Plasma images, filtered at characteristic carbon species emission wavelengths, were recorded, with 20 ns time resolution.

### 3. Experimental results

Figure 1 shows a time integrated spectrum obtained from a point 2.5 mm from the target surface, with 80 mTorr argon background. The spectrum shows emission lines associated with single ionized Carbon, single ionized Argon, and $C_2$ molecular bands, corresponding to vibrational states of the $d^3\Pi_g \rightarrow a^3\Pi_u$ Swan band transition [4]. The emission band around 405 nm has been assigned previously to transitions $A_1\Pi_u \rightarrow X_1\Sigma^+_g$, Swing band, of the neutral $C_3$ molecule [5]. A residual second harmonic laser emission line can also be identified, which is due to the optical configuration of the Nd:YAG laser cavity.

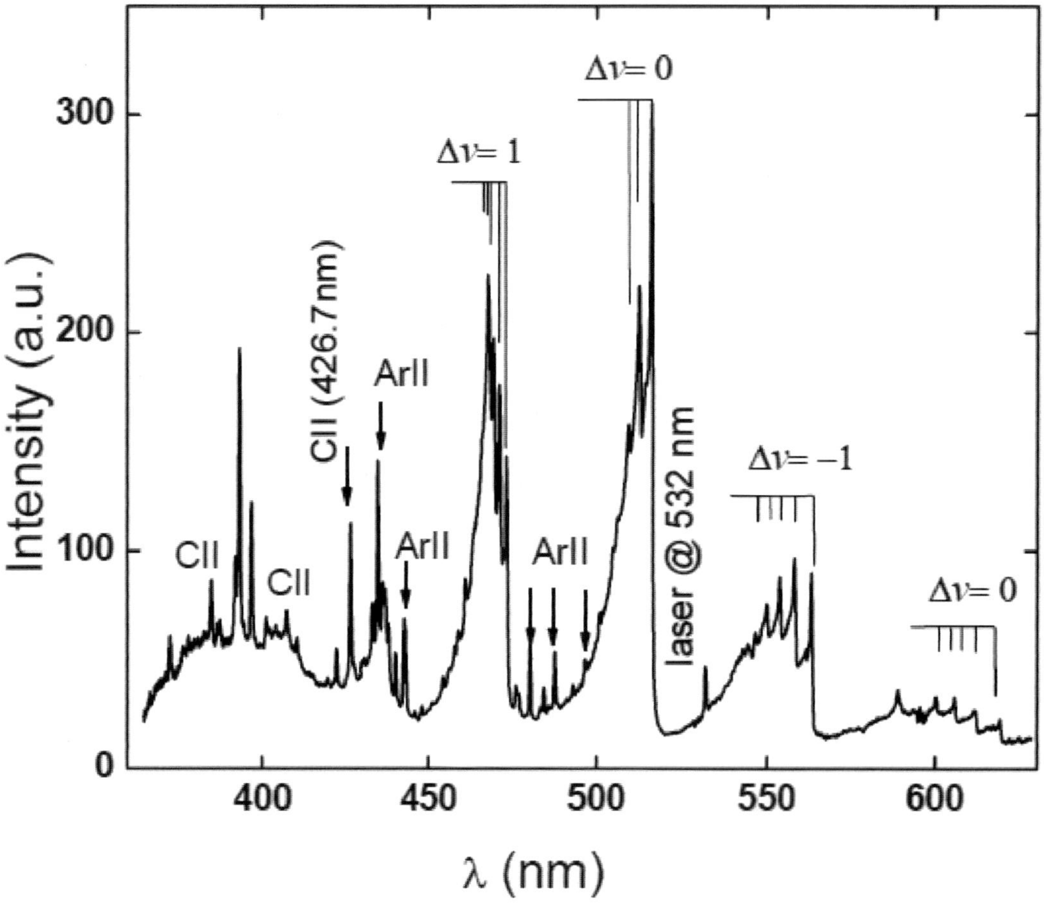


**Figure 1.** Time integrated spectrum, corresponding to light emission 2.5 mm $v = 0$ from the target surface, with 80 mTorr argon background. Emission from CII ions, ArII ions and $C_2$ molecular bands are identified.

Figure 2 shows the time evolution of light emission from a point 4.5 mm from the target surface, with 80 mTorr Argon pressure. The spectra correspond to emission from a $C_2$ molecular band ($\Delta v = 1$), centered at 467.5 nm, and CII ions at 426.78 nm. At early times, ArII emission lines are observed, which disappear when the CII and $C_2$ emission become prominent.

To investigate the dynamic features of the different carbon species in the expanding laser plasma we obtained plasma images using 10 nm band -pass filters centered at 470 nm, which correspond to the $C_2$, $\Delta v = 1$, Swan band. Figure 3 shows the dynamic evolution of the plasma emission from, (a) $C_2$ carbon molecules at the base pressure of 0.5 mTorr and, (b) $C_2$ carbon molecules at 80 mTorr Argon background. At the base pressure, the $C_2$ component is seen to remain nearly stationary. The dynamics is much more complex at 80mTorr Argon. In fact, two distinctive, expanding plasma fronts are seen to detach from the near target semi stationary plasma. The first one, seen in the 66 to 330 ns time interval, is a rather homogeneous front, but the second one, from 750 ns onwards, exhibits a polar dependence with respect to the normal to the target, with a noticeable decrease in brightness close to the cloud axis.

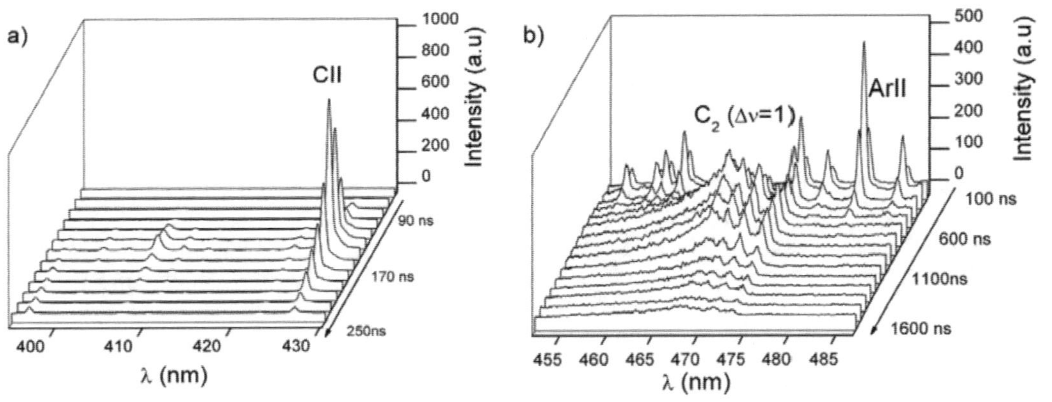


**Figure 2.** Time evolution of light emission from a point 4.5 mm from the target surface, with 80 mTorr Argon pressure. (a) $C_2$ molecular band ($\Delta v = 1$), centered at 467.5 nm, and (b) CII ions at 426.78 nm.

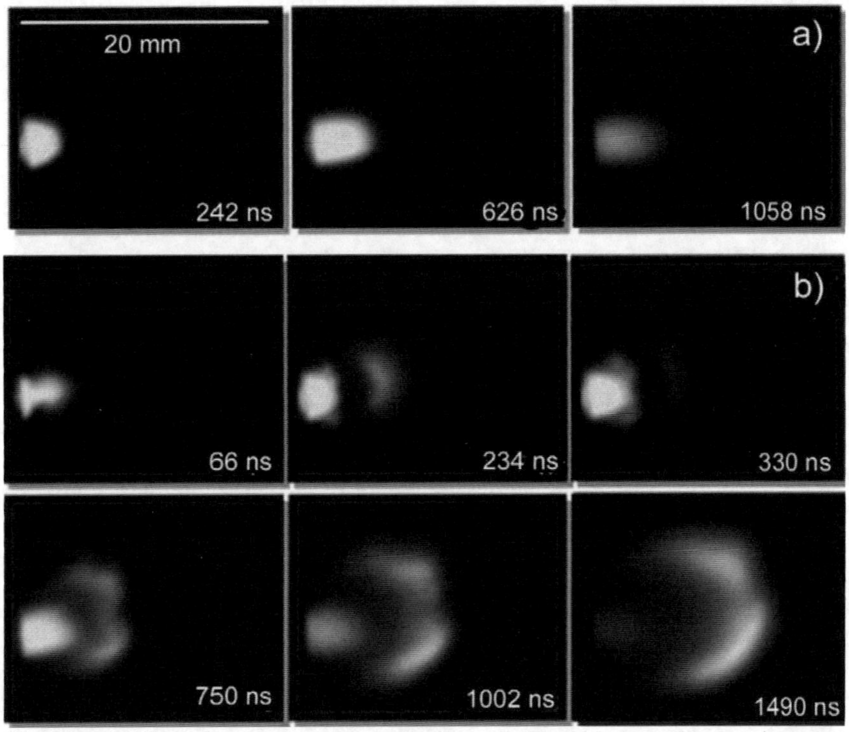


**Figure 3.** 20 ns exposure, spectrally resolved laser plasma images: (a) 0.5 mTorr base pressure, (b) 80 mTorr Argon.

## 4. Discussion

Several results have been reported on laser produced Carbon plasmas, using different laser fluences and gas backgrounds. At a high fluence of 50 J/cm$^2$, with Helium as background gas, the formation of $C_2$ molecules has been studied, using both, OES and plasma imaging [2]. It has been found that in vacuum $C_2$ formation is only seen near the target surface, but with Helium background, an expanding front of $C_2$ molecules is observed. In another experiment, at a higher fluence of 955 J/cm$^2$, two distinctive fronts, a fast and a slow one, have been identified in the temporal evolution of the $C_2$ rich laser Carbon plasma [1]. Ionization of the background neutral gas has been reported with a laser produced Tin plasma, in 100 mTorr Argon background [6]. Single ionized Argon background emission lines were found before Tin emission lines, due to laser plasma expansion, were observed. This might be caused by fast electrons produced at the early stages of the laser plasma, which propagate ahead of the laser plasma front.

Our time integrated spectrum in figure 1 shows single ionized Argon emission, from a point 4.5 mm from the target surface. Further time resolved observations indicate that the ionized Argon emission occurs before the Carbon molecules front is formed, as shown in figure 2(b). When the time resolved spectral emission seen in figure 2 is correlated with the time resolved images shown in figure 3, it can be concluded that the presence of ArII and CII ions coincide in space and time with the first plasma front, seen in figure 3(b).

## 5. Conclusions

We have investigated the role placed by Argon background in molecule formation in laser produced carbon plasmas. At the base pressure of 0.5 mTorr $C_2$ molecules are only observed near the target surface. At 80 mTorr Argon, two successive plasma fronts are observed. The first one is dominated by CII ions and propagates in a weakly ionized Argon background. The second one is dominated by the presence of $C_2$ molecules. It is evident that the Argon background favors Carbon molecule formation through three-body recombination processes. These observations contribute to understand morphological properties of thin Carbon films obtained by pulsed laser deposition, under identical experimental conditions, reported elsewhere.


## Acknowledgements

This work has been funded by FONDECYT project 1110380. H. M. Ruiz and F. Guzmán acknowledge doctoral studies scholarships from MECESUP and CONICYT, respectively.


## References


[1] Harilal S S, Issac R C, Bindhu C V, Nampoori V P N and Vallabhan C P G 1997 *J. Appl. Phys.* **82** 2140

[2] Al-Shboul K F, Harilal S S, Hassanein A and Polek M J 2011 *J. Appl. Phys.* **110** 053301

[3] Qian F, Singh R K, Dutta S K and Pronko P P 1995 *Appl. Phys. Lett.*. **67** 3120

[4] A Kushwaha and R K Thareja 2008 *Applied Optics* **47** G65

[5] Nemes L, Keszler A M, Parigger C G, Hornkohl J O, Michelsen H A and Stakhursky V 2007 *Applied Optics* **46** 4032

[6] Harilal S S, O'Shay B, Tao Y and Tillack M S 2006 *Journal of Applied Physics* **99** 083303

Journal of Physics: Conference Series **370** (2012) 012009 doi:10.1088/1742-6596/370/1/012009

# Adhesion improvement of carbon-based coatings through a high ionization deposition technique

**E Broitman and L Hultman**

Thin Film Physics Division, IFM, Linkoping University, SE 581 83 Linköping, Sweden

E-mail: esbro@ifm.liu.se

**Abstract**. The deposition of highly adherent carbon nitride ($CN_x$) films using a pretreatment with two high power impulse magnetron sputtering (HIPIMS) power supplies in a master-slave configuration is reviewed. SKF3 (AISI 52100) steel substrates were pretreated in the environment of a high ionized Cr+Ar plasma in order to sputter clean the surface and implant Cr metal ions. $CN_x$ films were subsequently deposited at room temperature by DC magnetron sputtering from a high purity C target in a $N_2$/Ar plasma discharge. All processing was done in an industrial-scale CemeCon CC800 coating system. A series of depositions were obtained with samples pretreated at different bias voltages (DC and pulsed). The adhesion of $CN_x$ films, evaluated by the Daimler-Benz Rockwell-C test, reaches strength quality *HF1*. Adhesion results are correlated to high resolution transmission electron microscopy observations confirming the formation of an optimal interfacial mixing layer of Cr and steel. The throwing power increase for HIPIMS coatings is associated to the higher ionization in the plasma discharge.

## 1. Introduction

During the last two decades, carbon-nitride ($CN_x$) coatings deposited by physical vapor deposition (PVD) techniques have enjoyed a growing interest in several disciplines. Amorphous and fullerene-like films exhibit very attractive properties, such as low friction and wear, high hardness, good chemical stability, and high resiliency to deformation, which make them suitable for replacing diamond-like carbon (DLC) films [1]. The first successful industrial application of this material has been the use of very thin (~2-5 nm) amorphous $CN_x$ films for the protection of hard disk drives [2-3]. However, commercial use of thicker (1-5 μm) coatings on steel substrates has been difficult because of the development of high compressive intrinsic stresses during deposition, which causes adhesion failures with delamination of films from the substrate surface [4].

The common practice to increase the adhesion of carbon-based coatings on steel substrates is to make a pretreatment of the substrates and the use of a metallurgical glue layer interposed between the surface substrate and the film. In this paper we present the deposition of highly adherent $CN_x$ films on martensitic steel substrates using a high plasma ionization pretreatment [5-6]. The plasma-processing methods used for improving adhesion of carbon-based coatings are reviewed in Section 2. A description of our original pretreatment and the experimental techniques are considered in Section 3. Section 4 presents the results and discussions, and Section 5 gives the conclusions.

Journal of Physics: Conference Series **370** (2012) 012009 doi:10.1088/1742-6596/370/1/012009

## 2. Review of plasma-processing pretreatments

### *2.1. Glow discharge cleaning*

*In-situ* surface preparation is an essential step in vacuum coating processing and is critical to assure a good film-substrate adhesion. For example, air-exposed stainless steel is covered by a native oxide layer (from 1 to 5 nm thick) composed of metal oxides and hydroxides [7]. Despite of pre-cleaning processes, other contaminants like organic and water molecules will also be present on the metal surface. Without the elimination of the oxide and contaminants, it is likely that any coating deposited on the surface will delaminate.

For many years, it has been known that cleaning of substrates using plasma techniques prior to physical vapor deposition processes can significantly improve coating adhesion. Such bombardment is expected not only to remove unwanted contaminant layers at the substrate surface, but also to create active dangling bonds, which may promote adhesion of the deposited film [8]. The first *in-situ* plasma cleaning technique was developed in 1935 by Strong for the cleaning of glass surfaces used in astronomical telescopes [9]. The equipment for this technique, included nowadays in basic commercial coating machines, consists basically of a DC high voltage "glow bar" that allows the formation of plasma. The surfaces in contact with the plasma acquire a negative voltage (several volts) with respect to the plasma. This sheath potential accelerates positive ions to the surface. When an ion contacts the surface it combines with electrons and releases its energy of ionization (5-20 eV). This low energy bombardment and heating desorbs adsorbed contaminants such as water vapor. This "ion scrubbing" can be done using argon or reactive gases like oxygen, chlorine, and hydrogen, the later are used to clean contaminants like hydrocarbons. In this technique, also known as "glow discharge cleaning" because of the light generated by the plasma, there is no bias voltage applied to the substrate (figure 1). The technique was popular in the 1960's and most of laboratory and commercial-scale PVD deposition systems commercialized in that decade were including this option [10].

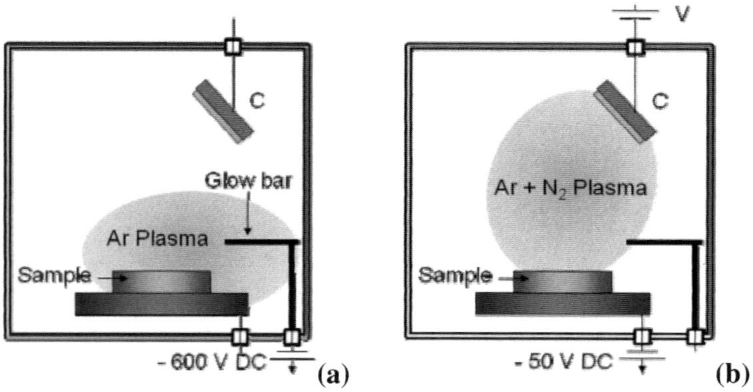


**Figure 1** (a) Glow discharge cleaning; a DC high voltage bar allows the formation of plasma. No bias voltage is applied to the substrate. (b) Deposition of carbon nitride films after substrate pre-treatment.

### *2.2. Sputter ion cleaning*

One of the most successful techniques for adhesion enhancement has been the pre-treatment of the substrate surface with a low energy (~500 eV) $Ar^+$ ions bombardment, *in-situ* before vacuum vapor deposition of the films [11]. The first report of this technique is back to 1955, when Farnsworth *et al* [12] reported using sputter cleaning with $Ar^+$ ions to prepare ultra-clean surfaces for low-energy electron diffraction studies. Later, sputter cleaning was adopted by D. M. Mattox [13] as a part of his *"Ion-Plating"* technique. Mattox's process, with a filament used to evaporate a metal and a DC voltage used to accelerate ions toward a biased substrate, evolved rapidly with the introduction of the DC magnetron sputtering technique for the deposition of the films. Before deposition, the plasma cleaning is performed by ions generated in the cathodes, while the acceleration of the ions toward the substrate to be cleaned can be performed by a negative DC voltage applied to the substrate [14]. Figure 2 shows a typical process where the ions are generated by a magnetron cathode with a Cr target and the bias is provided by a negative DC voltage: the magnetron generates mainly $Ar^+$ ions because the efficiency of Cr ion generation is very low. Radio frequency (RF) biasing has also been used since the 1970s to induce the energetic particle bombardment of insulating substrates. RF bias supplies, usually operating at 13.56 MHz, are connected to the substrates via a matching network [15].

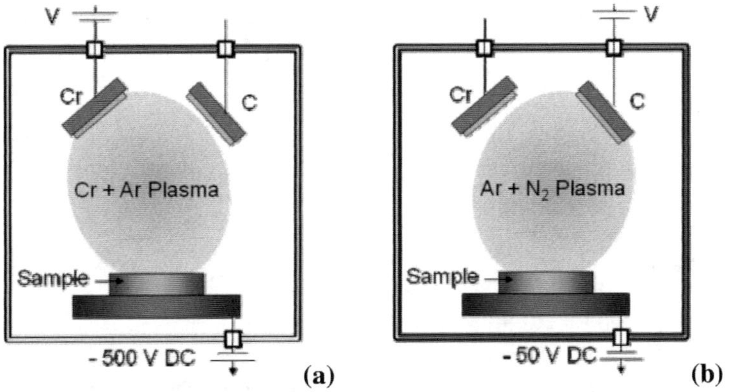


**Figure 2.** (a) Sputter ion cleaning; ions generated by a DC magnetron with a Cr target. A DC bias voltage is applied to the substrate. (b) Deposition of carbon nitride films after substrate pre-treatment.

In industrial production, where it is common to have a large deposition area, $Ar^+$ ions are usually provided by external sources. The "plasma booster" technology, introduced at the beginning of the 1990's by researchers from the company Leybold [16] has been evolving with the discovery of other ion sources. Anders [17] has recently reviewed some of these ion sources, such as hollow cathode discharges, linear sources of end-Hall type, and capacitively or inductively coupled plasma sources.

There are two drawbacks of this cleaning method. The first one is the substrate heating during ion bombardment that can change, per example, the properties of some temperature sensitive substrates like steels. Another disadvantage is the incorporation of the gas ions into the substrate. An inert gas like Ar can occupy interstitial sites and induce increased strains in the substrate lattice. These high strains can embrittle the substrate material by bringing it closer to its yield stress. Furthermore, when heated during pretreatment or under use, Ar could diffuse and agglomerate into bubbles, introducing porosity and weakening of the interface [18].

The use of unipolar pulsed substrate biasing in cathodic arc evaporation was introduced in 1991 by Olbrich et al [19] to reduce thermal input associated with ion bombardment. With the advent of pulsed-DC target sputtering, bias pulsing techniques with frequencies in the range 20-350 kHz become more popular in the late 1990s because it was then realized that it could provide improved control of substrate bombarding ion energy and flux, and be utilized to bias substrates during reactive deposition of insulating films [20]. Unipolar pulsed-DC substrate biasing is often carried out in combination with pulsed-DC target sputtering. In such situations, both target and substrate are pulsed at the same frequency (usually 100 kHz) in a synchronous "master-slave" configuration [21-23].

### *2.3. Ion-metal implantation*

An improvement in the adhesion of carbon-based coatings has been the pretreatment in metal arc discharge environment. Bombardment with metal ions is known to provide cleaner interfaces [24]: cathodic arc discharges produce a high flux of metal ions, which bombard the surface of the substrate and remove oxide layers. That was the base for the development in the 1990's of the so-called ABS (Arc Bond Sputtering) technique where a cathodic arc is used to generate metal ions that are accelerated by a DC bias voltage toward the substrate in order to produce an etching prior to the film deposition by magnetron sputtering [25]. As it has been pointed out by many researchers, the multiply ionized metal ions, which are produced in the vapor of the cathodic arc discharge, not only clean the substrate, but also produce low-energy implantation promoting the formation of a graded interface between the substrate and the coating. The formation of such interface provides a relief of film stress, increasing the adhesion of the coatings [26-28].

Journal of Physics: Conference Series **370** (2012) 012009 doi:10.1088/1742-6596/370/1/012009

The effects of DC bias voltage on the composition and structure of the graded interface for enhanced adhesion on steel has been the subject of many studies. For example, Schönjahn *et al* studied Cr arc etching of M2 mild steel, where an implantation/diffusion layer of Cr in the steel was observed at DC bias voltages of -800 to -1200V [29]. One of the problems of etching with cathodic arc is the production of molten particles which are deposited onto the surface to be cleaned. Münz *et al* investigated the generation of droplets by many target materials, including Cr [30]. They found that the number and size of the droplets depend on the melting point of the metal used during the ion etching, with a typical droplet size of 100 nm.

*2.4. HIPIMS technique*

Recently, the use of high power impulse magnetron sputtering (HIPIMS) technology has been proposed to increase the adhesion of coatings on metal substrates [31]. HIPIMS produces a high flux of ions with high metal content similar to an arc discharge, but without droplets. The ions (gas and metals) are accelerated to the substrate by the use of a DC bias, producing a high rate of substrate etching and also the implantation of metal atoms like in arc discharges.

The metal ion etching by HIPIMS using a DC bias voltage has been first reported by Ehiasarian *et al* [32], as in the scheme shown in figure 3. The highly ionized plasma conditions of HIPIMS with 80% metal ions resulted in a more efficient ion bombardment, improving the etching of the substrate and increasing the adhesion of the coatings. It has also been claimed that the presence of a DC bias voltage facilitates the formation of a shallow metal implantation on the surface, promoting the formation of an adherent interface between the substrate and the coating [33].

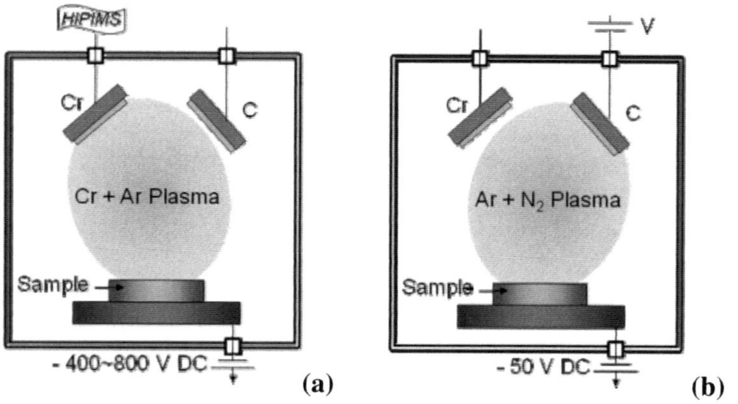


**Figure 3.** (a) Ion-metal cleaning and implantation; ions generated by a HIPIMS magnetron with a Cr target. A DC bias voltage is applied to the substrate. (b) Deposition of carbon nitride films after substrate pre-treatment.

Different ion metal etching prior to the deposition of coatings have been reported: Cr or Nb for CrN [18, 34, 35] , V for TiAlCN/VCN multilayers [36], Cr for CrAlYN/CrN multilayers [37, 38] , WC for W-DLC and C-DLC multilayers [39], and Cr for DLC and metal-DLC coatings [40]. Besides our previous work [5-6], to our knowledge there are not reports on the deposition of $CN_x$ (carbon nitride) coatings using the HIPIMS interface sputter cleaning, or on the use of two HIPIMS power supplies for etching (one connected to the cathode and one to the table) [41] .

The most important problem to solve during the substrate pretreatment by HIPIMS for the cleaning and interface formation is the high current flowing from the cathode to the substrates. In some DC power supplies used for biasing, this high current is interpreted by the bias power supply as the presence of arcing, which causes the power supply to interrupt the operation. This problem has been partially overcome installing large capacitor-resistors banks or special power supplies with filters that can stand the current pulse arriving to the substrates [42, 43]. However, the constancy of the bias during the discharge will depend on the size of the capacitors and resistors. From an industrial point of view, the drawback is that this device occupies a vital space in the chambers, especially if a nearly constant bias voltage is desired.

Based on our innovation [5] we have recently reported results for the deposition of highly adherent $CN_x$ films on steel substrates using a novel HIPIMS pretreatment in an industrial production chamber [6]. During the pretreatment, the substrates were pulse-biased in the environment of a HIPIMS Cr plasma in order to sputter clean the surface and to implant $Cr^+$ metal ions, as shown in figure 4. Subsequently, $CN_x$ films with superior adhesion to the steel substrates were obtained by DC unbalanced magnetron sputtering [6].

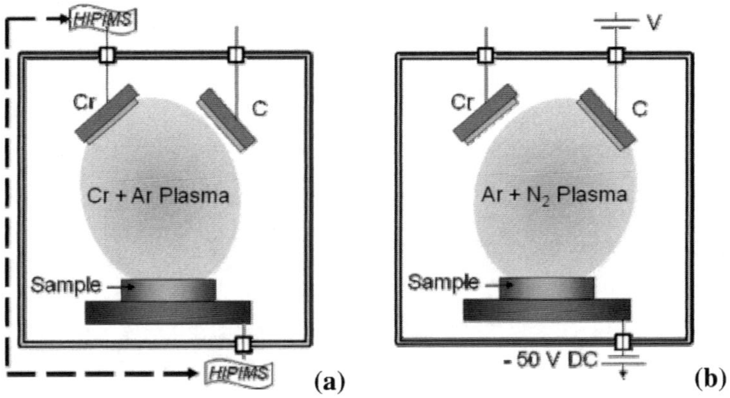


**Figure 4.** (a) Ion-metal cleaning and implantation; ions generated by HIPIMS magnetron sputtering with a Cr target. A HIPIMS bias voltage is applied to the substrate in a master-slave configuration.
(b) Deposition of carbon nitride films substrate after pretreatment.

## 3. Experiments

The experiments were carried out in a CC-800/9 industrial system manufactured by CemeCon AG in Würselen, Germany. The chamber (85 x 85 x 100 cm) is equipped with four rectangular magnetron sputtering cathodes (50 x 8.8 cm) facing a table holder that can provide up to three-fold rotation to the substrates. Due to the nature of our experiments only two out of four cathodes mounted in the system were used. The system is equipped with DC magnetron power supplies Pinnacle 3000 (Advance Energy) and HIPIMS power supplies SINEX 3 (Chemfilt Ionsputtering AB, Linköping, Sweden). The targets materials were Cr and C (99.8% pure). The substrates used in the experiments were Si (100) wafers and SKF3 (AISI 52100) bearing stainless steel.

Cross-sectional specimens for transmission electron microscopy (TEM) investigations were made by gluing slices into Ti disks followed by mechanical polishing and grazing incidence (4°) Ar+ ion thinning at 10 kV to electron transparency using a Technoorg Linda IV6 ion milling unit. The ion energy was decreased in the final period of the ion milling procedure down to 500 eV to decrease the ion damage. The high-resolution TEM (HRTEM) investigations were made in a Technai G2 and a JEOL 3010 electron microscope operated at 200 and 300 kV, respectively. EDS line scans were recorded in the Technai G2 microscope. Cross-sectional scanning electron microscopy (SEM) experiments were made using a LEO SEM equipment operated at 5 kV.

### *3.1. Substrate pretreatment and coating*

The metal substrates were mirror polished by mechanical polishing with a sequence of 6, 3, and 1 μm diamond pastes, and then were cleaned using a sequence of ultrasonic washing stages with alkaline detergents (Decon-90 at 5%), de-ionized water rinsing and air-drying. Prior to loading into the chamber, all the substrates were ultrasonic cleaned with acetone and isopropanol, and blow-drying with nitrogen gas.

Different substrate pre-treatments were carried out prior to the deposition of the CNx films (see figures 2, 3, and 4):

- a first group of substrates (figure 2) were biased to a DC voltage of – 1200 V while the Cr target was operated in a DC mode at a pressure of 3 mTorr of Ar and a current of 1 A during 30 minutes;

- a second group of substrates (figure 3) were biased to a DC voltage of -600 V while a HIPIMS source was operated at a pressure of 3 mTorr, with a peak target voltage of 1100 V, a frequency of 150 Hz, and a pulse width of 100 s, during 30 minutes;
- the third group of substrates (figure 4) were biased to a HIPIMS source with peak voltages of in the range 400-800 V, a frequency of 150 Hz, and a pulse width of 200 s. The target was also connected to a second HIPIMS source operated at a pressure of 3 mTorr, with a peak target voltage in the range 500-1100 V, a frequency of 150 Hz, and a pulse width of 100 s, during 30 minutes. Both HIPIMS power supplies were synchronized, being the one in the target the "master" and the one in the substrate the "slave". The voltage pulse in the target was programmed to start 10 s after the pulse in the substrate in order to assure the stability of the bias voltage (figure 5)

The coating step was carried out by DC magnetron sputtering of a C target in a reactive atmosphere of Ar and $N_2$ (partial pressure ratio of 84:16 and total pressure of 3 mTorr), target power of 1800 W, and a negative bias voltage of 25 V.

No heating treatment was used during the pretreatment or the deposition steps.

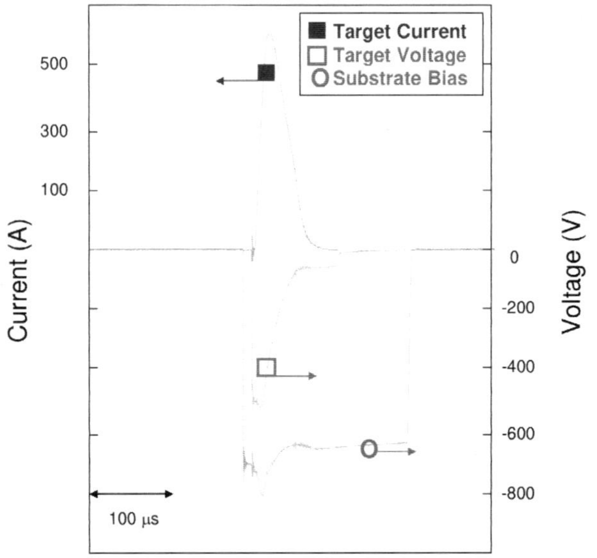


**Figure 5.** Target current (■), target voltage (□), and substrate bias voltage (O) waveforms recorded in one of the experiments of the configuration illustrated in figure 4.

*3.2. Cathode current and plasma characterization*

The target current was measured with a Tektronix CT-04 high current transformer together with a Tektronix TCP202 current probe mounted right at the connection to the vacuum chamber. A Tektronix P6015 high voltage probe was used to measure the cathode voltage. Current and voltage waveforms were recorded with the Picoscope 3000 real time digital storage PC oscilloscope from Pico Technology.

The optical emission from the plasma was measured in the light-of-sight geometry through a port window of the chamber and the probe was directed towards the race track. A spectrometer (Mechelle Sensicam 900) connected to a collimator via an optical fiber was used to record the emission from the plasma. The spectral range of the spectrometer was 300-1100 nm. The optical emission spectra were acquired at argon gas pressures ranging between 0.5 and 10 mTorr and for different target voltages (implying different peak power levels). The shutter speed of the spectrometer was set to 100 ms, meaning that all data were time-averaged over ca. 30 pulses (the pulsing frequency was fixed at 300 Hz).

### *3.3. Throwing power*
The throwing power of the Cr sputtered atoms was evaluated on Si substrates oriented at five different angles with respect to the cathode surface: 0°, 45°, 90°, 135°, and 180° from the horizontal. The Cr films were deposited by DC magnetron sputtering and HIPIMS, as shown in figure 2(a) with a DC bias of 60 V, and figure 4(a), using a HIPIMS bias of 60 V. After deposition, the coating thickness was measured by SEM on cross-sectional samples. The throwing power was calculated as a % of the film thickness relative to the film deposited in front of the surface cathode (0° from the horizontal).

### *3.4. Adhesion test*
The adhesion was measured by the Daimler-Benz Rockwell-C adhesion test. Developed by the Union of German Engineers (Verein Deutscher Ingenieure, VDI [44]), it uses a standard Rockwell hardness tester fitted with a Rockwell C-type diamond cone indenter with an applied load of 150 kg. The result is obtained by using an optical microscope and classifying the adhesion as *HF1* to *HF6* according to the level of cracking and coating delamination around the indent (figure 6). Only indents classified as *HF1* and *HF2* correspond to adequate adhesion [45].

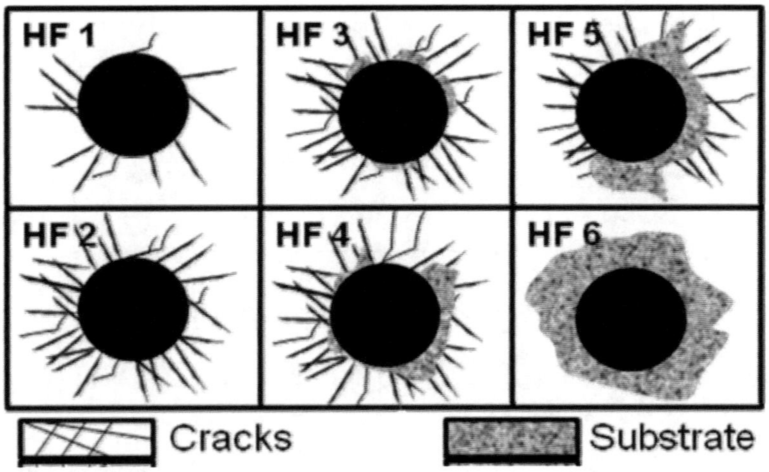


**Figure 6.** The Daimler-Benz Rockwell-C adhesion test chart (adapted from reference 44).

## 4. Results and discussion

### *4.1. Plasma chemistry during pre-treatment*
The chemistry of the ion flux during the HIPIMS pretreatment was studied by optical emission at different energies. In figure 7 the optical emission intensities of the strongest lines are plotted for $Cr^+$ ions and Cr neutrals, respectively, versus energy per pulse at an argon gas pressure of 3 mTorr [6]. The emission lines chosen for comparison are Cr-I at 399.2 nm and Cr-II at 336.9 nm. It is observed that there is a nearly linear relation between the intensity of optical emission of both $Cr^+$ ions and Cr neutrals and the energy per pulse. Especially, the slope of the linear function for $Cr^+$ ions is much larger than that of Cr neutrals. The detail quantification of the optical emission spectra is, however, complicated due to the dynamic nature of the discharge, lack of thermodynamic equilibrium (unknown electron temperature), and experimental setup (line-of-sight geometry), etc [6].

### *4.2. Interface characterization*
The microstructure of the interface between $CN_x$ films and steel substrates was studied by cross-sectional TEM in samples deposited with the HIPIMS biased substrate method of figure 4. Figure 8 shows the formation of a modified Cr layer between the steel substrate and the $CN_x$ film prepared by this method. The layer is characterized by a polycrystalline structure of approximately 100 nm thick.

The transition between the substrate and the modified layer, which appears to have more contrast, has been previously analyzed by HRTEM; a ~5 nm amorphous transition layer was found between the Cr and the steel [6]. The overall structure of the interface is dense, with no formation of bubbles or voids, pointing to a strong bonding between the steel substrate and the $CN_x$ coating.

The modified layer is formed during the pretreatment stage where $Ar^+$ and $Cr^+$ ions generated by the HIPIMS discharge on the Cr target bombard the substrate biased by the second HIPIMS power supply. Two counteracting processes compete during the formation of the polycrystalline layer: (a) the Cr concentration decreases due to sputter removal; and (b) the Cr concentration increases due to condensation on the surface.

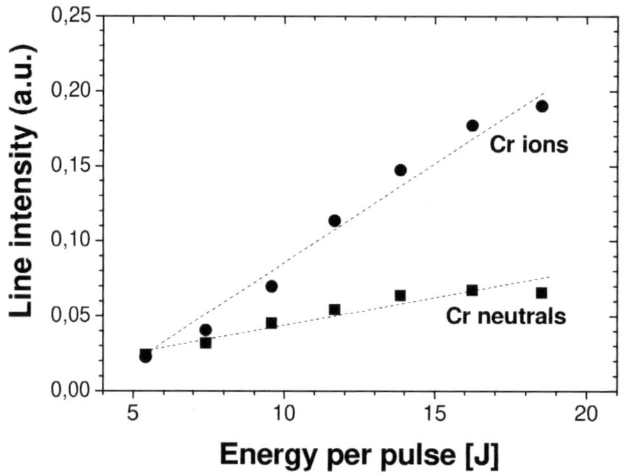


**Figure 7.** The dependence of emission intensity from Cr-I (339.1 nm) and Cr-II (336.8 nm) lines on the energy density per pulse

The effective sputter removal effect of $Ar^+$ plus $Cr^+$ ions bombardment was evaluated by a simple experiment on two Si(100) substrates using the configuration of figure 4(a). The Cr target was operated during 30 minutes at a pressure of 3 mTorr of Ar, with a peak target voltage of 1100 V, a frequency of 150 Hz, and a pulse width of 100 s. These conditions are similar to the sample shown in figure 5, however, one of the Si substrates had a floating bias. After the experiment, cross-sectional SEM observations of the Si substrates revealed a Cr layer of 1.2 μm thickness for the substrate held at floating bias, in contrast to the 0.1 μm of the biased sample grown under HIPIMS $Ar^+ + Cr^+$ ion bombardment.

We have previously shown that the amorphous transition layer is a mixture of Cr and Fe [6]. The presence of this mixed amorphous interface can be explained by the magnitude and composition of the ion flux. The introduction of Cr into the substrate is initiated by the direct physical implantation of Cr extracted from the plasma and accelerated by the high bias voltage. Additionally, the retention of Cr is facilitated by the high affinity of the metal ion, which could be incorporated at lattice sites of the substrate. In the system Cr-Fe there is a complete thermodynamic solubility and there is no limit to the expected concentration that can be incorporated [18, 35]. Recently, Ehiasarian *et al* [18] have simulated the depth of Cr incorporation resulting from implantation using the TRIDYN code. Assuming a steel substrate and bombarding ion flux concentration of 50 at. % $Cr^{3+}$, 30 at % $Cr^{4+}$ and 20 at. % $Ar^{1+}$, he obtained maximum range of implantation of 7 nm for a DC bias voltage of 600 V, similar to the depth profile we have measured previously [6]. This result contrasts with the ones from other authors [18, 35] where a mixed interface of 15-20 nm has been observed. The difference can be associated to thermal diffusion of Cr into the substrate. While these authors have been using substrate temperatures in the range 250-500 °C, our samples have been deposited at room temperature. It is well known that during DC bias cleaning of the substrates the surface temperature increases. In our case, we know the maximum substrate temperature during the dual-HIPIMS plasma etching (figure 4) was below 200 °C because we did not observe a change in the HV hardness of our SKF3 steel substrates [46].

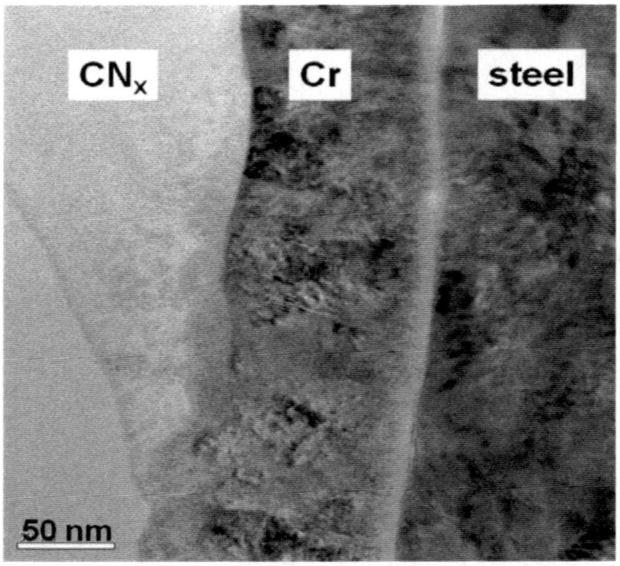


**Figure 8.** Cross-sectional TEM image showing the $CN_x$/steel interface. A polycrystalline Cr microstructure of about 100 nm thickness is observed

Cross-sectional and plan-view SEM images of the $CN_x$ films (not shown here) have revealed a coating with a dense microstructure, similar to that observed in amorphous $CN_x$ films deposited in laboratory-scale chambers [1, 47].

*4.3. Throwing power*

Figure 9 compares the throwing power of films deposited by DC magnetron sputtering and HIPIMS as a function of the substrate angle with respect to the surface of the cathode. It clearly shows that the films are more uniform when they are deposited by HIPIMS.

It is well known that in the vapor evaporation technique the transport of the vapor atoms is almost strictly line-of-sight and follows the Knudsen's cosine law $\cos\theta/r^2$, where r is the radial distance from the source and $\theta$ is the angle between the radial vector and the normal to the receiving surface. The main reason to be a line-of-sight deposition is due to the low mean free path of the evaporated atoms, in the order of the distance between the evaporating source and the substrate [48].

In the case of DC magnetrons sputtering, like in figure 3(a), Cr atoms are ejected from the target following "under-cosine" or "over-cosine" distributions for low and high bombarding energies, respectively. Two additional differences with vacuum evaporation will appear, originated from the higher deposition pressure and the substrate biasing possibility. For working pressures in the order of millitorrs, the mean free path of the evaporated atoms will be shorter than the distance between the cathode and the substrate. A large portion of these atoms will be deflected by the series of collisions with gas molecules and many of them will reach surfaces, which are not in the line-of-sight of the cathode. In addition, if the substrate is biased, the small amount of ionized Cr atoms generated in the discharge will be accelerated towards the substrate along the electric field lines produced by the bias and will also be deposited in non-line-of-sight regions of the substrate. Due to these two effects, dc magnetron sputtering is considered to have higher throwing power than vacuum evaporation [48].

When considering the case of HIPIMS magnetron sputtering, like in figure 4(a), the higher ionization in the discharge will increase the influence of the substrate bias: a higher amount of Cr ions will follow the electric field lines of the substrate and as a result, we will get a higher throwing power than DC magnetron sputtering, as shown in figure 9

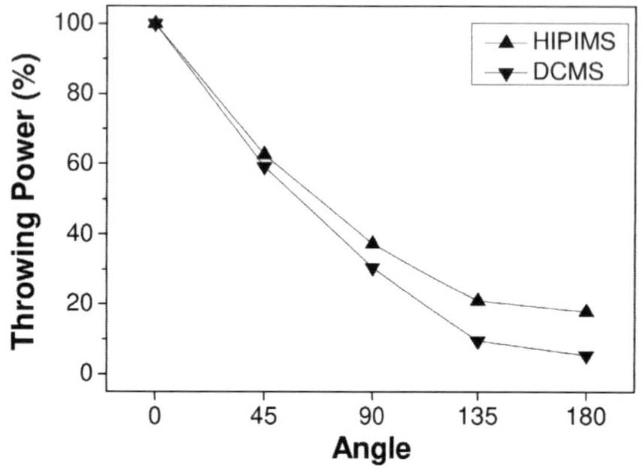


**Figure 9.** Throwing power of CrN films as a function of the angle between the surface of the target and the surface of the substrate for DC magnetron sputtering and HIPIMS magnetron sputtering. Both depositions have been done at the same pressure.

*4.4. Adhesion tests*

Figure 10 illustrates typical results from the Rockwell-C tests on samples deposited by the three different substrate pretreatment methods of figures 2, 3, and 4.

Figures 10(a) and 10(b) illustrates the resulting indentation when the samples were biased to a DC voltage of – 1200 V and the Cr target was operated in a DC mode, like in figure 2(a). The indent can be identified to the adhesion strength quality *HF5* shown in the chart of figure 6. The $CN_x$ coating shows radial cracks and large delaminations around the crater caused by the piling up of the substrate, and can be related to the poor adhesion of the film to the substrate.

Figures 10(c) and 10(d) show the resulting indentation of coatings when the samples were biased to a DC voltage of -600 V while the Cr target was operated with a HIPIMS source, like in figure 3(a). In this case the $CN_x$ coating has a typical failure that can be related to the adhesion strength quality *HF3*: cracks in the indentation area with small delamination in the vicinity of the crater.

Figures 10(e) and 10(f) show no visible delaminations around the indentation crater, and the presence of very few fine cracks, typical of adhesion strength quality *HF1*. This excellent adhesion was obtained when both substrate bias and Cr target were connected to two synchronized HIPIMS power supplies, as shown in figure 4.

The results of figure 10 can be correlated to the different pretreatments. We have shown [6] that at the optimum conditions of 1100 V Cr cathode voltage and 600 V bias voltage, a gradual interface between Cr and steel is formed (figure 8) that can withstand the high stresses of the $CN_x$ films. As we have discussed in section 4.2, at the beginning of the plasma pretreatment, the $Ar^+$ + $Cr^+$ ion bombardment have enough time to clean the substrate from oxides and other contaminants, and tends to sputter away the exposed steel substrate and implant Cr ions. As a result, the film adhesion with an optimal intermixing at the interface becomes very strong, with a high HF quality in the adhesion test (figure 10).

## 5. Conclusions

Carbon nitride films were deposited onto SKF3 (AISI 52100) bearing stainless steel substrates in an industrial chamber by DC magnetron sputtering with various interfacial treatments. A novel plasma HIPIMS pretreatment is employed where two HIPIMS power supplies are used in a master-slave configuration: one establishes the discharge in the cathode and the second produces a pulsed substrate bias. This configuration provides a more stable bias voltage during HIPIMS plasma pretreatment, resulting in an alternative way to achieve excellent results without the use of specially designed DC power supplies.

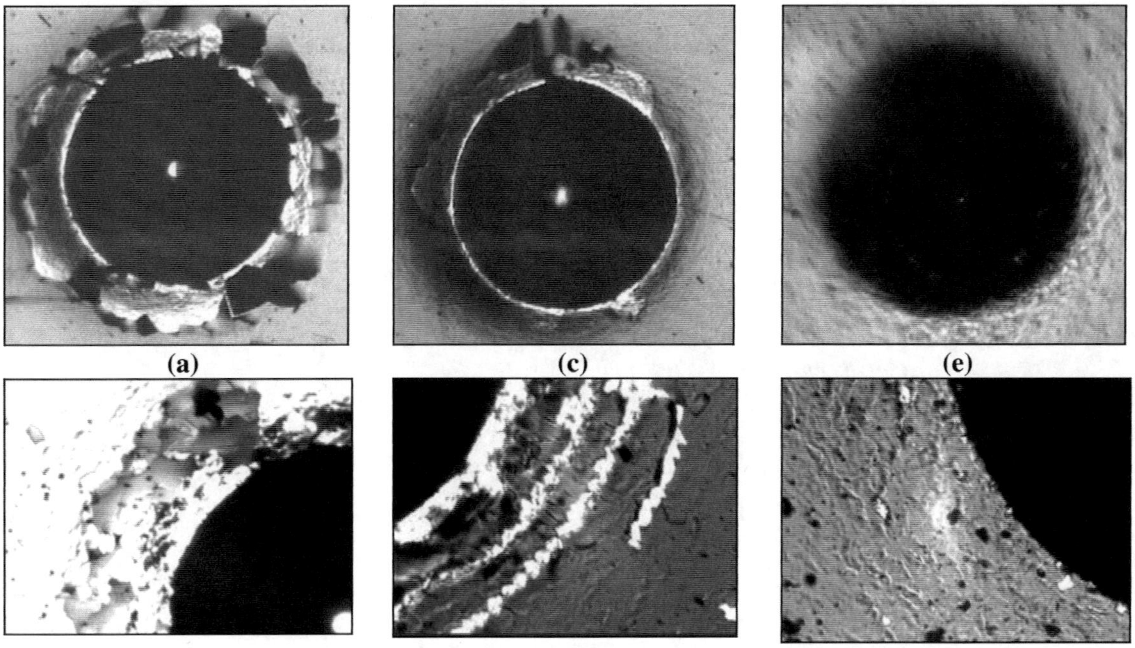


**Figure 10.** Adhesion strength HF of $CN_x$ films deposited by DC magnetron sputtering. Each optical micrograph represents a sample deposited by a different substrate plasma treatment. (a) & (b): *HF6,* when gas ions are generated by a DC magnetron cathode and the bias is provided by a negative DC voltage, like in figure 2. (c) & (d): *HF3*, when the gas and metal ions are generated by an HIPIMS discharge in the cathode and the bias is provided by a negative DC voltage, like in figure 3. (e) & (f): *HF1,* when two HIPIMS power supplies are used in a master-slave configuration, like in figure 4.

The throwing power of films deposited by DC magnetron sputtering and HIPIMS has been compared. We show that the higher throwing power of HIPIMS coatings can be associated to the higher ionization in the plasma discharge.

Adhesion of $CN_x$ films evaluated by the Daimler-Benz Rockwell-C and the scratch tests was remarkably increased to maximum value by our pretreatment through the formation of an optimal interfacial mixing layer of Cr and steel.

We demonstrate the feasibility of depositing thick (1-3 μm) adherent $CN_x$ coatings on steel substrates in an industrial scale chamber, opening the possibility for new tribological applications of carbon nitride films. The findings of this study are also expected to be of general validity and applicability in the industrial scale deposition of carbon-based coating materials.

**Acknowledgements**
E.B acknowledges the support from the Swedish Government Strategic Research Area Grant in Materials Science. Dr. G. Greczynski from Linköping University (Sweden) and Dr. Zs. Czigány from MFA (Hungary) are acknowledged for the help in acquiring the data for figures 7 and 8, respectively.

**References**

[1] Broitman E, Neidhardt J and Hultman L 2007 *Tribology of Diamond-Like Carbon Films :*

*Fundamentals and Applications* ed C Donnet and A Erdemir (New York: Springer) pp 620-53
[2] Broitman E, Gueorguiev G K, Furlan A, Son N T, Gellman A J, Stafström S and Hultman L 2008 *Thin Solid Films* **517** 1106-10
[3] Broitman E, Pushkarev VV, Gellman A J, Neidhardt J, Furlan A and Hultman 2006 L *Thin Solid Films* **513** 979-83
[4] Broitman E, Zheng W T, Sjostrom H, Ivanov I, Greene J E and Sundgren J E 1998 *Appl. Phys. Lett.* **72** 2532-4
[5] Broitman E, Hultman L and Cremer R 2008 *Beschichtungsverfahren und Vorrichtung zum Beschichten (Coating process and device for coating)*: Germany - Patent # DE102008021912A1 Date: 05112009
[6] Broitman E, Czigány Zs, Greczynski G, Böhlmark J, Cremer R and Hultman L 2010 *Surf. Coat. Technol.* **204** 3349-57
[7] Bertrand N, Drevillon B, Gheorghiu A, Senemaud C, Martinu L and Klemberg-Sapieha J E 1998 *J. Vac. Sci. Tech.* A **16** 6-12
[8] Chapman B N 1974 *J. Vac. Sci. Tech.* **11** 106-13
[9] Strong J 1935 *Rev. Sci. Inst.* **6** 97
[10] Mattox D M 1998 *Handbook of Vacuum Technology* ed DM Hoffman, B Sing and JH Thomas, (New York: Academic Press) ch 49
[11] Baglin J E E 1991 *Fundamentals of Adhesion* ed L-H Lee (New York: Springer) p 363
[12] Farnsworth H E, Schlier R E, George T H and Burger R M 1955 *J. App. Phys.* **26** 252-3
[13] Mattox D M 1965 *Electrochem. Technol.* **2** 295
[14] Ohring M 1992 *The Materials Science of Thin Films* (New York: Academic Press) 129-31
[15] Bunshah R F 1995*Handbook of Deposition Technologies for Films and Coatings* ed RF Bunshah (New York: Noyes Publications) 117-20
[16] Hofmann D, Beilsswenger S and Feuerstein A 1991 *Surf. Coat. Technol.* **49** 330-5
[17] Anders A 2005 *Surf. Coat. Tech.* **200** 1893-906
[18] Ehiasarian A P, Wen J G and Petrov I 2007 *J. App. Phys.* **101** 054301-10
[19] Olbrich W, Fessman J, Kampschulte G and Ebberink 1991 *J Surf Coat Technol* **49** 258-62
[20] Audronis M, Matthews A and Leyland A 2008 *J. Phys. D: App. Phys.* **3** 035309
[21] Arnell R D, Kelly P J and Bradley J W 2004 *Surf. Coat. Technol.* **188-189** 158-63
[22] Anders A 2004 *Surf. Coat. Technol.* **183** 301
[23] Cooke K E, Hamsphire J, Southall W and Teer D G 2004 *Surf. Coat. Technol.* **177-178** 789-94
[24] Håkansson G, Hultman L, Sundgren J E, Greene J E and Münz W D 1991 *Surf. Coat. Technol.* **48** 51-67
[25] Münz W D, Schulze D and Hauzer F J M 1992 *Surf. Coat. Technol.* **50** 169-78
[26] Anders A 2002 *Vacuum* **67** 673-86
[27] Brown I G 1998 *Annu Rev Mater Sci* **28** 243-69
[28] Helmersson U, Lattemann M, Bohlmark J, Ehiasarian A P and Gudmundsson J T 2006 *Thin Solid Films* **513** 1-24
[29] Schonjahn C, Ehiasarian A P, Lewis D B, New R, Munz W D, Twesten R D and Petrov I 2001 *J. Vac. Sci. Technol. A* **19** 4
[30] Münz W D, Smith I J, Lewis D B and Creasey S 1997 *Vacuum* **48** 473-81
[31] Münz W D 2008 *Vakuum in Forschung und Praxis* **20** 27-32
[32] Ehiasarian A P, Münz W D, Hultman L, Helmersson U and Petrov I 2003 *Surf. Coat. Technol.* **163-164** 267-72
[33] Ehiasarian A P, Hovsepian P E, Hultman L and Helmersson U 2004 *Thin Solid Films* **457** 270-7
[34] Paulitsch J, Schenkel M, Schintlmeister A, Hutter H and Mayrhofer P H 2010 *Thin Solid Films* **518** 5553-7
[35] Lattemann M, Ehiasarian A P, Bohlmark J, Persson P Å O and Helmersson U 2006 *Surf. Coat. Technol.* **200** 6495-9

[36] Hovsepian P E, Ehiasarian A P, Deeming A and Schimpf C 2008 *Vacuum* **82** 1312-7
[37] Hovsepian P E, Ehiasarian A P and Ratayski U 2009 *Surf. Coat. Technol.* **203** 1237-43
[38] Hovsepian P E, Reinhard C and Ehiasarian A P 2006 *Surf. Coat. Technol.* **201** 4105-10
[39] Munz W D, Schenkel M, Kunkel S, Paulitsch J and Bewilogua K 2008 *J. Phys.: Conf. Series* **100** 082001
[40] Donohue L A, Torosyan A, May P, Wolfe D E, Kulik J and T J Eden 2009 *Plating & Surface Finishing* 38-46
[41] *Databases used for research of scientific publications: SciFinder, Web of Science, and Google Scholar*
[42] *Bias power supply type BP12 manufactured by Huettinger Electronic*
[43] Ehiasarian A P, Tietema R, Hovsepian P E and Doerwald D 2007*Vacuum treatment apparatus with additional voltage supply - Patent Application GB20060007269 20060411*, Hauzer Techno Coating BV (NL); Univ Sheffield Hallam (GB): United Kingdom
[44] Verein-Deutscher-Ingenieure 1992 *Daimler Benz Adhesion Test VDI 3198* (Dusseldorf: VDI-Verlag) p 7
[45] Vidakis N, Antoniadis A and Bilalis N 2003 *J. Mat. Process. Technol.* **143-144** 481-5
[46] Guyer R A 1996 *Rolling bearings handbook and troubleshooting guide* (New York: CRC Press) pp 4-5
[47] Broitman E, Hellgren N, Czigany Z, Twesten R, Luning J, Petrov I, Hulttman L and Holloway B C 2003 *J. Vac. Sci. Technol. A* **21** 851-9
[48] Ohring M 1992*The materials science of thin films* (Boston: Academic Press) ch 3

# Nitriding of Ti substrate using energetic ions from plasma focus device

**A Henriquez[1], H Bhuyan[1], M Favre[1], B Bora[1], E Wyndham[1], H Chuaqui[1], S Mändl[2], J W Gerlach[2] and D Manova[2]**

[1]Department of Physics, Pontificia Universidad Católica de Chile, Santiago, Chile
[2]Leibniz-Institut für Oberflächenmodifizierung, Leipzig, Germany

E-mail: aphenriq@uc.cl

**Abstract.** Plasma Focus (PF) discharge is a pulsed plasma producing discharge that generates high temperature and high density plasma for a short duration. PF devices are known to emit intense ion beams pulses of characteristic energy in the keV to a few MeV range, in a time scale of tens of nanoseconds. We have previously investigated the ion flux and energy spectrum of ion beams emitted from a low energy PF, operating at 20 kV, with 1.8 kJ stored energy. It was observed that the ion beams have wide range of energy and intensity spectra with a clear angular anisotropy. Due to the wide range of ion energy and intensity spectra PF has become a subject of current interest for its applications in material sciences including surface modification and thin film deposition. The purpose of this study is the formation of titanium nitride (TiN) thin film and to investigate the structural properties of the TiN thin films in terms of PF angular positions. Substrates like Ti and Ti/Si were nitrided in a 1.8 kJ PF device at different angular positions with respect to the PF axis in order to correlate their surface properties with ion beam parameters. Preliminary characterizations of the ion implanted substrates have been conducted, using SEM, EDX and XRD. Our results indicate the formation of nanocrystalline TiN thin film only in certain angular positions. Angular dependency of the surface morphology was observed, which shows that the surface features strongly depends on ion beam energy and flux. With increasing angular positions, a reduction in the deposition rate and the sputter rate is observed. A pronounced nanostructured surface is only observed at the axis of the pinched plasma column, indicating the dominant role of sputtering and perhaps melting and fast re-crystallization of the surface in creating the nanostructures.

## 1. Introduction

In plasma focus (PF), a high density, high temperature, short duration plasma column is formed, following the radial compression phase that characterizes PF discharges [1]. In addition to x-ray and neutron pulses, the PF is a well-known source of energetic ion beams, of characteristic energy from hundreds of keV to tens of MeV [2]. This characteristic energy range exceeds up to several hundred times the maximum energy to be expected from the capacitor bank voltage. Several properties of PF emitted ion beams have been measured, including their angular distributions and energy spectra [3]. The use of PF for thin film deposition purposes has shown that this pulsed plasma device possesses interesting features of high deposition rates, energetic deposition and possible film deposition under a reactive background gas pressure [4]. Plasma focus based deposition of TiN thin films on titanium significantly improves its surface properties such as wear and corrosion resistance and hardness, making it attractive in many industrial applications including aeronautical, biomedical and nuclear fields.

## 2. Experimental apparatus

The experiment was conducted in a low energy, 1.8 kJ, Mather type [1] PF device. A schematic diagram of the experimental arrangements along with the discharge circuit is shown in figure 1. The electrode assembly of the PF chamber consists of a hollow central anode of length 10 cm and outer diameter 2.4 cm encircled by six cathode rods of length 9.7 cm and diameter 0.8 cm uniformly spaced coaxially at a diameter of 6.3 cm. The electrode material is copper and a titanium insert on the top of the anode is used. The two electrodes are separated by a 3 cm long Pyrex glass insulator sleeve fixed around the bottom of the anode as shown in figure 1. A hollow anode configuration was used in order to decrease plasma contamination with heavy ions resulting from anode ablation. The PF device is powered by a capacitor bank of 9 μF, at 30 kV maximum charging voltage. At 20 kV operating voltage the capacitor bank delivers a maximum current of 160 kA with 1.8 μs quarter period. The system inductance at the maximum current is around 146 nH.

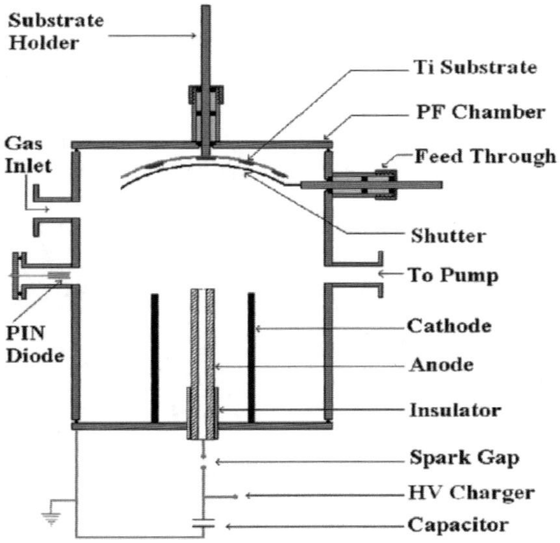


**Figure 1.** Experimental set-up of PF device with Ti, Ti/Si substrate at 0, 10, 15 and 25 degrees from the anode axis.

Nitrogen ions emitted from our PF device are implanted on different technologically important substrates like silicon/titanium (Ti/Si) and titanium (Ti) at different angular positions of 0, 10, 15 and 20 degrees with respect to the PF axis at a distance of 16 cm. The substrates were ion deposited by 20 numbers of PF discharges.

## 3. Experimental results and discussion

Figure 2 shows XRD patterns of nitrogen ions implanted titanium substrates at different angular positions. Formation of TiN (small peaks at around 36.5° [100] and 42.5°) appears to be observed in the case of the substrate placed at 0°. In addition, a comparative analysis between the XRD patterns for different angular positions probably indicates formation of TiN compound at 10 and 15 degrees, although TiN peaks are not clearly visible because of the superposition with other peaks. On the other hand, nitrogen ion implanted substrates at 25° angular positions indicate the presence of only α-Ti phase. This is because at 25° angular position the ion energy and ion flux is less in comparison to the 0°, 10° and 15° [5].

Figure 3 shows the XRD analysis of titanium substrate at different angular positions. The XRD spectra shows the appearance of small peaks of TiN at 0° and only small peaks due to α-Ti and β-Ti at larger angles.

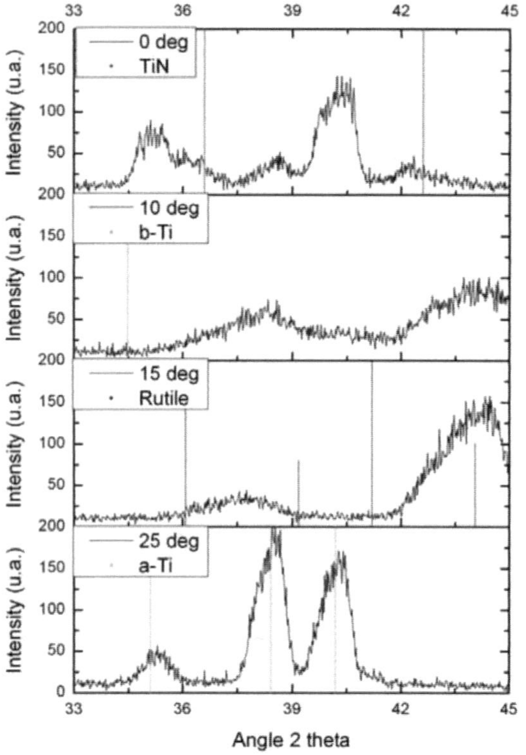


**Figure 2.** XRD analysis for titanium substrates at 0, 10, 15 and 25 degrees with XRD spectra for Ti, rutile and TiN.

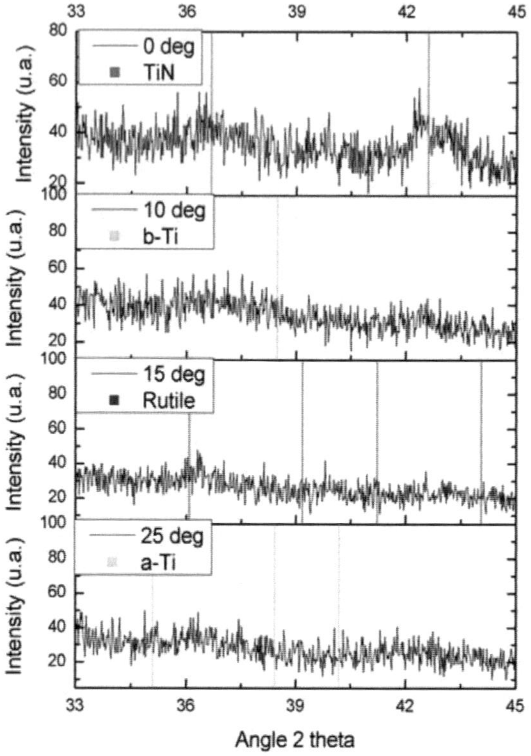


**Figure 3.** XRD analysis for Ti/Si substrate at 0, 10, 15 and 25 degrees.

A comparison between the ion implanted Ti/Si and Ti substrates point out that the incoming PF ion energy is higher than 100 keV for optimum ion flux. Owing to this high energy the ions overshoot the Ti layer on Ti/Si substrate and this why we had not observed the formation of TiN layer.

SEM images of nitrogen ion implanted Ti substrates at different angular positions are shown in figure 4. Clear changes on the surface morphology are observed at different angular positions. These changes are obviously due to the variation of ion energy and ion flux with different angular positions. Formations of nanostructures are observed on the surface. The appearance of layer structures are observed, where in some zones these layers are fractured due to the high energy of the ions. More cracks are observed on the substrates placed at 0°, where the ion energy was found maximum [5]. On the other hand at 10° and 15° the layers are almost regular with very few cracks where ion fluxes were found maximum with moderate ion energy.

The EDS spectra of these SEM micrographs indicate the presence of nitrogen, copper and oxygen on the substrate, irrespective of the angular positions. Copper seen in the EDS spectra is eroded from the electrode materials due to the ion bombardment. The oxygen observed was probably due to the absorption of oxygen by the unexposed Ti substrate before the irradiation and also the high Ti reactivity with residual water in the vacuum chamber.

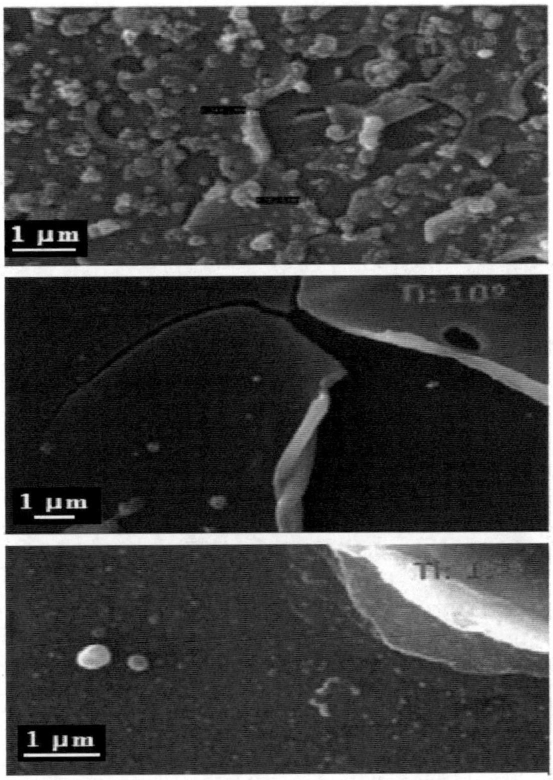


**Figure 4.** SEM images of the ion implanted titanium substrates at angular positions of 0°, 10° and 15°.

## 4. Conclusions

We have studied the interaction of nitrogen ions of a 1.8 kJ PF device with different kind of substrate materials like silicon and titanium which shows deposition of nanostructured thin films. Changes on the surface topology and in the composites formed on the substrate with the variation of the angular position of the substrate were found. In both substrates the XRD analysis shows that TiN could be found at 0 degrees but not at larger angles. SEM images shows nanostructure visible on all samples, independent of substrate and flaking or spaling of film during deposition more pronounced in centre and for Ti substrate. Further works are in progress.

## Acknowledgments

This work has been funded by Programa de Cooperación Científica Internacional CONICYT/DAAD and Proyecto Puente VRI N° 06/2011. University of Leipzig is acknowledged for providing RBS beamtime. The technical assistance of Dirk De Joung, David Alvarez and Juan Galvez is highly appreciated.

## References

[1] Mather J W 1965 *Phys. Fluids* **8** 366
[2] Sadowski M, Zebrowski J, Rydygier and Kucinski J 1988 *Plasma Phys. Control. Fusion* **30** 763
[3] Bhuyan H et al 2007 *J. Phys. D: Appl. Phys.* **40** 127
[4] Rawat R S, Chew W M, Lee P, White T and Lee S 2003 *Surf. & Coat. Technol.* **173** 276-84
[5] Bhuyan H, Favre M, Valderrama E, Chuaqui H and Wyndham E 2006 *Journal of Physics D: Applied Physics* **39** 3596

Journal of Physics: Conference Series **370** (2012) 012011 doi:10.1088/1742-6596/370/1/012011

# Electrical characterization of an air microplasma jet operated at a low frequency ac voltage

**L Giuliani[1,*], D Grondona[1,*], H Kelly[1,*] and F Minotti[1,*]**

[1]Instituto de Física del Plasma (CONICET), Departamento de Física, Facultad de Ciencias Exactas y Naturales (UBA) Ciudad Universitaria Pab. I, 1428 Buenos Aires, Argentina
[*]Member of CONICET

E-mail: leandro@tinfip.lfp.uba.ar

**Abstract** In this work a small plasma jet is generated applying an ac high voltage (kV) of low frequency (50 Hz) between two disk-shaped electrodes with a hole in the center and separated by a centrally perforated dielectric material. A plasma jet emerges from the electrode system to the room air when a large air flow rate is passing through the holes, for inter-electrode voltage drops around 1 ÷ 3 kV. The electrical characteristics of the discharge, voltage and current were studied varying the applied voltage amplitude and the gas flow rate. It was found that the microplasma jet was stable during a long period of time and the gas temperature remained almost at room temperature. These characteristics make this discharge suitable for biological applications.

## 1. Introduction

Non-thermal discharges generated at reduced pressures in vacuum reactors have a well established, broad spectrum of applications in material science. Nowadays, usage of non-thermal atmospheric plasmas is studied with growing interest, as these plasmas provide a cheaper and more convenient alternative to low-pressure plasma technology. Another important advantage of using atmospheric plasmas is the possibility of processing materials which are not resistant to high vacuum. An attractive application of these plasmas is its utilization in removal and sterilization of bio-films and planktonic bacterial

The plasma generated by an electric discharge within a gas contains free electrons and ions, several active species (i.e. atomic or molecular radicals, O, OH and excited molecules) and energetic UV photons with well known sterilization properties. Several applications of this kind of plasmas in the biology field were performed using a RF discharge called "plasma needle", with a great variety of studies including surface sterilization and bacterial removing and deactivation. *In vivo* applications were also studied, oriented to wound disinfection and dental cavities sterilization without drilling [1,2,3,4]. Due to the great importance and variety of applications a great amount of work appeared recently concerning biological applications of non-thermal plasmas, where it was studied the optimization of the discharge parameters [5], the dependency in the gas flow [6] and its capacity for bactericidal applications [7,8,9]. The majority of the devices employed for biological purposes work with noble gases or a mixture of them, due to the easiness of producing the discharge, with a small amount of $O_2$ or air which produces most of the active species. Nevertheless, Lu et al. [10] presented a stable discharge in pure air achieving a more reactive plasma. Recently, non-thermal plasmas at atmospheric pressure called "plasma jet" or "microplasma jet" that can be applied to biological applications were developed using a great variety of ac-sources, which are as efficient as RF sources and less expensive [11,12,13].

In this work we present a small plasma jet generated by applying an ac high voltage (kV) of low frequency (50 Hz) between two disk-shaped electrodes with a hole in the center and separated by a dielectric material. A plasma jet emerges from the electrode system to the room air when a large air flow rate is passing through the holes, for inter-electrode voltage drops around 1 – 3 kV. The electrical characteristics of the discharge, voltage and current were studied varying the applied voltage amplitude and the gas flow rate. The plasma jet appearance and temperature were also studied.

## 2. Experimental set–up

A schematic of the experimental device used to produce the atmospheric plasma jet is shown in figure 1. The device consists in two electrodes with a hole of 1 mm diameter, through which air is flowing. The two electrodes are made of stainless steel disks of 20 mm diameter and 3 mm thickness attached to the surface of a centrally perforated dielectric disk (Teflon) of 1 mm thickness. The hole in the center of the dielectric disk is 2 mm in diameter.

The ac power supply is a commercially available transformer for neon light (25 kV peak to peak voltage, 50mA, 50 Hz) which is connected to a variable autotransformer (Variac) in order to allow voltage control. The air used was commercial analytic air (H2O<5.0 ppm/v, CO2<10.0 ppm/v) and the flow could be varied up to 17 l/min.

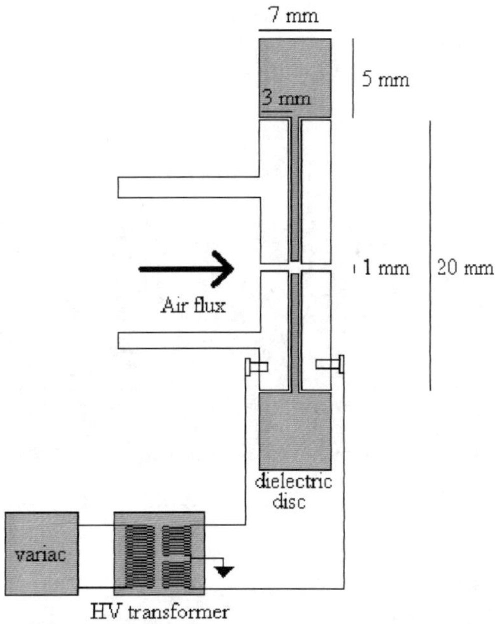


**Figure 1**. Schematic of the experimental set–up.

The electrical characteristics of the discharge and the appearance of the plasma jet were studied varying the applied voltage amplitude and the gas flow rate. The voltage $V_{ag}$ between one of the electrodes and the grounded middle point of the transformer (i.e. half the electrodes voltage) was measured using a high voltage probe (1000x) connected to a digital Tektronix oscilloscope (70 MHz, 1GS/s), and the current $I$ was measured using a commercial current transformer (2.5 V/A). Photographs of the microplasma jet were taken using a 10 Mpixel digital camera (exposure times of

Journal of Physics: Conference Series **370** (2012) 012011 doi:10.1088/1742-6596/370/1/012011

1/8 s) and the microplasma jet average temperature was measured using a k-type thermocouple located inside the plasma jet. The air flow was measured using a stainless steel float flow meter.

## 3. Experimental results and discussion

Figure 2 shows a photograph of the device with the plasma jet emerging from the electrode system.

**Figure 2.** Photograph of the experimental device.

Figure 3 shows images of the microplasma jet operated in air for different open circuit voltage amplitudes ($V_{ag\ op}$) and a flow rate of 9.2 l/min. It was found that as the voltage decreases, the jet becomes shorter and less luminous. The discharge could not be maintained at $V_{ag\ op}$ values smaller than 5.5 kV.

The flow rate was varied between 4.7 to 13.5 l/min. Figure 4 shows images of the microplasma jet operated in air for different flow rates and $V_{ag\ op}$ = 10 kV. The length of the plasma jet increase with increasing flow rate up to about 9 l/min, and with higher flow rates the jet becomes shorter. Jets longer than 1 cm were obtained for a flow rate of 9.2 l/min and $V_{ag\ op}$ = 10 kV. The temperature of the jet measured with the thermocouple is close to room temperature. The plasma jet can be touched without any harm. Figure 5 shows the jet in contact with a human finger.

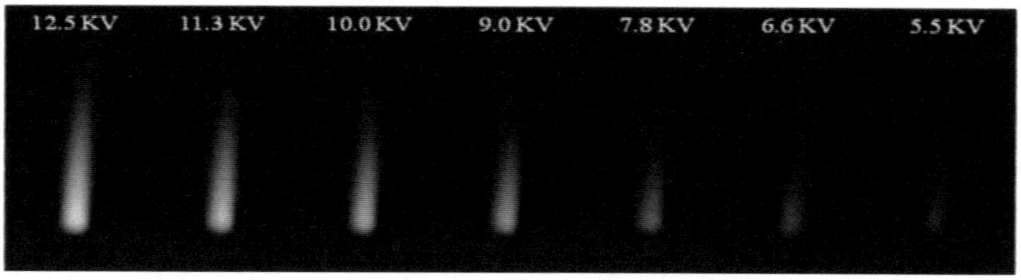


**Figure 3.** Jet photographs for different open circuit voltages of the transformer and a flow rate of 8.2 l/min.

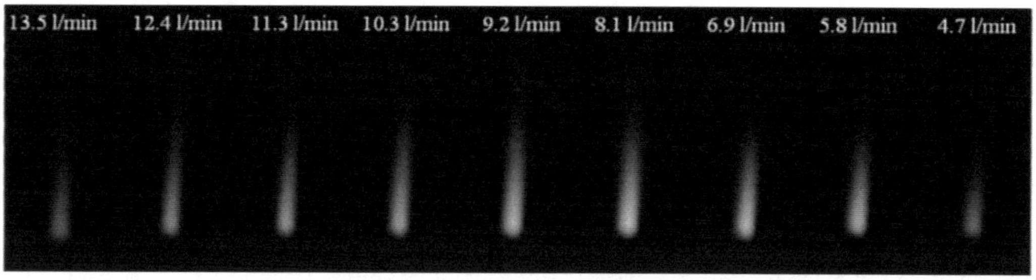


**Figure 4.** Jet photographs for different air flows and $V_{ag\ op}$ = 10 kV.

Typical signals of $V_{ag}$ and $I$ taken during the discharge are shown in figure 6. The voltage of the discharge has a large amount of pulses with amplitudes of about 3 kV and duration around 150 ns for a flow rate of 9.2 l/min and $V_{ag\ op} = 10$ kV, reaching minima of some hundreds of volts.

The voltage pulses are accompanied of very short current peaks which can reach some amperes and last about 20 ns at the base of the pulse.

**Figure 5.** Photograph of a finger touching the plasma jet transformer.

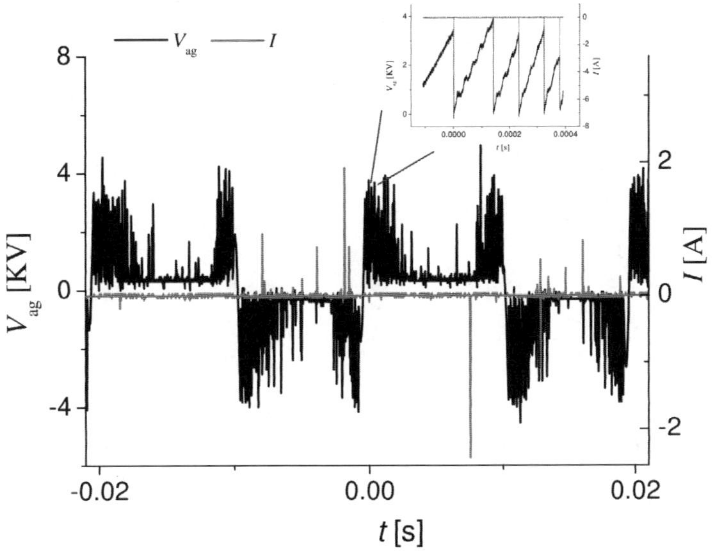


**Figure 6**. Typical voltage and current waveforms of the discharge.

Figure 7 shows a time scale expansion of the voltage signal for two different open circuit voltages of the transformer ($V_{ag\ op}$ = 12.5 kV and 5.5 kV) and for an air flow of 9.2 l/min. It can be seen that as $V_{ag\ op}$ increases the pulse amplitude does not change, but the pulse train becomes shorter and the pulse frequency increases. The increase of this frequency with increasing values of $V_{ag\ op}$, can be appreciated in figure 8.

Figure 9 shows a time scale expansion of the voltage signal for two different air flows ( air flow 13.5 l/min and 4.7 l/min) and a fixed open circuit voltage of $V_{ag\ op}$ = 10 kV. It can be seen that the frequency of the peaks does not change, but the pulse number and amplitude increases, as the flow increases. The increase of the pulse amplitude with the flow rate increasing can be appreciated in figure 10.

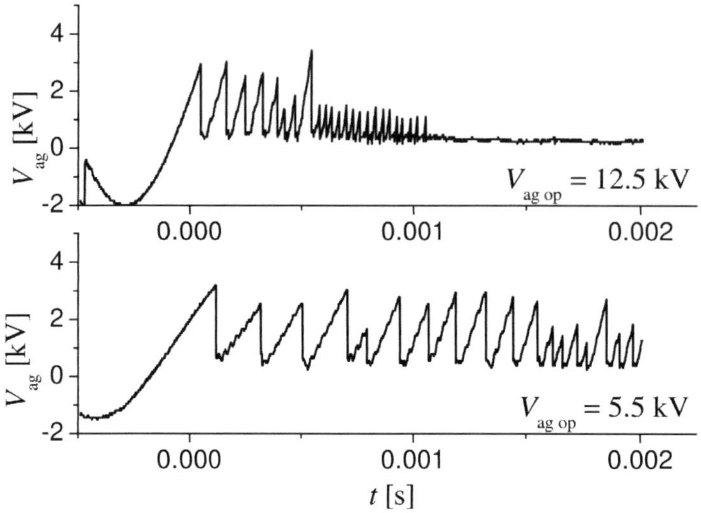


**Figure 7.** $V_{ag}$ for the two extreme applied open circuit voltage values of the transformer, for a fix air flow of 9.2 l/min.

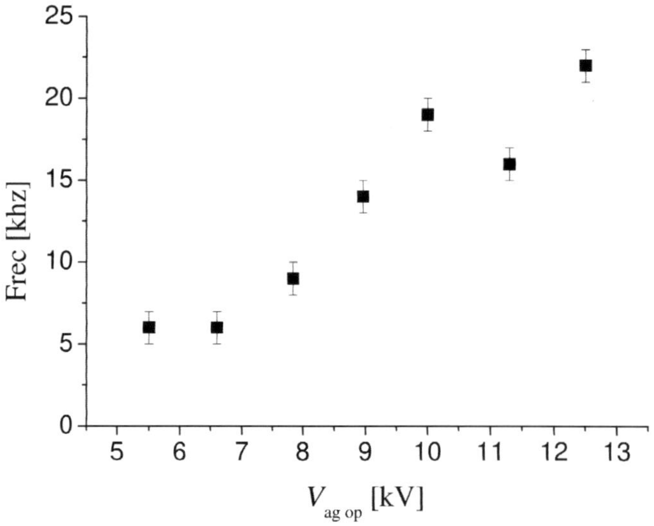


**Figure 8.** Peak frequency as a function of $V_{ac\ op}$.

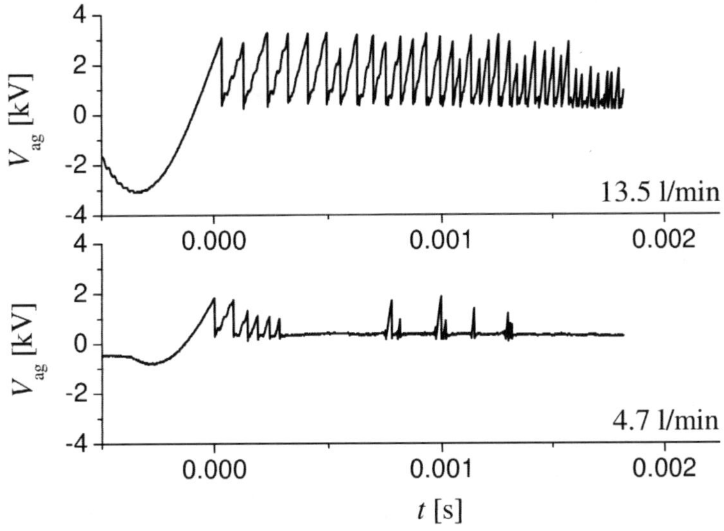


**Figure 9.** $V_{ag}$ for the two extreme air flows, for a fixed open circuit voltage of the transformer at 10 kV.

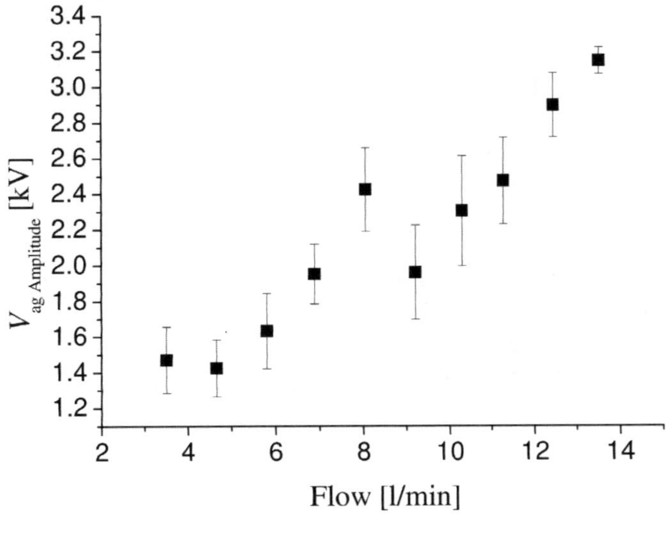


**Figure 10.** Pulse amplitude as a function of the air flow.

## 4. Conclusions

A stable microplasma jet of about one centimeter long was obtained within a pure air flow. The jet was formed at atmospheric pressure and room temperature and could be safely put in contact with human skin.

The voltage of the discharge is pulsed with pulse amplitude of a few kV, reaching minima of some hundreds of volts. The voltage drops are accompanied of very short current peaks which can reach some amperes and last about 20 ns at the base. The pulse amplitude was found to be independent of the power source voltage and to increase with the air flow due, quite probably, to an increment in the pressure.

The frequency of the discharge pulses was independent of the air flow, and increased with the power source voltage. This is appreciated in the jet as an increment of the length and luminosity. The length and luminosity of the jet varied with the air flow and there was an optimum for each value of the power source voltage.

All these studies are directed to the achievement of an appropriate discharge configuration for the technological development of devices used for sterilization, bacterial removal and deactivation, and also for in vivo applications as wound disinfection and decontamination of dental cavities without drilling.

**Acknowledgements**
This work was supported by grants from the Universidad de Buenos Aires (code: 20020090100141) and CONICET (PIP 11220090100219).

**References**

[1] Stoffels E, Flikweert A J, Stoffels W W and Kroesen G M W 2002 *Plasma Sources Sci. Technol.* **11** 383–8
[2] Sladek R E J and Stoffels E 2005 *J. Phys. D: Appl. Phys.* **38** 1716–21
[3] Sladek R E J, Stoffels E, Walraven R, Tielbeek P J A and Koolhoven R A 2004 *IEEE Transactions on Plasma Science* **32** 1540-3
[4] Stoffels E, Sakiyama Y and Graves D B 2008 *IEEE Transactions on Plasma Science* **36** 1441-57
[5] Simon A, Anghel S D and Papp J 2008 *Journal of Optoelectronics and Advanced materials* **10** 2077–81
[6] Goree J, Liu B and Drake D 2006 *J. Phys. D: Appl. Phys.* **39** 3479–86
[7] Lee M H, Park B J, Jin S C, Kim D, Han I, Kim J, Hyun S O, Chung K and Park J 2009 *New Journal of Physics* **11** 115022 (11pp)
[8] Zhang X, Huang J, Liu X, Peng L, Guo L, Lv G, Chen W, Feng K and Yang S 2009 *Journal of Applied Physics* **105** 063302 (5pp)
[9] Malovic G, Puac N, Lazovic S and Petrovic Z 2010 *Plasma Sources Sci. Technol.* **19** 034014 (7pp)
[10] Lu X, Xiong Z, Zhao F, Xian Y, Xiong Q, Gong W, Zou C, Jiang Z and Pan Y 2009 *Applied Physics Letters* **95** 181501 (3pp)
[11] Hong Y C and Uhm H S 2006 *Applied Physics Letters* **89** 221504 (3pp)
[12] Mohamed A -A H, Kolb J F and Schoenbach K H 2010 *Eur. Phys. J. D* **60** 517–22
[13] Anghel S D, Simon A, Papiu M A and Dinu O E 2011 *Rom. Journ. Phys.* **56** 90–4

Journal of Physics: Conference Series **370** (2012) 012012 doi:10.1088/1742-6596/370/1/012012

# Application of electron beam plasma for biopolymers modification

**T M Vasilieva**

Moscow Institute of Physics and Technology, Dolgoprudny, Moscow region, 141700, Russia

E-mail: tmvasilieva@gmail.com

**Abstract**. The effects of the Electron Beam Plasma treatment on natural polysaccharide chitosan were studied experimentally. Low molecular water-soluble products of chitosan and chitooligosaccharides were obtained by treating the original polymers in the Electron Beam Plasma of oxygen and water vapor. The molecular mass of the products varied from 18 kDa to monomeric fragments. The degradation of the original polymers was due to the action of active oxygen particles (atomic and singlet oxygen) and the particles of the water plasmolysis (hydroxyl radicals, hydrogen peroxides). The 95% yield of low molecular weight chitosans was attained by optimizing the treatment conditions. The studies of the antimicrobial activity of low molecular products showed that they strongly inhibit the multiplication of colon bacillus, aurococcus and yeast-like fungi. The EBP-stimulated degradation of polysaccharides and proteins were found to result from breaking β-1,4 glycosidic bounds and peptide bonds, respectively.

## 1. Introduction

The natural renewable biopolymers chitin and, especially, chitosan are very promising for technological and industrial applications such as agriculture, food processing, cosmetics production and others [1, 2]. Chitosan, linear heteropolymers of β-1,4-linked 2-amino-2-deoxy-D-glucopyranose and 2-acet-amido-2-deoxy-D-glucopyranose units, has many unique biological properties namely high biocompatibility with living tissues, biodegradability, ability to the complexation, and low toxicity. In medicine and pharmaceutics the water-soluble low molecular weight chitosans (less than 10 kDa) are usually required. These substances can be used as immune response-modulating or antibacterial agents, sorbents, radioprotectors, and for the production of microcapsules, thing films, and substrates for cell cultures [1, 2].

To produce the low molecular weight chitosans (LMWC) several techniques, including chemical, enzymatic, and radical treatment have been suggested [3]. Simple and rather low-cost chemical treatment is a conventional method, however toxic wastes and environment contamination are inherent in the chemical chitin and chitosan processing as well as in all techniques mentioned above. Besides, the chemical treatment is very time consuming and usually takes several hours. Thus, the development of the effective techniques for quick and environment friendly chitosan degradation is the burning issue of the day. The novel approach to the water-soluble low molecular weight chitosan production which is based on the Electron Beam Plasma (EBP) application is considered in the present paper.

The EBP is generated by injecting an electron beam (EB) into a gaseous medium. Under typical conditions of the EBP generation (medium pressure $0,1 < P_m < 10$ kPa and moderate EB power $N_b < 1$ kW) the plasma is strongly non-equilibrium and cold. Being injected into the gas the EB ionizes the gas, excites the gas molecules and is able to cause molecule dissociation. The generated EBP has a complex composition and contains a lot of chemically active particles that do not exist under equilibrium conditions. With respect to non-equilibrium plasmas generated in conventional ways (for instance, with respect to gas discharge plasmas) the EBP has the following advantages:

- the EB can be injected into any gases, vapors and gas-vapor mixtures;
- the EBP bulk does not contract even at very high gas pressures ($P_m$ ~10 kPa and higher);
- the solid powders and liquid droplets injected into the gas do not prevent the EBP generation; large-size bodies can be inserted into the plasma bulk;
- both solid powders and thing films can be treated in the EBP;
- very high concentrations of chemically active particles can be obtained even at low (up to room) temperatures;
- the process of the EBP-treatment is absolutely controllable and the treatment results are replicable.

The aims of the present study were as follows:

1) to experimentally prove the possibility of the EBP-stimulated hydrolysis of native chitosans and formation of water-soluble low molecular weight products as a result of the chitosans plasma chemical processing;
2) to obtain the high yield of the low molecular weight products by optimizing the treatment conditions;
3) to characterize both the structure of the low molecular weight products of the plasma chemical treatment and their bioactivity;
4) to compare the effects of the EBP-treatment on polysaccharides and some other natural biopolymers, e.g. proteins.

## 2. Materials and methods

### *2.1. Materials*

Crab shell chitosan with the degree of deacetylation and molecular weight of 95% and 500 kDa, respectively, was used as the original substance for the further EBP-treatment. The high molecular weight blood protein fibrin-monomer ($M_r$ ~ 340 kDa) was chosen as reference substance since its EBP-modification has been studied in detail previously [4].

### *2.2. Characterization of EBP-treated biopolymers*

*2.2.1. Solubility measurements.* 100±0.1 mg of the preliminary dried sample ($m_s$) were placed into a tube and 1.5 ml of distilled water were added to the sample. The resulting mixture was incubated for 24 h at room temperature under periodic mixing. After the incubation the mixture was centrifuged for 5 min and 1 ml of centrifuged was taken and dried. The mass of the dry residue ($m_{dr}$) was measured with an accuracy ±0.1 mg. The sample solubility was calculated as the $(m_{dr}/ m_s)\cdot 100\%$ ratio.

*2.2.2. Molecular mass characterization.* To characterize the molecular masses of the EBP-treatment products the exclusion chromatography was applied. The chromatograph *Staier* (Russia) and the chromatographic column *Phenomenex* BioSep-Sec-S-3000 (USA) with the efficiency of 30000 theoretical plates were used. The analysis conditions were as follows: the elutriating agent – 0.1 M phosphate buffer (pH 6.86) containing 0.05% $NaN_3$; the elution rate - 1 ml/min; temperature - 30°C; UV-detector with the wavelength 280 nm.

The effects of the EBP-treatment on fibrin-monomer molecular mass and structure were detected by means of the IR-spectroscopy, ion-exchange chromatography, horizontal electrophoresis in agar gel, immunoelectrophoresis and PAGE-electrophoresis as well [4].

*2.2.3. Biological activity of the EBP-produced LMWC.* The inhibition of the bacteria growth *in vitro* was measured to quantitatively characterize the bioactivity of LMCW obtained by the plasma treatment, gram-positive (*S. aureus*), gram-negative (*E. coli, Ps. aeruginosa*) microorganisms and yeast-like fungi (*C. albicans*) being used in these experiments.

*2.2.4. Biological activity of the EBP-treated fibrin-monomer.* The EBP-modification products of fibrin-monomer were tested as the platelet aggregation inhibitors [5].

## 3. Treatment procedure

For the controllable chitosan modification and LMWC production the special Electron Beam Plasma chemical Reactor (EBPR) was designed. The EBPR, its operation modes and optimization of the biomaterial treatment regimes were described in detail in [4].

Figure 1 illustrates the treatment procedure. The focused continuous electron beam (EB) 3 generated by the electron-beam gun 1 which is located in the high vacuum chamber 2 is injected into the working chamber 5 filled with the plasma-generating gas through the injection window 4 [6]. In passing through the gas the EB is scattered in elastic collisions and the energy of fast electrons gradually diminishes during various inelastic interactions with the medium (ionization, excitation, dissociation). As a result, the cloud 10 of the EBP is generated, all plasma parameters being functions of $x$, $y$, and $z$ coordinates ($z$ is the axis of the EB injection).

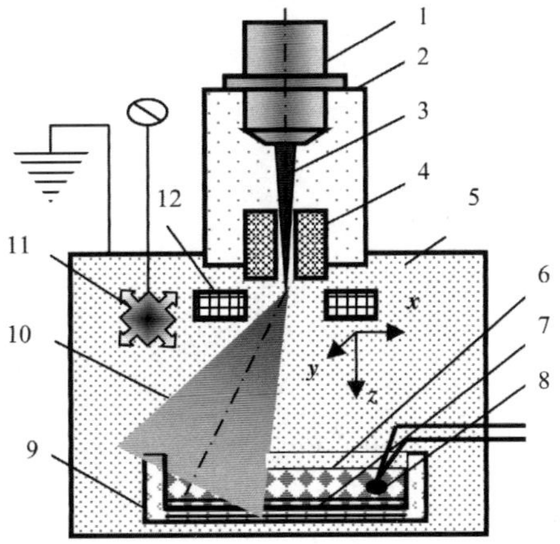


**Figure 1.** The design of the plasma chemical reactor and the treatment procedure. 1–electron beam; 2–high vacuum chamber; 3–EB; 4–injection window; 5–working chamber; 6–mixing layer of the powder to be treated; 7–piezoceramic plate; 8–temperature sensor; 9–glass container; 10–EBP cloud; 11–water evaporator; 12–scanning system.

The electromagnetic scanning system 12, which is placed inside the working chamber near the injection window is able to deflect the injected EB axis in $x$ and $y$ directions and, thereby, to control the spatial distribution of the plasma particles over the plasma bulk. The working chamber is preliminary evacuated to pressure ~$10^{-2}$ Torr and then filled with the plasma generating media (oxygen or water vapor). The water vapor was produced by means of the electrically heated evaporator 11 placed inside the working chamber. The samples to be treated were inserted into the EBPR reaction zone as solid powders with characteristic particle size ~ 100 mcm. The powder partially fills the glass container 9 over the thin plate 7 made of piezoelectric ceramics which is at the container bottom. Being fed with AC-voltage the plate vibrates, throws up the powder particles and forms the mixing layer 6 of the treated material inside the container. The miniature thermo-sensor 8 is inserted into the

Journal of Physics: Conference Series **370** (2012) 012012 doi:10.1088/1742-6596/370/1/012012

container to monitor the material temperature $T_s$ during the treatment. To prevent the thermal destruction of the biological material all samples were processed at $T_s$< 50 °C. In the case of proteins $T_s$ was ~ 37 °C. The temperature control was carried out by selecting the EB current $I_b$ ($1 < I_b < 100$ mA).

The experimental conditions were as follows:

- the pressure of the plasma generating gas was 4 Torr, 9 Torr, and 40 Torr for oxygen, water vapor, and helium, respectively;
- the distance between the injection window and sample surface was 250 mm;
- the EB scanning - square raster with 130 mm side length.

The optimal treatment time $\tau$ to achieve the required dose of the material irradiation was experimentally found by varying $\tau$ from 5 (for proteins) to 10 min (for chitosan).

## 4. Chitosan treatment in the electron-beam plasma. Production of LMWC

The original chitosan was not water-soluble while its EBP-treatment products became soluble and the effect increased with the prolongation of $\tau$. The maximum yield of water-soluble substances was obtained after 10 min and the solubility of these products was up to 95% up at room temperature.

The exclusion chromatography of the EBP-treated chitosan revealed the formation of a number of LMWC with molecular weight varied from 18 kDa to monomeric fragments (Figure 2). The majority of products formed with the EBP-treatment were oligosaccharides with the molecular mass 1 kDa (the elution time 11.3 min).

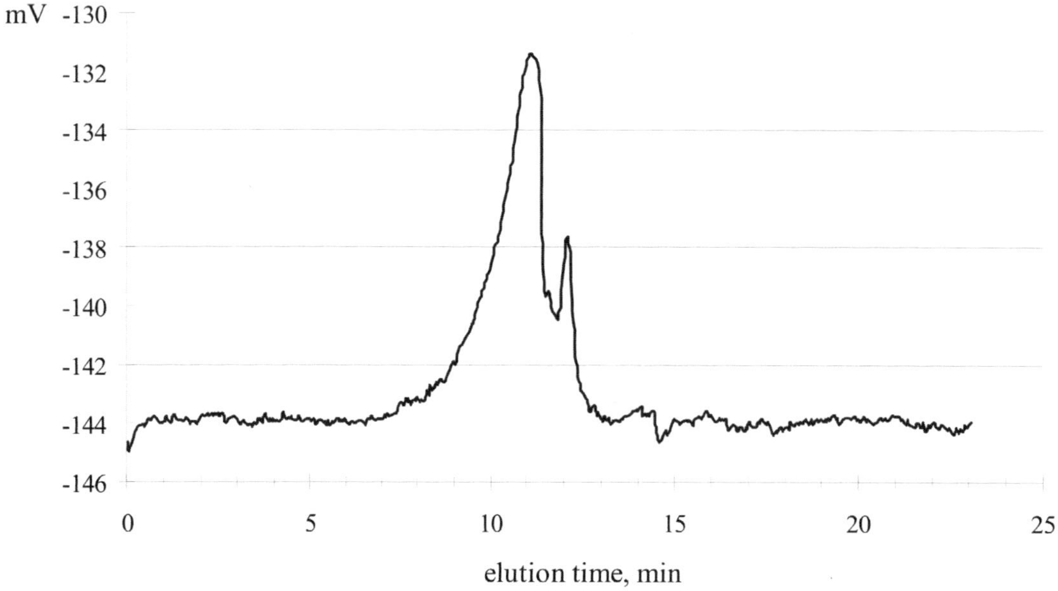


**Figure 2.** The exclusion chromatogram of chitosan treated in the EBP of water vapor for 10 min.

The degradation of the original polymer is due to the action of free radicals formed in the EBP. Active oxygen particles (O, O$^{\bullet}$, singlet oxygen) that are produced in plasma chemical processes and the products of the water plasmolysis (e.g. OH$^{\bullet}$) are likely to be of the most importance. These chemically active particles break the β-1,4 glycosidic bound and decrease the chitosan molecular weight Figure 3 illustrates the possible degradation mechanism [7].

The LMWC with molecular weight 30-180 kDa and chitooligosaccharide mixtures are known to possess antimicrobial properties [8, 9]. To quantitatively characterize the bioactivity of the EBP-produced LMCW the bacteria growth *in vitro* was studied (Table 1). The LMCW produced by the EBP-treatment in water vapor at concentration 1000 mcg/ml were found to completely suppress the multiplication of colon bacillus, aurococcus and yeast-like fungi. At lower doses the EBP-treatment products were also active and strongly inhibited the microorganism multiplication. We suppose that the antibacterial activity of the EBP-produced LMCW results from the LMCW interaction with the cell walls of microorganisms. This mechanism was considered in detail in [10].

**Table 1.** The microorganism growth under EBP-treated chitosan.

| Test microorganism | EBP-treated chitosan concentration, mcg/ml | | | | Control |
|---|---|---|---|---|---|
| | 1000 | 500 | 250 | 125 | |
| *E. coli* | ---- | ± | ± | + | + |
| *Ps. aeruginosa* | + | + | + | + | + |
| *S. aureus* | ---- | ± | + | + | + |
| *C. albicans* | ---- | ± | + | + | + |

--- the absence of microorganism growth; ± weak microorganism growth; + microorganism growth comparable with reference sample

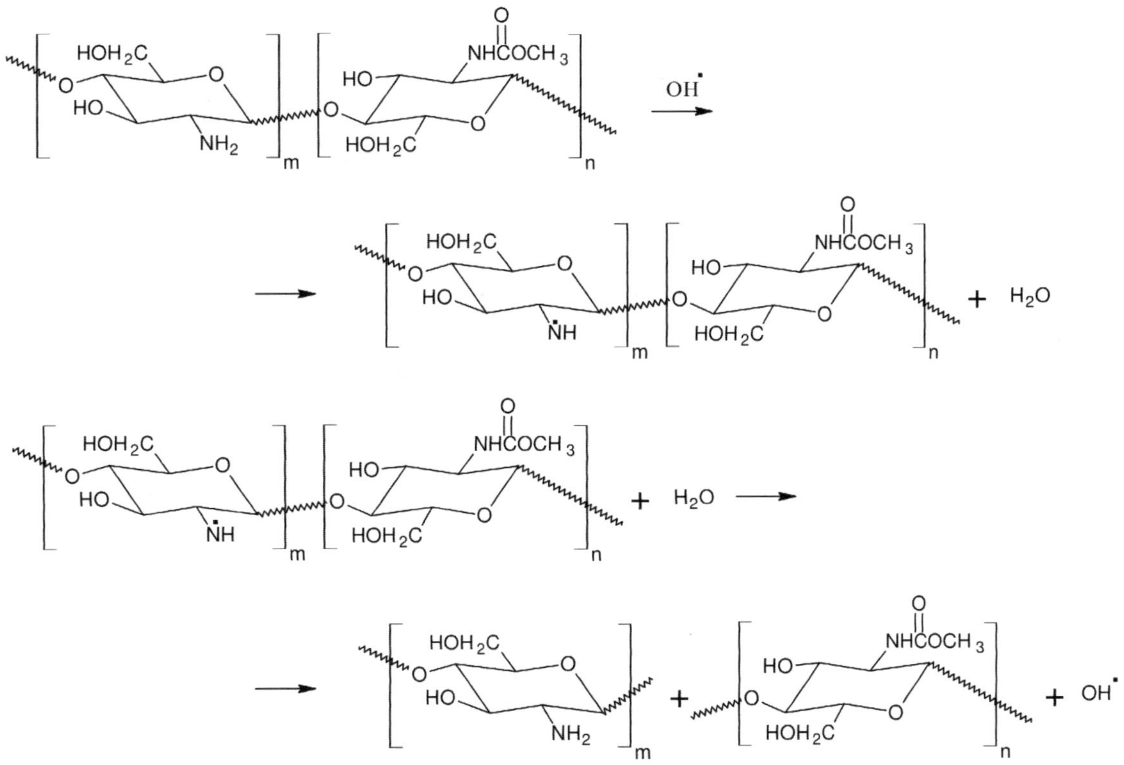

**Figure 3.** The scheme of chitosan degradation under hydroxyl radical action in the EBP of the water vapor [6].

## 5. The EBP-treatment of fibrin-monomer

The EBP-treatment of protein FM changed its structure radically due to the action of the chemically active plasma particles. Originally water-indissoluble native FM was found to become soluble at room temperature without bunching. The EBP-modified products did not exhibit the specific antigenic properties of the original FM and did not react with specific antibodies, while the native substance gave a specific precipitation line.

To reveal the changes in the primary and secondary structure of the EBP-treated FM the IR-spectroscopy was used. The EBR-treatment of the FM for $\tau = 5$ min caused the partial destruction of the peptide –CO-NH-bonds in the primary FM structure and the oxidation of the disulfide bonds responsible for the tertiary peptides structure. These resulted in the FM destruction and low molecular weight peptides formation.

To characterize the molecular masses of the EBP-treatment products the exclusion chromatography was applied. The peaks corresponding to 6 individual peptides with the elution times 12.30; 12.55; 13.17; 13.70; and 13.94 min were observed in the exclusion chromatograms of the FM modified by the EBP (figure 4).

The water-soluble products of the FM treated by EBP under specially adjusted conditions were found to decrease the platelet aggregation down to ≈ 33-35 % *in vitro* at concentrations $1\times10^{-4}$-1 mg/ml [5]. The peak corresponding to the elution time 12.3 min (molecular weight ≈ 650 Da) was observed at the exclusion chromatograms of the FM modified in the EBP of helium and in the EBP of water vapor. This peptide is likely to be responsible for inhibiting the platelet aggregation.

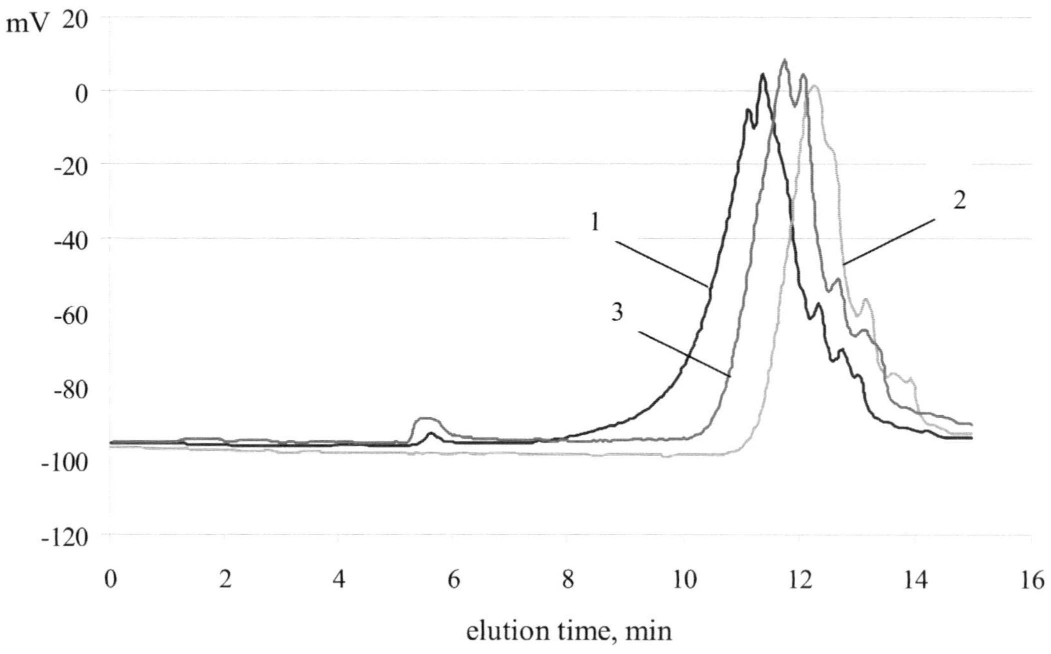


**Figure 4.** The exclusion chromatogram of fibrin-monomer treated in the EBP: 1– EBP of helium, $\tau = 10$ min; 2- EBP of water vapor, $\tau = 5$ min; 3- EBP of water vapor, $\tau = 15$ min.

## 6. Conclusions

- The possibility of the EBP-stimulated hydrolysis of native chitosan and formation of water-soluble low molecular weight products was proved experimentally.
- The 95% yield of low molecular weight EBP-treatment products was attained by optimizing the treatment procedure. The high yields of low molecular weight water soluble products are obtained at treatment time ~ 10 min whereas the traditional chitosan hydrolysis usually takes several days. The hazardous by-products and toxic wastes are not generated during the EBP-treatment.
- The low molecular water-soluble forms of the chitosan obtained by its treatment in the EBP of oxygen and water vapor were found to inhibit the multiplication of colon bacillus, aurococcus and yeast-like fungi.
- The EBP-stimulated degradation mechanisms occurring both in polysaccharides and high molecular weight proteins are similar. The active oxygen species produced in plasma chemical reactions and the products of water plasmolysis are responsible for the LMWC, chitooligosaccharides, and peptides formation.

Thus, our experiments have demonstrated that the EBP can be used for the effective and controllable modification of some natural biopolymers. The technique involved is likely to be promising to engineer the compounds with unique pharmacological activities and the EBPRs seem to be competitive with technologies conventionally used in the pharmaceutical industry.

**Acknowledgments**
This study was supported by Russian Foundation of Fundamental Research, grant 12-04-01189-a.

**References**

[1] Ray S D 2011 *Acta Pol. Pharm.* **68** 619-22
[2] Laurienzo P 2010 *Mar. Drugs* **8** 2435-65
[3] Aam B B, Heggset E B, Norberg A L, Sorlie M, Varum K M and Eijsink V G H 2010 *Mar. Drugs* **8** 1482-517
[4] Vasilieva T 2010 *IEEE Transac. Plasma Sci.* **38** 1903-7
[5] Vasilieva T 2008 *Peptide Science 2007*, ed. S Aimoto and S Ono S (The Japanese Peptide Society) 35-8
[6] Vasiliev M N 2001 *Encyclopedia of the low-temperature plasma*, ed. V E Fortov (Moscow: Nauka) **XI** 436-44
[7] Chang K L B, Tai M C and Cheng F H 2001 J. Agric. Food Chem. **49** 4845-51
[8] Fernandes J C, Tavaria F K, Fonseca S C, Ramos O S, Pintado M E and Malcata F X 2010 *J. Microbiol. Biotechnol.* **20** 311-8
[9] Wang Y, Zhou P, Yu J, Pan X, Wang P, Lan W and Tao S 2007 *Asia Pac. J. Clin. Nutr.* **16** Suppl. 1 174-7
[10] Kong M, Chen X G, Xing K and Park H J 2010 *Int. J. Food Microbiol.* **144** 51-63

Journal of Physics: Conference Series **370** (2012) 012013 doi:10.1088/1742-6596/370/1/012013

# Pulsed laser deposition of carbon films in low pressure neutral gas background

**F Guzmán[1], H M Ruiz[1], M Favre[1], S Hevia[1], H Bhuyan[1], H Chuaqui[1], E Wyndham[1] and M Flores[2]**

[1]Departamento de Física, Pontificia Universidad Católica de Chile, Av. Vicuña Mackenna 4860, Santiago, Chile
[2]Departamento de Física, Facultad de Ciencias Físicas y Matemáticas, Universidad de Chile, Av. Blanco Encalada 2008, Santiago, Chile

E-mail: faguzman@uc.cl

**Abstract**. We have investigated Pulsed Laser Deposition (PLD) of thin carbon films using a graphite target, in a low pressure Argon gas background. Raman spectroscopy based structural analysis of the films shows a correlation between films properties and pressure of the buffer gas background. The morphology of the resulting thin films was characterized with Raman spectroscopy. Time resolved Optical Emission Spectroscopy (OES) observations indicate that, at the substrate position, and for background pressures above 80 mTorr, the dominant species in the expanding carbon plasma are $C_2$ molecules. This is also consistent with Faraday cup observations, which show a strong decrease in the carbon ions content, as the pressure is increased. Our main result is the observation of a sharp transition in the film morphology, around 300 mTorr, from Diamond like Carbon (DLC) at lower pressures, to amorphous Carbon at higher pressures.

## 1. Introduction

Pulsed laser deposition (PLD) is a well established technique used for deposition of different materials, including metals, semiconductors and polymers [1]. Particular attention has been given to PLD of carbon coatings, due to their morphological, structural, mechanical and electrical properties [2], which lead to potential biomedical applications. In this context we report on the characterization of thin carbon films obtained by PLD in an argon background and discuss the resulting morphology in correlation with properties of the laser produced carbon plasma.

## 2. The experiment

The experiments were performed in a vacuum chamber at Argon background gas pressures between 0.5 and 700 mTorr. A Nd:YAG laser pulse, 370 mJ, 3.5 ns, at 1.06 μm, operating at 10 Hz was applied to a rotating graphite target, at approximately 45° to the normal, with a fluence of 6.7 J/cm$^2$. A Silicon (100) substrate was located 5 cm from the laser spot, and the exposure time was fixed at 1.5 min. To characterize the plasma composition we used time integrated and time resolved visible spectroscopy, using a spectrometer combined with a 15 ns gated OMA. Axial Faraday cup measurements were performed to characterize the ionic content and dynamics of the laser plasma plume. The morphology and composition of the resulting thin carbon films was investigated with Raman spectroscopy, atomic force microscopy (AFM) and secondary ion mass spectroscopy (SIMS).

## 3. Plasma properties

Figure 1 shows a 20 ms, time integrated spectrum of the Carbon plasma emission collected from a point on axis, 20 mm from the target surface, with 50 mTorr Argon background. The spectrum shows emission lines associated with single ionized Carbon, and molecular bands. Additional spectroscopic observations, closer to the target surface, show also single ionized Argon lines. These emission lines are either not observed or not resolved at this distance. Here the spectrum is dominated by emission corresponding to vibrational states of the $d^3\Pi_g \rightarrow a^3\Pi_u$ Swan band, of the neutral $C_2$ molecule [3]. Further observations of the laser plasma plume, with 15 ns time resolution, published elsewhere [4], show an increase in the molecular content of the laser plume, as pressure increases.

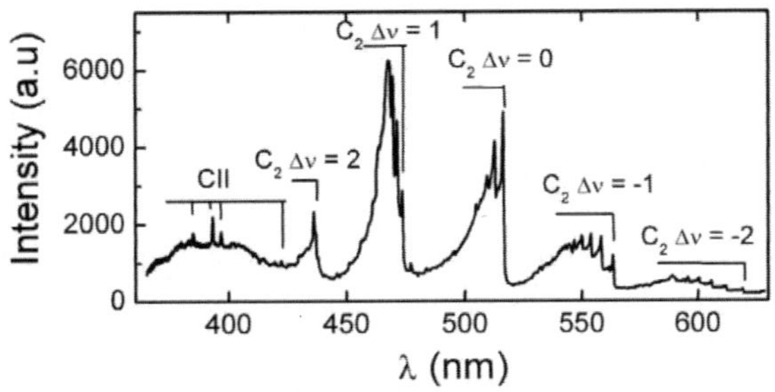


**Figure 1.** Time integrated spectrum of the carbon plasma emission from a point on axis, 20 mm from the target surface, at 50 mTorr Argon background.

Figure 2 shows signals from Faraday cups located at 7 and 10 cm from the graphite target, for three different pressures. The distinctive feature is the observation of two well defined plasma fronts, in agreement with previous observations [5]. Characteristic velocities of the order of $4{\bullet}10^4$ m/s and $3{\bullet}10^3$ m/s are inferred from the Faraday cup data for the fast and slow plasma fronts.

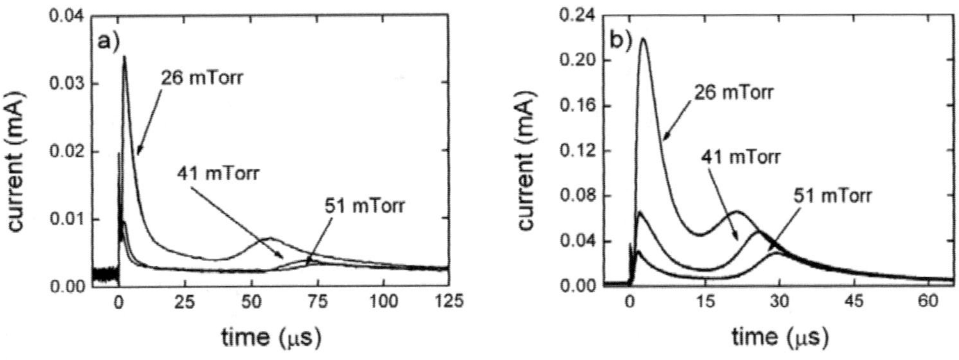


**Figure 2.** Characteristic signals of Faraday cups, as a function of pressure, located at (a) 10 cm, and (b) 7 cm, from the target surface.

On inspecting the leading edge or peak value of the fast component in the Faraday cup signals, an apparent reduction in the characteristic time of flight might be observed, as the Argon background pressure increases. This might imply an increase in the ions characteristic velocity, as the pressure increases, which is unlikely. The observed effect is explained by the fact that an increase in background pressure also increases the collision rate of lower energy ions. As a result of this process, a significant fraction of lower energy ions are scattered in the radial direction, thus resulting in a shift to higher energies of the peak value of the Faraday cup signals. The trend in the slow component, increase in the time of flight with increasing pressure, is also consistent with a global higher collision rate. In this case, the lower characteristic energy results in a global energy loss, which can also be seen along the axial direction, due to the much lower ion flux, in comparison with the fast component.

On the base of the spectroscopic observations [4] it is reasonable to assume that the dominant species in the Faraday cup signals is CII ions. Under these conditions, the characteristic energy of the carbon ions in the fast component is estimated to be around 250 eV. A species composition for the slow component can not be established, as it might be a mixture of slow CII ions and ionized $C_2$ molecules.

## 4. Raman analysis of carbon films

The Raman analysis of DLC-films can be performed following a general procedure proposed by Ferrari *et al.* [6]. Raman spectra of thin carbon films are characterized, in the 1000 to 1800 $cm^{-1}$ range, by two distinctive peaks, namely D and G peaks. Relative intensity and position of peaks depend on film morphology. Usually, the G peak is fitted whit a single skewed Lorentzian peak described by the Breit-Wigner-Fano (BWF) lineshape, and the D peak is fitted with a Lorentzian.

Figure 3 shows characteristic Raman spectra, with deconvoluted D and G peaks, at 170 and 380 mTorr Argon background pressure, together with the evolution of the peak intensity ratio, as a function of the Argon background pressure.

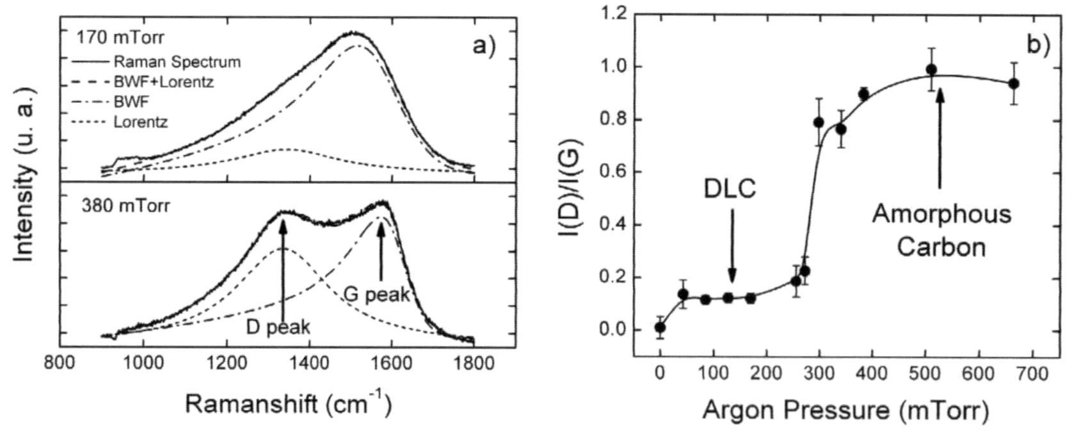


**Figure 3.** (a) Characteristic Raman spectra at two different pressures, showing D and G peaks deconvolution, (b) Intensity ratio of D and G peaks, as a function of pressure.

The intensity ratio, *I*(D)/*I*(G), is determined by the degree of graphitization of the films. The Raman spectrum at 170 mTorr is characteristic of a diamond like film (DLC), whereas that at 380 mTorr corresponds to amorphous carbon [6][7]. The *I*(D)/*I*(G) ratio was found to change with Argon pressure, in the ~0.1 to ~1.0 Torr range, thus indicating a transition from DLC to amorphous carbon, as pressure increases. The small peak in the Raman spectra around 900 $cm^{-1}$ is due to a vibrational state of the Silicon substrate. It can be used, together with peak G, to estimate the film thickness, if an independent thickness measure is available. This was done using SIMS, to find that as pressure increases, the characteristic thickness of the carbon films decreases from 90 to 20 nm.

## 5. Discussion

The evolution of the *I*(D)/*I*(G) ratio shows a sharp transition from DLC to amorphous carbon films. The OES observations [4] show that as the Argon background pressure is increased the laser plasma plume dominant species are $C_2$ molecules. On the other hand, the Langmuir probe measurements indicate that as pressure increases the relative intensity of the second plasma front, as compared with the first. On the base of these observations it can be conjectured that above the transition pressure of ~170 mTorr Argon, the main contribution to Carbon film growth is due to the second front, with low energy ions and dominant $C_2$ molecules, being the film growth process closer to a traditional deposition process. At lower pressure the dominant species are high energy CII ions. The characteristic high energy of the ions might lead to transient heating, followed by $C_2$ deposition.

## 6. Conclusions

We have characterized the carbon ions content of a laser produced Carbon plasma produced in low pressure Argon background, and the morphological properties of thin Carbon films resulting from laser Carbon plasma deposition on Silicon substrates. We have found that the structural properties of the Carbon films depend on the Argon background, with a sharp transition from DLC to amorphous Carbon around 300 mTorr. Further analysis and parameter scan is required to explain the observed transition, which will be reported elsewhere.

## Acknowledgements

This work has been funded by FONDECYT project 1110380. F. Guzmán and H. M. Ruiz acknowledge doctoral studies scholarships from CONICYT and MECESUP, respectively.

## References

[1] Eason R (Ed.) 2007 *Pulsed laser deposition of thin films* (New York: Wiley)
[2] Kumar Roy R and Lee K R 2007 *J. Bio. Mat. Res. B: Appl. Biomaterials* **83B** 72
[3] Kushwaha A and Thareja R K 2008 *Appl. Optics* **47** G65
[4] Ruiz H, Guzmán F, Favre M, Bhuyan H, Chuaqui H and Wyndham E 2011 *Plasma Sources Sci. Technol.* in press
[5] Budtz-Jørgensen C, Mond M, Doggett B and Lunney J 2005 *J. Phys. D: Appl. Phys.* **38** 1892
[6] Ferrari A and Robertson J 2000 *Phys. Rev. B* **61** 14095
[7] Ferrari A and Robertson J 2001 *Phys. Rev. B* **64** 075414
[8] Koivusaari K J, Levoska J and Leppävouri S 1999 *J. Appl. Phys.* **85** 2915

Journal of Physics: Conference Series 370 (2012) 012014 doi:10.1088/1742-6596/370/1/012014

# Degradation of tetrafluoroethane using three-phase gliding arc plasma

**J Pacheco[1,2], M García[2], M Pacheco[1], R Valdivia[1,2], C Rivera[1] and M Garduño[2]**

[1]Instituto Nacional de Investigaciones Nucleares Carretera Mexico-Toluca, La Marquesa Ocoyoacac, Mexico
[2]InstitutoTecnológico de Toluca, Av. Tecnológico S/N Ex Rancho la Virgen Metepec, Mexico

E-mail: marquidia.pacheco@inin.gob.mx

**Abstract.** The use of many chlorofluorocarbons (CFC´s) has negatively impacted the ozone layer. The Montreal Protocol was implemented, as a temporary solution for this problem by replacing CFC's by hydrofluorocarbons (HFC's). These kinds of gases have the propriety to be free of chlorine. However, in a next future, the Montreal Protocol also considers the replacement of HFC's because they have a high global warming potential when they enter in contact with the atmosphere. One of the methods to remove those compounds is the gliding arc plasma because it presents some advantages. The inlet system works near the atmospheric pressure and has a transition region from plasma at thermodynamic local partial equilibrium to non-thermal plasma; allowing high gas and electronic temperatures. Results present a promissory possibility to be scaled and to give an industrial service.

## 1. Introduction

The three-phase gliding arc plasma is a self-oscillating discharge formed in three divergent electrodes. The discharge is initially developed at the shortest distance between the two electrodes having the biggest electric potential difference. After that, the arc diverges along these electrodes. The discharge length grows until a critical length, determined by the power supply. Beyond this point, the energy obtained by the electrical supply can no longer sustain the equilibrium phase of the gliding arc and the electrical conductivity in plasma channel decreases (when the arc length increases) until the extinction of the discharge. Then, a restrike appears, leading to a new periodic cycle [1,2].

This work presents an alternative to treat, in a safety way, the HFC's. These gases are normally discarded by the refrigerant industry and their environmental impact has been amply measured. One of advantages to work with a three-phase gliding arc plasma is that the system works near to the atmospheric pressure and can treat a continuous gas flow at low power with the possibility to convert a considerable volume of HFC's [2-6].

The study of the degradation includes an electrical analysis considering the efficiency of HFC's conversion and gives preliminary parameters that will allow improving the system to be scaled in near future.

Results obtained in the present article, show that the gliding arc reactor proposed is effective to treat halogenated compounds, difficult to trait with conventional methods or non-thermal plasma [7].

Journal of Physics: Conference Series **370** (2012) 012014 doi:10.1088/1742-6596/370/1/012014

## 2. Experimental setup

The experimental setup consists in a plasma reactor, a power supply, flow controllers and measurement systems and is schematically depicted in figure 1. The plasma reactor contains three copper electrodes (see figure 2) disposed inside of a cylindrical quartz tube of 56 mm ID and length of 410 mm. The power supply is composed by three phase high voltage transformer, a variable transformer controls the ignition discharge and three inductances stabilize the discharge plasma. The exhaust gases are directly connected to an alkaline scrubber system employed as a safety measure in case of acid gas were formed, following this system, there are two ports to sample and to vent residual gases. One of the ports is used for gas chromatography and the other one is for FTIR analysis.

The gases from the cylinders were controlled by three mass flow rate controllers (MKS, πMFC), in order to add water steam to the mixture, gases were bubbled in water and the mixture was then injected to the reactor. The discharges were formed at the closest points of the copper electrodes, having a very short contact time. Then they are spread by gliding axially along the electrodes by the gas force action. The arc disappeared at the end of the electrodes and a new discharge immediately is formed at the initial section. The copper electrodes were not cooled; as a result, the electrical energy was directly transferred to the treated gas mixture.

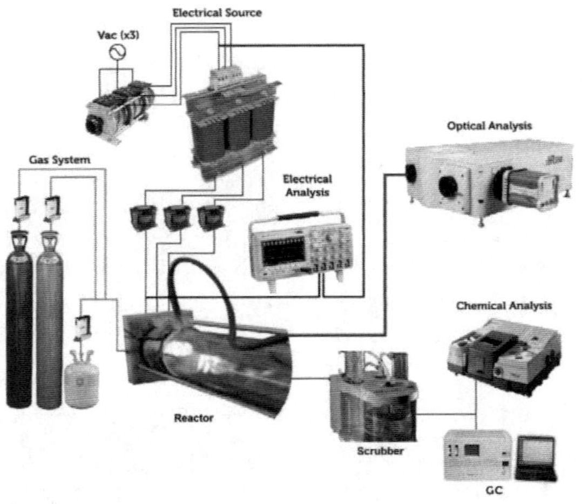


**Figure 1.** HFC's degradation system.

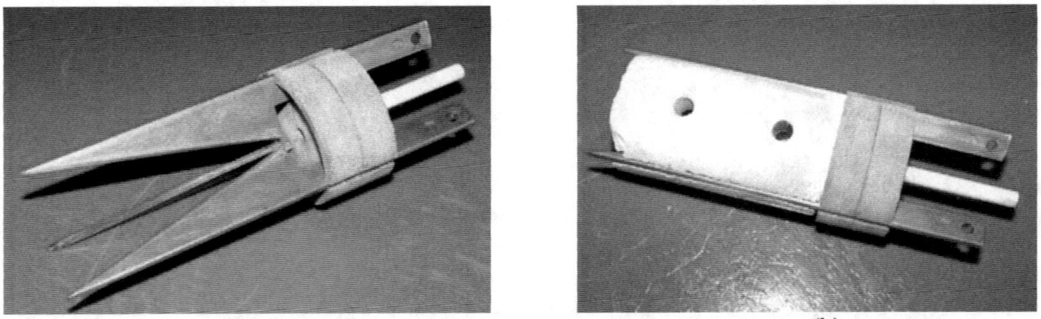

(a) (b)

**Figure 2.** (a) Triangular electrodes, (b) triangular electrodes with ceramic pieces.

## 3. Chemical background

The tetrafluoroethane was diluted in argon and air in a concentration of 20489 ppm in volume and a proportion given in table 1. The mixture of gases was bubbled to add water and injected to the reactor to be treated. The total flow rate applied to the reactor was 14.1 l.p.m.

**Table 1.** Mixture of gases to be treated by plasma.

| Gas | **Concentration** |
|---|---|
| R134-a | 0.289 l.p.m. |
| Ar | 12.7 l.p.m. |
| Air | 1.1 l.p.m. |

The mechanism of degradation developed inside of the reactor was described by two main reactions; the first reaction considers a mixture of tetrafluoroethane and water steam described in the reaction (1)

$$CH_2FCF_3 + H2O \rightarrow CO + C + 4HF \quad (1)$$

When this reaction was developed in the degradation system, the reactor walls were covered with a carbon layer (see figure 3(a)) who worked like a conductor forcing to stop the degradation process.

The second mechanism of degradation (2) includes water steam and oxygen in the mixture of gases. It presents an enhancement, because the presence of oxygen from air avoids the carbon formation over the inner walls (see figure 3(b)). [4,6].

$$CH_2FCF_3 + H_2O + 1/2O_2 \rightarrow 2CO + 4HF \quad (2)$$

One possibility to reduce the presence of CO is by oxidizing it with oxygen to reduce its negatives effects as is shown by reaction (3) [5, 6]

$$CO + 1/2O_2 \rightarrow CO_2 \quad (3)$$

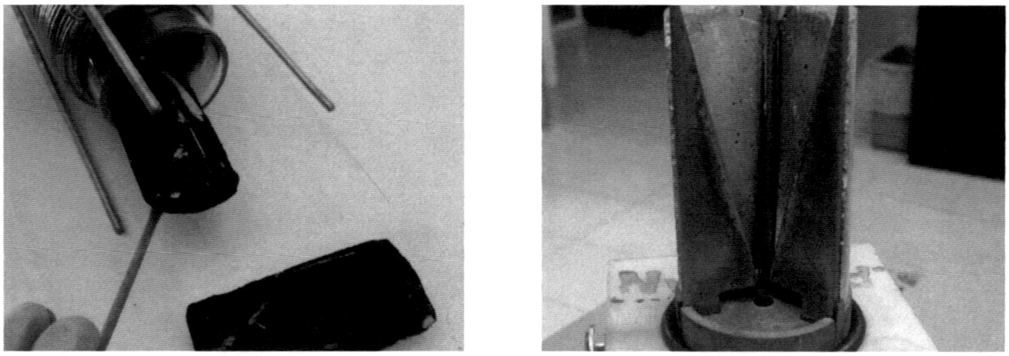

(a) (b)

**Figure 3.** (a) Internal walls of the reactor covered of carbon, mechanism of degradation (1), (b) internal walls of the reactor carbon free mechanism of degradation (2).

## 4. Chemical analysis

To determinate the conversion factor of the system was necessary to determinate the initial and final concentration of tetrafluoroethane. The technique used to quantify was gas chromatography. The detector used was a FID (Flame Ionization Detector), a packed column of Porapak Q and Argon like a carrier gas was used in the analysis.

Journal of Physics: Conference Series **370** (2012) 012014 doi:10.1088/1742-6596/370/1/012014

The detector has a temperature of 200° C and the oven has a temperature of 70° C isothermal, the carrier gas flow was 30ml min-1.

The initial concentration of gas was 20489ppm and the final concentration was 1390 ppm, therefore the conversion rate obtained is around 93.21%. All the chemicals measures were verified three times.

The figure 4 is a chromatogram that shows the conversion of tetrafluoroethane before and after the plasma degradation.

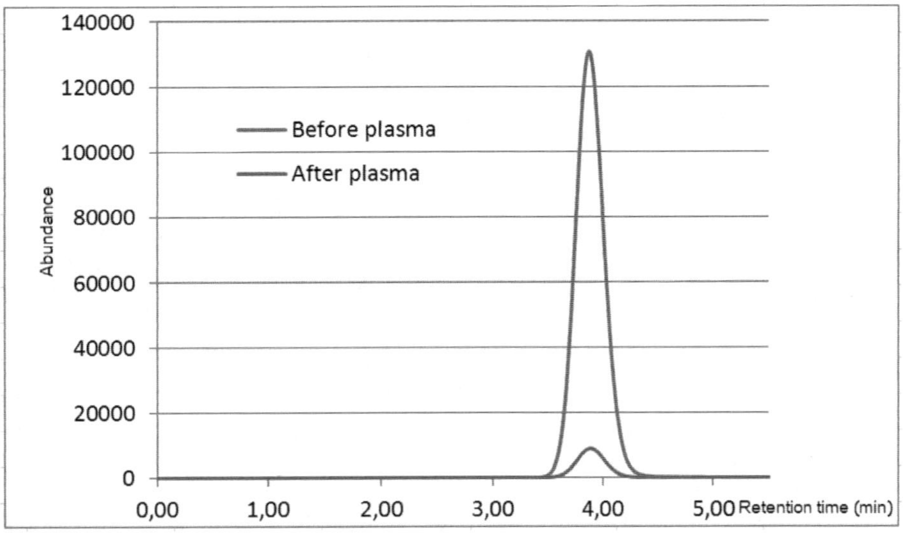


**Figure 4.** Chromatogram of tetrafluoroethane degradation, employing fid detector, argon as carrier gas 30ml $min^{-1}$.

## 5. Electrical analysis

Voltages and currents were measured with an oscilloscope Tektronics MSO 2024, these signal can be observed in figure 5. It is worth to note that the value represents only one phase of the system, complementing it with other two electrical phases. The results here presented correspond to electrical signals during the tetrafluoroethane treatment with gliding plasma. These signals are not modified in a considerable way compared to these obtained when only argon is present in the reactor. This is because the very low HFC's concentrations employed.

With the instantaneous voltage and current reported in figure 5, the corresponding instantaneous power is reported in figure 6 considering one and three phases, according with equation (4), the total RMS power can be calculated.

$$Three - phase\ power\ RMS = \sqrt{\frac{3}{T}\int_0^T [v(t)\times i(t)]^2\, dt} \tag{4}$$

From the RMS power values and the conversion of HFC's the destruction efficiency can be obtained as follows:

$$Destruction\ efficiency = \frac{Converted\ HFC's}{Supplied\ power} \tag{5}$$

The destruction efficiency was calculated in 33.36 l $kW^{-1}$ $hr^{-1}$

Results obtained by [10] to treat the same molecule with another plasma technology (i.e. plasma torch with water) show efficiencies near to 100% at 0.16 kg/h of C2H2F4 mass yield rate. In this

Journal of Physics: Conference Series **370** (2012) 012014 doi:10.1088/1742-6596/370/1/012014

work efficiencies of 93% at 0.074 kg/h were obtained, however the input energy in the system here proposed is low (around 519W). [7,9]

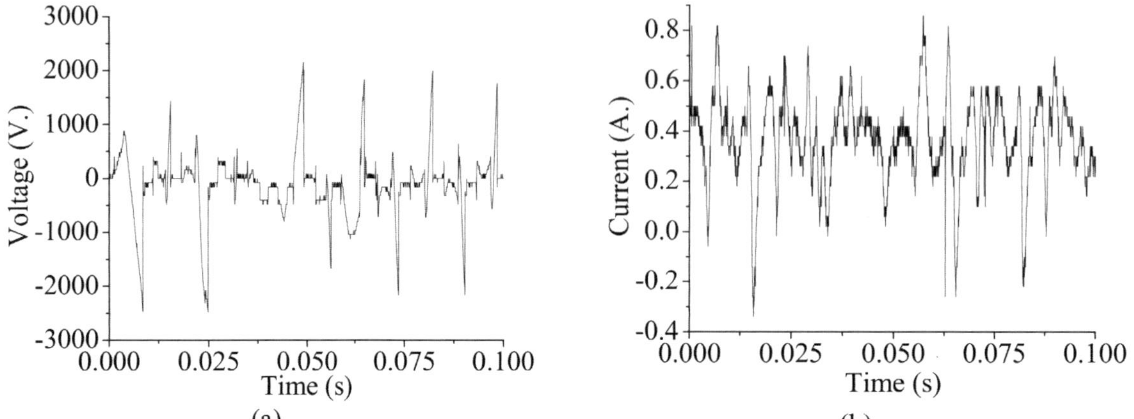


**Figure 5.** (a) Plasma voltage discharge, (b) plasma current discharge.

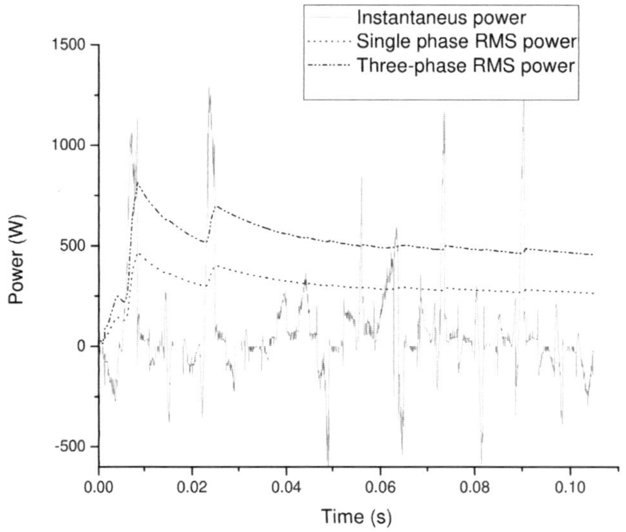


**Figure 6.** Instantaneous and RMS plasma power.

## 6. Conclusions

The degradation by three-phase gliding arc plasma is an effective way to reduce the gases that produce greenhouse effect like the tetrafluoroethane. The plasma reactor has a very high energy density media, the active species accelerate chemical reactions, besides the plasma reactor is compactness and has a low energetic cost.

This system can be used to detoxify other halogenated compounds [7-9]. Like a future work we will analyze the formation of species and radicals of NOx formed in the reactor by the presence of nitrogen and oxygen in the reactor and the improvement of HFC's treatment efficiency at low power consumption.

**Acknowledgments**
The authors would like to thank CONACyT for the financial founds and for the scholarship 11557, as well as the LAP-ININ collaborators for the technical support and assistance during the experimental tests.

**References**

[1] Fridman A A, Petrousov A, Chapelle J, Cormier J-M, Czermichowski A, Lesueur H and Stevefelt J 1994 *J. Physique III* **4** 1449–65
[2] Garduño M, Estrada N, Pacheco J, Pacheco M, Garcia M, Valdivia R, Rivera C and Gonzalez J 2011 *IEEE Transactions on Plasma Science* **39** 2890–1
[3] Montreal Protocol on Substances That Deplete the Ozone Layer 1999 *HFC and PFC Task Force of the Technology and Economic Assessment Panel* (Nairobi, Kenya: UNEP)
[4] Takita Y, Tanabe T, Ito M, Ogura M, Muraya T, Yasuda S, Nishiguchi H and Ishihara T 2002 *Ind. Eng. Chem. Res* **41** 2585-90
[5] Technology of Decomposition of Ozone Layer Depletion Compounds 1991 *Committee on the Decomposition of Ozone Layer Depletion Compounds, Ministry of Trade and Industry: Japan* p 43
[6] Simmonds P, O'Doherty S, Huang J, Prinn R, Derwent R G, Ryall D, Nickless G and Cunnold D 1998 *J. of Geophysical Research* **103** 029-37
[7] Indartoa A, Yangb D, Azharic C, Mohtarc W, Choia J, Leea H and Songa H 2007 *Chem. Engineering J.* **131** 337–41
[8] Labelle C and Gleason K 2001 *J. of App. Polymer Science* **80** 2084–92
[9] Jasin M, Dors M and Mizeraczyk J 2009 *Plasma Chem Plasma Process* **29** 363–72
[10] Watanabe T and Tsuru T 2008 *Thin Solid Films* **516** 4391–6

# Low temperature deposition of low stress silicon nitride by reactive magnetron sputtering

**A P Mousinho[1,2], R D Mansano[1], L S Zambom[3] and A Passaro[2]**

[1]Departamento de Engenharia de Sistemas Eletrônicos
Escola Politécnica da Universidade de São Paulo - Brazil
[2]Laboratório de Engenharia Virtual - Divisão de Física Aplicada
Instituto de Estudos Avançados, IEAV-CTA, Brazil
[3]Departamento de Sistemas Eletrônicos, Fatec/SP – Centro Paula Souza, Brazil

E-mail: mousinho@lsi.usp.br

**Abstract.** Silicon nitride thin film is a well-know kind of material in semiconductor industries. These films are used in important applications such as protective layers; insulators for optical devices, as dielectric material for thin film transistors (TFTs), etc. In this work, silicon nitride thin films were deposited at room temperature through of the reaction of sputtered silicon by argon ions and nitrogen gas with relatively high deposition rates in sputtering reactive processes. We have obtained silicon nitride thin films with low stress for MEMS (Micro Electro Mechanical Systems) fabrication and in post-processing of sensors and actuators devices. Silicon nitride thin films obtained in this work showed stoichiometric ($Si_3N_4$) and/or rich in silicon with low mechanical stress and good dielectric properties. The films were analyzed by scanning electron microscope, micro-Raman spectroscopy, profilometry, current-voltage (I-V) and capacitance-voltage (C-V) measurements, laser-based measurement for determination the internal stress and Rutherford Backscattering.

## 1. Introduction

Silicon nitride ($Si_3N_4$ or $SiN_l$) is a very important ceramic due to its chemical, mechanical and electronic properties. It is chemically inert and show superior mechanical properties at high temperatures [1]. Silicon nitride is an important material in microelectronics due to its high resistivity, higher dielectric constant compared to silicon dioxide, mechanical strength, and chemical inertness. It is used as a gate insulator in thin film transistors that are in flat panel displays. It is also used in non-volatile memory devices because of its trapping ability. It can be applied as an oxidation mask and diffusion barrier in integrated circuit technology [2].

Silicon nitride films are mainly prepared by chemical vapor deposition (CVD) and plasma enhanced chemical vapor deposition (PECVD) [2-5]. Even though chemical vapor deposition is widely used for deposition of these films, the main disadvantages of this technique are the incorporation of hydrogen in the films and high substrate temperatures [6]. The entrapped hydrogen in the films can deteriorate the properties of silicon nitride, and the high substrate temperature is not generally desired in microelectronic applications [7].

Reactive Magnetron Sputtering [8-10] at low substrate temperature can provide silicon nitride films with extremely low hydrogen content [3]. In this work, we have obtained silicon nitride films

Published under licence by IOP Publishing Ltd

deposited by Reactive Magnetron Sputtering with low temperature deposition, low stress and low hydrogen content for MEMS (micro electro mechanical systems) fabrication [11-12]. Thus, flexibility in controlling the flux and energy of the ions will facilitate the growth of thin films with desired properties.

In this article we discuss the properties of silicon nitride films prepared by Reactive Magnetron Sputtering. The properties are studied using micro-Raman spectroscopy [13], profilometry, laser-based measurement for determination the internal stress [14-15], scanning electron microscopy [16] and Rutherford Backscattering spectrometry [17] and current-voltage (I-V) and capacitance-voltage (C-V) measurements [18].

## 2. Materials and methods

### *2.1. Deposition system*
The silicon nitride films were deposited in a magnetron sputtering plasma system. We used a silicon target and nitrogen and argon as the process gases. The pressure was fixed in 5 mTorr while the applied power was varied (250 W and 350 W) and the process time was varied too (from 10 min until 120 min). For obtaining a step in a silicon nitride films for measuring the thickness a mechanical mask was used (protecting a small area in the surface samples).

### *2.2. Samples*
The substrates used to deposit the films were three-inch diameter silicon wafers, 380 μm thick and orientation (100). They were submitted to a Piranha clean followed by a diluted HF dip, before the films deposition. The silicon nitride films were deposited at room temperature with relatively high deposition rates.

### *2.3. Techniques of analyses*
For evaluating the deposition rates, we measured the silicon nitride thickness using a Dektak 3030 profilometry from Sloan-Veeco. For analyzing the surface topography and the profiles of the samples we used a FEI NOVA nano400 scanning electron microscope (SEM).

We analyzed the films with a vis-UV Raman spectrometer system (Renishaw micro-Raman 2000 spectrometer on a 40x objective with a photomultiplier). Unpolarized Raman spectra were acquired at $\lambda$=514.6 nm, the spectral resolution was about 4-6 $cm^{-1}$ and the power on the sample was kept well below 1 mW.

The Si/N ratio was analyzed by Rutherford Backscattering Spectrometry (RBS). The total stress of the silicon nitride films was obtained using a KLA-Tencor FLX stress measuring system. For electrical measurements metal/nitride/silicon capacitors were formed by thermal evaporation of 300 nm thick aluminium films. The capacitors were patterned with a mask composed of 700 μm pads. On the wafer backside was evaporated a 300 nm thick Al film. C-V measurements at 1 MHz were made to determinate the dielectric constant and I-V measurements were made to determinate the dielectric breakdown electric field and conductivity. The results obtained are shown in this work.

## 3. Results and discussion
Firstly, we have made a plasma composition effects study to verify the influence of this parameter in the deposition rate, in the Si/N ratio and in the electrical properties. For this, the deposition time (30 min and 60 min) and the applied RF power (250 W and 350 W) were varied. The plasma composition effects in these parameters can be observed in figures 1, 2 and 3.

Observing figure 1, we can notice that the deposition rate decreases with the nitrogen addition in the argon plasma. This effect occurs due the lesser collision section of the nitrogen ions, when we compared with the collision section of the argon ions. Thus, lesser material removal of the target occurs when more nitrogen is present. In general, the thickness increases with the increase of deposition time and the applied RF power (as we can see in figure 2).

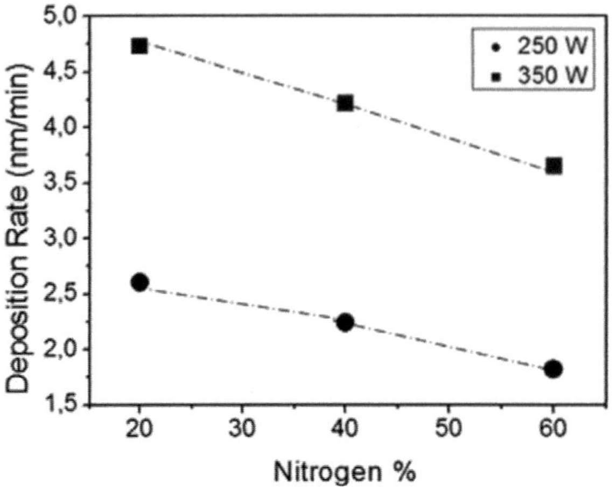


**Figure 1.** Deposition rate.

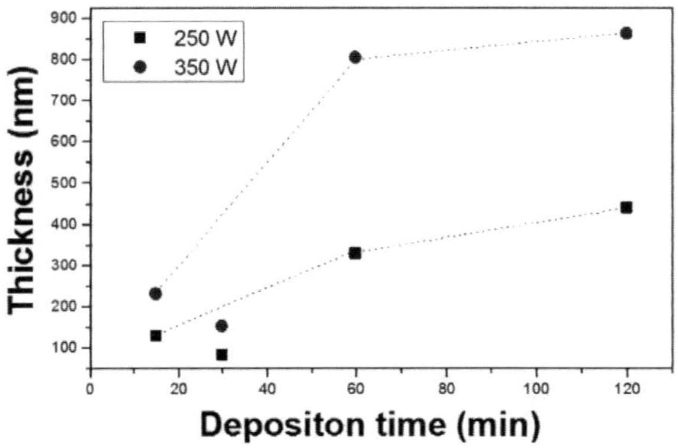


**Figure 2.** Thickness variation.

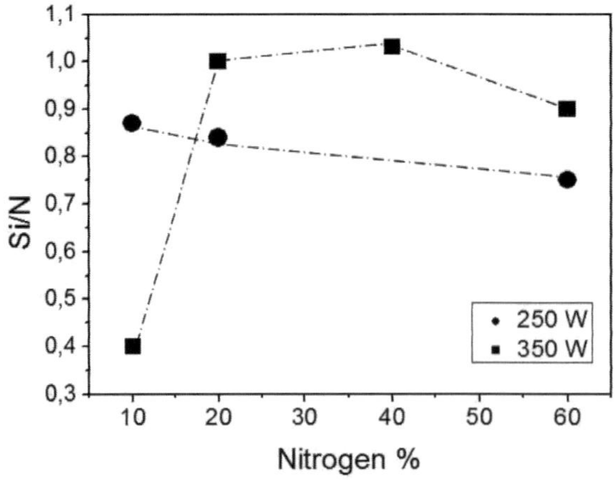


**Figure 3.** Si/N ratio.

The most stoichiometric silicon nitride films ($Si_3N_4$) were obtained for plasma processes with nitrogen content of 40% and content argon of 60%, and with applied FR power of 250W. The process parameters effect in the dielectric constant and the stress of the silicon nitride is presented in figures 4 and 5, respectively.

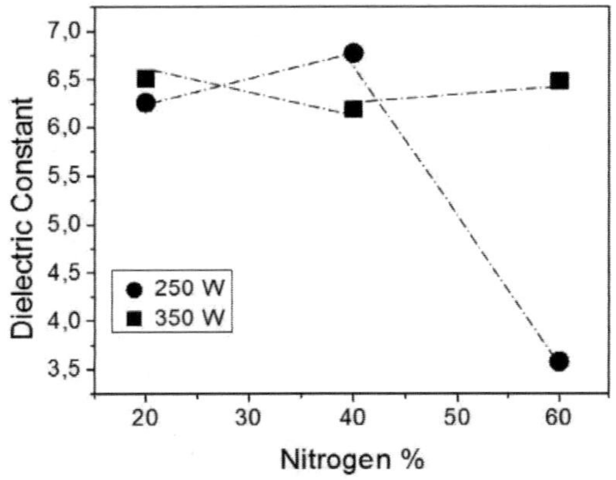


**Figure 4.** Dielectric constant.

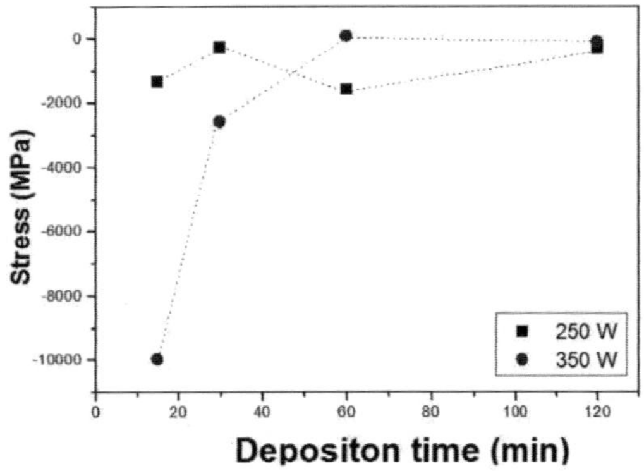


**Figure 5.** Stress of the silicon nitride films for 40% Nitrogen content.

The most stoichiometric silicon nitride films ($Si_3N_4$) presented a dielectric constant of 7.0 (see figure 5). In general, silicon nitride films obtained with high nitrogen content present low mechanical stress. Thus, for measuring the mechanical stress of the silicon nitride films we have selected films with high nitrogen content and the nitrogen content was not varied (fixed at 40%) and the deposition time and the applied RF power were varied (see figure 5).

Observing the figures, we can notice that the deposition rate decreases when the deposition time is increased. This effect occurs due the increase of the density of the silicon nitride films and the better accommodation of the atoms in the films. So, the mechanical stress decreases.

In such a way, these films can be applied in MEMS fabrication and as mask to KOH etching, how we can see in figure 6 that shows a silicon structure with silicon nitride rich in silicon as mask. The silicon nitride films obtained in this work showed low mechanical stress, high nitrogen content, good dielectric constant and high deposition rate at room temperature.

**Figure 6.** Silicon structure with silicon nitride rich in silicon as mask.

## 4. Conclusions

In this work, we have obtained silicon nitride thin films at room temperature at relatively high deposition rates with sputtering reactive plasma processes. The films showed good stoichiometry ($Si_3N_4$) and/or rich in silicon with low mechanical stress and good dielectric properties. Consequently, the silicon nitride films can be used for MEMS fabrication, electrical insulators, and gate dielectric in thin films transistors (TFTs), diffusion mark and KOK mask, etc. Thus, the silicon nitride films obtained in this work can be used in different technological areas.

**Acknowledgments**

The authors are appreciative to Mr. Nelson Ordonez, Mr. Alexandre M. Camponucci for technical input and support, and FAPESP, CAPES and CNPq for financial support.

## References

[1] Alvarez F and Valladares A 2002 *Revista Mexicana de Física* **528** 48
[2] Vargheese K and Rao G 2000 *Indian J. of Technology* **560012** 8
[3] Nowling G, Babayan S, Jankovic V and Hicks R 2002 *Plasma Source Science and Technology* **97** 11
[4] Huang H, Winchester K, Liu Y, Hu X, Musca C, Dell J and Faraone L 2005 *J. Microlectronics and Microengineering* **608** 15
[5] Bagaturyants A 2001 *J. Surface Science* **213** 486
[6] Wiederhorn S, Hockey B and French J 1999 *J. European Ceramic Society* **2273** 19
[7] Barbottin G and Vapaille A 1999 *Instabilities in Silicon Devices* (Amsterdam: Springer)
[8] Howson R 1994 *Pure & Appl, Chem.* **1311** 66
[9] Safi I 2000 *Surface and Coatings Technology* **203** 127
[10] Musil J 2005 *Thin Solid Films* **208** 475
[11] Judy J 2001 *Smart Material and Structures* **1115** 10
[12] Dai C 2005 *Tamkang J. Science and Engineering* **197** 8
[13] Wolf I 1996 *Semicond. Sci. Technology* **139** 11
[14] Zecchino M 2004 *Thin Film Stress Measurement Using Dektak Profiles* (Veeco Inc)
[15] Zecchino M 2004 *Thin Films Stress Measurement* (Tencor Inc)
[16] Dulap M and Adaskaveg J 1997 *Introduction to the SEM* (Instrumentation UC Davis)
[17] Mayer M 2003 *Rutherford Backscattering Spectrometry* (Trieste)
[18] Srisvastara V 2009 *International J. Recent Trends in Engineering* **4** 1

Journal of Physics: Conference Series **370** (2012) 012016 doi:10.1088/1742-6596/370/1/012016

# Study of the efficiency of magnetic island macroparticle filters for different vacuum arc configurations

**A Arias Duran[1,3], M Fazio[2], A Kleiman[2,3], L Giuliani[2,4], A Marquez[2,4] and D Grondona[2,4]**

[1]LP&MC, Dept. de Física, FCEyN-UBA, and IFIBA-CONICET, Cdad. Universitaria, Pab. I, C1428EGA Cdad. Autónoma de Buenos Aires, Argentina
[2]INFIP, UBA-CONICET, Cdad. Universitaria, Pab. I, C1428EGA Cdad. Autónoma de Buenos Aires, Argentina.
[3]Fellow of CONICET
[4]Member of CONICET

E-mail: grondona@df.uba.ar

**Abstract**. In the present work, we studied the efficiency of a magnetic island filter for two vacuum arc designs, pulsed and continuous. The magnetic island filter consisted in a straight duct with an external solenoid and a magnet enclosed in a housing (magnetic island) located inside the duct on its axis, and both magnetic fields are in opposite direction. The housing of the magnetic island obstructs the line of sight between the cathode and the substrate. In this arrangement, the charged plasma components move along the curved magnetic field lines, around the magnetic island, but the macroparticles move in straight paths and deposit on the wall of the magnetic island housing. The performance of the filter was characterized for different external and internal field strengths. The plasma transport efficiency was analyzed by measuring the ion saturation current with Langmuir probes and the deposited mass rate. The ion transmission efficiency around the system axis achieved values around 25%. Observation of the coating surface morphology with optical microscopy determined that the macroparticles were effectively removed.

## 1. Introduction

Coatings produced by vacuum arc deposition have macroparticles incorporated; therefore they have not the quality required for some applications. Magnetic filtering techniques are used to remove the macroparticles from the plasma jet emitted in cathodic arc discharges. In this kind of filters a magnetic field with intensity higher than that required for electron magnetization is applied. The magnetized electrons are forced to spiral along the magnetic field lines. The ions are not magnetized but follow the electrons in order to maintain a quasineutral state of the plasma jet. However, the macroparticles trajectories are not considerably affected by the magnetic field, so they travel in straight trajectories and impact the filter walls.

A great variety of magnetic filters with different geometries has been implemented [1,2]. The simplest magnetic filter is the straight filter. This filter consists in a duct with a coil wrapped which generates an axial magnetic field. In this kind of filter, although the ratio of the plasma to macroparticles in the coating increases due to the ion confinement, the filter is not very efficient in macroparticle reduction because the substrate is in the line of sight of the cathode.

The ''magnetic island filter'' is a straight filter in which a central shield is inserted. The shield is a magnetic structure that helps guiding the plasma around it. This filter does not have a line of sight between cathode and substrate [1,3]. The island is an electromagnetic coil or a permanent magnet located inside the duct, on its axis. The magnetic fields generated by the straight filter and the island are in opposite direction on its axis. In this arrangement, the charged plasma components move along the curved magnetic field lines, around the magnetic island to the treated surface, but the macroparticles move in straight paths and deposits on the case of the magnetic island (figure 1). This kind of filter is well suited for small substrates and shows a great efficiency removing macroparticles [4,5].

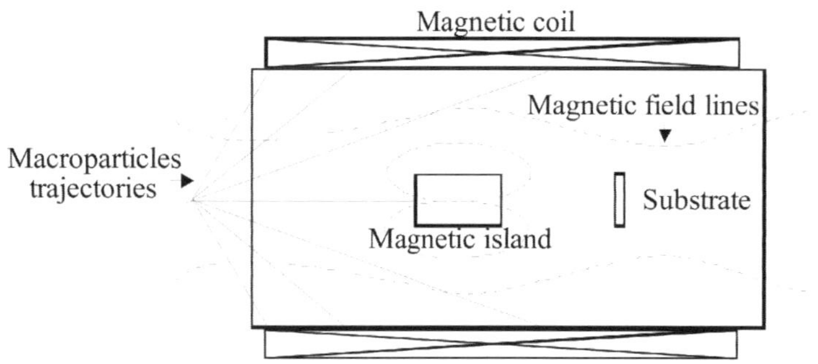


**Figure 1.** Schematic diagram of a magnetic island filter.

In this work, the performance of magnetic island filters for two vacuum arc designs, pulsed and continuous, were studied. The plasma transport efficiency and macroparticle reduction in Titanium coatings were compare with those obtained with a straight filter for different magnetic field configurations.

## 2. Experimental set-up

The experimental study of the magnetic filter was carried out in a dc vacuum arc and in a pulsed vacuum arc discharge. In figure 2 schematic drawings of both vacuum arcs are shown.

In the pulsed device (figure 2(a)), the arc was produced by discharging an electrolytic capacitor bank with C = 0.075 F, connected to a series inductor-resistor (L = 2 mH, R = 0.33 Ω), which critically damped the discharge. The arc duration is about 35 ms, with a peak current of (450 ± 20) A and an interelectrode voltage of (45 ± 5) V. The charging voltage was 280 V and the arc was ignited with a mechanically controlled tungsten rod. The chamber pressure was maintained at a base pressure $< 10^{-2}$ Pa during the whole arc discharge with an oil diffusion pump. A grounded Ti cathode (5 cm in length and 1 cm in diameter) was located 1 cm in front of an annular anode with an aperture of 5 cm and a thickness of 2 cm. The entrance of the straight magnetic filter was placed at the end of the anode, separated by a 1 cm insulating ring, at 4 cm from the cathode frontal surface. The magnetic field is generated by an external coil wrapped around a stainless steel tube (22 cm long, 5 cm inner diameter). The coil (3 layers of 30 turns each) was fed with dc current from an independent power source. The magnetic field strength was measured with a calibrated Hall probe with a precision of 10%, and was characterized with the value measured at the duct center ($B_{ext}$). The range of $B_{ext}$ values investigated was 3 – 43 mT. The magnetic island filter consisted of cylindrical permanent magnets enclosed in an aluminum jacket (1 cm diameter and 2 cm long). The amount of permanent magnets inside the jacket was modified. Two different magnetic island field strengths ($B_{isl}$) were employed, one of 10 mT and the other of 60 mT. These values were measured at the duct axis on the jacket frontal surface. The magnetic island was placed at the duct axis at 11 cm from the cathode frontal surface.

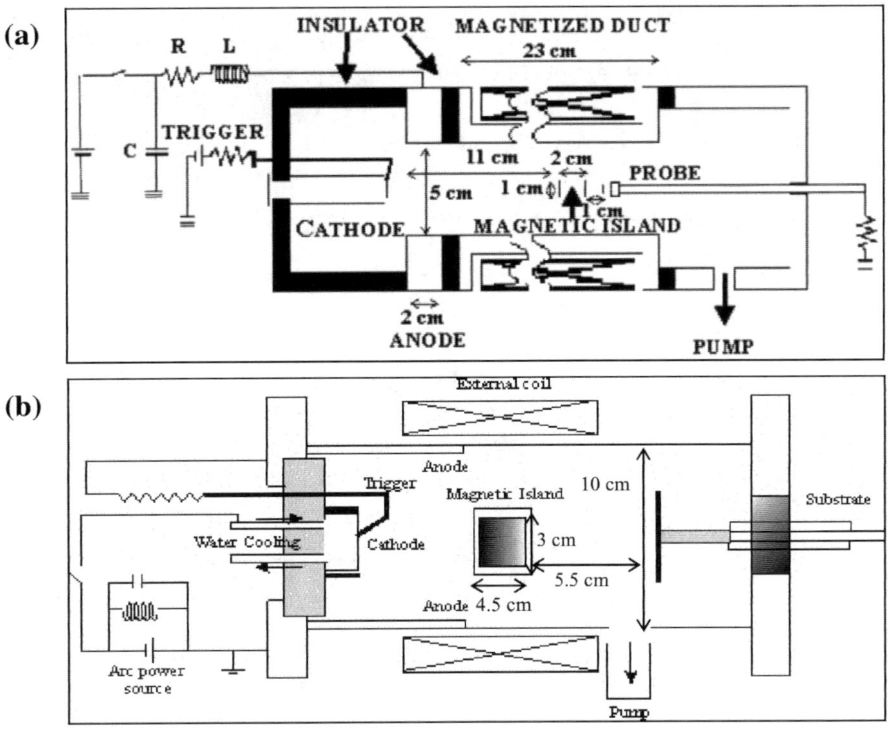


**Figure 2.** Schematic diagrams of the magnetic island in the vacuum arcs: (a) pulsed discharge and (b) dc discharge.

In the dc cathodic arc (figure 2(b)) the discharge circuit consisted in a current supply (18 kW, 150 A) in parallel with a capacitor bank (165 mF) connected to the electrodes through a series inductor (2.8 mH) in order to provide arc stability. The arc was run with a 125 A current. A Ti cathode of 5 cm in diameter and an annular copper anode of 8 cm inner diameter were employed. Both electrodes were refrigerated during the arc discharge by a water cooling system by means of a centrifuge pump with a nominal discharge of 50 l/min. The substrate chamber was electrically insulated and connected with the main discharge chamber through a straight duct (25 cm long, 10 cm inner diameter), which operates as the straight magnetic filter. One vacuum system, composed of a mechanical and two diffusion pumps, pumps the chamber to a base pressure of less than $10^{-2}$ Pa. The field of the straight magnetic filter was generated by an external coil of 20 cm in length with 30 turns of 18 cm in diameter). The range of $B_{ext}$ values investigated was 0 – 8 mT. The jacket of the magnetic island filter was a cylinder of 3cm diameter and 4.5 cm long with cylindrical permanent magnets inside. Two different magnetic island field strengths were employed, one of 11 mT and the other of 36 mT. The magnetic island was placed at the duct axis at 20 cm from the cathode frontal surface.

Plots of the field lines on a longitudinal plane (from the axis to the external coil radius) for two field configurations applied to the dc and pulsed arcs are shown in figure 3. Figure 3(a) and (b) correspond to the pulsed vacuum arc system with $B_{ext}$ = 43 mT and $B_{isl}$ = 60 mT and 10 mT, respectively. Figure 3(c) and (d) correspond to the dc vacuum arc with the same value of $B_{ext}$ = 7mT and with $B_{isl}$ = 36 mT and $B_{isl}$ = 11 mT, respectively.. Values of the field component parallel to the axis registered with the Hall probe are also indicated in figure 3.

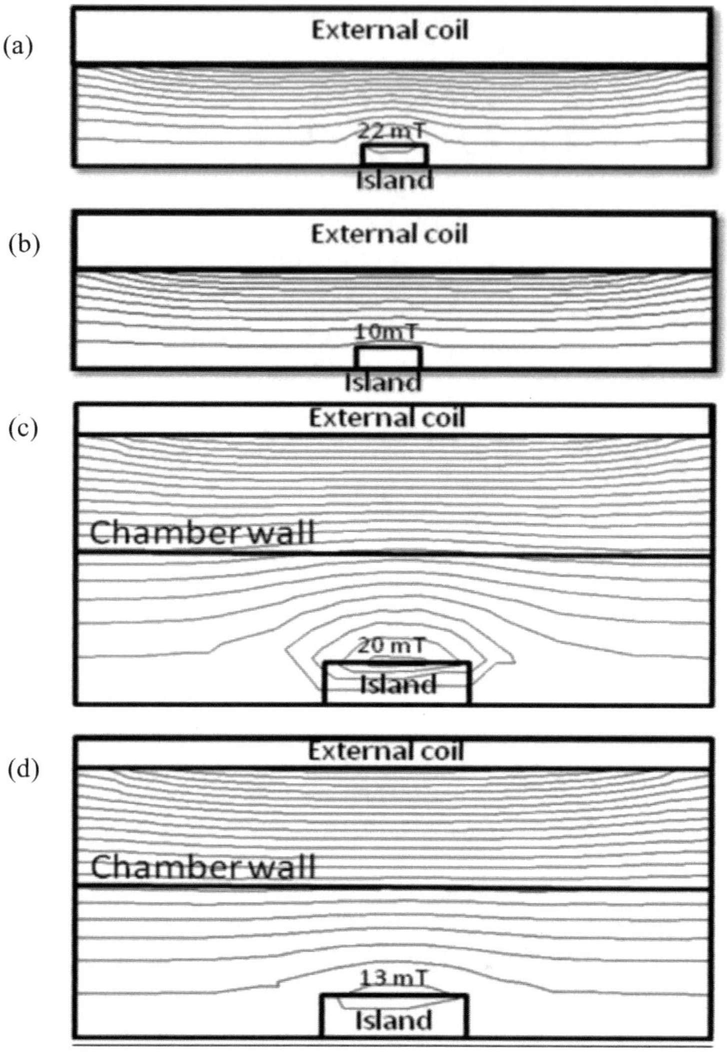


**Figure 3.** Magnetic field configuration for the island filter: (a) pulsed arc, $B_{ext}$ = 43 mT and with $B_{isl}$ = 60 mT; (b) pulsed arc, $B_{ext}$ = 43 mT and with $B_{isl}$ = 10 mT; (c) dc arc, $B_{ext}$ = 7mT and with $B_{isl}$ = 36 mT; and (d) dc arc, $B_{ext}$ = 7mT and with $B_{isl}$ = 11 mT.

In order to evaluate the efficiency of the plasma passage through the island, the ion flux was measured with a probe in the pulsed vacuum arc and the growth rate was determined in the dc arc for different field configurations. In the pulsed arc, a circular plane copper probe (0.6 cm diameter) was placed at 14 cm from the cathode frontal surface at the tube axis and at 1 cm behind the magnetic island jacket. For the measurement of the ion saturation current, taking into account the relatively high plasma potential [6], the probe was simply biased to ground through a resistance of 300Ω to ensure that the probe voltage was well below the floating potential during the discharge. The electrical signals were registered using a four-channel digitizing oscilloscope (100MSs$^{-1}$ sampling rate, 60MHz analogical bandwidth). In the dc arc, steel circular samples of 3 cm diameter placed 5.5 cm behind the magnetic island have been exposed to the discharge for 180 s. The growth rate was assessed by weighting the samples before and after the deposition process using an Ohaus scale, model AS200 and assuming that the film had bulk Ti density ($\rho_{Ti}$ = 4.5 g/cm$^3$).

The surface morphology of coatings deposited on silicon substrates was observed using an optical microscope (Olympus BX60M). In the pulsed arc the silicon substrates were exposed to 5 discharges, while in the dc arc they were exposed in the same conditions as the steel samples.

## 3. Results

In figure 4 the ion saturation current ($I_{ion}$) measured in the pulsed arc (figure 4(a)) and the growth rate ($r$) measured in the dc arc (figure 4(b)) as function of $B_{ext}$ without and with the presence of the magnetic island inside the chambers are presented.

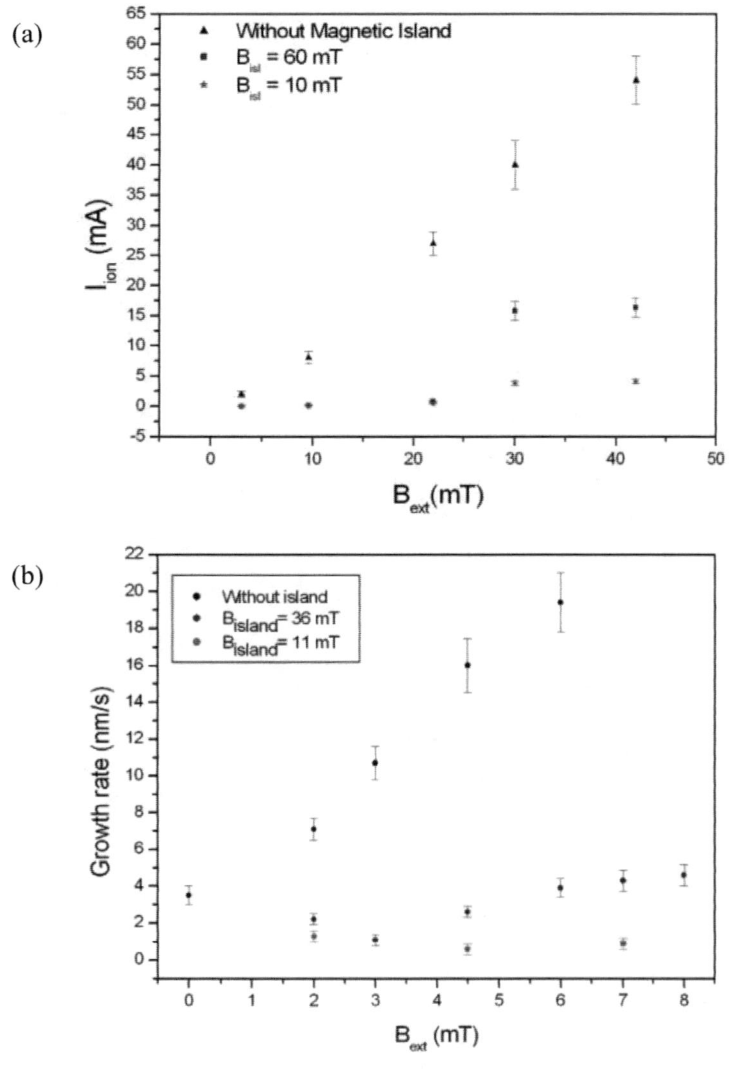


**Figure 4.** Ion flux as function of the magnetic field strength: (a) ion current measured in the pulsed arc and (b) growth rate ($r$) measured in the dc arc.

In figure 4(a), it can be observed that without the island $I_{ion}$ increased as $B_{ext}$ increased, varying from 2 mA for the lowest $B_{ext}$ to 55 mA for $B_{ext}$ = 43 mT. With the island, when $B_{ext}$ was lower than 25 mT the registered ion flux was not significant independently on the $B_{isl}$ value, while for higher $B_{ext}$

$I_{ion}$ augmented, $I_{ion}$ for $B_{isl}$ = 60 mT being a factor 4 higher than $I_{ion}$ for $B_{isl}$ = 10 mT. The measurements of $r$ as function of the magnetic filed in the dc arc (figure 4(b)) showed a similar behavior to $I_{ion}$. Without the island $r$ augmented from 3.5 nm/s up to 19 nm/s corresponding to $B_{ext}$ = 6 mT. For higher values of $B_{ext}$ $r$ could not be determined because the ion flux was so concentrated on the axis that a burned spot appeared on the surface after the deposition process. With the island, for the case $B_{isl}$ = 11 mT $r$ varied slightly with $B_{ext}$ taking a value of approximately 1 nm/s, while for the case $B_{isl}$ = 36 mT $r$ increased with $B_{ext}$ achieving a value of 4.5 nm/s.

In figure 5 optical micrographs (with 500x magnification) of Ti coatings show the surface morphology typically observed without and with the magnetic island filter in both vacuum arcs. The pictures presented in figure 5 correspond to samples deposited in the following conditions: (a) with the pulsed arc without the island, with $B_{ext}$ = 43 mT; (b) with the dc arc without the island, with $B_{ext}$ = 6 mT; (c) with the pulsed arc with the island with $B_{ext}$ = 43 mT and $B_{isl}$ = 60 mT; and (d) with the dc arc with the island with $B_{ext}$ = 7 mT and $B_{isl}$ = 36 mT. An estimation of the coating thickness of the pulse vacuum arc was obtained from the charge deposited, which was calculated from the probe ion current integration. The coating thicknesses obtained with $B_{ext}$ = 43 mT and without the island was ~ 4 nm (figure 5(a)), and with $B_{isl}$ = 60 mT a thicknesses of ~ 1 nm (figure 5(c)). The coating thickness with the dc arc was obtained from the deposited mass. The values obtained with $B_{ext}$ = 6 mT was ~ 3.2 μm (figure 5(b)); and with $B_{ext}$ = 6 mT and $B_{isl}$=36 mT was ~ 0.7 μm (figure 5(d)).

Without the island many macroparticles of different sizes could be distinguished onto the substrate for all $B_{ext}$ values, in the case of the dc arc the number of macroparticles was higher and macroparticles with larger diameters were observed. Using the magnetic island filter it can be seen that practically no macroparticles were visible for both arcs.

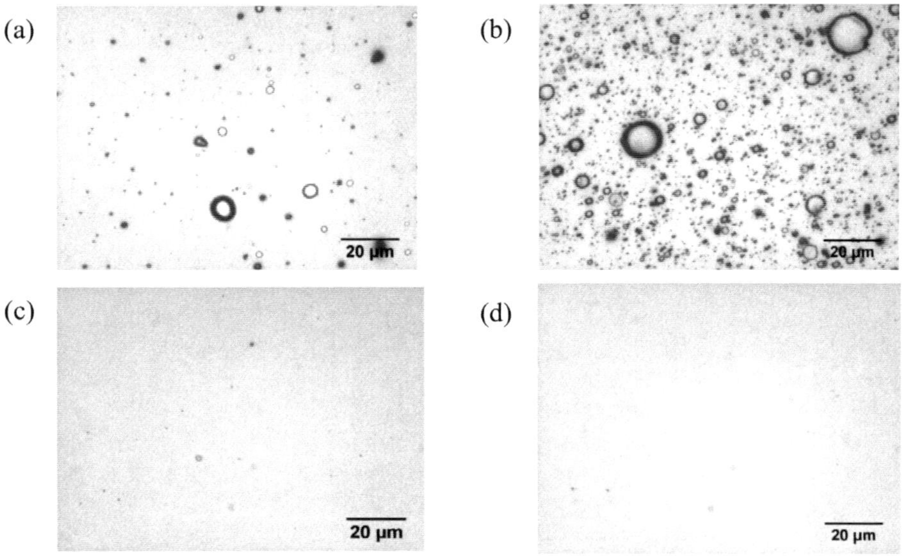


**Figure 5.** Optical micrographs of Ti coating surfaces: (a) pulsed arc without the island, with $B_{ext}$ = 43 mT; (b) dc arc without the island, with $B_{ext}$ = 6 mT; (c) pulsed arc with the island with $B_{ext}$ = 43 mT and $B_{isl}$ = 60 mT; (d) dc arc with the island with $B_{ext}$ = 7 mT and $B_{isl}$ = 36 mT.

## 4. Conclusions

In the present work, we studied the efficiency of the magnetic island filter for two vacuum arc systems, one pulsed and the other continuous. The geometrical dimensions of the filter for both systems were chosen according to the cathode and the straight filter diameters with similar proportions. The performance of the filter was characterized for different external and internal field strengths. The ion transmission efficiency was characterized by comparing the ion flux for the maximum $B_{ext}$ with and without the island on the axis. The maximum axial ion flux that passed around the magnetic island was found for the maximum $B_{isl}$ and achieved values around 25% for both systems. The fact that the higher ion transmission was found for the field configurations with higher strengths could be associated to decrease of the magnetic diffusion diminishing the ion losses to the chamber wall. The amount of macroparticles present on the film obtained with the magnetic island filter was compare with those present on the film obtained with the simple straight filter for different magnetic field configurations. Observation of the coating surface morphology with optical microscopy determined that the macroparticles were effectively removed.

**Acknowledgments**
This work was supported by grants from Universidad de Buenos Aires and CONICET.

**References**

[1] Karpov D A 1997 *Surf. Coat. Technol.* **96** 22-33
[2] Anders A 2008 *Cathodic Arcs* (New York: Springer)
[3] Aksenov I I, Belous V A, Padalka V G and Khoroshikh V M 1982 Swedish Patent N° 82188-8
[4] Kleiman A, Márquez A and Boxman R 2008 *Plasma Sources Science and Technology* **17** 015008
[5] Bolt H, Koch F, Rodet J L, Karpov D and Menzel S 1999 *Surf. Coat. Technol.* **116–119** 956-62
[6] Giuliani L, Grondona D, Kelly H and Minotti F O 2007 *J.Phys. D: Appl. Phys.* **40** 401-8

Journal of Physics: Conference Series **370** (2012) 012017 doi:10.1088/1742-6596/370/1/012017

# Thin $TiO_2$ films deposited by implantation and sputtering in RF inductively coupled plasmas

**R Valencia-Alvarado[1], A de la Piedad-Beneitez[2], R López-Callejas[1,2], S R Barocio[1], A Mercado-Cabrera[1], R Peña-Eguiluz[1], A E Muñoz-Castro[1], B G Rodríguez-Méndez[1] and J M de la Rosa-Vázquez[3]**

[1]Instituto Nacional de Investigaciones Nucleares, Plasma Physics Laboratory, AP 18-1027, 11801 México DF, Mexico
[2]Instituto Tecnológico de Toluca, AP 890, Toluca, Estado de México, México
[3] ESIME-Zacatenco-IPN, AP 07738, México DF, Mexico

Email: raul.valencia@inin.gob.mx

**Abstract:** The achievement of titanium dioxide ($TiO_2$) thin films in the rutile crystalline phase is reported. The samples result from the implantation of oxygen ions of Ti in argon/oxygen plasma generated by inductively coupled RF at a commercial 13.56 MHz frequency. Simultaneously, a sputtering process is conducted on the titanium target in order to produce $TiO_2$ thin films in the anatase phase over silicon and glass substrates. Both implantation and sputtering processes shared the same 500 W plasma with the target, polarized between 0 and -3 kV. The substrates were placed between 2 and 3 cm from the target, this distance being found to be determinant of the $TiO_2$ deposition rate. The rutile phase in the target was obtained at temperatures in the order of 680 °C and the anatase (unbiased) one at about 300 °C without any auxiliary heating. The crystalline phases were characterized by x ray diffraction and Raman spectroscopy. The morphology and average roughness were established by means of scanning electronic and atomic force microscopy, whereas the reaction products generated during the oxidation process were analyzed by mass spectrometry. Finally, the stoichiometric composition was measured by means of X-ray photoelectron spectroscopy.

## 1. Introduction

Today, thin titanium dioxide ($TiO_2$) films attract considerable interest of multiple research groups due to a vast number of useful properties, particularly in photocatalytic applications, water purification, air treatment and the development of biocompatible coatings of human implants [1-4]. In the biomedical sector, $TiO_2$ films are considered an especially biocompatible material, some of whose effects on the human body have recently been compiled by Ellingsen and Lyngstadaas [5], and a decisive factor in the development of bioengineering materials, as described by Ratner and Oshida [6-7]. The ecological value of these films arises from their photo-catalytic qualities, as described by, e.g., Sakai and Tavares [8-9].

In their investigations, Anpo [10] has found that the photo-catalytic activity of $TiO_2$ increases when the particle diameter is reduced under 10 nm. Liu [11] has pointed out that an increase in the few nm thickness of a titanium oxidized coat, naturally found over titanium samples, clearly improves the $TiO_2$ bioactivity. The photo-catalytic properties of $TiO_2$ in the anatase and rutile phases were initially reported by Fujishima and Honda in 1972 [12]. Among the plasma processing techniques connected with film deposition, one finds chemical vapor deposition [13], electron-beam evaporation [14], ion-beam assisted deposition [15], RF reactive magnetron sputtering [16], plasma source ion implantation [17], and inductively coupled plasma-assisted sputtering [18-20].

This work describes the simultaneous growth of anatase and rutile by means of an implantation/diffusion and sputtering process in inductively coupled radio frequency plasma of calibrated argon/oxygen mixtures without post annealing.

## 2. Setup

The implantation/diffusion and sputtering process has been conducted within a Pyrex-like glass cylindrical vessel, 50 cm long and 20 cm in diameter, around which a 3 mm diameter copper pipe antenna is wound helicoidally. The plasma is generated from calibrated argon/oxygen (80%/20%) gas mixtures by transferring 500 W to the antennae through an automatic coupling network from a 1.2 kW RF generator where induction works as in a transformer with the plasma playing the role of a secondary coil. Figure 1 exhibits this experimental set up, with its substrate holder, reported elsewhere [18-19].

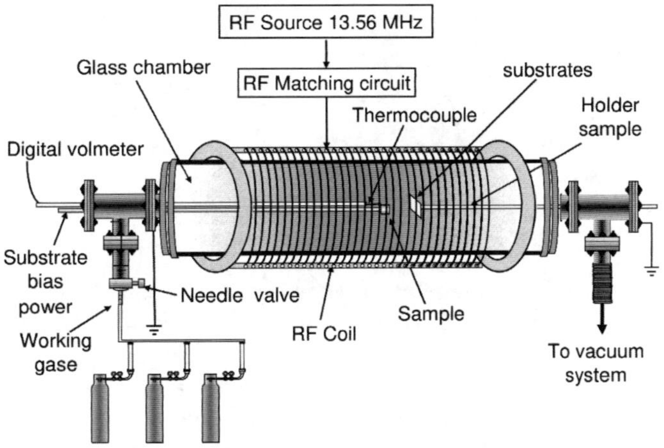


**Figure 1.** Diagram of the plasma reactor.

The vessel was evacuated at pressures $<5\times10^{-6}$ mbar by a turbomolecular pump operating at 500 l/s. After attaining the base pressure, the work argon/oxygen mixture was admitted through a precision leak valve up to $5\times10^{-2}$ mbar pressure at which the highest plasma density is obtained. Commercially pure titanium (CpTi) cylindrical samples, 9 mm in diameter and 5 mm long, were introduced to the vessel and oxidized by the plasma during periods up to 5 hours by means of a sputtering and implantation/diffusion process. The substrate temperature was kept about 250 °C by the action of the plasma alone and when the Ti sample is biased, this temperature was raised to 300 °C by the sputtering action. When the samples were biased at -3000 $V_{DC}$, the ion bombardment increased the sample temperature to around 680 °C, as measured by a commercial K type thermocouple. Insulation between the biased piece and the thermocouple-reading meter was provided by an isolation amplifier. Electrically floated silicon and glass substrates were placed on the holder 2 and 3 cm away from the CpTi sample, becoming thermalized between 250 °C (by the plasma alone) and 300 °C (during the oxidation process).

Both the titanium targets and the glass/silicon substrates were finally taken out and cut in cross sections in order to measure their $TiO_2$ film depth. XRD and Raman spectroscopy techniques, used to characterize the oxidized samples, indicate the growth of the anatase and rutile phases. XRD spectra were used to estimate the mean crystallite size from the Scherrer equation. The surface morphologies and titanium/oxygen atomic percentage on sample and silicon sputtering films were characterized by scanning electron microscopy (SEM) and atomic force microscopy (AFM). The stoichiometric composition was measured thanks to X-ray photoelectron spectroscopy (XPS) and the reaction products during the oxidation process by means of a residual gas analyzer (RGA)

## 3. Results and discussion

Figure 2 presents SEM images of the cross section of titanium and the growth dioxide ($TiO_2$) film resulting from 5 h processing. Figure 2(a) corresponds to silicon substrate coating with $TiO_2$ by sputtering. The layer is in the order of 2 μm and the rate lies between 5 and 7 nm/min. Figure 2(b) shows the growth of the crystalline anatase phase on the silicon substrate.

(a) (b)

**Figure 2.** Layer depth: (a) implantation/diffusion, (b) growth of the anatase phase.

Looking for an estimate of its chemical composition, the coating was analyzed using 15 kV electron dispersive spectroscopy (EDS) in conjunction with the SEM. In a typical silicon substrate, the atomic percentages found were 75 at % of oxygen, 23 at % of titanium and 2 at % silicon; similar percentages were identified in the sample (73 at % oxygen and 27 at % titanium), indicating the chemical composition of the film albeit not the stoichiometric ratio of two parts of oxygen per one of titanium. The sample surfaces and substrates changed in color during the oxidation process according to the temperature, gas pressure and bias voltage of the treatment. The colors appear to be directly related to the composition of the oxide formed as well as to the thickness of the oxide layer. Initially, the titanium sample surface acquires a dark blue shade which turns silvery and, by the end of the process, goes into lead grayish. This evolution suggests oxidation in the rutile phase. The case of the glass and silicon substrates involves the appearance of a two color circular pattern ranging from purple red to olive green, pointing to the existence of the anatase phase. The multiplication of these concentric strips can be related to the increase of the deposited film depth as the processing time goes by.

The new crystalline phases of the titanium sample surface modified by implantation and diffusion have been analyzed by X-ray diffraction using $CuK_{\alpha}$ radiation within a $2\theta$=20-90° scan range. Figure 3(a) displays the X-ray diffraction patterns of the $5\times10^{-2}$ mbar sample, this spectrum reveals that the rutile phase is present at $2\theta$=27.8°, 36.5°, 41.6°, 44.4°, 54.8°, 56.9°, 62.9°, 69.3° and 70.0°. Given the Scherrer equation, the grain size of the $TiO_2$ film in the rutile phase implanted with the preferential crystalline orientation (110) was 58 nm. Figure 3(b) shows the Raman spectrum from the same titanium implanted and diffused displaying the rutile peaks.

The thin films grown on the glass/silicon substrates have been characterized by Raman spectroscopy. The spectrum from glass (figure 4(a)) reveals the presence of the anatase phase, interestingly enough, grown at temperatures below 350 °C. The Raman amplitudes at wavelengths 149 and 399 $cm^{-1}$ enabled to calculate the relative amount of anatase in the $TiO_2$ films. The resulting ~5 nm/min deposition rates are considered acceptable in this kind of oxidation process. Figure 4(b) describes a similar growth on silicon.

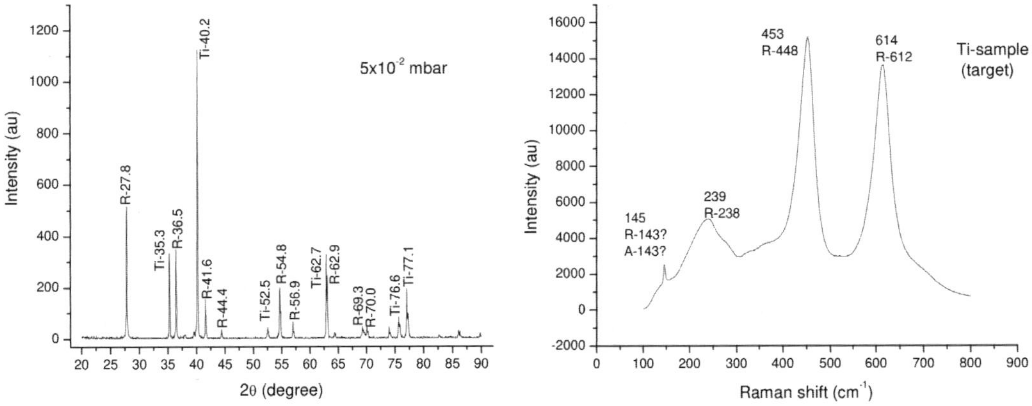


(a) (b)

**Figure 3.** (a) X-ray diffraction patterns at $5\times10^{-2}$ mbar and (b) the Raman spectrum from oxidized sample.

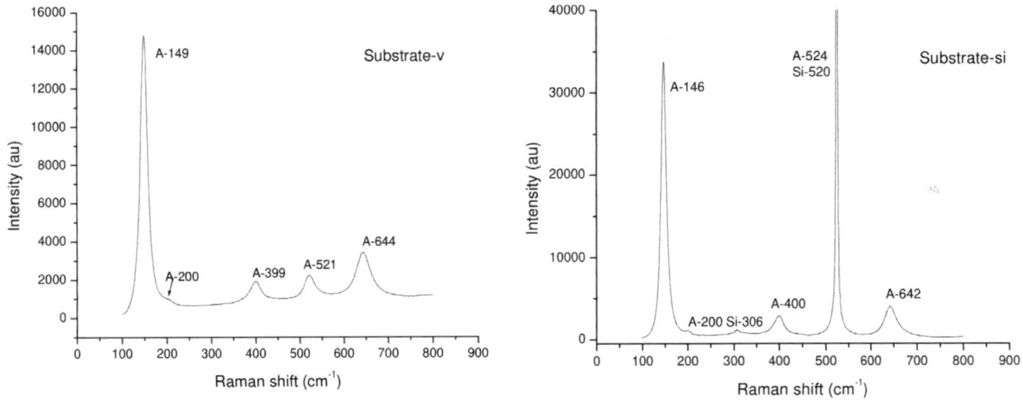


(a) (b)

**Figure 4.** Raman spectra from the films deposited on glass (a) and silicon (b).

Figure 5 displays the XRD spectrum from the film deposited on a silicon substrate, showing the anatase phase peaks, which validates the Raman spectrum in figure 4.

Figure 6 displays the AFM image of $TiO_2$ film deposited on the silicon substrate corresponding to the anatase phase. This film presents root mean squared (RMS) values of 1.6 nm indicating the existence of a smooth surface.

Figure 7(a) show typical resolution XPS spectra of $Ti_{2p}$ and $O_{1s}$, respectively, from the anatase $TiO_2$ film. The $Ti_{2p}$ region is split in $2p_{3/2}$ and $2p_{1/2}$ doublet peaks located at 458 and 464 eV binding energy values. This doublet corresponds to Ti(IV). The figure 7(b) shows the spectra from the $O_{1s}$ region taken from the anatase $TiO_2$ film surface. The XPS spectrum binding energy of the main $O_{1s}$ peak appeared at 530 eV. This peak can be associated to the typical stoichiometric oxide of $TiO_2$.

The analysis of the reaction products generated by the oxidation process has been carried out by means of a mass spectrometer coupled to a CUADREX 200 Leybold quadrupole gas analyzer sensor and a turbomolecular vacuum system, resulting in a mass resolution of 1 atomic mass unit. The array was connected through an adaptor to the discharge reactor outlet. The atomic masses of the reaction products generated by the plasma implantation and sputtering were detected as: $Ti^+$ with mass 48, $TiO_2^+$ with mass 80, Ar with mass 40, as well as masses 16 and 32 corresponding to atomic ($O^+$) and molecular ($O_2^+$) oxygen.

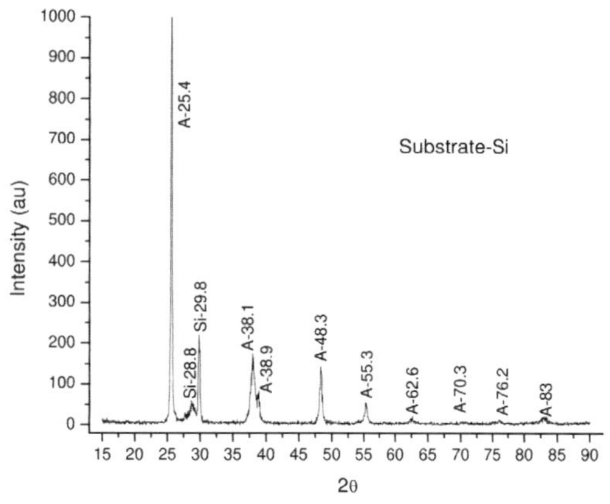


**Figure 5.** XRD-Spectrum from the film grown on silicon revealing the anatase phase film deposited on the silicon substrate.

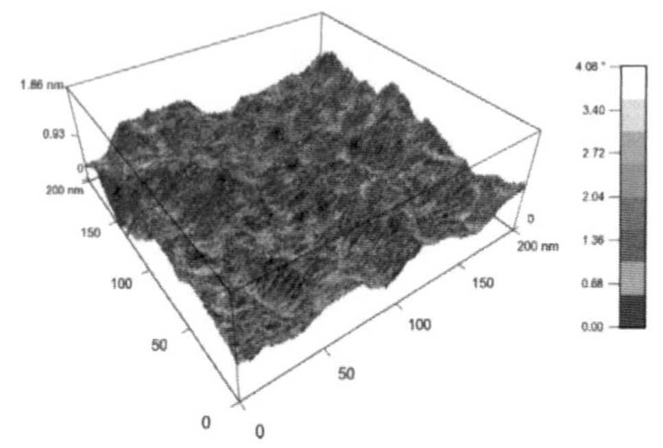

**Figure 6.** AFM-image of the surface of $TiO_2$.

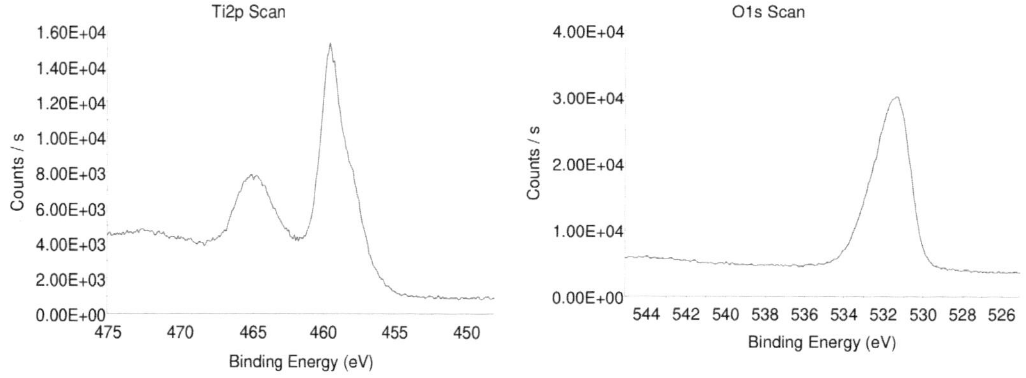


(a) (b)

**Figure 7.** (a) $Ti_{2p}$ from the anatase phase deposited on silicon substrate, (b) $O_{1s}$ from the anatase phase deposited on silicon substrate.

## 4. Conclusions

An inductively coupled plasma source supplied by a specifically designed and built RF generator has allowed obtaining $TiO_2$ films in the rutile phase by means of the implantation/diffusion of oxygen atoms into a titanium target while, at the same time, films in the anatase phase were developed by titanium sputtering with argon ions. Atomic concentration rates above 2 are due to the presence of TiO-like oxides in addition to the $TiO_2$ An oxygen atomic percentage of 55 at %, versus a 35 at % of titanium leads to a stoichiometry well below to 2. By contrast, percentages on the glass are 70 at % vs 23 at %, respectively, which can be attributed to the silicon oxide composition of the glass itself. Color changes of the samples during their implantation/diffusion processing suggest the evolution of their composition and structure. Thus, a transition from deep blue to silvery white is typically achieved by the end of the process, frequently accompanied by iridescent concentric circular patterns whose number could be related with the deposited film depth. The deposited anatase phase did not require any additional annealing after the oxidation. The RGA technique can be considered as an efficient methodology in the optimization of crystalline growth of $TiO_2$ films on both titanium samples and silicon or glass substrates.

**Acknowledgements**
This project has been partially funded by CONACyT, México. The authors are grateful to Dr. Manuel Espinoza and Mr. Jorge Pérez, for their technical support in the XRD, EDS, AFM diagnostics; and XPS diagnostic to Cuauhtémoc Gutiérrez Rivera and Ma Eugenia Camacho López. The technical assistance received from Pedro Ángeles Espinoza, María Teresa Torres M. and Isaías Contreras Villa is much appreciated.

**References**

[1] Baroch P, Musil J, Vlcek J, Nam K H and Han J G 2005 *Surf. & Coat. Tech.* **193** 107-11
[2] Li Z G, Wu Y X and Miyake S 2009 *Surf.& Coat. Tech.* **203** 3661-8
[3] Chu P K 2007 *IEEE Trans. on Plasma Sci.* **35-2** 181-7
[4] Lee S H, Kim H W, Lee E J, Hi L H and Kim H E 2006 *J. of Biomaterials Applications* **20** 195-208
[5] Ellingsen J E and Lyngstadaas S P (editors) 2003 *Bio-Implant Interface* (Boca Raton: CRC Press LLC)
[6] Ratner B D, Hoffman A S, Schoen F J and Lemons J E (editors) 2004 *Biomaterials Science, An Introduction to Materials in Medicine* 2nd Edition (Elsevier Academic Press)
[7] Oshida Y 2007 *Bioscience and Bioengineering of Titanium Materials* (Oxford: Elsevier)
[8] Sakai T, Kuniyoshi Y, Aoki W, Ezone S, Endo T and Hoshi Y 2008 *Jpn. J. Appl. Phys.* **47** 6548
[9] Tavares C J, Vieira J, Rebouta L, Hungerford G, Coutinho P, Teixeira V, Carneiro J O and Fernandes A J 2007 *Mater. Sci. Eng. B* **138** 139
[10] Anpo M, Shima T, Kodama S and Kubokaa T 1987 *J. Phys. Chem.* **91** 4305
[11] Liu X, Chu P K, and Ding C 2004 *Mater. Sci. Eng.* **R47** 49
[12] Fujishima A and Honda K 1972 *Nature* **238** 37
[13] Sun H, Wang C, Pang S, Li X, Tao Y, Tang H and Liu M 2008 *J. Non-Cryst. Solids* **354** 1440
[14] Habibi M H, Talebian N and Choi J H 2007 *Dyes and Pigments* **73** 103
[15] Yang C, Fan H, Xi Y, Chen J and Li Z 2008 *Appl. Surf. Sci.* **254** 2685
[16] da Silva L L G, Ueda M, da Silva M M and Codaro E N 2006 *IEEE Trans. on Plasma Sci.* **34-4/1** 1141
[17] Okimura K and Furumi T 2004 *Jap. J. Appl. Phys.* **43-5B** L655
[18] Valencia-Alvarado R, de la Piedad-Beneitez A, López-Callejas R, Barocio S R, Mercado-Cabrera A, Peña-Eguiluz R, Muñoz-Castro A E and de la Rosaq-Vázquez J 2009 *Vacuum* **83** S264
[19] Valencia-Alvarado R, de la Piedad-Beneitez A, de la Rosaq-Vázquez J, López-Callejas R, Barocio S R, Mercado-Cabrera A, Peña-Eguiluz R and Muñoz-Castro A E 2010 *Int. J. Nanomanufacturing* **5** 62-8

Journal of Physics: Conference Series **370** (2012) 012018 doi:10.1088/1742-6596/370/1/012018

# Time effect and aliquot concentration in *Streptococcus mutans* elimination by plasma needle

**E García-Alcantara[1], R López-Callejas[1,2], R Peña-Eguiluz[2], S Lagunas-Bernabé[3], R Valencia-Alvarado[2], A Mercado-Cabrera[2], S R Barocio[2], A E Muñoz-Castro[2], B G Rodríguez-Méndez[2] and A de la Piedad-Beneitez[1]**

[1]Instituto Tecnológico de Toluca, AP 890, Toluca, Estado de México, México
[2] Instituto Nacional de Investigaciones Nucleares, AP 18-1027, CP 11801 México DF
[3]Centro de Investigación y Estudios Avanzados en Salud Animal,
Facultad de Medicina Veterinaria y Zootecnia, Universidad Autónoma del Estado de México, Autopista Toluca-Atlacomulco km 15.5, Toluca, Estado de México, 50200, México

E-mail: regulo.lopez@inin.gob.mx

**Abstract:** Atmospheric plasma needle systems are being intensively studied with a view to potential applications in medicine. The aim of this *in vitro* study is the improved elimination of *Streptococcus Mutants* (*S. mutans*) bacteria. A 5 ml volume of Luria-Bertani culture medium has been inoculated with a test bacterial population and incubated during 24 hours, followed by ten dilutions producing aliquots at 20, 50 and 100 micro l per dilution. Each aliquot is deposited on a paper filter and then exposed to a 2 W RF room pressure helium plasma needle discharge at a 1.5 l.p.m. rate for 1, 3, 5 or 7 minutes. Each sample paper is placed in a test tube, again containing Luria-Bertani fluid, in order to develop a new bacterium colony after a 24h incubation period. The plasma needle lethality has been evaluated from absorbance studies by means of a 6305 Jeway spectrophotometer at a 600 nm wavelength, indicating a clear correlation with exposure time. These studies validate the high disinfection efficacy of the plasma needle.

## 1. Introduction

The plasma medicine concept is emerging worldwide with promising applications such as bacterial inactivation [1,2] dental cavity [3], surface sterilization [4,5], live skin treatment [6,7], blood coagulation [8] or tissue cauterization [9]. Such applications require a wide variety of plasma sources: plasma needle [10], atmospheric-pressure plasma plume [11], floating-electrode dielectric barrier discharge (FE-DBD) [12], atmospheric-pressure glow discharge torch [13], microhollow cathode discharge air plasma jet [14], microwave plasma torch [15], helium plasma jets [3], dielectric barrier discharge (DBD) [16] and nanosecond plasma gun [17], among many others. Provided that all these plasma devices are operated in room air, their use can easily overcome the limitations imposed by the currently available vacuum-based plasma facilities. More importantly, atmospheric plasma can generate active chemical species, which can be propelled toward the surface to be treated. The short life of these species is desirable as they should not remain after the treatment is completed.

Journal of Physics: Conference Series **370** (2012) 012018 doi:10.1088/1742-6596/370/1/012018

The plasma produced by a plasma needle is restricted to the tip of the jet which cannot contact directly the samples. The plasma needle consists of a thin wire (~μm in diameter) with a sharpened tip powered by radio frequency (RF 13.56 MHz). Helium (He) is typically used as a source gas to ignite and sustain stable plasma. The needle has various advantageous features: no pump is required, the discharge is localized into a region in the order of hundreds of micrometers to millimeters, and the neutral gas temperature is usually not far from the room one. Another important factor is that the process is operated in an open environment.

Dental caries is an infectious pathologic process that destroys irreversibly the dental hard core tissue. Several micro-organisms are associated with caries production, of which the *Streptococcus mutans* or *S. mutans* [18] is the most active, given its ability to adhere to the tooth surface while being the most acidic. *S. mutans* can also produce extracellular polysaccharides out of sugars and break down glycoproteins naturally found in saliva, factors of primordial importance in the development of tooth decay, as the usual *S. mutans* population in the mouth reaches $10^5$ CFU/ml [2], [10], [11] [19].
It follows the relevance of designing preventive policies directed to control this bacteria. Several researchers have worked on the inactivation of different bacteria by means of non-thermal plasma [19], [20], [21] identifying fundamental parameters such as gap distance, applied voltage, source gas flow rate, etc.

In this paper a low temperature open air helium plasma needle has been applied to the gram-positive oral *S. mutans* bacteria in order to study the effects of plasma deactivation as a function of time and aliquot concentration.

## 2. Experimental set-up

### *2.1 Plasma source*

The plasma needle possesses a coaxial configuration constituted by a copper axial powered filament electrode surrounded by a ceramic tube that provides both mechanical rigidity and electric insulation. The components are assembled on a Nylamid SL® stainless steel frame which includes the gas jet and the electrode connection to a female BNC. It produces a plasma spot of ~1 mm in diameter at power levels in the order of a few hundred mW (figure 1). Previous to the treatment, a full diagnostics of the needle plasma, based on optical emission spectrometry (OES) indicated substantial UV emission in the range 300-400 nm. Active oxygen radicals (O and OH) were also detected [22].

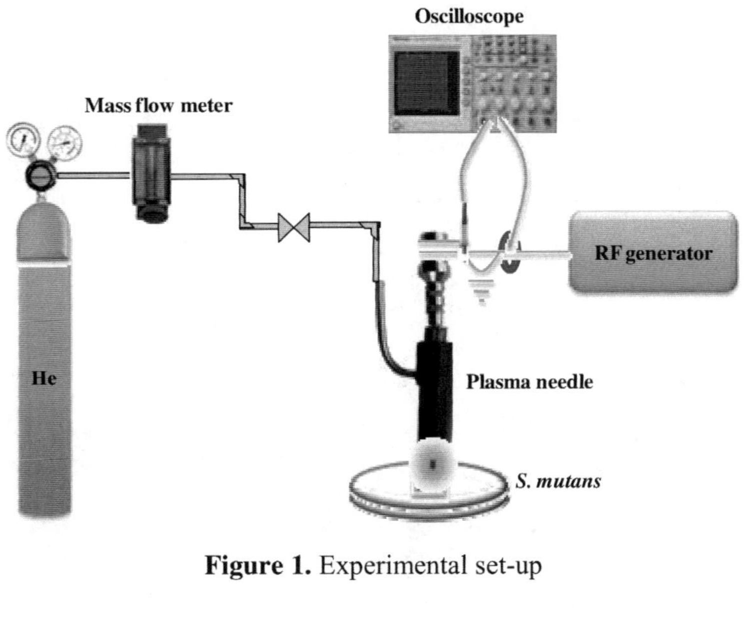


**Figure 1.** Experimental set-up

*2.2 Sample preparation*
A strain of ATCC3358 *S. Mutans* was inoculated in 5 ml of Luria-Bertani (LB) culture stock, continuously stirred and then incubated for a 24 h period. Then, 1:10 dilutions of it were made, taking 20, 50 and 100 microlitre aliquots respectively, each of them being deposited on previously sterilized filter paper and subjected to the helium plasma treatment at a 2 W RF generator power level. The experiment was carried out with a previously optimized 1.5 l.p.m. gas flow for 1, 3, 5 or 7 minute periods. Afterwards, the filter papers were placed in test tubes in order to form new bacterial colonies. To this purpose, each paper was stirred for 10 minutes at 5000 rpm and then incubated, again for a 24 h period.

*2.3 Colony count up*
The counting up process of the new colonies was conducted by means of turbidimetry on a 6305 Jaway spectrophotometer at a 600 nm wavelength which determined the bacteria/ml content of the original undiluted sample. Every sample was diluted down to $10^{10}$ bacteria/ml and then its optical absorbance was measured against a calibrated standard. This process requires producing a turbidity plot in the 540-660 nm wavelength range on the assumption that the proportion of light scattered by the sample is directly proportional to the concentration of cells in the culture.

## 3. Results and discussion
In the interest of applying the plasma needle technology to dental treatments, the end jet temperature has been established around 28°C at 3 mm of the sample approximately. These results can guarantee not to cause necrosis to the dental pulp whose temperature is in the order of 37°C [24].

Some of the general results of the treatment are showed in figure 2. The control sample absorbance value lies between 0.225 and 0.187. All the treated samples showed some degree of bacterial elimination, considering the initial (control) $10^{10}$ bacteria/ml bacterial population density.

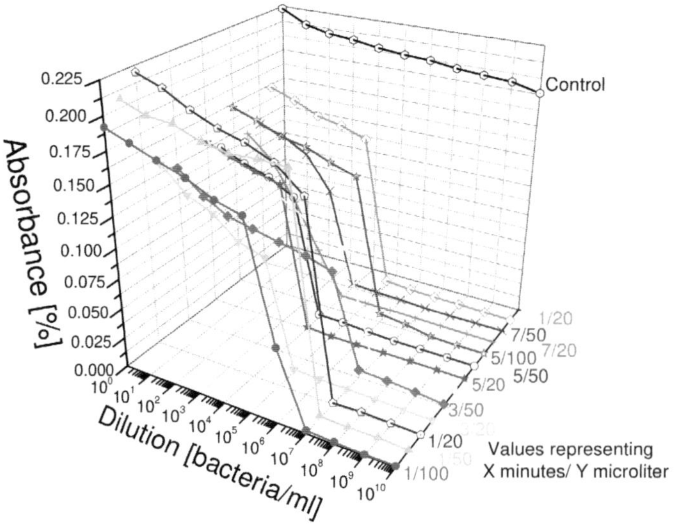


**Figure 2.** Absorbance curves corresponding to 20, 50 and 100 microlitre aliquots of the culture after a 1, 3, 5 or 7 minute exposure to helium plasma

As to the 20 microlitre volume, the population density descends two orders of magnitude, $10^7$ to $10^5$ bacteria/ml, in 1 and 3 treatment minutes, respectively. After 5 and 7 minutes the density decayed one order, to $10^4$ bacteria/ml. In the case of the 50 microlitre volume, the rhythm of elimination was maintained. However in a 100 microlitre concentration volume, the original $10^7$ bacteria/ml population remained unaltered after a 3 minute exposure. Only after 5 minutes did the density declined in one order of magnitude and, at 5 minutes, it reached $10^5$ bacteria/ml. In all, as expected, the bacterial elimination is favored by smaller volumes (20 and 50 microlitre) and longer exposure times (5 and 7 minutes) although the quantitative relationship among these variables is far from being monotonic, as seen in figure 2. Previously reported data [19, 20] have established typical elimination rates for *S. mutans* as $10^5$ bacteria/ml in 5 minutes.

## 4. Conclusions

Typical operational conditions for the elimination of *S. mutans* up to a 99.9% by means of a low temperature open air helium plasma needle have been experimentally established as a 1.5 l.p.m. gas flow ionized by a 2 W RF generator applied during 5 minutes in order to achieve a population density reduction from $10^{10}$ bacteria/ml to $10^4$ bacteria/ml. Thus, this technique can be successfully applied to higher *S. mutans* concentrations, which entails considerable expectations for the plasma needle as an effective tool in the sterilization of dental cavities.

**Acknowledgements**

This work has received financial support from CONACYT, Mexico. The technical support of María Teresa Torres Martínez, Pedro Angeles Espinoza and Isaías Contreras Villa is highly appreciated by the authors.

## References

[1] Morfill G E, Shimizu T, Steffes B and Schmidt H-U 2009 *New J. Phys.* **11** 115019 (10 pp)
[2] Zhou X, Xiong Z, Cao Y, Lu X and Liu D 2010 *IEEE Trans. Plasmas Sci.* **38** 3370-4
[3] Rupf S, Lehmann A, Hannig M, Schäfer B, Schubert A, Feldmann U and Schindler A 2010 *J. Med. Microbiol.* **59** 206-12
[4] Moisan M, Barbeau J, Crevier M-C, Pelletier J, Philip N and Saoudi B 2002 *Pure Appl. Chem.* **74** 349–58
[5] Cooper M, Fridman G, Staack D, Gutsol A F, Vasilets V N, Anandan S, Cho Y I, Fridman A and Tsapin A 2009 *IEEE Trans. Plasmas Sci.* **37** 866-71
[6] Kilmer S, Semchyshyn N, Shah G and Fitzpatrick R 2007 *Laser Med. Sci.* **22** 101–9
[7] Fridman G, Friedman G, Gutsol A, Shekhter A B, Vasilets V N and Fridman A 2008 *Plasma Proc. Polym.* **5** 503–33
[8] Grund K E, Straub T and Farin G 1999 *Bailliere Clin. Gastr.* **13** 67–84
[9] Martines E, Zuin M, Cavazzana R, Gazza E, Serianni G, Spagnolo S, Spolaore M, Leonardi A, Deligianni V, Brun P, Aragona M, Castagliuolo I and Brun P 2009 *New J. Phys.* **11** 115014 (8 pp)
[10] Stoffels E, Flikweert A J, Stoffels W W and Kroesen G M W 2002 *Plasma Sources Sci. Technol.* **11** 383-88
[11] Laroussi M and Lu X 2005 *Appl. Phys. Lett.* **87** 113902 (3 pp)
[12] Fridman G, Peddinghaus M, Ayan H, Fridman A, Balasubramanian M, Gutsol A, Brooks A and Friedman G 2006 *Plasma Chem. Plasma Process.* **26** 425-42
[13] Coulombe S, Léveilleé V, Yonson S and Leask R L 2006 *Pure Appl. Chem.* **78** 1147-56
[14] Kolb J F, Mohamed A-A H, Price R O, Swanson R J, Bowman A, Chiavarini R L, Stacey M and Schoenbach K H 2008 *Appl. Phys. Lett.* **92** 241501 (3 pp)
[15] Shimizu T, Steffes B, Pompl R, Jamitzky F, Bunk W, Ramrath K, Georgi M, Stolz W, Schmidt H-U, Urayama T, Fujii S and Morfill G E 2008 *Plasma Process. Polym.* **5** 577-82
[16] Kuchenbecker M, Bibinov N, Kaemling A, Wandke D, Awakowicz P and Viöl W 2009 *J. Phys. D: Appl. Phys.* **42** 045212 (10 pp)
[17] Robert E, Barbosa E, Dozias S, Vandamme M, Cachoncinlle C, Viladrosa R and Pouvesle J M

2009 *Plasma Process. Polym.* **6** 795–802
[18] Hillman J D, Brooks T A, Michalek S M, Harmon C C, Snoep J L and van Derweijden C C 2000 *Infect. Immun.* **68** 543-9
[19] Goree J, Liu B and Drake D 2006 *J. Phys D: Appl. Phys* **39** 3479-86
[20] Sladek R E J, Baede T A and Stoffels E 2006 *IEEE Trans. Plasma Sci.* **34** 1325-9
[21] Yang B, Chen J, Yu Q, Li H, Lin M, Mustapha A, Hong L and Wang Y 2011 *J. Dent.* **39** 48-56
[22] Pérez-Martínez J A, Peña-Eguiluz R, López-Callejas R, Mercado-Cabrera A, Valencia-Alvarado R, Barocio S R and de la Piedad-Beneitez A 2008 *Plasma Proc. Polym.* **5** 593-8
[23] Ulusoy Ç, Irmak Ö, Bağiş Y H and Ulusoy Ö İ A 2008 *Eur. J. Orthodon.* **30** 413–7
[24] Lazović S, Puač N, Miletić M, Pavlica D, Jovanović M, Bugarski D, Mojsilović S, Maletić D, Malović G, Milenković P and Petrović Z 2009 *New J. Phys.* **12** 083037 (22 pp)

# Zinc oxide thin films deposited by magnetron sputtering with various oxygen/argon concentrations

**L R Damiani and R D Mansano**

Departamento de Engenharia de Sistemas Eletrônicos - Universidade de São Paulo
Av. Professor Luciano Gualberto, 158, trav.3, - São Paulo/SP - Brazil

E-mail: larissar@lsi.usp.br

**Abstract**. Deposition of zinc oxide thin films onto silicon and thermally oxidized silicon substrates was performed by radio frequency magnetron sputtering. The obtained films were studied in function of percentage of oxygen and argon flow as process gases. Stoichiometric films were obtained in oxygen rich process conditions, which are more suitable to be p-type doped. Growth of zinc oxide films over thermally oxidized substrates resulted in better structural characteristics.

## 1. Introduction

Zinc oxide (ZnO) is a kind of semiconductor classified as a transparent conductive oxide (TCO), as it may exhibit electrical conductivity and optical transparency. It is a promising substitute to indium-tin oxide (ITO), widely used in optical devices as transparent electrode. ZnO, by nature, is an n-type material due to intrinsic defects as interstitial zinc atoms ($Zn_i$) and oxygen vacancies ($V_o$) on its structure [1]. However, it has been recognized that it is very important to grow high-quality both p- and n-type ZnO thin films for the development of ZnO-based optoelectronic devices [2]. Even with poor conductivity, formation of p-type ZnO has been observed when the films were deposited on $O_2$-rich conditions, without the presence of dopants [3]. These films have excess of oxygen in contrast to normal n-type ZnO films, which have excess of zinc [4].

For the deposition of ZnO, silicon substrates have many advantages, such as large wafer size and the possibility of integrating nitride semiconductors with the silicon technology. For research purposes, they are especially suitable because they are easy to cleave, have low surface roughness and present good chemical purity, facilitating the characterization processes of samples. However, for applications where transparency in the visible range is needed, amorphous materials (such as glass or polymers) are often used as substrates. Besides, the large mismatch of the lattice and thermal expansion coefficients between Si and ZnO makes the direct growth of ZnO on a Si substrate difficult, and frequently results in amorphous or polycrystalline films [5].

Magnetron sputtering is a physical vapor deposition method that creates plasma in the chamber to sputter the target material, commonly using argon as process gas. In this work, radio frequency (RF) magnetron sputtering was used for the deposition of the thin layers. The process is initiated in a glow discharge, produced in the vacuum chamber, under controlled gas flow. ZnO thin films were deposited onto silicon (Si) substrates. The main goal of this study is to find a process gas mixture that results in stoichiometric ZnO thin films, reducing intrinsic defects that impose n-type conductivity. In addition,

Published under licence by IOP Publishing Ltd

thermally oxidized silicon ($SiO_2/Si$) substrates were also used in order to study the difference in the structural characteristics occasioned by the substrate type where films were grown.

## 2. Experiment

Silicon (p-type, crystallographic orientation (100)) substrates were firstly cleaned with an ammonium hydroxide solution ($H_2O + H_2O_2 + NH_4OH$ - 5:1:1, for 10 minutes) to clean organic contamination and particles. They were then rinsed in with deionized water and immersed in a hydrochloric acid solution ($H_2O + H_2O_2 + HCl$ – 4:1:1, for 10 minutes) in order to remove metallic contaminants. The wafers were rinsed again and immersed in fluoridric acid ($H_2O + HF$ - 20:1), until the wafers seemed to be dry when taken out of the solution, in order to clean off the native silicon dioxide ($SiO_2$).

Half of the clean Si substrates were thermally oxidized in a wet process for 210 minutes at 1150ºC, resulting in a 288nm layer of $SiO_2$.

Thin films of ZnO were deposited over the previously cleaned silicon and thermally oxidized silicon substrates by RF magnetron sputtering, using a 4" high-purity (99.999%) zinc oxide ceramic target. After placing each substrate in the chamber, the system was evacuated to a vacuum of the order of $5x10^{-7}$ Torr using a turbomolecular pump. A distance of 10cm was maintained between target and substrate. Argon and oxygen were used as sputtering gases, with 10±0.5mTorr of process pressure (controlled by gas flow) and 200W of RF power. The percentage of oxygen flow in the process gas varied over the range 7.3 – 12.5% (deposition time: 90 min), and an additional film without oxygen in the gas composition was fabricated (using 100% argon, with 60 min of deposition time). Table 1 summarizes the deposition characteristics, both for Si and $SiO_2/Si$ substrates.

In order to obtain the I-V curves of the ZnO/Si samples, aluminum contacts were deposited by thermal evaporation. The Al top electrodes were evaporated through a shadow mask, resulting in various Al cylinders of approximately 1 mm² each. On the back of each sample, a whole Al layer was deposited.

**Table 1**. Deposition parameters used for the deposition of ZnO thin films onto Si and SiO2/Si substrates, with 200W of RF power and 10mTorr of process pressure.

| Process: | Gas flow (sccm) | Percentage of gas flow | Deposition time (min) |
|---|---|---|---|
| a) | 83,4 (Ar) | 100% (Ar) | 60 |
| b) | 76,0 (Ar) - 6 ($O_2$) | 92,7% (Ar) - 7,3% ($O_2$) | 90 |
| c) | 74,6 (Ar) - 8,1 ($O_2$) | 90,2% (Ar) - 9,8% ($O_2$) | 90 |
| d) | 69,5 (Ar) - 9,9 ($O_2$) | 87,5 % (Ar) - 12,5% ($O_2$) | 90 |

## 3. Results and discussion

The thicknesses of the samples were obtained by surface profilometry. The deposition rates of the ZnO/Si films (figure 1) tend to decrease as oxygen concentration increases. According to the literature, with oxygen content, the amount of Ar ions is decreased in the sputtering ambient, changing the deposition rate. Also, oxygen chemisorbs on the target and forms a surface layer of adsorbed oxygen. This layer inhibits the sputtering of the atoms [6].

The spectra of the ZnO/Si samples obtained by rutherford backscattering spectrometry (RBS) are shown in figure 2. In order to find out the stoichiometry of the films, simulation was performed using the RBS analysis package built on Genplot (RUMP), and simulated spectra were compared to the experimental data. The results shows that films with more oxygen concentration in the process gas tend to incorporate more oxygen in film composition, as expected. When the oxygen flow gets to 9.8%, the films present good stoichiometry (same proportion of Zn and O in the films). According to the RBS simulation, introduction of oxygen in the process gas results in films with higher roughness values, as the FUZZ command, which simulates the roughness of surface or interface, had to be used so the simulation would fit the experimental spectra. It was observed that roughness tends to increase proportionally to the oxygen concentration, until 9.8% of oxygen flow. With a further increase in the oxygen flow, the roughness value kept constant. The values of surface roughness were not measured in this work, only estimated by the RBS spectra.

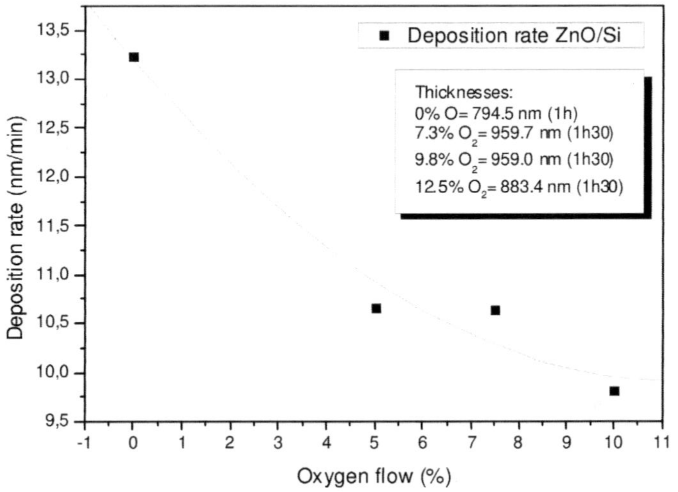


**Figure 1.** Deposition rate of ZnO/Si films in function of oxygen flow percentage in the process gas, along with the thicknesses obtained by profilometry.

The cristallinity of the $ZnO/SiO_2$ films (figure 3(a)) and ZnO/Si films (figure 3(b)) was studied by x-ray diffraction (XRD). The films obtained with oxygen in gas composition are strongly c-axis oriented along the (002) direction, resulted from the hexagonal (wurtzite) ZnO crystal structure. Both samples obtained with 100% Ar as process gas also presented other peaks of low intensity, indicating polycrystalline characteristics. The ZnO/Si with 100% Ar sample also showed an amorphous phase along with low intensity peaks. From the literature, it is known that the usage of $SiO_2$ buffer layer decreases the stress in the film. Stress in the ZnO thin film deposited on the bare Si substrate mainly results form the mismatch of lattice constants and thermal expansion coefficients between ZnO and Si substrate, resulting in a degradation of cristallinity [7].

An improvement of the cristallinity with the increase in oxygen concentration for both substrates can be observed, probably due to the suppression of defects on the structure. Optimum results were found for films with 7.3 and 9.8% of oxygen in gas composition, as they resulted in highest (002) peaks.

The I-V curves (figure 4) of the samples ZnO/Si were obtained by applying voltage from the top Al electrodes (deposited on ZnO) to the bottom electrodes (deposited under the silicon substrates). The results suggest that the oxygen incorporation influence the resistivity behavior. Although all the samples are poorly conductive, as they are not doped, the oxygen vacancies (specially present on samples obtained with less oxygen in gas composition) act as electron donors, increasing the probable n-type conductivity.

Scanning electron microscopy (SEM) images of $ZnO/SiO_2/Si$ (figure 5) and ZnO/Si samples (figure 6) show that the structures of the films change drastically with the addition of oxygen in gas composition. Figure 5(a) shows clusters of mean size of 110 nm, formed by smaller grains, and figure 6(a) shows grains of different shapes and sized, agreeing with XRD spectra, which presented many small peaks and amorphous phase. Both samples obtained with no oxygen in gas composition appear to be flatter than the other samples, suggesting a lower surface roughness. These observations agree with the RBS spectra. For the films obtained with oxygen partial pressure ((b), (c) and (d) of figure 5 and figure 6), uniformly distributed spherical grains compose the surfaces.

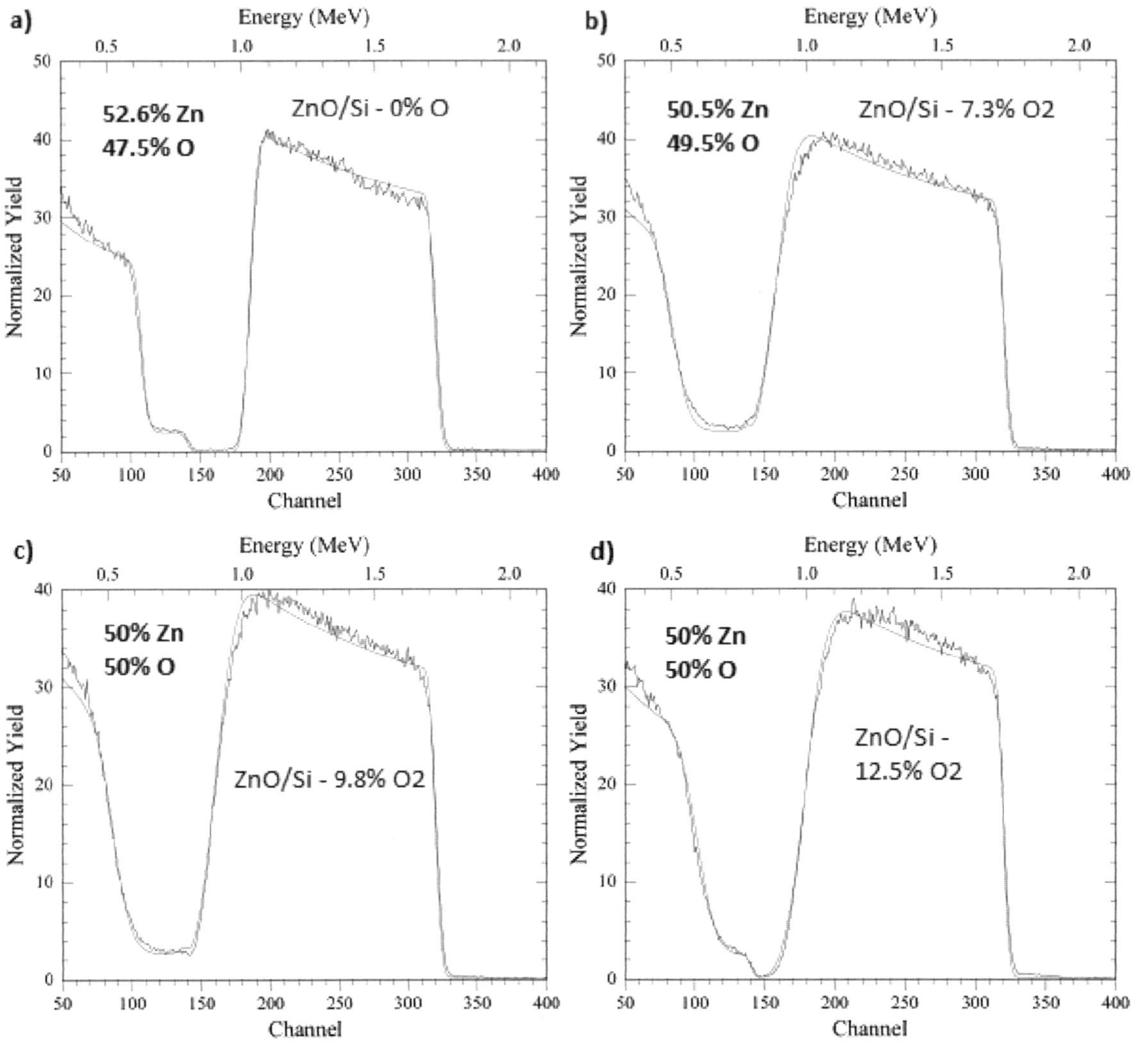


**Figure 2.** RBS spectra of ZnO/Si, showing atomic percentage of zinc and oxygen in film composition for samples obtained with (a) 0% $O_2$, (b) 7.3% $O_2$, (c) 9.8% $O_2$ and (d) 12.5% $O_2$.

## 4. Conclusions

Stoichiometric ZnO films were prepared by RF magnetron sputtering using a high-purity ZnO ceramic target. The layers were deposited onto silicon and thermally oxidized silicon substrates using argon and oxygen as sputter gases, 200W of RF power and 10mTorr of total process pressure. Substrate type has influenced the microstructure and cristallinity of the ZnO films, as they show better characteristics when deposited onto $SiO_2$/Si substrates. According to RBS measurements, films with higher oxygen partial pressure tend to incorporate more oxygen on the films, while films fabricated with low or no oxygen in gas composition presented oxygen vacancies. These defects are responsible for increasing n-type conductivity and reducing cristallinity. Best results of chemical composition and structural characteristics were found for samples produced with 7.5% of oxygen in the gas composition. They show good stoichiometry, improved cristallinity and reduced n-type conductivity. Therefore, this composition is more suitable for p-type doping of the films.

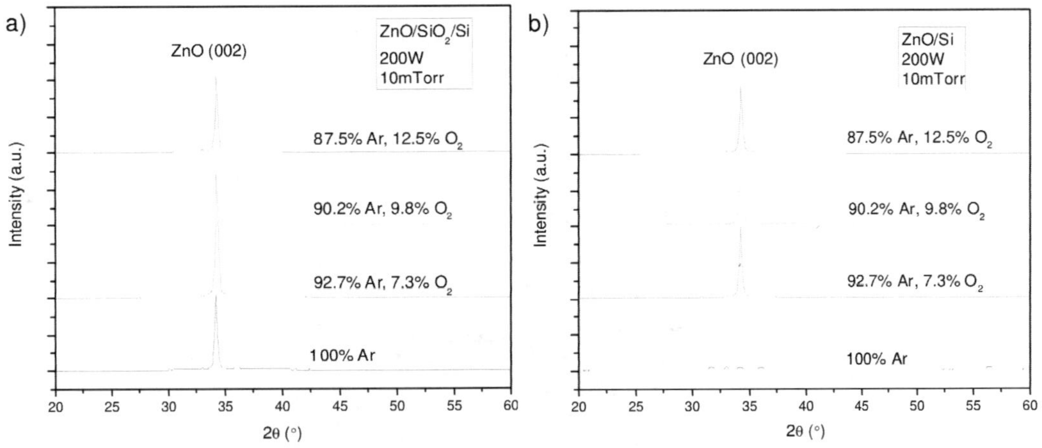


**Figure 3.** XRD spectra obtained for (a) ZnO/SiO2/Si films and (b) ZnO/Si films.

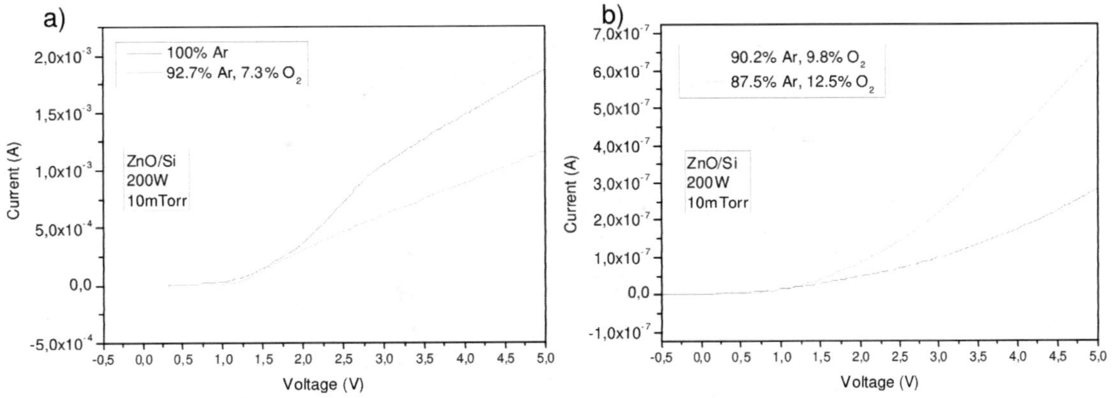


**Figure 4.** I-V curves of ZnO/Si films with (a) 0% $O_2$ and 7.3% $O_2$ and (b) 9.8% $O_2$ and 12.5% $O_2$.

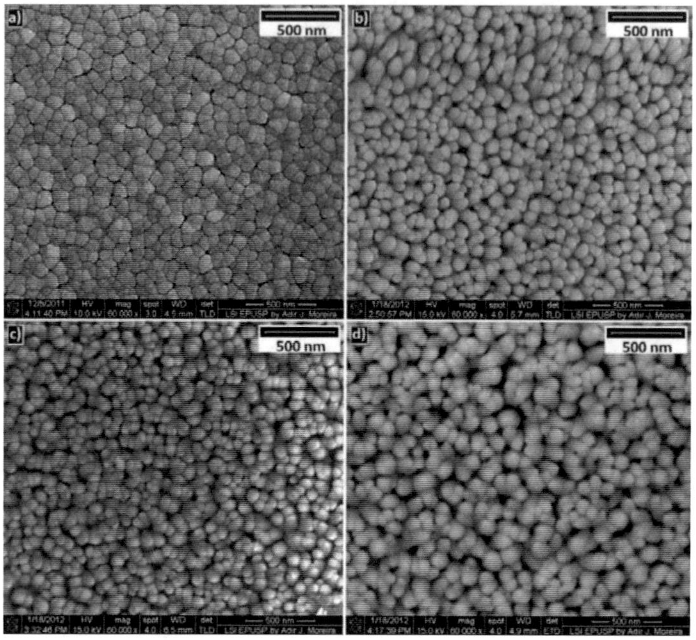


**Figure 5.** SEM images obtained for ZnO/SiO2/Si samples with (a) 0% $O_2$, (b) 7.3% $O_2$, (c) 9.8% $O_2$ and (d) 12.5% $O_2$.

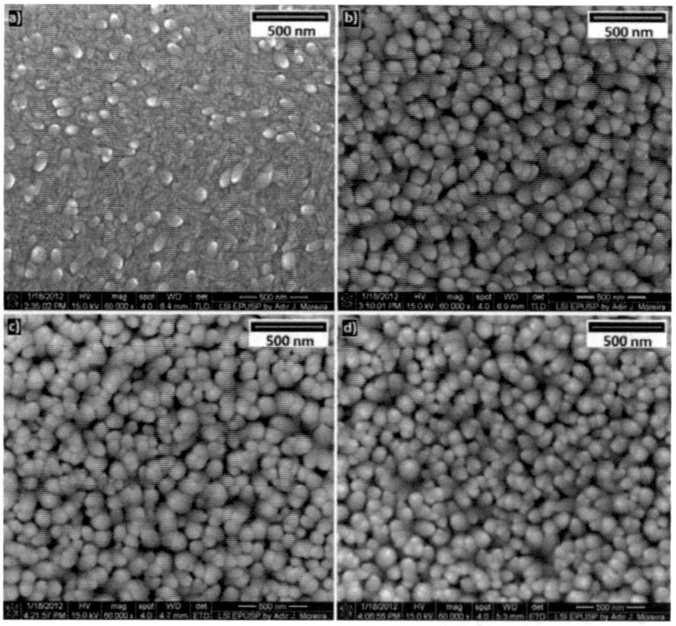


**Figure 6.** SEM images obtained for ZnO/Si samples with (a) 0% $O_2$, (b) 7.3% $O_2$, (c) 9.8% $O_2$ and (d) 12.5% $O_2$.

**Acknowledgements**
The authors would like to thank Dr. Luís da Silva Zambom and Mr. Nelson Ordonez for technical support and FAPESP, CNPq and CAPES for financial support.

**References**

[1] Yamamoto T et al. 2001 *Physica B* **302-303** 155-62
[2] Xiaodan Z et al. 2007 *Applied Surface Science* **253** 3825-7
[3] Lee K H et al. 2010 *Applied Surface Science* **256** 4241-5
[4] Ding R et al. 2007 *Vacuum* **82** 510-3
[5] Shen W J et al. 2006 *Journal Of Physics D: Applied Physics* **39** 269–73
[6] Sundaram K B et al. 1997 *Thin Solid Films* **295** 87-91
[7] Xu L H et al. 2011 *Journal of Physics: Conference Series* **276** 012051

Journal of Physics: Conference Series **370** (2012) 012020 doi:10.1088/1742-6596/370/1/012020

# Growth evolution of ZnO thin films deposited by RF magnetron sputtering

**A M Rosa[1], E P da Silva[1], E Amorim[1], M Chaves[1], A C Catto[2], P N Lisboa-Filho[2] and J R R Bortoleto[1]**

[1] Laboratory of Technological Plasmas, São Paulo State University-UNESP, Av. 3 de Março, 511, Sorocaba, Brazil
[2] Grupo de Materiais Avançados, São Paulo State University-UNESP, Av. Eng. Luiz Edmundo Carrijo Coube, 14-01, Bauru, Brazil

E-mail: dre_rosa_16@hotmail.com

**Abstract**. We study the surface morphology evolution of ZnO thin films grown on glass substrates as a function of thickness by RF magnetron sputtering technique. The surface topography of the samples is measured by atomic force microscopy (AFM). All AFM images of the films are analyzed using scaling concepts. The results show that the surface morphology is initially formed by a small grains structure. The grains increase in size and height with growth time resulting in the formation of a mounds-like structure. The growth exponent, $\beta$, and the exponent defining the evolution of the characteristic wavelength of the surface, p, amounted to $\beta = 0.76 \pm 0.08$ and $p = 0.3 \pm 0.05$. From these exponents, the surface morphology is determined by the nonlocal shadowing effects, that is the dominant mechanism, due to the incident deposition particles during film growth.

## 1. Introduction

Zinc oxide (ZnO) has gained much attention due to its versatile nature. Actually, ZnO is a wide band gap semiconductor (3.37 eV at room temperature) that can be used in numerous applications such as solar cells [1], flat displays, heat mirrors, TFTs and chemical sensors [2]. In particular, it is a promising alternative to indium tin oxide (ITO) in transparent conducting oxide (TCO) applications, due to its low cost, non-toxicity and the stability under hydrogen plasma [2,3]. ZnO thin films have been synthesized by a wide variety of techniques including chemical and physical routes, such as pulsed laser deposition [4], thermal evaporation [5], chemical vapor deposition [6,7], electron beam evaporation [5], spray pyrolysis [8], sol-gel method [9] and magnetron sputtering on a variety of substrates [10,11]. The advantages of magnetron sputtering are high deposition rate, low substrate temperature, good adhesion of the film to the substrate and the simple apparatus [10].

In thin film growth, the morphologies of growth fronts and their evolution reflect the microscopic growth dynamics. Under the common conditions, the growth of thin films is a process of far from-equilibrium that can lead to the evolution of nanostructures on front surfaces [12]. Thus, many of the properties of thin films, from the point of view of the applications, depend on their surface properties, particularly on the surface roughness [4,12,13].

The knowledge of the mechanisms that determine the thin film morphological evolution has motivated a large volume of research last years. A great emphasis has been given to the theoretical and experimental study of thin film growth evolution [14]. Two typical lengths are still employed to study the growth dynamics [15]: the surface roughness, $\sigma$ (the rms height fluctuation around the mean height value) and the lateral correlation length, $\xi$, is linked with the characteristic distance between mounds on the surface. It is assumed that $\sigma$ and $\xi$ evolves with growth time, t, as $\sigma \sim t^{\beta}$ and $\xi \sim t^{p}$, respectively [15]. The exponent $\beta$ is known as the growth exponent and p is known as the exponent defining the evolution of the characteristic wavelength of the surface [12,14,16]. These exponents depend on the growth mechanisms operating during deposition, which resulting the film structure [4].

In this paper, we investigate the growth dynamics of ZnO thin films via the RF magnetron sputtering technique. This research consist the determination of the growth regime of the thin film through the analysis of the exponents scaling. Thus, the experimental determination of the values of the exponents $\beta$ and p and their comparisons with the existing values of the literature allow identifying the main growth mechanisms involved in the film deposition.

## 2. Experiment

The experimental setup used in this work consists of a cylindrical stainless steel chamber, with 330 mm diameter and 200 mm high. Inside the reactor there are two stainless steel circular plates (electrodes) with 100 mm diameter, separated by distance of 50 mm. Gases were controlled through needle valves (Edwards, model LV-10K) and by flow meter and pressure was monitored by Pirani gauge. The system was evacuated by a vacuum rotary vane pump (Leybold D25B) lowering pressure to $1.0\times10^{-3}$ Torr. The plasma was activated by coupling a source of RF (13.56 MHz, 300W with impedance matching Tokyo Hy-Power Labs) capacitively coupled to the internal electrodes of the chamber.

ZnO films were deposited on $10\times30\times3$ mm$^3$ rectangular glass substrates by RF magnetron sputtering technique. The glass substrates were ultrasonically cleaned in detergent diluted in deionized water, rinsed in new ultrasonic bath in deionized water and a final ultrasonic bath in isopropyl alcohol. Afterwards, the substrates were dried in a hot air flow and positioned in the sample holder (top electrode). A metallic zinc target with 50 mm diameter and 99.99% purity was used as zinc source. The plasma was activated by a 13.56 MHz RF power of 70 W in argon pressure of $1.0\times10^{-2}$ Torr and flow of oxygen of 0.10 sccm. In the lower electrode was mounted a set of magnets of Samarium-Cobalt (SmCo), with circular geometry, and on top of the set of magnets was placed the metallic zinc target. In the top electrode was placed the glass substrate and this was grounded while the lower electrode was applied of RF power. The substrate temperature was measured after the deposition by a thermocouple and it was maintained around 353 K. The ZnO thin film growth rate was approximately 5 nm/ min. In order to study the growth evolution, different ZnO films with thickness ranging from 0 up to 547 nm were deposited.

The thickness was measured in a profilometer Veeco Dektak 150. Surface morphology was measured with atomic force microscopy (AFM, XE-100, Park Systems) operating in non-contact mode. Silicon cantilevers were employed (nominal radius of 10 nm). The scan area was 1 μm×1 μm with a resolution of 512×512 pixels. The AFM images were acquired; the surface roughness was characterized in terms of root mean square roughness (RMS), $\sigma$, and lateral correlation length, $\xi$.

The parameters $\sigma$ and $\xi$ are obtained from the surface morphology measured by AFM. $\sigma$ is mathematically defined by equation [17]:

$$\sigma = \sqrt{\frac{1}{N}\sum_{i=1}^{N}\left(hi - \bar{h}\right)^{2}} \tag{1}$$

The value of $\sigma$ was obtained directly from the 1 μm×1 μm AFM images using the software XEI (Park Systems). The lateral correlation length $\xi$ is related to the characteristic distance between mounds on the surface. The average value of $\xi$ was calculated from a Lorentz fitting of the power spectral density (PSD) calculated from 1 μm×1 μm AFM images. The PSD curves are provided by software XEI. The peak of the PSD curves for each thickness was used to select the wavelength $\lambda$ on the surface. The

Journal of Physics: Conference Series **370** (2012) 012020 doi:10.1088/1742-6596/370/1/012020

average value of $\xi$ was considered the inverse of $\lambda$, $\xi = 1/\lambda$. These measurements were used to determine the growth exponent $\beta$ and the exponent defining the evolution of the characteristic wavelength of the surface p. These exponents were derived fitting in graphics of roughness and correlation length, respectively. The crystalline degree and crystal orientation of ZnO thin films was measured using an x-ray diffractometer system (D/MAX-2100/PC, Rigaku) in grazing incidence geometry. X-ray diffraction patterns confirm the proper phase formation of the material.

## 3. Results and discussion

Figure 1 shows AFM images of the ZnO thin films grown on glass substrate as a function of thicknesses. The surface morphology is mainly formed by small grains. As the deposition goes further increases the density of large structure, which grows continuously in lateral size and height, tends to exhibit mounds-like structure. As shown AFM images in figure 1, the grain size spatial distribution is very homogeneous for all films. Vasco et al. [4] reported the presence of two growth regimes for ZnO/InP(100) films grown by PLD technique. Similar behavior is found in the first regime in which the surface morphology is mainly formed by a granular structure as well. In second regime, there are an increase of the large structure, which grows both in width and height, tending to display a stepped pyramid-like structure.

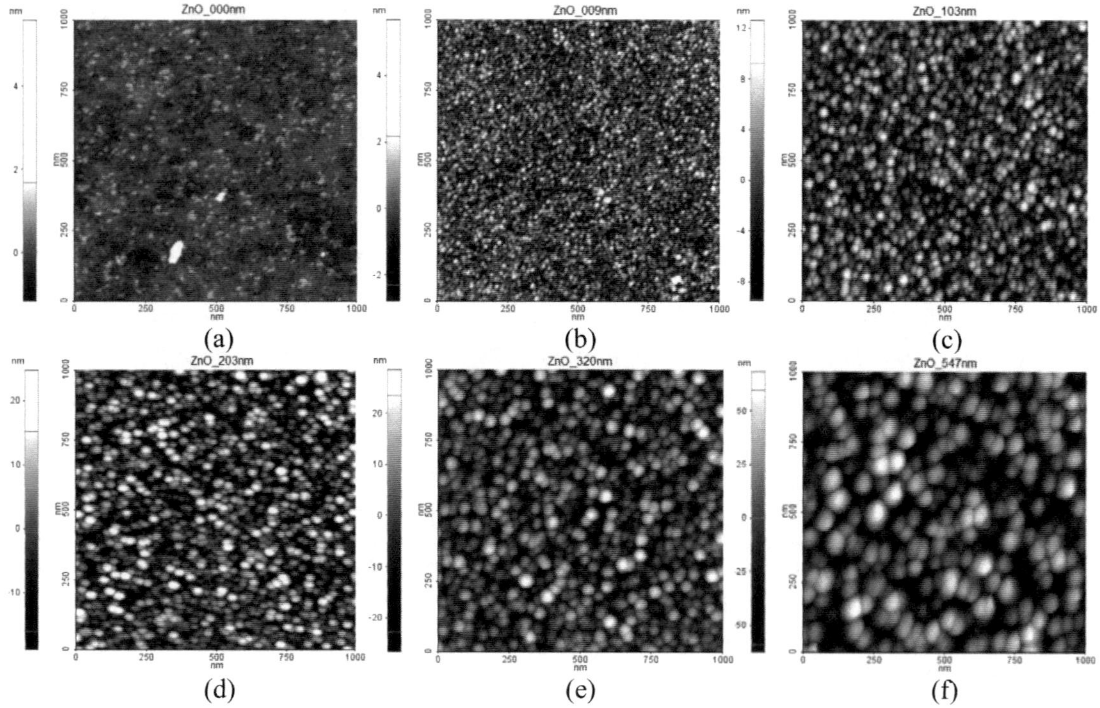


**Figure 1**. 1 μm×1 μm AFM images of glass substrate and ZnO thin films with different thickness. (a) glass substrate, (b) 9 nm, (c) 103 nm, (d) 203 nm, (e) 320 nm and (f) 547 nm.

Figure 2 shows the profile height of the surface topography of the (a) glass substrate and ZnO thin films with (b) 203 and (c) 547 nm thick along the fast scan direction. The growth evolution of the granular structures and the formation of mound-like structure are evident. From figure 2 a rough surface is apparent, i.e., the increase along of vertical direction of surface irregularities. The same behavior has been reported by Fu and Shen for thin Al/Si(100) [16] films sputter-deposited. They reported the increase of mounds in size and height with growth time.

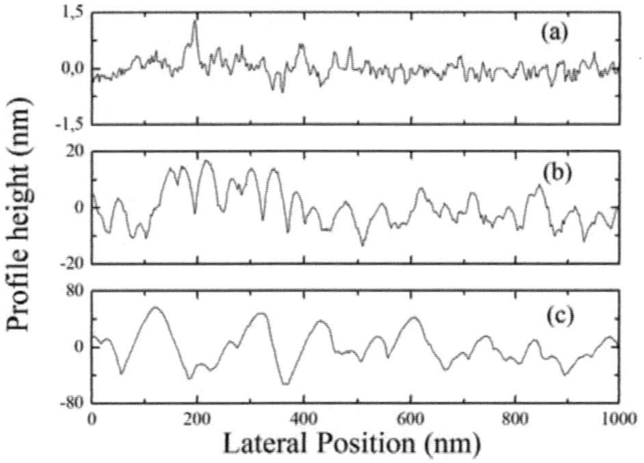


**Figure 2.** Surfaces profiles taken from 1 μm×1 μm AFM images of glass substrate and ZnO thin films. (a) glass substrate, (b) 203 nm thick with $\xi = 79.5$ nm and (c) 547 nm thick with $\xi = 179$ nm.

In figure 3 it is shown the evolution of $\sigma$ with films thicknesses and the calculation of exponent $\beta$. From the measurements of AFM images is determined the $\sigma$ data. The growth exponent $\beta$ is calculated from the slope of the log-log representation of the surface roughness as a function of thickness. By fitting the roughness data, the $\beta = 0.76 \pm 0.08$ while Vasco et al. [4] and Álvarez et al [12] reported $\beta = 1.0 \pm 0.1$ and Fu and Shen [16] related $\beta = 0.56 \pm 0.03$. So, the process of film growth is characterized by a growth unstable, i.e., $\beta \geq 0.5$ [14]. The growth of structures along the vertical direction is faster for larger $\beta$ value, i.e., for the films ZnO/InP(100) [4] and $ZrO_2$ [12] compared with our films.

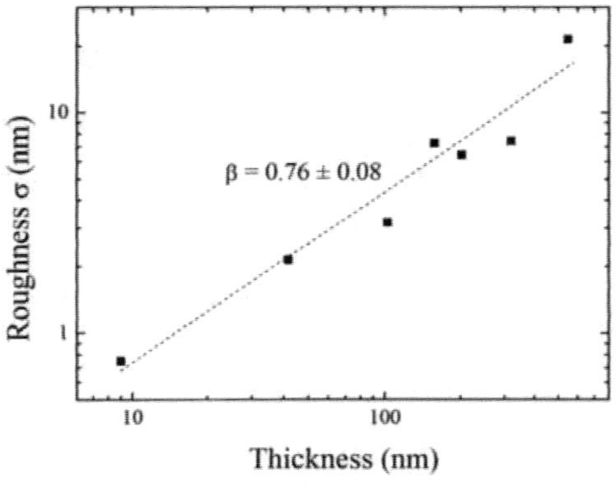


**Figure 3.** The roughness surface evolution with the ZnO films thicknesses as measured on 1 μm×1 μm AFM images and the calculation of the exponent $\beta$ for the films grown. The dash line is the best fit obtained for the calculation.

Figure 4 shows the evolution of $\xi$ with film thicknesses. The lateral correlation length $\xi$ is derived from the peaks $\lambda$ of the 2D power spectrum density (PSD) of the AFM images of the grown films, i.e., $\xi = 1/\lambda$. The p is calculated from the slope of the log-log representation of the lateral correlation length as a function of thickness. By fitting of the $\xi$ data, $p = 0.30 \pm 0.05$ while Vasco et al. [4] p =

Journal of Physics: Conference Series **370** (2012) 012020 doi:10.1088/1742-6596/370/1/012020

0.33 ± 0.05, Álvarez et al. [12] p = 0.34 ± 0.03 and Fu and Shen [16] reported p = 0.36. In a simplified way, AFM data show that ξ evolves from 43 up to 179 nm with film thickness.

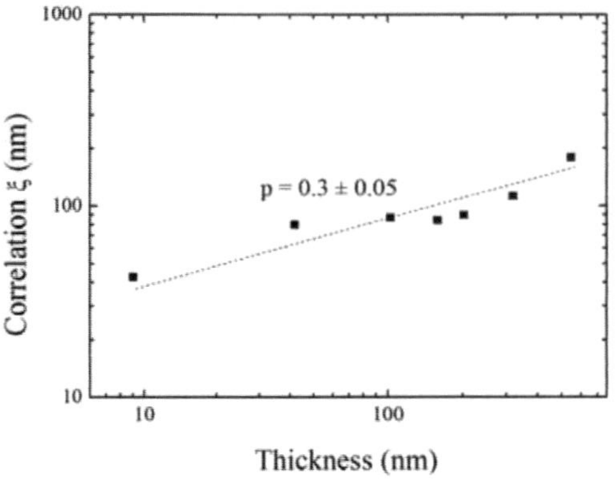


**Figure 4.** The correlation length ξ evolution as calculated from a Lorentz fitting of 2D power spectrum density of 1 μm×1 μm AFM images. The calculation of the exponent p for the grown films and the dash line is the best fit obtained.

Figure 5 displays (a) the x-ray diffraction pattern, and (b) correlation length ξ, solid square, and grain size, empty triangle, for ZnO films with different thicknesses. In the spectrum of figure 5(a) is possible to observe increasing the intensity of the (002) and (201) peaks with thickness. No peak is observed in ZnO film with 9 nm thick, because it is very thin it is impossible to detect the diffraction peaks. The presence of the (100), (002) and (201) peaks in ZnO film with 547 nm thick indicates that it is a polycrystal, this fact is less evident in smaller thicknesses. A strong (002) peak can be seen with a preferential orientation along c-axis perpendicular to the substrate surface according to the literature [18]. Lin et al. [10] reported that ZnO film deposited with thickness below 500 nm were polycrystalline with a c-axis preferential orientation. This preferential orientation in sputtered ZnO films is due to the lowest surface free energy of (002) plane according to Fujimura et al. [19].

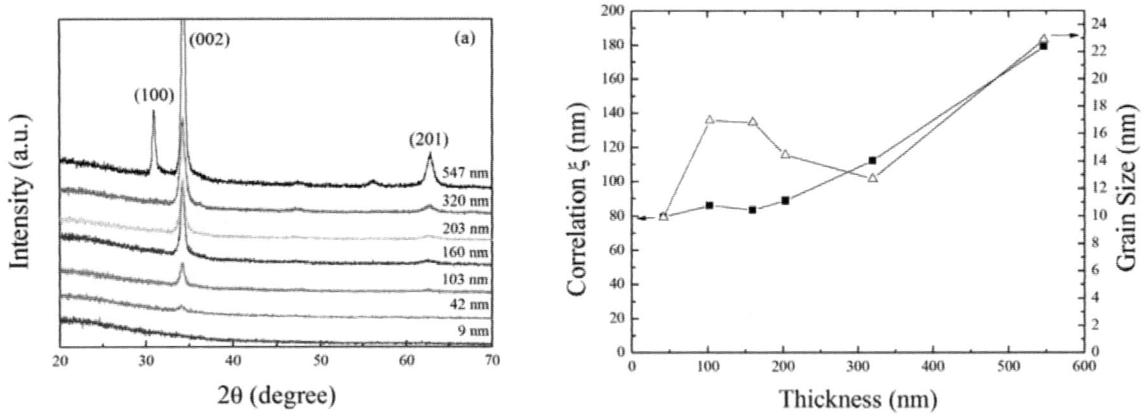


**Figure 5.** (a) The x-ray diffraction pattern of ZnO films and (b) correlation length ξ (solid square) and the grain size (empty triangle) for ZnO films with different thicknesses.

Figure 5(b) shows similar behavior of correlation $\xi$ and size grain that tend to increase as a function of thicknesses. The grain size of ZnO films is calculated using Scherrer's formula [10,20]:

$$D = \frac{0.9\lambda}{\beta \cos \theta} \tag{2}$$

where $\lambda = 1.54$ Å, $\beta$ the value FWHM in radians and $\theta$ Bragg angle. The grain size increased with the increase of film thickness. The film with 203 nm thick has about 14 nm of grain size and 79.5 nm of $\xi$ while the film thickness of 547 nm is observed about 23 nm and 179 nm of $\xi$. It is clear in figures 2(b) and 2(c) that shown surfaces profiles of ZnO films.

The formation of mounds-like structures on the surface of ZnO films is related to the dynamic growth far from equilibrium ($\beta \geq 0.5$) and to nonlocal effects. The nonlocal effects are growth mechanisms that during the deposition affect the film morphology evolution such as step-edge barriers, the shadowing, diffusional instability, sputtering, etc. [4,12,16]. Among them, two mechanisms must be taken into account: the shadowing of incident deposition particles, which becomes important when random incidence, and the diffusion mechanism tends to smooth a growing surface [12,14,21].

The shadowing has a strong effect on the formation of a mounds-like structure [14]. This occurs due to the vapor atoms, that incoming from all directions, stick where are first deposited [14,21]. The grass model illustrates nicely the effect of shadowing [14,21]. In this model, the growth of each "blade of grass" is related to the amount of light arriving at their tips. The initial configuration is blades of grass with small but random differences in heights. The shortest are shadowed by their taller neighbors, and stop growing. This also happens during the deposition when the valleys are shadowed by the peaks. This creates instability in the growth of the film that leads to the increases of the structures along the vertical direction causing the roughness. As a competing mechanism, diffusion erodes peaks and fills in the valleys, i.e., creates a more energetically stable structure [12,21]

By considering the experimental results obtained for the ZnO thin films grown on glass substrates is possible to identify the dominant growth mechanism that operates during deposition. The growth exponent $\beta = 0.76$ is found in our work while Vasco et al. [4] and Álvarez et al. [12] found a higher $\beta$ value, $\beta = 1.0$. They related that the dominant mechanism is the shadowing. Sometimes it is not possible to identify the dominant mechanism as reported by Fu and Shen [16]. They related that the nonlocal shadowing effect work together with diffusion mechanism in the film growth resulting in a lower $\beta$ value, $\beta = 0.56$. From these results it can be concluded that the growth dynamic behavior observed when the mounds-like structures developed is consistent with that expected when shadowing effects determine the growth mode. This behavior is consistent with the images of AFM in the figure 1 that show the development both in lateral size and height of the small grains as a function of thicknesses. Together with shadowing effect, diffusion mechanism is present in the growth dynamic, but at a lower magnitude. It must be noted that the shadowing effects dominates over diffusion mechanism.

## 4. Conclusions

In conclusion, it has been observed that ZnO thin films grown on glass substrates by RF magnetron sputtering technique show that the surface morphology is initially formed by small grains. As the deposition goes further increases the grains both lateral size and height resulting in the formation of a mounds-like structure. We have determined the exponents that characterize the dynamic growth and found that $\beta = 0.76 \pm 0.08$ and $p = 0.30 \pm 0.05$. From these results compared with the existing values in the literature can be concluded that the mechanism dominant in the surface morphology evolution is the nonlocal shadowing effects. However, together with shadowing effect, the diffusion mechanism acts in the growth dynamic at a lower magnitude. The growth behavior observed for the mounds-like structure is consistent with the existence of nonlocal shadowing effects.

**Acknowledgments**
The financial support from Fapesp and CNPq are gratefully acknowledged.

**References**

[1] Li Y L, Lee D Y, Min S R, Cho H N, Kim J S and Chung C W 2008 *Japan. J. Appl. Phys.* **47** 6896
[2] Lee S, Bang S, Park J, Park S, Jeong W and Jeon H 2010 *Phys. Status Solidi* **207** 1845-9
[3] Jiang M, Liu X and Wang H 2009 *Surface & Coatings Technology* **203** 3750-3
[4] Vasco E, Zaldo C and Vázquez L 2001 *J. Phys.: Condens. Matter* **13** L663-72
[5] Mtangi W,. Auret F D, Janse van Rensburg P J, Coelho S M M, Legodi M J et al. 2011 *J. Appl. Phys.* **100** 094504
[6] Kasuga M and Ishihara S 1976 *Japan. J. Appl. Phys.* **15** 1835
[7] Choi Y S, Hwang D K, Kwon B J, Kang J W, Cho Y H and Park S J 2011 *Japan. J. Appl. Phys.* **50** 10550
[8] Sonmez E, Aydin S, Yilmaz M, Yurtcan M T, Karacali T and Ertugrul M 2012 *Journal of Nanomaterials* **2012** 950793
[9] Mamata M H, Khusaimib Z, Musaa M Z, Maleka M F and Rusopa M 2011 *Sensor and Actuadors A* **171** 241-7
[10] Lin S-S and Huang J-H 2004 *Surface & Coatings Technology* **185** 222-07
[11] Shin J W, No Y S, Lee J Y, Kim J Y, Choi W K and Kim T W 2011 *Appl. Surf. Sci.* **257** 7516-20
[12] Álvarez R, Palmero A, Pietro-López L O, Yubero F, Cotrino J and Cruz W 2010 *J. Appl. Phys.* **107** 54311
[13] Liu Y, Yang S, Wei G, Song H, Cheng C, Xué C and Yuan Y 2011 *Surface & Coatings Technology* **205** 3530-4
[14] Barabási A L and Stanley H E 1995 *Fractal Concepts in Surface Growth* (Cambridge: Cambridge University Press)
[15] Smilauer P and Vvedensky D 1995 *Phys. Rev.* B **52** 14263
[16] Fu T and Shen Y G 2008 *Phys. B* **403** 2306-11
[17] Gadelmawala E S, Koura M M, Maksound T M, Elewa I M and Soliman H H 2002 *Journal of Materials Processing Techonology* **123** 133-45
[18] Rolo A G, Campos J A, Viseu T, Arôso T L and Cerqueira M F 2007 *Superllatices and Microstructures.* **4 2** 265-009
[19] Fujimura N, Nishihara T, Goto S, Xu J and Ito T 1993 *J. Cryst. Growth* **130** 269
[20] Ramana C V, Vemuri R S, Fernandez I and Campbell A L 2009 *Appl. Phys. Lett.* **95** 231905
[21] Karunasiri R P U, Bruinsma R and Rudnick J 1989 *Physical Review Letters* **62** 788-91

Journal of Physics: Conference Series **370** (2012) 012021 doi:10.1088/1742-6596/370/1/012021

# Nanocrystalline nitride coatings deposited by vacuum arc plasma-assisted method

**O V Krysina[1], N N Koval[1], Yu F Ivanov[1], N A Timchenko[2], T Baumbach[3], S Doyle[3] and T Slobodskyy[3]**

[1]Institute of High Current Electronics, 2/3 Akademichesky ave., Tomsk, 634055, Russia
[2]National Research Tomsk Polytechnic University, IHTP, 30 Lenin ave., Tomsk, 634050 Russia
[3]Institute for Synchrotron radiation, Forschungszentrum Karlsruhe, 76344 Eggenstein-Leopoldshafen, Germany

E-mail: krysina_82@mail.ru

**Abstract.** In the given work, experiments on research of formation of titanium nitride doped with copper (≤12 at %) produced by plasma-assisted vacuum arc deposition by evaporation of sintered Ti-Cu cathodes were carried out. It was revealed that Ti-Cu-N coatings have high hardness (≈40 GPa), high elastic recovery (≥50%), low friction coefficient (≈0.2) and high adhesion to a substrate compared with typical TiN coatings. By methods of transmission electron microscopy of thin foils and x-ray diffraction, it was showed that the coating crystallites consist of δ-TiN with the average crystallite size of 10-30 nm and the sheath of doping elements (copper) with thickness of 2-3 monolayers is formed around of TiN crystallites.

## 1. Introduction

Alongside with extensive studies on the plasma-assisted synthesis of nanocrystalline coatings, investigations of their structure-phase composition, properties and of the effect of the doping element on the way by which nanostructuration of the coating occurred are carried out. Two basic models of the structure formation in these coatings depending on their phase composition have been proposed. A detailed description of the model classification is given elsewhere [1,2]. In the first case, a coating of the nc-MeN/hard phase type is formed with a-$Si_3N_4$, BN, etc., playing the part of the hard phase. For instance, in Ti–Si–N coatings, nanosized crystallites of the basic phase (TiN) are embedded in an amorphous silicon nitride ($Si_3N_4$) matrix. In the second case, when metals which do not form compounds with nitrogen, such as Cu, Ni, Y, Au, etc., are added to MeN coatings, atoms of the doped element surround crystallites of the basic element nitrides, thus restricting their growth on the nanometer scale, and a coating of the nc-MeN/soft phase system (Me = Ti, Zr, Cr, Ta, etc.) is formed. Unfortunately, the capabilities of the existing research techniques restrict the scope of investigation on nanostructured and amorphous materials. Therefore, direct verification of the proposed models of nanostructuration of multi-component coatings with a certain coating is not always possible, and the use of complex procedures, such as those based on the properties of synchrotron radiation, is required.

This paper presents a study on the structure, phase composition, and element composition of nitride coatings synthesized on metal substrates by evaporating Ti cathode and Ti-Cu composite cathode. The goal of the study was to elucidate the effect of the doping elements on the features of the formation of nanocrystalline structure, on the basic phase crystallite size, and on the properties of the coatings.

## 2. The test material and the experimental procedure

The deposition of nitride coatings in low-pressure arc discharges was carried out on a plasma-ion set-up equipped with a standard arc evaporator and PINK, an original gas-plasma generator developed at the Institute of High Current Electronics, SB RAS (Tomsk) [3]. Additional ionization of the working gas by means of a plasma source with filament cathode made it possible to increase the efficiency of preliminary cleaning of the specimen surface and to realize the formation of nitride compounds under the conditions of plasma assistance.

The generation of multi-component plasma and the subsequent condensation of coatings were carried out by evaporating Ti cathodes containing 12at.%Cu in a nitrogen medium. To compare the obtained coatings and the widely used two-component coatings by their structure-phase and element composition and by mechanical properties, conventional TiN coatings were produced and investigated. Beryllium foil of thickness 0.5 mm, WC-8%Co hard alloy, and AISI 304 steel were used as substrates for x-ray diffraction analysis, mechanical and transmission electron microscopy (TEM) investigation, correspondingly. After mechanical grinding, polishing, and washing in an ultrasonic bath, the specimens were placed, on a substrate holder, in a vacuum chamber at a distance of 300 mm from the cathode. Immediately before the deposition of coatings, the surface of specimens was cleaned and activated by bombardment with accelerated argon ions at a negative substrate potential of ~1 kV. During the bombardment, the specimen surface heated up to ~300 °C.

To improve the adhesive characteristics of a nitride coating, its formation was preceded by deposition of a titanium sublayer of thickness about 100 nm by evaporating the cathode material in an argon medium. Synthesis of all Ti-Cu-N coatings was carried out in the following parameter ranges: bias voltage $U_b = -(100\text{–}300)$ V, gas pressure $p = 0.3\text{–}0.4$ Pa, discharge current $I_d = 50\text{–}100$ A, and deposition temperature $T = 300\text{–}400$ °C. The coating thickness was 3–5 μm at a coating growth rate of 1–3 μm/h.

Investigations of the deposited coatings were carried out by the following methods: optical microscopy (OLYMPUS GX71), transmission (EM-125) and scanning electron microscopy (Philips SEM 515 equipped with EDAX ECON IV, an element composition microanalyzer), micro- and nanoindentation (PMT-3, NHT-S-AX-000X Nano Hardness Tester), scratch testing (MST-S-AX-000 Micro-Scratch Tester), and the Calotest method. The phase and element composition was investigated on powder diffraction stations, on the synchrotron radiation (SR) channels of the VEPP-3 (RFA-SR, PRD SR) energy store and of the ANKA (XRPD) SR source.

## 3. Results and discussion

Measurements of the micro- and nanohardness of the test coatings were performed in order to investigate the influence of crystallite size on mechanical properties of the samples. The measurements performed at a normal load of 500 and 50 mN, respectively, have shown that the hardness of the coatings produced by evaporating composite cathodes is greater than the hardness of titanium nitride (≈25 GPa) by a factor of ≈1.5-2. The coatings obtained with the use of Ti-12 at. % Cu cathodes possess superhardness (≈40 GPa). Analysis of the loading-unloading curves obtained by the nanoindentation method has given the elastic strain of the test coatings. The greatest residual strain of 75% was observed for a titanium nitride coating and the least of 50% for Ti-Cu-N coatings. The degree of elastic recovery of the surface shape for the coatings formed by evaporating powder cathodes was 2 times greater than that for the TiN coatings. The Young modulus of the Ti-Cu-N coatings was in the range of 500-550 GPa.

The scratch-test method was used to determine the critical load at which destruction of a coating begins. For TiN, this quantity was about 3.6 N, whereas the destruction of Ti-Cu-N coatings of ≈10 N begins at a critical load two times larger than that for TiN coatings.

Journal of Physics: Conference Series **370** (2012) 012021 doi:10.1088/1742-6596/370/1/012021

With the use of transmission electron diffraction microscopy of thin foils it has been found that the coatings formed by evaporation of composite cathodes consist of nanosized crystallites oriented randomly relative to each other. This follows from the strongly pronounced ring structure of the electron diffraction patterns (figure 1(b)). Electron diffraction analysis has shown that the crystallites of the coatings consist of δ-TiN. The measurements of size of coating crystallites were carried out on dark-light images, the average size of crystallite was estimated by the methods of statistical analysis [4]. The size of crystallites in the coatings containing copper is equal to 10-30 nm (figure 1(a), figure 1(c)); the coatings produced by evaporating pure titanium have crystallites of average size 100 nm.

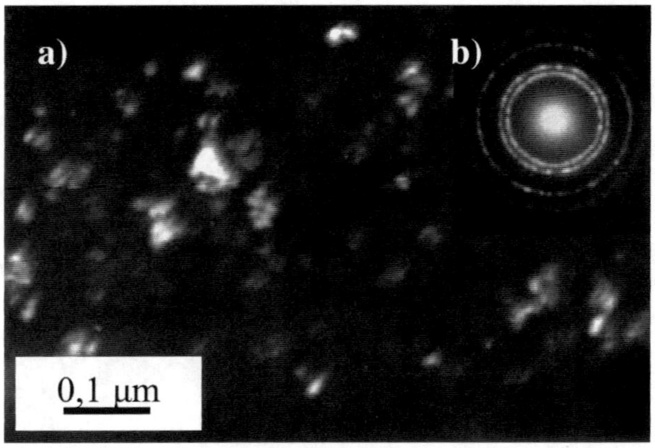


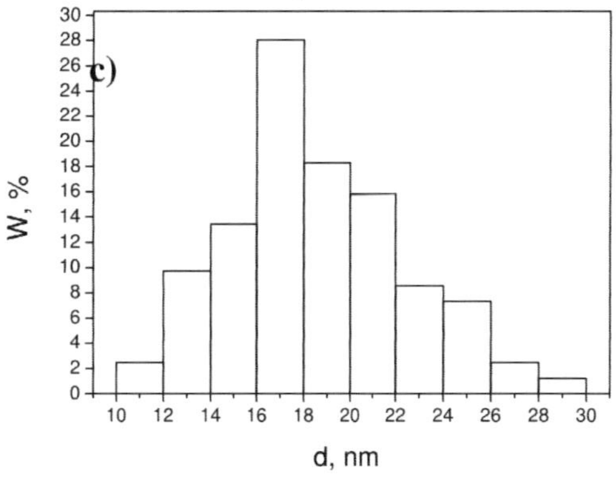


**Figure 1.** Image of the structure of a Ti-Cu-N coating formed by evaporating Ti-12 at.% Cu composite cathode in nitrogen: dark field in the reflection of a type (111) ring of TiN (a), electron diffraction pattern (b). TiN crystallite size distribution in the Ti-Cu-N coating obtained by evaporation of Ti-12at.%Cu cathode (c).

The investigations performed by the above methods gave insufficient information to construct a model of the processes responsible for the variations of the coating properties and to answer the major questions of which is the role of the impurity atoms in the formation of the nanocrystalline structure of synthesized coatings, where they are localized, and whether they form their own crystallographic phase.

It is necessary to notice that the composite cathode material was investigated earlier [5]. It was showed that copper in cathode material is situated uniformly mainly on boundary of main phase (α-Ti) or in compound with Ti ($CuTi_2$). The size of titanium based phases is in the range of 3-20 μm for Ti-12at.%Cu cathode. The size of cathode spots of vacuum arc is in range of 50-300 μm. This fact is evidence of uniformly evaporation of composite cathode elements.

Investigations of crystalline structure and element composition of coatings based on titanium nitride with additional elements have been carried out on the stations of the store powder diffraction and x-ray fluorescence element analysis (VEPP-3) [6] and of the ANKA synchrotron radiation source [7]. We rely on the data for the construction of a model of the coatings synthesis. The results of these investigations are given in figure 2 and figure 3.

In figure 2 the beryllium substrate shows up in all diffraction patterns in the form of very narrow strongly pronounced reflections. The reflections from the synthesized coating structure are wider owing to their nanocrystalline structure.

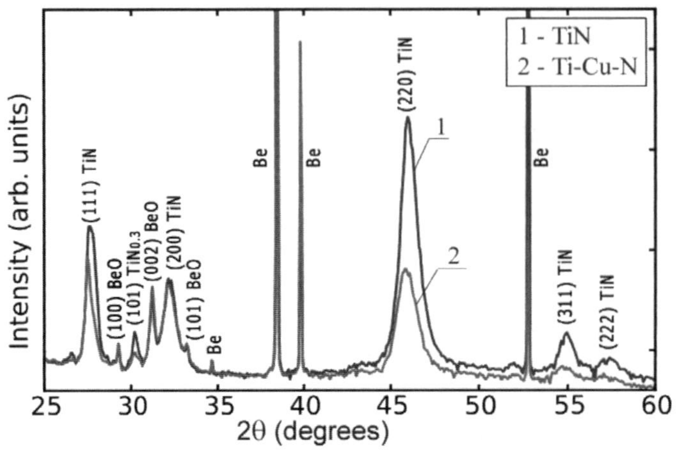


**Figure 2.** Diffraction patterns of TiN coatings without doping (TiN) and doped with Cu (Ti-Cu-N) obtained for the photon energy of 10.5 keV.

The diffraction patterns of the layers of titanium nitride and of titanium nitride doped with copper are identical [see figure 2]. The location of reflections testifies to the presence of TiN, $TiN_{0.3}$, and Ti phases in the coatings. The reflections from (111), (200), (220), (311), (222) TiN planes were revealed. The reflection from (101) plane corresponds to $TiN_{0.3}$ phase. The rest peaks belong to Be and BeO phases. Shifts of reflections and new reflections are not observed in the diffraction patterns obtained for specimens doped with copper. These data allow the conclusion that copper atoms do not form compounds with titanium or nitrogen and, equally, do not form their own crystalline phase. As the data of x-ray fluorescence analysis presented in figure 3 confirm the presence of copper in the test coating in amounts corresponding to its content in the composite cathode evaporated in an arc discharge (Ti-12at.%Cu), it can be concluded that copper is in the amorphous state at the crystallite boundaries. No signal from heavy elements is present in fluorescence data. A stand alone peak at 85 keV is attributed to Be windows. (The presence of Fe-peak is connected with insignificant evaporation of cathode holder material (stainless steel) during ignition of arc discharge). Therefore, we conclude that the only time it takes for copper atoms to form a closed sheath around a growing TiN crystallite determines the time of growth of the crystallite and, hence, its size.

In order to explain the observed phenomenon we developed an intuitive model of the growth process. The model assumes that a layer of finite thickness should be formed around a growing grain in order to prevent further growth of the grain. Spherical form of the grain is chosen as a statistical approximation. For a given grain radius $R$ the volume of the grain is calculated as $V = (4/3)\pi R^3$, similarly the volume of a cover layer with thickness $dR$ is calculated as $dV = (4/3)\pi((R+dR)^3 - R^3)$. The concentration of doping material can be deduced as:

$$x = 1 - \frac{R^3}{(R+dR)^3} \tag{2}$$

Plot of the grain size dependence on the doping material concentration required to completely cover a growing grain for different cover layer thicknesses shown in figure 4. From the figure it is easy to see that the grain size will grow with decrease in the dopant concentration until other processes start to dominate limiting the size of the grain. In the case of grain sizes below 5 nm the dependence might be slightly modified by triangulation correction.

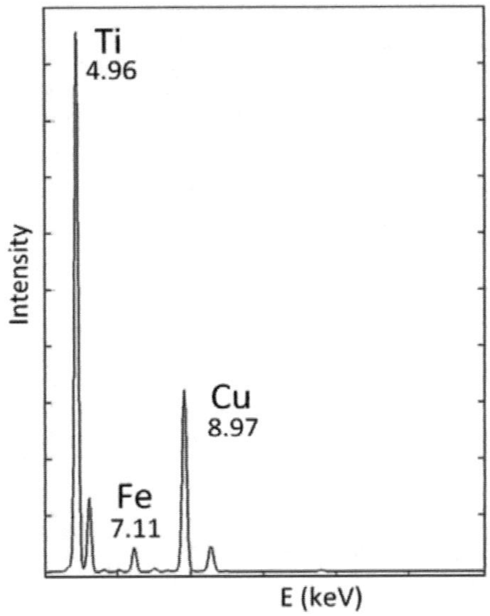


**Figure 3.** X-ray fluorescence spectra of Ti-Cu-N specimens near the K-edges of Ti and Cu obtained for the excitation with 20 keV photons.

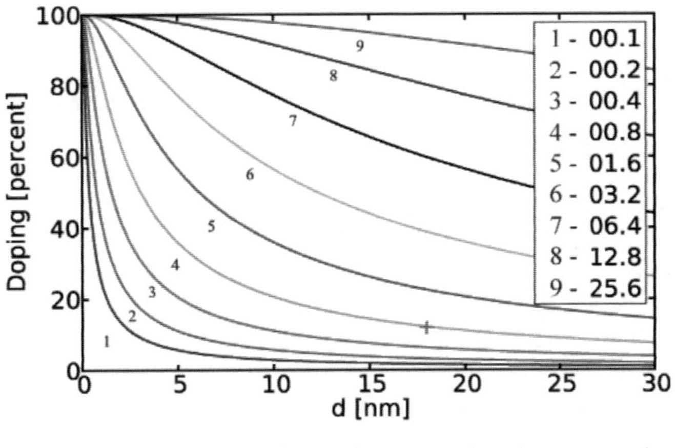


**Figure 4.** Simulation of maximum grain size at a given doping concentration. The lines correspond to different coverage needed to prevent further crystallite growth in [nm]. Our experimental conditions are indicated by a cross.

If the grain sizes and doping concentrations are known the cover layer thickness is expressed as:

$$dR = R\left(\frac{1}{\sqrt[3]{1-x}} - 1\right) \tag{3}$$

For our experimental conditions, with 18 nm average grain size and 12at.%Cu concentration we find 0.74 nm cover layer thickness (cross in the figure 4) corresponding to 2.8 single atomic layers coverage of the grain (Atomic radius of copper is equal to 132 pm).

## 4. Conclusion

By method of transmission electron microscopy of thin foils in diffraction mode the formation of the nanocrystalline coatings produced by evaporating Ti-Cu composite cathodes in low-pressure arc discharges in the presence of nitrogen plasma was revealed. The coating crystallites consist of δ-TiN. For the copper-containing coatings the average crystallite size is 10-30 nm; the coatings produced by evaporating pure titanium have crystallites of average size 100 nm.

With the powder diffraction method using synchrotron radiation it has been found that copper does not participate in the formation of crystalline phases in the coating; that is, they are in the amorphous state. It has been revealed that the copper concentration in the Ti-Cu-N coatings synthesized by the vacuum-arc method coincides with that in the evaporated cathode (12at.%).

The obtained data confirm the following model of nanostructuration of coatings based on titanium nitride. The nanocrystallization of these materials occurs when an amorphous sheath of doping elements (copper) is formed around of TiN crystallites, thereby preventing their further growth.

**Acknowledgments**

The work was partly supported by RFBR under Grant No. 11-08-00625-a.

## References

[1] Veprek S and Reiprich S 1995 *Thin Solid Films* **268** 64

[2] Musil J, Baroch P and Zeman P 2008 in: Wei R (Ed) *Handbook of plasma surface engineering research and its practical applications* (India: Research Singpost) Chapter 1

[3] Vintizenko L G, Grigoriev S V, Koval N N, Tolkachev V S, Lopatin I V and Schanin P M 2001 *Russian Physics Journal* **44** 927

[4] Tomas Gareth and Goringe Michael J 1979 *Transmission electron microscopy of materials* (New York: John Wiley & Sons)

[5] Pribitkov G A, Korosteleva E N, Psakhie S G, Goncharenko I M, Ivanov Yu F, Koval N N, Schanin P M, Gurskih A V, Korjova V V and Mironov Yu P 2004 in: Korovin S, Ryabchikov A (Eds.), *Proceedings of 7$^{th}$ International Conference on Modification of Materials with Particle Beams and Plasma Flows* (Tomsk: Publishing house of the IAO SB RAS) p. 163

[6] Tolochko B P, Kochubey D I , Shmakov A N, Ten K A, Sharafutdinov M R and Erenburg S B 2007 in: *Proceedings of the 5$^{th}$ International specialized exhibition "LaboratoryExpo'07"*, Expodesign Moscow Russia p. 92

[7] Heinrich J (Ed) 2007 *ANKA Instrumentation Book*, ISS Institute for Synchrotron Radiation, Forschungzentrum Karlsruhe GmbH, a member of the Helmholtz Association, Germany,

[8] Lipson G and Stiple G 1972 *Interpretation of Powder Patterns* [Russian translation], Mir, Moscow

# Lubricating coating prepared by PIIID on a forming tool

**J F Martinatti[1], L V Santos[2], S F Durrant[1], N C Cruz[1] and E C Rangel[1]**

[1]Paulista State University, Av. Três de Março, 511, 18087-180, Sorocaba, SP, Brazil
[2]Technological Institute of Aeronautics, São José dos Campos, SP, Brazil

E-mail: elidiane@sorocaba.unesp.br

**Abstract.** In this work, the performance of a-C:H films produced by the hybrid Plasma Immersion Ion Implantation and Deposition technique as lubricating layers for a steel forming tool has been investigated. Hardened steel (AISI M2, 64 HRC) plates coated with a commercial TiN layer were used as substrates and the films were deposited in a vacuum chamber fitted with two parallel-plate electrodes. The discharges were generated in atmospheres composed of 91% $C_2H_2$ and 9% Ar by the application of radiofrequency power (13.56 MHz, 100 W) to the upper electrode while the lower one, also used as the sample holder, was biased with high voltage negative pulses (3.6 kV, 30 μs, 300 Hz). A deposition time of 840 s was used. The effects of the gas pressure, $p$, on thickness, molecular structure, wettability, surface morphology and topography, hardness and friction coefficient of the films were investigated. Film thickness increased from 0.3 to 0.5 μm when $p$ was increased from 2.7 to 16.5 Pa. Generally, the films were slightly hydrophilic, with contact angles of around 84°, and the deposition decreased the roughness of the steel. A polymer-like structure was detected in high pressure depositions and an amorphous carbon structure derived from the low pressure procedures. Hardness decreased from 8.2 to 7.0 GPa with increasing $p$. Improvement in tribological performance was indicated by the fall in the friction coefficient from 0.5 to 0.2 as the deposition pressure was reduced. Operating at the latter value (of μ) would lead to a significant reduction in wear and hence to significant economy in diverse industrial applications.

## 1. Introduction

Hydrogenated amorphous carbon films, a-C:H, are part of a class of organic materials with very attractive properties such as high hardness, optical transparency and low coefficient of friction. The characteristics of the a-C:H films are strongly dependent on chemical composition and the proportion of $sp^3$ hybridized carbon atoms. That is, the higher the fraction of tetrahedrally bonded carbon of $sp^3$ atoms the higher the mechanical resistance of such materials, whose hardness can reach values close to that of diamond. It is well known that the formation of $sp^3$ groups during the deposition is enhanced when the film grows under ion bombardment [1]. More specifically, when the energy per deposited carbon ion is around 150 eV, the energy transferred by the ion impact promotes the breakage of $sp^2$ bonds and enhances the formation of $sp^3$ groups. Therefore, by controlling the energy of the ions it is possible to modify the microstructure of the bombarded material. On the other hand, the higher the fraction of $sp^3$ bonds, the higher the internal stress, which may limit the physical stability of the material. When the forces associated with the internal stress overcome the cohesive forces between film and substrate, the coating peels off. Although a-C:H coatings show great potential in applications, these are still limited by high levels of internal stress [2]. Specific investigations have to be conducted for each particular application to adjust the levels of internal stress (for each particular substrate) to provide good coating adhesion without deleterious effects on the desirable properties.

In this context, Plasma Immersion Ion Implantation and Deposition (PIIID) [3] is a very convenient technique. Similarly to the conventional Plasma Enhanced Chemical Vapor (PECVD) process [4,5], the deposition is enhanced by the plasma. In PIIID experiments, however, the substrate holder is biased with high voltage negative pulses which accelerate the positive ions towards the sample. Plasma immersion ion implantation and deposition is an environmentally friendly and cost effective technique that is able to eradicate the need for intermediary layers to improve the physical stability of the stressed a-C:H films.

Aiming at the improvement of physical resistance and the reduction of the friction coefficient, a-C:H films have been deposited by PIIID on AISI M2 and TiN coated AISI M2 steel forming tools. The influence of the plasma pressure on chemical composition, roughness, hardness, friction coefficient, morphology and topography of the coated tools has been studied.

## 2. Experiment

Heat treated AISI M2 steel samples with and without a 9 μm- thick TiN top layer were obtained from forming tools used in the automotive part stamping process. Prior to the depositions the samples not coated by the TiN film were mirror polished and all the samples were sonicated in detergent, distilled water and isopropyl alcohol baths and then dried using a hot air gun. The samples were further cleaned by the exposure for 600 s to plasmas generated in atmospheres composed of 50% Ar and 50% $H_2$ at a total pressure of 4.1 Pa. The plasmas were excited by the application of radiofrequency power (13.56 MHz, 150 W) to two parallel plate electrodes fitted inside a cylindrical stainless steel vacuum chamber fully described elsewhere [6]. After the cleaning procedure, without exposing the samples to air, the atmosphere was changed to 91% $C_2H_2$ and 9% Ar for PIIID experiments. In all the experiments, the background pressure was 2.8 Pa while the total pressure owing to the gas feed ranged from 2.7 to 16.5 Pa. The radiofrequency power was kept at 100 W and for ion implantation the substrates were biased with -3600 V pulses at a repetition rate of 300 Hz and pulse length of 30 μs. The chemical composition of the films has been investigated by Infrared Reflectance Absorbance Spectroscopy (IRRAS) using a Jasco FTIR-410 spectrometer. The thickness and the roughness of the coatings have been determined by profilometry using a Veeco Dektak D150 while the morphology has been assessed by scanning electron microscopy with a Quanta Inspect S – FEI Company microscope. Surface wettability was evaluated by measuring the water contact angle using a Ramé-Hart 100-00 goniometer. Contact angle and thickness measurements have been performed using glass substrates to avoid any artifact associated with tool roughness. A Hysitron TriboIndenter was used to evaluate film hardness through nanoindentation experiments with a Berkovich tip and a CETR-UMT micro tribometer was employed for the tribological characterizations.

## 3. Results and discussions

The film thickness, $h$, as function of the total pressure, $p$, is shown in figure 1. In the range of pressures investigated, a continuous growth of $h$ as $p$ was increased was observed. This tendency can be understood considering that the higher the pressure the higher the availability of film precursor species.

Figure 2 shows the mean roughness, $R_a$, of the samples prepared directly on polished M2 steel and on TiN-M2 steel as a function of the pressure of the gases. It is observed that when the substrate has a high roughness (TiN-M2 steel), film deposition leads to a reduction in $R_a$, independently of the pressure employed. When the substrate has a low surface roughness (M2 steel) the deposition of the film tends initially to an increase in $R_a$ with an increase in pressure up to 5.5 Pa. For depositions made at greater pressures, there is a tendency for $R_a$ to fall. According to a study by Salvadori [7], the roughness tends to increase beyond its initial value owing to the greater deposition rate on peaks than in valleys. In agreement with the BD model [8], with an increase in film thickness, the roughness decreases owing to the filling of the valleys by the lateral growth of the film surface. For both of the substrates, the lower roughnesses were obtained using plasmas at greater pressures.

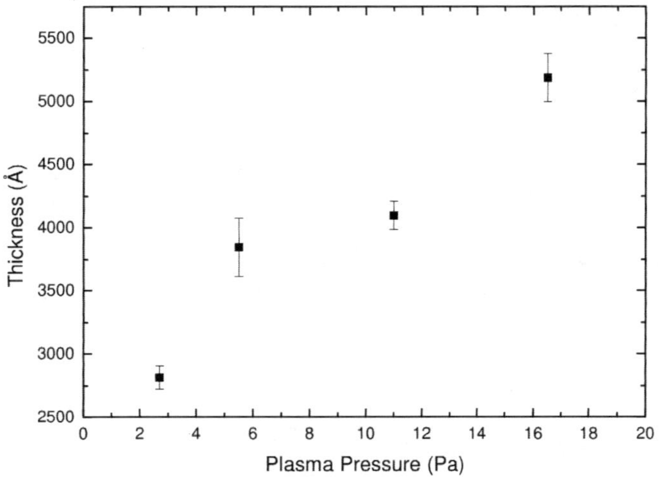


**Figure 1.** Film thickness as a function of the total gas pressure.

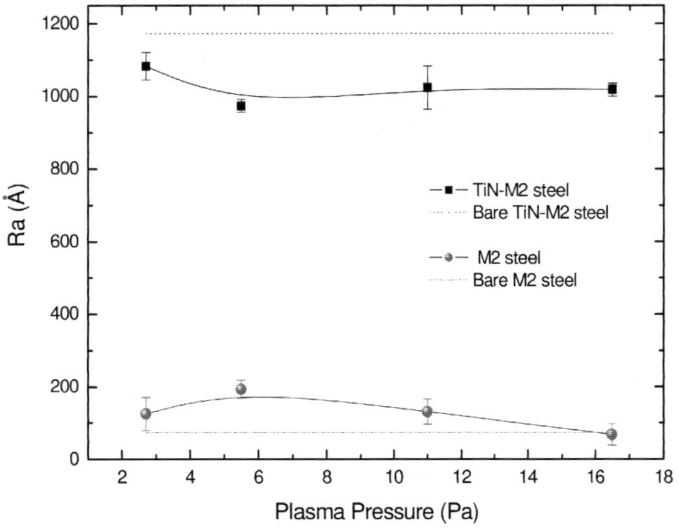


**Figure 2.** Average roughness of the films as a function of the plasma pressure. Dotted lines represent the $R_a$ values for the uncoated substrates.

Figure 3(a) shows a topographic image of the TiN-M2 steel surface, obtained by AFM, where the peaks and valleys owing to the surface treatment (hard turning) before the application of the TiN layer are evident. Figure 3(b) shows the surface topography of the system containing the film deposited at 2.7 Pa, where the pattern produced by the working of the surface is still clearly observable. Figure 3(c) shows the topography of the sample that received the film from a plasma at $p$ = 5.5 Pa. Comparing this image with the previous one, it can be seen that the valleys have been filled-in by the film, resulting in a more regular surface with a reduced surface roughness. Figures 3(d, e) show the topography of surfaces prepared at 11.0 and 16.5 Pa plasmas, respectively. Triangular structures are observed, which indicate pointed defects in the surface and the continuous disappearance of the marks left by the machining of these samples by the film coverage. Holes are also identified more frequently in the last two images.

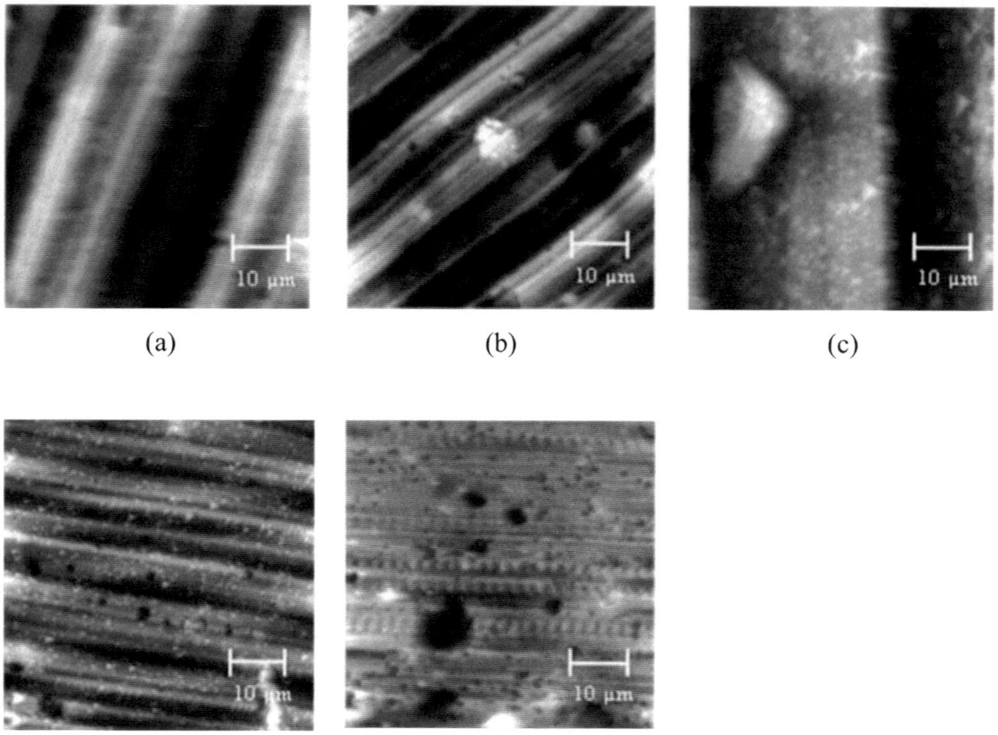


**Figure 3.** Atomic force microscopy images of the bare TiN-M2 substrate (a) and of the same material after exposure to plasma immersion ion implantation and deposition at 2.7 Pa (b), 5.5 Pa (c), 11.0 Pa (d) and 16.5 Pa (e). Image area: 50 X 50 $\mu m^2$.

Figure 4 shows a line profile of the surface of the TiN-M2 steel sample after the deposition of the a-C:H film at a total gas pressure of 2.7 Pa. It is clearly observed that the profile produced by machining of the surface is not completely covered by the deposition of the a-C:H film.

Figure 5 shows the infrared spectra of films deposited on polished M2 steel for various values of $p$. The spectra of the films deposited at lower gas pressures (2.7 and 5.5 Pa) do not present bands characteristic of plasma-deposited organic films. There is a slight indication that OH groups are present, namely the absorption at around 3620 $cm^{-1}$. Film thickness cannot be the only factor to be considered here since spectra characteristic of polymer-like films were obtained from even thinner films in a previous study [9]. For films deposited at higher values of $p$, there are bands related to the stretching and the deformation of C-H groups (2920 and 1434 $cm^{-1}$, respectively) and the stretching of C=C groups (1615 $cm^{-1}$). These alterations in the spectra may be ascribed to the diminution in the energy delivered to the material per unit of area, since the deposition rate grows, shifting the a-C:H to a polymer-like structure.

From the above results it is clear that the film characteristics depend on the plasma pressure employed for deposition. A structure typical of hydrogenated amorphous carbon, which is transparent in the infrared part of the electromagnetic spectrum, is obtained at lower pressures while a polymeric structure is obtained when the pressure is increased [10].

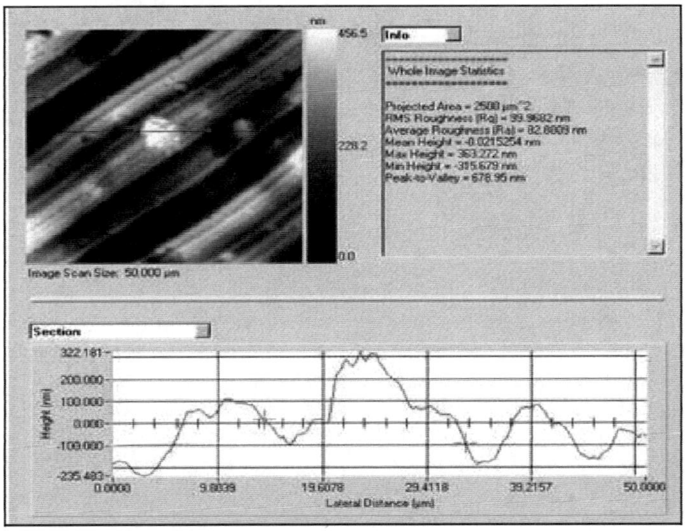


**Figure 4.** Line profile in the surface of the steel sample AISI M2 with TiN layer after deposition of the a-C:H film using a total gas pressure of 2.7 Pa.

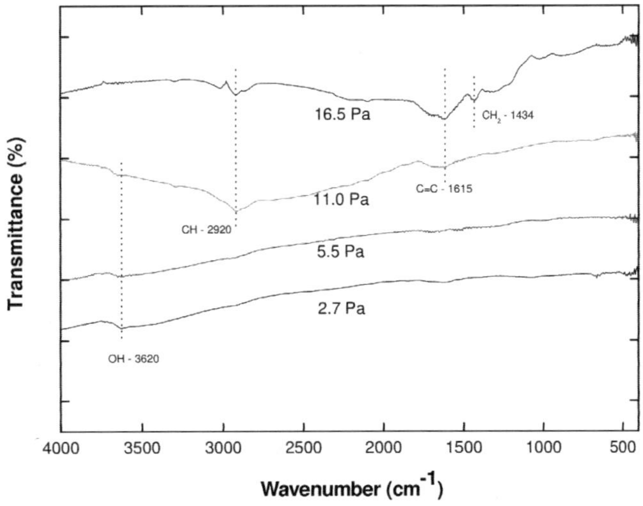


**Figure 5.** Infrared spectra of films produced by PIIID using different plasma pressures. Spectra were acquired from films deposited onto polished stainless steel plates.

Figure 6 shows the water contact angle, θ, of the samples as a function of *p*. These measurements were undertaken on the same day as the deposition of samples prepared on glass substrates. Considering that the uncertainty in the measurements is about 4°, the contact angle remains almost constant (~ 85°) as *p* varies; that is, the films are slightly hydrophilic. This result is very similar to that obtained in a previous study in which the pulsed polarization voltage was varied [9] and may be also interpreted in terms of oxygen incorporation in the film (figure 5).

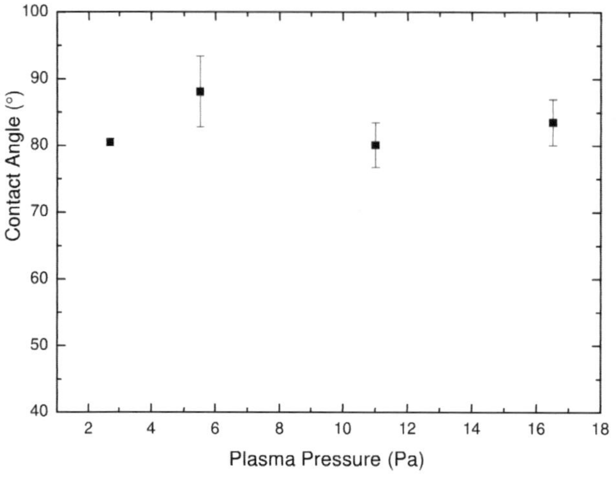


**Figure 6.** Contact angle of the films deposited onto glass substrates as a function of the plasma pressure.

Figure 7 shows the hardness, H, as a function of depth for films deposited on polished M2 steel, at pressures ranging from 2.7 and 16.5 Pa. The hardness of the as-received substrate and of quartz, the standard used for calibration of the indenter tip area, are also shown. It is important to note that, the hardness measurements were realized on the film prepared directly on M2 steel since the results obtained from samples prepared on TiN-M2 steel are influenced by the greater surface irregularities presented by these.

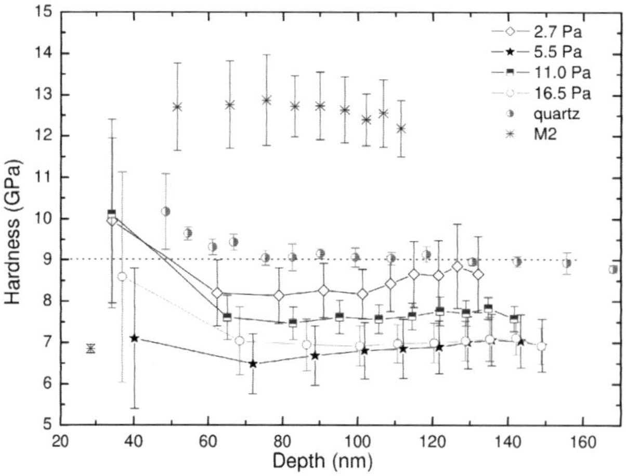


**Figure 7.** Hardness of the films as a function of depth for the samples prepared on M2 steel in plasmas of different pressures.

For depths greater than 60 nm, for which the hardness of quartz stabilizes at ~ 9 GPa, the hardest film was that deposited at the lowest pressure (2.7 Pa). It is interesting to note that for the samples deposited at the lowest pressures (2.7 and 5.5 Pa), there is an increase in hardness at increasing tip penetration, indicating the growing interference of the mechanical properties of the substrate with the measurements. For the remaining films this effect is not observed owing to the greater film thicknesses. All the films are softer than the substrate.

Figure 8 shows the hardness of the films as a function of *p*, determined from the data in figure 7, in regions that correspond to 13 until 22% of the total film thickness. Generally, there is a fall in H for higher *p*, the lowest hardness being obtained at a pressure of 5.5 Pa. The film with the greatest mechanical resistance was that prepared at the lowest pressure. This result is consistent with an increase in the mean energy of the plasma species when the pressure is reduced. The increase in the mean free path of the ions with a reduction in pressure results in an increase in the energy supplied to the growing layer by ion bombardment. Both phenomena may favor the precipitation of $sp^3$ nuclei in the structure, thus increasing the hardness. The higher H values for the samples prepared at 11.0 and 16.5 Pa (compared to that prepared at 5.5 Pa) may be a consequence of the interference of the mechanical properties of the substrate since in these samples hardness was taken at a depth greater than 15% of the film thickness. It is believed from Raman analyses (not shown), that the hardness values for such samples are even lower than 7.0 GPa.

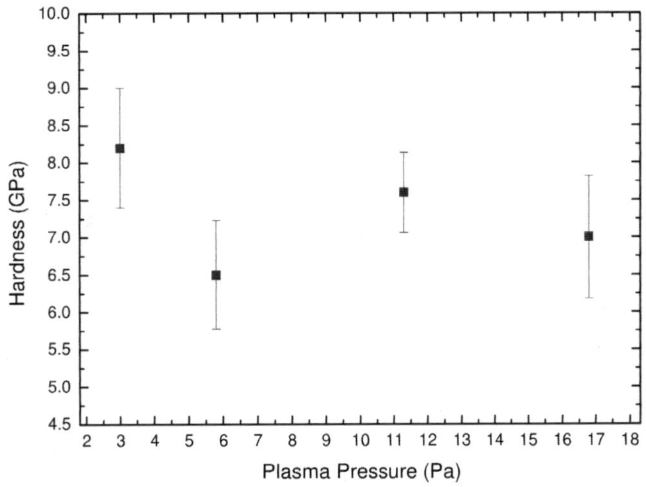


**Figure 8.** Hardness of the films as a function of the plasma pressure at a constant depth.

Figure 9 shows the friction coefficient, $\mu$, as a function of the friction time, t, for samples prepared directly onto M2 steel at different pressures. The results obtained for the bare substrate are also included for comparison. In the sample prepared with $p$ = 2.7 Pa, the curve begins at 0.22 but rises continually with increasing test duration, reaching around 0.7 at 300 s. The initial $\mu$ value was as high as that of the film that was deposited in 5.5 Pa plasmas. Moreover, its growth is faster indicating an earlier failure in this case. For the films prepared at the highest pressure plasmas (11.0 and 16.5 Pa), the loss of protection occurs in the first stages of the tribological test. This behavior is ascribed to the rise in the deposition rate with increasing *p*, since this affects the film structure in two different ways: it enhances the internal stress as a consequence of the higher proportion of defects incorporated into the structure and reduces the implantation of carbon ions on the steel surface, a process that has a beneficial effect on the connection of amorphous carbon to steel substrates via carbide formation [5,6]. Carbon ion implantation also reduces structural mismatches between film and substrate, reducing the interfacial stress and improving connectivity [11]. The ion bombardment process is also decisive in influencing the mechanical resistance of the layers.

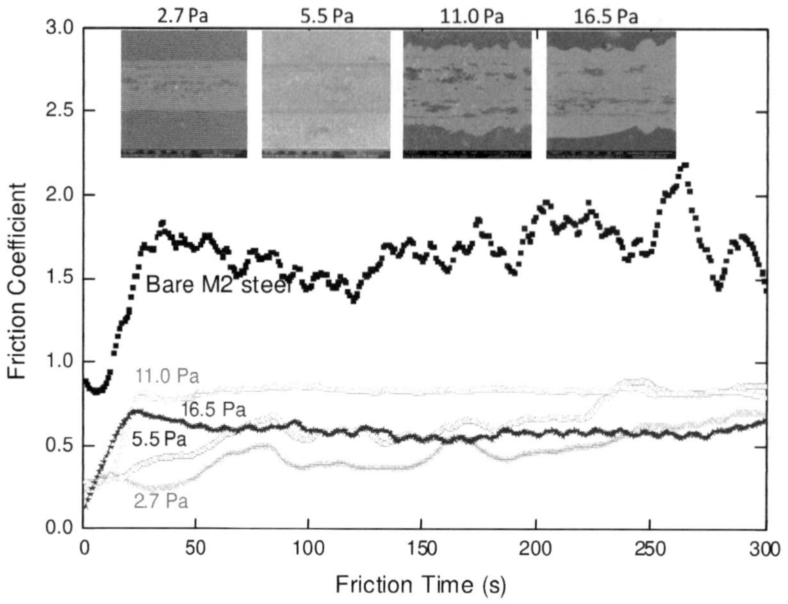


**Figure 9.** Friction coefficient as a function of the experiment time for the pair AISI 404 steel sphere and samples prepared on M2 steel by PIIDI with pressure varying from 2.7 to 16.5 Pa. Friction coefficient for the pair sphere-bare M2 steel is also presented. Micrographs of the tracks formed during the friction measurements are presented as insets. The bar scale in the pictures corresponds to 200 μm.

The insets in figure 9 are micrographs of the samples detailing the track regions. Wear tracks containing sparsely distributed tribolayers are clearly detected in all the samples but remarkable differences are noticed. First of all, the dimensions of the tracks generated in the samples treated at the lowest pressure plasmas (2.7 and 5.5 Pa) are considerably lower. Secondly, less film detachment occurred in the latter. Finally, the formation of a tribolayer was also more frequent for the steel treated at the lowest pressure. On the other hand, the film is completely removed from the track as depositions are conducted in plasmas of 11.0 and 16.5 Pa. In these cases the track dimensions are substantially higher, revealing a weaker connection between film and substrate. Therefore, the little damage produced in the coating deposited in 2.7 Pa plasmas is ascribed to an association of the low friction coefficient, stronger adhesion and superior mechanical resistance of this film.

Figure 10 shows the friction coefficient results for the samples prepared on TiN-M2 steel as well as for the bare TiN-M2 substrate. For deposition carried out in plasmas of 2.7 Pa, the curve initiates around 0.22 reaching 0.28 at the end of the experiment. Since the direct contact of the tribometer sphere with the TiN coating would result in friction coefficients of around 0.6 [12], the strong improvement provoked by this film on the system tribological performance is readily noticed. For the sample which received the film in plasmas of 5.5 Pa, the $\mu$ value starts around 0.24 and increases slowly during the test duration reaching 0.5 in 300 s. It is interesting to observe how $\mu$ reaches the 0.5 plateau in successively shorter times as one further increases $p$ (11.0 and 16.5 Pa).

The characteristics of the tracks produced during the risk tests can be evaluated by the inset of figure 10. Just a slight mark is generated in the sample exposed to 2.7 Pa plasmas. The waviness of the substrate is also clearly detected. According to the micrographs and EDS (not shown) data, wear occurred just in the top regions with no tribolayer formation, while the depressions were kept unchanged. An increment in the track dimensions is the only modification observed as the test was conducted in the sample treated in 5.5 Pa plasmas. But upon further increasing $p$, there is complete removal of the organic layer from the track. Evidence of tribolayer formation is obtained only for the highest pressure prepared sample (16.5 Pa).

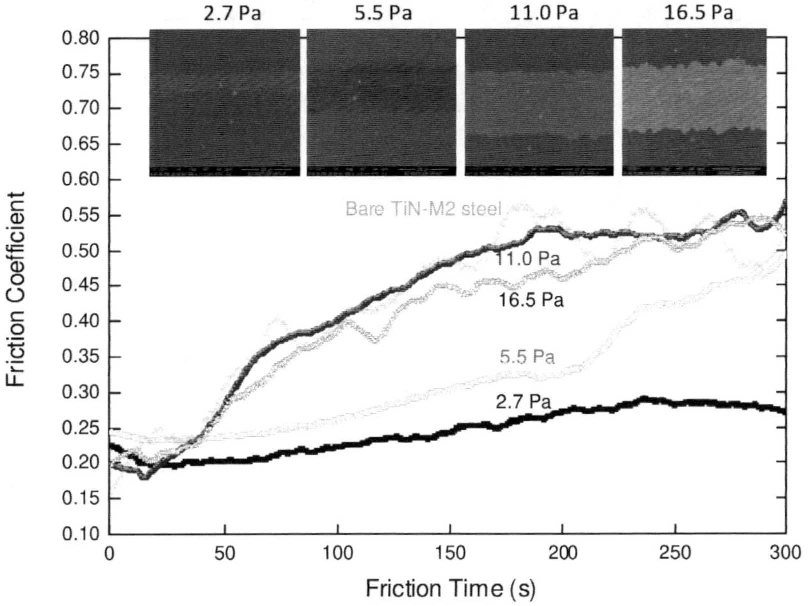


**Figure 10.** Friction coefficient as a function of the experiment time for the pair AISI 404 steel sphere and samples prepared on TiN-M2 steel by PIIDI at different plasma pressures. The friction coefficient for the pair sphere-bare TiN-M2 steel is also presented. Micrographs of the tracks formed during the friction measurements are presented as insets. The bar scale in the pictures corresponds to 200 μm.

Therefore, considering the results presented here, it was possible to verify that the system tribological properties improve with decreasing deposition pressure, independently of the employed substrate. The higher proportion of carbon atoms incorporated into the surfaces by ion implantation is one of the responsible factors. Film hardness is also important. Despite being rougher, the samples prepared on TiN M2 steel presented a better result than that prepared on the smooth M2 steel. This fact confirms that adhesion and mechanical resistance are primordial aspects of the tribological performance of hydrogenated amorphous carbon.

## 4. Conclusions

It was observed that the pressure of the depositing plasma has a significant influence on the tribological response of hydrogenated amorphous carbon deposited on M2 and on TiN-M2 steel. The substrate chemical composition and topography are also decisive for the final film properties. Moreover, the mechanical resistance of the coating is an important factor since it is directly related to the film durability. New studies should be developed to investigate the effect of the film hardness on the tribological properties of these specific systems. It should be kept in mind, however, that extremely hard coatings may compromise the friction coefficient and wear rate since they may become brittle. Adhesion of the film to the substrate was observed to be another issue for the system tribological behavior. The formation of carbides, especially in the TiN substrate, substantially increased the connectivity of both materials. Lower deposition pressures also improved the film connectivity. The initial topography of the substrate was not the most important factor. The association of greater mechanical resistance, adhesion and lower friction coefficient is responsible for the lower wear rates produced in the optimized sample. Indeed, in this sample, the friction coefficient was reduced by more than 50%, which can represent a significant cost reduction for a series of industrial applications. Finally it should be pointed out that the friction conditions employed here were severe considering the applied load. Certainly this compromised the film durability. New studies will be addressed to evaluating the tribological behavior of the same systems under milder friction conditions.

## Acknowledgements

The authors acknowledge the financial support of FAPESP (São Paulo State Research Foundation) and CNPq (Brazilian Scientific and Technological Development Council).

## References

[1] Anders A 2000 *Handbook of Plasma Immersion Ion Implantation and Deposition* (John Wiley & Sons) chapter 1
[2] Peng X L and Clyne T W 1998 *Thin Solid Films* **312** 219
[3] Anders A 2000 *Handbook of Plasma Immersion Ion Implantation and Deposition* (John Wiley & Sons) pp 281
[4] Martinu L, Klemberg-Sapieha J E, Kuttel O M, Raveh A and Wertheimer M R 1994 *J. Vac. Sci. Technol. A* **12** 1360
[5] Martinu L and Wertheimer M R 1993 *Materials Science Forum* **140-142** 405
[6] Rangel R C C, Souza M E P, Schreiner W H, Freire C M A, Rangel E C and Cruz N C *Surf. & Coat. Technol.* **204** 3022
[7] Salvatori M C, Martins D R and Cattani M 2006 *Surf. & Coat. Technol.* **200** 5119
[8] Barabasi A L and Stanley H E 1995 *Fractal Concepts in Surface Growth* (Cambridge University Press)
[9] Martinatti J F, Santos L V, Cruz N C and Rangel E C 2011 *Eur. Phys. J. Appl. Phys.* **56** 24014
[10] Silva S R P 2003 *Properties of Amorphous Carbon* EMIS data views series, n$^{o}$. 29, (INSPEC publication) pp 128-34
[11] Shum P W, Zhou Z F, Li K Y and Chan C Y 2004 *Thin Solid Films* **458** 203
[12] Yuhara D A 2000 *Aplicação de PVD em Ferramentas de Corte* (Brasil: Brasimet Com. Ind. S.A) pp. 1-9

# Removal of main exhaust gases of vehicles by a double dielectric barrier discharge

**M Pacheco[1], E Alva[2], R Valdivia[1], J Pacheco[1], C Rivera[1], A Santana[3], J Huertas[3], B Lefort[4] and N Estrada[5]**

[1]Instituto Nacional de Investigaciones Nucleares, Carretera México-Toluca s/n, La Marquesa, Ocoyoacac, Estado de México, Mexico
[2]Universidad Nacional Autónoma de México, Ciudad Universitaria, Coyoacan, Mexico
[3]Instituto Tecnológico y de Estudios Superiores de Monterrey, Campus Toluca, Eduardo Monroy Cárdenas 2000, Toluca, Mexico
[4]Institut Supérieur de l'Automobile et des Transports, 49 Rue Demoiselle Bourgeois, Nevers, France
[5]Instituto Tecnológico de Toluca, Av. Instituto Tecnológico, Metepec México

E-mail: marquidia.pacheco@inin.gob.mx

**Abstract.** Because the health effects and their contribution to climate change, the emissions of toxic gases are becoming more controlled. In order to improve the diminution of toxic gases to the atmosphere, several techniques have been developed; here it will be focus only to automotive emissions. This work deals about the treatment of toxic gases emitted from vehicles by a non-thermal plasma. Several tests were done in a 4-cylinder 2002/Z16SE motor to characterize the vehicle emissions. With these results gas mixture simulating the exhaust gases vehicles, was used in experiments at different conditions employing a double dielectric barrier reactor for their treatment. The removal efficiencies superior to 90% show the competence of the non-thermal plasma reactor to treat these gases. Experimental results are explained with the aid of a simple chemical model that suggests a possible mechanism of degradation of toxic gases. The plasma reactor employed could works at 12V supplied without difficulty by a vehicle battery.

## 1. Introduction

Pollutants like nitrogen oxide (NOx), particulate matter (PM), sulfur dioxide ($SO_2$), volatile organic compounds (VOCs) and carbon monoxide (CO) are mostly generated by vehicles. The situation becomes more difficult because the inexistence of catalytic converter in around 30% of the vehicle fleet [1] and because the poisoning of catalytic converters with sulfur still present in Mexican gasoline. In January 2012 the Mexican Center for Environmental Rights [2] reports 38,000 death produced by atmospheric pollution between 2001 and 2009.

European regulations [3,4] are now focused on reducing greenhouse gases, above all carbon dioxide ($CO_2$) and, subsequently, modifications in vehicles have been done in order to diminish it; particularly the consumption of diesel is enhanced. The diesel, successfully, diminishes $CO_2$, nevertheless the NOx and PM emissions increase and the use of conventional catalytic converters is not adequate [5].

The problem of toxic gases emission persists and the use of alternative technologies becomes essential. One promising alternative is the use of non-thermal plasma as it can be depicted in next paragraph.

In the branch of automotive pollution control several reactor configurations have been proposed (i.e. cylindrical pulsed corona [6,7], packed pellet cylindrical configuration [8], dielectric discharge with flat electrodes [9]). All these configurations show a good removal of $NO_x$ and PM; the packed cylindrical configuration has a relative better efficiency certainly due to a major plasma-polluted gas interaction, however in this type of reactors the exhaust gas pressure diminishes when the gas collides with a barrier (pellets), this fact implies an augment in consumption of gasoline in vehicles.

In this work a double dielectric barrier in a cylindrical configuration (2DBD) reactor is proposed to reduce gaseous emission of vehicles. The double dielectric allows a better distribution of plasma in the reactor; therefore a higher residence time of toxic gas was expected without diminishing the gas exhaust pressure.

## 2. Basic concepts

The chemistry in plasmas in a non-thermodynamic equilibrium state or commonly named cold plasmas are mostly based on radicals reactions, obtained from electronic collisions. For the chemistry model in a double dielectric barrier reactor, a detailed work was done in [10]. In this case the kinetic model used to represent the exhaust gas treatment was divided in two phases.

During the first step, it is considered an energetic (500 Td [11]) and quasi instantaneously ($1\times10^{-8}$ s) streamer head is formed, with an initial electron density of 1 $cm^{-3}$ [12-14]. In this phase two mechanisms for active species formation were considered: dissociation by direct electron impact with gas molecules and quenching of the excited states such as $O(^1D)$ and $N_2(A)$. Some primary radicals ($OH^\bullet$, $O^\bullet$, $N^\bullet$) appear during this phase. The radicals production is supposed uniformly distributed within the whole volume. The formation of a streamer channel connected to the streamer head is considered as the second phase; where NOx, propane and $CO_2$ react with active species.

The reaction mechanisms of the NOx-$C_3H_8$ and $CO_2$ degradation are respectively showed in figure 1(a) and (b). These mechanisms takes into account thirty three species ($e^-$, $N_2$, $O_2$, $H_2O$, $O^\bullet$, $N^\bullet$, $OH^\bullet$, $H^\bullet$, $O_2(a^1\Delta g)$, $O(^1D)$, $N_2(A)$, $O^-_2$, $O^-$, $O^+_2$, $H^-$, $O_3$, $NO_2$, $NO_3$, $N_2O_5$, $N_2O$, $HNO_3$, $HNO_2$, $C_3H_8$, i-$C_3H_7$, $(CH_3)_2CHO_2$ , $(CH_3)_2CHO$, $CH_3C(O)CH_3$, $CO_2$, CO, $CO^{2-}_3$,COOH, HCOOH,C).

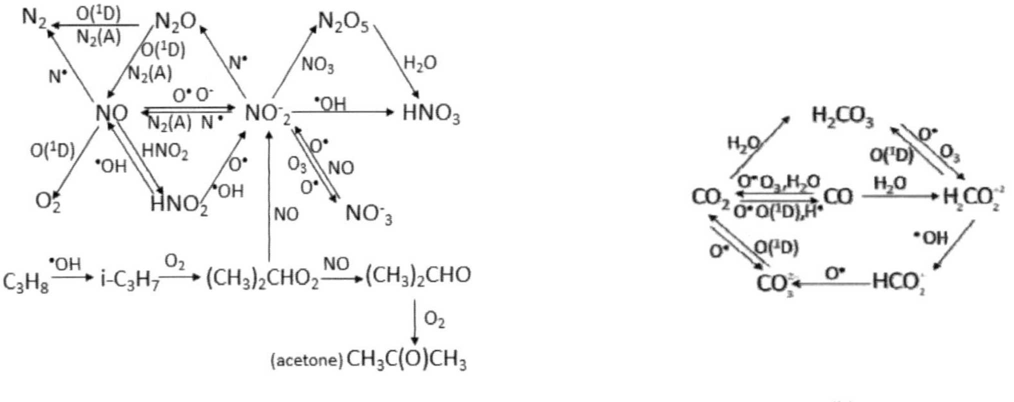


(a) (b)

**Figure 1.** Chemical mechanism proposed (a) NOx-$C_3H_8$ conversion, (b) $CO_2$ conversion.

### 3. Experimental setup

The analysis of the gas exhaust in a 4-cylinder 2002/Z16SE motor (figure 2(a)) was realized under several conditions of acceleration (expressed in revolutions per minute- rpm) directly bounded with the opening of the butterfly valve and the weight transfer, denoted by the change in the load borne during acceleration. These conditions simulate different situations during vehicle driving, notably the acceleration and the increasing load during a hill climb, for example.

The gas exhaust analysis was realized with PG-250 and IR-FGA400XDS sensors that respectively detect NOx, $SO_2$, $CO_2$, CO, $O_2$ and Total Hydrocarbons. In relation to the quantification of propane, it cannot be effectuated; a qualitative analysis was only completed by employing Fourier transform infrared spectroscopy (FTIR) equipment.

Concerning the gas treatment by plasma, in figure 2(b) the experimental setup is schematized, it consists in a tube of Pyrex glass with an internal radius of 11.1 mm, a length of 75 mm and a thickness of 1.2 mm. A stainless steel concentric electrode of radius of 2.54 mm is set inside the Pyrex tube and a tube of alumina is wrapping it. A metallic mesh covering the Pyrex tube plays the role of the external electrode. The ideal mixture simulating the gas exhaust of the motor studied is continuously introduced into the 2DBD reactor at a gas flow of 3 liters per minute. An applied power ($P_A$) of less than 30W was applied to maintain the plasma. The outlet gases from the reactor were analyzed with the PG250 and a mass spectrometer RGA. An electrical diagnosis was also realized in order to calculate the consumed power (Pc) and power efficiency ($P_E$).

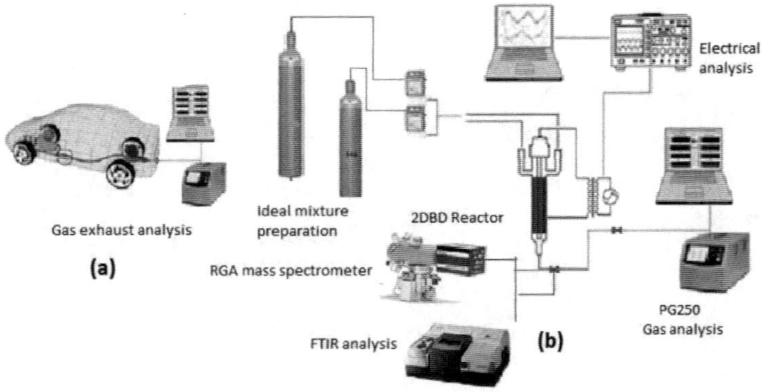


**Figure 2.** Experimental setup. (a) Exhaust gas characterization (b) Characterization of toxic gases treatment by a 2DBD.

The composition of the gas mixture to be treated with the plasma reactor is detailed in table 2. This composition was taken from results obtained during the analysis of the motor exhaust gases (reported in next section).

**Table 1.** Mixture of gas treated with plasma.

| Compound | Concentration |
|---|---|
| $C_3H_8$ | 1200ppm |
| CO | 4% |
| $CO_2$ | 12% |
| NO | 3000ppm |

## 4. Results

### *4.1. Motor emissions characterization*

Several driving conditions in the motor are reported in figure 3. Generally, an augmentation of noxious emissions is obtained with the increase of rpm. This could be explained by the fact that a higher number of rpm is the result of opening the butterfly valve and then, a higher air alimentation producing the oxidation of the gases due to higher temperatures and higher concentrations of oxygen.

In general, the concentration of NOx varies from 30 ppm to 2000 ppm, the CO augments from 2500 to 5000 ppm, the total hydrocarbons (HC) concentrations vary from 500ppm to 1200ppm and the rang covered by $CO_2$ concentrations is from 5%v.v to 15% v.v. Experimental errors fluctuates from 2 to 4.5%. All these concentrations are more important at higher rpm and at higher weight transfer.

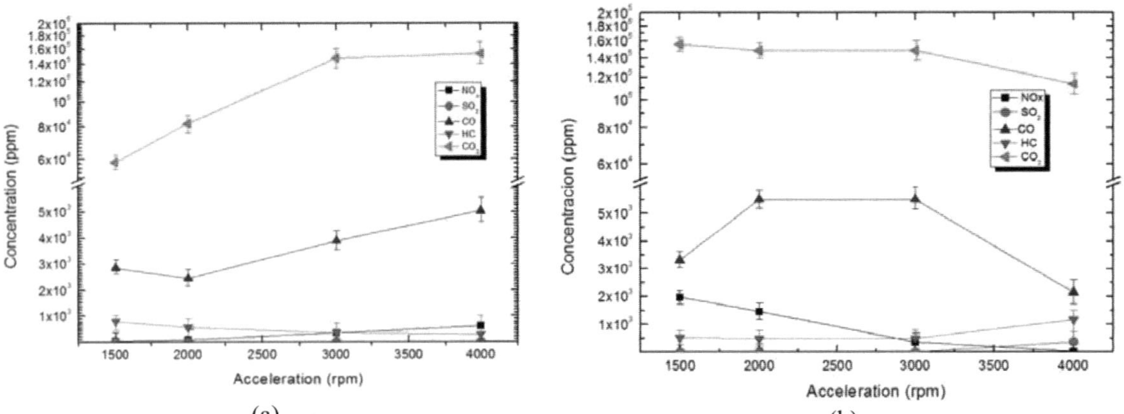


**Figure 3.** Motor conditions (a) without load, (b) with load.

When weight transfer is bigger (figure 3(b)) at 4000 rpm an unusual phenomena can be observed; the $CO_2$ and CO concentrations diminish, the oxidation of nitrogen and sulfur compounds to $NO_x$ and $SO_2$, seems to be favored. At higher temperatures the Zeldovich mechanism could be dominant in chemical reactions giving preference to the NOx formation [15]. For the $SO_2$ formation it is shown that the increase in temperature corresponds to a decrease of equilibrium constant values as result of reactions in direction of the reactants formation [16].

In both graphics presented in figure 3, the $SO_2$ concentration is not as much as 15ppm, except at 4000 rpm and higher weight transfer, where the concentration attains 372ppm. In order to simplify the experiences the $SO_2$ was not taken into account.

### *4.2. Plasma reactor to treat noxious emissions*

Several experimental conditions were tested; the better frequency of operation fluctuates at around 16 kHz. The applied power varied from 10 W (1 A, 10 V) to almost 30 W (1 A, 27 V).

A characteristic diagram of current and voltage in function of time is reported in figure 4. A relatively stable and continuous discharge with some microdischarges could be appreciated in the current curve. The discharge is a bit more stable than these found in a single dielectric barrier as was reported by Valdivia et al [17].

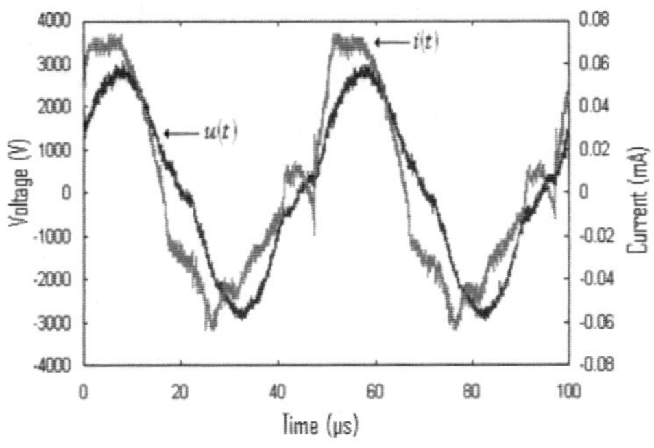


**Figure 4.** Voltage and current signals.

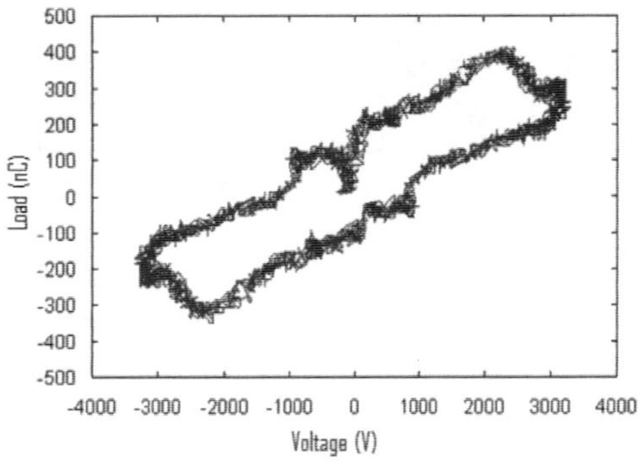


**Figure 5.** Lissajous diagram.

The Lissajous diagram depicted in figure 5 is representative of the dielectric barrier discharge that includes both the phase shift between the voltage and current signals and the accumulated charge in the dielectric of the reactor glass wall. This dielectric material prevents both the effect of electric arc and the microdischarges in the plasma. The electrical power $P_C$, was calculated; this oscillates between 7 and 23 W. After obtaining these values, and considering the power $P_A$, can finally achieve the efficiency $P_E$, as reported in table 2.

**Table 2.** Influence of electrical parameters into gas treatment

| Frequency (kHz) | $P_A$(W) | $P_C$ (W) | $P_E$ (W) | $NO_x$ removal (%) | CO removal (%) | $CO_2$ removal (%) |
|---|---|---|---|---|---|---|
| 16 | 10.30 | 8.18 | 79% | 94.54 | 43.99 | 30 |
| 16.03 | 10.40 | 9.43 | 83% | 96.9 | 15 | 25 |
| 16.8 | 27.26 | 23.7 | 86% | 96.24 | 37.64 | 15 |
| 16.82 | 20.85 | 17.95 | 86% | 98.03 | 15 | 30 |

From table 2, it can be observed that $NO_x$ removal is quite good (superior to 94%) at any condition, nevertheless the best removal was obtained at higher power efficiency.

For CO and $CO_2$ removal, the influence of the electrical conditions studied seems to be not conclusive; a possibly explanation of the oscillation in both concentrations is the competition for $O^•$ radical as observed in figure 1(b).

The FTIR spectra obtained are reported in figure 6; here, when plasma is applied (figure 6(b)), a diminution of the propane can be qualitatively observed at around 3000 $cm^{-1}$. Another characteristic band that appears in the products at around 1700 $cm^{-1}$ is characteristic of nitric acid. The formation of acids is corroborated by a conductometer; a concentration of 20-40 ppm of $HNO_3$ was found.

From mass spectrometry study (figure 7), several compounds in products were formed after the plasma treatment; for example, in figure 7(b) the line situated at 17 uma, corresponds to $O^{-2}$, $O^+$ and $OH^-$. Line at 30uma depicts the $N_2$ and NO. It could be observed that acetone is also formed, reinforcing the mechanism schematized in figure 1(b) and also proposed by Dorai [18], however the propionaldehyde founded in [20] is not obtained at our experimental conditions, possibly because the concentrations of radicals $^•OH$ are not enough.

## 5. Conclusions

The degradation of the chemical compound gas: NO, CO, $CO_2$, and $C_3H_8$ is successfully through chemical reactions and mainly free radicals like: $O^•$, $NO^•$ and $^•OH$, this compounds are produced basically through electronic collision with neutral molecule of water stream, oxygen and nitrogen. The degradation removal efficiency results for the automotive mixture depicts the feasibility to use in the future this technology of non-thermal plasma for the treatment of automotive toxic gas pollutants by taking care of remove the acids formed by using simply means, like activate charcoal.

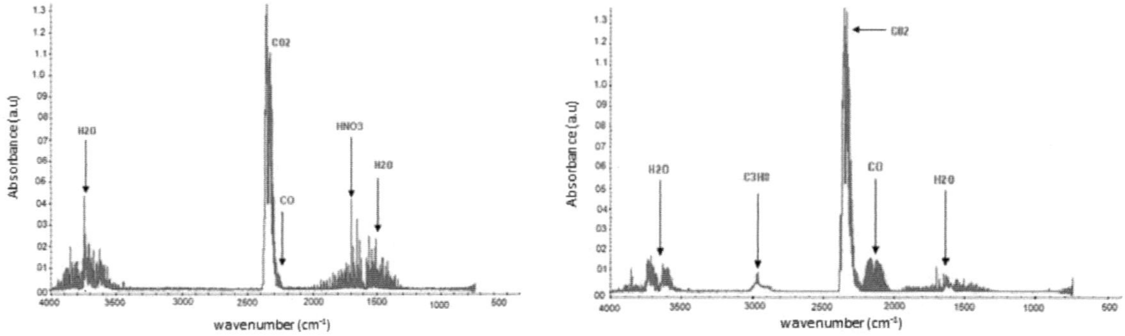


**Figure 6.** FTIR analysis of (a) non treated gas mixture and (b) after treatment products.

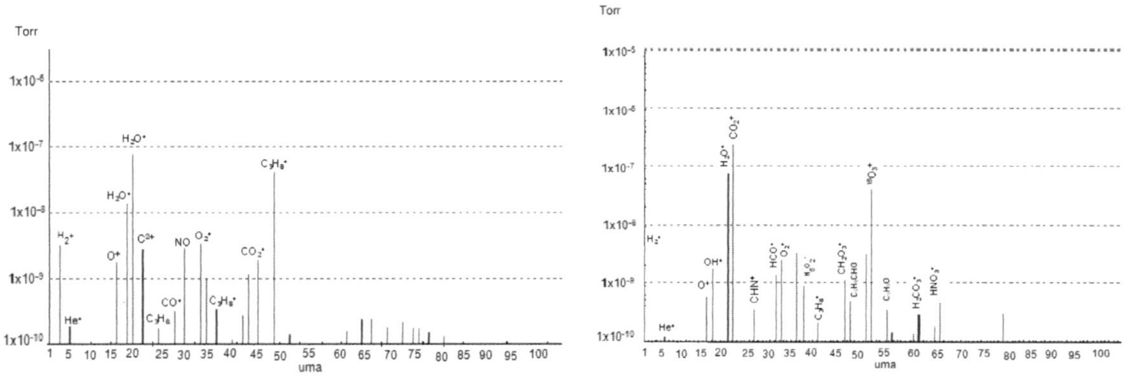


(a) (b)

**Figure 7.** Mass spectra of automotive emissions (a) without plasma treatment (b) with plasma treatment.

## Acknowledgments

We are very indebted to F. Ramos, M. Durán, M. Hidalgo, M. García, G. Soria, for the experimental support. The presentation of this work was supported financially, COMECYT–Gobierno del Estado de México 2009CO2-131982 and by CONACYT for the scholarship CVU 225878.

## References

[1] SEM 2008 *Inventario de emisiones de contaminantes tóxicos de la ZMVM* 49-50 SMA Gobierno del DF Mexico
[2] INE 2011 *Cuarto almanaque de datos y tendencias de la calidad del aire en 20 ciudades mexicanas 2000-2009* SEMARNAT Mexico
[3] Kaspar J, Fornasiero P and Hickey N 2003 *Catalysis Today* **77** 419-49
[4] Mierlo J, Maggetto G and Lataire P 2006 *Energy Conversion and Management* **47** 2748-60
[5] Twigg M 2007 *Appl Cat. B: Env.* **70** 2-15

[6] Qianli W, Chaochen M, Lei Z and Jing H 2008 *SAE Technical Paper* 2008-01-1793
[7] Xing Y, Liu Z, Couttenye R, Willis W, Suib S, Fanson P, Hirata H and Ibe M 2008 *J. Cat.* **253** 28-36
[8] Matsui Y, Sato S, Takashima K, Katsura S and Mizuno A 2003 *SAE Technical Paper* 2003-01-1185
[9] Kim Y, Naito K, Fujikawa H, Ogawa T, Tan I, Hasegawa K and Tanaka H 2007 *SAE Technical Paper* 2007-01-0236
[10] Estrada N 2010 PhD Thesis *Estudio de un reactor de doble barrera dieléctrica y su aplicación en el tratamiento de efluentes gaseosos* Instituto Tecnológico de Toluca Mexico
[11] Kim Y, Woo S, Hong S and Young H 2000 *Annual Report*, Seoul National University 151-742
[12] Fridman A 2008 *Plasma Chemistry* (Cambridge University Press) **1** 23-63
[13] Lowke J and Morrow R 1995 *IEEE Trans. Plasma Sci.* **23** 661-70
[14] Pacheco M, Pacheco J, Moreno H, Mercado A, Díaz J and Yousfi M 2007 *Plasma Sci.Tech.* **9** 682-5
[15] Amneus P, Mauss F, Kraft M, Vressner A and Johansson B 2005 *SAE Technical Paper* 2005-01-0126
[16] Kuburovic M, Duric S, Jovovic A and Karan M 2002 *Thermal Sci.* **6** 71-9
[17] Valdivia R, Pacheco J, Pacheco M, Benítez J and López R 2006 *Plasma S. Sci. Tech.* **15** 237-45
[18] Dorai R 2000 PhD Thesis *Modeling of plasma remediation of NOx using global kinetic models accounting for hydrocarbons* University of Illinois USA

Journal of Physics: Conference Series **370** (2012) 012024 doi:10.1088/1742-6596/370/1/012024

# Pulsed dielectric barrier discharge for *Bacillus subtilis* inactivation in water

**A N Hernández-Arias[1], B G Rodríguez-Méndez[2], R López-Callejas[1,2], R Valencia-Alvarado[2], A Mercado-Cabrera[2], R Peña-Eguiluz[2], S R Barocio[2], A E Muñoz-Castro[2] and A de la Piedad Beneitez[1]**

[1] Instituto Tecnológico de Toluca, AP. 890, Toluca, México
[2] Instituto Nacional de Investigaciones Nucleares, AP. 18-1027, 11801, México

E-mail: benjamin.rodriguez@inin.gob.mx

**Abstract**. The inactivation of *Bacillus subtilis* bacteria in water has been experimentally studied by means of a pulsed dielectric barrier discharge (PDBD) in a coaxial reactor endowed with an alumina dielectric. The plasma source is capable of operating at atmospheric pressure with gas, water or hybrid gas–liquid media at adjustable 25 kV pulses, 30 μs long and at a 500 Hz frequency. In order to evaluate the inactivation efficiency of the system, a set of experiments were designed on the basis of oxygen flow control. The initial data have showed a significant bacterial rate reduction of $10^3$-$10^7$ CFU/mL. Additional results proved that applying an oxygen flow for a few seconds during the PDBD treatment inactivates the *Bacillus subtilis* population with 99.99% effectiveness. As a reference, without gas flow but with the same exposure times, this percentage is reduced to ~90%. The analysis of the relationship between inactivation rate and chemical species in the discharge has been carried out using optical emission spectroscopy as to identifying the main reactive species. Reactive oxygen species such as atomic oxygen and ozone tuned out to be the dominant germicidal species. Some proposed inactivation mechanisms of this technique are discussed.

## 1. Introduction

Bacteria are omnipresent. They are found in all natural habitats and some species have been found in boiling hot spring as well as extreme cold condition [1]. A number of aerobic endospore-forming bacterial species are opportunistic pathogens of animals (including humans). These bacteria are widely distributed in the environment where their habitats are soils of all kinds, ranging from alkaline to acid, cold to hot, as well as bottom deposits of water. Thus, the term "aerobic endospore-forming bacteria" is used to embrace *Bacillus* species and related genera, capable of forming resistant endospores. *Bacillus subtilis* (*B. subtilis*) is an endospore-forming Gram-positive bacterium, extremely dormant metabolically. This dormancy is the key to their resistance to many agents, including heat, radiation, pressure, desiccation and chemicals, hence their survival over long periods [2].

Some pathogen bacteria contaminate solid or liquid food, and particularly drinking water, originating diseases to the point of being the main cause of human death [3]. Clearly, applying some proportionately effective disinfection methods becomes indispensable. Conventional water disinfection techniques generally involve the use of hazardous chemicals that lead to large waste streams that must be eventually treated, in order to avoid adverse environmental impacts. By contrast, non-thermal plasmas used for inactivation of harmful bacteria and pollutants are recently drawing

Journal of Physics: Conference Series **370** (2012) 012024 doi:10.1088/1742-6596/370/1/012024

increasing attention in the study of biological and chemical decontamination [4]. There have been an increasing number of reports on the capability of non-thermal plasmas to inactivate a wide range of microorganisms, including Gram-negative and Gram-positive bacteria [5,6]. Most of the early investigations have used arc or corona discharges where the abrupt release of energy through a narrow plasma channel causes a so-called electrohydraulic shock wave. Apart from these high pressure waves, discharges in aqueous solutions initiate a number of effects that cause microbial inactivation: formation of chemical reactive radicals by electron impact ionization, photo dissociation as well as direct interaction of UV photons with the DNA of living microorganism [7]. In this case, the presence of dielectrics plays an important role in arcing prevention generating a uniform room pressure plasma discharge.

The *B. subtilis* species has received particular attention due to its highly resistive nature, endospore-forming capability and are believed to be among the most resistant microorganism. As a result, this type of bacteria has often been chosen for inactivation experiments due to its comprehensive characterization as a model organism for laboratory studies. Evidence of inactivation and/or complete removal of *B. subtilis* in water with $10^3$, $10^5$ and $10^7$ bacterial cells/mL has been found in the present study, using pulsed dielectric barrier discharge (PDBD) plasma at atmospheric pressure, with oxygen ($O_2$) as work gas, thus proving the efficacy of this process to inactivate bacteria.

## 2. Methods

### *2.1. Pulsed dielectric barrier discharge system*

Efforts have been specifically focused on bacterial inactivation by direct plasma application in a water environment via PDBD at atmospheric pressure. Figure 1 shows the experimental layout including the adjustable 100-1000 Hz power supply whose pulse width is 1-50 μs, operating at 1-30 kV. The reactor has been made out of a stainless steel tube of 1.25 mm inner diameter, operating as a cathode, and a 0.1 mm diameter tungsten wire which has been adapted as anode. The wire has been covered with alumina ceramic as dielectric. Total length of the reactor is 300 mm, including a work gas inlet.

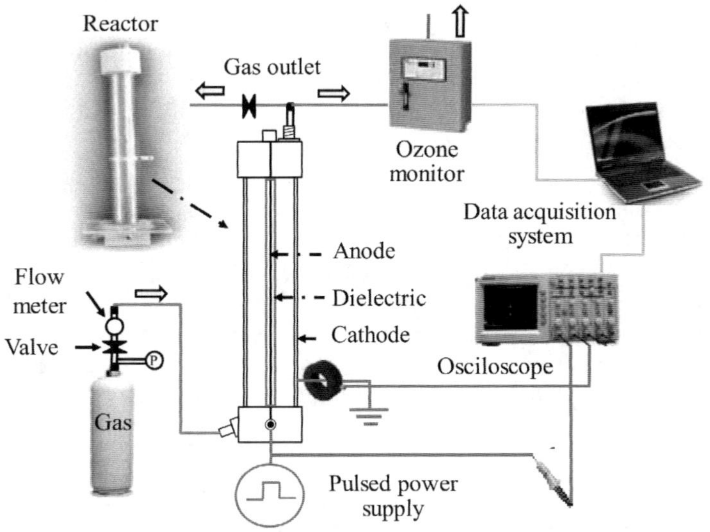


**Figure 1.** Experimental layout for pulsed dielectric barrier discharge in a water environment.

Voltage, current and power are measured using an oscilloscope (Tektronix TDS-2024) connected to a voltage divider (Tektronix P6015) and a current transformer (Stangenes 2-0.1W). Main plasma

Journal of Physics: Conference Series **370** (2012) 012024 doi:10.1088/1742-6596/370/1/012024

diagnostics include spectroscopy of the light emission from the plasma (Jaz OceanOptics) and ozone monitoring (Teledyne Instruments 460L) interfaced by means of a personal computer.

### 2.2. *Bacteria culture*

5 mL of Tryptone Soya Broth (TSB) solution was inoculated with a strain of *B. subtilis* and incubated for 24 h at 37°C until the bacterial concentration reached approximately $10^7$ bacterial cells/mL. Then, the solution was twice centrifuged for 10 minutes at 5000 rpm. The sedimented bacteria were put in a 5 mL suspension of distilled and sterilized water. The final concentration was determined by direct phase contrast microscopy and then diluted to concentrations of $10^3$, $10^5$ and $10^7$ bacteria/mL.

### 2.3. *Plasma treatment*

The plasma treatment of the culture was carried out placing 15 mL of water containing $10^3$, $10^5$ and $10^7$ bacteria/mL controlled concentrations into the reactor, namely, between the dielectric and the cathode. The power supply was adjusted to 25 kV, 500 Hz and to a ~30 μs pulse width. Experiments were conducted in water during periods dependent on each concentration and in two modes: with and without gas flow. In the second case, the reactor gas inlet was closed generating a discharge in a heterogeneous media of air and water. In the first case, and in order to produce reactive oxygen species (ROS), oxygen gas was injected at 0.5 L/min flow rate in continuous mode for each treatment period, producing gas bubbles and allowing a continuous generation of ozone. Data acquisition of ozone concentration and analysis by means of optical emission spectroscopy (OES) along the process was performed in every experiment, after which, the microbiological analysis was completed.

Immediately after the PDBD treatment, the samples were removed from the reactor and transferred to sterile test tubes. A standard culturing technique was followed in order to practice the microbiological evaluation of the *B. subtilis* survivability. Culture in Petri dishes containing Tryptone Soya Agar (TSA) was implemented where a three dish set per sample was compounded by inoculating 100μL aliquots that were taken from the serially diluted bacterial suspension. The inoculated dishes, once labeled, were incubated for 24 h periods at 37°C. After incubation, colony forming units (CFU) were counted, and concentrations of CFU/mL were determined by colony counts and plotted as a function of the treatment time. The surviving fraction is determined by comparison between the number of surviving CFUs from the treated samples and those of the untreated sample.

## 3. Results

The electric discharges produce physical phenomena and reactive chemical species acting synergistically and thereby accelerating the elimination rate of microorganisms. In particular, molecular breaking down in water and gas media can generate powerful oxidation species such as the hydroxyl radical (OH) with a high potential to inactivate the bacteria [8], [9]. In order to investigate this possibility, we identify the chemical species present in the discharge applying OES.

Figure 2 shows the discharge emission bands with and without oxygen flow in the wavelength range of 270-870 nm. Figure 2(a) displays the respective emission spectra produced by the PDBD in $H_2O$ without $O_2$ flow. The main products appear as follows: hydroxyl radical (OH) in the 306.3 nm band ($A^2\Sigma^+ \rightarrow X^2\Pi$, 0-0) and the bands of the $N_2$ second positive system (337.1 nm, $C^3\Pi_u \rightarrow B^3\Pi_g$, 0-0), the $N_2^+$ first negative system (358.2, $B^2\Sigma_u^+ \rightarrow X^2\Sigma_g^+$, 1-0), the $N_2$ second positive system (380.4 nm, $C^3\Pi_u \rightarrow B^3\Pi_g$, 0-2), $H_\alpha$ (656.3 nm, $3d^2D \rightarrow 2p^2P^o$), and O (777.4 nm, $3p^5P \rightarrow 3s^5S^o$). Figure 2(b) shows the effects of the addition of $O_2$. The emission intensities of $H_\gamma$ (434 nm, $5d^2D \rightarrow 2p^2P^o$), $H_\beta$ (486.1, $4d^2D \rightarrow 2p^2P^o$), $H_\alpha$ (656.3 nm, $3d^2D \rightarrow 2p^2P^o$), O (777.4 nm, $3p^5P \rightarrow 3s^5S^o$), O (844.6 nm, $3p^3P \rightarrow 3s^3S^o$), are evident. A strong emission is observed in the $O_2^+$ first negative system (529.5, $b^4\Sigma_g^- \rightarrow a^4\Pi_u$, 2-0) and the OH radical in the band of 613.7 nm ($X^2\Pi$, 5-0) with a significant difference from the oxygen flow free case.

The ozone concentration was monitored during each experiment using an ultraviolet (UV) absorption ozone meter (Ozone Monitor model 460L, Teledyne Instruments™) tuned in to 253.7 nm. Data acquisition from the ozone monitor was processed on LabView™ software and plotted versus treatment time. Similar curves of ozone concentration in ppm were obtained from all experiments as shown in figure 3. The ozone concentration was saturated within 140-155 ppm in continuous oxygen flow while the ozone level, measured in the process without gas flow, remained under 10 ppm.

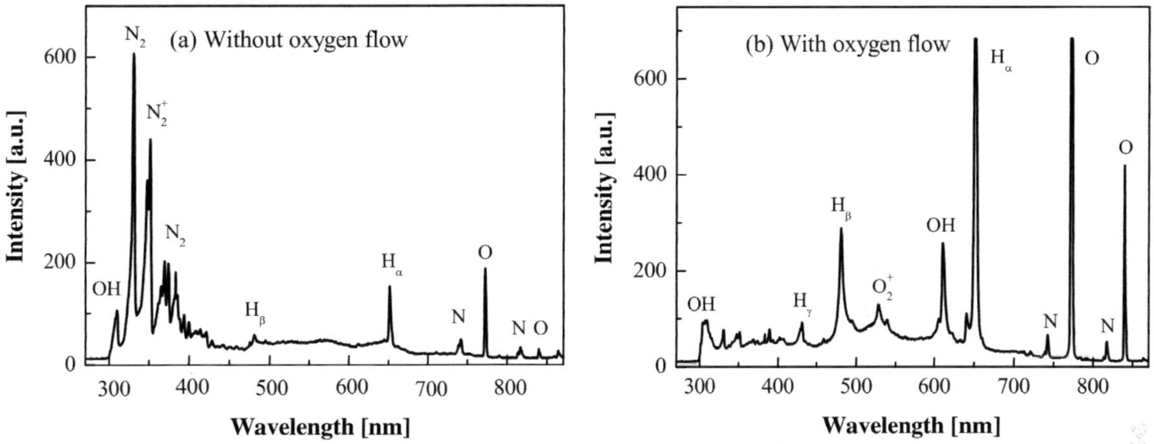


**Figure 2.** Optical emission spectroscopy of the in PDBD in a water environment: (a) without oxygen gas flow and (b) with continuous oxygen gas flow into the reactor.

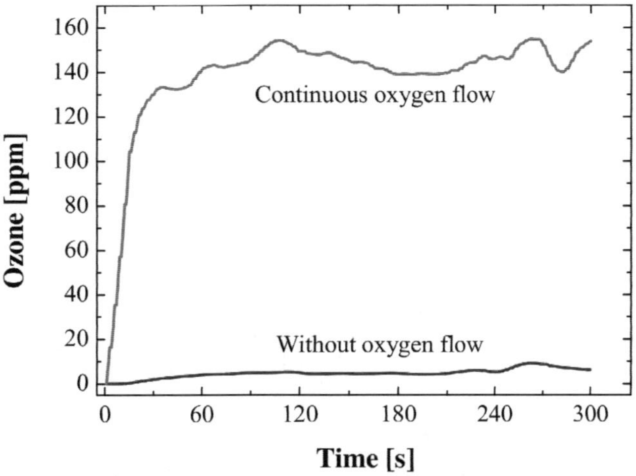


**Figure 3.** Curves of ozone concentration without oxygen flow and with continuous oxygen flow.

The aim of the experiment has been to determine the way in which the gas flow mode or the concentration of the ROS should shorten the inactivation time. A characterization of the inactivation phases of *B. subtilis* during the plasma treatment versus time was carried out. Initial experiments of *B. subtilis* inactivation were performed at $10^3$ CFU/mL for different treatment periods with and without oxygen flow. The quantitative results are presented in figure 4.

The survival curves in figure 4 indicate that the gas flow increased the efficacy of the process. In particular, figure 4(a) shows the survivability percentage of *B. subtilis* at a $10^3$ CFU/mL, where the pure discharge yields, without gas flow, a 6.785% of survivability after 120 s, whereas the addition of $O_2$ leads, over the same time interval, to 0.08% of survivability equivalent to a 99.92% inactivation. By contrast, without gas flow, the inactivation percentage is only 99.85% after 360 s. Figure 4(b) shows the survivability percentage at a $10^5$ CFU/mL concentration. In this case, 99.99% inactivation was achieved at 360 s in the gas flow mode, while 90.38% of inactivation was obtained in the same time and 99.30% after 600 s, both, without gas flow in the reactor. Samples at elevated concentrations as high as $10^7$ CFU/mL reached 99.85% inactivation after 1200 s with oxygen flow whereas 90% of inactivation was obtained during a similar period without gas flow, as shown in figure 4(c).

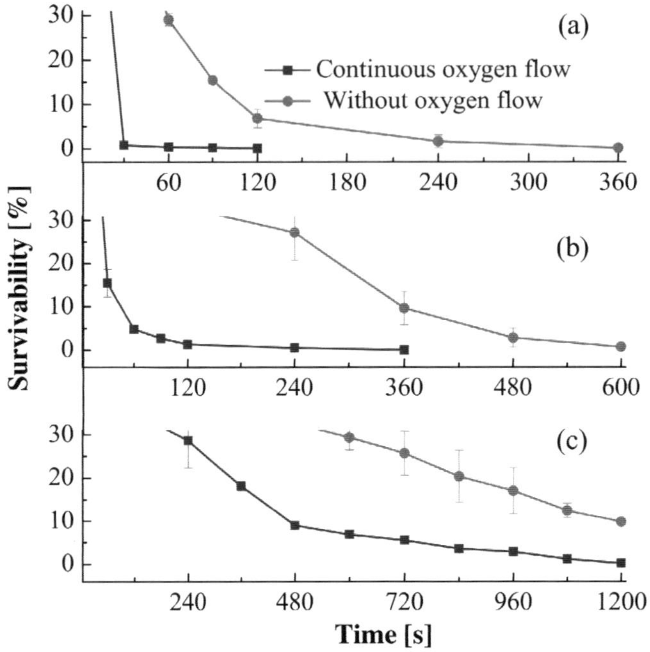


**Figure 4.** Survivability percentage of B. *subtillis* at (a) $10^3$, (b) $10^5$ and (c) $10^7$ CFU/mL versus exposure time.

We estimate that the pure discharge survival curve reflects inactivation by UV photons, mainly since no enough radicals from the gaseous phase are available. In contrast, when adding $O_2$, which partially dissociates to yield O, $O_2^+$, $O_3$, then both UV and ROS contribute to the inactivation. A logarithm plot of the number of surviving microorganisms as a function of exposure time is presented in figure 5.

According to figure 5, the process with $O_2$ gas flow injection raises the elimination efficacy in all the cases. Thus, a reduction ~ 4 $\log_{10}$ is attained at concentrations $10^3$ CFU/mL, and similar reductions are achieved at $10^5$ CFU/mL. At $10^7$ CFU/mL, a ~3 $\log_{10}$ reduction is obtained when $O_2$ gas is present. Finally, if concentrations of $10^7$ CFU/mL are chosen, around 1 $\log_{10}$ reduction demands 1200 s of treatment without oxygen gas flow. As expected, higher reduction requires even longer treatment periods.

## 4. Discussion

The inactivation of *B. subtilis* can be caused either by an individual or synergistic effect of plasma components on the microorganism. We believe that the difference between processes presented above is due to the effect of the reactive species generated by the addition of $O_2$ to the discharge. The reactive species can be atomic and molecular radicals, for instance O and $O_3$, and excited molecules or ions, for example, the $O_2^+$. Then, the subsequent gas discharge radicals, $O_3$, high-energy electrons, and UV radiation act synergistically to inactivate the bacteria. On its part, in the absence of reactive species, only intrinsic photodesorption and rupture of the cytoplasmic membrane due to the electrostatic force can take place. Therefore, a longer treatment time is required for bacterial inactivation, according to the mechanism presented in [7].

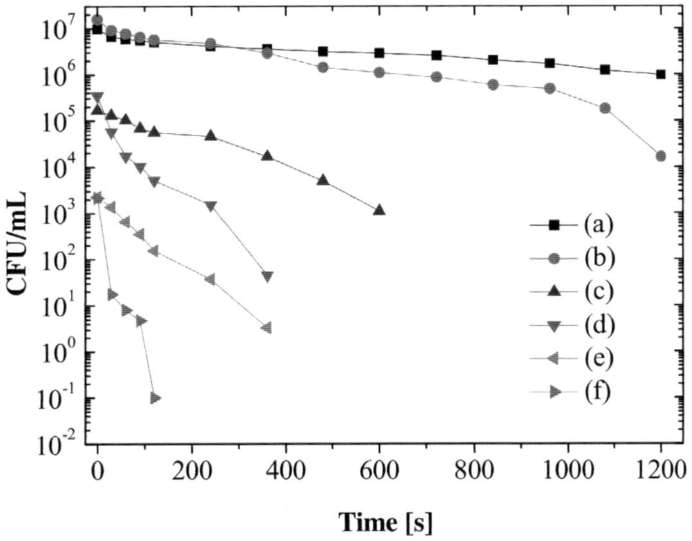


**Figure 5.** Inactivation kinetics curve of the *Bacillus subtilis* by PDBD at: (a) $10^7$, (c) $10^5$, and (e) $10^3$ CFU/mL without gas flow; (b) $10^7$, (d) $10^5$, and (f) $10^3$ CFU/mL in continuous oxygen flow.

The experimental results suggest that the addition of $O_2$ increased the effect of inactivation by ROS albeit generated OH radicals have an inactivation effect too. In order to understand this, we observed the difference between spectra in figure 2 in order to discuss the production processes of O and OH radicals under the added oxygen condition. The additional $O_2$ can be dissociated through reactions (1)-(3), while considerable proportions of O ($^1$D) and O atoms can be produced according to [8]:

$$e + O2 \rightarrow O- + O \quad (1)$$

$$e + O2 \rightarrow e + O + O\ (1D) \quad (2)$$

$$e + O2 \rightarrow O+ + O + 2e \quad (3)$$

The main pathway leading to OH radical formation by discharge is thought to be the result of the direct dissociation of water molecules by electron impact, leading to the generation of H and O as:

$$e + H2O \rightarrow e + H + OH \quad (4)$$

$$e + H2O \rightarrow 2H + O \quad (5)$$

However, it has also been reported that the generation of OH radicals increases along with the concentration of $O_2$ molecules [9]. Likewise, a lot of OH radicals can be produced by the interaction of O($^1$D) and $H_2O$:

$$O\ (1D) + H2O \rightarrow 2OH \quad (6)$$

$O_3$ is the primary species produced by an electrical discharge occurring in either air or oxygen. It is the only active species that can diffuse into the liquid phase and react with aqueous contaminants that are susceptible to degradation by direct attack [10] e.g., as molecular $O_3$. Ozone can also act indirectly, via radical-intermediates formed during $O_3$ decomposition in the aqueous media. OH radical is the most important species formed during $O_3$ decomposition:

$$H + O3 \rightarrow OH + O2 \quad (7)$$

$$O3 + OH \rightarrow HO2 + O2 \quad (8)$$

$$HO2 + O3 \rightarrow OH + 2O2 \quad (9)$$

These findings can be interpreted as active chemical species, being formed under gas flow, reacting with the environment and interacting with the cytoplasmic membrane of *B. subtillis*. Similar observations have been reported by Sato *et al.* [11] and Chen *et al.* [12]. Likewise, ROS and OH radicals are suggested to play a fundamental role in the bacterial inactivation and to cause physical destruction of the cytoplasmic membrane by oxidation after plasma exposure.

## 5. Conclusions

This report has proven that the pulsed dielectric barrier discharge technique can be successfully used in the inactivation of microorganisms in water, and, in particular, that its application is lethal to *Bacillus subtilis* even at high concentrations. The obtained results revealed the possible physical and chemical mechanisms of bacterial inactivation. Through optical emission spectroscopy and inactivation kinetics studies, it has been established that PDBD is effective to produce radicals in such a way that observed bacteria inactivation was most probably induced by the radicals, UV photons, and the electric field, acting synergistically. According to this, a relationship between the presence of the OH radical and the bacterial inactivation efficiency has been suggested. This finding is highly relevant for future applications of electric discharges to bacterial inactivation. Future experiments and simulations will focus on evaluating all contributors, and thus bringing forth more understanding on the mechanisms of bacteria inactivation in water.

## Acknowledgments

This work has received financial support from DGEST and CONACYT, Mexico. The authors are much obliged to I. A. Rojas O., M. T. Torres M., P. Angeles E., I. Contreras V. and M. Pacheco P. for their technical support.

## References

[1] Harder T, Lau S C K, Tam W Y and Qian P W 2004 *FEMS Microbiology Ecology* **47** 93
[2] Trivedi P C, Pandey S and Bhadauria S 2010 *Text Book of Microbiology* (Jaipur: Aavishkar Publishers) pp 92-4
[3] Ford T and Hamnere S 2010 *Environmental Microbiology*, ed R Mitchell and J D Gu (Hoboken: Wiley-Blackwell) pp 33-45
[4] Fujii T, Arao Y and Rea M 2008 *J. Phys.: Conf. Ser.* **142** 012070
[5] Gweon B, Kim D B, Moon S Y and Choe W 2009 *Current Applied Physics* **9** 625
[6] Guivan M M, Kamikozawa T, Kurokawa H, Motomura H, Kadowaki K and Jinno M 2010 *IEEE Trans. Plasma Sci.* **38** 1972
[7] Laroussi M 2005 *Plasma Process. Polym.* **2** 391
[8] Feng L, Wenchun W, Su W, Chunsheng R and Younian W 2005 *Plasma Sci. & Tech.* **7** 2851
[9] Ono R and Oda T 2008 *J. Phys. D: Appl. Phys.* **41** 035204
[10] Lukes P, Appleton A T and Locke B R 2004 *IEEE Trans. Ind. Appls.* **40** 60
[11] Sato M 2008 *Plasma Sources Sci. Technol.* **17** 024021
[12] Chen C W, Lee H M and Chang M B 2009 *Journal of Electrostatics* **67** 703

# Surface characteristics analysis of polypropylene treated by dielectric barrier discharge at atmospheric pressure

**T M C Nishime[1], A Toth[2], L R O Hein[1] and K G Kostov[1]**

[1] Engineering Faculty in Guaratinguetá – FEG, Universidade Estadual Paulista – UNESP, 12516-410, Guaratinguetá, SP, Brazil
[2] Institute of Materials and Environmental Chemistry, Hungarian Academy of Sciences, H-1025, Budapest, Hungary

E-mail: mayumi.nishime@gmail.com

**Abstract**. Polypropylene (PP) samples were treated by Dielectric Barrier Discharge (DBD) in order to modify their surface characteristics. The XPS analysis reveals that the DBD treatment added oxygen atoms to the PP surface. These polar groups cause increase in the wettability as shown by water contact angle measurements. The formation of low-molecular-weight oxidized materials (LMWOMs) in the form of small nodules on the PP surface was observed by atomic force microscopy (AFM). The presence of oxygen polar groups on the PP surface was also confirmed by infrared spectroscopy (FTIR). All analysis were performed before and after rinsing the treated samples in water and showed that the LMWOM can be removed from the surface by polar solvents.

## 1. Introduction

The polypropylene (PP) is a common engineering polymer widely employed in industry because of its good mechanical properties and low processing costs. It is used in several industrial fields, such as food packaging, protective films, conventional adhesives, printing, coating, etc. Many of these industrial applications require good adhesion properties [1,2]. However, due to its non-polar nature and low surface energy the PP exhibits poor adhesion and wetting [3]. Thus, in order to improve these characteristics many processes using different kind of plasmas have been applied. One type of discharge that is broadly investigated is so-called dielectric barrier discharge (DBD) [4].

DBD is a kind of electrical discharge, which occurs between two metal electrodes, providing that the space between them is partially filled with a dielectric material. It generates large amount of ions, energetic electrons, and exited reactive species under low temperature and high-pressure conditions [5]. Unlike other plasma reactors, DBD can treat materials in air without costly vacuum systems. This discharge demonstrates a remarkable flexibility with respect to its geometrical shape, operation parameters and working gas [6].

When polymers are treated by DBD, the samples are placed in the discharge gap where are exposed to the reactive species of the plasma. The ultraviolet (UV) photons and energetic electrons collide with the material surface breaking covalent bonds [2,3]. Afterwards, the free radicals created on the polymers surface can combine with oxygen atoms in the air thus increasing the polymer surface energy. However, the incorporation of new functional groups onto the materials surface has some limitations, one of them is the polymer surface degradation which results in the production of low-

Published under licence by IOP Publishing Ltd

molecular-weight oxidized materials (LMWOMs) [7]. LMWOMs are short highly oxidized oligomers weakly bonded to the polymers surface that agglomerate into globular features on the polymer surface and can be easily removed by washing with polar solvents [3,8].

This work reports the effect of DBD on polypropylene samples focusing on the surface modification and production of LMWOM during the treatment. The surface chemistry and structural modification of PP were analyzed with contact angle measurements, atomic force microscopy (AFM), x-ray photoelectron spectroscopy (XPS) and infrared spectroscopy (FTIR) before and after rinsing the treated films with water.

## 2. Experimental setup

### *2.1. Material*

A transparent 0.3-mm-thick PP film, provided by Goodfellow, Brazil, was cut into small pieces with 1.0x1.5 $cm^2$ size. Before plasma treatment the samples were cleaned using an ultrasonic cleaner with distilled water and isopropyl alcohol. After treatment some samples were rinsed 1 minute in distilled water, air dried for 1 hour and then subjected to further analysis.

### *2.2. Plasma treatment*

The treatments were performed in a 11.5-cm-diam parallel plate reactor. Both reactor electrodes were covered by 0.5-mm-thick polyester film and the distance between them was fixed at 3 mm. The upper electrode was grounded while the lower one was employed as high-voltage electrode. The high-voltage power supply is consisted of a step-up high-voltage transformer driven by a variac autotransformer. A load resistor of 1 kΩ protected the transformer in case of DBD to arc transition. All plasma treatments were carried out in air at atmospheric pressure and ambient temperature. The experimental arrangement is sketched in figure 1.

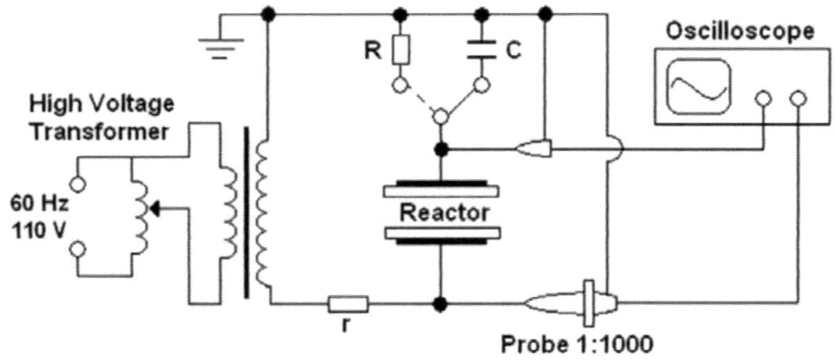


**Figure 1.** Schematic view of the DBD experimental setup.

The applied AC voltage was fixed at 35 kV peak to peak and was measured by a 1000:1 high-voltage probe (Tektronix – P6015A) and monitored on a digital oscilloscope (Tektronix – TDS 2024B). The treatments were performed at the line frequency of 60Hz. For displaying the waveform of the discharge current on the oscilloscope the reactor upper plate was grounded through a current measuring resistor of 68 Ω. The charge Q on the reactor plates was obtained from the voltage drop across a serial capacitor of 0.91 μF.

To obtain the mean electric power we employed the so-called Lissajous figure. It is obtained when plotting the transported electric charge as a function of the applied periodical voltage. In most cases the Lissajous figure is almost a parallelogram, whose area is equal to the electric energy consumed per voltage cycle. The power of DBD reactor is obtained by multiplying the consumed electric energy to the discharge frequency [5].

Treatments of PP films were performed for different time intervals to analyze the effects of plasma exposure on polymer surface.

### *2.3. Characterization techniques*

The treated samples were later analyzed with contact angle measurements, x-ray photoelectron spectroscopy (XPS), infrared spectroscopy (FTIR) and atomic force microscopy (AFM) to verify the changes on the PP surface. Some treated and washed (with distilled water) PP films were also analyzed to compare their structural modification with that of unwashed ones.

The contact angle between a test liquid and the polymer surface was obtained by the sessile drop method on a standard Rame-Hart goniometer, model 300-F1 using the DROPImage software. The measurements were carried out in controlled environment with 60% relative humidity and temperature of 21 °C. The volume of distilled water drop was 2 μℓ. At least five different drops were deposited on the polymer surface to obtain the average value of the contact angles.

Surface chemical characterization was carried out by x-ray photoelectron spectroscopy by a Kratos XSAM 800 spectrometer, using Mg K$\alpha$ radiation and fixed analyzer transmission mode (80 and 40 eV pass energies for the wide scan and detailed spectra, respectively). The spectra were referenced to the C 1s line (binding energy, BE=285.0 eV) of the hydrocarbon type carbon.

AFM, was used to analyze surface morphology of the polymer samples. The AFM analysis was performed in air with a Shimadzu, model spm9600-j3 equipment. During the analysis, the microscope was operated in tapping mode with a scanning rate of 0.5 Hz for all scanning sizes using etched silicon probe (k = 50 N/m). From the AFM images on 10.0x10.0 $\mu m^2$ areas the root mean square roughness ($r_{rms}$) of the topographic profiles was evaluated. For each sample several images on different spots were acquired and after that the average value of the roughness was calculated.

To analyze the changes of the surface chemical composition and the kind of chemical bonding Fourier transform infrared spectrometer PerkinElmer, model spectrum 100 was used. The spectrometer is equipped with a universal ATR accessory. The internal reflection element is a diamond crystal. The FTIR spectra were recorded using a resolution of 4 $cm^{-1}$ and 16 scans were taken for each measurement.

## 3. Results and discussion

### *3.1. XPS analysis*

The untreated PP sample exhibited a relatively narrow and symmetric C 1s peaks. In addition, a low intensity O1s peak was also detected at about 532.3 eV, probably due to slight surface oxidation, resulting in surface O-content of about 2 at%. XPS surface analysis performed on the DBD-treated PP samples revealed an elevated surface oxidation – intense O 1s photoelectron and O KLL Auger peaks could be detected in addition to the C 1s and C KLL peaks of carbon. Quantification of the detailed spectra showed a surface composition: C 80.4 at% and O 19.6 at%, resulting in an O/C ratio = 0.244.

Figure 2 presents details of C 1s and O 1s peaks of the PP sample treated for 10 min. The C 1s peak of the untreated PP is also shown for comparison. Synthesis of XPS C 1s and O 1s peak deconvolution is listed in table 1.

The XPS reveals that the DBD treatment added oxygen atoms to the PP surface in the form of single, double and triple oxidized carbon bonding.

### *3.2. AFM analysis*

The AFM images of PP surface, shown in figure 3, indicate the formation of globular structures whose size increases with the treatment time. The nodules are constituted by highly oxidized short polymer fragments, called in the literature LMWOMs [3]. These structures can be easily removed by washing in water, resulting in changes in the surface morphology and roughness.

In table 2 is presented the PP roughness before and after water rinsing the treated samples.

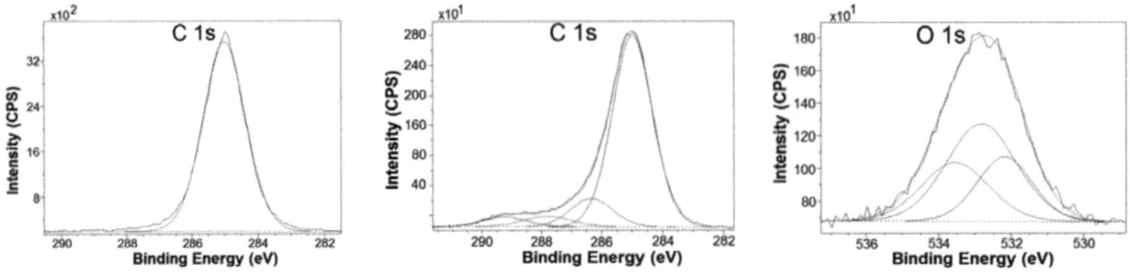


**Figure 2.** C 1s and O 1s peaks of DBD treated and untreated PP.

**Table 1.** Results from peaks synthesis for air DBD-treated PP.

| Peak | Component | Bonding mode | BE applied eV | Percentage at% |
|---|---|---|---|---|
| C 1s | C1 | C-C, CH | 285.0 | 62 |
| | C2 | C-OH, C-O-C | 286.4 | 10 |
| | C3 | C=O, O-C-O | 287.8 | 4 |
| | C4 | C(=O)OH | 289.3 | 4 |
| O 1s | O1 | C=O | 532.2 | 5 |
| | O2 | C-O-C, C-OH | 532.8 | 9 |
| | O3 | C(=O)OH | 533.6 | 5 |

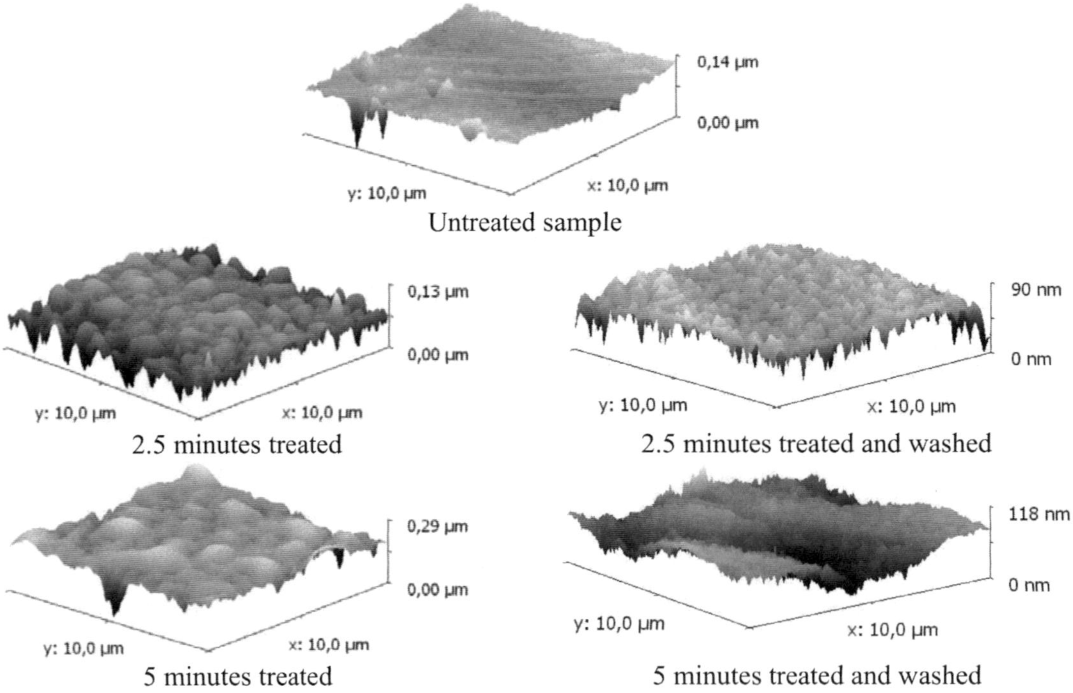


**Figure 3.** AFM images of polypropylene samples.

**Table 2.** Comparison of the PP roughness between washed and unwashed samples.

| | | Untreated | 2.5 minutes | 5 minutes |
|---|---|---|---|---|
| Rms (nm) | Unwashed | 5,1±0,9 | 13±0,4 | 20±0,6 |
| | Washed | | 10±0,2 | 6,1±0,4 |

### *3.3. Contact angle measurements*

The figure 4 shows that plasma treatment enhanced the PP hydrophilicity due to the introduction of polar oxygen-containing groups to the surface as evidenced by XPS analysis. Couple of minutes DBD treatment converted the initially hydrophobic PP surface into hydrophilic with contact angle around 65°. Further increase of the treatment time did not lead to significant changes in the PP wetting properties. This finding can be explained by the accumulation of LMWOMs on the surface that prevents the oxidation of deeper atomic layers. However, after washing the samples, the LMWOM was removed which resulted in smaller oxygen content on the surface and correspondingly higher water contact angle.

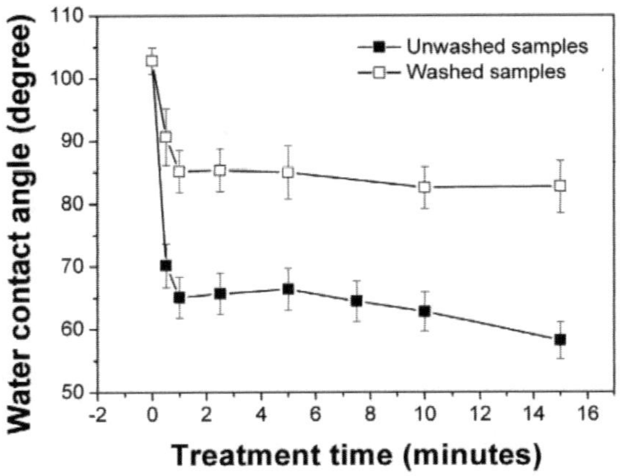


**Figure 4.** Reduction of the contact angle as a function of the treatment time.

### *3.4. Infrared analysis – FTIR*

The ATR-FTIR spectra of DBD-treated PP samples shown in figure 5 clearly show the appearance of some new features [9]. A very broad bond between 3100 and 3650 $cm^{-1}$ is attributed to the OH stretching vibration. A broad pick at 1750 $cm^{-1}$ appears owing to the presence of the carbonyl groups (C=O). The peak at 1280 $cm^{-1}$ is associated with COOH groups. After the surface washing in water these new features disappear turning the FTIR spectrum of the washed sample identical to the one of the untreated PP. These changes are corroborated by the AFM analysis, which showed the existence of nodular structures on the surface that were removed by washing. The spectrum of washed samples in figure 5 exhibits lost of oxygen containing chemical groups, which confirms the presence of LMWOM in these structures.

## 4. Conclusion

The DBD treatment of PP modifies both the surface chemical composition (shown by XPS and FTIR) and the surface morphology (AFM). When working in air, the DBD produces oxygen-containing functional groups, which are introduced in the PP chain turning the material more hydrophilic and increasing its roughness.

The AFM analysis of PP samples reveals an increase of the surface roughness with increasing the time of plasma exposure. This is caused by the formation of short highly oxidized polymer fragments (LMWOMs) that tend to agglomerate into nodule-like structures. However, the PP samples washed in water did not fully recover their original wetting properties. This suggests that oxidation also occurred at deeper and more permanent levels of the samples.

There was also observed that after washing the treated samples in water they lost part of their hydrophobicity because the LMWOMs were dissolved, which also affected the surface roughness.

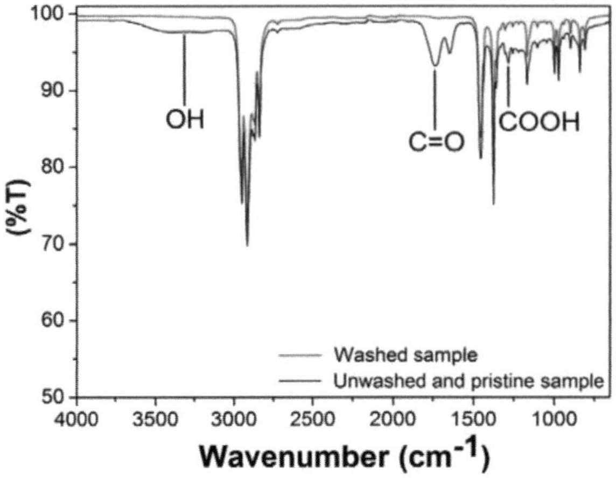


**Figure 5.** Comparison of the PP infrared spectra.

**Acknowledgements**

The authors acknowledge financial support from CNPq under the grants 146411/2010-2 and 507445/2010-4. Partial financial support from the OTKA-NKTH K-67741 project (Hungary) is acknowledged.

**References**

[1] Leroux F, Campagne C, Perwuelz A and Gengembre L 2008 *J. Colloid Interface Sci.* **328** 412
[2] Kwon O J, Tang S, Myung S W, Lu N and Choi H S 2005 *Surface & Coatings Technology* **192**
[3] Sellin N and Campos J S C 2003 *Materials Research* **2** 163
[4] Fang Z, Lin J, Yang H, Qiu Y and Kuffel E 2009 *IEEE Transactions on Plasma Science* **37** 5
[5] Kostov K G, Honda R Y, Alves L M S and Kayama M E 2009 *Brazilian Journal of Physics* **39** 2
[6] Wagner H E, Brandenburg R, Kozlov K V, Sonnenfeld A, Michel P and Behnke J F 2003 *Vacuum* **71** 417
[7] Guimond S and Wertheimer M R 2004 *J. of Applied Polymer Science* **94** 1291
[8] Hare L A O, Leadley S and Parbhoo B 2002 *Surf. Interface Anal.* **33** 335
[9] Morent R, Geyter N, Leys C, Gengembre L and Payen E 2007 *Surf. Interface Anal.* **40** 597

Journal of Physics: Conference Series **370** (2012) 012026 doi:10.1088/1742-6596/370/1/012026

# Deactivation of *Escherichia coli* in a post-discharge chamber coupled to an atmospheric pressure multi-electrode DBD plasma source

**V H Pérez-Ruiz[1], R Peña-Eguiluz[2], R López-Callejas[1,2], A Mercado-Cabrera[2], A E Muñoz-Castro[2], S R Barocio[2], R Valencia-Alvarado[2], A de la Piedad Beneitez[1] and B G Rodríguez-Méndez[2]**

[1] Instituto Tecnológico de Toluca, Av. Tecnológico S/N, A.P. 890, Toluca, Mexico
[2] Instituto Nacional de Investigaciones Nucleares, Plasma Physics Laboratory, A.P. 18-1027, C.P. 11801, México D.F., Mexico

E-mail: rosendo.eguiluz@inin.gob.mx

**Abstract.** Experimental results from applying a room pressure RF multi-electrode DBD plasma source to the inhibition of the population growth of Gram negative Escherichia coli (*E. coli*) within a post-discharge reactor are reported. The sample to be treated is deposited in the post-discharge chamber at about 50 mm from the plasma source outlet. Thus, the active species generated by the source are conveyed toward the chamber by the working gas flow. The plasma characterization included the measurement of the axial temperature at different distances from the reactor outlet by means of a K-type thermocouple. The resulting 294 K to 322 K temperature interval corresponded to distances between 10 mm to 1 mm respectively. As the material under treatment is placed further away, any thermal damage of the sample by the plasma is prevented. The measurement and optimization of the ozone $O_3$ concentration has also been carried out, provided that this is an active specie with particularly high germicide power. The effectiveness treatment of the *E. coli* bacteria growth inhibition by the proposed plasma source reached 99% when a $10^3$ CFU/mL concentration on an agar plate had been exposed during ten minutes.

## 1. Introduction

Sterilization by means of a room pressure glow plasma discharge is considered a highly effective method given its multiple advantages. Among them, one finds: the short duration required by its application, its great accessibility to almost any surface topology, as well as its target friendly operation temperature despite its microbial lethality [1].

Several studies have shown that room temperature RF plasmas are generally constituted by ions, free electrons, neutral particles as well as multiple reactive species such as molecular or atomic radicals. It has been long known that the reactive oxygen species (ROS) play an important role in microbial elimination [2]. Thus, in order to enhance the production of ROS such as ozone $O_3$, which actively interferes with the normal cell respiration process, some proportion of $O_2$ is added to the discharge working gas [3].

In the relative literature, the concept of bioburden refers to the relative number of viable individuals to be found at any time on the surface of an object or medium [4]. In the particular case of sterilized

Journal of Physics: Conference Series **370** (2012) 012026 doi:10.1088/1742-6596/370/1/012026

surgical instruments, the bioburden oscillates typically within $10^1$-$10^4$ UFC, once the instrument has been used. Furthermore, the bioburden levels after washing remain under $10^2$ UFC [4, 5]. The present study describes the capacity of a room pressure non thermal glow plasma discharge system to inhibit the growth of *E. coli* bacteria cultured on Petri dishes. This, with a view to decontaminating surfaces and, in due course, sterilizing surgical implements.

## 2. Experimental setup

Figure 1 displays the experimental array assembled around a portable cylindrical multi-electrode reactor, specifically designed to generate room pressure non thermal plasma discharges. The reactor is coupled to a stainless steel 135×140×95 mm box post-discharge chamber were the samples to be treated are placed. An ENI ACG-3B RF power supply, operating at 13.56 MHz, sustains the discharge through a PI matching network. A PC data acquisition system is associated to a 460L Teledyne Instruments monitor intended to measure the $O_3$ concentration in the discharge, while a type-K thermocouple provides a temperature profile with respect to the distance to the reactor outlet.

The cylindrical reactor is structured as an alternate succession of coaxial Pyrex glass tubes and copper mesh covers acting as electrodes, as seen in figure 1, resulting in a multiple capacitive system. Then, the injected 2 SLPM He and 0.066 SLPM $O_2$ gas mixtures follow a longitudinal flow between every pair of glass tubes with a return trajectory along every second pair of tubes, so as to play the role of a serpentine path. The tube lengths are 60 mm, 65 mm and 70 mm, their external diameters being 14 mm, 22 mm and 30 mm respectively. The effective distance between mesh electrodes is ~4 mm.

The material to be sterilized is placed at 50 mm from the reactor outlet, within the post-discharge chamber, in order to be better exposed to plasma reactive species. In order to validate the capability of the generated plasma to inhibit bacterial growth, an *E. coli* ATCC8739 bacterial strain was inoculated in 5 mL of Luria-Bertani (LB) medium and then incubated for 24 hours at 37°C. Later serial dilutions were carried out on a decimal basis until a $10^4$ CFU/mL concentration was achieved. Finally, sterile Petri dishes containing solid LB agar were inoculated with 0.1 mL of the solution and immediately placed in the post-discharge chamber to be exposed to the plasma species for 2, 4, 6, 8 and 10 minute periods. The treated dishes were incubated at 37°C during 24 hours before counting up the surviving bacterial colonies. Naturally, one untreated dish was included in every set for control purposes.

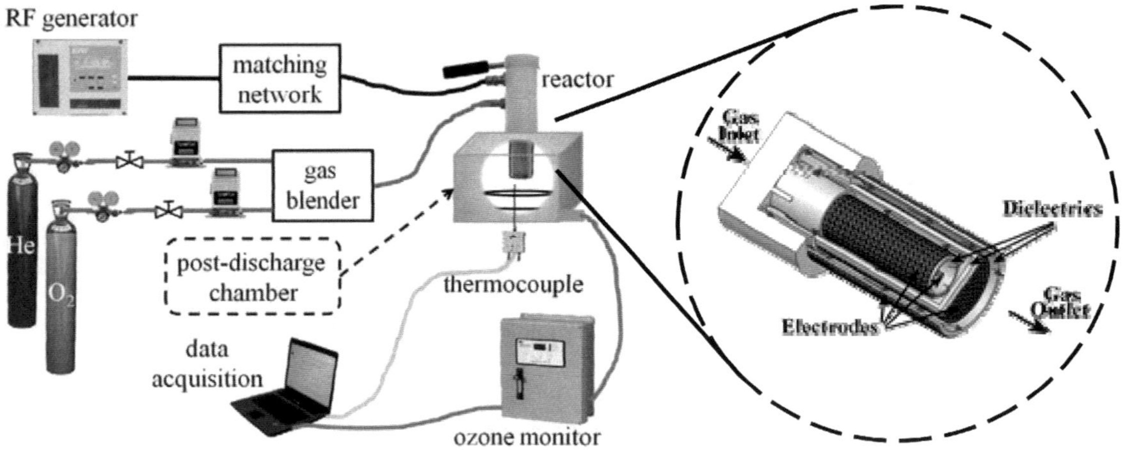


**Figure 1.** Experimental setup.

## 3. Results and discussion

*3.1. Ozone generation*

In order to maximize the $O_3$ volume produced within the reactor with a view to eliminating microbes from surfaces and instruments, ozone concentration was measured at different gas flows. The optimal combination, seen in figure 2, was identified to be 0.066 SLPM of $O_2$ added to 2 SLPM of He, namely, 3.3% of the helium flow, which comes closer to the 97 % He + 3% $O_2$ results published in [6].

*3.2 Thermal temperature at the reactor outlet*

A thermal temperature profile, shown in figure 3, has been constructed by measuring this variable outside the reactor at several distances from its outlet. To this purpose, a K-type thermocouple revealed a range going from 294 K at 10 mm to 322 K at 1 mm, proving that the proposed plasma facility is exempt from producing any thermal damage to the treated objects.

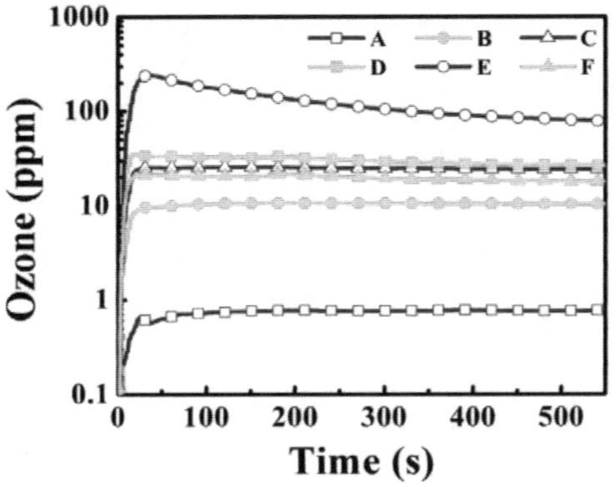


**Figure 2.** $O_3$ concentration generated by (A) He (2 SLPM), and its mixtures with: (B) $O_2$ (0.016 SLPM), (C) $O_2$ (0.033 SLPM), (D) $O_2$ (0.05 SLPM), (E) $O_2$ (0.066 SLPM), (F) $O_2$ (0.083 SLPM).

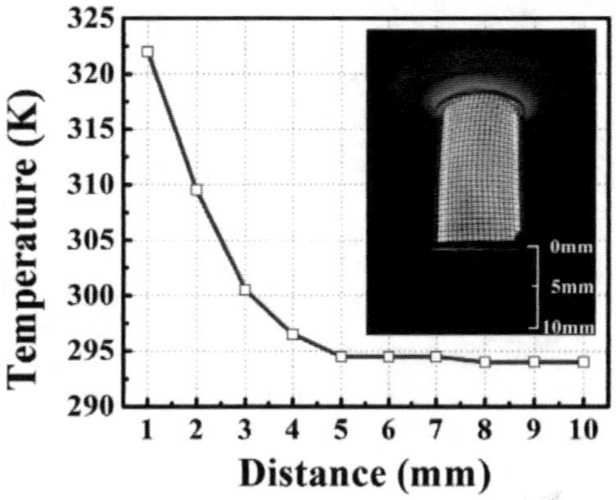


**Figure 3.** Thermal temperature profile with respect to the distance to the reactor outlet.

*3.3. Inhibition of E. coli colony growth*

Figure 4 exhibits the effects on the treated samples of the plasma discharge at several power levels and exposure times, as specified on table 1. Clearly, the number of surviving CFU diminishes as the exposure time increases.

## 4. Conclusions

The capacity of a room pressure glow plasma discharge produced in a multi-electrode DBD plasma source coupled to a post-discharge chamber to inhibit the growth of *E. coli* bacteria in the order of $10^3$ CFU/ml has been proven. Thus, the sterilization of surfaces without a direct contact with the plasma, and therefore at no risk of material structural damage, is feasible. The proposed plasma facility shows that thermal temperature is not necessarily involved in the elimination of the bacteria provided that samples can be successfully treated at 50 mm from the reactor outlet where the exposure temperature is ~294K, practically the same as the room temperature. Given the high oxidative power of ozone resulting in immediate cell damage, the $O_3$ concentration produced in the discharge has been maximized with purposes of biological decontamination.

**Table 1.** Results from inhibition of *E. coli* population growth.

| Applied power | 25W | | 30W | |
|---|---|---|---|---|
| Exposure time (min) | CFU | % elimination | CFU | % elimination |
| 0 | 1635 | 0 | 1537 | 0 |
| 2 | 519 | 68.25 | 667 | 65.77 |
| 4 | 434 | 73.45 | 41 | 97.33 |
| 6 | 111 | 93.21 | 29 | 98.11 |
| 8 | 92 | 94.37 | 7 | 99.54 |
| 10 | 20 | 98.77 | 0 | 100.00 |

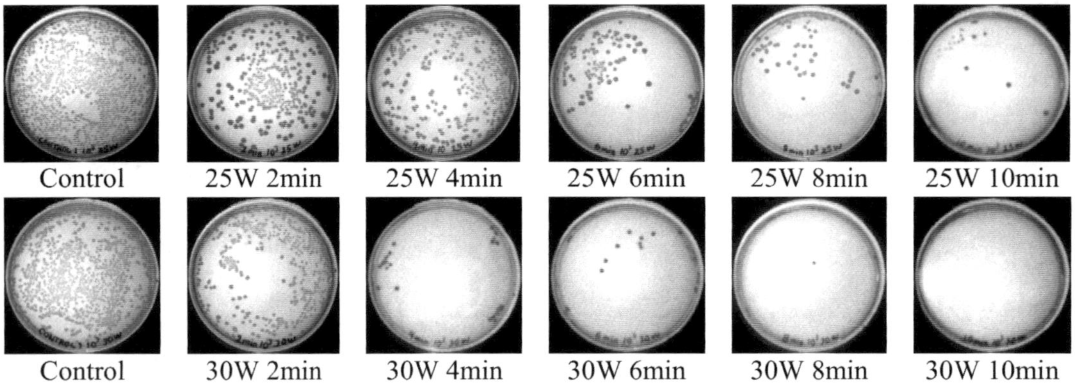


**Figure 4.** Deactivation of *E. coli* in agar by a plasma discharge at different power levels and exposure times.

**Acknowledgements**
This work has received partial financial support from CONACYT and COMECYT, México. The authors do appreciate the technical collaboration received from P. Angeles-Espinosa, M. Lugo-Hernández, M. T. Torres-Martínez and I. Contreras-Villa.

**References**

[1] Laroussi M 2009 *IEEE T. Plasma Sci.* **6** 714-25
[2] Deng X T, Shi J J and Kong M G 2007 *J. Appl. Phys.* **101** 074701
[3] Moisan M, Barbeau J, Moreau S, Pelletier J, Tabrizian M and Saoudi L'H 2001 *Int. J. Pharm.* **226** 1-21
[4] Rutala W A, Gergen M F, Jones J F and Weber D J 1999 *Infect. Control Hosp. Epidemiol.* **20** 514-6
[5] Pinto F G, De Souza R, Da Silva C, Mimica L M and Graziano K 2010 *Am. J. Infect. Cont.* **38** 229-33
[6] Laroussi M 2002 *IEEE T. Plasma Sci.* **30** 1409-15

Journal of Physics: Conference Series **370** (2012) 012027 doi:10.1088/1742-6596/370/1/012027

# Rutile titanium dioxide films deposited with a vacuum arc at different temperatures

**L Franco Arias[1,3], A Kleiman[1,3], E Heredia[2] and A Márquez[1,4]**

[1]Instituto de Física del Plasma, CONICET-UBA, Ciudad Universitaria, Pab. 1, 1428 Buenos Aires, Argentina
[2]Centro de Investigaciones en Sólidos (CINSO), CONICET-CITEDEF, J.B. de La Salle 4397, B1603ALO Villa Martelli, Argentina
[3]Fellow of CONICET
[4]Member of CONICET

E-mail: linafranco1@gmail.com

**Abstract**. Rutile crystalline phase of $TiO_2$ has been one of the most investigated materials for medical applications. Its implementation as a surface layer on biomedical implants has shown to improve hemocompatibility and biocompatibility. In this work, titanium dioxide coatings were deposited on glass and steel 316L substrates using cathodic arc deposition. The coatings were obtained at different substrate temperatures; varying from room temperature to 600°C. The crystalline structure of the films was identified by glancing angle X-ray diffraction. Depending on the substrate material and on its temperature during the deposition process, anatase, anatse+rutile and rutile structures were observed. It was determined that rutile films can be obtained below 600 °C with this deposition method.

## 1. Introduction

Titanium dioxide has been widely studied due to its outstanding physical and chemical properties. This compound is known to exhibit three main polymorphs: rutile, anatase and brookite. Rutile is the only thermodynamically stable phase at all temperatures, while anatase phase can be retained when it is nanostructured. Since its high refractive index, rutile films are used as optical coating in lenses and optical fibers. In the last few years, the most investigated topic on rutile thin films has been its implementation as a surface layer on biomedical implants; it has been found that these coatings improve hemocompatibility as well as biocompatibility [1,2]. It was demonstrated that when the structure of a $TiO_2$ layer changes from anatase to rutile, the dissolution rate of metallic ions in a simulated body fluid can be substantially decreased [3]. Rutile has been shown to be more chemically stable than anatase at both low and high pH levels [4].

$TiO_2$ films have been prepared by a great variety of deposition techniques such as sol-gel [5], chemical vapor deposition [6], evaporation [7], various reactive sputtering techniques [8], ion beam assisted process [9], atomic layer deposition [10], pulsed laser deposition [11] and cathodic arc deposition (CAD) [12]. In cathodic arc discharges, a metallic plasma jet generated from the cathode surface flies away from it, penetrating into the chamber volume, with energies typically in the range 30-100 eV. If a substrate is located facing the plasma jet, its surface is coated with a metallic film; the production of compound films being achieved when a reactive neutral gas flows into the discharge chamber. Phase formation during deposition by CAD is mainly determined by the substrate temperature and the ion energy, yielding amorphous $TiO_2$ at low temperature and low ion energy. Therefore, synthesizing crystalline $TiO_2$ by CAD requires either heating or biasing the substrate during deposition. According to Löbl diagram, rutile phase can be achieved when in the process the particle energy is higher than 10 eV at room temperature or beyond 800°C with thermal energies [13]. Crystalline $TiO_2$ coatings can also be attained by annealing amorphous films, requiring an annealing temperature higher than 800ºC for rutile transformation. However, this sort of treatment can generate tensions in the film that may cause cracking or even delamination.

The purpose of this work was to obtain rutile $TiO_2$ films crystallized directly in the growth stage, avoiding the subsequent annealing. The crystalline structure of films grown at different temperatures on steel 316L and glass substrates was studied.

## 2. Experimental details

The vacuum arc device used to synthesize the films, described in detail in a previous work [14], is schematically shown in figure 1. The arc was run with a current of 120 A between a Ti cathode and a grounded vacuum chamber serving as anode. The system was operated in an oxygen atmosphere with a continuous incoming gas flux of 63 sccm and a working pressure in the range of 2–5 Pa. Substrates were located at 30 cm from the cathode surface on a heater, which allowed fixing the temperature with an accuracy of 5%. $TiO_2$ samples were obtained at different temperatures ranging from room temperature to 570 °C. The temperature was measured with a thermocouple located on the front surface of the sample that was coated. The heater was electrically isolated of the chamber. Samples were exposed to the discharge during 2 min. Substrates were square pieces of 3×3 $cm^2$ of glass and of stainless steel 316L.

In order to evaluate the deposited mass, the samples were weighed before and after deposition using a Ohaus analytical balance, model AS200. The crystalline structure of the films was identified by X-ray diffraction (XRD) using a Philips PW 3710 diffractometer with a $CuK_\alpha$ radiation source. It was operated with glancing angle geometry by using a Philips thin film attachment, with an incidence angle of 1°.

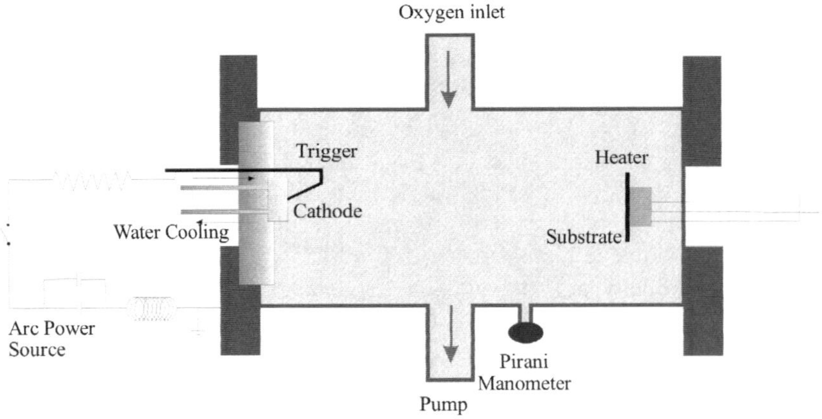


**Figure 1.** Schematic diagram of the employed CAD device.

## 3. Results and discussion

The deposited mass on the substrates was ~ 1.1mg. The film thickness was estimated to be ~ 300 nm, assuming a film density of 3.8 g/$cm^3$, this value corresponding to the density reported for amorphous and anatase films [15].

Figure 2 shows XRD patterns obtained from $TiO_2$ films deposited on glass substrates at 25, 360, 400, 450, 500 and 560°C. At room temperature the film grew amorphous. The peak associated to the

(101) plane of the anatase phase was visibly detected in the diffractograms corresponding to samples deposited at 360ºC, while those deposited at temperatures in the range from 400°C up to 500ºC presented both anatase and rutile phases. As the temperature increased the relative height of anatase peaks respect to rutile peaks diminished. In the case of samples obtained at 560°C only the presence of rutile structure was clearly registered. The broad elevation observed in all patterns for 2θ in the range 20-30 deg corresponds to amorphous glass.

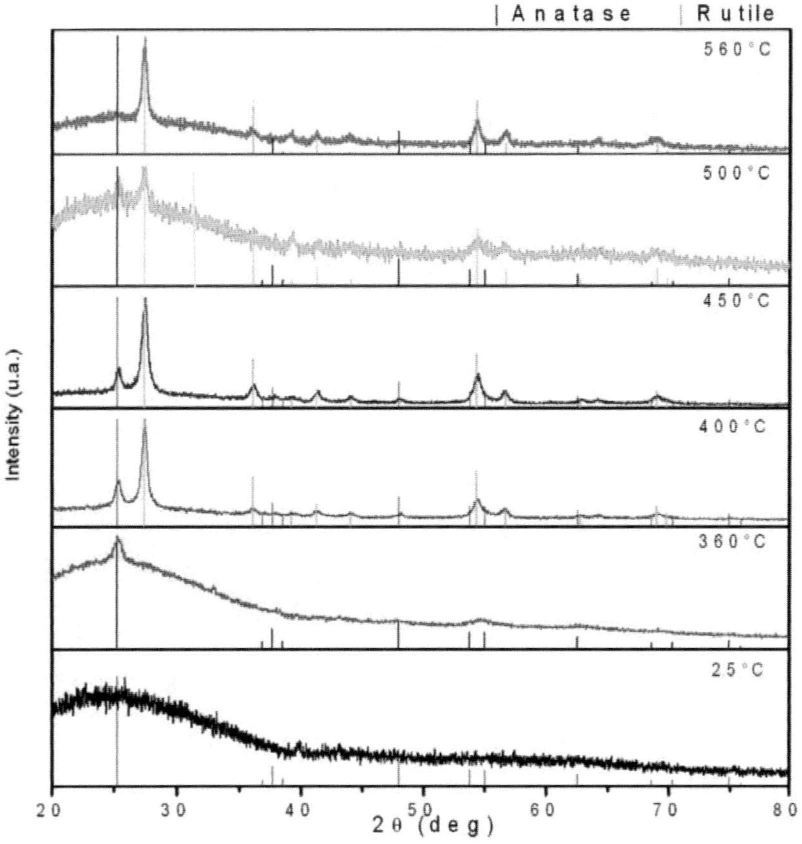


**Figure 2.** XRD patterns of $TiO_2$ films grown on glass substrates at different temperatures.

XRD patterns corresponding to films deposited on steel 316L are plotted in figure 3. In the same figure the diffractogram obtained from a non coated substrate is also shown. The phase transformations as functions of the process temperature observed from the diffractograms were very similar to the behavior described for the glass substrates. At room temperature just the substrate peaks were registered, indicating an amorphous film; at intermediate temperature the presence of both anatase and rutile phases was observed; and for the film grown at 570°C only the presence of rutile phase was evident. However, a difference came out comparing with glass substrate for films grown at 360ºC, on steel substrate not only anatase peaks appeared in the diffractogram, but also rutile peaks were visible.

Figure 4 shows the phase diagram proposed by Löbl *et al.* [13] for the synthesis of different $TiO_2$ structures depending on growth conditions: ion energy and substrate temperature. The working region corresponding to this work - taking into account the employed temperature range and plausible values for the ion energy - is indicated in the diagram. XRD patterns obtained from samples grown at 360°C revealed different structures for films deposited on different substrates.

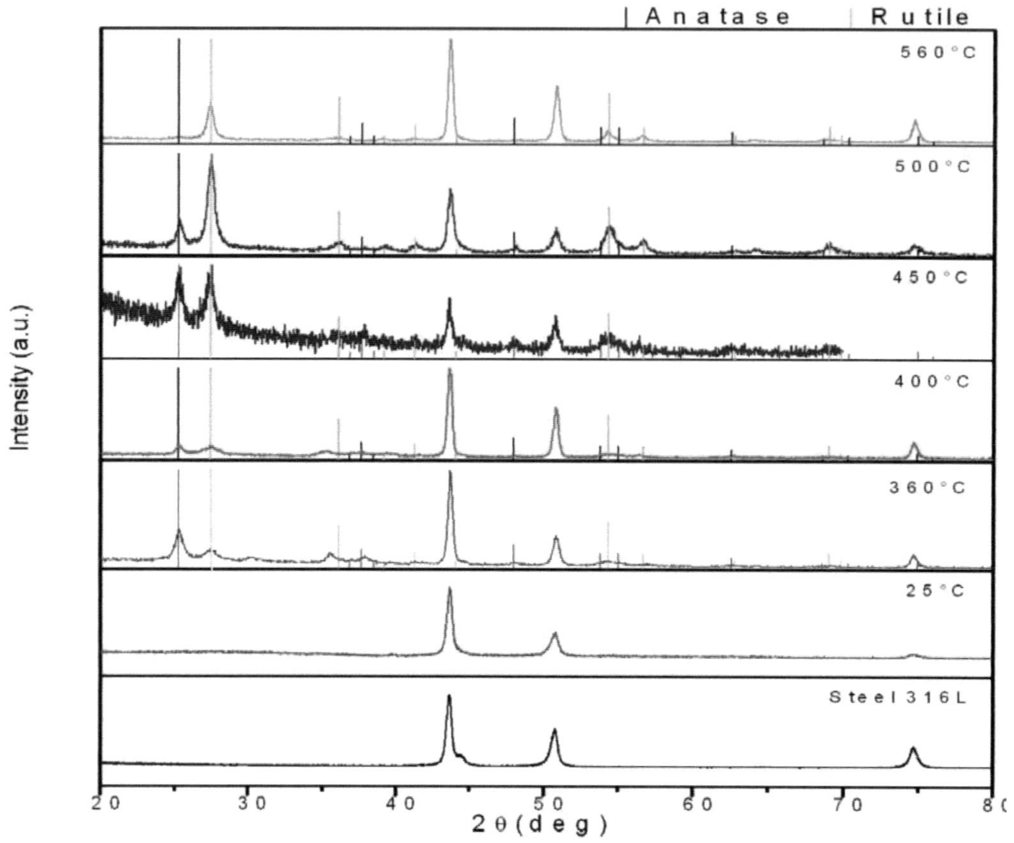


**Figure 3.** XRD patterns of $TiO_2$ films grown on steel 316L substrates at different temperatures.

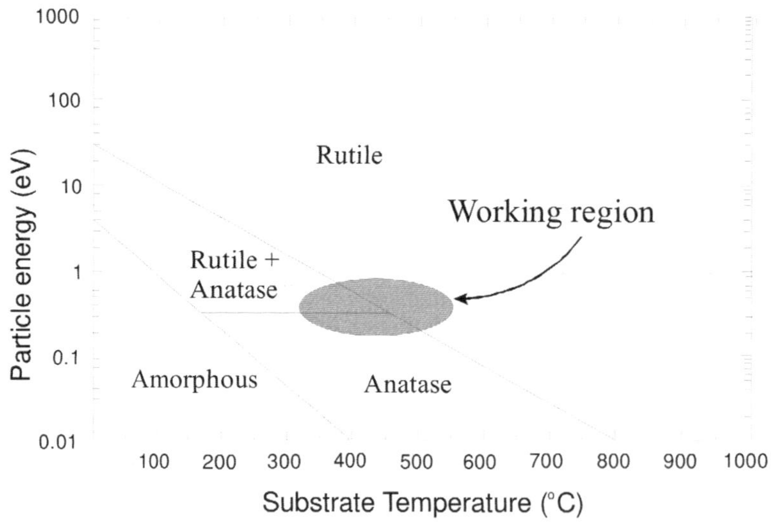


**Figure 4.** Phase diagram proposed by Löbl *et al.* [13]. The working region (temperature and energy ranges) is indicated.

Samples grown on glass presented only anatase phase, while rutile was also detected in films obtained on steel. This is in agreement with Löbl's diagram considering that the energy range of involved ions is close to – and likely includes – the threshold for obtaining anatase or anatase+rutile. Nonetheless, it seems rutile phase can be obtained at a lower temperature on steel than on glass substrates. This fact suggests that the electrical properties of the substrate affect on the energy of the ions arriving at the surface.

## 4. Conclusions

The obtained results indicate that cathodic arc deposition is a capable technique to synthesize rutile $TiO_2$ films on glass and steel substrates. Energetic particles provided by the plasma jet ejected from the cathode allow the growth of this phase without biasing and with in-situ heating at relatively low temperatures. By changing the substrate temperature employed in the process is possible to control the proportion of rutile to anatase in the film. A subsequent annealing that may cause delamination or cracking on the films was avoided.

**Acknowledgments**

This work was supported by grants from Universidad de Buenos Aires and CONICET.

## References

[1] Huang N, Chen Y, Luo J, Yi J, Lu R, Xiao J, Xue Z and Liu X 1994 *Biomater. J.Appl.* **8** 404-12
[2] Larsson C, Thomsen P, Lausmaa J, Rodahl M, Kasemo B and Ericson L 1994 *Biomaterials* **15** 1062-74
[3] Tsyganov I, Maitz M F and Wieser E 2004 *Appl. Surf. Sci* **235** 156-63
[4] Forberg S 1986 *Adv. Ceram.* **20** 321-7
[5] Sabate J, Anderson M A, Kikkawa H, Xu Q, Cervera-March S and Hill Jr C G 1992 *J. Catal.* **134** 36-46
[6] Williams L M and Hess D W 1983 *J. Vac. Sci. Technol.* **A 1** 1810
[7] Fujii T, Sakata N, Takada J, Miura Y, Daitoh Y and Takano M 1994 *J.Mater. Res.* **9** 1468-73
[8] Okimura K, Shibata A, Maeda N, Tachibana K, Noguchi Y and Tsuchida K 1995 *J. Jpn. Appl. Phys.* **34** 4950-5
[9] Gilo M and Croitoru N 1996 *Thin Solid Films* **283** 84-9
[10] Aarik J, Aidla A, Kiisler A-A, Uustare T and Sammelselg V 1997 *Thin Solid Films* **305** 270-3
[11] Yoon H S, Kim S K and Im H S 1997 *Bull. Kor. Chem. Soc.* **18** 640-3
[12] Kleiman A, Márquez A and Lamas D G 2007 *Surf. Coat. Technol.* **201** 6358-62
[13] Löbl P, Huppertz M and Mergel D 1994 *Thin Solid Films* **251** 72-9
[14] Márquez A, Blanco G, Fernandez de Rapp M E, Lamas D G and Tarulla R 2004 *Surf. Coat. Technol.* **187** 154-60
[15] Prasai B, Cai B, Kylee Underwood M, Lewis J P and Drabold D A 2012 Properties of amorphous and crystalline titanium dioxide from first principles *J. Mater. Sci.*, DOI 10.1007/s10853-012-6439-6

# Thin films generated by plasma immersion ion implantation and deposition of hexamethyldisilazane mixed with nitrogen in different proportions

**F V P Kodaira**[1], **R P Mota**[1], **V A Hills**[1,2], **R I Honda**[1], **M E Kayama**[1], **K G Kostov**[1] **and M A Algatti**[1]

[1]UNESP – Campus Guaratinguetá, Av. Dr. Ariberto Pereira da Cunha, 333, 12516-410 – Guaratinguetá - SP, Brazil
[2]School of Physics and Astronomy, The University of Nottingham, University Park, Nottingham, NG7 2RD, UK

E-mail: fevpk@yahoo.com.br

**Abstract.** In this work, it was used a plasma system composed of a cylindrical stainless steel reactor, a radio-frequency (13.56MHz) power source fixed at either 25 W or 70 W, a power source with a negative bias of 10kV and a 100Hz pulse. The system worked at an operational pressure of 80mTorr which consisted of varying concentrations of the monomer HMDSN and gaseous nitrogen in ratios: HMDSN (mTorr)/nitrogen (mTorr) from 70/10 to 20/60 in terms of operational pressure. The structural characterization of the films was done by FTIR spectroscopy. Absorptions were observed between 3500 $cm^{-1}$ to 3200 $cm^{-1}$, 3000 $cm^{-1}$ to 2900 $cm^{-1}$, 2500 $cm^{-1}$ to 2000 $cm^{-1}$, 1500 $cm^{-1}$ to 700 $cm^{-1}$, corresponding, respectively, to OH radicals, C-H stretching bonds in CH2 and CH3 molecules, C-N bonds, and finally, strain C-H bonds , Si-CH3 and Si-N groups, for both the 70 W and the 25 W.The contact angle for water was approximately 100° and the surface energy is near 25mJ/$m^2$ which represents a hydrophobic surface, measured by goniometric method. The aging of the film was also analyzed by measuring the contact angle over a period of time. The stabilization was observed after 4 weeks. The refractive index of these materials presents values from 1.73 to 1.65 measured by ultraviolet-visible technique.

## 1. Introduction

Material processing using low-temperature, low-pressure RF excited plasmas is essential in many current scientific and technological issues encompassing microelectronics, optical, electrical, biomaterial and others industries [1-12]. This feature of such plasmas gives rise to a very reactive chemistry in a relative cold environment whose kinetics is not easily controlled [13]. Plasma Immersion Ion Implantation & Deposition (PIIID) is a highly efficient technique for surface modification and thin film deposition. It can produce or modify metals, semiconductors and polymeric materials [14-18]. In this case, the materials are produced in a gas or vapor discharge and it is polarized simultaneously with high voltage pulses. Plasma ions are accelerated toward the films and implanted into their surfaces while the films are deposited.

In this work, PIIID process is used to obtain thin films from mixtures in various proportions of hexamethyldisilazane (HMDSN) and nitrogen at two power conditions, 25 and 70 Watts excited by

13.56 MHz and polarized by pulses of negative 10 KV in 100 Hz. The samples were analyzed by infrared, uv-visible and goniometric techniques.

## 2. Experimental setup

The PIIID occurs inside a cylindrical stainless steel reactor 21.5 cm in diameter and 24.5 cm in height, with pipes to evacuation, controlled gas admission, and optical observation. Inside the reactor, there are two parallel stainless steel discs shaped plain electrodes with a gap of 3cm between them. The upper electrode is powered by a 13.56 MHz radiofrequency power source for the deposition; the lower one is fed by a 100 Hz and 10 kV high voltage power source for the ion implantation. In this lower electrode is where the glass and aluminum substrates are placed.

The system is initially evacuated by mechanical and turbomolecular vacuum pump, reaching the pressure of $10^{-5}$Torr, cleaning the system. After that, the turbomolecular pump is switched off and the mechanical pump keeps the system at $10^{-3}$ Torr. Then the HMDSN enters the reactor by a precise needle valve until it reaches the required pressure, so the nitrogen enters the same way and completes the pressure. The aim is to vary the proportion of the HMDSN and the nitrogen from 10mTorr to 10 mTorr so that the sum of them both is equal 80 mTorr in each deposition and try them all in both 25 W and 70 W of RF power for 20 minutes. The molecular structure of the samples was investigated by FTIR analysis using a Perkin Elmer Spectrum 100. The refractive index was measured by a Perkin Elmer Lambda 25 UV/VIS Spectrometer. The contact angle, surface energy and aging of the films were analyzed by a Hamé-Hart 300 – F1 equipment. The aging of the film was also analyzed in a period of 4 weeks.

## 3. Results and discussions

Figures 1 and 2 illustrate the infrared spectra of PIIID – HMDSN/Nitrogen films deposited at 70 W and 25 W radiofrequency power supply respectively. However, the HMDSN and nitrogen concentration was modified in each implantation/deposition process. The most important vibration modes of the molecular structure of these thin films are presented in table 1.

**Table 1.** PIIID – HMDSN/nitrogen film infrared spectra vibration modes.

| Vibration modes | Absorption peak ($cm^{-1}$) |
|---|---|
| O-H stretching | 3400 |
| N-H stretching | 3350 |
| C-H stretching in $CH_3$ groups | 2960 |
| C-H stretching in $CH_2$ groups | 2900 |
| Si-H stretching | 2130 |
| C=O stretching | 1700 |
| N-H bending in NH2 groups / C=C stretching | 1650 |
| C=N / C=O bending | 1600 |
| CH3 bending in Si-$(CH_3)_x$ groups | 1230 |
| Si-O stretching | 1050 |
| Si-N stretching | 940 |
| Si-$(CH_3)_x$ rocking – stretching | 850 |
| Si-C stretching | 800 |

For 70 W or 25 W PIIID, the same vibrational modes were observed in the samples, independently of the HMDSN-nitrogen concentration in the plasma, as shown in the figures 1 and 2. On the other hand, the intensity of the same bands like Si-$(CH_3)_x$, Si-O, Si-N in the range 1500-900 $cm^{-1}$ was changed. This implies that the chemical structure of the films was modified by PIIID process. Although the HMDSN does not contain oxygen, it was observed in the molecular structure of the films. It can be attributed to the atmospheric or residual oxygen present inside the reactor during PIIID process that reacts with the free radicals in the samples structure. Si-O and Si-H vibration modes in the

films are due to three concomitants effects of ion implantation processing: chain breakage, unsaturation and crosslinking. It is known that the impact of energetic ions with a target material causes chemical bond breakages and promotes emission of atomic and molecular species [19].

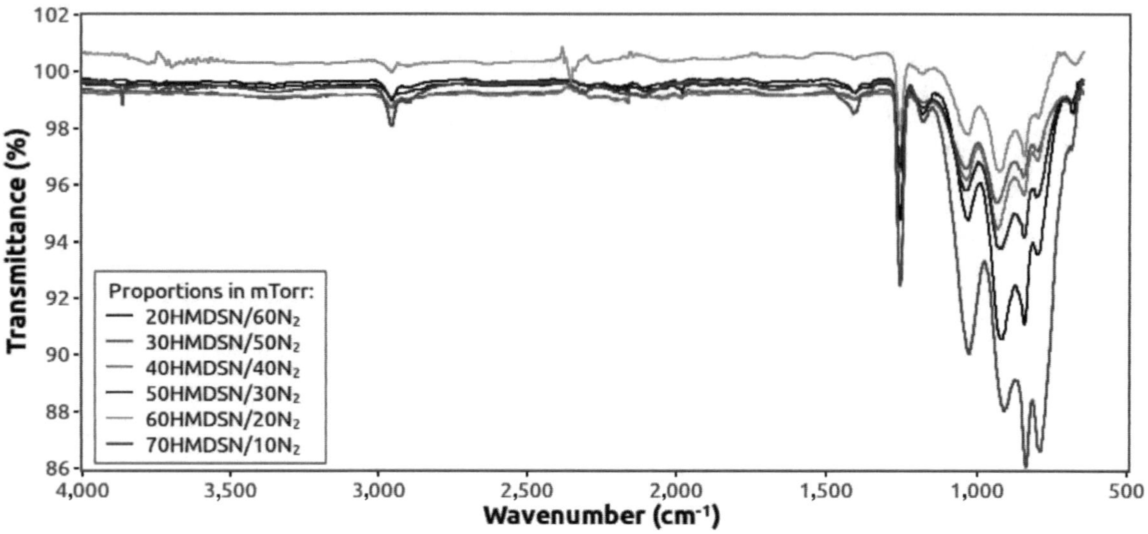


**Figure 1.** IR spectra of HMDS/Nitrogen PIIID films at 70 W.

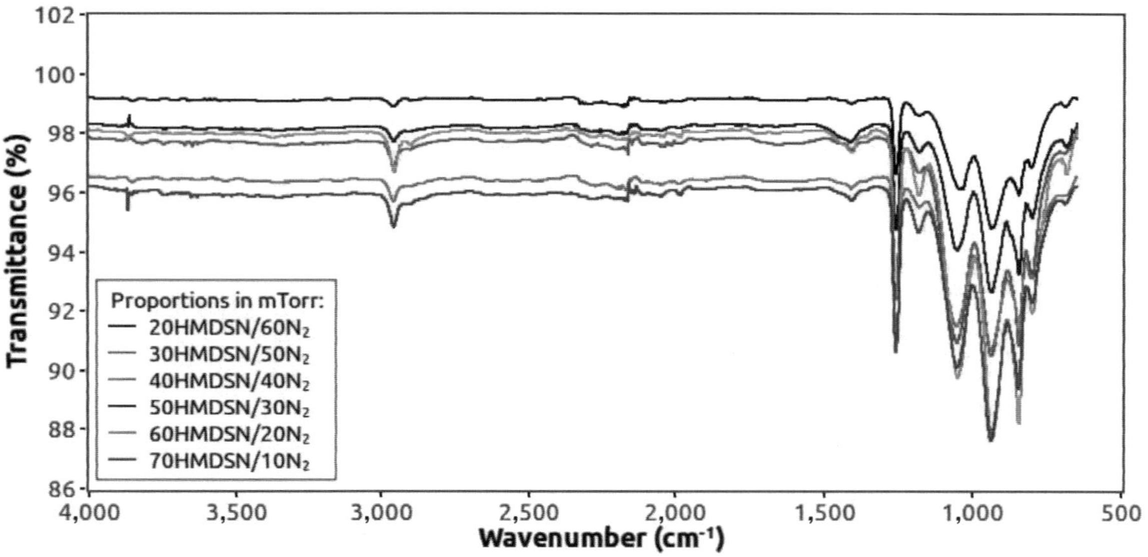


**Figure 2.** IR spectra of HMDS/Nitrogen PIIID films at 25 W.

In polymeric materials, hydrogen bonds are preferentially lost because they belong to chain termination and hydrogen atom progressive emission induces the appearance of dangling bonds. The dangling bonds can recombine through unsaturation process such as double bond between carbon atoms or between carbon atoms and other species present in the polymeric structure.

Figure 3 shows the behavior of the refractive index of the films obtained at 70 W and 25 W. At 70 W the rcfractive index decreases form 1.73 to 1.70 while it decreases from 1.69 to 1.65 for the samples obtained at 25 W.

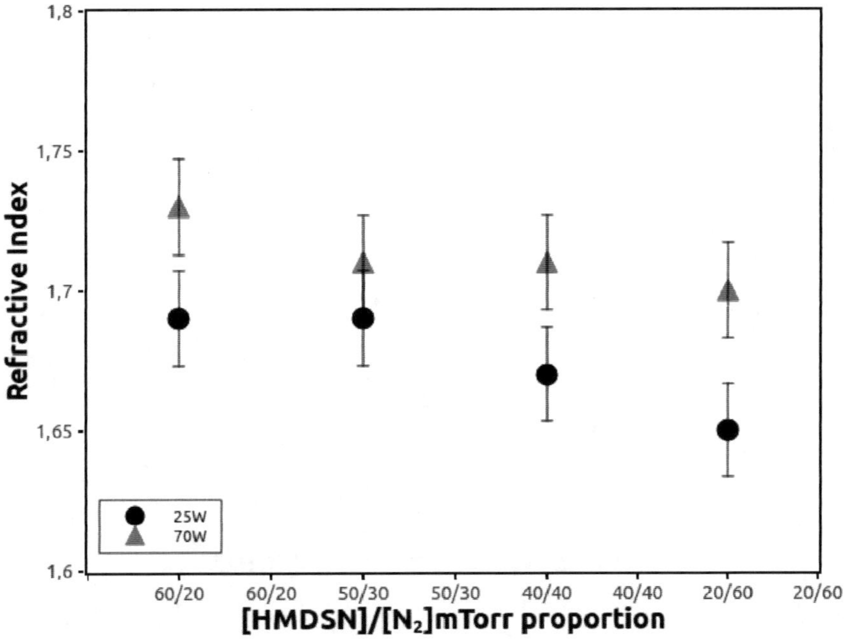


**Figure 3.** The refractive Index of the films obtained by PIIID at 70 W and 25 W in some different proportions of HMDS/Nitrogen.

This slight tendency in the behavior of the refractive index is the same in these cases, 70 W or 25 W. The PIIID process at 25 W produces films more colorless than at 70 W, which are slightly yellow. These alterations in the optical properties of the samples can be due to the random structure of the chains in a plasma polymer. Because the structures are random it makes theoretical interpretation of the experimental results rather difficult. However, it seems reasonable to suppose that some trends in the physical behavior of a "conventional" polymer must be followed by a HMDSN-Nitrogen PIIID polymer film. For example, the density of $\pi$ bonds is influenced by the degree of hydrogenation that depends in turn on the applied power and different concentration of HMDSN and nitrogen in the PIIID process. The detachment of hydrogen, for example, from the polymer also affects bond lengths, thus changing the refractive index. In both cases (70 and 25 W) the films thicknesses decrease from 100 to 55nm while the concentration of nitrogen was higher than HMDSN. In the PIIID process nitrogen can contribute to make chemical bonds to form the film and it can promote ablation to destroy the polymer and so to reduce the film thickness.

Figures 4 and 5 present the contact angle and surface energy as a function of HMDSN and nitrogen concentrations at 70 and 25 W PIIID process respectively.

Independently of the HMDSN and Nitrogen concentration, in both cases of the PIIID process at 70 or 25 W, the contact angle and surface energy presented the same behavior with very close values. As the contact angle is near 100 degree, the samples present hydrophobic character and it was not changed after aging effect over the films. A representative aging effect can be observed in the figure 6. The films obtained in this work result of a competitive process of polymerization and ablation. As was discussed first, these films are formed mainly by free radicals within their structures and surfaces. So, rapid recombination of atoms and molecules involving oxygen contributed to create polar groups and to saturate the surface bounds responsible for the hydrophobic character of the samples for a long time.

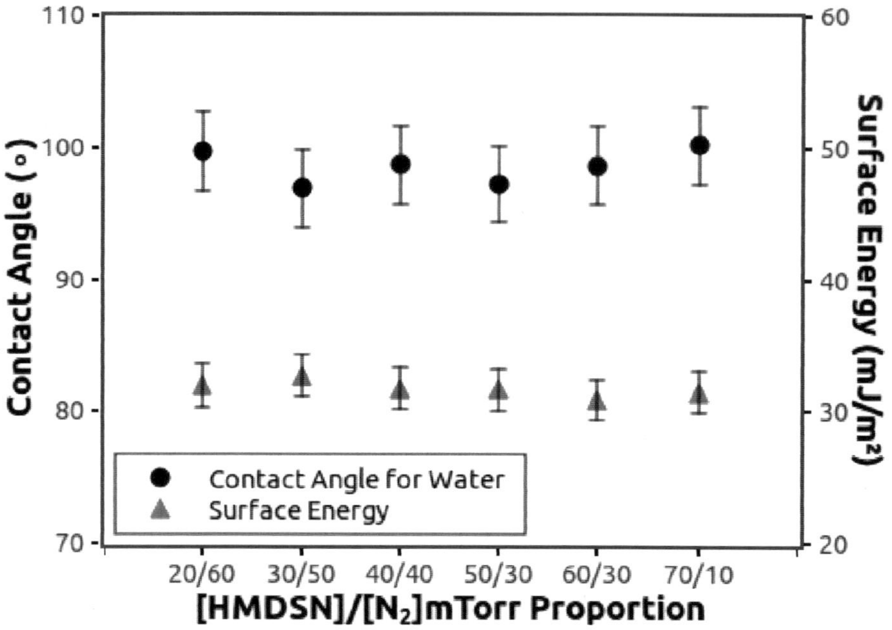


**Figure 4.** Contact angle and surface energy as a function of HMDSN and nitrogen concentration at 70 W PIIID process.

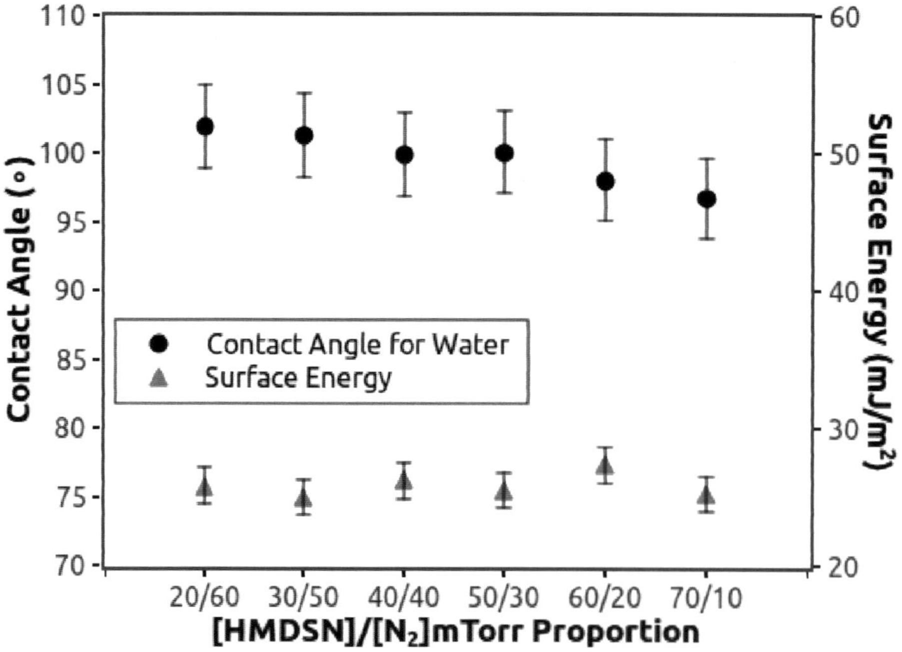


**Figure 5.** Contact angle and surface energy as a function of HMDSN and nitrogen concentration at 70 W PIIID process.

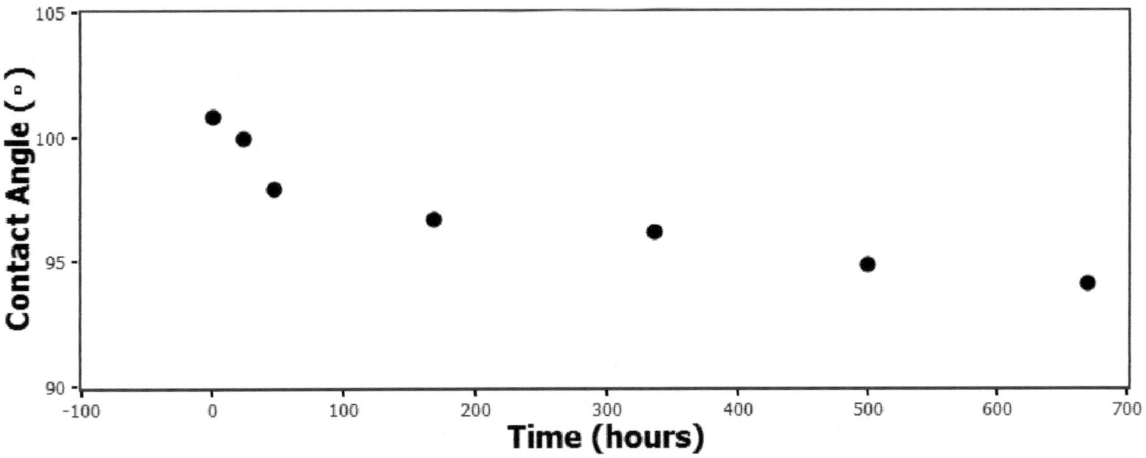


**Figure 6.** Aging effect represented by sample deposited at 60/20mTorr HMDSN/Nitrogen concentration at 25 W radiofrequency power.

## 4. Conclusions

The results of infrared measurements show groups not present in the mixture HMDSN-Nitrogen, but present in the films as a consequence of the PIIID to produce films with crosslinking structure and formed by free radicals.

The PIIID process is a competitive one characterized by polymerization and ablation simultaneously responsible of the formation of nanofilms that present hydrophobic character.

## Acknowledgements

The Authors thank CNPq for financial support.

## References

[1] d'Agostinho R, Favia P, Oehr C and Wertheimer M R 2005 Low-temperature plasma processing of materials: past, present and future *Plasma Process. Polym.* **2** 7-15

[2] d'Agostinho R, Favia P, Kawai Y, Ikegami H, Sato N and Arefi-Kosari F 2008 (eds) *Advanced Plasma Technology* (Weinheim, Germany: Wiley-VCH)

[3] Hippler R, Kersten H, Schmidt M and Shoenbach K H 2008 (eds) *Low Temperature Plasmas: Fundamentals, Technologies and Techniques* (Weinheim, Germany: Wiley-VCH)

[4] Biederman H 2004 (ed) *Plasma Polymer Films* (London: Imperial College Press)

[5] Bogaerts A, Neyts E, Gijbels R and van der Muller J 2002 *Spectrochim. Acta* **B 57** 609

[6] Chu P K, Chen J Y, Wang L P and Huang N 2002 *Mat. Sci. Eng. Rep.* **36** 143-206

[7] Shul R J and Pearton S J 2000 (eds) *Handbook of Advanced Plasma Processing Techniques* (Berlin: Springer)

[8] Inagaki N 1996 *Plasma Surface Modification and Plasma Polymerization* (Technomic Publishing Company, Inc., USA)

[9] Roth J R 1995 *Industrial Plasma Engineering: Principles* v.1 (UK: IOP Publishing)

[10] Roth J R 2001 *Industrial Plasma Engineering: Applications* v.2 (UK: IOP Publishing)

[11] d'Agostino R 1990 (ed.) *Plasma Deposition, Treatment and Etching of Polymers*, (New York: Academic Press)

[12] d'Agostino R, Favia P and Fracassi F 1997 (eds.) *Plasma Processing of Polymers*, NATO ASI *series E: Applied Sciences* **vol 346** (Dordrecht, The Netherlands: Kluwer Academic Publishers)

[13] Yasuda H 1985 *Plasma Polymerization* (New York: Academic Press, Inc.)

[14] Poon R W Y, Ho J P Y, Liu X, Chung C Y, Chu P K, Yeung K W K, Lu W W and Cheung M C 2005 Improvements of anti-corrosion and mechanical properties of NITi orthopedic materials by acetylene, nitrogen and oxygen plasma immersion ion implantation *Nucl. Instrum. Meth. B* **237** 411

[15] Fu R K Y, Cheung I T L, Mei Y F, Shek C H, Siu G G, Chu P K, Yang W M, Leng Y X, Huang Y X, Tian X B and Yang S Q 2005 Surface modification of polymeric materials by plasma immersion ion implantation *Nucl. Instrum. Meth. B* **237** 417

[16] Gan B K, Bilek M M M, Kondyurin A, Mizuno K and McKenzie D R 2006 Etching and structural changes in nitrogen plasma immersion ion implanted polystirene films *Nucl. Instrum. Meth. B* **247** 254

[17] Sui J and Cai W 2006 Mechanical Properties and anti-corrosion behavior of the diamond-like carbon films *Surf. Coat. Tech.* **201** 1323

[18] Yeung K W K, Chan Y L, Lam K O, Liu X M, Wu S L, Liu K Y, Chung C Y, Lu W W, Chan D, Luk K D K, Chu P K and Cheung K M C, 2008 New plasma surface-treated memory alloys: towards a new generation of "smart" Orthopedic Materials *Mat. Sci. Eng. C* **28** 454

[19] Santos D C R, Rangel R C C, Mota R P, Cruz N C, Schreiner W H and Rangel E C 2004 Modification of plasma polymer films by ion implantation *Mater. Res.* **7-3** 493

# Tribological behavior of DLC films deposited on nitrided and post-oxidized stainless steel by PACVD

**E L Dalibon**[1], **S P Brühl**[1] **and D Heim**[2]

[1]Surface Engineering Group, Universidad Tecnológica Nacional, Facultad Regional Concepción del Uruguay, Ing. Pereira 676, E3264BTD - C. del Uruguay, Argentina
[2]Upper Austria University of Applied Sciences, Stelzhamerstraße 23, 4600 Wels, Austria

E-mail: dalibone@frcu.utn.edu.ar

**Abstract**. In this work, the tribological behavior and adhesion of DLC films deposited by PACVD on AISI 420 martensitic stainless steel was evaluated. Prior to DLC deposition, the samples were nitrided and some of them also post-oxidized. The films were characterized by Raman and EDS, microhardness was assessed with Vickers indenter and the microstructure was analyzed by OM, SEM, FIB. Fretting and linear reciprocating sliding tests were performed using a WC ball as counterpart, and the adhesion of the DLC films was characterized using the Scratch Test and Rockwell C indentation. Corrosion behavior was evaluated by the Salt Spray Fog Test. The film showed a hardness of only about 1500 HV but it was about 15-20 microns thick. The results of the mechanical tests showed that pre-treatments (nitriding and oxidizing) of the substrate did not have a big influence in the tribological behavior of the coating. However, the nitriding treatment before the DLC coating process reduced the interface stress and enhanced the adhesion. Additionally, all the films evidenced good corrosion resistance in a saline environment, better than the AISI 420 itself.

## 1. Introduction

DLC "Diamond like carbon" coatings are known to be very hard and chemically inert and they also show a low friction coefficient and a high wear resistance. These properties can be useful in a wide range of applications such as drills, seals, bearings, dry cutting operations, medical implants and surgical instruments [1]. However, these films have a problem; they may easy peel off from the substrate in certain situations involving friction. The failure of the film is related to a bad adhesion to the substrates, and other characteristics of the film itself, for instance internal stresses, hardness, surface roughness, topography and elastic modulus [2]. In order to overcome this problem and improve the mechanical and tribological behavior of the films according to various requirements, different interlayer systems have been proposed.

The deposition of interlayers such as titanium, silicon, chromium, CrN or tungsten, have been developed to enhance the adhesion on metal substrates [3,4], compositionally graded coatings have been deposited to minimize the stress concentration and improve the coating adhesion [5], and also duplex process including diffusion treatments previous to the coating have been proposed. Plasma nitriding of a steel substrate is a good possibility because it improves the tribological behavior, increases the load capability of the coating and extends its lifetime [6]. But the characteristics of the nitride interlayer are very important because they can influence not only the adhesion but also the mechanical behavior of the system, for example, the presence of a compound layer may deteriorate the load bearing capacity in low alloyed steels [7]. Also in stainless steels, a top porous and brittle nitrided layer affects the adhesion and it is necessary to remove it prior to the DLC deposition [6].

Besides, a combination consisting of plasma nitriding followed by a post-oxidizing treatment has been applied in low alloyed steels to achieve superior corrosion resistance; nevertheless there are few studies about this topic in stainless steels [7].

In this work, DLC films were deposited by the Plasma Assisted Chemical Vapor Deposition technique (PACVD) on AISI 420 martensitic stainless steel. Prior to DLC deposition, the samples were nitrided and some of them also post-oxidized. Two duplex systems and a single coating are then compared in wear and corrosion situations.

## 2. Experiment

AISI 420 martensitic stainless steel samples were cut from a plate 1.5 mm in thickness and they were prepared as discs, 22 mm in diameter. The chemical composition in mass percent of AISI 420 is 0.38% C, 13% Cr, 0.44% Mn, 0.72% Si, 0.02% P and Fe as balance. All samples were previously heat treated (quench and tempering) to get the martensitic structure and they reached hardness values around 550 HV (50 g load). Before nitriding, they were grounded and polished in one face.

The nitriding treatment was performed in a standard industrialized DC pulsed process in the firm Rübig (Austria) for 14 hours in with a $N_2$ flow rate of 90 sl/h and $H_2$ flow rate of 30 sl/h at a temperature of 500°C. The mean voltage between electrodes was 550 V in a pulsed mode with a 200μs on-time period and 300 μs off-time period. Oxidizing was carried after nitriding in the same chamber using water steam. The DLC coatings are in fact a:C-H-Si films (silicon containing amorphous hydrogenated carbon) and they were deposited by the Plasma Assisted Chemical Vapor Deposition technique (PACVD) with the same DC pulsed discharge as used for nitriding with HMDSO and acetylene as gas precursors. The deposition rate was 2 μm/h. The process was carried out in a semi industrial facility at the University of Applied Sciences in Wels, Austria.

Microhardness was assessed with a Vickers indenter, 50g load. Microstructure was observed by optical microscopy, scanning electron microscopy (SEM), focused ion beam (FIB). The DLC films were characterized by Raman spectroscopy and energy dispersive spectroscopy (EDS) microanalysis. X-ray diffraction analyses were performed in a PANalytical diffractometer with Cu-Kα radiation and graphite monochromator. Diffraction patterns were collected in the Bragg-Brentano configuration. Fretting and linear reciprocating sliding tests were performed in a self-made machine using a WC ball 5mm in diameter as counterpart and a frequency of 11.7 Hz. Fretting tests were performed with a track length of 84 microns, a duration of 30 min and at a load of 8.49 N. Reciprocating sliding tests were carried out with a track length of 953 microns, a duration of 90 min and using loads of 4.42 N and 12 N. The worn surface or wear scars were analyzed with SEM, White Light Interferometry and a mechanical profilometer. The adhesion of the DLC films was characterized using the methods of Scratch Test and Rockwell C indentation with loads of 2 kg and 150 kg respectively. Finally, corrosion behavior was evaluated by $CuSO_4$ spot and the Salt Spray Fog Test (ASTM B117).

## 3. Results

### *3.1. Coating*

The samples were named: DLC, N+DLC and N+O+DLC, where "N" and "O" mean Nitriding and Oxidizing, respectively. The EDS spectrum of the film showed Si, C and O, as expected (figure 1), and the Raman spectrum of the film showed only a peak around 1500 $cm^{-1}$ of amorphous carbon [8], containing hydrogen but without graphitization (figure 1). A thin amorphous silicon interlayer was deposited to improve the film adhesion.

### *3.2. Microstructure and hardness*

The hardness of the DLC films reached 1500 $HV_{0,05}$ This is considered a "soft a-C: film", it results from a high hydrogen content and it is known to have also a low elastic modulus, 80-90 MPa. Its thickness varies between 15 and 20 microns (figure 3) and in the case of the N+O+DLC sample, it has an irregular interface with the previous layer, a thin and porous oxide layer of 3 microns which can be observed in the FIB image, in a place where the film was detached (figure 4). It can be noticed that the oxidized surface has imperfections which could affect adhesion. It is composed of magnetite ($Fe_3O_4$) and hematite ($Fe_2O_3$) phases according to the XRD analysis (figure 5). The first phase (magnetite) is known to be compact and homogeneous but the second (hematite) is porous and brittle and it could affect the surface roughness and the adhesion [9].

Underneath the oxide layer, a nitrided diffusion zone of about 90 microns in thickness could be detected by optical microscopy (OM) and it is showed in figure 6. The XRD pattern reveals that it is composed of α Fe (martensite) and the nitrides $Fe_4N$ and CrN, which were formed because nitriding was carried out at high temperature. Only the α peaks were identified in the XRD diffractogram of the DLC sample (figure 5) because the film is amorphous and transparent to x ray radiation.

### *3.3. Fretting test*

In the fretting tests, no big difference regarding the different pretreatment could be observed according to the calculated volume loss from the wear scar analysis (table 1 and figure 7). It could also be demonstrated that the film deformed plastically without breaking in all cases, and in general the tribological behavior of the film was far better than in the untreated samples, as it can be observed comparing figure 7 and 8.

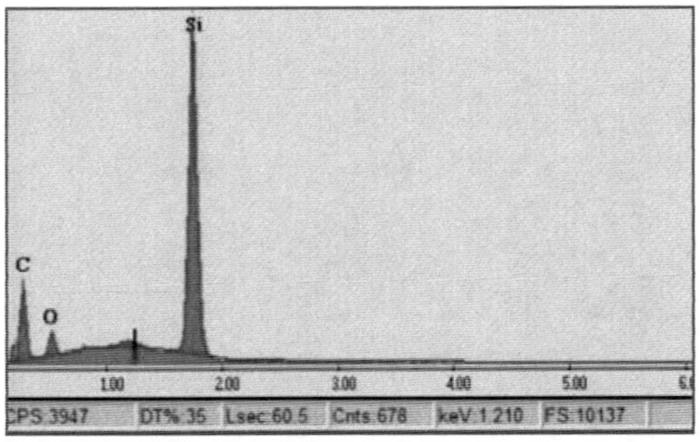


**Figure 1.** EDS spectrum N+O+DLC sample.

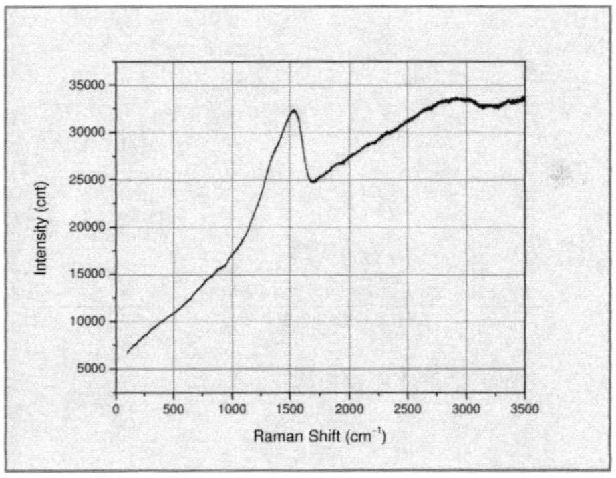


**Figure 2.** Raman spectrum DLC sample.

**Figure 3.** SEM image of the DLC film.

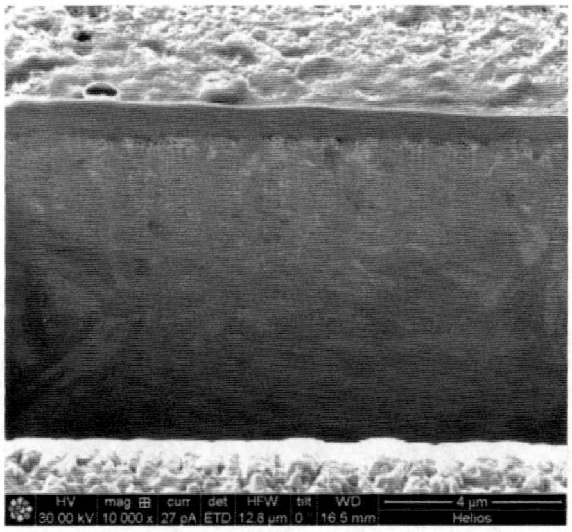

**Figure 4.** SEM-FIB view of the oxide layer.

### *3.4. Reciprocating sliding tests*

In both wear tests using 4.42 N and 12 N loads, the depth of the wear tracks was lower in the treated samples than in the untreated samples (figure 9). However, there is not a big difference between the different pretreatments. The thick DLC film resulted mechanically resistant and underwent the test with the highest load without breaking, regardless of the substrate or the prior surface pre-treatment.

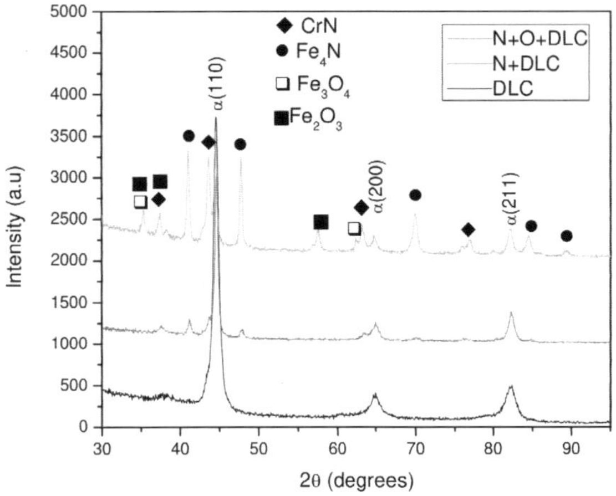


**Figure 5.** XRD diffractograms of the three samples.

**Figure 6.** OM micrograph of N+DLC sample.

**Table 1.** Quantitative wear results.

| Samples | Relative Wear volume loss | Wear scar depth | Roughness |
|---|---|---|---|
| DLC+O+N | 0.39 | 1.46 ± 0.03 | 0.2551 |
| DLC+N | 0.40 | 1.53 ± 0.02 | 0.064 |
| DLC | 0.41 | 1.54 ± 0.03 | 0.028 |
| Untreated | 1.00 | 2.34 ± 0.02 | 0.0043 |

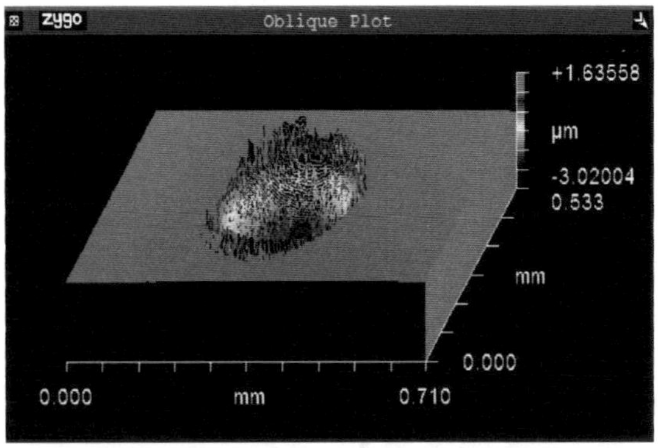


**Figure 7.** WLI 3D plot of the fretting wear scar in an untreated sample.

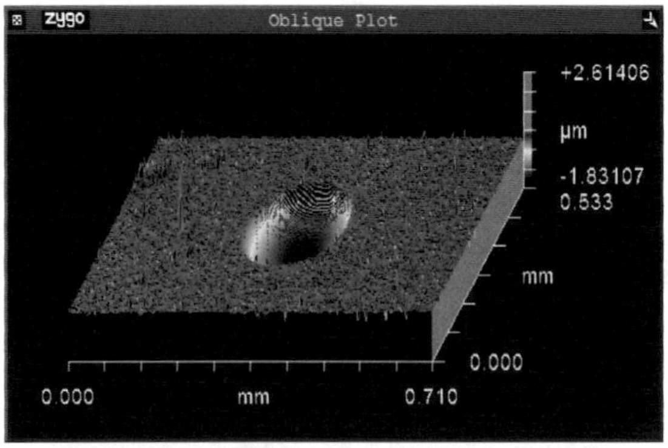


**Figure 8.** WLI 3D plot of the wear scar in a DLC coated sample.

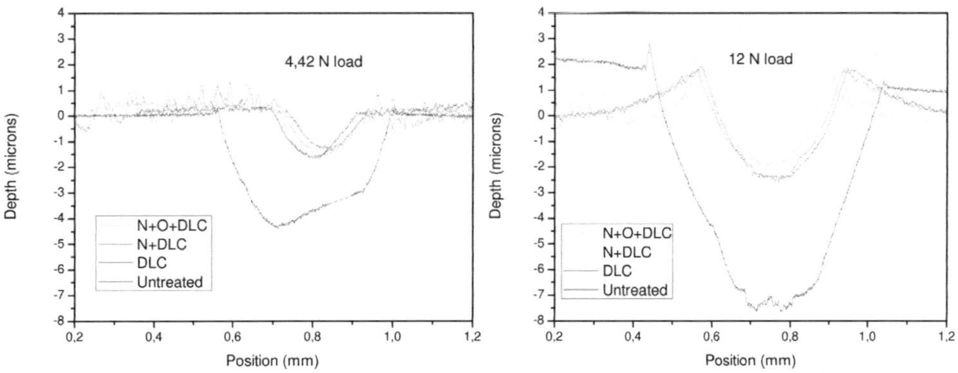


**Figure 9.** Depth profile of wear tracks.

*3.5. Adhesion*

The adhesion quality of the DLC pretreated samples is better than that of the sample without any pretreatment. In this case a significant spalling and delamination at the vicinity of the indentation can be observed (figure 10). In the N+O+DLC sample, a part of the coating surrounding the indentation was detached, indicating that the presence of hematite deteriorated the adhesion strength as it was also observed by other authors [10]. The best behavior was achieved in the sample which had only the nitriding treatment prior to the DLC coating. This diffusion modified layer, even with nitrides precipitations, resulted in a good interlayer to reduce the internal stress between the DLC film and the steel substrate [6].

In the scratch test the same results were obtained, the N+DLC sample had better adhesion than the other samples tested with 20 N load. In the untreated sample small areas of apparent delamination could be observed along the edge of the track. In the N+O+DLC, the high roughness could have influenced in the film adhesion.

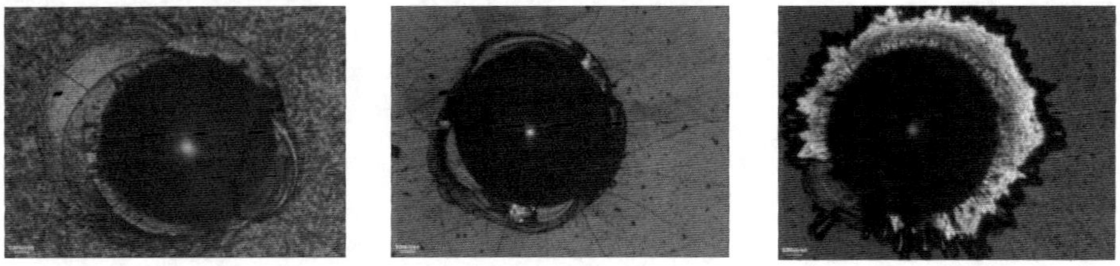

N+O+DLC N+DLC DLC

**Figure 10.** Optical micrographs 100x of samples after Rockwell C indentation.

*3.6. Corrosion behavior*

All treated samples did not show sensitization after 6 min of $CuSO_4$. The untreated sample did not pass this test, and cupper deposited immediately, showing the absence of passivity.

After 100 hours exposure in the salt spray chamber, the coated samples were clean with no sign of general or local corrosion (figure 11). The untreated sample had some pits and beginning of general corrosion.

## 4. Conclusions

The results of the mechanical tests show that the different pretreatments of the steel substrates do not have a big influence in the tribological behavior of the samples in the evaluated conditions; since the DLC films are thick enough and they have a good mechanical resistance. They underwent the fretting and reciprocating wear test with a good performance. However, the adhesion quality of the DLC film over the nitrided sample without oxidation (N+DLC) was better than in the post-oxidized one (N+O+DLC) and the sample without any pre-treatment. The nitride layer provided a graded interface between the DLC coating and substrate; which reduced the internal stress and enhanced the adhesion. On the contrary, the oxide layer consisting of hematite and magnetite did not promote a good adhesion considering nitrided stainless steel as a substrate.

The films also evidenced a good corrosion resistance in a saline environment, far better than the AISI 420 stainless steel itself, without any treatment but quench and tempering.

N+O+DLC N+DLC DLC

**Figure 11.** Surface of coated samples aspect after the salt spray fog test.

**Acknowledgements**
The authors would like to thank to Mg. Eng. Agustina Guitar and the Saarland University (Germany) for the SEM-FIB analysis; to Dr. Adriana Márquez (INFIP-UBA) for the use of the Scratch Tester, to Dr. E. B: Halac (CNEA) for the Raman measurements and to Dr. Mariela Desimone (INTEMA) for XRD measurements. Also to the Materials Departament of CNEA-CAC (Argentina) for the use of SEM and EDS, to the firm Rübig (Austria) for performing the diffusion treatments. Finally to UTN and ANPCyT (Argentina) for the Doctoral Fellowship of Eugenia Dalibon.

**References**
[1] Li K Y, Zhou Z F, Bello I, Lee C S and Lee S T 2005 *Wear* **258** 1577
[2] Wu X, Ohana T, Nakamura T and Tanaka A 2010 *Wear* **268** 329
[3] Glozman O, Halperin G, Etsion I, Berner A, Shectman D, Lee G H and Hoffman A 1999 *Diamond and Related Materials* **8** 859
[4] Trava-Airoldi V J, Bonetti L F, Capote G, Fernandes J A, Blando E, Hübler R, Radi P A, Santos L V and Corat E J 2007 *Thin Solid Films* **516** 272
[5] Choy K L and Felix E 2000 *Materials Science and Engineering* **A278** 162
[6] Chicot D, Puchi-Cabrera E S, Decoopman X, Roudet F, Lesage J and Staia M H 2011 *Diamond & Related Materials* **20** 1344
[7] Forsich C, Heim D and Mueller T 2008 *Surface & Coatings Technology* **203** 521
[8] Baba K, Hatada R, Flege S and Ensinger W 2009 *Surface & Coatings Technology* **203** 2747
[9] Rovani A C, Fischer R R, Cemin F, Echeverrigaray F G, Basso R L O, Amorim C L G, Soares G V, Baumvol I J R and Figueroa C A 2010 *Scripta Materialia* **62** 863
[10] Esfahani A, Sohi M H, Rassizadehghani J and Mahboubi F 2008 *Vacuum* **82** 346

# Platinum nanoparticle deposition on polymeric membranes for fuel cell applications

**A J Moreira, S Lopera, N Ordonez and R D Mansano**

Electronic System Engineering, University of São Paulo, São Paulo, Brazil

E-mail: adir@lsi.usp.br

**Abstract.** This work aimed to show an alternative to produce platinum nanoparticles directly on a polymeric membrane using plasma technique, in order to make these nanoparticles adhere to the membrane, in size, shape and homogeneity controlled by the process without damaging the polymeric material. In this manner the cell's production time is reduced since the catalyst is directly deposited on the polymeric membrane; the time of the process is approximately five minutes for each side of the membrane, and the total time for each membrane is 10 minutes. With this exposure time, and the advantage of controlling the other parameters such as pressure, RF power, gas flow rate and temperature of the electrode, it was possible to obtain platinum nanoparticles with dimensions of about 50 nm scattered homogenously on the membrane, without damaging the structure of the polymeric material and, consequently, affecting its performance. Together with platinum nanoparticles were also deposited carbon nanoparticles, so that these acted as catalyst support, avoiding self poisoning. Electrochemical activity tests were performed to test the efficiency of the cell where it was exposed to different pressures and flow rates of $O_2$ and $H_2$, reaching open-circuit voltage of 750 mVolts.

## 1. Introduction

Nowadays there is a great concern about the generation of energy due to the emission of pollutants that greatly contribute to air pollution and environmental disasters caused by fossils and nuclear fuels [1]. A technology that has been studied in depth recently is the process that involves fuel cells, which not only is a source of renewable energy but also contributes to reducing pollution and threats to nature. An example of fuel cell is one that uses the proton exchange membrane technology. A catalyst (usually a platinum nanoparticle) is added to this membrane. The platinum is a rare metal and tends to become increasingly scarce due to its various applications. Thus, studies are conducted to reduce the amount of platinum used in fuel cells. One way to reduce the quantity of the metal and not affect the performance of the cell is to also reduce the size of these nanoparticles. Just as the nanoparticles are reduced the effects of size and surface are increasingly noticeable. Researchers have observed in metallic and semiconducting nanoparticles for example that by reducing the size of nanoparticles is a confinement of electrons is achieved. This confinement causes a quantum effect in the material. This is observed through the magnetic and optical properties, as well as the electronic conduction which are different in a larger size nanoparticle. The size also affects the thermodynamic properties such as heat capacity [2]. As to the surface effects, there is an increase in the surface free energy, which is caused by incomplete chemical bonds of atoms at the surface, since these are less tied than the atoms of the nanoparticles. These effects affect surface properties such as melting temperature, chemical reactivity and optical absorption. The melting temperature of platinum for example, falls from 1,773 °C in solid volume to 600 °C, for nanoparticle size 8 nm [3].

The particle shape also changes as it approaches the nanometer scale. This makes the knowledge of the properties of these nanoparticles very important for the application of this technology whatever the area it is applied to [1,4,5,6].

The techniques used to obtain nanoparticles vary according to the availability of resources, as well as the desired final application, and range from simple processes as the dissolution of the liquid phase wet via to a process more expensive and complicated processes.

This present work aimed to show the attainment of platinum nanoparticles using a physicochemical process, in which plasma equipment was employed.

The plasma technique for obtaining carbon nanotubes and nanoparticles in general has increased significantly due to its reproducibility, uniformity of size and shape of particles and nanotubes generated as well as the quickness of the procedures. In catalysts, studies have been developed to generate metal nanoparticles directly on carbon particles, allowing an increase in catalyst efficiency [7,8]. Studies using capacitive-coupled plasma equipment showed that by to submitting a platinum target to argon plasma it is possible to extract the platinum particles from the raw metal and deposit them on a support inside the reactor. And also if these platinum nanoparticles mixed with a gas containing carbon, it is possible to generate carbon particles decorated with platinum nanoparticles, which in turn can significantly increase the performance of a fuel cell [9].

In the case of proton exchange fuel cell, the carbon support is critical for the proper functioning of the cell because it will cause this exchange to take place efficiently and it will avoid the self poising of platinum nanoparticle. The polymer membrane fuel cell operates at low temperature between 60 and 100 °C. Its operation requires highly pure hydrogen in order to avoid the poisoning of the electrolyte. It consists of two electrodes (hydrogen and oxygen) separated by an electrolyte and electrically connected to an external circuit through which the electrical current flows. Electrolytes are coated by the catalyst, which is platinum in the majority of cases. This serves to accelerate the reactions in electrolytes and break the hydrogen molecules [10,11]. These functions may vary depending on the nanoscale size, as stated previously, and this can directly affect the catalytic activity [2]. One of the materials commonly used as electrolyte is *Nafion*®, which is obtained by copolymerization of perfluorinated vinyl ether, which has as its terminal group the $SO_2F$, with tetrafluoroethylene. The performance of this electrolyte will depend on their physicochemical characteristics like: high ionic conductivity, chemical stability, low permeability to the driver and oxidant, low electro-osmotic flow and thermal stability. The proton conductivity through the *Nafion*® membrane, is one of most important properties for the application in fuel cells, because stopping the flow of proton across the membrane, would cause the cell's circuit to be interrupted and thus the cell to stop working. Another important factor is the transport of water through the membrane while keeping the cell always in optimal humidity condition. If this condition is not satisfactory, the cell's performance can be affected in three ways: 1- Dehydratin of the regions of the membrane on the cathode side and affecting the proton conductivity through the membrane. 2- Accumulation of water in the cathode hinders the diffusion of oxygen to the catalyst. 3- Minimization of the negative electro-osmotic effects by the diffusion of the water returning from the cathodes to the anode. These considerations are important when dealing with material processed at low pressure plasma, because the conditions within the chamber such as humidity, pressure, temperature, U.V. radiation and ion bombardment can modify physicochemical or even structural properties, compromising the performance of the cell [11].

The most common methods of coating these electrolytes involve using sheets of carbon as support, where platinum nanoparticles are then deposited. When ready, these catalysts are hot pressed to the *Nafion*® membrane on both sides, then coating the electrolyte [12].

The purpose of this study was to produce platinum nanoparticles directly on the *Nafion*® membrane, so that these nanoparticles were homogeneously scattered, within the nanoscale and homogeneous shape without compromising the electrolyte in both its structure and its performance.

## 2. Material

The plasma equipment used to obtain the nanoparticles was parallel plate plasma with arc system, with a volume chamber of approximately 3.5 liters. The plasma was generated by a radio frequency generator operating at 13.56 MHz. The catalysts were produced using a 99.98 % pure platinum wire, and it was assembled in the form of a coil with five turns. The diameter of each turn was 10 mm approximately and spaced at approximately 0.5 mm. The gases used for plasma generation were industrial acetylene ($C_2H_4$) and argon (Ar). The acetylene gas was the source used to produce the carbon support and the argon gas was used to increase the ion bombardment on the platinum wire pulling particles and directing them towards the catalytic support. 99,999% pure nitrogen gas was used for opening and closing the chamber. *Nafion*® was used as the electrolyte; it was 50×50 $mm^2$ and 1mm thick. The gases used as fuel were hydrogen ($H_2$) 5.0 (99.999% pure) and oxygen ($O_2$) 4.0 (99.99%). The membrane was prepared after obtaining the catalysts, using sulfuric acid ($H_2SO_4$) MOS grade, hydrogen peroxide ($H_2O_2$) MOS grade and deionized water level 18MΩ.cm. The samples were observed through a scanning electron microscope with field emission gun *Nova NanoSEM 400* FEI Company. The metal deposition was observed through an Energy Dispersive Spectroscopy (EDS) *Noran System Six* of *Thermo Electron Corporation*, by SEM 6460LV of JEOl and an x-ray diffractometer *PANalytical XRD Empyream* and the electrical resistance of the material deposited on the membrane was measured by four point probe *FPP 5000* of *VEECO Instruments INC.*

## 3. Methodology

### *3.1. Assembling the platinum coil inside the plasma reactor*

The previously molded platinum coil was connected to the plasma electrode through at the ends of the rods which acted as a support sustaining the coil at a high of approximately 10 mm (figure 1). This way, the membrane can be placed between the coil and the electrode causing the metal particles to be directed towards the electrolyte during the process.

Additionally it was also inserted between the coil and the membrane an aluminum mold with an opening that corresponds to the cell's work area in the center of the mold, causing the released particles to reach only the cell's work area, leaving the sides of the electrolyte free for assembling.

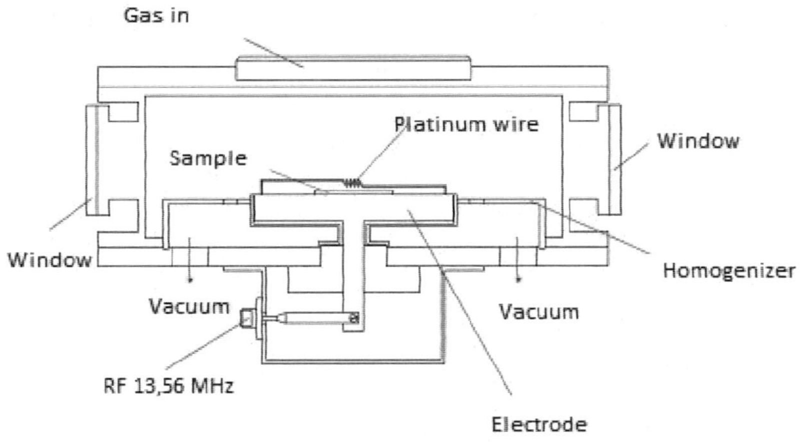


**Figure 1.** Assembling of platinum coil inside plasma reactor.

### *3.2. Obtention of nanoparticles deposition by plasma process*

After assembling the platinum coil on the electrode, the polymeric membrane was inserted in the reactor, between the electrode and the coil. The aluminum mold was placed on the membrane so that the nanoparticles were deposited exclusively in the cell's work region.

After assembling the cell in the chamber, the lid of the reactor was closed, the valves were opened and the system was evacuated until 1mTorr of pressure. So, with the reduced pressure, the gases of the processes were inserted at flows of 18 sccm for the acetylene and 45 sccm for the argon. Then the total system work pressure was controlled in 500 mTorr.

Once the flow and pressure were controlled the RF power was adjusted to 200 Watts. These conditions provided the heating of the electrode, which can harm the membrane [11], so the electrode's temperature was reduced to 10°C and after five minutes, the power was turned off. After that the valves were closed and the gas pressure was reduced again in order to clean the chamber.

With the pressure close to 1 mTorr, the vacuum pump valves were closed and then the nitrogen gas was inserted to open the chamber. With the chamber opened the membrane was moved for deposition on the other side.

### *3.3. Preparation of polymeric membrane*

After the deposition of nanoparticles in the polymer membrane, the following step was the preparation of the membrane to application in the fuel cell. For the preparation of the cell, chemical cleaning processes were carried out to activate the cell. The following steps were taken: Bath in a hydrogen peroxide solution 3 vol at 80 °C for 1 hour, then three deionized water baths were performed for three hours each at 80 °C and then one sulfuric acid bath also for one hour at 80 °C. Finally more three baths in deionized water for one hour each at 80 °C.

### *3.4. Assembling of the fuel cell*

After preparing the membrane, it was then assembled in an experimental parker fuel cell. After fixing the membrane in half-cells, these were closed and proceeded to be tested for leakage. The cell was inserted into a water container and the system was pressurized. After observing that there was no leakage in the system, the electrolyte was moistened with deionized water through the method of dragging for 15 minutes.

### *3.5. Tests of electrochemical activities*

The oxygen and hydrogen gases were connected to the fuel cell and monitored by pressure gauges, so that it was possible to observe the effects of gas pressure variation in the voltage measurements achieved. This voltage was taken as a function of pressure up to a point where it reached the maximum for the designed cell.

## 4. Results and discussions

In order to evaluate the shape of the platinum nanoparticles and carbon deposition, as well as the coverage area of the deposition of these particles, tests were performed on $10\times10$ mm$^2$ glass laminas. Thus it was possible to adjust the parameters of RF power, pressure, gas flow rate and exposure time to reach homogeneous shape and sizes of platinum nanoparticles and carbon deposition. These tests also served to check the influence of the plasma temperature on the sample, which would hinder the application of this process on polymeric material. The distance between the coil and the substrate was set as this is also an important factor in heating the substrate, that is: during the process, the coil heats up due the effects of generating the plasma, such as current and ionic bombardment. This temperature can cause irreversible damages to the membrane. On the other hand, if the distance is too large, the dispersion of the particles may not be enough to ensure good homogenization, and may not generate the plasma within the coil, especially if the coil is out of the sheath region of the plasma, which would hinder the release of the platinum particles. The process time was 5 minutes, so that the substrate temperature was near to room temperature and there was a deposition of homogeneously shaped and sized nanoparticles scattered in the substrate (figure 2).

In these images it was possible to observe that the platinum nanoparticles did not overlap each other as they were deposited on the carbon film, which led to infer that in a polymeric membrane these nanoparticles would be well scattered and would contribute to good cell performance. Observing these images it was also possible to calculate by extrapolating the amount of platinum nanoparticle deposited, and in this way a result of 200 nanoparticles/mm² was reached. Some of these samples in glass were analyzed by EDS, and the results obtained showed the presence of platinum (figure 3). With this information, the plasma process on the polymeric material was then carried out.

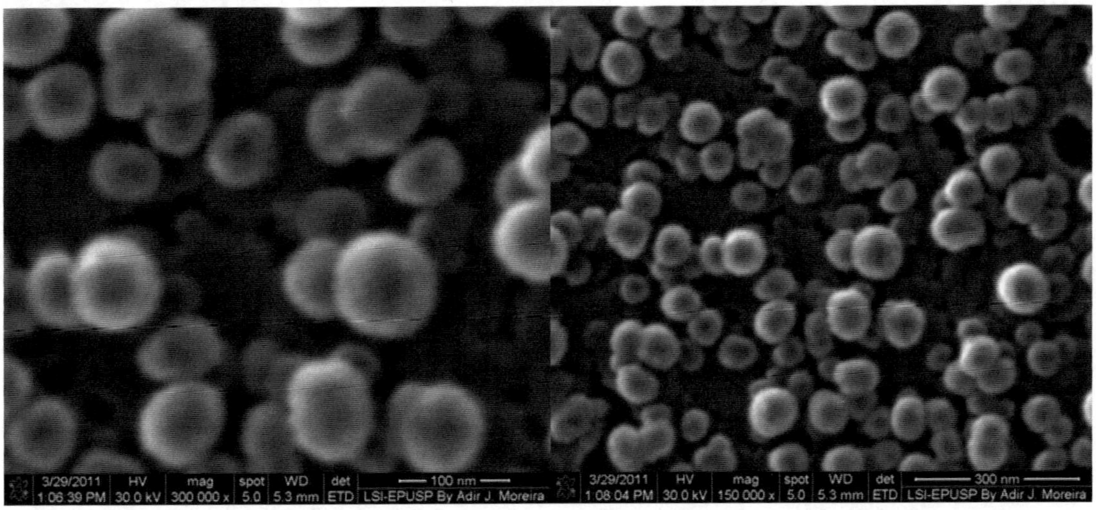


**Figure 2.** Deposition of platinum nanoparticles and carbon on glass substrate.

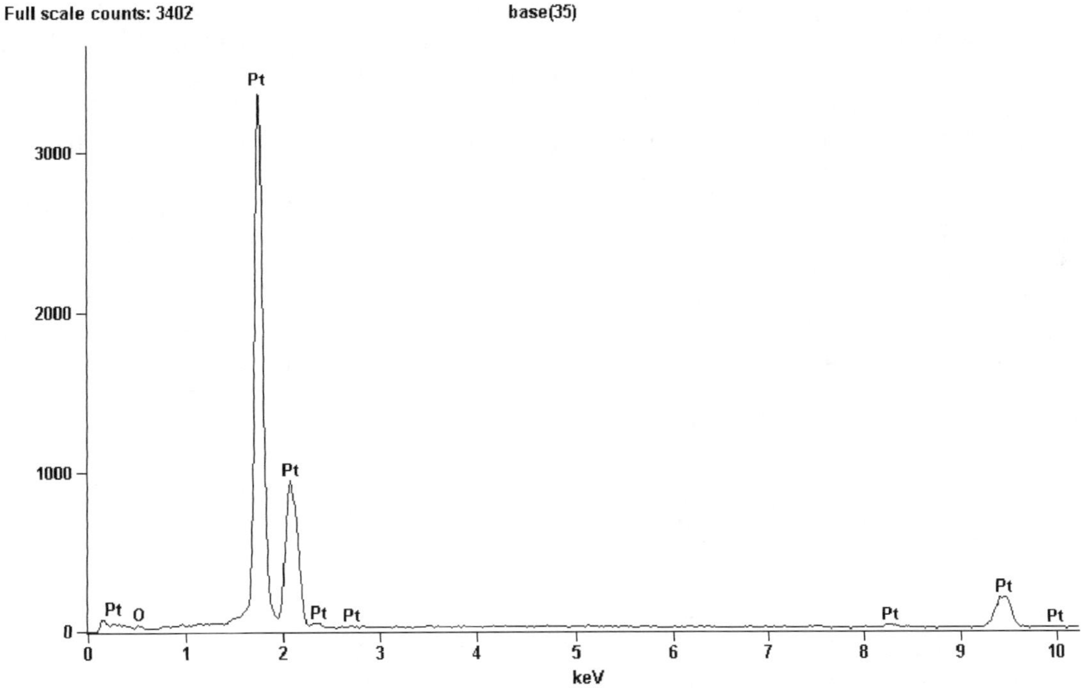


**Figure 3.** Analysis by EDS showed the presence of platinum on the substrate.

The next step was to submit the polymeric membrane to the plasma process in order to evaluate if the platinum nanoparticle generated and the carbon film would be reproducible in this material, and if the material would support the process conditions. The *Nafion*® membrane was then placed in the reactor and exposed to the process under the same conditions of pressure, gas flow, power and process time, and also the cooling of the electrode. In the case of the membrane the difference was in the process time, because this had to be done in two stages, in a total of 10 minutes (5 minute each side of the membrane). However, by adjusting the temperature of the electrode to 10 °C, the opening time of the chamber already cools the electrode enough so that when the process is restarted on the other side of the membrane, the temperature inside the reactor does not damage the material. After the process was performed on both sides, it was possible to observe the fuel cell's work area filled with the deposited material (figure 4).

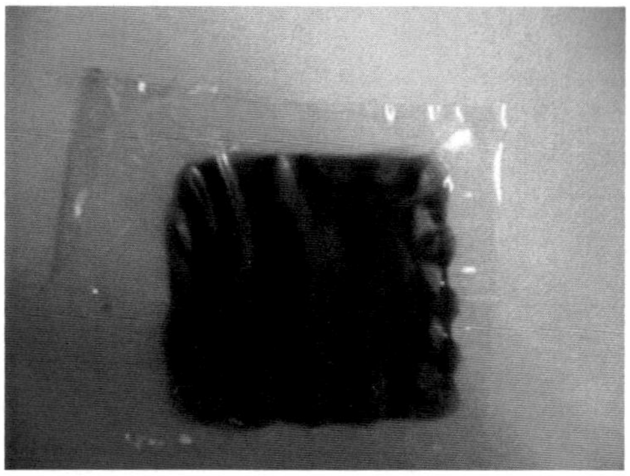

**Figure 4.** Polymeric membrane with the work area filled with nanoparticle of platinum and carbon.

The polymeric membranes have properties that may be altered when exposed to certain conditions of pressure, humidity, temperature, etc. This issue was observed when the sample was exposed to the low plasma process. Although the temperature was controlled, a dry membrane was found. This was due to the lack of moisture inside the chamber. To avoid this problem, the membrane was fixed in the aluminum support being stretched and enabling the deposition in the entire region of the polymeric membrane. The images obtained by SEM-FEG show that the parameters used both in the glass and in the polymeric material, were not in a matter of reproductive size, but the shape and homogeneity of the dispersion on the substrate was satisfactory. The nanoparticle size increased to approximately 250 nm, but the format remained rounded (figure 5). The increase in size of the nanoparticles can be explained by the relief of the substrate, which makes the smaller nanoparticles grains accumulate until the end of the process generating the larger grains.

An indication that the substrate relief influences in nanoparticles dispersion, is that the larger nanoparticles were founded in the furrows of the material, and in the same sample were also found smaller nanoparticles on the surface (figure 6). It can also be observed that the shape of nanoparticles generated on the surface does not correspond to what was obtained in the glass sample. This may be due to the fact that during the plasma process very small grains are stripped off the platinum wire and deposited on the material, because of the energy and amount of grains removed, the grains join together with other grains already deposited in the cavities of the membrane, thus forming larger and rounded grains. The smaller rod-shaped grains have a smaller amount of added platinum grains, since there is a greater chance of dispersion of these grains on the surface of the sample. This can be clearly observed on the glass, as the surface is flat and the grains are dispersed in small sizes. But this apparently not so homogeneous distribution of nanoparticles did not cause a discrepancy in the measurement of electric resistance of the sample, it was measured an average of 13.45 Ω, and the difference between the five points measured in the sample was 0.67 Ω, which proves that the concentration of particles was well distributed on the work region of the fuel cell.

As mentioned earlier, the humidity of the membrane is an important factor. During the process low level of moisture may cause the pores of the membrane create furrows.

These furrows cause the nanoparticles to deposit unevenly, with higher concentrations in some regions and more scattered and in others. But, in general, the whole work area of the fuel cell was covered with the material of the catalysis, enabling its application in the fuel cell. It is noteworthy that these furrows were not caused by the plasma process exclusively, but that the membrane already has these cavities in its construction. What happens is that during the plasma process these furrows tend to get bigger due to lack of moisture in the chamber. To try and solve this problem, the membrane was stretched, and the area of furrow formation was reduced and the nanoparticles deposition was more evenly distributed.

**Figure 5.** Platinum nanoparticles on the *Nafion*® membrane.

**Figure 6.** Nanoparticles on *Nafion*® (surface region).

The x-ray analysis performed on *Nafion*® processed samples by plasma confirmed the presence of platinum (figure 8), among other constituents of the material, and of other materials present in the construction of the reactor. Comparing the plasma process on the polymeric membrane with platinum wire (raw material) and platinum deposition plasma process on a glass sample, it was possible to see the difference between the elements present in the two sample (glass and polymeric membrane), besides to separate the position platinum point. So, the figure 9 shows the analysis of platinum wire

Journal of Physics: Conference Series **370** (2012) 012030 doi:10.1088/1742-6596/370/1/012030

without plasma exposition. This material was put directly on x-ray support, for platinum position definition.

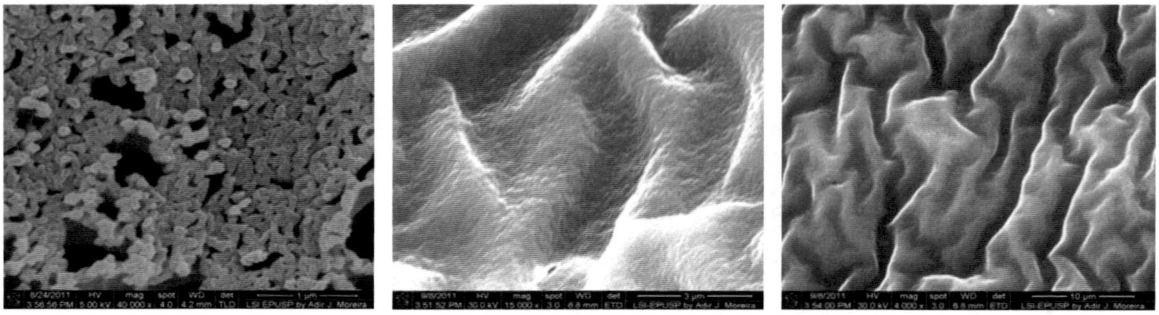

**Figure 7.** Polymeric membrane furrows.

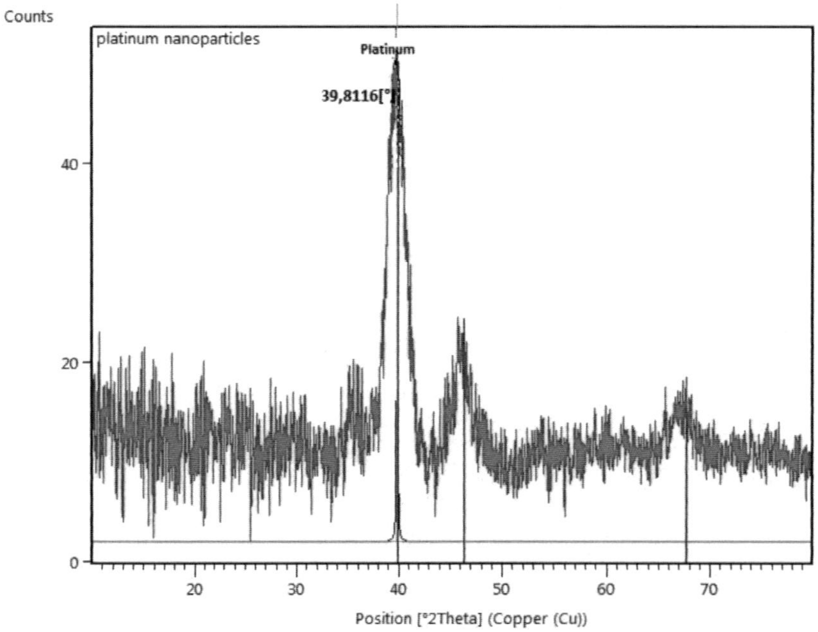


**Figure 8.** X-ray analysis showing the presence of platinum in *Nafion®* membrane.

Figure 10 shows the glass sample after platinum plasma process deposition in the same condition of polymeric membrane process. The noise shows the chemical elements present on the sample.

Not all chemical elements observed on the noise are contaminants, and almost all belong to the material construction. It is possible to observe in the figures that the platinum peaks were higher than noise which represent the contaminants, which leads to inferring that the interference in the cell's performance is very low, further showcasing the viability of this process directly on polymeric membrane for obtaining nanoparticles. These results of the plasma process on the membrane were tested for electro-chemical activity. Firstly, the membrane was chemically activated, for it to come into the appropriate operating system, enabling the pores and the channels, responsible for the electro-osmotic flow. Moreover, it was observed that during the production of nanoparticles there was strong membrane dehydration which could potentially affect its proton conductivity, so watering was necessary.

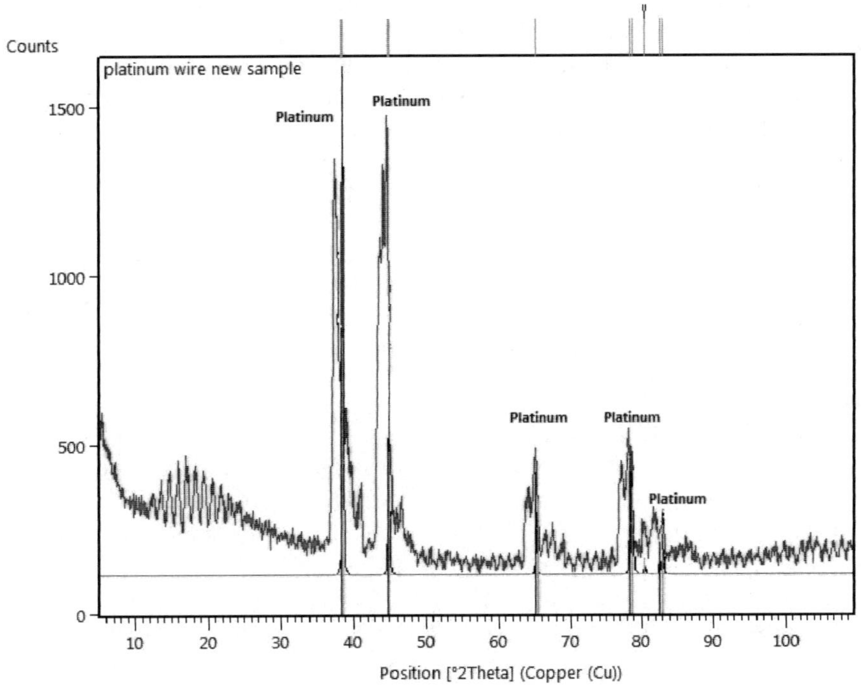


**Figure 9.** X-ray analysis on wire platinum showing the platinum peak.

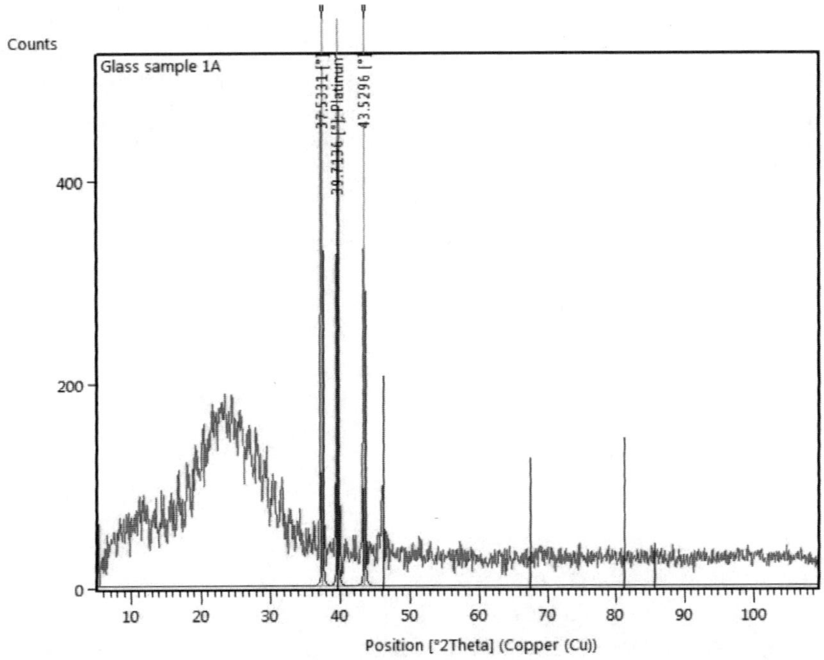


**Figure 10.** X-ray analysis on glass sample after plasma process.

During the activation process and hydration of the membrane the roughness of the membrane's surface disappeared, showing a good recovery of the material when hydrated.

It is known that the water flow affects the performance of fuel cells. So dehydration of the membrane on the side of the cathode and the accumulation of water were avoided.

The results of electro-chemical activity of the cell were obtained by altering the pressure of the oxygen and hydrogen (table 1), the maximum voltage was 750 mVolts, until the circuit was open. The purpose of this was to achieve the maximum voltage before the membrane was fully compressed, thereby reducing the pores and channels and expelling water from the polymeric matrix.

**Table 1.** Voltage obtained on polymeric matrix as a function of gas pressure.

| Voltage (mVolts) | | 380 | 700 | 750 | 680 | 400 | 300 |
|---|---|---|---|---|---|---|---|
| Pressure | Oxygen (bar) | 0 | 0.5 | 1.0 | 2.0 | 2.0 | 3.0 |
| | Hydrogen (bar) | 0.5 | 1.0 | 1.0 | 1.0 | 2.0 | 2.0 |

The results show that maximum voltage was reached for pressures of hydrogen and oxygen in equal proportions, and that the cell's performance levels drop as more oxygen than hydrogen is provided. This is due to the compression cell undergoes in the anode region, preventing $H^+$ ions move and form water. With higher pressure of hydrogen than oxygen $H^+$ ions were expected to move and the cell's performance to be of the highest voltage. Although every effort has been made to maintain good hydration in the membrane, and avoid external contamination, the result was not completely satisfactory and the voltage has not been high enough. Another factor to be considered is in the amount of platinum deposited on the sample it was higher than expected (200 particles/$\mu m^2$) and this larger quantity of nanoparticles may have caused a self poisoning that caused the decrease of performance. As noted in previous tests, this problem can be avoided by reducing the process time. This way the prevention of possible damage to the cell is also obtained since it is exposed to the effects of plasma for less time.

## 5. Conclusions

Although the maximum voltage achieved is still relatively low, the results obtained in this work showed the possibility of using plasma for direct platinum nanoparticles deposition on polymeric membranes. It was observed that the temperature does not damage the polymeric matrix and that the nanoparticles are deposited along its entire length. This is an important factor for good catalytic action. Nevertheless the amount of nanoparticles deposited should be reduced, thus preventing self the poisoning of the cell.

Another possibility observed in this work was the formation of nanoparticles of other material along with platinum, in order to be used as a support for platinum. In this way, platinum can be used in smaller quantities, saving material without losing efficiency of the catalyst.

## Acknowledgements

The authors thanks FAPESP and CNPQ for the financial support, the Laboratory of Integrated System of Polytechnic School for supplying all support to realize this work; We are grateful to the Laboratório de Filmes Finos do IFUSP for the SEM facility (FAPESP Proc. #00/8231-1) for EDS equipment and Alexandre Marques Camponnuci, Julio Cesar dos Santos and Rubens de Alcantara Pereira for technical support

## References

[1] Taylor J M 2006 *New dimensions for Manufacturing a UK Strategy for Nanotechnology – Report of the UK advisory Group on Nanotechnology Applications* submitted to Lord Sainsbury Minister for Science and Innovation

[2] Klabunde K J 2001 Nanoescale Materials in Chemistry *Introduction to Nanotechnology* (John Willey & Sons Inc.)

[3] Wang H Z, Gao L, Gui LH and Guo J K 1998 Preparation and properties of intragranular Al2O3 –SiC nanocomposites *Nanostructures Materials* **10** 947-53
[4] Su F, Poh C K, Tian Z, Hu Z and Lin J 2010 Electrochemical Behavior of Pt Nanoparticle Supported on Meso and Microporous Carbons for Fuel Cells *Energy Fuel* **24** 3727-32
[5] Wei K, Wang Y, Lai C, Ning C, Wu D, Wu G, Zhao N, Chen X and Ye J 2004 Synthesis and characterization of hidroxyapatite nanobelts and nanoparticles *Material Letters* **59** 220-5
[6] Wing C E G 2006 Las nanopartículas: pequeñas estructuras com gran potencial, El Inin hoy, *Contato Nuclear* 24-29
[7] Perles C E 2008 Propriedades Fisico-quimicas Relacionadas ao Desenvolvimento de Membranas de Nafion® para Aplicação em Células a Combustível do Tipo PEMFC *Polímeros Ciência e Tecnologia* **18** 281-8
[8] David V P, Sanches, Huing P, Koslov M E, Baughman R, Vidic R D and Yun M 2010 Carbon Nanotube/Platinum (Pt) Sheet as an Improved Cathode for Microbial Fuel Cells *Energy Fuel*, vol 24 5897–902
[9] Ferreira H S and Rangel M C 2009 Nanotechnology: General Aspect and Potential Application in Catalysis *Quimica Nova* **32** 7
[10] Traça A A 2005 *O Hidrogênio como combustível* Faculdade de Ciências e Tecnologia da Universidade de Coimbra
[11] Richardson H H, Hickman Z N, Govorov A O, Thomas A C, Zhang W and Kordesh M E 2006 Thermooptical Properties of Gold Nanoparticles Embedded in Ice: Characterization of heat Generation and Melting *Nano Letters* **6** 783-8
[12] Dilonardo E, Milela A, Palumbo F, Capitani G, D'Agostino R and Fracassi F 2010 One Step Plasma Deposition of Platinum Contanining Nanocomposite Coating *Plasma Processes and Polimers* **7** 51-8

Journal of Physics: Conference Series **370** (2012) 012031 doi:10.1088/1742-6596/370/1/012031

# Chemical characterization of 4140 steel implanted by nitrogen ions

**E D V Niño[1,3], J L Pinto[2], V Dugar-Zhabon[1] and J A Henao[2]**

[1] Grupo de Investigación en Física y Tecnología del Plasma y Corrosión – FITEK, Universidad Industrial de Santander, A.A. 678, Bucaramanga, Santander, Colombia
[2] Grupo de Investigación en Química Estructural – GIQUE, Universidad Industrial de Santander, A.A. 678, Bucaramanga, Santander, Colombia
[3] Grupo de Investigación en Tecnología del Plasma – GINTEP, Universidad Pontificia Bolivariana, A.A. 2932, Bucaramanga, Santander, Colombia

E-mail: ely.valbuena@upb.edu.co

**Abstract.** AISI SAE 4140 steel samples of different surface roughness which are implanted with 20 keV and 30 keV nitrogen ions at a dose of $10^{17}$ ions/cm$^2$ are studied. The crystal phases of nitrogen compositions of the implanted samples, obtained with help of an x-ray diffraction method, are confronted with the data reported by the International Centre for Diffraction Data (ICDD) PDF-2. The implantation treatment is realized in high-voltage pulsed discharges at low pressures. The crystal structure of the implanted solid surfaces is analyzed by the x-ray diffraction technique which permits to identify the possible newly formed compounds and to identify any change in the surface structure of the treated samples. A decrease in the intensity of the plane (110), a reduction of the cell unity in values of 2-theta and a diminishing of the crystallite dimensions in comparison with non-implanted samples are observed.

## 1. Introduction

It is well known that the plasma immersion methods of ion implantation which do not change the sample geometric dimensions are used to modify the physical and chemical properties of solid surfaces [1]. One of these methods is the 3DII method reported by Khvesyuk and Tsygankov [2]. In this method, the high voltage low pressure discharge is ignited without a pre-discharge [3]. It is important that the cathode drop zone of the 3DII discharge should be very stable to guarantee a small energy dispersion of the ions accelerated in the cathode zone and their normal incidence on the treated surface. This suggests that the electron emission remains constant during the treatment process due to the ion bombardment of the surface, thus simplifying the implanted ion dose calculations [4]. In the Industrial University of Santander (Colombia), the 3DII method is implemented in the JUPITER (Joint Universal Plasma and Ion Technologies of Experimental Reactor) [5,6] equipment which is used for modifying the superficial metal layers.

In this paper, the surface characteristics of the AISI SAE 4140 steel samples exposed to a high voltage pulse discharge in the JUPITER reactor are presented. Their study includes four stages. The first step consists in choosing the sample material, designing, manufacturing and sample preparation. On the second stage the chemical composition of the substrate material is analyzed. The third stage is devoted to the sample implantation treatment. On the fourth stage the chemical characteristics of the sample surface phases and compounds are qualitatively analyzed by comparing the obtained diffraction profiles with those reported in the PDF-2 database of the International Centre for Diffraction Data (ICDD).

## 2. Chemical composition of the samples

The chemical composition analysis of samples is made through an optical emission spectroscopy (EEO) technique which permits to determine precisely the sample material type. The spectroscopy results indicate that the sample material correspond to an AISI SAE 4140 alloy steel whose chemical composition corresponds (see table 1) to the data reported in ASTM A 322 standard [7]. The steel AISI SAE 4140 [8,9] is chosen because it is widely used in industry.

**Table 1.** Chemical composition of the simple tested AISI SAE 4140 steel.

| Element | % | Element | % |
|---|---|---|---|
| C | 0.40 | Cu | 0.17 |
| Cr | 0.85 | Sn | 0.019 |
| Mo | 0.14 | V | 0.026 |
| Mn | 0.85 | Al | 0.002 |
| Si | 0.24 | Ni | 0.09 |
| S | 0.018 | Nb | 0.007 |
| P | 0.012 | Cu | 0.17 |

## 3. Sample preparation

The samples of a cylindrical geometry of 1.8 cm in diameter and 0.5 cm thick are polished with abrasive SiC paper of numbers of 320, 600, 1200 and 6μ and with a 1μ cloth according to ASTM E 3-95 standard [10,11]. This procedure also eliminates the surface impurities which are formed because of the contact with the environment.

The information on the structure characteristics of the tested material is obtained with the metallographic analysis which is fulfilled using an Olympus microscope OBM X3. In order to compare the pre- and post-nitrogen ion implantation surface roughness, the roughness of every prepared surface is measured thrice with the help of a T500 Hommel roughness tester [12,13]. Table 2 reports the roughness parameters $R_a$, $R_z$ and $R_{max}$ which evidence that the nitrogen ion dose rise reduces the roughness significantly due to the pickling effect that occurs in the sputtering process and the collisions of the incident ions with the crystal structure of the material [12,13].

**Table 2.** Roughness for implanted and without implant samples.

| Voltage Discharge | Finish Superficial | $R_z$ (μm) | | $R_a$ (μm) | | $R_{max}$ (μm) | |
|---|---|---|---|---|---|---|---|
| | | Without Implant | Implanted | Without Implant | Implanted | Without Implant | Implanted |
| 30kV | Cloth 6μ | 0.48 | 0.15 | 0.08 | 0.02 | 1.12 | 0.21 |
| 20kV | Cloth 6μ | 0.48 | 0.36 | 0.08 | 0.03 | 1.12 | 0.82 |
| 30kV | Cloth 1μ | 0.51 | 0.31 | 0.03 | 0.03 | 1.73 | 0.78 |
| 20kV | Cloth 1μ | 0.51 | 0.46 | 0.03 | 0.05 | 1.73 | 1.00 |

## 4. Surface treatment samples

The data obtained in this work together with the previous studies results[14-16] permit to establish the range of discharge voltages, the dose of implanted ions and the gas pressures for optimal condition for the 3DII implantation technology. The samples are located in the discharge chamber (see figure 1) on the surface of the cathode that renders the sample into cathode part. The chamber walls act as the anode. In the high voltage discharge, the sample is surrounded by the cathode drop zone where the ions are accelerated and incident onto the surface sample. The potential which accelerates the ions is almost equal to the applied cathode-anode voltage. The nitrogen ion implantation is performed in pulsed discharges of 20 kV and 30 kV, at pulse duration of 0.40 ms, pulse repetition frequency of 30 Hz and a 60-minute exposure treatment time at pressures of 0.40 Pa and 0.45 Pa.

**Figure 1.** Container with a sample holder on the cathode surface.

The implanted dose was determined by taking into account the value of the secondary ion-electron emission coefficient ($\gamma$) and the total current value obtained during the sample surface treatment. The secondary emission coefficient for a discharge of 30 kV is taken $\gamma = 1$, taking into consideration that the sample holders are made of 304 stainless steel [17]. In table 3, the total current values for the samples polished with the clothes of 1μ and 6μ and at discharge voltages of 20 y 30 kV are presented.

**Table 3.** Total current for the sample treatment.

| Voltage Discharge | Polished Surface | Total Current [A] |
|---|---|---|
| 20kV | Cloth 6μ | 0.359 |
| 20kV | Cloth 1μ | 0.324 |
| 30kV | Cloth 6μ | 0.678 |
| 30kV | Cloth 1μ | 0.648 |

With the data on total current and secondary emission coefficient values, the total ion flow incident on the sample surfaces are obtained. The dose of nitrogen ions implanted into the surface is calculated in accordance with the following equation [12,13]

$$D = 2\Phi f d t \left[\text{ions cm}^2\right] \quad (1)$$

where $D$ is the dose of implanted ions, $\Phi$ is the ion flux incident on the sample, $f$ is the pulse repetition frequency, $d$ is the pulse duration, $t$ is the time of implantation and number two is due to the fact that the ions in the nitrogen discharge are molecular ions $N_2^+$ which are dissociated into two atomic ions $N^+$ being penetrated into a solid.

Table 4 data show the 20 keV and 30 keV nitrogen ion doses [12,13] (with the accuracy not worse than 0.0015) obtained for the samples of different roughness.

**Table 4.** Nitrogen doses implanted at energies of 20 keV and 30 keV.

| Voltage Discharge | Polished Surface | Nitrogen Doses [ions/cm$^2$] |
|---|---|---|
| 20kV | Cloth 6μ | 4.4409*10$^{17}$ |
| 20kV | Cloth 1μ | 4.0089*10$^{17}$ |
| 30kV | Cloth 6μ | 7.5501*10$^{17}$ |
| 30kV | Cloth 1μ | 7.1269*10$^{17}$ |

## 5. XRD characterization

The structural characterization of the sample surfaces is fulfilled with the method of x-ray diffraction on polycrystals (XRDPD) by using Bragg-Bentano geometry equipment. The samples are mounted on its aluminum sample holder of a Rudaku diffractometer, model D/MAX IIIB, and the tests are fulfilled under the working conditions reported in table 5. The qualitative and comparative analysis of phases and compounds of the implanted and non-implanted samples of different roughness is made by comparing the diffraction profiles obtained experimentally with the profiles registered in PDF-2 data of the International center for Diffraction Data (ICDD). In the process of nitrogen implantation into steel, the formation of iron nitride FexN is theoretically possible, but in the obtained spectra the peaks corresponding to this compound have not been observed, which can be attributed to a relatively low content of the nitrogen atoms in the steel substrate. The nitrogen atoms may reside in the crystal interstice spaces and are screened or are overlapped by some element or compound.

**Table 5.** Working conditions of the XRD.

| Parameter | Value |
|---|---|
| Voltage | 40kV |
| Current | 20mA |
| Slit DS | 1° |
| Slit RS | 0.3mm |
| Slit SS | 1° |
| Sampling | 0.02° 2-Theta |
| Measuring Range | 30°-120° 2-Theta |
| Radiation | $CuK_{\alpha 1}$ |
| Monochromator | Graphite |
| Type of measure | Continuous |
| Scanning speed | 1.2°/min |

Nevertheless, the superposition of the obtained profiles for each of the samples (implanted and non-implanted) evidences that nitrogen causes some changes in the AISI SAE 4140 steel surface: the crystal cell size decreasing, some increase in the 2-theta positions of the diffraction maxima and diminishing of the intensity of the (110) diffraction plane. The higher deformation is observed in the samples which get a higher nitrogen dose; these are the samples with 6 μ polishing and implanted by

30 keV ions. In this sample, the cubic crystal structure is transformed into a pseudocubic (see figures 2(a) and 2(b)). The crystallite dimensions for each of the samples are calculated in accordance with the Braker TOPAS program [18] which determines that the small displacements of the (110), (200), (211), (220) and (310) are generated by micro-tensions provoked by the implanted ions. The values presented in table 6 show that the crystallite dimensions are reduced by implantation due to the nitrogen ions collisions with the crystal structure.

## 6. Conclusions and recommendations

In this article, it is demonstrated that the high voltage pulse discharges excited at very low pressures are well suited to implantation treatment of steels. It is observed that the implantation of 20 keV and 30 keV nitrogen ions into the AISI SAE 4140 steel sample decreases significantly the superficial roughness. It is determined that the implantation of high energy nitrogen ions also causes some decrease in the (110) plane intensity, reduces the cell in 2-theta values and diminishes the crystallite dimensions in respect to the cases without implantation.

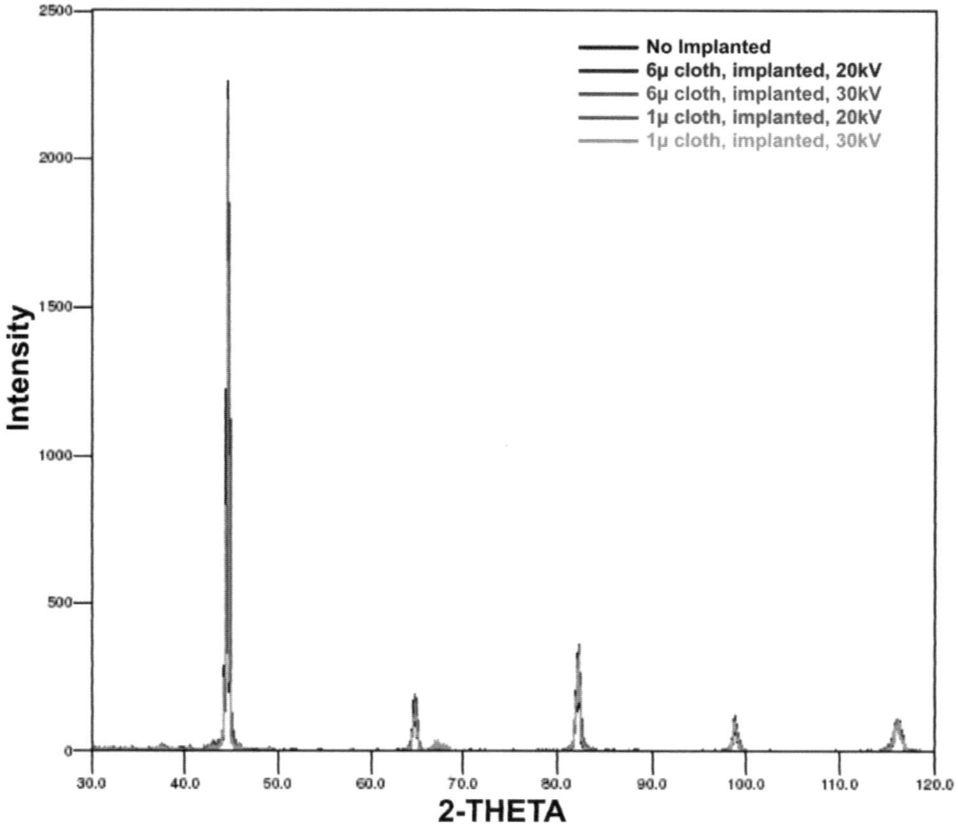


**Figure 2(a).** Overlapping XRD pattern. Superposition of the diffraction profiles for the implanted and non-implanted samples of different roughness.

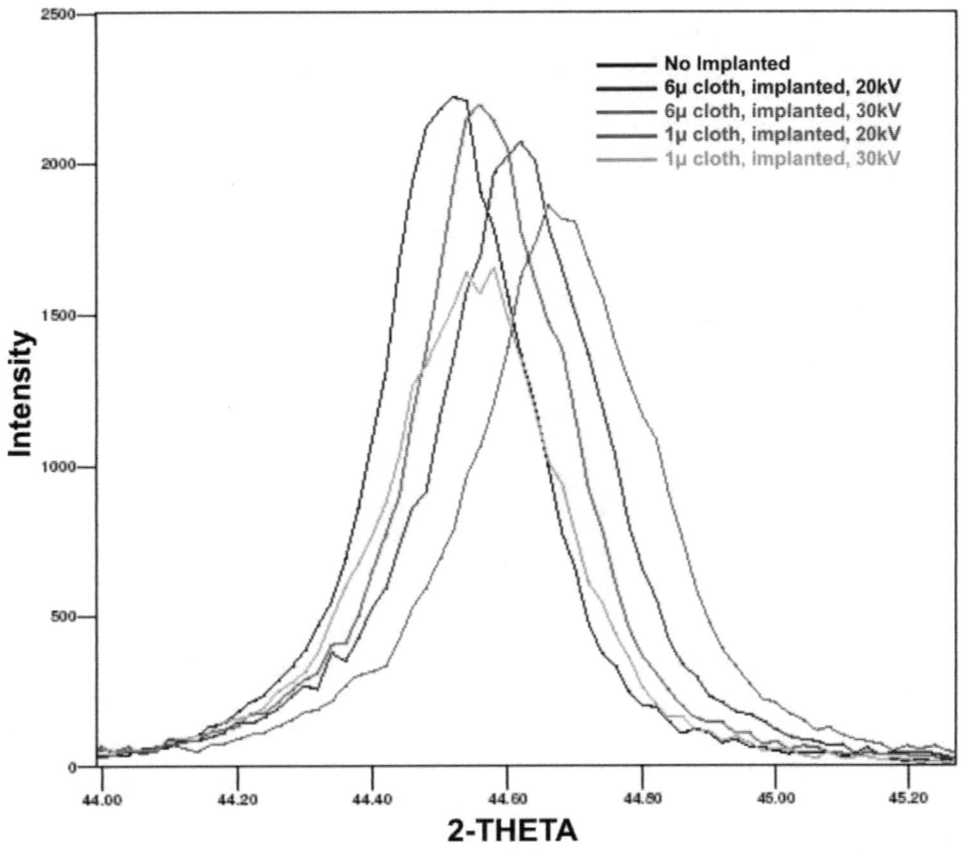


**Figure 2(b).** Overlapping XRD pattern. Enlargement of superposition of the diffraction profiles for the implanted and non-implanted samples of different roughness in the plane (1 1 0).

**Table 6.** Crystallite size calculated for the plane (1 1 0).

| Sample surface smoothing | Plane(1 1 0)Crystallite Size (nm) |
|---|---|
| No Implanted | 90.0 |
| 6μ cloth, implanted, 30kV | 58.6 |
| 6μ cloth, implanted, 20kV | 68.7 |
| 1μ cloth, implanted, 30kV | 66.5 |
| 1μ cloth, implanted, 20kV | 69.9 |

## Acknowledgments

This work was carried out with financial support under the Colciencias code 1102-06-17623.

## References

[1] Current M I, Yamada I and Cheung N W 1992 *New Technologies for Ion Implantation* (Ion Implantation Technology) ed J F Ziegler (North-Holland)
[2] Khvesyuk V I and Tsygankov P A 1997 *Surface and Coatings Technology* **96** 68
[3] Conrad J R 1989 *Materials Science and Engineering* **A116** 197
[4] McClure G W 1961 *Physical Review* **124-4** 969

[5] Dougar-Jabon V D, Castro B J, Dulce Moreno H J and Tsygankov P A 1999 *Revista Colombiana de Física* **31** 181
[6] Dougar-Jabon V D, Dulce Moreno H J and Tsygankov P A 2003 *Revista Respuestas* **8** 19
[7] ASTM A322 2007 *Standard Specification for Steel Bars, Alloy, Standard Grades*
[8] Valvuena Niño E D and Dougar-Jabon V D 2006 *Revista Colombiana de Física* **38** 61
[9] Machado J, Brühl S, Feugeas J N and Gomez B J 1999 *Surface Engineering* **15** 313
[10] ASTM E 3-95 1995 *Standard Practice for Preparation of Metallographic Specimens*
[11] ASTM E3-01 2007 *Standard Practice for Preparation of Metallographic Specimens*
[12] Peña D, Fontalvo P, Estupiñan H, V Niño E D, Vesga W 2009 *Revista Dyna* **76** 43-52
[13] Valbuena Niño E D, Dulcé H J and Dugar-Zhabon V 2010 *Revista Colombiana de Física* **42** 387-92
[14] Dougar-Jabon V D, Dulce Moreno J and Tsygankov P A 2002 *Review of Scientific Instruments* **73** 828
[15] Rueda A, Dulce H, Peña G and Torres P 2006 *Revista Colombiana de Física* **38** 894
[16] Dulce Moreno J H, Tsygankov P A, Dugar-Zhabon V D, V Niño E D and Parada F F 2011 *Puente Revista Científica* **5** 21-6
[17] Shamim M, Scheuer J T, Fetherston R P and Conrad J R 1991 *Journal of Applied Physics* **70-9** 4756
[18] Coelho A 2007 *Topas Academic Version 4.1. Computer Software* (Topas Academic Coelho Software Brisbane)

Journal of Physics: Conference Series **370** (2012) 012032 doi:10.1088/1742-6596/370/1/012032

# Structural and tribological properties of Ti-TiN coatings obtained with a vacuum arc

**L S Vaca[1], A Márquez[2,3] and S P Brühl[1]**

[1]Grupo de Ingeniería de Superficies – Facultad Regional Concepción del Uruguay, Universidad Tecnológica Nacional, Argentina
[2] INFIP, UBA-CONICET, Cdad. Universitaria, Pab. I, C1428EGA Cdad. Autónoma de Buenos Aires, Argentina
[3]Member of CONICET

E-mail: laurasvaca@gmail.com

**Abstract.** In this work TiN and Ti-TiN bilayer coatings obtained with a vacuum arc on AISI 316 samples were proposed in order to improve hardness and tribological properties. The coatings were deposited with a DC cathodic arc running a 130 A current discharge between a Titanium cathode and the vacuum chamber employed as anode. The samples were placed 20 cm away from the cathode surface. TiN films were obtained employing a continuous nitrogen flux of ~ 20 sccm and a working pressure in the range of 3 $10^{-2}$ Pa. The structural and tribological properties were characterized by means of an optical microscope, scanning electron microscopy (SEM), x-ray diffraction (XRD), scratch test and fretting. The film thickness was about 0.5µm. Ti-TiN coatings presented better adhesion than TiN coatings and improved the fretting fatigue life compared to the untreated steel.

## 1. Introduction

The AISI 316L stainless steel is widely used in chemical, nuclear and food industry as well as in medical implants due to its excellent corrosion resistance and good biocompatibility. However, its low hardness, wear resistance and high friction coefficient limit the potential application of this material [1-3].

A way of improving the corrosion resistance and the adhesive wear resistance or the friction coefficient could be to coat the surface with a ceramic film like TiN. These kinds of coatings are widely used for hardening cutting tools made of high speed steels [4-6]. In particular, cathodic arcs have proved to be very attractive to deposit TiN coatings because they promote the formation of dense nanostructured films with a good adhesion and a very high deposition rate. In cathodic arc devices, an electrical discharge of high current (~100 A) and low voltage (~20 V) runs between two electrodes in a vacuum chamber. From the surface of the metallic cathode, a beam of metallic ions is ejected with energies between 20 – 100 eV; the ion current being approximately a 10% of the total discharge current. When a sample is exposed to the plasma, the surface is coated with a metallic film. If a reactive gas flows in the chamber, the substrate is covered with a metallic compound of the cathode material [7-9].

Journal of Physics: Conference Series **370** (2012) 012032 doi:10.1088/1742-6596/370/1/012032

In this work TiN and Ti-TiN bilayer coatings on AISI 316L samples were obtained with a Titanium vacuum arc. Their mechanical properties were investigated and their corrosion resistance was also qualitatively assessed.

## 2. Experimental procedure

The samples were cut out with a size of 26 mm in diameter from a plate 2 mm in thickness, of AISI 316L whose certified composition is presented in table 1. The surfaces were polished and cleaned using an ultrasonic cleaner in acetone during 10 minutes.

The coatings were deposited with a DC cathodic arc discharge employing a Titanium cathode and the grounded vacuum chamber acting as anode. The arc current was set at 125 A. The samples were placed 30 cm away from the cathode surface and connected electrically to ground.

**Table 1.** Chemical composition AISI 316L.

| **C** % | **Si** % | **Mn** % | **P** % | **S** % | **Cr** % | **Mo** % | **Ni** % | **Cu** % | **Ti** % | **N** **ppm** |
|---|---|---|---|---|---|---|---|---|---|---|
| 0.017 | 0.500 | 1.37 | 0.027 | 0.002 | 16.65 | 2.04 | 10.03 | 0.140 | 0.006 | 216 |

A group of samples was coated with TiN films and a second group of samples with a Ti-TiN bilayer. The samples coated with TiN were exposed to four discharges of three minutes each (a total time of 12 min) with 20 sccm $N_2$ flow at a work pressure of $3.10^{-2}$ Pa. The Ti-TiN samples received a discharge in vacuum of 1.5 minutes at a lower working pressure, $10^{-2}$ Pa, and then three discharges of three minutes each with a 20 sccm $N_2$ flow, achieving a working pressure of $3.10^{-2}$ Pa, adding up a total time of exposure to the discharge of 10.5 min. In all cases, the waiting time between two consecutive discharges was 15 minutes.

The deposited mass was determined by weighting the samples before and after the discharges using an Ohaus analytical balance, model AS200. The coating adhesion was evaluated by means of the Tape Test performed with a commercial scotch tape and the Scratch Test employing a Hand Operated Scratch Tester (model 706, Sheen Instrument Ltd.) with a 1 mm tungsten carbide ball with a needle point and with constant loads varying from 5 to 20N. The coating surfaces were observed with a scanning electron microscope Philips SEM 505 (with Soft Imaging System ADDA II (SIS)). Also, the samples were cut and embedded in resin, so as to observe the lateral profile of coatings with a scanning electron microscope Philips 515. The coating structure was studied with X ray diffraction (XRD) employing a Philips diffractometer 1732/10.

Qualitative corrosion tests were performed in order to evaluate the corrosion resistance of the samples. First, the $CuSO_4.5H_2O$ drop test was performed according to ASTM A967 in practice "D" to evaluate the passivity of the material, then the samples were attacked with Marble reagent during 10 s. Marble reagent contains chlorhydric acid (4g $CuSO_4$, 20 ml HCl, 20 ml $H_2O$) and it is normally used to reveal structure in austenitic stainless steels.

Wear resistance was evaluated by means of the linear reciprocal sliding and fretting tests. The samples were subjected to a reciprocal sliding wear test, with 2.85 N load, WC counterpart ball of 5 mm in diameter, 953 μm in amplitude and half hour duration. The reciprocal test wear tracks were observed with 50x magnification with an optical microscope Mikoba with Digital Camera DCM500 (USB2.0) Resolution 5 Mpixels and the wear scar depth was measured with a roughness tester acting as a profilometer (Mitutoyo Surftest SJ-210). The fretting test was carried out with the same load, counterpart ball and time, but only with 84 μm amplitude, in order to analyze another kind of wear mechanism, mainly plastic deformation and less abrasion. The wear scars were observed with scanning electron microscope (SEM).

## 3. Results

The deposited mass in the case of TiN coatings was (0.7 ± 0.1) mg while the gained mass in the samples coated with Ti-TiN was (1.6 ± 0.1) mg. The big difference in the deposited mass for similar exposure times can be attributed to the fact that the erosion rate of the cathode is higher in vacuum than in the presence of nitrogen. This could be the consequence of the fact that the contamination of the surface not only diminishes the erosion rate, but also changes the conditions of the discharges turning them less repetitive.

The thickness of the TiN coatings can be estimated assuming a density for the film as the tabulated for the bulk material ($\rho_{TiN}$= 5.4 g/cm$^3$), then the TiN film thickness was ≈ 250 nm. In the case of the bilayers, composed by Ti ($\rho_{Ti}$= 4.5 g/cm$^3$) and TiN, both density values were employed and the film thickness was estimated in the range (630 – 750) nm. SEM micrographs with 5000x magnification of lateral cuts of a TiN film and a bilayer film are shown in figure 1. In the SEM images the film cannot be observed along the whole surface; because part of the film was broken when the sample was cut previously to the inclusion in resin. However, the film thickness could be measured and was about 700 nm for the Ti-TiN coating and 300 nm for the TiN coating, agreeing with the calculated values.

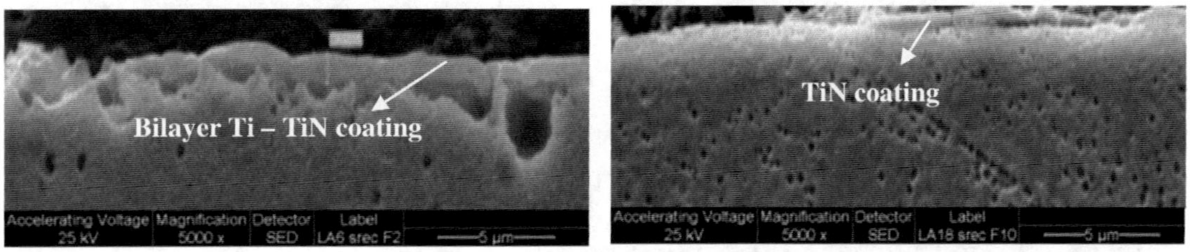


**Figure 1**. SEM micrograph of lateral cut of samples: (a) Ti-TiN coating and (b) TiN coating.

In figure 2, XRD patterns of the bilayer and TiN coatings are shown. In figure 2(a) the characteristic diffraction peaks of austenite and TiN corresponding to planes (111) and (200) can be distinguished clearly; and the small peak located between TiN peaks at 38.5 degrees can be associated to the Ti (101) plane. In the case of the TiN coating, the same TiN peaks were registered in the diffractogram and also appeared the third TiN peak, corresponding to the (220) plane.

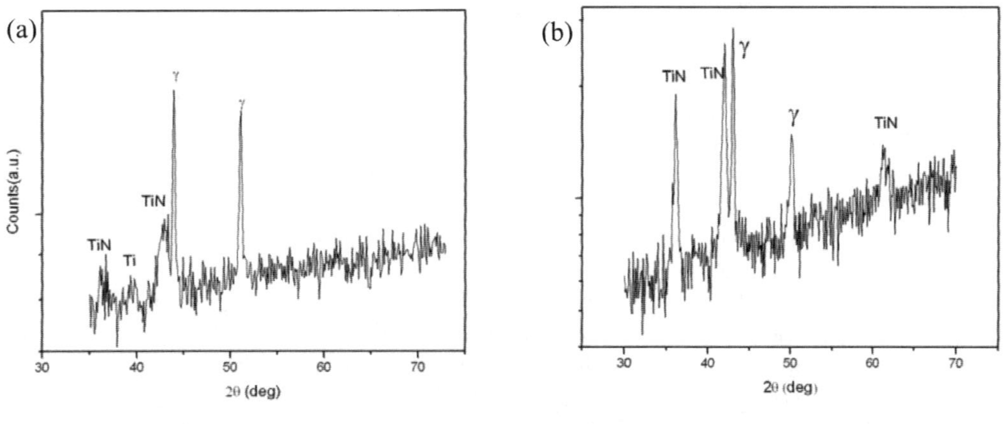


**Figure 2**. X-ray diffractograms for: (a) Ti-TiN coating and (b) TiN coating.

The surface morphology of the coatings can be observed in the SEM micrographs presented in figure 3. The Ti-TiN bilayer coatings exhibited an irregular surface with the presence of a few macroparticles while the TiN coatings surface was smoother. Small holes, probably left by the detachment of macroparticles, can also be noticed on both surfaces. However, SEM images with 100kx magnification of lateral cuts of the coatings showed compact films without holes, as shown in the SEM image of figure 4.

The coating roughness was (0.224 ± 0.020) μm for the Ti –TiN bilayer and (0.088 ± 0.007) μm for the TiN film.

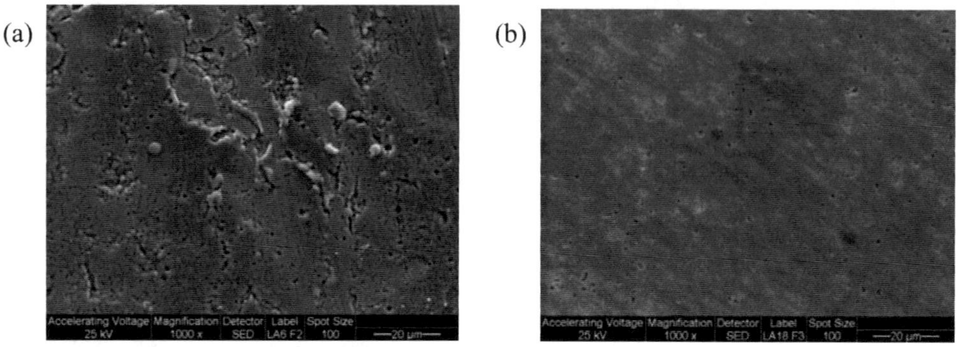


**Figure 3.** SEM images 5000x magnification of surface coatings (a) Ti-TiN and (b) TiN.

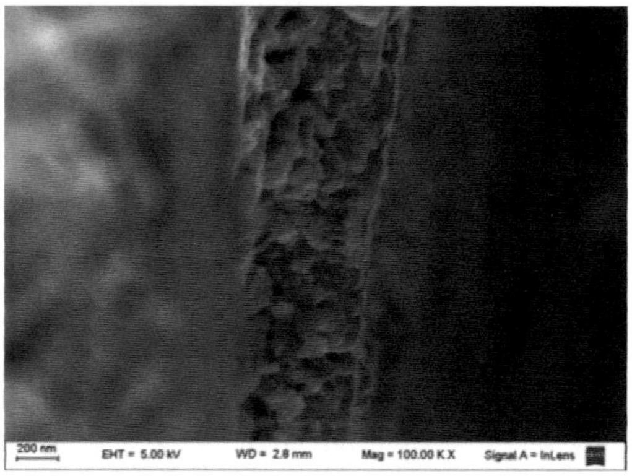


**Figure 4**. SEM image with 100kx magnification of the lateral profile of a Ti-TiN coating.

The Scratch Test allowed determining that the bilayer coatings had better adhesion to the substrates than the TiN coatings. Pictures of the tracks for the different films are shown in figure 5. For the TiN films some delamination failures on the track were detected with a load of 10 N, while for the bilayer coating no failures were observed in the tested load range, until 20 N. Probably, the better adhesion behavior of the Ti – TiN film could be attributed to its greater thickness but it could be also a consequence of the presence of the Ti interlayer, which generates a graduated hardness profile between the film and the substrate [10-12].

The $CuSO_4$ $5H_2O$ drop test and the Marble reagent attack yielded negative results in both types of coatings. Even though these are qualitative experiments; they could indicate that the films are chemically inert like other authors reported for these kind of films [13,14].

SEM images of the scars after performing the fretting test are shown in figure 6. The Ti-TiN coating (figure 6(b)) demonstrated a better behavior than the untreated sample and the TiN coating. Anyway, the scars in both coated samples are of smaller size than in the untreated material. In spite of the surface imperfections, it can be noted that in the Ti-TiN coating the fretting test only produced a small deformation while in the TiN coating the scar were bigger, a small amount of abrasion was observed such as in the untreated sample and also the film was delaminated around the scar.

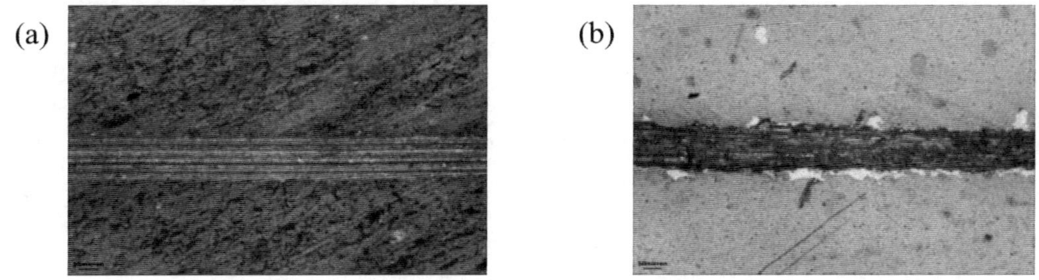


**Figure 5.** Scratch track (a) obtained with a 20N load on Ti-TiN and (b) obtained with 10N load on TiN.

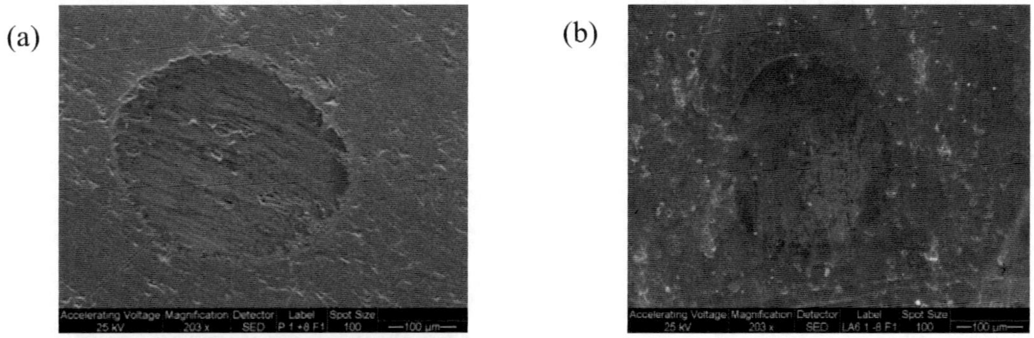


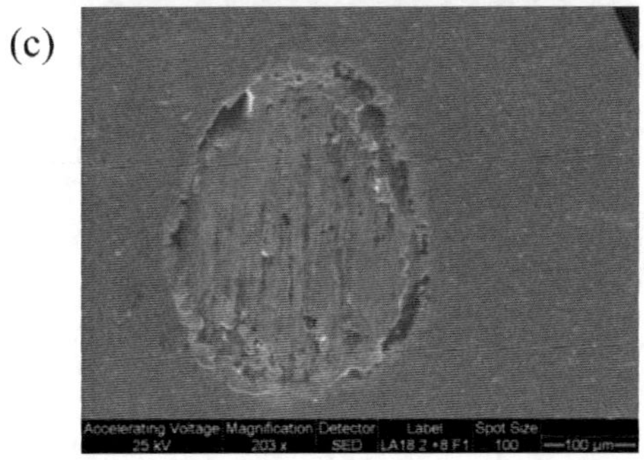


**Figure 6.** SEM images with 200x magnification of fretting scars: (a) untreated sample, (b)Ti-TiN coating and (c) TiN coating.

In the case of the linear reciprocating tests, it can be seen in figure 7 that in the untreated sample the wear track was bigger and deeper, and severe abrasion could be observed, together with a large amount of deformation. Such as in the previous test, both coated samples showed a better wear resistance than the untreated material, but the wear scar in the Ti-TiN coated sample was the smaller. The TiN film presented again some delamination around the scar.

The scar profiles were obtained with the profilometer. For a better comparison, the profile points were copied in an Origin worksheet, they were all aligned and centered, and the final profiles were depicted all together in one graphic (figure 8).

The scar profiles sections have been measured and the values were compared to that of the untreated sample. Assuming this result as the volume loss in the wear test, then in the TiN coated sample the loss was about 57% respect to the value measured for the untreated sample and in the Ti-TiN, it was only 19%. In both films the wear track was deeper than the film thickness, meaning that the films were lost throughout the test, and as the TiN film was thinner than the Ti – TiN one, its wear resistance was lower.

**Figure 7.** Micrograph with 50x magnification of reciprocal sliding scars: (a) untreated sample, (b)Ti-TiN coating and (c) TiN coating.

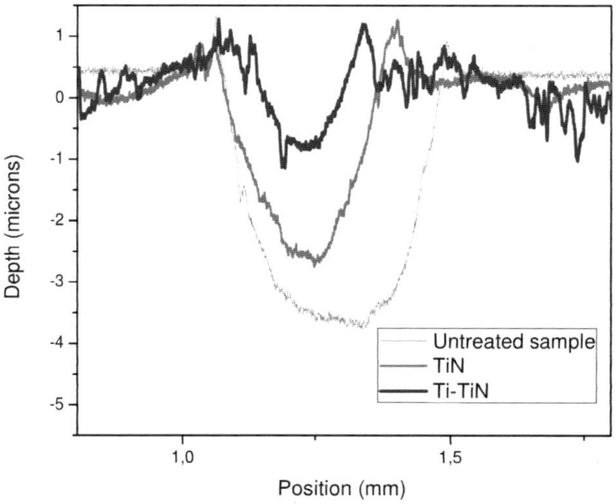


**Figure 8.** Reciprocal sliding wear scars.

## 4. Conclusions

The qualitative corrosion tests showed that both coatings were chemically inert, quantitative corrosion tests should be carried out in the future. In all mechanical tests the scars were greater in the TiN film than in the Ti – TiN film and some delamination was observed around the scars. Also the scratch test revealed the failure of the TiN coating for 10 N load when the Ti-TiN was not broken with 20 N. Both films protected the steel in the tested conditions but it should notice that in the case of the Ti-TiN film, all the scars were very shallow indicating a better wear resistance, especially in the linear reciprocating test, where it was increased 75% with respect to the untreated material.

Finally, it is generally accepted that the presence of macroparticles is a drawback of the cathodic arc process; although in these work the performance of the coatings did not seem to be strongly affected by them, they could be eliminated with the incorporation of filters like a magnetic island.

**Acknowledgments**
The authors would like to thank J. Tara and M. Sánchez (CINDECA-UNLP) for their technical assistance in DRX and SEM; J. Cimetta, N. García, N. Pecina, G. Schierloh and E. Dalibón (FRCU-UTN) for the collaboration in the wear tests. They would also like to thank the financial support from FRCU and CONICET.

**References**

[1] Kottel C M, Sprague J A and Smidt F A 1994 *ASM Handbook* vol 5 (Ohio: ASM International)
[2] ASM International Handbook Committee 1990 *Metals Handbook* vol 1 (Ohio: ASM International)
[3] Sendriks A J 1996 *Corrosion of Stainless Steel* 2nd Ed. (New York: Wiley & Sons)
[4] Rybiak R, Fouvry S, Liskiewicz T and Wendler B 2007 *Surf. Coat. Technol.* **202** 1753–63
[5] Wu P-Q, Tang B and Celis J-P 2006 *Surf. Coat. Technol.* **201** 413-7
[6] de Wit E, Froyen L and Celis J–P 1998 *Wear* **221** 124–33
[7] Bruzzone H, Kelly H, Márquez A, Lamas D, Ansaldi A and Oviedo C 1996 *Plasma Source Sci. and Technol.* **5** 582–7
[8] Boxman R L, Martin P J and Sanders D M (eds) 1995 *Handbook of Vacuum Arc Science and Technology* (New Jersey: Noyes)
[9] Krella A 2009 *Surf. Coat. Technol.* **204** 263-70
[10] Chen L, Wang S Q, Zhou S Z, Li J and Zhang Y Z 2008 *Surf. Coat. Technol.* **26** 456-60
[11] Thao Y, Lin G, Xiao J, Du H, Dong C and Gao L 2011 *Appl. Surf. Sci.* **257** 2683-8
[12] Puchi-Cabrera E S, Martínez F, Herrera J and Berríos J A 2004 *Surf. Coat. Technol.* **182** 276-86
[13] Liu C, Yang D, Lin G and Qi M 2005 *Mat. Lett.* **59** 3813-9.
[14] Liu C, Lin G, Yang D and Qi M 2006 *Surf. Coat. Technol.* **200** 4011-6

Journal of Physics: Conference Series **370** (2012) 012033 doi:10.1088/1742-6596/370/1/012033

# Simulation of plume-plasma expansion with one-dimensional Particle-in-Cell

**C A Gonzalez[1], J A Arteaga[1],Y H Gomez[2], J Osorio[2], J A Jaramillo[2] and H Riascos[1]**

[1] Grupo Plasma Láser y Aplicaciones, Universidad Tecnológica de Pereira A.A 097, Pereira, Colombia

[2] Grupo Sirius, Universidad Tecnológica de Pereira A.A 097, Pereira, Colombia

E-mail: caangonzalez@utp.edu.co

**Abstract.** In this work we present the analysis of the dynamic of the expansion of Al Plasma produced by Nd:YAG laser (1064 nm, 500mJ, 9 ms, 10 Hz) in vacuum. To study the Coulomb interaction between the particles of the initial states of the plasma expansion, we use the one dimensional Particle-in-Cell method (PIC) and finite difference method. We considered an ideal model, that is, we assume that the plasma is in a local thermal equilibrium, the ablated particles have a fixed temperature and a constant evaporation flux ($J$) from the aluminium surface. To obtain more accurate results we use high computing exploiting the parallelization of this kind of algorithms. The mean velocity and particles densities are determined for different times of the expansion.

## 1. Introduction

Studies of plasma produced by pulsed laser have been taking great importance due to its multiple applications in different scientific and industrial fields. One of its most important applications is the deposition of thin films by laser ablation commonly called Pulsed Laser Deposition (PLD) [1, 2, 3, 4].

PLD involves the incidence of high-energy laser in a solid that absorbs its radiation. This energy is transmitted to the ejected species in the form of kinetic energy [5, 6]; Ablation process is generated when the kinetic energy absorbed by the species is greater than its binding energy in the material. The result of this process is a cloud of particles composed of ions, electrons and neutrals, called plasma plume. In this cloud there is a collective interaction between the ablated particles through the fields generated from these particles. The representation of this interaction is a complex and nonlinear problem, which includes the use of a wide range of physical concepts. Research about pulsed laser deposition is made through theoretical and experimental models.

Many important details of the plasma are highly difficult to obtain experimentally. Therefore the numerical simulation plays an important role in understanding the phenomena that occurs there. In this way the theoretical models can be checked and additionally, get more detailed information about what happened in the process.

In this paper we used the PIC computational method for studying the one-dimensional initial stage of the plasma plume expansion in vacuum. It was accomplished by analyzing the change of kinetic momentum produced by the Coulomb interaction between the plasma particles. To make the results more approximate to those predicted by analytical models it is required to define a finer discrete model. However, the use of fine discrete models increases exponentially the algorithms simulation times. It is for this reason that is required to use a strategy of high performance computing, where we can take advantage of the latest parallelization platforms such as GPUs.

Here we present the acceleration of ions and electrons of Al plume plasma in vacuum as function of interaction time between laser pulse and the target using PIC method and high performance, assuming Coulomb interaction for species in plasma only. It was determined that some instabilities were generated while the expansion plume occurred.

## 2. Model

The study of the expanding plasma plume in vacuum consists of two stages. An initial stage where the particles are ejected from the solid surface, during the interaction between laser pulse and the solid in a finite time $t_0$. And then, a second stage, where no additional particles are ejected allowing for the Coulomb interaction between particles in the cloud.

Particles emitted in the initial stage are principally composed of ions (charge +e, mass M) and electrons (charge -e, mass m). These particles were ejected in the normal direction to the surface ($x - axis$) through a constant evaporation flux $J$. The generated cloud of ions and electrons as a whole has a neutral electric charge, because the evaporation flows of both particles are assumed to be equal and constant. In addition, these particles are ejected from the surface with a thermal energy $kT_0$ that depends on both the material and the characteristics of the laser, in our case we assumed that they reach thermal equilibrium.

In the second stage it is assumed that the particles are ejected with initial velocities that satisfy a half-space Maxwell distribution function

$$f_i(x = 0, v_x, t) = J\left(\frac{M}{kT_0}\right) exp\left(-\frac{Mv_x^2}{2kT_0}\right) \tag{1}$$

where $J = n_0 v_0$ is the flow of ions or electrons immediately after the incidence of the laser, $k$ is Boltzmann's constant and $v_x$ is the velocity perpendicular to the surface. Immediately after the expulsion of particles, the expansion process of the cloud begins due to the Coulomb interaction between particles. This interaction is based on the potential difference generated by the charge distribution in certain regions of the cloud described by the equation:

$$\nabla^2 \phi = -e\,\frac{n_i - n_e}{\epsilon_0} \tag{2}$$

where $n_{i,e}$ represents the density of particles of both species. The fundamental basis for describing the evolution of collisionless plasma is given by the Vlasov equation:

$$\partial_t f + v_x \partial_x f + \frac{eE}{M}\partial_{v_x} f = 0. \tag{3}$$

This equation determines the time evolution of a statistical system described by a distribution function $f$. In this way it achieves the kinetic model of particles in phase space ($x$,$v$). From (3) it is possible to obtain different conservation laws, i.e. the conservation of the particle numbers (zero order momentum)[7].

## 3. Simulation

### *3.1. PIC description*

The PIC method is a technique that allows the tracking of particles in a space. In order to develop a computational model, a discretization of the space is needed as well as the definition

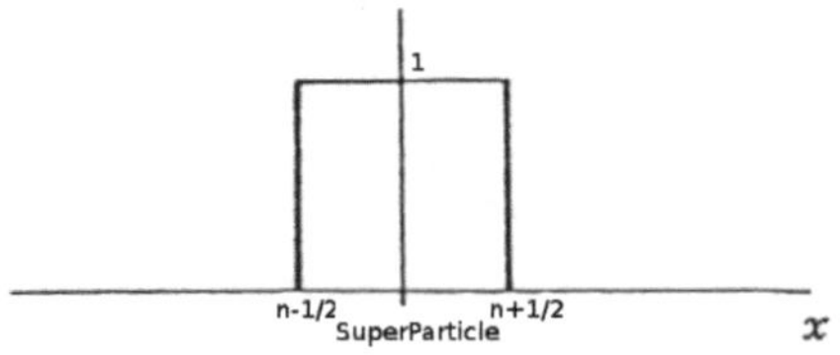


**Figure 1.** Zeroth order weight.

of the initial values of the system. Later, the forces are calculated by using those initial values, and the next position of each particle in the discretized space. Using next state we can close the loop and make the calculations during the time in order to complete the path of the movement of each particle.

First, a charge is assigned to the group of particles in a cell, called 'superparticle'. Since the charge is not uniformly distributed, a potential difference can be generated between the neighboring cells that is assumed to be a field of zero order weight[8, 9],that is, the n-th cell particles are affected from the difference of charge between both the $n-1$ and $n+1$ cell, no matter the real position of the center of the charge of the $n-th$ cell as it is shown in figure 1. Once the charge density at each cell is known, the Poisson equation is solved to find the potential difference between the cell and its neighbors, subsequently the electric field due to the potential difference given by the equation: $E_n = -\nabla\phi_n$ is evaluated.

By knowing the n-th cell's electric field, its effect on each particle of that cell is weighted. Then, their new position and velocity are found, according to Newton's second law, for each discretized time step $\Delta t$. The summary of the PICs method is shown in figure 2.

### *3.2. Discrete space*

Since the equations describing the system are differential equations, the most appropriate method to solve them in the discrete space is the one-dimensional finite difference method [10, 11], this is,

$$\frac{\partial^2\phi}{\partial x^2} \approx \frac{\phi_{n+1} - 2\phi_n + 2\phi_{n-1}}{\Delta x^2} \tag{4}$$

where n represents the $n-th$ cell. To integrate the equations of motion we use the $leap-frog$ algorithm which creates a coherent calculation cycle, given by the following formulas [12]:

$$\begin{aligned} x_{n+1} &= x_n + v_{n+1/2}\Delta t \\ v_{n+3/2} &= v_{n+1/2} + \frac{eE_n}{M}\Delta t \end{aligned} \tag{5}$$

### *3.3. Stability criteria*

To apply the PIC method in the expansion of the plume, the simulation space must be divided into cells smaller than the Debye length $\Delta x \ll x_0$, this in order to neglect the collisions among particles [13].

The time $t_0$ is the time that the particles emitted take to go through a Debye length with an ejection speed $v_0$, ($t_0 = x_0/v_0$). According to these conditions $\Delta t$ should be less than the

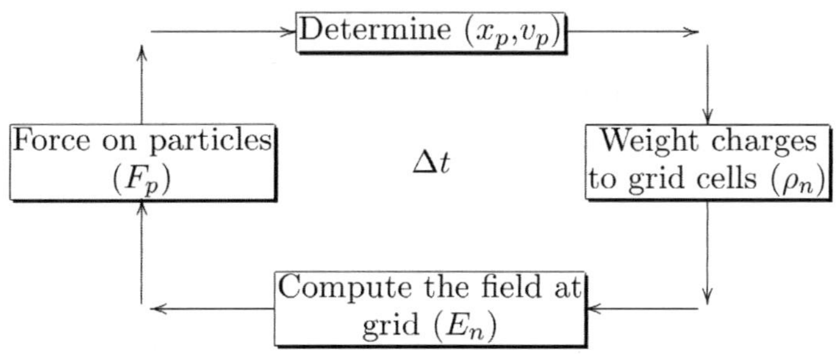


**Figure 2.** Basic cycle of the Particle In Cell simulation.

plasma wave period($\Delta t \ll 2\pi/\omega_e$) to satisfy the Courant criteria where $\omega_e$ is the electron plasma frequency.

For dimensionless physical quantities, all the equations that govern the system were normalized using $x_0$, $t_0$, ie, $\Delta x/x_0$, $\Delta t/t_0$. For the distribution function, the kinetic energy of each particle is normalized with the average energy $kT_0$. Finally, the normalized field takes the form $E/(v_0 M/et_0)$.

### *3.4. Simulation details*

Since all particles in the system get an initial velocity greater than zero, the boundary condition $\phi(x=0)=0$ is considered in order to ensure that all the particles that return to the surface are reflected [14]. When the simulation space $x_0$ is overflown by the particles ($x_p > x_0$) , these are re-injected at, $x_p = x_0 - x_p$ in order to satisfy the conservation of the charge.

As a consequence of the previous conditions, the electric potential must be periodic $\phi(x) = \phi(x+x_0)$. According to previous authors [15], the approximate parameters to simulate Al plasma produced by Nd: YAG laser (1064nm, 500mJ, 9ms, 10Hz) are presented in table 1. The initial velocity for the ions given by $v_{0,i} = \sqrt{2kT_o/\pi M}$; in our case determined a temperature $T_0 = 10000K$ and for electrons $v_{0,e} \approx 7v_{o,i}$.

### *3.5. Algorithm optimization*

In order to improve the performance of the algorithm, a data dependence analysis between functions must be made. In this case, the force particle function and electric field grid function are independent of each other, thus allowing for the parallel execution of both processes in different CPU cores. That is why the high performance tools, based on floating point operations, known as GPUs, were used in the present work. GPUs act as coprocessors to the CPU's data bus using a high-transfer [16]. In order to further reduce simulation times and take the best advantage of the GPU, the use of OpenCL framework from the KRONOS consortium [17] was used. This framework allows the creation of calculus kernels using a programming language like C, C++, or FORTRAN and it's also costless.

**Table 1.** Simulation parameters for Al plasma.

| $N_p$ | $N_g$ | $\Delta x/x_0$ | $\Delta t/t_0$ |
|---|---|---|---|
| 1x$10^5$ | 5x$10^3$ | 5x$10^{-3}$ | 1x$10^{-5}$ |

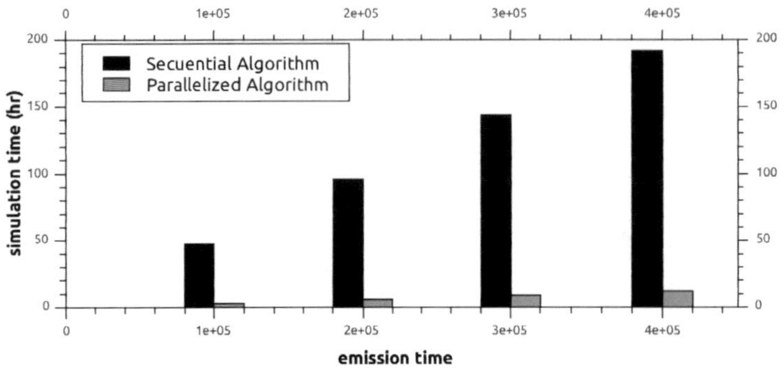


**Figure 3.** Execution times for the sequential and parallelized algorithms.

Parallelized and sequential algorithm was run on a computer with the following characteristics: GTX580 Video Card that has 512 processing cores, 16 GB of RAM Intel Core i7 processor, 2 solid state drives and a motherboard P6T7 Axus SuperComputer.

Figure 3 shows the linear behavior of both processes where the GPU's efficiency is evidenced, having an improvement of 16x. Also, the execution time is doubled at each increment of emission times shown for the two algorithms.

## 4. Results and discussions

To investigate the dynamics of expansion of the plasma plume, we construct the phase space of the system, which consists of identifying each of the particles according to its velocity ($v_p$) and position ($x_p$) in an instant of time t. Figure 4 shows the phase space ($x_p$,$v_p$) system for the emission time $t_0$, where the black dots indicate the presence of an ion, and the blue ones the presence of an electron. From figure 4 it is observed that the electrons achieve greater distance and speed compared with ions, this due to the ratio of their masses ($m \ll M$). In the region where both species coincide in position, it shows clearly the oscillation of the electrons around the ions, the electrons are trapped in the potential that they generate. After this region, a certain number of electrons that exceed their kinetic energy and thus manage to move freely ($mv_e = cte$) can be found. As for the dynamics of ions it is obvious the linear collective behavior, that can be known thoroughly studying its density. To find the particle density we turn to the solution of the expression, $n(x,t) = \int f(x, v_x, t)dv_x$ divided into intervals of equal length, the simulation space; in our case we took ten points for each $x_0$ ($x/x_0 = 10$) as shown in Equation(1), and

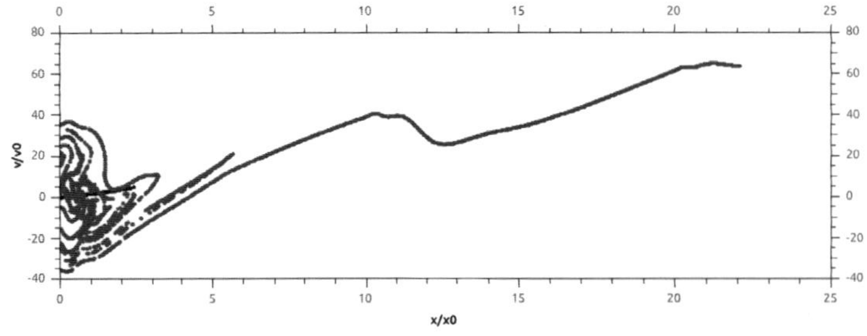


**Figure 4.** Ion phase-space (black) and electron phase space (blue).

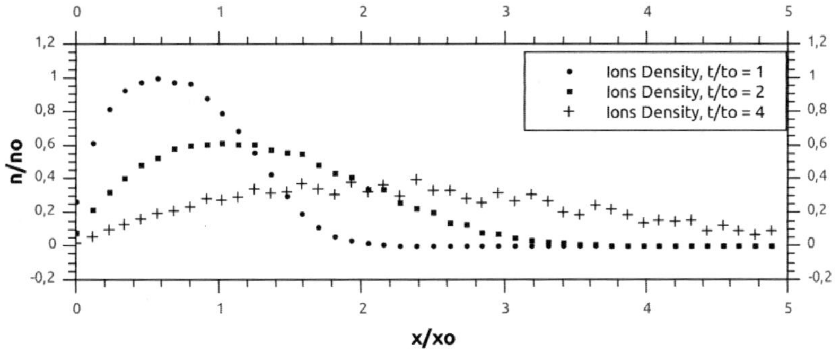


**Figure 5.** Ion density vs ablated surface distance at different times.

finally, we integrated the velocity of each particle that was found at each interval. Figure 5 shows that for an emission time ($t/t_0 = 1$), the ion density decreases exponentially for $x > 0.5x_0$, just as the two fluid model predicts. As long as the post-emission time ($t/t_0 > 1$) increases, density is distributed homogeneously in a larger space and, therefore showing the plume expansion. On the other hand, the electron density in a small region coincides with the ion density, according with the quasi-neutral plasma condition plasma.

The evolution of the mean velocity of ions $u(x,t) = \frac{1}{n(x,t)} \int v_x f(x, v_x, t) dv_x$ as these were far from the surface was found, as shown in figure 6, by the same method that the figure 5 was done, with the distribution function $f(x, v_x, t)$ we have resolved the equation.

Figure 6 shows the ions acceleration as they move away from the surface, this acceleration may be caused by the potential difference between these ions. As shown in figure 5, the ion density decreases exponentially down to zero, after this, the ions do not experience a change in their momentum. Therefore, the distance to the surface grows linearly with the momentum they have.

Unfortunately our results do not agree with those of other authors [15, 18], probably due to the fact that the previous results are related to certain factors, such as initial conditions, boundary conditions, and other factors inherent to the computational model. On the other hand, the consideration of the field weight in our case was zero-order, which has major implications on the final results, since the potential of a cell affects not only the neighbor cells but also the entire

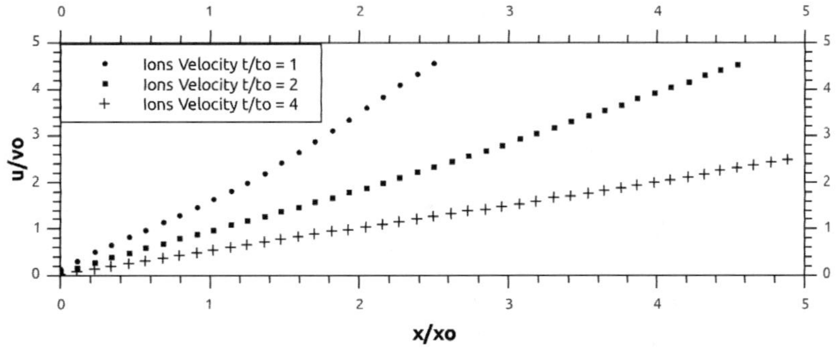


**Figure 6.** Ion mean velocity vs. distance to wall for different times of the expansion.

system itself.

## 5. Conclusions

We have calculated the density for ions and electrons of a plasma plume in expansion finding that these densities decrease exponentially down to zero and then the movement was not change. While their mean velocities increase linearly. For an evaporation time larger than $t_0$ two stream instabilities were generated. An important aspect of this work was to have made the acceleration of the code using high performance computing with the use of graphic accelerator cards, as the results were obtained in a short time in comparison to the simulation of a sequential code. This acceleration enabled us to make changes to our source code (initial conditions) without having to wait long.

## References

[1] Chrisey D and Hubler G 1994 *Pulsed laser deposition of thin films* Wiley-interscience publication (J. Wiley) ISBN 9780471592181
[2] Eason R 2007 *Pulsed laser deposition of thin films: applications-led growth of functional materials* (Wiley-Interscience) ISBN 9780471447092
[3] Riascos H, Neidhardt J, Randnóczi G, Emmerlich J, Zambrano G, Hultman L and Prieto P 2006 *Thin Solid Films* **497** 1–6
[4] Franco L, Pérez J and Riascos H 2008 *Microelectronics Journal* **39** 1363–1365 ISSN 0026-2692
[5] GBekefi 1976 *Principle of Laser Plasmas*
[6] Miller J and Haglund R 1998 *Laser ablation and desorption* Experimental methods in the physical sciences (Academic Press) ISBN 9780124759756
[7] Kruer W 2003 *The physics of laser plasma interactions* Frontiers in Physics (Westview Press) ISBN 9780813340838
[8] Hockney R and Eastwood J 1981 *Computer simulation using particles* (McGraw-Hill International Book Co.) ISBN 9780070291089
[9] Birdsall C and Langdon A 1991 *Plasma physics via computer simulation* Adam Hilger series on plasma physics (Adam Hilger) ISBN 9780750301176
[10] Henrici P 1962 *Discrete variable methods in ordinary differential equations* (Wiley)
[11] Appert K, Succi S, Vaclavik J and Villard L 1987 *Computer Physics Reports* **6** 335–349 ISSN 0167-7977
[12] Cartwright K, Verboncoeur J and Birdsall C 2000 *Journal of Computational Physics* **162** 483–513
[13] Chen F 1984 *Introduction to Plasma Physics and Controlled Fusion: Plasma physics* Introduction to Plasma Physics and Controlled Fusion (Plenum Press) ISBN 9780306413322
[14] Fehske H, Schneider R and Weisse A 2008 *Computational many-particle physics* Lecture notes in physics (Springer) ISBN 9783540746850
[15] Nedelea T and Urbassek H M 2002 *Phys. Plasmas* **9** 3209–16
[16] Macedonia M 2003 *IEEE Computer* **36** 106–8
[17] Munshi A, Gaster B, Mattson T, Fung J and Ginsburg D 2011 *OpenCL Programming Guide* OpenGL Series (Addison Wesley Professional) ISBN 9780321749642
[18] Ellegaard O, Nedelea T, Schou J and Urbassek H 2002 *Applied Surface Science* **197–198** 229–238 ISSN 0169-4332

# Drift instability grow rates in non-ideal inhomogeneous bi-dust plasmas

**J Puerta, E Castro, P Martin and F Blanco**

Departamento de Física, Universidad Simón Bolívar, Apdo: 8900, Caracas - Venezuela

E-mail: jpuerta@usb.ve

**Abstract.** Dust Acoustic Wave propagation in unmagnetized inhomogeneous dusty plasmas has been extensively studied, considering non-ideal Padé type state equation. In this paper, the drift instability is analyzed in a weakly magnetized dusty plasma including inhomogeneities of the type called gradient instabilities considering two kind of dust particles. A fully kinetic approach is achieved using a dispersion relation in a polynomial form through a multipole approximation. The drift instability for the dust acoustic mode is extensively analyzed including the non-ideal effect in dusty plasmas, using the state equation introduced by Ree and Hoover. These unstable modes are well discriminated and treated as a function of the density gradient and dust grain radius. Temperature gradients and charging processes will be ignored.

## 1. Introduction

Plasma inhomogeneities across the magnetic field in the presence of finite - size charged grains cause a wide class of instabilities of an inhomogeneous dusty plasma called gradient instabilities. Such instabilities can be studied in the approximations of magnetic fields with parallel, straight field lines in order to simplify our treatment. We look for instabilities in the very low frequency regime where a new spectrum of instabilities and waves appears, induced by the dust collective dynamics: Dust-Acoustic-Waves (DAWS), Dust-Ion-Acoustic-Waves (DIAWS), etc. The frequency of DAWS are around 10 Hz as determined in the laboratory, and even below that in astrophysical plasmas [1],[2]. In the case that grains are in the micron range we expect a non-ideal behavior due to the fact that the particulate are highly charged and intermolecular forces play certainly an important role. In order to discuss this problem we compare the ideal properties with the simple hard-core model and in a following work we will use a better model by considering a square-well model and the Padé rational approximant to the state equation [3]. The hard-sphere gas model, to our knowledge, is more realistic than the simple application of the Van der Waals state equation [4] . In this paper we show an analysis of the electrostatic waves and instabilities growth rates in a weakly non-ideal, magnetized dusty plasma with density gradients, but with temperature gradients and charge fluctuations neglected. As introduced before, the non-ideal behavior is characterized by the square-well model from Ree and Hoover, defined by a new state equation

$$p = nk_bT(1 + b_0n\frac{1 + a_1b_0n + a_2b_0^2n^2}{1 - b_1b_0n + b_2b_0^2n^2}). \tag{1}$$

Here $b_0 = 2\pi\sigma^3/3$ with $\sigma$ the grain radius.

## 2. Theoretical model

In this paper we introduce an new numerical treatment in combination with a more realistic formulation of the state equation to simulate weak non ideal effects in order to analyze inhomogeneous Vlasov-Dusty Plasma systems where a linearized dispersion relation is obtained. Due to the low frequency range enough energy can be transferred from the particle to the wave and instabilities can be generated. In order to get an adequate linear dispersion relation with a magnetic field given by $\vec{B} = B_0\hat{k}$ for maxwellian multi-species plasmas $(electron, ion\ and\ dust)$, we introduce the well known and very accurate multipolar approximation for the **Z** dispersion function.

In the presence of a magnetic field we have the distribution function of the specie $\alpha$, as solution of the kinetic equation.

$$\frac{df_\alpha}{dt} = \frac{q_\alpha}{m_\alpha}\nabla\phi \cdot \frac{\partial f_{0\alpha}}{\partial \mathbf{v}} \tag{2}$$

given by

$$f(\mathbf{r},\mathbf{v},t) = \frac{q_\alpha}{m_\alpha}\int_{-\infty}^{t} \exp[i\omega(t-t')]\nabla\phi(r(t'))\cdot\frac{\partial f_{0\alpha}}{\partial \mathbf{v}}\,dt' \tag{3}$$

where $\alpha = e, i, d1, d2$. Now, the dispersion relation in terms of the dielectric susceptibilities in the low frequency approximation $(\omega,\ k_z v_T \ll \omega_c)$ is

$$1 + \sum_\alpha \chi_\alpha = 0 \tag{4}$$

where,

$$\chi_{0\alpha} = \frac{1}{(\kappa\lambda_{D\alpha})^2}\left[1 + l_\alpha \frac{\omega}{\sqrt{2}k_z}\mathbf{Z}(\xi_\alpha)\mathbf{I}_{0\alpha}\,\mathbf{e}^{-\mathbf{z}_\alpha}\right] \tag{5}$$

with :

$$l_\alpha = 1 - \frac{k_y T_\alpha}{m_\alpha\omega_{0\alpha}^2}\left(\frac{d}{dx}\ln n_{0\alpha} + \frac{dT_\alpha}{dx}\frac{\partial}{\partial T_\alpha}\right); \quad z_\alpha = \frac{k_y^2 T_\alpha}{m_\alpha\omega_{0\alpha}^2}; \quad \xi_\alpha = \frac{\omega}{\sqrt{2k_z v_{T_\alpha}^2}}$$

Taking into account the low frequency range, enough energy could be transferred from the particle to the wave inducing instabilities. As an adequate linear dispersion relation with a magnetic field defined by $\mathbf{B} = B\,\hat{\mathbf{k}}$ for Maxwellian multi-species plasmas, we introduce the well known and very accurate multipolar approximation for the **Z** dispersion function neglecting temperature gradients ($i.e.$ $\frac{\partial T}{\partial x} = 0$).

Further, in order to simplify our expressions, we need:

$$\frac{d}{dT\alpha}\left(\frac{1}{v_\alpha}\right) = -\frac{m_\alpha}{2\,T_\alpha^{3/2}}; \quad \frac{dz_\alpha}{dT_\alpha} = \frac{k_y^2}{m_\alpha\omega_{0\alpha}}; \quad \frac{d\xi_\alpha}{dT_\alpha} = -\frac{\omega}{\sqrt{2}\,k_z v_{T_\alpha^2}\sqrt{m_\alpha T_\alpha}} \tag{6}$$

Now, using the following identity for the dispersion function **Z**

$$\mathbf{Z}' = -\mathbf{2}[\mathbf{1} + \xi_\alpha\mathbf{Z}(\xi_\alpha)], \tag{7}$$

we obtain after several cumbersome algebraic manipulations the general dielectric susceptibilities in the form

$$\chi_{0\alpha} = \frac{1}{(\kappa\lambda_{D\alpha})}\left[1 + \frac{\omega\mathbf{Z}I_{0\alpha}e^{-z_\alpha}}{\sqrt{2}\,k_z v_{T_\alpha}}\left\{1 - \frac{k_y T_\alpha}{m_\alpha}\left(\frac{n'_{0\alpha}}{n_{0\alpha}}\right)\right.\right.$$
$$\left. + T'_\alpha\left(-\sqrt{\frac{m_\alpha}{T_\alpha^3}}\frac{v_{T\alpha}}{2} + \frac{\mathbf{Z}'\xi'}{\mathbf{Z}} + \frac{I'_0 z'_\alpha}{I_0} - z'_\alpha\right) \tag{8}$$

with $\alpha = e, i, d1, d2$
In this work we neglect temperature gradients, as was mentioned before, then we have in such cases only the following expression

$$\chi_\alpha = \frac{1}{(\kappa\lambda_{D\alpha})}\left[1 + \frac{\omega \mathbf{Z} I_{0\alpha} e^{-z_\alpha}}{\sqrt{2}\, k_z v_{T_\alpha}}\left\{1 - \frac{k_y T_\alpha}{m_\alpha}\frac{n'_{0\alpha}}{n_{0\alpha}}\right\}\right] \tag{9}$$

In order to put our dispersion relation in a dimensionless form, we introduce the following suitable definitions:

| $\lambda_{D\alpha}$ | $K$ | $\xi_\alpha$ | $\mu_\alpha$ | $\Theta_\alpha$ |
|---|---|---|---|---|
| $\sqrt{\frac{T_\alpha}{n_{0\alpha} Z_\alpha e^2}}$ | $\kappa\lambda_{Di}$ | $\frac{\omega}{\sqrt{2} k_z v_{T\alpha}}$ | $\frac{n_{0\alpha}}{n_{0i}}$ | $\frac{T_\alpha}{T_i}$ |

| $\omega_{0\alpha}$ | $\kappa\lambda_{D\alpha}$ | $\Omega$ | $\Omega_{0\alpha}$ | $U_\alpha$ |
|---|---|---|---|---|
| $\frac{z_\alpha e B}{m_\alpha}$ | $K\sqrt{\frac{\Theta_\alpha}{\mu_\alpha}}$ | $\frac{\omega}{\omega_{pi}}$ | $\frac{\omega_{0\alpha}}{\omega_{pi}}$ | $\frac{v_{T\alpha}}{c_{si}}$ |

Now, using these results and assuming that $\omega \ll \omega_{0i} \ll \omega_{0d}$ we can write down (9) as

$$1 + \chi_{0e} + \chi_{0i} + \chi_{0d1} + \chi_{0d2} = 0 \tag{10}$$

with

$$\chi_{0\alpha} = \frac{\mu_\alpha}{K^2\, Theta_\alpha}\left[1 + \frac{\Omega\, \mathbf{Z} I_{0\alpha} e^{-z_\alpha}}{\sqrt{2}\, K_z U_\alpha}\left\{1 - \frac{K_y U_\alpha^2}{\Omega\, \Omega_{0\alpha}}\frac{1}{\Lambda_{n_\alpha}}\right\}\right] \tag{11}$$

Now it is very convenient to define the following relations [4]:

$$\frac{1}{L_p} = \frac{\nabla p_d}{p_d};\ \frac{1}{L_n} = \frac{\nabla n_d}{n_d} \tag{12}$$

$$L_p = \frac{p_d}{L_n} \equiv \frac{\alpha_1 n_d + \alpha_2 n_d^2 + \alpha_3 n_d^3 + \alpha_4 n_d^4}{1 + \beta_1 n_d + \beta_2 n_d^2}\frac{1}{L_n} \tag{13}$$

where, $d = d1, d2$ and the coefficients of equation(13) are given in table 1.

**Table 1.** Coefficients of equation (13).

| Nr. | Coefficients | Value |
|---|---|---|
| 1 | $\alpha_1$ | $1.000000\ k_B\ T_d$ |
| 2 | $\alpha_2$ | $0.438507\ k_B\ T_d\ b_0^2$ |
| 3 | $\alpha_3$ | $0.14482\ k_B\ T_d\ b_0^3$ |
| 4 | $\alpha_4$ | $0.017329\ k_B\ T_d\ b_0^4$ |
| 5 | $\beta_1$ | $-0.561493\ b_0$ |
| 6 | $\beta_2$ | $0.081313\ b_0^2$ |

Now, introducing the multipolar approximation to for the function $\mathbf{Z}$ we can get a rational polynomial expression in the form [5]

$$\sum_j a_i\ \Omega_i \,/ \sum_j b_j\ \Omega_j = 0 \tag{14}$$

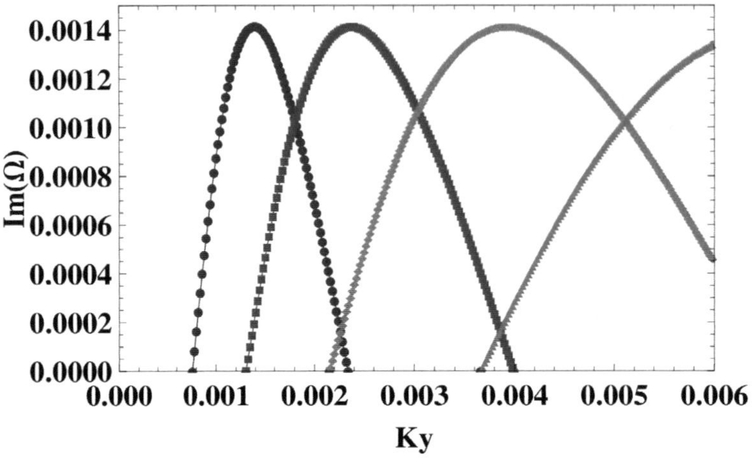


**Figure 1.** Normalized maximum growth rate as a function of normalized perpendicular wavelength for slightly inhomogeneous plasma ($\Lambda_p = 1 \times 10^2$) and for a relatively inhomogeneous one ($\Lambda_{p1} = 5 \times 10^1$), $r_d = 1.0, 0.7, 0.5, 0.3$.

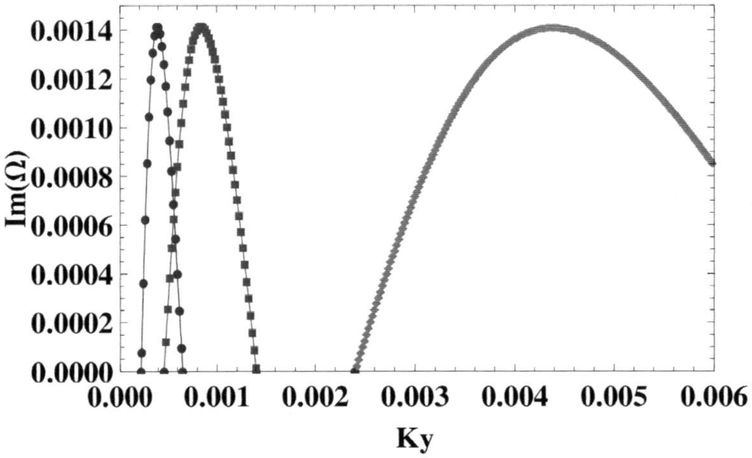


**Figure 2.** Normalized maximum growth rate as a function of normalized perpendicular wavelength for slightly inhomogeneous plasma ($\Lambda_p = 1.5 \times 10^2$) and for a relatively inhomogeneous one ($\Lambda_p = 2.5 \times 10^1$) $r_d = 1.0, 0.7, 0.5$.

where coefficients $a_i$ and $b_i$ of the system parameters. To solve this relation is very easy to find roots of the numerator. Analysis of of the solutions region permit us to find the gain $\gamma = Im(\Omega)$ in function of $1/K_y$ or simply versus $K_y$ as shown in figures 1 and 2.

## 3. Results and conclusions

The quasi-neutrality equation for dusty plasmas can be simplified due to the high state of charge of the dust particulate to the more simple relation

$$n_{oi} = Z_D n_{od} + Z_D n_{od1} + n_{oe} \tag{15}$$

and the electron susceptibility can be neglected in the dispersion relation. The range of the main parameters in the study of the low frequency oscillation of dust particulate is established by the approximations that conduced to the simplified dispersion relation

$$\Omega, K_z U_d \ll \Omega_{cd} \tag{16}$$

Non negligible unstable dust oscillations ($Im(\Omega) > 0$) are found for $\Omega_{cd} \simeq 10^{-1}$, $K_z U_d \simeq 10^{-2}$. For slightly inhomogeneous plasmas, the shape of the dust instability ($Im(\Omega)_{\max}$) curve as function of the perpendicular to magnetic field wavelength ($1/K_y$) is similar to that for ions, previously studied [6]. The maximum value of the instability increases slightly with the particle radius $r_d$ and decreases towards smaller radius shifting to the ideal region. A widening of the growth rate profile is to observe in the region of smaller radius. For typical laboratory weakly dusty plasma ($n_{d\sim}10^4 m^{-3}$, $Z_D \sim 10^3$) the instability of dust acoustic or electrostatic waves is narrower and smaller than that for ions. In figure 1 the maximum of the instability corresponds to the typical shape of instability of weakly inhomogeneous dusty plasmas and density gradients lengths of $\Lambda_p \equiv \Lambda \leq 1.5 \times 10^1$), while for the figure 2 the region of instability appears for density gradient lengths of the order of ($\Lambda_p \equiv \Lambda \leq 2.5 \times 10^1$). The introduction of of a new dust particle reduce the the growth rate change and the achieved maxima are very similar by changing the radius.

$$r_d \ll \lambda_{Di} \tag{17}$$

necessary condition for the exhibition of dust acoustic waves.

## Acknowledgments

Work supported by grant DID-G22, Universidad Simón Bolívar, Caracas, Venezuela

## References

[1] Chen J H, Du J B and Lin I 1994 *J. Phys. D: Appl. Phys.* **27** 296
[2] Barkan A Merlino R L and D'Angelo N 1995 *Phys. Plasmas* **2** 3563
[3] Ree F H and Hoover W G 1964 *J. Chem. Phys.* **40** 939
[4] Rao N N 2000 *Frontier in Dusty Plasmas* ed Y. Nakamura, T. Yakota and P. K. Shukla (Elsevier Science) pp 13-20
[5] Martin P 1980 *J. Math. Phys.* **21** 280
[6] Puerta J and Cereceda C 1996 *Plasma Physics Proc. ICPP 96* ed H. Sugay and T. Hayashi pp 94-7

Journal of Physics: Conference Series **370** (2012) 012035 doi:10.1088/1742-6596/370/1/012035

# Numerical modeling of the gas breakdown development in the space–charge layer inside the nozzle of a transferred arc torch

**B Mancinelli[1], L Prevosto[1,*] and F O Minotti[1,2,*]**

[1]Grupo de Descargas Eléctricas, Departamento Ing. Electromecánica, Universidad Tecnológica Nacional, Regional Venado Tuerto, Laprida 651, 2600 Venado Tuerto, Santa Fe, Argentina
[2]Instituto de Física del Plasma (CONICET), Departamento de Física, Facultad de Ciencias Exactas y Naturales (UBA) Ciudad Universitaria Pab. I, 1428 Buenos Aires, Argentina
[*]Member of the CONICET

E-mail: bmancinelli@arnet.com.ar

**Abstract** Double–arcing is a phenomenon that occurs when a transferred arc, flowing inside an electrically insulated nozzle, breaks into two separate arcs: one that connects the cathode with the nozzle, and another that connects the nozzle with the anode. Experimental evidence suggests that the reason for double–arcing is a Townsend like breakdown occurring in the thin space–charge layer, which separates the plasma from the metallic nozzle, due to the high voltage drop across it. Breakdown phenomena in a gas between metallic electrodes have been extensively studied; however the present case involves breakdown of a high–temperature gas between one electrode (the nozzle) and a plasma boundary. A 1–D model of the gas breakdown development in the space–charge layer contiguous to the nozzle of a cutting arc torch operated with oxygen is reported. The dynamics of the discharge is analyzed. The kinetic scheme includes processes of ionization of heavy particles by electron impact, electron attachment, electron–ion recombination and ion–ion recombination.

## 1. Introduction

The plasma cutting process is characterized by a transferred electric arc that is established between a cathode, which is a part of the cutting torch, and a work–piece (the metal to be cut) acting as the anode [1]. In order to obtain a high–quality cut, the plasma jet must be as collimated as possible. To this end, the transferred arc is constricted by a narrow nozzle (bore of the order of one millimeter).

In the normal mode of torch operation the nozzle is a floating conductor. However, since the metallic nozzle itself is at a constant floating voltage, the zero current balance is fulfilled globally along the whole collecting area of the nozzle, but not locally. This means that one part of the nozzle will collect electrons while the other part will collect ions. Furthermore, due to the fact that the electric current is mostly carried by the electrons, the electron collecting section of the nozzle will be very small as compared to the ion collecting region. Hence, the nozzle floating potential must be close to the arc voltage at the nozzle inlet. In consequence, the voltage drop between the metallic nozzle and the plasma at the nozzle exit reaches a value very close to the total arc voltage drop inside the nozzle [2]. Under certain operating conditions the level of arc stabilization provided by the vortex flow can be insufficient, and thus the arc, following the path of smallest electrical resistance (the electrical conductivity of the nozzle is much higher than that of the confined arc column), creates two separated arc channels: one from the cathode to the nozzle and another from the nozzle to the anode. Such type of arc instability is called double–arcing. Experimental results [3] suggest that the reason for double–arcing is a Townsend like breakdown occurring in the thin space–charge layer (which separates the plasma from the nozzle wall) at the nozzle exit.

Breakdown phenomena in a gas between metallic electrodes have been extensively studied (e.g., [4]); however, double–arcing involves breakdown of a high–temperature gas between one electrode (the nozzle) and a plasma boundary. A 1–D model of the gas breakdown development in the space–charge layer contiguous to the nozzle of a cutting arc torch operated with oxygen is reported. The dynamics of the discharge is analyzed. The kinetic scheme includes processes of ionization of heavy particles by electron impact, electron attachment, electron–ion recombination and ion–ion recombination.

## 2. Numerical modeling

### 2.1. Model geometry

As the Townsend discharge in the space–charge layer likely occurs in the vicinity of the nozzle exit, the 1–D domain is restricted to this region. The model corresponds to a collisional space–charge layer contiguous to the nozzle wall of a 30 A oxygen cutting torch. Since the layer remains thin (about 21 μm [3]) as compared to the nozzle orifice radius (0.5 mm), a planar geometry was used ($x$ is anti-parallel to the radial coordinate at the nozzle exit). The domain is shown in figure 1.

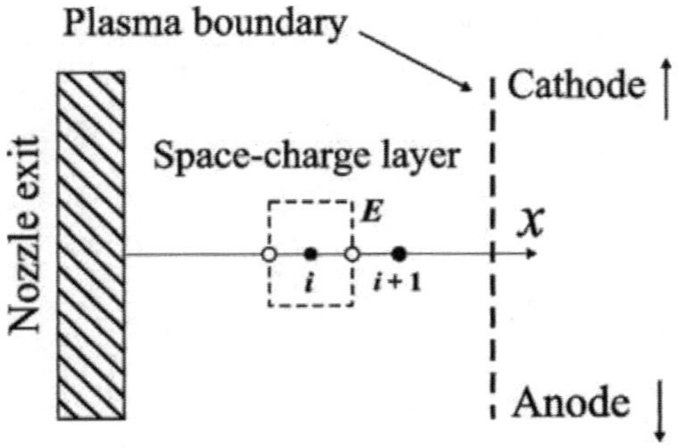


**Figure 1**. Schematic of the domain.

### 2.2. Main equations of the layer hydrodynamic approximation

The numerical model includes the following balance equations for charged particles:

$$\frac{\partial N_e}{\partial t} + \nabla \cdot \left(N_e \bar{v}_e - D_e \nabla N_e\right) = k_{ion} N_e N - k_{att} N_e N^2 - k_{rec}^{ei} N_p N_e, \tag{1}$$

$$\frac{\partial N_p}{\partial t} + \nabla \cdot \left(N_p \bar{v}_p - D_p \nabla N_p\right) = k_{ion} N_e N - k_{rec}^{ii} N_n N_p N - k_{rec}^{ei} N_p N_e, \tag{2}$$

$$\frac{\partial N_n}{\partial t} + \nabla \cdot N_n \bar{v}_n = k_{att} N_e N^2 - k_{rec}^{ii} N_n N_p N. \tag{3}$$

Here, $N_e$, $N_p$, $N_n$ and $N$ are the numerical densities of electrons, positive ($O_2^+$) and negative ($O_2^-$) ions, and neutral ($O_2$) particles; respectively. $v_e$, $v_p$ and $v_n$ are the drift velocity of electrons, positive and negative ions, respectively. $D_e$ and $D_p$ are respectively the diffusivity of electrons and positive ions. $k_{ion}$, $k_{rec}^{ei}$, $k_{rec}^{ii}$ and $k_{att}$ are, respectively, the rate constants of ionization, electron–ion and ion–ion recombination, and electron attachment. The equation of state for the pressure ($p$) was also used

$$\frac{p}{k} = \left(N_p + N_n + N\right)T_h + N_e T_e, \tag{4}$$

$k$ being Boltzmann's constant and $T_e$ and $T_h$ the electron and heavy particle temperatures, respectively. The electric field ($\overline{E}$) in the discharge gap is described in terms of the electrostatic potential $\varphi$

$$\overline{E} = -\nabla\varphi \tag{5}$$

whose distribution over the gap was determined by solving the Poisson's equation

$$\nabla^2\varphi = -\frac{e}{\varepsilon_0}\left(N_p - N_e - N_n\right), \tag{6}$$

($e$ is the absolute value of the electron charge and $\varepsilon_0$ the permittivity of free space).

*2.3. Kinetic scheme*

The following processes were incorporated to the model [4, 5]

Direct electron impact ionization

$$e + O_2 \rightarrow e + e + O_2^+, \tag{7}$$

$$k_{ion} = \exp(-13.2 - 21.8/\vartheta)\, m^3 s^{-1},$$

where $\vartheta \equiv E/N$ (in units of $10^{-16}$ V cm$^2$).

Electron-ion recombination

$$e + O_2^+ \rightarrow O + O, \tag{8}$$

$$k_{rec}^{ei} = 2.0\times10^{-13}\left(\frac{300}{T_e}\right) m^3 s^{-1}$$

Electron attachment

$$e + O_2 + O_2 \rightarrow O_2 + O_2^-, \tag{9}$$

$$k_{att} = 2.0\times10^{-41}\left(\frac{300}{T_e}\right) m^6 s^{-1}.$$

Ion-ion recombination

$$O_2^+ + O_2^- + O_2 \rightarrow O_2 + O_2 + O_2, \tag{10}$$

$$k_{rec}^{ii} = 2.0\times10^{-37}\, m^6 s^{-1}.$$

*2.4. Initial particle distributions in the layer and boundary conditions*

The results were obtained for a gas pressure $p$ = 0.1 MPa, a gas temperature of $T_h$ = 1000 K, and an electron temperature of $T_e$ = 5400 K. [3] The initial profiles of the electron, positive ion, and neutral densities together with the initial voltage distribution were taken from a previously published numerical model of the layer [6]. At the initial time zero level of negative ions was assigned. The nozzle and plasma boundary voltages are held constant at values of -155 and -22 V, respectively. [6]

The nozzle wall was considered as an absorbing surface for the positive ions. The secondary emission caused by positive ions and UV photons was also considered at this surface [7]. The plasma border was modeled as an open boundary for the charged particles.

*2.5. Numerical aspects*

The governing equations were solved by solving them using a finite difference discretization technique of first-order accuracy in space and time. The domain was discretized into 50 cells. To ensure stability a semi-implicit method was used in which the Poisson equation is first solved using explicitly predicted values of the densities, and the transport equations are then solved implicitly for the densities [8].

The drift–diffusion flux of charged particles was discretized using the Scharfetter–Gummel scheme [9]

$$J_E = \frac{v_E}{\exp\left(\frac{v_E \Delta x}{D_E}\right) - 1}\left[\exp\left(\frac{v_E \Delta x}{D_E}\right) N_i - N_{i+1}\right], \tag{11}$$

where the sub–index $E$ indicates east boundary of the cell (see figure 1). $\Delta x$ is the cell dimension.

## 3. Numerical results and discussion

The dynamics of the Townsend breakdown development in the space–charged layer is illustrated in figures 2 to 5. The results show that the breakdown is produced on time scales of the order of $10^{-6}$ s. The cathode layer of the discharge (with a length scale about 5 μm) in front of the negatively biased nozzle wall is clearly recognized. In this region the positive ion density exceeds the negative charge density, strongly shielding the electric field outside it. The remaining gap region contains quasi–neutral plasma with a weak electric field.

The electrons are lost mostly by electron attachment to $O_2$ molecules in accordance with the reaction (9).

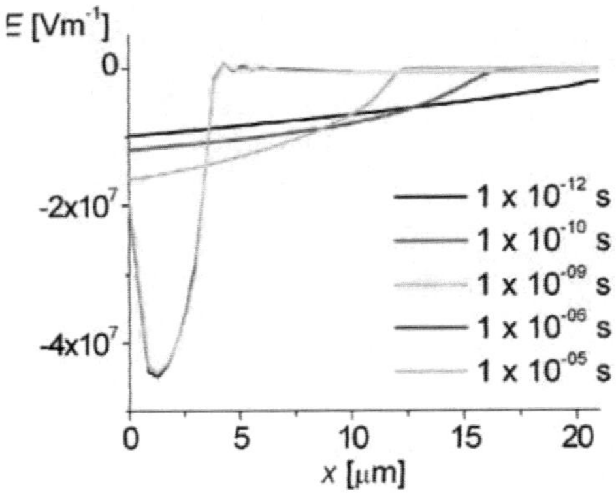


**Figure 2**. Calculated electric field.

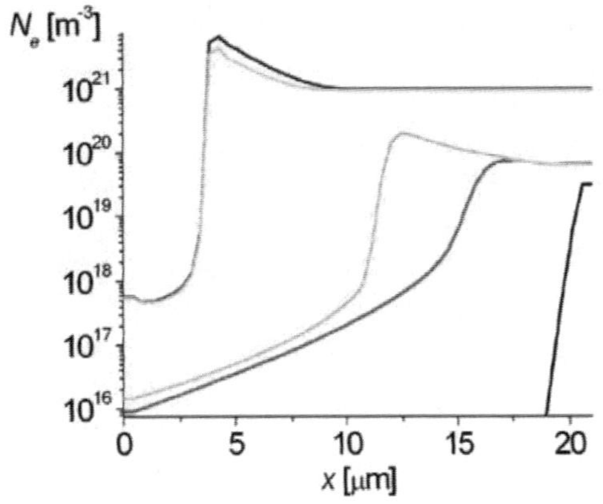


**Figure 3**. Calculated electron density.

## 4. Conclusions

A 1–D model of the gas breakdown development in the space–charge layer contiguous to the nozzle of a 30 A cutting arc torch operated with oxygen was reported. The dynamics of the discharge was analyzed. The kinetic scheme includes processes of ionization of heavy particles by electron impact, electron attachment, electron–ion recombination and ion–ion recombination. The numerical results show that the Townsend breakdown is produced on time scales of the order of $10^{-6}$ s. The so–called cathode layer (with a length scale about 5 μm) in front of the nozzle wall was clearly recognized. The remaining gap region contains quasi–neutral plasma. It has been also shown that electron attachment to $O_2$ molecules plays an important role in the discharge dynamics.

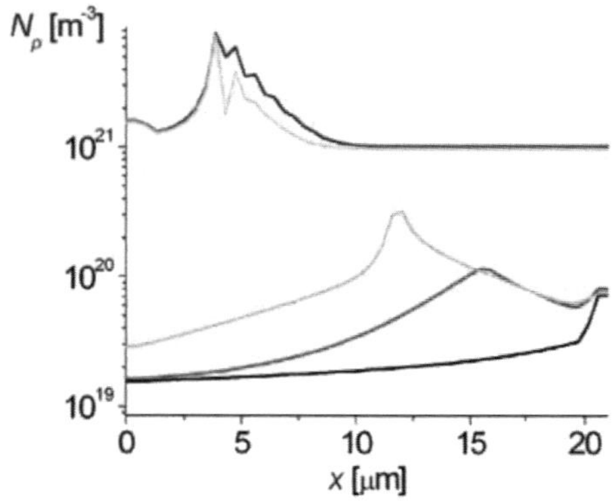


**Figure 4**. Calculated positive ions density.

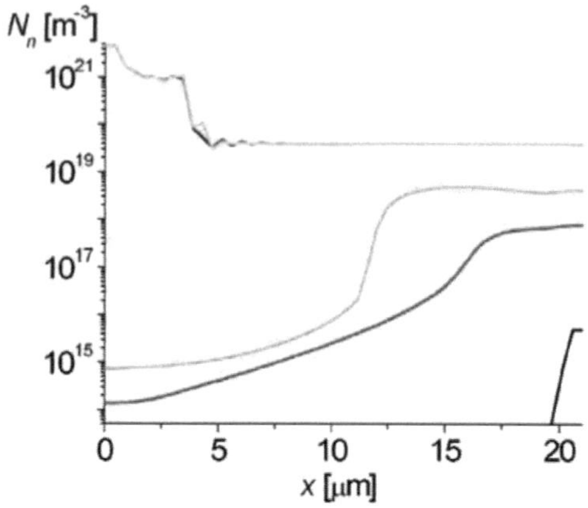


**Figure 5**. Calculated negative ions density.

**Acknowledgements**
This work was supported by grants from the Universidad de Buenos Aires (PID X108), CONICET (PIP 5378) and Universidad Tecnológica Nacional (PID 1389).

**References**

[1] Boulos M, Fauchais P and Pfender E 1994 *Thermal Plasmas, Fundamentals and Applications Vol. 1* (Plenum Press)
[2] Nemchinsky V A and Severance W S 2006 *J. Phys. D: Appl Phys* **39** R423
[3] Prevosto L, Kelly H and Mancinelli B 2009 *J. Phys. D: Appl. Phys.* **105** 013309
[4] Pancheshnyi S V and Starikovskii A Yu 2003 *J. Phys. D: Appl Phys* **36** 2683
[5] Kossyi I A, Kostinsky A Yu, Matveyev A A and Silakov V P 1992 *Plasma Sources Sci. Technol.* **1** 207
[6] Prevosto L, Kelly H and Mancinelli B 2009 *J. Phys. D: Appl. Phys.* **105** 123303
[7] Raizer Y P 1991 *Gas Discharge Physics* (Berlin, Germany: Springer)
[8] Ventzek P L G, Sommerer T J, Hoekstra R J and Kushner M J 1993 *Appl. Phys. Lett.* **63** 605
[9] Kulikovsky A A 2001 *J. Comput. Phys.* **173** 716

Journal of Physics: Conference Series **370** (2012) 012036 doi:10.1088/1742-6596/370/1/012036

# Modeling of fast capillary-discharge for soft x-ray lasers

**C P Teh[1,2], B R Wong[2] and K H Kwek[1]**

[1]Department of Physics, University of Malaya, 50603 Kuala Lumpur, Malaysia
[2]Institute Mathematical Sciences, University of Malaya, 50603 Kuala Lumpur, Malaysia

E-mail: tcp5441@um.edu.my

**Abstract**. In this work we present the results of the numerical investigations of the one-dimensional single-fluid magneto-hydrodynamics (MHD) model, with separate electron and ion temperatures, of a capillary-discharge collisional soft x-ray laser. The MHD equations are solved by the Lagrangian cylindrical geometry approach. The effects of the gas filling pressure on the plasma densities and temperatures, and implosion-pinch phase for soft x-ray lasing conditions have been analyzed. The results are compared with experimental measurements of the operating pressure ranges.

## 1. Introduction

Soft x-ray lasers are invaluable tools for studying high-density plasmas and permitting one to see smaller features in microscopy, to write finer patterns in lithography and to generate shorter pulses. These lasers have applications for the materials science community, both inside and outside the laboratory, by supplying detailed information about the atomic structure of new and existing materials. These in turn lead to new scientific understandings, perhaps through surface science, chemistry and physics, providing feedback to the enabling technologies. Development of soft x-ray spectral region is presently in a rapid growth and there is much interplay between the science associated technologies.

In 1984, soft x-ray lasers were first demonstrated in plasmas generated by powerful laser drivers in large laboratories [1, 2]. However, the large pump lasers used for this kind of soft x-ray laser experiments are only available at a few large laser facilities in the world due to their large size, complexity, and extremely high cost. Therefore the demonstration of a capillary discharge soft x-ray laser operating in the transition of Neon (Ne)-like ions by Rocca *et al* [3] in 1994 opened the possibility to develop compact, efficient and simpler soft x-ray laser. This similar demonstration had then been only repeated by Ben-kish *et al* [4] in 2001, followed by the work of Niimi *et al* [5] and Kukhlevsky *et al* [6] in the following year. The quest for improving these systems still continues by parallel efforts involving theory and computer simulations of the experiments to understand the laboratory observation and to optimize the resulting energy density. Even after more than a decade, systems with improved performance and efficiency are still being reported [7].

In this paper, the capillary discharge soft x-ray laser described by Tan and Kwek [8] is modeled using two-temperature, one dimensional single-fluid magneto-hydrodynamics (MHD) equations. The details of the governing equations and the computational simulation will be applied and shown. The model was then utilized to numerically evaluate the time and radial dependences of capillary plasma characteristics of fast capillary discharges with current rise time of around 50 ns and peak currents 16 kA in a capillary 3 mm in diameter filled with pre-ionized argon (Ar) gas. The results were then

Journal of Physics: Conference Series **370** (2012) 012036 doi:10.1088/1742-6596/370/1/012036

compared with experimental data [8] from different initial argon filled pressures. The effects of the gas filling pressure on the plasma densities and temperatures, during implosion-pinch phase for soft x-ray lasing conditions were analyzed.

## 2. Magneto-hydrodynamics model

The basic MHD equations used are described below. The continuity equation (mass conservation) is given by

$$\frac{\partial \rho}{\partial t} + \nabla \cdot (\rho \mathbf{u}) = 0,$$

where $\rho$ is mass density and $\mathbf{u}$ is the plasma fluid velocity. Generally, in any steady state process, the rate at which mass enters a system is equal to the rate at which mass leaves the system. In this numerical scheme, the continuity equation is automatically satisfied.

The momentum conservation equation is solved in the one-fluid approximation where the plasma electrons and ions are assumed to flow together as a single fluid. The momentum equations is governed by

$$\rho \frac{du}{dt} = -\nabla (P_e + P_i + q) - \mathbf{J} \times \mathbf{B},$$

where $P_e$ and $P_i$ are the thermal pressure due to electrons and ions, respectively, $\mathbf{B}$ denotes the magnetic field induced by the axial current and $\mathbf{J}$ is the axial component of the electric current density. Here, $q$ is the von Neumann artificial viscosity. The artificial viscosity, which in included in the equation of motion to handle shocks, effectively smoothes the shock over a small number of zones.

The energy balance for electrons and ions is described by

$$C_{v,e} \frac{dT_e}{dt} = V\nabla \cdot (\kappa_e \nabla T_e) + \psi_{joule} - \left( \frac{dE_e}{dV} + P_e \right) \frac{dV}{dt} - \frac{(T_e - T_i)}{t_{eq}},$$

$$C_{v,i} \frac{dT_i}{dt} = V\nabla \cdot (\kappa_i \nabla T_i) - \left( \frac{dE_i}{dV} + P_i \right) \frac{dV}{dt} + \frac{(T_e - T_i)}{t_{eq}}.$$

where $C_{v,\alpha}$, $T_\alpha$, $E_\alpha$, $P_\alpha$ and $\kappa_\alpha$ are, respectively, the specific heat, temperature (in eV), specific internal energy, pressure and thermal conductivity, of the electrons ($\alpha = e$) and ions ($\alpha = i$). While $V$ is the specific volume, $\psi_{joule}$ is joule heating term and $t_{eq}$ is the electron-ion collisional coupling term.

The difference between the temperature diffusion equations for the electrons and ions is the absence of the Joule heating term in the ion equation. Since the currents are mainly due to the lighter electrons, the energy is transferred to the ions through the collision specified by the term involving $t_{eq}$.

The magnetic field transport and diffusion equation reads

$$\frac{d\mathbf{B}}{dt} = \frac{\eta}{4\pi} (\nabla \cdot \nabla) \mathbf{B} - (\nabla \cdot \mathbf{u}) \mathbf{B},$$

where $\eta$ is defined as the plasma electrical resistivity. This equation is obtained using Maxwell equations and generalized Ohm's law neglecting the time variation of the electric field [9].

These MHD equations have been adopted in the Lagrangian reference frame using the specific volume expression. The position of each cell and the velocity on cell boundary are determined by

$$\frac{dr}{dt} = u.$$

All dynamical values in the equations evolve as a function of position $r$ from the capillary axis at a time $t$.

## 3. The requirement characteristic of populations and gains in neon-like argon

As a first estimation of absolute gain of this transition, the fractional abundance of charge states is evaluated using steady-state ionization model.

The atomic physics PrismSPECT code [10] solves the equation of state and provides radiative properties for plasma in local thermodynamic equilibrium (LTE) state. The ionization population fractions in the range $T_e \in 1 - 1000$ eV for a given electron density $N_e \sim 2.0 \times 10^{18}$ cm$^{-3}$ are depicted on the figure 1. This value has been chosen because of the value of the pinch electron density of the case study will be studied in next section. The maximum ionization population fraction of neon-like $Ar^{8+}$ ion, $f^{8+} \approx 0.95$ is accomplished at temperature $T_e \approx 40$ eV. The temperature range of Ne-like ionization state is approximately between 10 – 200 eV. It should be kept in mind that the steady-state ionization model does not depend on ionization dynamics, namely, ionization plasma history.

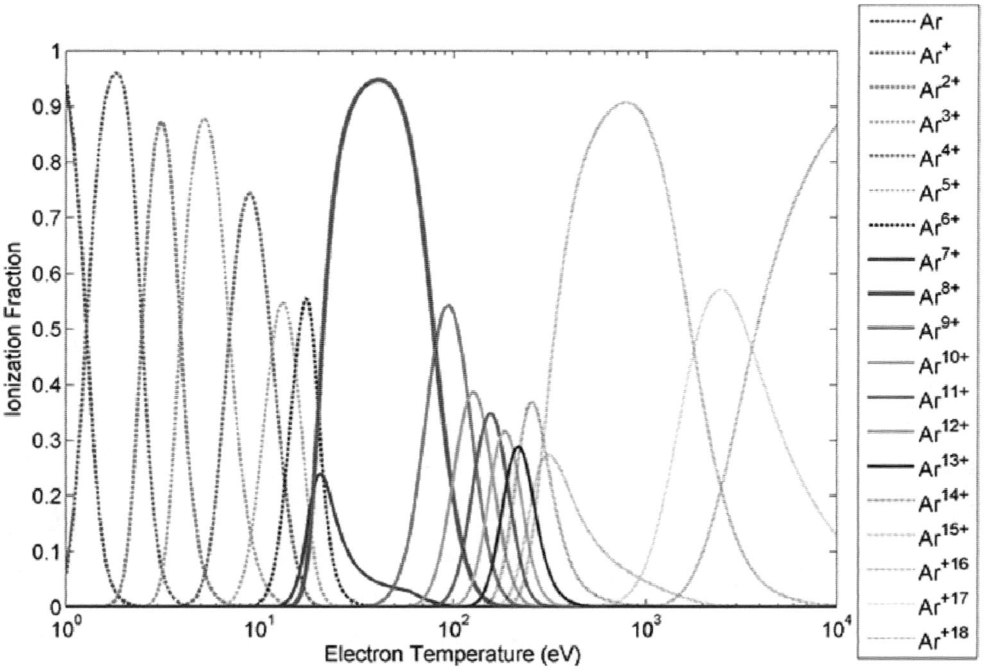


**Figure 1.** Argon ionization fractions dependences on plasma electron temperature for atom density $N_0 = 2 \times 10^{18}$ cm$^{-3}$.

Figure 2 presents a contour plot of gain against $N_e$ and $T_e$ computed using steady-ion ionization model. The effect of reabsorption of the resonance radiations on the distribution of population density of excited-states has been neglected (optically-thin plasma assumption). It has been assumed that $T_e = T_i$. This assumption generally sets the upper limit of gain in capillary-discharge.

Figure 2 indicates that a large gain formed for $N_e \approx 2 \times 10^{18}$ cm$^{-3}$. For the regime $N_e > 10^{20}$ cm$^{-3}$, collisional depopulation exceeds the radiative decay rates. Hence, the populations become distributed according to the Boltzmann distribution function and the collisional quench of the population inversion occurs (i.e., unsuitable gain regime). Although, the fractional abundance of Ne-like Ar becomes a maximum at $T_e \approx 40$ eV using the steady-state approximation for ionization model, the optimum regime of gain is approximately at $T_e \approx 60$ eV, as shown in figure 2. This fact shows a mismatch between electron temperature that maximizes gain and fractional abundance of Ne-like Ar ion, in which is a general problem of Ne-like x-ray lasers [11]. The mismatch arises because the excitation energy into the $n = 3$ levels is typically three-quarters of the ionization energy and thus a temperature which gives large ground-state excitation rates will lead to large ionization rates if the electron density is sufficient (the relaxation time for the ionization balance $\sim 10^{11}\, n_e^{-1}$) [12].

The collisions between particles in the plasma transfer energy from particle to particle in a random fashion, thereby heating the plasma and ionizing the atoms. Generally, multiple ionization states are formed, each with its own characteristic emission lines, leading to a rich spectrum of lines. By carefully controlling the temperature and density of the plasma, the population of specific ionization species can be preferentially established. Generally, such high temperatures and densities are established at or close to pinch radius.

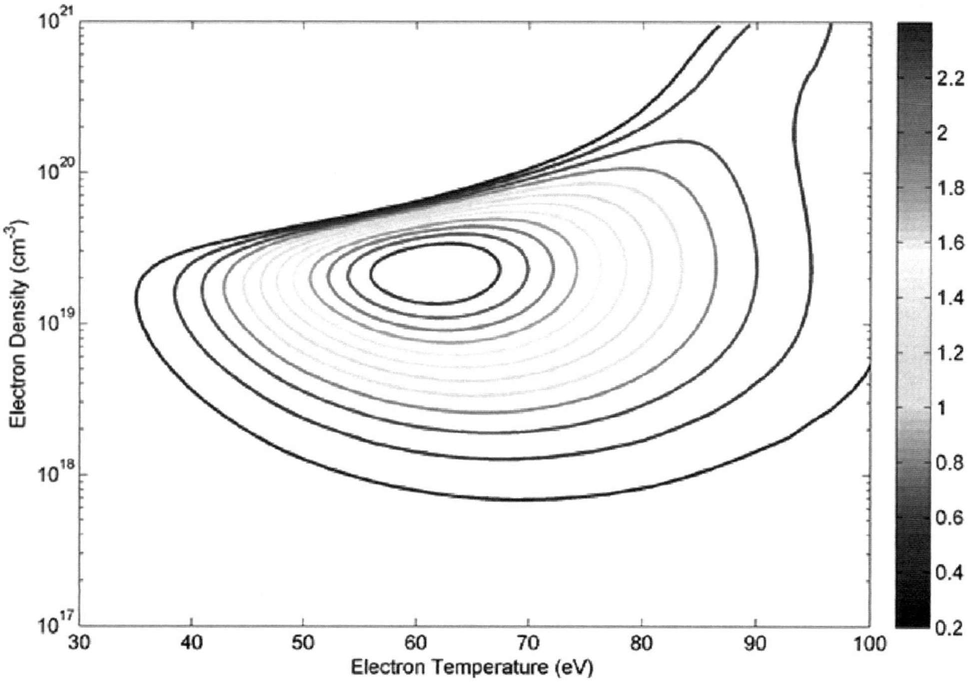


**Figure 2.** The contour of gain in 3p $^1S_0$ – 3s $^1P_1$ transition as function of the electron temperature and density.

Under certain plasma conditions (electron density of 0.1 – 4 x $10^{19}$ cm$^{-3}$, and electron temperature ≈ 40 – 90 eV, as in results from figure 2), collisional electron impact excitation of the ground state Ne-like ions produces a population inversion between the 3p ($J$=0) and the 3s ($J$=1) levels resulting in amplification at the 46.9 nm wavelength.

## 4. Results and discussion

Using the general physical model outlined in section 2, the discharge channel can be simulated with specific values of the controlling parameters. The parameters of a working discharge x-ray laser reported in reference [8] are used. Here the radius of the capillary has been fixed at 0.15 cm, and current rise time is about 50 ns. For this basic situation, the filling pressures and the current profiles are varied to investigate the impact of these variations on the plasma. Initially, the pre-ionized plasma is assumed to be uniformly distributed with $T_e = T_i = 1\ eV$. For the simulations, we use the HELIOS-CR code described in reference [13]. It is a modeling platform that can handle LTE and non-LTE plasmas in one dimension. The peak discharge current 16 kA is used in this paper. These are the typical values used in the laser of reference [8].

Figure 3 presents the calculated results for a capillary discharge with a peak current of 16 kA and 0.15 mbar gas filling pressure. The temporal evolution of the trajectories of argon plasma elements inside the channel of this configuration is illustrated in figure 3(a). The blue line represents the discharge current. Figure 3(b) shows contour plots of the logarithm of the electron density measured in $cm^{-3}$ while figure 3(c) shows the electron temperature measured in eV. These figures also provide quantitative information about the distribution of the plasma parameters inside the channel at different times. Some characteristics features are discussed below.

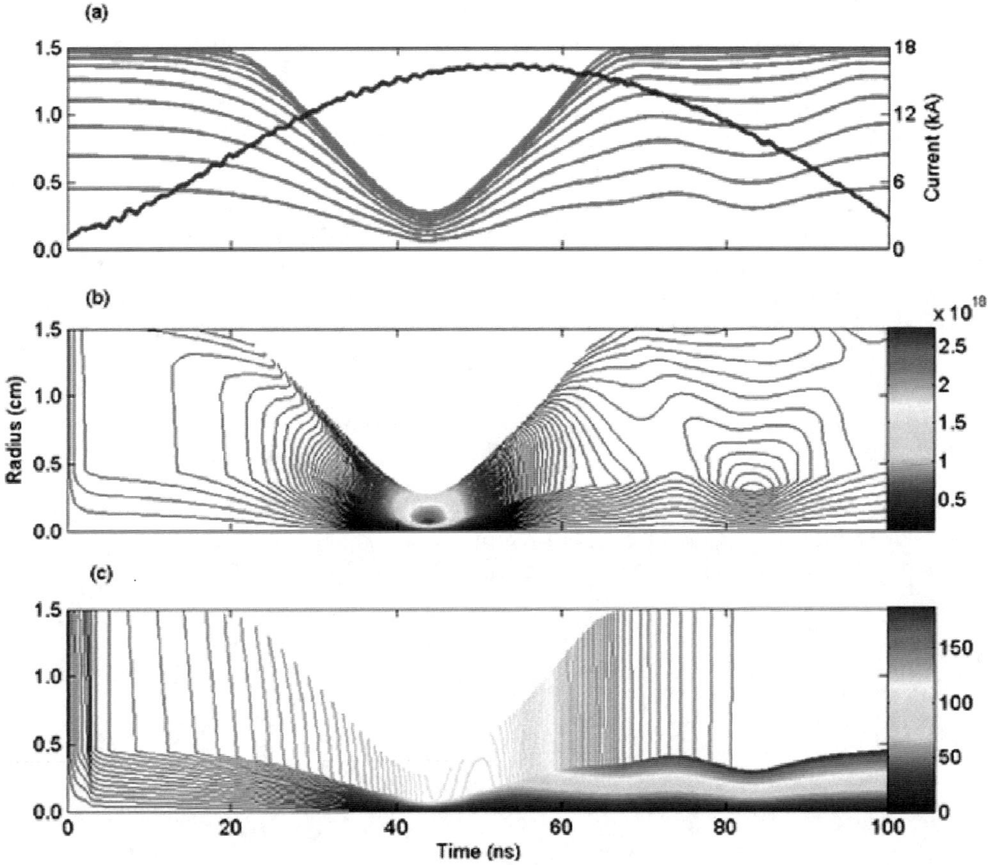


**Figure 3.** Basic parameters of discharge in capillary with diameter of 3 mm filled with argon at a initial gas filling pressure of 0.15 mbar for $I_0 = 16$ kA and rise time 50 ns. (a) The plasma time-space flow diagram; blue line corresponds to the discharge current at outer boundary; (b) contour lines of the decimal logarithm of the electron density (measured $cm^{-3}$) on the $(t, r)$ plane; (c) contour lines of the electron temperature (measured in eV).

The dynamics of a Z-pinch plasma can be described by the balance between the thermal pressure and the magnetic pressure. In the early stages ($t < 20$ ns) the plasma is not compressed immediately but actually tries to expand due to the fact that the plasma pressure exceeds other forces. However, this expansion is constrained by the rigid wall of the capillary. The pressure in the outer zones of the plasma is due to both the initial filling pressure and Joule heating. Generally, the current tends to flow in the outer part of the cylindrical plasma, and Joule heating occurs due to current passing through the plasma. This heating results in an increase in pressure which can be sufficiently high for the outer regions of the plasma to initially expand before being accelerated inward by the magnetic piston.

When the discharge current has reached a value such that the magnetic pressure is larger than the thermal pressure ($t > 20$ ns), the plasma is compressed inward onto the axis. At the same time, a vacuum region appears near the wall and a converging shock wave is formed which propagates towards the channel axis. The shock waves eventually meet at the axis, and are subsequently reflected. During compression, the plasma becomes hotter due to Joule and shock heating, and the thermal pressure increases. Although, before the time 50 ns, when the total current has been increasing towards the maximum value, the discharge current is high, and the Lorentz force dominates. Thus, the compression stops. When the thermal pressure balances the magnetic pressure, the plasma stagnates ($t \approx 43.5$ ns) and then expands ($t > 43.5$ ns). During the compression, a high density and temperature plasma on the axis is produced, as shown in figure 3(b) and 3(c). The temperature $T_e$ reaches the value of 70 eV and continues to heat up to 160 eV during the expansion phase. This is because the inner zones of plasma continue to compress again after the first pinch. The value of electron density $N_e$ at the pinch is about 3 x $10^{18}$ cm$^{-3}$. A hot plasma core with the radial dimension $r_{core}$ = 0.266 mm and central mass density $\rho_{core}$ = 1.22 x $10^{-4}$ g/cm$^3$ is formed.

At the stage of expansion the MHD instabilities may break the uniformity of the plasma column, leading to turbulent mixing, enhanced heat transport and resistance. However, from the theory of the classical pinch effect, MHD instabilities begin to evolve when the expansion velocity begins to decrease [14]. If the lasing occurs during the few first nanoseconds after the reflection of the shock wave, MHD instabilities have no time to evolve. The validity of this description of the plasma behavior in the capillary discharge ends at the moment when the diverging shock wave reaches the capillary wall.

The time dependence of small signal gain factor $G$ is determined by the time dependences of both laser level populations $n^{8+}_{3p\,3S^1}$ and $n^{8+}_{3s\,3P^0}$. In this case above (a non-ablating alumina capillary with radius $r_0$ = 1.5 mm filled with initial gas filling pressure of 0.15 mbar, current pulse with $T_{1/4}$ = 50 ns and current peak value $I_{max}$ = 16 kA) the gain factor has a shape of very short peak with maximum value $G(t_G)$ = 0.8 cm$^{-1}$ at the time $t_G$ = 44.5 ns (see figure 4). The peak value of gain is achieved during the pinch time.

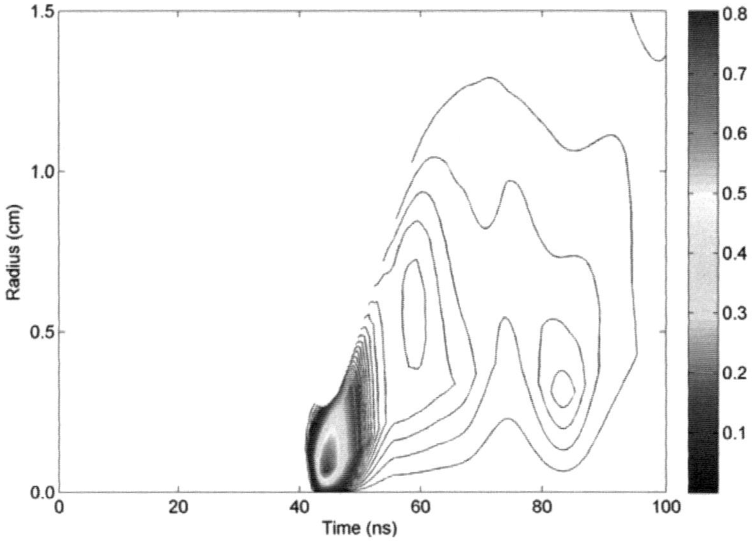


**Figure 4.** The spatial and temporal evolution of gain

## 5. Comparison with experiments

The simulations of the capillary discharge dynamics for $r_0$ = 1.5 mm, $T_{1/4}$ = 50 ns, $I_{max}$ = 16 kA and various initial gas filling pressure $p_0$ of argon have been performed here. For given capillary radius $r_0$ and current waveform, the peak value of gain factor $G(t_G)$ may be expressed as a function of the initial gas filling pressure (see figure 5). The gain curve shows that the measurable gain factor for $r_0$ = 1.5 mm, $T_{1/4}$ = 50 ns, $I_{max}$ = 16 kA is found in the pressure range 0.11 mbar < $p_0$ < 0.29 mbar.

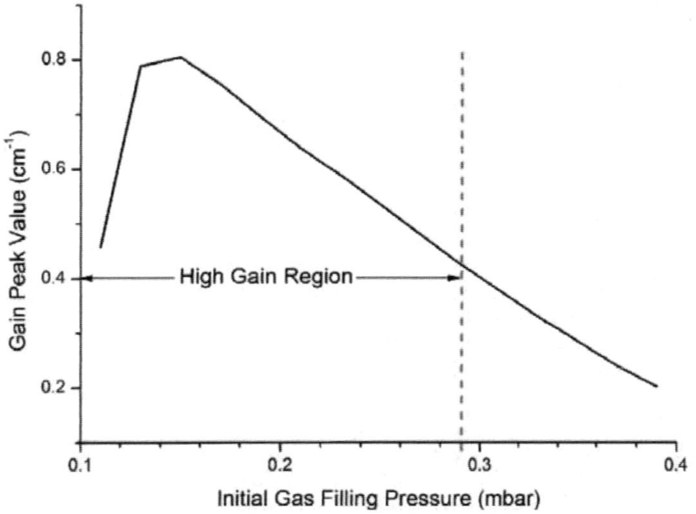


**Figure 5.** Dependence of gain factor peak value on the initial gas filling pressures.

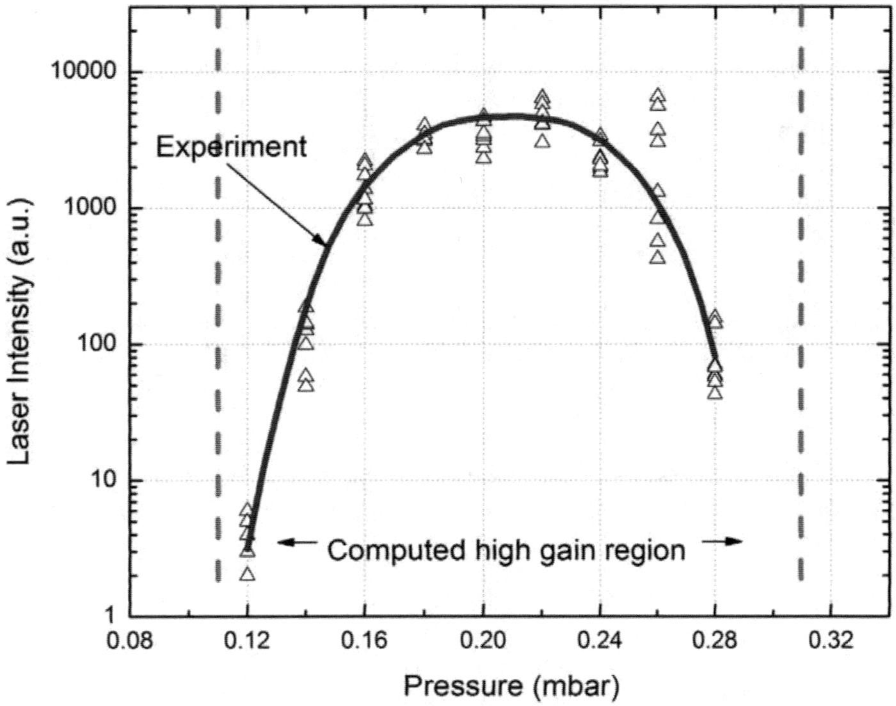


**Figure 6.** Comparison between experimental data and simulation data. The main discharge current is 16 kA.

From the MHD simulation, the electron densities and temperatures that fulfilled the plasma conditions and high gain factors obtained are in between 0.11 to 0.29 mbar. These results have been further computed which included refraction effect due to electron density gradient. The simulation output of the intensity are compared with the experimental data [8] shown in figure 6. The dotted lines are the range of gas filling pressures that are computed by MHD simulation. The soft x-ray lasers (experimental data) have been produced in the range of 0.12 to 0.28 mbar (the solid line) which show good agreement if compare to simulation data.

## 6. Conclusions

The results from an MHD modeling of a fast capillary-discharge for soft x-ray lasers have been compared with experiment results. The effects of the gas filling pressure on the plasma densities and temperatures, and implosion-pinch phase for soft x-ray lasing conditions resulting in high gain factors have been analyzed. Our findings indicate that the simulations performed gave reasonably good agreement with experiment.

**Acknowledgements**

This work has been supported in part by the Postgraduate Research Fund (PPP) Project No. PS330/2009C and in part by University of Malaya Grant RG031/09AFR.

## References

[1] Matthews D L *et al.* 1985 Demonstration of a soft x-ray amplifier *Phys. Rev. Lett.* **54** 110

[2] Suckewer S, Skinner C H, Milchberg H, Keane C and Voorhees D 1985 Amplification of stimulated soft x-ray emission in a confined plasma column *Phys. Rev. Lett.* **55** 1753

[3] Rocca J J, Shlyaptsev V, Tomasel F G, Cortázar O D, Hartshorn D and Chilla J L A 1994 Demonstration of a Discharge Pumped Table-Top Soft-X-Ray Laser *Phys. Rev. Lett.* **73** 2192

[4] Ben-Kish A, Shuker M, Nemirovsky R A, Fisher A, Ron A and Schwob J L 2001 Plasma Dynamics in Capillary Discharge Soft X-Ray Lasers *Phys. Rev. Lett.* **87** 015002

[5] Niimi G, Sakamoto N, Nakajima M, Hayashi Y, Watanabe M, Okino A, Horioka K and Hotta E 2002 Study of low current capillary discharge for compact Soft x-ray laser. *AIP Conf. Proc.*: AIP) pp 103-6

[6] Kukhlevsky S V, Ritucci A, Kozma I Z, Kaiser J, Shlyaptseva A, Tomassetti G and Samek O 2002 Atomic Model Calculations of Gain Saturation in the 46.9 nm Line of Ne-like Ar *Contrib. Plasm. Phys.* **42** 109-18

[7] Heinbuch S, Grisham M, Martz D and Rocca J J 2005 Demonstration of a desk-top size high repetition rate soft x-ray laser *Opt. Express* **13** 4050

[8] Tan C A and Kwek K H 2007 Influence of current prepulse on capillary-discharge extreme-ultraviolet laser *Phys. Rev. A* **75** 043808

[9] Burnett N H and Offenberger A A 1974 Magnetohydrodynamic behavior of a laser-heated solenoid *J. Appl. Phys.* **45** 2155-62

[10] MacFarlane J J, Golovkin I E, Woodruff P R, Welch D R, Oliver B V, Mehlhorn T A and Campbell R B 2003 *Inertial Fusion Sciences and Applications* (American Nuclear Society)

[11] Holden P B and Pert G J 1996 Long-wavelength, prepulsed driving as a means to greatly increase the gain in low-Z Ne-like XUV lasers *J. Phys. B: At. Mol.Phys.* **29** 2151

[12] Whitten B L, London R A and Walling R S 1988 Scaling of neonlike lasers using exploding foil targets *J. Opt. Soc. Am. B* **5** 2537-47

[13] MacFarlane J J, Golovkin I E and Woodruff P R 2006 HELIOS-CR - A 1-D radiation-magnetohydrodynamics code with inline atomic kinetics modeling *J. Quant. Spectrosc. Radiat. Trans.* **99** 381-97

[14] Vrba P, Vrbová M, Bobrova N and Sasorov P 2005 Modelling of a nitrogen x-ray laser pumped by capillary discharge *Cent. Eur. J. Phys.* **3** 564-80

# Acid gas degradation by non-thermal plasma and energy estimation

**R Valdivia-Barrientos[1,2], M Ibañez-Olvera[2], J Pacheco-Sotelo[1,2], M Pacheco-Pacheco[1], N Estrada-Martinez[2], J Silva-Rosas[2] and C Rivera-Rodríguez[1]**

[1] Instituto Nacional de Investigaciones Nucleares, Carretera México-Toluca s/n, Ocoyoacac, CP 52750, Mexico
[2] Instituto Tecnológico de Toluca, Av. Instituto Tecnológico s/n, Metepec, CP 52140, Mexico

E-mail: ricardo.valdivia@inin.gob.mx

**Abstract.** This paper describes a method to estimate the energy efficiency in the system performed to treat acid gases by plasma discharges. The electrical energy consumed by the plasma is evaluated by an electrical diagnosis, taking into account the experimental voltage and current applied to the power source. The estimation of the electrical energy generated by solid oxide fuel cells is based on the method of modeling the energy produced by the species generated by the plasma discharge and taking also into account the temperature of the gases.

## 1. Introduction

Actually, the necessity to treat the mixture of gases like carbon dioxide ($CO_2$) and hydrogen sulfide ($H_2S$) has a great interest and some conventional techniques which are employed for this purpose are based on the incineration process and pyrolisys or gasification processes [1-5]. The hybrid plasma offers at least three fundamental properties that make it attractive for this application: The temperature and energy density far exceeds that of conventional techniques, Ability to produce active species with much lower reaction times than those achieved by conventional techniques, The active species have high concentrations of energy, even when the temperature of the plasma remains varied [6]. Consequently, high energy density and the low inertia of plasma are used to break the molecular bonds of the gases in question, obtaining as a final product constituent particles and gases with high value-added energy.

The main strategies for the waste management are the increase of material recovery, which can reduce the landfill disposal and the improvement of energy recovery from waste. In the plasma process the organic compounds are thermally decomposed in their constituent elements and converted into a synthesis gas (*syngas*), which mainly consist of hydrogen and carbon monoxide, while the inorganic material are melted and converted into a dense, inert, non-leachable vitrified slag [7-10]. This *syngas* is cleaner than that produced by conventional gasification processes [7] due to the high temperatures involved, which allow breaking down all the tars, char and dioxins.

In [11] Galeno *et. al.* proposed the integration between plasma gasification unit and a solid oxide fuel cell (SOFC) and established the energy suitable by using a numerical model based on the combination of a thermochemical model and a electrochemical model for the SOFC, and they found that the *syngas* has a lower heating value of 9.20 MJ/kg and it is composed by the species shown in table 1.

**Table 1.** *Syngas* composition.

| Specie | Composition (*vol* %) |
|---|---|
| $H_2$ | 28.65 |
| CO | 37.37 |
| $CO_2$ | 1.41 |
| $H_2O$ | 14.19 |
| $N_2$ | 17.12 |
| $H_2S$ | 0.22 |
| others | 1.04 |

Considering a previous plasma model presented by Hirsch *et. al.* [12], in this paper the specific energy consumption is computed and the main species generated by the plasma discharge are estimated by mass balance. It is worth mentioning that the output species represent only an estimate of the main products generated during the treatment of acid gases by plasma discharge.

## 2. Model

The method to estimate the energy efficiency in the system performed to treat acid gases by plasma discharges is performed in two steps. The first step is to estimate the efficiency of degradation depending on the plasma temperature as it is described in figure 1.

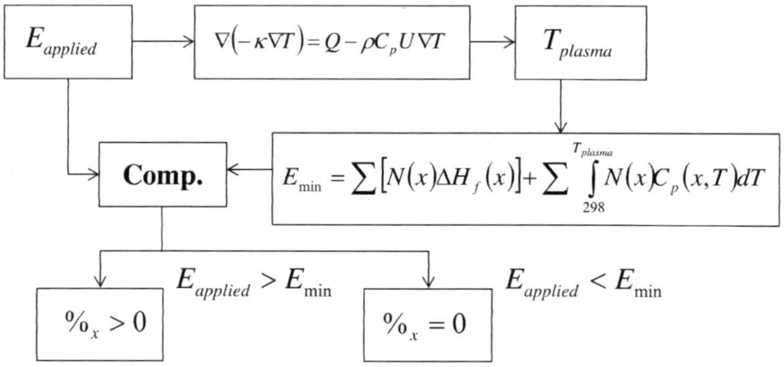


**Figure 1.** Structure of the method.

In figure 1, the applied energy $E_{applied}$ to the plasma is used to compute the distribution of temperature in the plasma $T_{plasma}$ by using some transport coefficients depending on the temperature $T$ of the gases (thermal conductivity $\kappa$, heat source density $Q$, mass density $\rho$, thermal capacity $C_p$ and $U$ is the vector of velocity) [12]. The term source $Q$ is calculated from

$$Q=\sigma E^2-q \tag{1}$$

by considering the electrical field $E=-\nabla V$ and the electrical conductivity $\sigma$ depending on the temperature $T$. The temperature dependent quantity $q$ is defined by an interpolation function and introduced to account for radiation losses [13, 14]. The mass density is calculated from the ideal gas law

$$\rho = \frac{pM}{RT} \tag{2}$$

where $p$ is the static pressure of the gas, $M$ the molar mass of the gas and $R$ = 8.32 $JK^{-1}mol^{-1}$ the perfect gas constant.

The plasma modeling described in this paragraph is quite simple and approximate. Considering the plasma at LTE, without any sheath, does not describe the fine mechanisms involved, as arc discharge plasma. However, it provides a satisfactory temperature profile in a relatively short computing time and with reasonable memory requirement for a desktop computer. This temperature profile as well as the mol number $N$ of the species $x$, the molar formation enthalpy $\Delta H_f$, and the degradation fraction of the species $x$ $\%_x$ are then used as the input parameter for the computing of the specific energy consumption $E_{min}$ [15].

## 3. Mass balance

Having estimated the value of the percentage of degradation according to the total power applied to the plasma reactor, the temperature and the inward species, the mass balance is proposed as shown in figure 2.

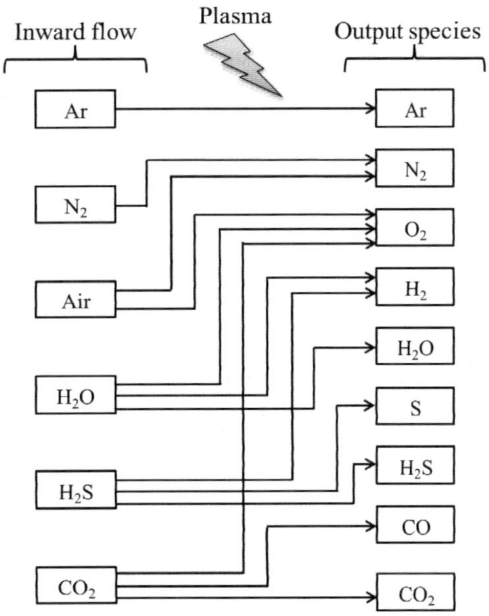


**Figure 2.** Schema of the mass balance.

The concentration of the output species depends on the species in the inward flow and power applied, since the latter determines the percentage of degradation for the waste gases ($CO_2$ and $H_2S$). To estimate the concentration of output species, we performed a mass balance as follows. If the concentration of each input atom is equal to every output atom, we have:

$$C_x^E = C_x^S \tag{3}$$

Where $C_x^E$ is the concentration of the inward species $x$ and $C_x^S$ is the concentration of the output species $x$. In the case of argon, being neutral species, the equivalent balance is:

$$C_{Ar}^{S} = C_{Ar}^{E} \tag{4}$$

Relating the degradation percentages for input species $H_2S$, $CO_2$ and $H_2O$, we have the following outlet concentrations corresponding to the portion not degraded by plasma:

$$C_{H_2S}^{S} = \left(1 - \frac{\%_{H_2S}}{100}\right) C_{H_2S}^{E} \tag{5}$$

$$C_{CO_2}^{S} = \left(1 - \frac{\%_{CO_2}}{100}\right) C_{CO_2}^{E} \tag{6}$$

$$C_{H_2O}^{S} = \left(1 - \frac{\%_{H_2O}}{100}\right) C_{H_2O}^{E} \tag{7}$$

On the other hand, the $H_2S$ portion degraded is converted to species that determines the following reaction:

$$1H_2S \rightarrow 1H_2 + 1S \tag{8}$$

The probability of formation of $H_2$ and S is proposed to be 50%. In the case of CO2, the reaction that determines the species generated from its degradation is:

$$CO_2 \rightarrow CO + \frac{1}{2}O_2 \tag{9}$$

where the probability of generation of CO is 75% and $O_2$ is 25%. For $H_2O$, the reaction that determines the degradation is:

$$H_2O \rightarrow H_2 + \frac{1}{2}O_2 \tag{10}$$

where the probability of generation of $O_2$ is proposed 25% and 75% for $H_2$.. In the case of air, it is known that its general composition is 79% $N_2$ y 21% $O_2$. In that case, the concentrations of these output species are:

$$C_{N_2}^{S} = (0.79) C_{Air}^{E} \tag{11}$$

$$C_{O_2}^{S} = (0.21) C_{Air}^{E} \tag{12}$$

In the same way to the case of Argon, the balance for each atom is realized. The addition of each output concentration for every species gives their total output concentration (see figure 2).

## 4. Results

Figure 3 presents the image of the hybrid plasma discharge [16]. The plasma model was implemented in the free software Saturne with the approximated geometry of the discharge and 2 kW of applied power; the result of the temperature profile is shown in figure 4.

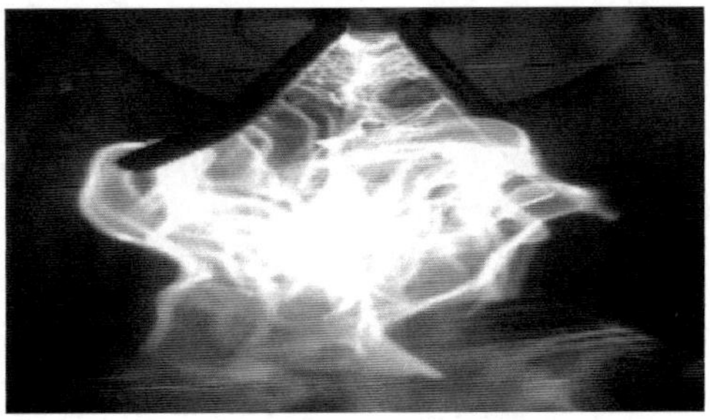

**Figure 3.** Hybrid plasma discharge.

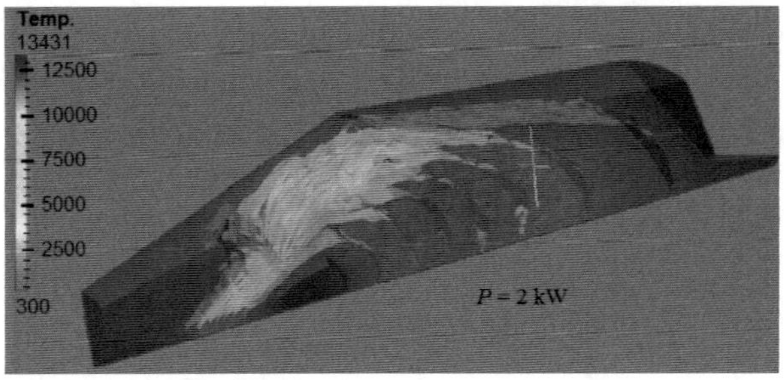


**Figure 4.** Temperature profile of plasma discharge.

The temperature profile presents high temperature values, more than 10000 K in the arc region. This temperature is sufficient to crack the hydrogen sulfide and carbon dioxide molecules; the non-thermal plasma at the end on the electrodes does not permit the recombination process of these molecules. Figure 5 presents the estimation of the minimum power to degrade the $CO_2$ and $H_2S$ molecules.

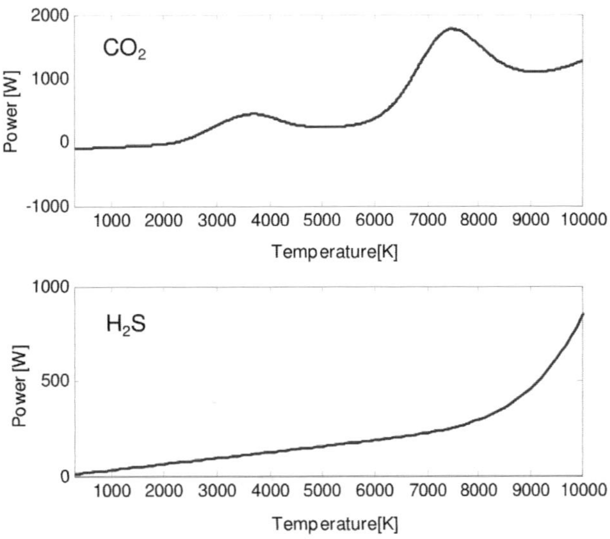


**Figure 5.** $P_{min}$ for $CO_2$ and $H_2S$.

According to this, the degradation percentage values are estimated as follow. The $\%_{H_2S}$ value is estimated to be 95%; the molecule of carbon dioxide is stronger than $H_2S$, then the $\%_{CO_2}$ value is estimated to be 50%; in the case of $H_2O$, the value $\%_{H_2O}$ is 30% due to the recombination in the steam phase. A numerical test of the mass balance was run taking into account the % values of degradation, supposing 10 minutes of the duration of the experiment, and 20 l.p.m. inward gas flow with the composition described in table 2.

In the case of the species $N_2$ and $O_2$, their concentration is the equivalent value to introducing 8.46% of air. In this mixture of inward species, the total concentration is 31.7899 g/min. The mass balance was computed according to the schema presented in figure 2 and the equations (4) to (12). The result is presented in table 3.

**Table 2.** Composition of the inward gas flow.

| Specie IN | Concentration (g/min) | Composition (*vol* %) |
|---|---|---|
| Ar | 13.7842 | 41.27 |
| $N_2$ | 1.5839 | 6.68 |
| $O_2$ | 0.4811 | 1.78 |
| $H_2O$ | 1.9900 | 9.95 |
| $H_2S$ | 3.8976 | 13.44 |
| $CO_2$ | 10.0531 | 26.88 |

**Table 3.** Result of the mass balance. The final mass accumulation is after 10 minutes of treatment duration.

| Specie OUT | Concentration (g/min) | Composition (vol %) | Final mass (g) |
|---|---|---|---|
| Ar | 13.7842 | 43.36 | 137.842 |
| $N_2$ | 1.5839 | 4.98 | 15.839 |
| $O_2$ | 1.8869 | 5.94 | 18.869 |
| $H_2$ | 2.2991 | 7.23 | 22.991 |
| $H_2O$ | 1.3930 | 4.38 | 13.930 |
| S | 1.8514 | 5.82 | 18.513 |
| $H_2S$ | 0.1949 | 0.61 | 1.948 |
| CO | 3.7699 | 11.86 | 37.699 |
| $CO_2$ | 5.0266 | 15.81 | 50.265 |

The total concentration of the output species is 31.7899 g/min, the same value of the total concentration of the inward species; as a result the mass balance presents a correct agreement. The mixture gas of the output species $H_2$, CO, $H_2O$ and $N_2$ represents mainly the composition of *syngas*. Table 4 presents the composition of *syngas* in the output gas flow, having a good agreement with the composition reported in [11] (see table 1). In this case, to total mass of *syngas* is 0.09046 kg; If the lower heating value of *syngas* in a fuel cell is 9.20 MJ/kg, then the total energy produced is estimated to be 0.8322 MJ.

In the total output gas flow there are also species like $CO_2$ and $H_2S$ that are not totally degraded. In this case, these gases can be extracted and feed backed to the inward flow. Otherwise, it is recommended to increase the applied power to reach higher degradation values for these molecules and reduce their concentration in the output.

The energy consumption to degrade the mixture of gases for 10 minutes and 2kW of applied power is calculated to be 1.2 MJ. Relating the energy consumption and production of electrical energy, the resultant efficiency of the system is 69.35%.

## 5. Conclusions

From gases like $CO_2$ and $H_2S$, the plasma process produces *syngas* and it can be applied to fuel cells to generate electrical energy. The efficiency estimated can reach 69.35%, making the plasma process a quasi-self sustained system.

The remaining gas species can be separated from the total flow and feed backed the system, especially in the case of Ar, $CO_2$ and $H_2S$.

The plasma model and the mass balance are quite simple and approximate. In the mass balance only the main species are considered but it is well known that plasma discharges produces a great quantity of free radical that can interact with other species creating chain reactions. However, the model here depicted presents a good approximation to the results mentioned in section 1.

**Table 4.** Output *syngas* composition.

| Specie | Composition (*vol* %) |
|---|---|
| $H_2$ | 25.41 |
| CO | 41.68 |
| $H_2O$ | 15.40 |
| $N_2$ | 17.51 |

**Acknowledgments**
The authors are very indebted to F. Ramos-Flores, M. Durán-García, G. Jiménez-Aviles, A. Salazar-Torres, G. Soria-Arguello, M. García-Ramírez, E. Alva-Rojas, R. Reyes-Gutiérrez and M. Hidalgo-Pérez for the experimental support. This work was effectuated and financed by ININ and CONACyT.

**References**

[1] Consonni S, Giugliano M and Grosso M 2005 *Waste Manage.* **25** 123-35
[2] Chang Y H, Chen W C and Chang N B 1998 *J. Hazard. Mater.* **58** 33-45
[3] Caputo A C and Pelaggage P M 2002 *Appl. Therm. Eng.* **22** 439-48
[4] Belgiorno V, De Feo D, Della Roca C and Napoli R M A 2003 *Waste Manage.* **23** 1-15
[5] Altafini C R, Wander P R and Barreto R M 2003 *Energ. Convers. Manage.* **44** 2763-77
[6] Lange H, Huczko A, Chojecki G, Golimowski J, Plotczyk W, Ostrega B, Dziadko D, Pacheco J, Ishigaki T and Tanaka H 2002 *Acta Agrophysica* **80** 327-34
[7] Mountouris A, Voutass E and Tassious D 2006 *Energ. Convers. Manage.* **47** 1723-37
[8] Vaidyanathana A, Mulhollanda J, Ryua J, Stuart S M and Circeo L J 2007 *J Environ. Manage.* **82** 77-82
[9] Leal-Quiros E 2004 *Braz. J. Phys.* **34(4B)** 1587-93
[10] Huang H and Tang L 2007 *Energ. Convers. Manage.* **48** 1331-7
[11] Galeno G, Minutillo M and Perna A 2011 *Int. J. Hydrogen Energ.* **36** 1692-1701
[12] Hirsch J, Baumann B, Wolff M, Bhosle S and Valdivia-Barrientos R 2010 *J. Phys. D: Appl. Phys.* **43** 234002 (7pp)
[13] De Groot J J and Van Vliet J A 1986 The high pressure sodium lamp (Philips Technical Library)
[14] Club EDF Arc Électrique 1984 L'arc électrique et ses applications (Tome 1, Etude physique de l'arc électrique, Édit. CNRS)
[15] Bo Z, Yan J, Li X, Chi Y and Cen K 2009 *J. Hazard. Mater.* **166** 1210-6
[16] Garduño-Aparicio M, Estrada-Martinez M, Pacheco-Sotelo J, Pacheco-Pacheco M, Garcia-Ramírez M, Valdivia-Barrientos R, Rivera-Rodríguez C and Gonzalez J J 2011 *IEEE Trans. Plasma Sci.* **39** 2890-1

# Magnetothermal instabilities in magnetized anisotropic plasmas

**M S Nakwacki**[1] **and J Peralta-Ramos**[2]

[1]Instituto de Astronomía, Geofísica e Ciências Atmosféricas, Universidade de São Paulo, Rua do Matão 1226, Cidade Universitária, 05508-090 São Paulo, Brazil
[2]Instituto de Física Teórica, Universidade Estadual Paulista, Rua Doutor Bento Teobaldo Ferraz 271 - Bloco II, 01140-070 São Paulo, Brazil

E-mail: sole@astro.iag.usp.br

**Abstract.** Using the 16–moment transport equations for an ideal anisotropic collisionless plasma we analyze the influence of pressure anisotropy on the magnetothermal (MTI) and heat–flux–driven buoyancy (HBI) instabilities. We calculate the dispersion relation and the growth rates for these instabilities in the presence of a background heat flux and for configurations with static pressure anisotropy, finding that when the frequency at which heat conduction acts is much larger than any other frequency in the system (i.e. weak magnetic field) the pressure anisotropy has no effect on the MTI/HBI, provided the degree of anisotropy is small. In contrast, when this ordering of timescales does not apply the instability criteria depend on pressure anisotropy.

## 1. Introduction

In many magnetized dilute astrophysical plasmas thermal electron conduction occurs almost exclusively parallel to magnetic field lines. In this regime, the equations of ideal magnetohydrodynamics (MHD) that describe the plasma dynamics must be supplemented with anisotropic transport terms for energy and momentum due to the near free-streaming motions of particles along magnetic field lines [1]. [2, 3] have shown that anisotropic thermal conduction can fundamentally alter the Schwarzschild convective instability criterion in stratified atmospheres, in that convection sets in when a temperature gradient (as opposed to an entropy gradient) is present. This new convective instability is termed the magnetothermal instability (MTI) [2] or the heat–flux–driven buoyancy instability (HBI) [4] when the temperature decreases or increases with height, respectively. We note that the HBI only takes place if a background heat–flux is present. Essentially, the MTI and the HBI occur because the heat flux must follow the perturbed magnetic field lines, and thus there are regions in the plasma which are locally heated or cooled (see [2–6]).

The MTI and HBI couple the magnetic structure of the plasma to its thermal properties and can have important implications for galaxy clusters (see e.g.[5–8]). For example, in a weakly magnetized atmosphere local simulations have demonstrated that the MTI can amplify magnetic field acting as a magnetic dynamo and also drive strong turbulence.

Dilute magnetized plasmas not only exhibit anisotropic thermal conduction but also an anisotropic pressure tensor, which results from different kinetic temperatures for electron motion in the parallel and perpendicular direction to the magnetic field. For collisionless plasmas the pressure does not isotropize, and hence if we insist on using a fluid description the standard magnetohydrodynamics theory (MHD) must be modified [9–13].

The purpose of this work is to study the MTI and the HBI in a dilute magnetized plasma including the effect of pressure anisotropy. To this end, we employ the theory of kinetic MHD (KMHD) as derived by the 16–moment method from the Vlasov equation [10–12]. This formalism extends the well–known double–adiabatic theory of Chew, Goldenberg and Low (CGL) [9] in that it allows for nonvanishing (parallel and perpendicular) heat fluxes, which are neglected in the CGL theory. Although we keep the discussion quite general throughout the paper, we have in mind an astrophysical magnetized collisionless plasma, and as a typical example we consider the ICM in section 5. We note that our study is limited to the case where pressure perturbations are zero, i.e. we only treat the case of a static background pressure anisotropy [5, 14].

## 2. Basic equations

In this section we describe the equations and set the stage for the analysis of wave instabilities given later on (More complete accounts of KMHD and similar approaches can be found in [11–13]).

The inadequacies of the CGL approach (e.g. in the criterion and nonlinear evolution of the fire hose and mirror instabilities [13, 15, 16]) stem from the unwarranted neglect of the third–order moment of Vlasov equation. The 16–moment method provides a consistent way of deriving from the Vlasov equation the correct fluid equations for a heat–conducting anisotropic magnetized plasma. If the Larmor radius is much smaller than the other characteristic lengths of the plasma the equations can be considerably simplified [10] (see also [11, 12]):

$$\frac{d\rho}{dt} + \rho\vec{\nabla}\cdot\vec{v} = 0 \tag{1}$$

$$\begin{aligned}\rho\frac{d\vec{v}}{dt} + \vec{\nabla}\left(p_\perp + \frac{B^2}{8\pi}\right) - \frac{1}{4\pi}(\vec{B}\cdot\vec{\nabla})\vec{B} = \rho\vec{g} + \hat{b}\nabla_\parallel\Delta p \\ +\Delta p\left[\hat{b}(\vec{\nabla}\cdot\hat{b}) + \nabla_\parallel\hat{b}\right]\end{aligned} \tag{2}$$

$$\frac{d\vec{B}}{dt} + \vec{B}(\vec{\nabla}\cdot\vec{v}) - (\vec{B}\cdot\vec{\nabla})\vec{v} = 0 \tag{3}$$

$$\rho T_\parallel \frac{d}{dt}s_\parallel = -\vec{\nabla}\cdot\vec{Q}_\parallel \tag{4}$$

$$\rho T_\perp \frac{d}{dt}s_\perp = -\vec{\nabla}\cdot\vec{Q}_\perp \tag{5}$$

where

$$\frac{d}{dt} = \frac{\partial}{\partial t} + (\vec{v}\cdot\vec{\nabla}) \ , \tag{6}$$

$\Delta p = p_\perp - p_\parallel$, $\hat{b} = \vec{B}/B$, $\nabla_\parallel = \hat{b}\cdot\vec{\nabla}$ and $\vec{Q}_{\parallel,\perp}$ are the parallel and perpendicular heat fluxes, which we assume for simplicity to be given by Braginskii's approximation [1] $\vec{Q}_{\parallel,\perp} = -\hat{b}\chi\nabla_\parallel T_{\parallel,\perp}$ where $\chi$ is the electron thermal diffusivity (assumed to be a constant [4]).

The specific entropies associated with parallel and perpendicular motion are given by [15]

$$s_\parallel = \frac{c_V}{3}\ln\left(\frac{p_\parallel B^2}{\rho^3}\right) \quad \text{and} \quad s_\perp = \frac{2c_V}{3}\ln\left(\frac{p_\perp}{\rho B}\right) \tag{7}$$

where $c_V$ is the specific heat. Note that the total entropy

$$s = s_{\|} + s_{\perp} = c_V \ln\left(\frac{p_{\|}^{1/3} p_{\perp}^{2/3}}{\rho^{5/3}}\right) \tag{8}$$

reduces to the ordinary MHD expression for the specific entropy in the isotropic case $p_{\|} = p_{\perp}$.

## 3. MTI/HBI in the presence of pressure anisotropy

We consider a thermally stratified plasma in the presence of gravity $\vec{g} = -g\hat{z}$, so that in equilibrium $dp/dz = -\rho g$. The magnetic field of the equilibrium state is assumed to be homogeneous and in the $(x,z)$ plane $\vec{B} = B_x\hat{x} + B_z\hat{z}$ so that there is a background heat flux given by $\vec{Q}_{\|,\perp} = -\chi(b_x b_z \hat{x} + b_z^2 \hat{z}) dT_{\|,\perp}/dz$. As in [4], we assume a steady–state initial equilibrium so that $\vec{\nabla}\cdot\vec{Q} = 0$, which implies a linear temperature profile in $\hat{z}$. Note that $p_{\perp} = p_{\|} + \text{const.}$ and that we consider a vanishing initial velocity.

Under these circumstances, putting $\delta v_z \sim \exp(-i\omega t + i\vec{k}\cdot\vec{r})$ and similarly for other quantities the linearly perturbed equations can be written as

$$\omega\delta\rho - \rho k_v = 0 \tag{9}$$

$$\omega[k_v\vec{k} - k^2\delta\vec{v}] - (v_A^2 - \Delta p/\rho)k^2\tilde{b}\frac{\delta\vec{B}}{B} + [\vec{k}\tilde{b} - k^2\hat{b}]\tilde{b}(\delta p_{\perp} - \delta p_{\|}) + i(k_z\vec{k} - k^2\hat{z})g\delta\rho/\rho = 0 \tag{10}$$

$$\omega\delta\vec{B} - \vec{B}k_v + (\vec{k}\cdot\vec{B})\delta\vec{v} = 0 \tag{11}$$

$$i\frac{c_V}{3}\omega\left[\frac{1}{p_{\|}}(\delta p_{\|} + 2\delta p_{\perp}) - \frac{\delta\vec{B}}{B}(1-\alpha) + \frac{\delta\rho}{\rho}(3+2\alpha)\right] + \left(\frac{ds_{\|}}{dz} + \alpha\frac{ds_{\perp}}{dz}\right)\delta v_z - i\chi\tilde{b}\frac{D}{\rho}\left(\frac{d\ln T_{\|}}{dz} + \alpha\frac{d\ln T_{\perp}}{dz}\right) + \chi\frac{\tilde{b}^2}{\rho}\left(\frac{\delta T_{\|}}{T_{\|}} + \alpha\frac{\delta T_{\perp}}{T_{\perp}}\right) = 0 \tag{12}$$

$$i\frac{c_V}{3}\omega\left[\frac{1}{p_{\|}}(\delta p_{\|} - 2\delta p_{\perp}) - \frac{\delta\vec{B}}{B}(1+\alpha) + \frac{\delta\rho}{\rho}(3-2\alpha)\right] + \left(\frac{ds_{\|}}{dz} - \alpha\frac{ds_{\perp}}{dz}\right)\delta v_z - i\chi\tilde{b}\frac{D}{\rho}\left(\frac{d\ln T_{\|}}{dz} - \alpha\frac{d\ln T_{\perp}}{dz}\right) + \chi\frac{\tilde{b}^2}{\rho}\left(\frac{\delta T_{\|}}{T_{\|}} - \alpha\frac{\delta T_{\perp}}{T_{\perp}}\right) = 0 \tag{13}$$

where we have used that

$$\delta\vec{Q}_{\|,\perp} = -\chi\delta\hat{b}\nabla_{\|}T_{\|,\perp} - \chi\hat{b}(\delta\hat{b}\cdot\nabla T_{\|,\perp}) - i\chi\hat{b}\tilde{b}\delta T_{\|,\perp}\ , \tag{14}$$

$v_A = B/(4\pi\rho)^{1/2}$ is the Alfvén speed, and we have defined $\tilde{b} = \vec{k}\cdot\hat{b}$, $k_v = \vec{k}\cdot\delta\vec{v}$, $D = (\delta B_z - 2\delta B\big)/B$, and $\alpha = T_{\perp}/T_{\|}$.

## 4. Neglecting pressure perturbations

We will now consider the simpler case in which pressure perturbations are neglected. The growing modes of interest have growth times much longer than the sound crossing time of the perturbation, so it is sufficient to work in this approximation. From (9)–(13) we get the following dispersion relation

$$iA_3\omega^3 + A_2\omega^2 + (iA_1 + \tilde{A}_1)\omega + A_0 = 0 \tag{15}$$

where

$$A_0 = (\omega_A^2 - \omega_s^2)A_2 - gKp_{\parallel}\omega_{c,\parallel}(1+\alpha)\frac{d\ln T}{dz} \tag{16}$$

$$A_1 = \frac{2}{5}\rho c_V T_{\parallel}(1+\frac{2}{3}\alpha)(\omega_A^2 - F) - \rho N^2\frac{k_\perp^2}{k^2} \tag{17}$$

$$\tilde{A}_1 = \frac{4}{15}gT_{\parallel}c_V(1-\alpha)\tilde{b}\frac{k_\perp}{k^2}(b_x k_z - b_z k_x) \tag{18}$$

$$A_2 = p_{\parallel}(1+\alpha)\omega_{c,\parallel} \tag{19}$$

$$A_3 = \frac{2}{5}\rho c_V T_{\parallel}\left(1+\frac{2}{3}\alpha\right) \tag{20}$$

and $\omega_A = \vec{k}\cdot\vec{v_A}$, $N^2 = (2/5)gd(s_{\parallel} + \alpha s_\perp)/dz$, and $\omega_{c,\parallel} = (2/5)\chi\tilde{b}^2 T_{\parallel}/p_{\parallel}$ are the Alfvén, Brunt–Väisälä and conduction frequencies, respectively, suitably modified by the anisotropy parameter, and $F = (c_{s,\parallel}^2 - c_{s,\perp}^2)\tilde{b}^2$ with $c_{s,\parallel,\perp}^2 = p_{\parallel,\perp}/\rho$. In (15) we have put $K = \frac{1}{k^2}[(1-2b_z^2)k_\perp^2 + 2b_z b_x k_z k_x]$ for notational simplicity. To derive (15) we have used that $\delta\rho/\rho = -\delta T_{\parallel}/T_{\parallel} = -\delta T_\perp/T_\perp$ (see [2, 4]), and also that $d\ln T_{\parallel}/dz = d\ln T_\perp/dz = d\ln T/dz$, where $T = (T_{\parallel} + 2T_\perp)/3$ is the total temperature, which follows since $\alpha$ is assumed to be a constant.

It is interesting to analyze the special case of a weak magnetic field, in which there exists an ordering in frequencies given by [4]

$$\omega_c \gg \omega_d = \left(\frac{g}{H}\right)^{1/2} \gg \omega_A \tag{21}$$

where $H$ is the local scale-height of the system and $\omega_d$ its dynamical frequency. Besides, if $\alpha$ is sufficiently close to 1 then one can safely put $|F| < \omega_c$. In this limit the dispersion relation becomes

$$\omega^2 \simeq gK\frac{d\ln T_{\parallel}}{dz} = gK\frac{d\ln T}{dz} \tag{22}$$

This expression is identical to the result of [4], which shows that when the magnetic field is sufficiently weak so that the timescale ordering given in (21) holds and $\alpha \sim 1$ (as occurs e.g. in the ICM), the MTI and the HBI become independent of the pressure anisotropy.

Following [3], we will now analyze the stability of solutions to (15) using the Routh–Hurwitz criteria. It should be noted that the polynomial (15) has complex coefficients and therefore the generalized Routh–Hurwitz theorem must be used (for details see e.g.[17]). Applying this procedure to the polynomial in (15) we find the stability criteria $A_0 > 0$, $A_2 > 0$, and $A_2(A_1A_2 - A_0) > \tilde{A}_1^2$. Comparing this result to the isotropic case studied in [3], we see that the stability criteria get modified by pressure anisotropy in two ways. Not only do the expressions for the coefficients $A_n$ contain $\alpha$, but also an extra term $\tilde{A}_1$ appears which vanishes if $\alpha = 1$ or if $b_x k_z = b_z k_x$. The first effect does not change in a significant way the quantitative analysis of [3], so we shall not pursue it here. However, the appearance of $\tilde{A}_1$ in the third stability criteria implies that when pressure anisotropy is taken into account $A_1A_2 - A_0$ cannot be arbitrarily small, otherwise an instability develops. This represents a qualitative change with respect to the isotropic case, where $A_1A_2 - A_0 > 0$ for stable solutions. Note that this modification only affects transverse perturbations, since $\tilde{A}_1 = 0$ if $k_\perp = 0$.

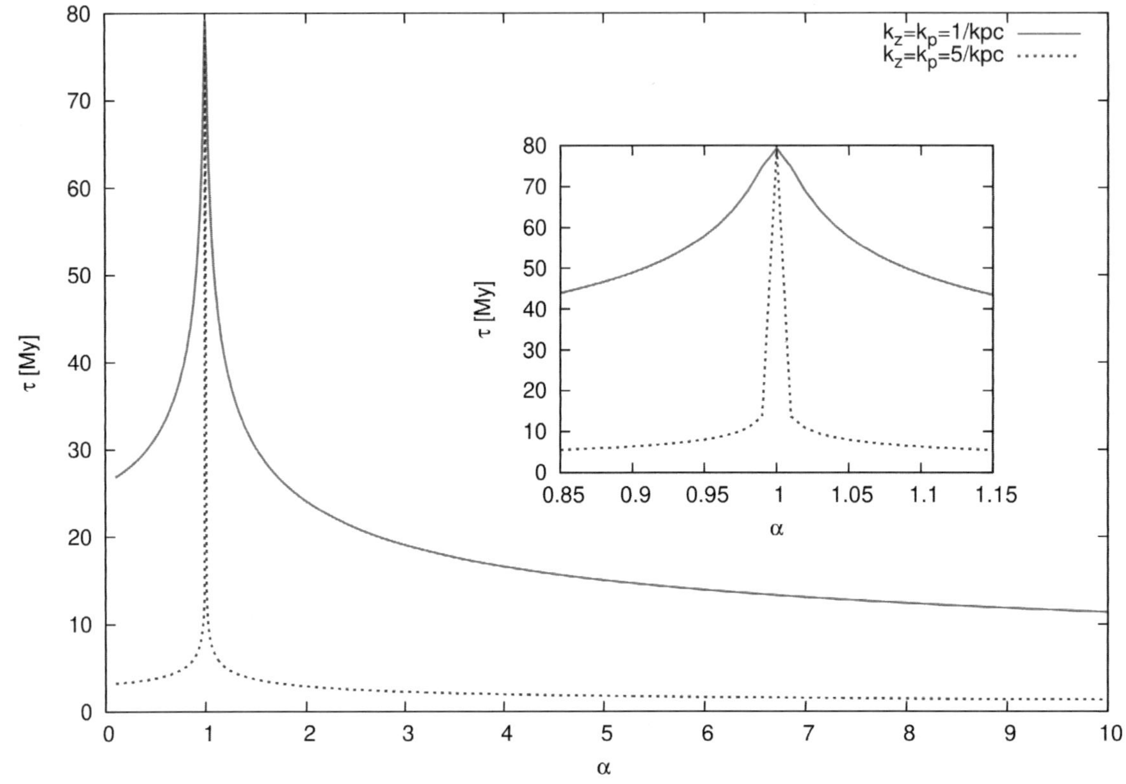


**Figure 1.** (Color online) Growth time for the MTI/HBI as a function of $\alpha$ for $k_z = k_\perp = 1$/kpc and $k_z = k_\perp = 5$/kpc. The inset is a zoom for $0.85 < \alpha < 1.15$, which is a closer range to the values expected in the ICM.

## 5. Application to the ICM

As an illustrative case, we will now present numerical results corresponding to the ICM. As typical values we take $\rho \sim 10^{-3}$ cm$^{-3}$, $B \sim 10^{-6}$ G, $T \sim 10^7$ K, and $P \sim 10^{-10}$ ergs cm$^{-3}$. The value of $g$ is estimated as the average of $g(r) = GM/r^2$ from $r = r_c$ to $r = 2r_c$, where $G$ is the gravitational constant, $M = 10^{14} M_\odot$ is the mass of the ICM and $r_c = 290$ kpc. This gives $g \sim 7 \times 10^{-7}$ cm s$^{-2}$. For the entropy gradient we take $3d\ln s/dz = -d\ln T/dz$, and taking the Brunt–Väisälä frequency to be $\sim 10^{-12}$ s$^{-1}$ [8] we get $d\ln T/dz \sim 10^{-23}$ cm$^{-1}$. The anisotropy parameter in the ICM can be written as $\alpha = 1/(1 \pm \varphi)$, where $\varphi = |p_\perp - p_\parallel|/p_\perp \sim 1/\sqrt{\mathrm{Re}}$ and Re is the Reynolds number [13]. To estimate the lower and upper bounds for the values of $\alpha$ we will consider Re $> 50$, so we get $\alpha \in [0.87, 1.16]$. Note that in the ICM Re $\gg 50$ for convective motions, so that actually even lower values for $|1-\alpha|$ are expected in this environment. However, in order to show the physical behavior of this instability as a function of the anisotropy parameter it is convenient to consider a broader, though unphysical, range for $\alpha$ (as we do in figure 1). In what follows we take a tangled magnetic field corresponding to $b_x = 0.3$.

In order to illustrate the dependence of the stability of solutions to (15) on pressure anisotropy, we show in figure 1 the growth time $\tau = 1/|\mathrm{Im}(\omega)|$ as a function of $\alpha$ for $k_z = k_\perp = 1$/kpc and $k_z = k_\perp = 5$/kpc.

It is seen from figure 1 that in the case of $\alpha = 1$ the growth time, $\tau$, is of the order of $\sim 100$ My. On the other hand, the presence of anisotropy causes a strong decrease in $\tau$, which

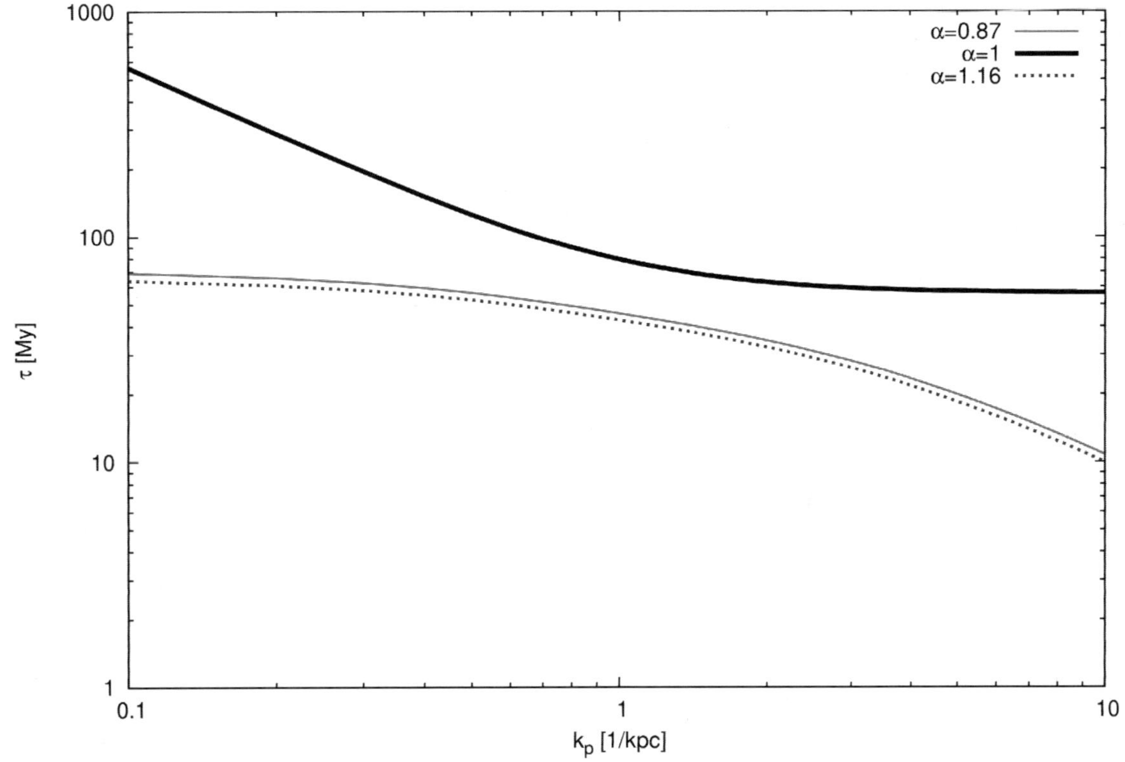


**Figure 2.** (Color online) Growth time for the MTI/HBI as a function of $k_\perp$ for $k_z = 1/\text{kpc}$ and $\alpha = 0.87; 1; 1.16$.

is considerably larger for smaller scales. The dependence of $\tau$ with $\alpha$ is similar in both cases ($\alpha < 1$ or $\alpha > 1$). The inset shows the behavior of $\tau$ in the range $0.85 < \alpha < 1.15$, which as mentioned is a closer range to the values expected in the ICM. It is seen that for scales $\sim 0.2$ kpc the growth time in the anisotropic case can be almost one order of magnitude smaller than the one corresponding to $\alpha = 1$. For larger scales ($> 1$ kpc) $\tau$ decreases by a factor of $\sim 2$ or less.

We now go over to analyze the dependence of $\tau$ with wave vector in the range $[0.1, 10] \times 1/\text{kpc}$ (magnetic tension stabilizes the instabilities on shorter scales). Figure 2 shows $\tau$ as a function of $k_\perp$ for $k_z = 1/\text{kpc}$ and for three values of $\alpha$. We note that for other values of $k_z$ in the studied range $[0.1, 10] \times 1/\text{kpc}$ the results are similar. It is seen that for any value of $k_\perp$ the growth time of the isotropic case is larger than the corresponding ones to $\alpha < 1$ and $\alpha > 1$. The difference in $\tau$ can reach one order of magnitude. As happened in the previous case, the behavior of $\tau$ is similar for both anisotropic cases. Interestingly, the decrease in $\tau$ with the anisotropy is significantly smaller in the kpc range. We will now briefly discuss some possible implications of our findings to the ICM (in the case where the timescale ordering of (21) does not apply). In this connection, the most important results of this work are that: (i) even with small pressure anisotropy the growth time of the MTI/HBI can become almost an order of magnitude smaller than the one corresponding to the isotropic case; (ii) the decrease of $\tau$ with anisotropy is smaller for scales $\sim$ kpc; and (iii) the effect of anisotropy on $\tau$ is larger at smaller scales. As mentioned in the Introduction, the MTI/HBI can be a significant source of magnetic field amplification

Journal of Physics: Conference Series **370** (2012) 012038 doi:10.1088/1742-6596/370/1/012038

[5, 6]. Our results imply that in the ICM the magnetic field amplification due to the MTI/HBI could be faster than what is expected in the isotropic case. Although we expect this effect to be small in the ICM because the magnetic field is weak and thus the timescale ordering holds, it could have some important consequences in the subsequent plasma dynamics following the linear regime.

## 6. Conclusions

We have found that if the magnetic field is sufficiently weak so that the dynamical frequency is larger than any other frequency of the system, the MTI and HBI do not depend on pressure anisotropy (provided the latter is small). On the other hand, if this timescale ordering does not apply the stability criteria for the MTI and the HBI will depend on pressure anisotropy. For the ICM, the growth time of the instability in the anisotropic case can be almost one order of magnitude smaller than the isotropic one.

The linear analysis of the MTI/HBI in anisotropic plasmas presented in this work is largely idealized, since we neglected viscosity and radiative cooling and we considered static pressure anisotropy configurations. In spite of this shortcomings, we believe that our analysis provides some insight into the effect of pressure anisotropy on the MTI/HBI, as well as on some possible implications for the dynamics of the ICM and similar astrophysical environments.

## Acknowledgments

We thank Matthew Kunz and Michael McCourt for useful comments. M.S.N. and J.P.R. are funded by Fundação de Amparo a Pesquisa do Estado de São Paulo (FAPESP, Brazil).

## References

[1] Braginskii S I 1965 *Reviews of Plasma Physics* **1** 205

[2] Balbus S A 2000 *The Astrophysical Journal* **534** 420–427 (*Preprint* arXiv:astro-ph/9906315)

[3] Balbus S A 2001 *The Astrophysical Journal* **562** 909–917 (*Preprint* arXiv:astro-ph/0106283)

[4] Quataert E 2008 *The Astrophysical Journal* **673** 758–762 (*Preprint* 0710.5521)

[5] McCourt M, Parrish I J, Sharma P and Quataert E 2011 *MNRAS* **413** 1295–1310 (*Preprint* 1009.2498)

[6] Parrish I J, Quataert E and Sharma P 2010 *The Astrophysical Journal* **712** L194–L198 (*Preprint* 0912.1851)

[7] Bogdanović T, Reynolds C S, Balbus S A and Parrish I J 2009 *The Astrophysical Journal* **704** 211–225 (*Preprint* 0905.4508)

[8] Ruszkowski M and Oh S P 2010 *The Astrophysical Journal* **713** 1332–1342 (*Preprint* 0911.5198)

[9] Chew G F, Goldberger M L and Low F E 1956 *Royal Society of London Proceedings Series A* **236** 112–118

[10] Oraevskii V, Chodura R and Feneberg W 1968 *Plasma Physics* **10** 819–828

[11] Barakat A R and Schunk R W 1982 *Plasma Physics* **24** 389–418

[12] Ramos J J 2003 *Physics of Plasmas* **10** 3601–3607

[13] Rosin M S, Schekochihin A A, Rincon F and Cowley S C 2011 *MNRAS* **413** 7–38 (*Preprint* 1002.4017)

[14] Kunz M W 2011 *MNRAS* **417** 602–616 (*Preprint* 1104.3595)

[15] Abraham-Shrauner B 1967 *Journal of Plasma Physics* **1** 361

[16] Dzhalilov N S, Kuznetsov V D and Staude J 2008 *Astron. Astrophys.* **489** 769–772 (*Preprint* 0805.2405)

[17] Gantmacher F R 1959 *Applications of the theory of matrices* (Netherlands: Interscience Publishers)

Journal of Physics: Conference Series **370** (2012) 012039 doi:10.1088/1742-6596/370/1/012039

# Plasma effects in harmonic emission in relativistic laser interactions with overdense plasma targets examined from single particle radiation models

**R Ondarza-Rovira[1] and T J M Boyd[2]**

[1]Instituto Nacional de Investigaciones Nucleares, A.P. 18-1027, México 11801, Distrito Federal, Mexico

[2]Centre for Theoretical Physics, University of Essex, Wivenhoe Park, Colchester CO4 3SQ, United Kingdom

E-mail: `ricardo.ondarza@inin.gob.mx`

**Abstract.** Intense femtosecond laser pulses incident on supercritical plasmas can generate reflected high-order harmonics of the fundamental frequency that extend over the xuv spectral region, characterized by a power decay $P_m \sim m^{-p}$, where $m$ is the harmonic number. It has been shown in earlier reports that plasma effects can modify the emission spectrum producing decay indexes $p$ significantly weaker than the *universal* value 8/3 predicted by the similarity theory, these being within the range $4/3 \leq p \leq 8/3$. In this paper we employ two single particle radiation models to demonstrate that plasma density effects substantially modify the emission radiated by a relativistic electron accelerated by the electric field of the laser pulse, and show that the harmonic power decay is commonly characterized by the indexes 4/3 and 5/3, with remarkable similitudes to those particle-in-cell results recently reported in ultrarelativistic laser-plasma interactions.

## 1. Introduction

Multiple harmonic generation at relativistic energies from laser irradiated solid targets has become a subject of enduring interest for the diversity of potential technological applications in nonlinear optics, and more recently as sources that are promissory for generating intense attosecond light pulses at short wavelengths [1]. High harmonic generation could provide extremely short pulses at the attosecond scale in the xuv regime, much shorter that the conventional x-ray sources, and appropriate for exploring the ultrafast nature of the dynamical processes in matter.

The spectral power decay of laser-generated harmonics in relativistic interactions with overdense plasmas has been characterised by various authors [2,3] according to laser-plasma parameters. In the mildly relativistic regime Gibbon [3] proposed an empirical relation from PIC simulations for the harmonic efficiency $\eta$, and showed this to be scaled as $\eta \sim 17.2\, a_0^4\, m^{-p}$, where the decay index $p = 5$, $a_0$ is the normalized quiver momentum $a_0 \sim 0.85\left(I_{18}\lambda_L^2\right)^{1/2}$, with $I_{18}$ denoting the laser intensity in units of $10^{18}$ W/cm$^2$, and $\lambda_L$ the wavelength of the incident light measured in microns. Relativistic effects arise for $a_0 \gtrsim 1$. The spectral power density was estimated in experiments of laser pulses of a few ps duration to span over decreasing values of $p$ for increasing laser intensities, from instance $p = 5.5$ at $a_0 \sim 0.64$ to $p = 3.38$ at $a_0 \sim 2.85$ [4].

More recently, Baeva *et al.* [5] proposed a model on the grounds of the similarity model, that asserts that the interaction physics remains similar solely for constant values of the parameter $S = (n_e/n_c)/a_0$, with $n_e/n_c$ being the ratio of the ambient electron plasma density to the critical density. This restriction lead to associate a "universal" value 8/3 to the harmonic power decay. Furthermore, another feature predicted by the $S$-similarity analysis states a cut-off in the harmonic spectra at $m = \sqrt{8\,\alpha}\,\gamma_s^3$, where $\gamma_s$ is the relativistic factor of the plasma surface and $\alpha$ is of order one. No experimental support was found in experiments of the spectrum cut-off, though the spectral power decay was showed to be fitted by indexes in the range $2.3 \lesssim p \lesssim 2.7$, with the upper limit in consonance with the predicted "universal" value.

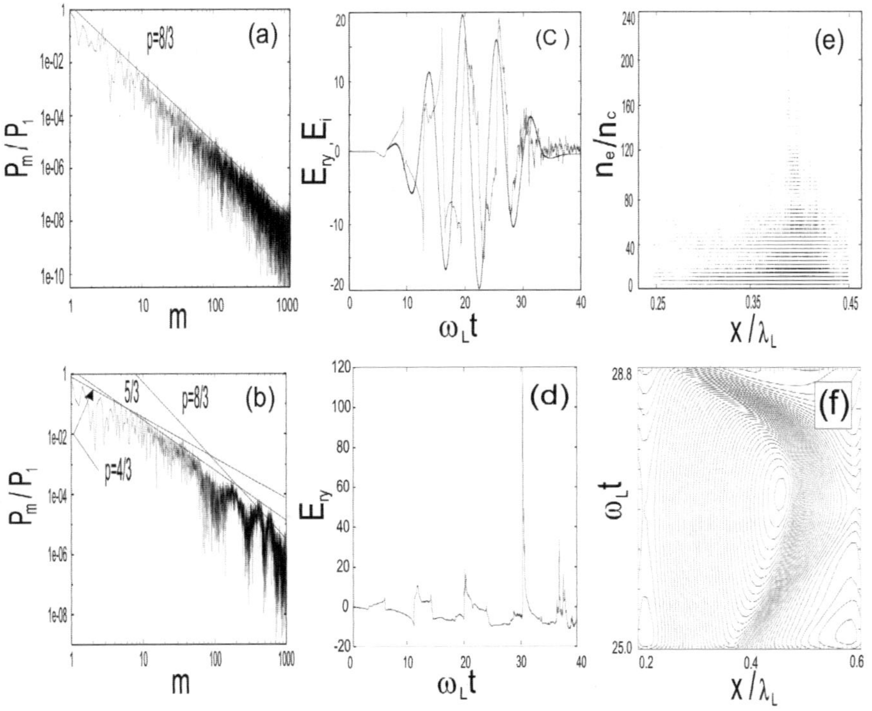


**Figure 1.** (a) PIC simulation results for the harmonic spectra from $s$-polarized laser pulses incident on a plane target plasma and (b) for $p$-polarized light; $a_0 = 20, n_e/n_c = 40$, pulse length $\tau_p = 17$ fs, $\lambda_L = 1054$ nm, angle of incidence $\theta = 23^{\rm o}$, $m = \omega_m/\omega_L$; (c),(d) show the incident $E_i$ and reflected fields $E_{ry}$ for these two polarizations, (e) shows the peak Langmuir condensate density, and (f) the contours of the corresponding Langmuir electric field, $E_x$.

## 2. Plasma effects

In mildly relativistic interactions, Boyd and Ondarza-Rovira (BOR) found evidence of emission at the plasma line and at its harmonics [6]. In the UR regime, the ponderomotive force of the Langmuir field is strong enough to govern plasma behavior over time scales $t > \omega_p^{-1}$ and highly localized electrostatic fields are generated in correspondence with density structures inside the plasma. These structures were identified as sources of emission at the plasma frequency and at its harmonics.

Particle-in-cell simulations where used to explore the harmonic power decay in ultrarelativistic laser-plasma interactions by BOR [7-9] and spectral decays characterized by prevalent indexes $p = 5/3$ were found and, less commonly, $p = 4/3$ in $p$-polarized light incident on dense plasma targets. When the incident laser pulse is normal to the plasma surface, the spectrum is invariantly characterized by a power law described by the index 8/3, regardless of the combination of parameters $(a_0, n_e/n_c)$, as shown in figure 1(a) [9]. Contrasting this, plasma effects are evident in the power harmonic decay for oblique incidence, since the excitation of plasma oscillations enable the enhancement of emission with lower harmonic power decay indexes, resulting in much flatter spectra, as shown in figure 1(b). The cases here shown correspond to the combination $a_0 = 20$ and $n_e/n_c = 40$, for a pulse length $\tau_p = 17$ fs, angle of incidence of $\theta = 23^o$, and $m = \omega_m/\omega_L$. For $p$-polarization field amplification is also observed (figure 1(d)), being an order of magnitude greater than the $s-$polarized case, figure 1(c), with an estimated width of about 10 as. These reflected attosecond spikes correlate with the electron

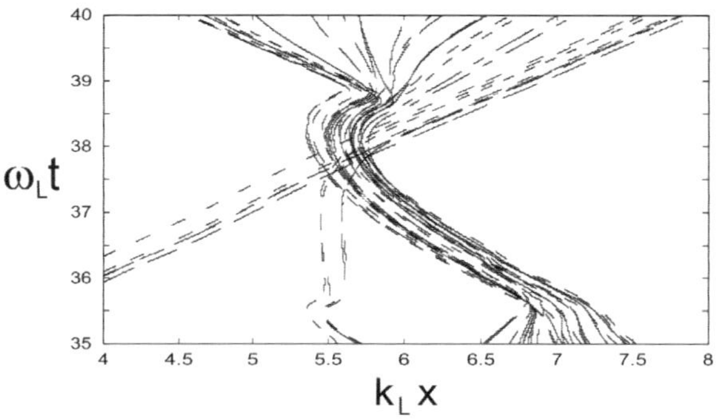


**Figure 2.** PIC simulation results for the electron trajectories around the region in the plasma where the emission source is correlated from the $E_x$ contours and the reflected field.

bunching observed within the plasma (figure 1(e)), and were identified as the sources of the emission.

Beside this, the generation of localized high electrostatic fields are found to be generated at sites of electron density bunching, as depicted in figure 1(f). Additionally, as figure 2 shows, the electron trajectories were followed around the regions in the plasma where these intense electrostatic fields were observed, and for which the emission of a reflected attosecond spike was correlated.

## 3. Single particle radiation models

In what follows we show that from simple dynamical models, the harmonic radiation emission from a charged particle accelerated by a strong electromagnetic pulse and perturbed by an external electrostatic field or by the plasma restoring force, is characterized by power decay indexes 5/3 or 4/3, in similitude with PIC results, as discussed in the previous section.

### *3.1. Plasma and harmonic emission from relativistic electrons perturbed by a soliton field*

Radiation from electron bunching in large-amplitude Langmuir waves excited in electron beam-plasma interactions has been studied in the past. Weatherall and Hobbs [10] showed that electrostatic bunching was an effective means of generating plasma harmonics. Subsequently, Weatherall and Benford (WB) [11] reported a spectrum showing strong emission at the plasma line and its harmonics with a power spectrum below a cut-off at $\omega \sim \omega_c$,

$$P(\omega) \sim \omega^{-5/3}; \qquad \omega \lesssim \omega_c = 2\,\gamma_b^2\, c/D\ , \tag{1}$$

where $\gamma_b$ is the relativistic factor for the electron beam and $D$ denotes the scale length of the localised electrostatic structures described by the Zakharov model of turbulence, [12]. Carrying this over to our PIC spectra, the cut-off frequency identified in (1) corresponds to $m_c = (\gamma_b^2/\pi)(\lambda_L/D)$. For $\gamma_b \sim 6, (\lambda_L/D) \sim 20, m_c \sim 200$. We note in passing that all three signatures of this spectrum, namely plasma emission at $\omega_p$ and its harmonics, spectral decay with $p = 5/3$ and a cut-off in a range typically $100 \lesssim m_c \lesssim 300$ are consistent with the BOR spectra reported in references [6] and [9]. Furthermore, the localized electrostatic structures in their strongly turbulent state, the so-called soliton, is consistent with the trajectory crossings, as shown in figure 2.

The electron dynamics was studied from the numerical integration of the equations of energy and motion

$$\frac{d(m\boldsymbol{v}\gamma)}{dt} = e\left(\boldsymbol{E} + \frac{\boldsymbol{v}}{c} \times \boldsymbol{B}\right) - \frac{2}{3}\,\pi\, e\, D\, \rho_0\, e^{-i\omega_p t}\ \hat{\boldsymbol{p}}\,, \tag{2}$$

where $e$ and $m$ are the electric charge and mass of the electron, respectively. The density oscillation of amplitude $\rho_0$ is localized within a scale length $D$, oscillates at the plasma frequency $\omega_p$ and has a dipole moment in the direction of the unit vector $\hat{\boldsymbol{p}}$. In figure 3(a) the beam radiation power obtained from the WB model is shown, characterized by a strong feature in the spectrum at the plasma frequency. The harmonic power decay is governed by the 5/3 index. In figure 3(b) the single particle radiation spectrum obtained from (2) is shown for the case $(a_0, n_e/n_c) = (30, 100)$, exhibiting a power spectrum decay $p \sim 5/3$. For other two combinations of parameters, figures 4(a) and 4(b) show that the power decay is described by $p \sim 5/3$ for (30, 200) and by $p \sim 4/3$ for (40, 100). We performed single particle simulations for a large number of combinations $(a_0, n_e/n_c)$ and found that indexes 5/3 and 4/3 remained predominant in the observed spectra.

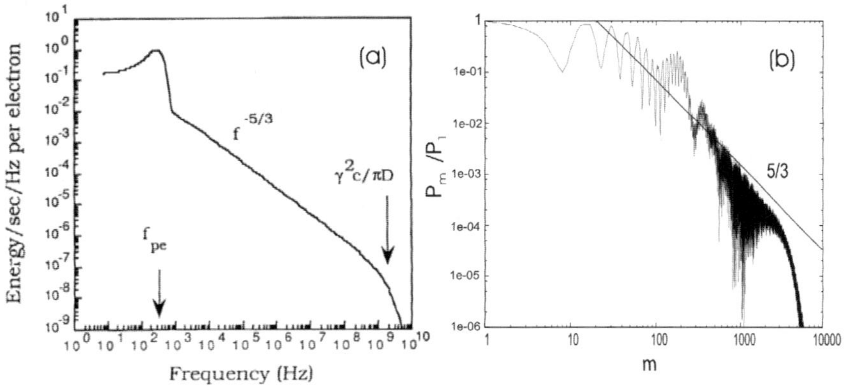


**Figure 3.** (a) Beam radiation power spectrum from WB model [11], and (b) single particle radiation from a laser-accelerated electron perturbed by a soliton field (2), for $a_0 = 30, n_e/n_c = 100$, both having power decays $p \sim 5/3$.

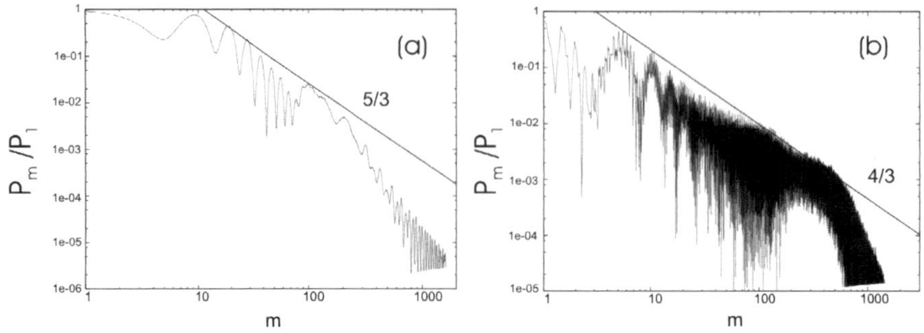


**Figure 4.** Harmonic spectra from (2) for (a) $a_0 = 30, n_e/n_c = 200$, and (b) $a_0 = 40, n_e/n_c = 100$, having respectively power decays $p \sim 5/3$, and 4/3.

*3.2. Plasma and harmonic emission from relativistic electrons perturbed by plasma potential*

The radiation from electrons accelerated by an intense laser pulse and perturbed by the restoring force of the plasma was studied by solving numerically the equations of energy and motion

$$\frac{d(m\boldsymbol{v}\gamma)}{dt} = e\left(\boldsymbol{E} + \frac{\boldsymbol{v}}{c} \times \boldsymbol{B}\right) - m\,\omega_p^2\,\boldsymbol{r}\,, \tag{3}$$

where the second term on the right represents the restoring force that oscillates at $\omega_p$. Different aspects of the particle dynamics were studied for a number of combinations of the laser amplitude and plasma density. In particular, the electron energy gain during and after the action of the laser pulse was explored. It was found that a similarity effect for the electron energy exists as a function of the parameter $S$. In figure 5(a) the similarity effect for the electron energy is shown for fixed amplitudes $a_0$ and varying densities, while figure 5(b) shows this effect for fixed plasma densities and varying laser intensities.

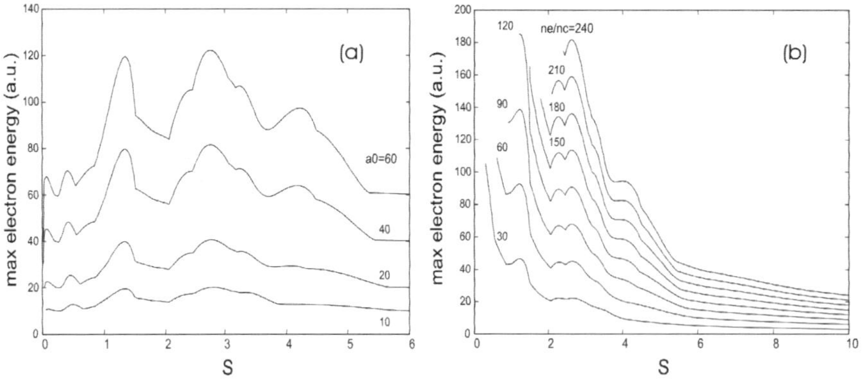


**Figure 5.** Similarity effects: maximum electron energy during the pulse vs $S = (n_e/n_c)/a_0$, for (a) $a_0$=10, 20, 40 & 60 and (b) $n_e/n_c = 30, 60, ..., 240$.

In figures 6(a-d) the radiation spectra for various combinations of the parameter $S$ are shown, and where the power decay indexes 5/3 are observed for these cases. Figure 6(c) shows for comparison a PIC harmonic spectrum from an ultrarelativistic laser-plasma interaction, characterized by a ~4/3 power decay index, reported by other authors [13].

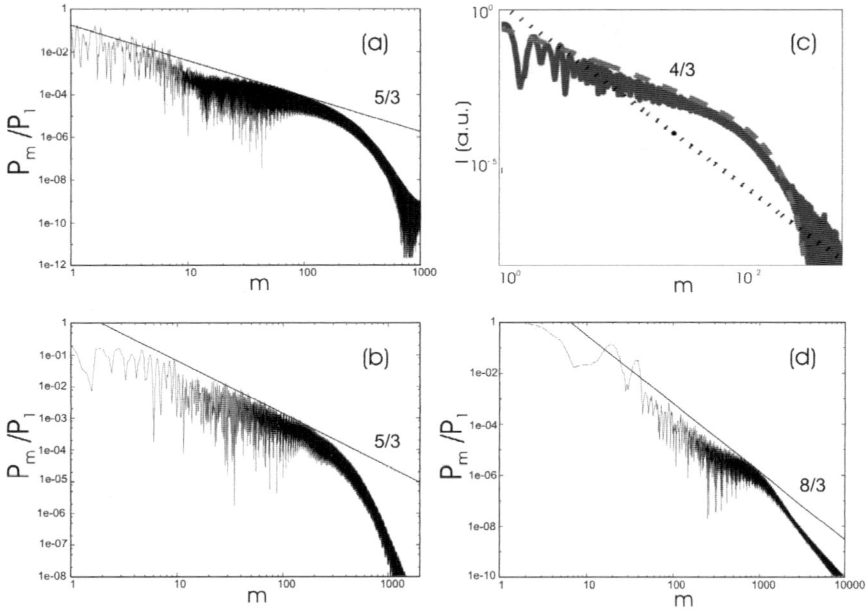


**Figure 6.** Harmonic spectra for (a) $(a_0, n_e/n_c; S) = (4, 70; 4.25)$, (b) for (20, 100; 5), (c) PIC harmonic spectrum from [13] showing a 5/3-4/3 decay for (60, 95; 1.6), and (d) for an $s$-polarization -like case condition, (15, 450; 30), showing an 8/3 decay.

## 4. Conclusions

In this paper we demonstrate the influence of plasma effects on harmonic spectra for relativistic laser-plasma interactions, from both PIC simulations and single particle models. Two possible particle radiation mechanisms were explored consistent with the highly localized electrostatic fields generated inside the plasma during the interaction and with the electron trajectory crossings that correlate with both the electron density bunching and the reflected attosecond electric field spikes observed from PIC simulations. Radiation from perturbed plasma electrons either by a soliton field or by a plasma potential is reminiscent of the 5/3 UR power law decay observed in PIC emission spectra.

## Acknowledgments

One of us (R.O.R.) acknowledges support from CONACyT under Contract No. 167185.

## References

[1] Teubner U and Gibbon P 2009 *Rev. Mod. Phys.* **81** 445
[2] Norreys P A *et al.* 1996 *Phys. Rev. Lett.* **76** 1832
[3] Gibbon P 1996 *Phys. Rev. Lett.* **76** 50; Teubner U and Gibbon P 2009 *Rev. Mod. Phys.* **81** 445
[4] Dromey B *et al.* 2006 *Nat. Phys.* **2** 456
[5] Baeva T, Gordienko S and Pukhov A 2006 *Phys. Rev. E* **74** 046404
[6] Boyd T J M and Ondarza-Rovira R 2008 *Phys. Rev. Lett.* **101** 125004; 2008 *Laser-Driven Relativistic Plasmas, AIP Conf. Proc.* **1024** 233
[7] Boyd T J M and Ondarza-Rovira R 2000 *Phys. Rev. Lett.* **85** 1440
[8] Boyd T J M and Ondarza-Rovira R 2010 *Phys. Lett. A* **374** 1517
[9] Boyd T J M and Ondarza-Rovira R 2010 *Phys. Plasmas* **17** 080701
[10] Weatherall J C and Hobbs WE 1986 *Phys. Fluids* **29** 2292

[11] Weatherall J C and Benford G 1991 *Astrophys. J.* **378** 543
[12] Zakharov V E 1967 *Soviet Phys. JETP* **24** 455
[13] an der Brügge D and Pukhov A 2010 *Phys. Plasmas* **17** 033110

Journal of Physics: Conference Series **370** (2012) 012040 doi:10.1088/1742-6596/370/1/012040

# One-dimensional hydrodynamical kinetics model of a cylindrical DBD reactor with $N_2$

**M Flores-Moreno[2], S Barocio-Delgado[1], A Mercado-Cabrera[1], R López-Callejas[1,2], R Peña-Eguiluz[1], A De la Piedad-Beneitez[2], B Rodríguez-Méndez[1] and A Muñoz-Castro[1]**

[1]Instituto Nacional de Investigaciones Nucleares, Plasma Physics Laboratory AP 18-1027, 11801 Mexico D. F., Mexico

[2]Instituto Tecnológico de Toluca, AP 890, Toluca, Mexico

E-mail: antonio.mercado@inin.gob.mx

**Abstract.** A numerical 1-D model of the chemical kinetics related hydrodynamics of room pressure $N_2$ plasma at 25°C is reported. This generic discharge is assumed to take place between two cylindrical concentric electrodes, coated in a dielectric material, biased between 1 kV and 10 kV at 60Hz - 3kHz. The model includes the integration of particles conservation and the momentum equations as well as the local field approximation and the Poisson equations for the sake of completeness. The outcome shows that an accumulation of electrons takes place in the close vicinity of the higher voltage electrode, due to the electric field convergence to the internal electrode. Thus, this is a region of intense ionization whereas the generation of free radicals would occur away from the internal electrode. The model predicts no significant influence of the electric field on the heavier particles whose density remains practically constant.

## 1. Introduction

The dielectric barrier discharge (DBD) is one of the most popular techniques for non-thermal plasma industrial production, given its versatile configuration and the convenience of operating at room pressure. DBDs have been applied to ozone production [1], treatment of live tissue [2], biomaterial modification for implant production [3], bacteria elimination [4], surface conditioning [5], atmospheric pollution control [6] and so on.

The cylindrical geometry being the most applied, e.g., in the experimental degradation of toxic gas [7],[8], waste water treatment [9], it is also the less theoretically studied geometry, especially from the point of view of hydrodynamics, albeit some advances in its electrodynamics have been recently achieved. In some particular cases, a few cylindrical geometry models have been proposed [10],[11]. Nevertheless, there is a need for understanding the chemical behavior and generation mechanisms of the chemical species, including the effects of the electric field, both in time and space.

The model presented here is structured on the continuity and momentum equations. The closure of the equation system is provided by the Poisson equation. Despite its simplifications, the model is able to predict the space-time distribution of electrons, ions, and neutrals as well as the evolution of the electric field.

## 2. The model

The model has been conceived to apply within the $1-10kV$ and $60Hz-3kHz$ power supply ranges. The AC power source is connected to the electrodes while keeping the external one grounded. The reactor is considered as an open system as it is continuously circulated by a $1L/min$ nitrogen flow of nitrogen at atmospheric pressure.

### *2.1. DBD cylindrical reactor*

A generic reactor has been assumed with both electrodes insulated with Pyrex$^{TM}$ glass, $8\times10^{-4}m$ thick at the inner electrode and $1\times10^{-3}m$ at the external one. Thus, the gap discharge becomes $3\times10^{-3}m$, figure 1.

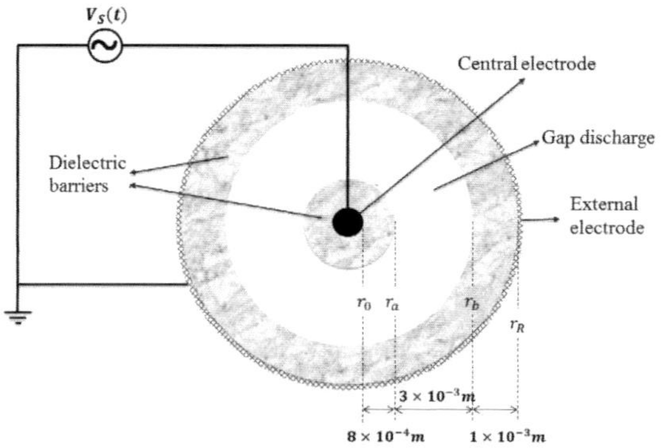


**Figure 1.** Cross sectional view of the cylindrical double dielectric barrier reactor.

The applied potential behaves sinusoidally and can be represented by the expression:

$$V_S(t) = V_0 sen(\omega t), \tag{1}$$

where $\omega$ is the angular frequency, $t$ time in seconds and $V_0$ is the maximal amplitude provided by the power supply. The potential difference through the dielectric barriers ($V_W$) and through the discharge gap ($V_G$) are expressed respectively as [12]:

$$V_W(t) = V_{DI} + V_{DE} = -\int_{r_0}^{r_a} E_r(r,t)dr - \int_{r_b}^{R} E_r(r,t)dr, \tag{2}$$

$$V_G(t) = -\int_{r_a}^{r_b} E_r(r,t)dr, \tag{3}$$

where $V_{DI}$ is the voltage through the internal dielectric, $V_{DE}$ through the external one and $E_r(r,t)$ is the electric field as a function of radial distance and time. The discharge current ($i_T$) has been calculated as the algebraic sum of the conduction ($i_C$) and displacement ($i_D$) currents [12]:

$$i_T(t) = i_C(t) + i_D(t), \tag{4}$$

where the conduction one is given by [12]

$$i_C(t) = \frac{eS}{R}\int_0^R [\Gamma_+(r,t) - \Gamma_e(r,t)]\,dr, \tag{5}$$

Here $e$ is the elementary charge, $S$ is the discharge cross section, $\Gamma_{+,e}(r,t)$ are the ion and electron fluxes, as functions of the radial position and time. The displacement current can be expressed as [13]:

$$i_D(t) = \frac{\sum i_C(r,t)dr}{d_G + d_D} + \frac{\epsilon_0 S}{d_G + d_D}\frac{dV_S}{dt}, \tag{6}$$

where $i_C$(r,t)=$eS[\Gamma_+(r,t)$-$\Gamma_e(r,t)]$ is the net current, $\epsilon_0$ is the permittivity of vacuum, $d_G$ is the gap space between insulated electrodes, $d_D$=($d_{DI}/\ \kappa$)+($d_{DE}/\kappa$ ) is the electrode dielectric thickness, $\kappa$ being the dielectric constant of the barrier.

### 2.2. Equations of the model

The main body of equations results from simplifying the moments of the Boltzmann equation. Thus, from the first moment one obtains the species continuity equation, (7), and the equation of momentum, derived from the second Boltzmann moment and neglecting inertia. The simultaneous solution to both equations enables to specify the species distribution in time and space. The general 1-D expressions of the equations of continuity and momentum in cylindrical coordinates are, after [14],

$$\frac{\partial n_i(r,t)}{\partial t} + \frac{1}{r}\frac{\partial}{\partial r}(r\Gamma_{i_r}(r,t)) = S_i, \tag{7}$$

$$\Gamma_{i_r}(r,t) = n_i(r,t)v_r - D_{i,eff}\left[\frac{\partial n_i(r,t)}{\partial r}\right], \tag{8}$$

where $n_i$ is the density of species $i$ (electrons, ions and neutral particles), $S_i$ is the source term, $v_r$ is the radial flux velocity of the neutrals and $D_{i,eff}$ is the effective diffusion coefficient of species $i$ and can be calculated from the gas kinetic theory. The radial velocity is assumed to be zero and only the second term relating the diffusion and gradient of density were considered. In the case of charged particles, the equation of motion is given by [14]:

$$\Gamma_{+,e_r}(r,t) = \pm n_{+,e}(r,t)\mu_{+,e}E_r(r,t) - D_{+,e}\left[\frac{\partial n_{+,e}(r,t)}{\partial r}\right], \tag{9}$$

where $\mu_{+,e}$ is the (ion, electron) mobility and $D_{+,e}$ is the diffusion coefficient of the charged species and ($\pm$) refers to the respective polarity of the charge.

The transport coefficients $\mu_e$ and $D_e$ are evaluated from BOLSIG+ [15] whereas $\mu_+$ and $D_+$ are available in [16],[17]. The electric field of equation (9) results from:

$$E_r(r,t) = -\frac{d\phi(r,t)}{dr}, \tag{10}$$

where the radial potential $\phi$ can be obtained from the Poisson equation

$$\nabla^2\phi(r,t) = -\frac{\rho(r,t)}{\epsilon_0}, \tag{11}$$

### 2.3. Reaction processes

Ten chemical species have been considered relevant for the model, namely $e^-$, $N^+$, $N_2{}^+$, $N_4^+$, $N(^2D)$, $N(^2P)$, N, $N_2(A)$, $N_2(a')$ and $N_2$ involved in 11 main reactions summarized in table 1.

### 2.4. Numerical method

The solution to equations (7) to (8) was achieved by the finite volume technique [23]. A 200 cell mesh has been considered. $1.0\times10^{-11}$ to $1.0\times10^{-9}s$ time increments were applied bearing in mind the Courant-Friedrich-Lewy condition [24]. The initial density profile was specified on the Townsend discharge criterion. The solution to the Poisson equation adopted the form of a three point finite difference [25].

**Table 1.** Chemical reactions of $N_2$ and their respective reaction rate coefficient values

| N° | Reaction | $k$ $[cm^3s^{-1}][cm^6s^{-1}]$ | Reference |
|---|---|---|---|
| | Electron-molecule interaction | | |
| 1. | $e^- + N_2 \rightarrow 2e^- + N^+ + N$ | $4.29 \times 10^{-12}\sqrt{T_e}(1 + 6.76 \times 10^{-6})exp(-2.96 \times 10^5 T_e^{-1})$ | [18] |
| 2. | $e^- + N_2 \rightarrow 2e^- + N_2^+$ | $5.05 \times 10^{-11}\sqrt{T_e}(1 + 1.10 \times 10^{-5})exp(-1.82 \times 10^5 T_e^{-1})$ | [18] |
| 3. | $e^- + N_2 \rightarrow e^- + N_2(A)$ | $6.21 \times 10^{-11}\sqrt{T_e}(1 + 2.46 \times 10^{-5})exp(-8.21 \times 10^4 T_e^{-1})$ | [18] |
| 4. | $e^- + N_2 \rightarrow e^- + N_2(a')$ | $log(-8.8 - 16.7/(E/N)) + log(-8.5 - 17.4/(E/N))$ $+log(-8.7 - 17.5/(E/N))$ | [19] |
| | Electron-ion recombination | | |
| 5. | $e^- + N^+ \rightarrow N$ | $3.5 \times 10^{-12}$ | [20] |
| 6. | $e^- + N_2^+ \rightarrow N_2$ | $4.0 \times 10^{-12}$ | [20] |
| 7. | $e^- + N_2^+ \rightarrow N(^2D) + N$ | $2.0 \times 10^{-7}(300T_e^{-1})^{0.50}$ | [19] |
| | Transference of charge | | |
| 8. | $N_4^+ + N_2 \rightarrow N_2^+ + 2N_2$ | $2.4 \times 10^{-15}$ | [21] |
| | Reactions with excited species | | |
| 9. | $N_2(A) + N_2(a') \rightarrow N_4^+ + e^-$ | $5.0 \times 10^{-11}$ | [22] |
| 10. | $N_2(a') + N_2(a') \rightarrow N_4^+ + e^-$ | $2.0 \times 10^{-10}$ | [22] |
| 11. | $N(^2D) + N(^2P) \rightarrow N_2^+ + e^-$ | $2.4 \times 10^{-14}$ | [22] |

### *2.5. Boundary conditions*

For the continuity equation (7):

$$n_e(r_a, t) = n_+(r_a, t) = 0, \tag{12}$$

$$n_e(r_b, t) = n_+(r_b, t) = 0, \tag{13}$$

The condition for Poisson equation (11) is set by the Gauss law [26] whereby

$$V_{r_b} = \frac{\lambda}{2\pi\kappa\varepsilon_0}\int_{r_R}^{r_b}\frac{dr}{r}, \tag{14}$$

where $\lambda$ is the charge density per length unit (C/m) given by

$$\lambda = \frac{2\pi\varepsilon_0 V_0 sen(\omega t)}{\frac{1}{\kappa}\ln(\frac{r_b}{r_R}) + \ln(\frac{r_a}{r_b}) + \frac{1}{\kappa}\ln(\frac{r_o}{r_a})}, \tag{15}$$

## 3. Results

Figures 2 to 5 shown below have been obtained in the following conditions: 3kHz, 10kV power supply.

### *3.1. Voltage and current in the discharge*

The waveforms corresponding to $V_S(t)$, $V_W(t)$, $V_G(t)$ and $i_T(t)$, all shown in figure 2.

In a cylindrical configuration discharge, contrary to a flat parallel plate one [12], the potential through the barriers $V_W(t)$ is not out of phase with respect to the supply $V_S(t)$. The voltages $V_W(t)$ and $V_G(t)$ follow the same sinusoidal wave and the gap potential $V_G(t)$ is the main component of $V_S(t)$. This fact implies a greater potential fall, $V_G(t)$, and a smaller dielectric one, $V_W(t)$, for the cylindrical discharge case.

Figure 2 presents the $V_S(t)$ and $i_T(t)$ curves with a mutual $\omega t = \pi/2$ shift, which confirms the capacitive property of a DBD reactor. Clearly the potential reaches its maxima at $\pi/2$, $3\pi/2$, $5\pi/2$ and $7\pi/2$, while the current curve shows no distortions, suggesting a discharge more homogeneous than filamentary.

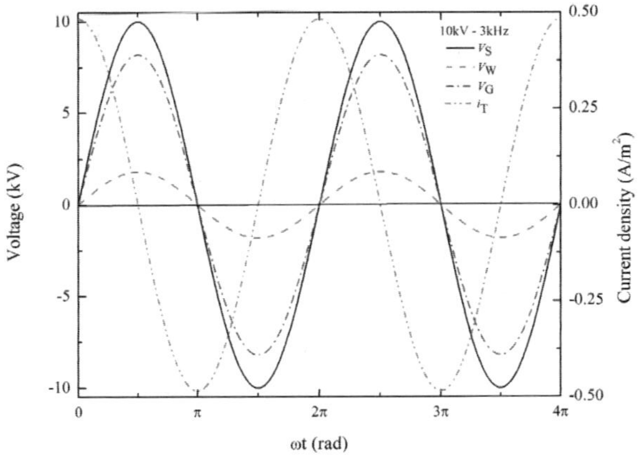


**Figure 2.** Evolution of $V_S(t)$, $V_W(t)$, $V_G(t)$ and $i_T(t)$.

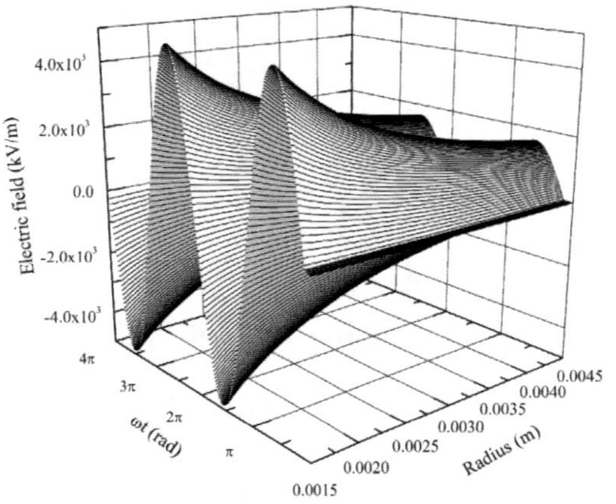


**Figure 3.** Radial and temporal distribution of the electric field in the DBD gap.

The evolution of the electric field can be observed in figure 3. The extent of the field variation reaches $\pm 4000$ kV/m. The field value at the left side extreme of the radial domain, r $= 1.5\times 10^{-3}$ m, is the maximal one, corresponding to the point of greatest potential, namely, the energized electrode.

*3.2. Space-time distribution of the species density and electric field*

Figures 4 and 5 describe the behavior, in time and space, of the electrons and neutrals species respectively. The neutral species have been added together and can be seen in figure 5. All the distribution functions have been considered on the discharge gap only.

The electron radial distribution is displayed in figure 4 to $\omega t = 4\pi$. The illustrated time-space evolution differs considerably from that of the flat plate case [12]. At $\pi/2$, $3\pi/2$, $5\pi/2$ and $7\pi/2$, when the potential reaches its maxima, the greatest electron concentration is found near the inner electrode. When the voltage goes to zero ($\pi$, $2\pi$, $3\pi$ and $4\pi$) the electron concentration is in the order of $10^7$ $m^{-3}$. This rather low value can be attributed to the lack of an electric field that promotes the generation of the species referred to in table 1. Nevertheless, the residual electron concentration facilitates the ionization process during the following voltage increase. The high density of this species suggests the existence of a negative charge concentration towards the inner electrode. Electrons presents a minimal density close to the external electrode when the potential vanishes.

Figure 5 displays the neutral species density, with the exception of molecular nitrogen. The neutral species are unaffected by the electric field, their only density variations within the inner electrode vicinity are due to ionization processes derived from electron impacts.

## 4. Conclusions

The 1-D model of an AC nitrogen plasma discharge in a double dielectric barrier reactor with cylindrical configuration has been developed with satisfactory results. The evolution in time and space of the chemical species densities, along with the respective electric potential and field distributions, has been obtained.

In a cylindrical DBD reactor with an active inner electrode, the chemical processes tend to take place in the close proximity to it, whereby the species concentration exhibits an exponentially increasing profile towards the inner electrode. These profiles are clearly distinct from those of the flat parallel plate DBD case.

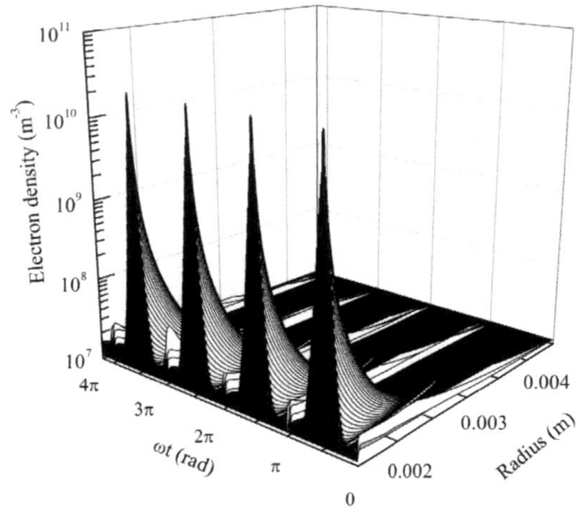


**Figure 4.** Radius and time electron distribution within the DBD reactor.

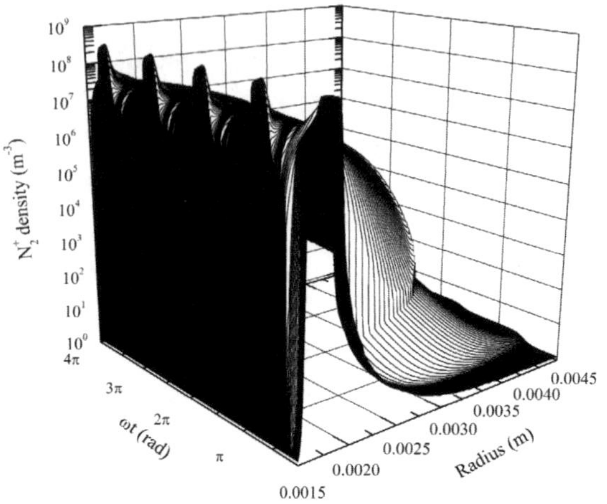


**Figure 5.** Space-time neutral species distribution in the DBD discharge.

Within its own region in the variable space, the model constitutes not only an effective approximation to the experimental behavior of DBD cylindrical plasma but also a solid basis for the development of more complex models including characteristics such as higher dimensionality or additional species and chemical reactions.

## Acknowledgments

The authors are grateful to the support of CONACyT, Mexico, through the graduate studies programme grant No. 227590. The technical consultation received from Ph.D. Ricardo Valdivia is highly appreciated.

## References

[1] Kogelschatz U 2003 *Plasma Chem. Plasma Proc.* **23** 1
[2] Bibinov N, Rajasekaran P, Mertmann P, Wandke D, Viöl W and Awakowicz P 2011 *Biomedical Engineering, Trends in Materials Science* (Intech Publisher) 123
[3] D'Sa R A, Burke G A and Meenan B J 2011 *Acta Biomater* **6** 2609
[4] Vojkovská H, Slámová J, Kozáková Z and Krčma F 2010 *Publ. Astron. Obs. Belgrade* **89** 331-4
[5] Kawakami R, Niibe M, Fukudome T, Takeichi A, Inaoka T and Tominaga K 2011 *Japan J. Appl. Phys.* **50** 01BE02
[6] Jun H, Kim H, Sakawuchi Y and Hong Y 2008 *J. Phys. D. Appl. Phys.* **41** 205213
[7] Panousis E, Papageorghiou L, Spyrou N, Loiseau J-F, Held B and Clément F 2007 *J. Phys. D. Appl. Phys.* **40** 4168
[8] Flores-Moreno M, Mercado-Cabrera A, Valencia-Alvarado R, López-Callejas R, Barocio-Delgado S, Peña-Eguiluz R, Muñoz-Castro A and Jaramillo-Sierra B 2009 *Int. J. of Plasma Environmental Sci. and Technol.* **3** 43
[9] Sato M, Soutome T, Mii S, Ohshima T and Yamada Y 2007 *Int. J. of Plasma Environmental Sci. and Technol.* **1** 71
[10] Dorai R and Kushner M J 2003 *J. Phys. D. Appl. Phys.* **36** 1075
[11] Godyak V and Sternberg N 2008 *Plasma Sources Sci. Technol* **17** 025004
[12] Oda A, Sakai, Y, Akashi H and Sugawara H 1999 *J. Phys. D. Appl. Phys.* **32** 2726
[13] Fridman A and Kenneedy L A 2004 *Plasma Physics and Engineering* (New York: Taylor and Francis) 853
[14] Libermann A M and Lichtenberg J A 1994 *Principles of Plasma Discharges and Materials Processing* (USA: John Wiley and Sons) 572
[15] Hagelaar G J M and Pitchford L C 2005 *Plasma Sources Sci. Technol.* http ://www.bolsig.laplace.univ-tlse.fr/ **14** 722

[16] Ellis H W, Pai R Y, McDaniel E N, Mason E A and Viehland L A 1976 *Atomic Data and Nuclear Data Tables* **17** 177
[17] Ellis H W, McDaniel E N, Albritton D L, Viehland L A, Lin S L and Mason E A 1978 *Atomic Data and Nuclear Data Tables* **22** 179
[18] Takaki K, Chang J-S and Kostov K G 2004 *IEEE Trans. Dielectrics Electrical Insulation* **11** 481
[19] Kossy I A, Kostinsky A Y, Matveyev A A and Silakov V P 1992 *Plasma Sources Sci. Technol.* **1** 207
[20] Eichwald O, Yousfi M, Hennad A and Benabdessadok M D 1997 *J. Appl. Phys.* **82** 4781
[21] Choi Y H, Kim J H and Hwang Y S 2006 *Thin Solid Films* **506-507** 389
[22] Lebedev Y A and Epstein I L 2007 *Plasma Phys. Rep.* **33** 63
[23] Versteeg H K and Malalasekera W 1995 *An Introduction to Computational Fluid Dynamics. The finite volume method* (Prentice Hall) 257
[24] Lymberopoulos D P and Economou D J 1995 *J. Res. Nat. Inst. Stand. Technol.* **100** 473
[25] Birdsall C K and Langdon A B 2004 *Plasma Physics via Computer Simulation* (New York: Taylor and Francis) 305
[26] Halliday D and Resnick R 2007 *Fundamentals of Physics* vol 2 (USA: Wiley) 125

# Studies of the cathode sheath of a low pressure hollow cathode discharge

**G Petraconi[1], A B Guimarães Neto[1], H S Maciel[1,2] and R S Pessoa[1,2]**

[1]Technological Institute of Aeronautics, S. J. dos Campos, SP, Brazil
[2]IP&D, University of Vale do Paraíba, S. J. dos Campos, SP, Brazil

E-mail: petra@ita.br

**Abstract.** In this work, a theoretical model for the cathode sheath using experimental data of a low pressure hollow cathode discharge (HCD) is derived considering it to be non-collisional. Secondary electrons emitted from the cathode surface are taken into account on the discharge model and their influence on the theoretical sheath potential profile is investigated. The plasma parameters and the floating potential profile along the discharge axis were inferred from the current-voltage characteristics of a single Langmuir probe positioned at the inter-cathode space of the HCD. For a low pressure HCD, typical values of the electron density and electron temperature are $n_e \approx 10^{16}$ m$^{-3}$ and $T_e = 4$ eV, respectively. By using the probe data the floating potential profile was determined to verify the position of the plasma-cathode sheath interface and to promote a qualitative discussion between theoretical and experimental results.

## 1. Introduction

The more important effects in a glow discharge occur next to the cathode, i.e. in the cathode sheath and negative-glow regions. The cathode sheath supports almost all the discharge voltage and is the main responsible for the processes of production of neutral and charged species. The electrons crossing the cathode fall and streaming into the negative glow acquire enough energy to promote more intense ionization and excitation rates as compared with the positive-column plasma. This makes the negative glow based discharges a useful tool in laser, lighting, and surface-processing applications [1-3].

HCD are the most common examples of discharges in which the negative glow plasma is generated. The hollow cathode device consists of a cathode with hollow structure (hole, aperture, etc.) and an arbitrarily shaped anode [4]. Compared to a conventional glow discharge between parallel plates, the hollow cathode geometry is characterized by higher current densities and larger concentrations of high-energy electrons. These properties are consequences of the phenomenon known as *hollow cathode effect* (HCE) [3,4].

A simplified theoretical model of non-collisional cathode sheath of a plane hollow cathode discharge is usually made by only considering the contribution of electrons and ions from the plasma volume. From this model we conclude that the sheath region should have a net positive electrical charge [5]. However, in real situation, we must take into account the contribution of secondary electrons produced as a result of the impact of ions, photons or neutral species or metastable against the cathode surface.

Journal of Physics: Conference Series **370** (2012) 012041 doi:10.1088/1742-6596/370/1/012041

## 2. Description of the model

We consider a plane of abscissa $x = 0$ coincident with the flat surface of the cathode. The sheath region is confined between the plane $x = 0$ and another one of abscissa $x_b > 0$. The electric potential in the sheath, as well as in the plasma, is described by the function $V = V(x)$, and at any point in the plasma is attributed $V(x) = 0$ to be consistent with the quasi-neutral conditions. The potential of the cathode in $x = 0$ is $V(0) = V_c < 0$ because, in this case, the plasma is the region of highest potential [5]. According to the condition of quasi-neutrality, the plasma density $n_o$ is equal to the ions density $n_{oi}$ and to the electron density $n_{oe}$. The secondary electrons are supposed to travel as a beam, leaving the cathode at $x = 0$. The electron beam is accelerated through the cathode sheath towards the plasma. The ions from plasma are accelerated by the cathode field, and the electrons from the plasma are decelerated by this field.

We consider that the secondary electrons are generated by the impact of ions, photons and neutral species - both stable and metastable - remembering that in the cathode region the emission by electron impact is not very significant [5]. Thus, we can make use of a dimensionless coefficient denoted by the effective (or apparent) electron secondary emission coefficient [6, 7], since it includes the effect of all relevant mechanisms in electron emission. This effective secondary electron emission coefficient is given by:

$$\gamma = \left.\frac{J^-}{J^+}\right|_{cathode} \tag{1}$$

Based on (1), and with the knowledge of the effective coefficient, we can determine the current density of secondary electrons $J^-$. Therefore, it is necessary to determine $J^+$. In particular, by the continuity equation, we have $J^+ = n_o e v_o$, so that (1), by applying the continuity equation, results in:

$$J^- = \gamma e n_o v_o = e n_{ose} v_{se} \tag{2}$$

where $v_o$ is the ion velocity at the sheath edge, the $n_{ose}$ and $v_{se}$ are the secondary electron density and secondary electron velocity, respectively. Thus, we have:

$$n_{ose} = \gamma n_o v_o \left(\frac{-2eV_C}{m}\right)^{-\frac{1}{2}} \tag{3}$$

The number density of secondary electrons in the plasma remains constant, but the same does not occur in the sheath. Such density can be denoted by $n^-$ and also calculated based on the equations of energy conservation and continuity:

$$n^- = \gamma n_0 v_0 \left[\frac{2e(V - V_C)}{m}\right]^{-\frac{1}{2}}, V \neq V_C \tag{4}$$

Also, the electron density in the plasma volume is given by the difference between $n_o$ and $n_{ose}$, so:

$$n_{oep} = n_o \left[1 - \gamma v_o \left(\frac{-2eV_C}{m}\right)^{-\frac{1}{2}}\right] \tag{5}$$

The electrons from the plasma are assumed to be in thermal equilibrium, called thermal electrons. The electrons density ($n_e$) in the sheath region satisfies the Boltzmann relation given by:

$$n_e(x) = n_{oep} \exp\left(\frac{e\,V(x)}{k\,T_e}\right), \tag{6}$$

where $n_{oep}$ is the electron density of plasma without the contribution from the electron beam, $V$ is the electric potential in the sheath, $e$ is electron charge, and $kTe$ is the electron temperature in eV. The ions are cold and monoenergetic. Using the continuity equation and the energy conservation, the ion density $n_i$ in the sheath can be obtained by:

$$n_i(x) = n_o\left(1 - \frac{2eV(x)}{Mv_o^{\,2}}\right)^{-1/2}, \tag{7}$$

where $M$ is the ion mass. The potential in the sheath is given by Poisson's equation

$$\frac{d^2V}{dx^2} = -\frac{\rho}{\varepsilon_o} = -\frac{1}{\varepsilon_o}\left(-n_e e - n^- e + n_i e\right), \tag{8}$$

where $\varepsilon_o$ is the vacuum permittivity and $\rho$ is the total charge density. By using (4), (5), (6), and (7), we obtain

$$\frac{\varepsilon_0}{en_0}\frac{d^2V}{dx^2} = \left[1 - \gamma v_0\left(\frac{-2eV_C}{m}\right)^{-\frac{1}{2}}\right]\exp\left(\frac{e\,V}{k\,T_e}\right) + \gamma v_0\left[\frac{2e(V - V_C)}{m}\right]^{-\frac{1}{2}} - \left(1 - \frac{2eV}{Mv_0^{\,2}}\right)^{-1/2}. \tag{9}$$

Introducing the normalized quantities

$$\eta = -\frac{e\,V}{k\,T_e},\quad \eta_C = -\frac{e\,V_C}{k\,T_e},\quad \xi = \frac{x}{\lambda_D},\quad \mu = \frac{M}{m},\quad \nu = \frac{v_o}{c_s}, \tag{10}$$

where $\lambda_D$ is the electron Debye length at the sheath edge and $c_s=(kTe/\mathrm{M})^{1/2}$ is the ion sound speed [8, 9], we obtain

$$\frac{d^2\eta}{d\xi^2} = -\left[1 - \gamma\nu(2\mu\eta_C)^{-\frac{1}{2}}\right]\exp(-\eta) - \gamma\nu(2\mu\eta_C)^{-\frac{1}{2}}\left(1 - \frac{\eta}{\eta_C}\right)^{-\frac{1}{2}} + \left(1 + 2\eta\nu^{-2}\right)^{-\frac{1}{2}}. \tag{11}$$

Equation (12) has as inconvenience the fact that ν is an unknown. To circumvent this, it is necessary to establish a relationship involving ν and other parameters of this equation, namely, γ, μ and $\eta_C$. By imposing equality of charge density at $\eta \to 0$ we have ensured one of the matching conditions as ξ becomes large. Also if η is to evolve smoothly to zero as ξ increases, then it may be supposed to vary asymptotically as [10]:

$$\eta = k\xi^{-n} \tag{12}$$

The substitution of (12) in (11), and after performing the Taylor series expansions around $\eta = 0$, which is particularly valid in the limiting region between the sheath and the plasma, results in an equation which can be satisfied only if the terms that multiply $\xi^{n}$ vanish. This condition leads to the following third-degree polynomial equation in ν:

$$k\nu^3 - \nu^2 + 1 = 0 \tag{13}$$

By solving (13) we obtain:

$$k = \gamma\left(2\mu\eta_C\right)^{-\frac{1}{2}}\left(1+\frac{1}{2\eta_C}\right) \tag{14}$$

The solution of (14) that we should adopt is one which, considering the fact that $k$ is typically much smaller than unity, gives the value of ν such that the Bohm criterion is satisfied and that tends to unity when $k$ approaches zero. It can be shown that ν is a positive real number if $0 < k < \frac{2}{3\sqrt{3}}$. This condition is satisfied by a wide range of typical values of γ, μ and $\eta_C$. The electric field and potential in the sheath can be obtained by integrating numerically (12) using $\eta = 0$ and $\frac{d\eta}{d\xi} = 0$, so:

$$\frac{1}{2}\left(\frac{d\eta}{d\xi}\right)^2 = \left[1-\gamma\nu\left(2\mu\eta_C\right)^{-\frac{1}{2}}\right]\left[\exp(-\eta)-1\right] + 2\gamma\nu\left(2\mu\eta_C\right)^{-\frac{1}{2}}\eta_C\left(1-\frac{\eta}{\eta_C}\right)^{\frac{1}{2}} - \\ -2\gamma\nu\left(2\mu\eta_C\right)^{-\frac{1}{2}}\eta_C + \nu^2\left(1+2\eta\nu^{-2}\right)^{\frac{1}{2}} - \nu^2 \tag{15}$$

The boundary condition for determining where the sheath edge is located, at ξ=s, is

$$\eta\left(s/\lambda_D\right) = -\frac{eV_b}{kT_e} = \frac{\nu^2}{2} \tag{16}$$

because the sheath potential ($V_b$) in this position is given by:

$$V_b = -\frac{1}{2e}Mv_0^{\,2} = -\frac{\nu^2 kT_e}{2e} \tag{17}$$

The mathematical problem constituted by (11) and condition $\eta(0) = \eta_C$ must be solved numerically to determine the theoretical profile of the electrical potential in the cathode sheath. The solution of the problem without the use of dimensionless variables depends on the knowledge of plasma parameters according to (10).

## 3. Experimental setup

A schematic diagram of the plane hollow-cathode discharge apparatus is illustrated in figure 1. The discharge resembles a glow discharge regime extending from the hollow anode (earth potential) up to flat cathodes made of aluminum covered by Teflon. The hollow anode is made of titanium of 7 mm external diameter which is covered by a ceramic tube terminated by an orifice of 2.0 mm diameter. The internal diameter of the hollow anode cavity is 4 mm. The orifice constricts the anode discharge to form electrostatic layers which contribute to increase the ionization rate in the anodic region. The vacuum glass chamber of 130 mm internal diameter and 300 mm length was preliminary evacuated to a pressure below $10^{-2}$ Pa. The plane hollow-cathode discharge was operated with argon gas for pressures ($P$) in the range of (10 – 50) Pa. The gases were fed through the anode hole at a flow rate in the range of (1 – 40) sccm. Thus, the gas flows through the hollow anode and becomes partially ionized in the process before it reaches the vacuum chamber. The self-sustained cathode-anode voltage ($V_d$) during operation was in the range of (350 – 900) V with a discharge current ($I_d$) in the range of (10 – 800) mA.

A cylindrical tungsten probe with 4.0 mm in length and 0.1 mm in diameter was positioned at the axis of the discharge between the cathodes. The axis of the probe is perpendicular to the normal to the cathode surface. Standard technique [5] was used to extract the argon plasma parameters from the current-voltage characteristic of the probe. The plasma generated between the cathodes is characterized by the presence of impurity metal-ions and gas ions due to intensive ion sputtering of the cathode. A relatively high content of atoms of the cathode material in the plasma is explained by a high kinetic energy (1 – 10 eV) [5] of the sputtered atoms. This effect promotes the contamination of the chamber walls and of the electrostatic probe. The combined effect of the ion bombardment by polarizing the probe with the cathode potential promotes control over the incandescence intensity of the probe to clean it.

## 4. Results and discussion

The resolution of several computational problems constituted by (15), with the variation of the parameters involved, shows that the model developed for the cathode sheath is not very sensitive to variations of the coefficient of secondary electron emission, $\gamma$, within the typical range of this parameter. Figure 2 shows the result from numerical integration of the (15). For a certain set of parameters we can conclude that the dimensionless potential profile in the cathode sheath is little influenced by variations of the $\gamma$ coefficient. On the one hand, this low sensitivity may indicate that the same effect is valid in practice due to the relative low densities of secondary electrons. Only an analysis including values of the coefficient $\gamma$ of the order of hundreds shows significant changes in the profiles of the electric potential for the same parameters adopted in figure 2. A relevant theoretical analysis refers to the variation of thickness of the cathode sheath with the plasma density, which can be easily varied by changing some discharge parameters as gas pressure, axial magnetic field, inter-cathode distance and discharge voltage. Figure 3 shows that the thickness of the cathode sheath decreases monotonically with increasing of the plasma density, which is a consequence of the direct dependence of such thickness on the Debye length.

If the electron temperature $T_e$ is expressed in electron volts, and the number density ($n_o$) in electrons per cubic meter, the Debye length ($\lambda_D$) is given by:

$$\lambda_D \approx 7434\left(\frac{T_e}{n_o}\right)^{1/2} \tag{18}$$

The Debye length is the characteristic thickness of sheaths which form between a plasma and an electrode, or a plasma and a wall. Equation (18) predicts a Debye shielding distance of approximately 0.15 mm for plasmas with an electron temperature of 4 eV and a number density of $10^{16}$ electrons/m$^3$. The increase of the discharge current and the plasma density gives lower Debye lengths, consequently, the thickness of the cathode sheaths decreases. This effect generates a greater plasma volume and, consequently, it is possible to enhance the discharge current by lowering the inter-cathode distance. In our experimental apparatus the inter-cathode distance was kept fixed at 19 mm. The discharge was operated by using argon as working gas. The mean values of the plasma parameters summarized in table 1 were obtained at different positions inside of the negative glow formed between the flat cathode surfaces for two operational conditions of the hollow cathode discharge:

- Condition I with $P = 15$ Pa, $V_d = 431$ V and $I_d = 15$ mA.
- Condition II with $P = 25$ Pa, $V_d = 428$ V and $I_d = 19$ mA.

The position of the interface region between the negative glow and the cathode sheath is visually identified by intense brightness difference between these two regions. Experimentally we can evaluate this interface region by moving the cylindrical Langmuir probe with respect to the cathode surface along the axis of the discharge. The mean values of the interface positions also are presented in table 1.

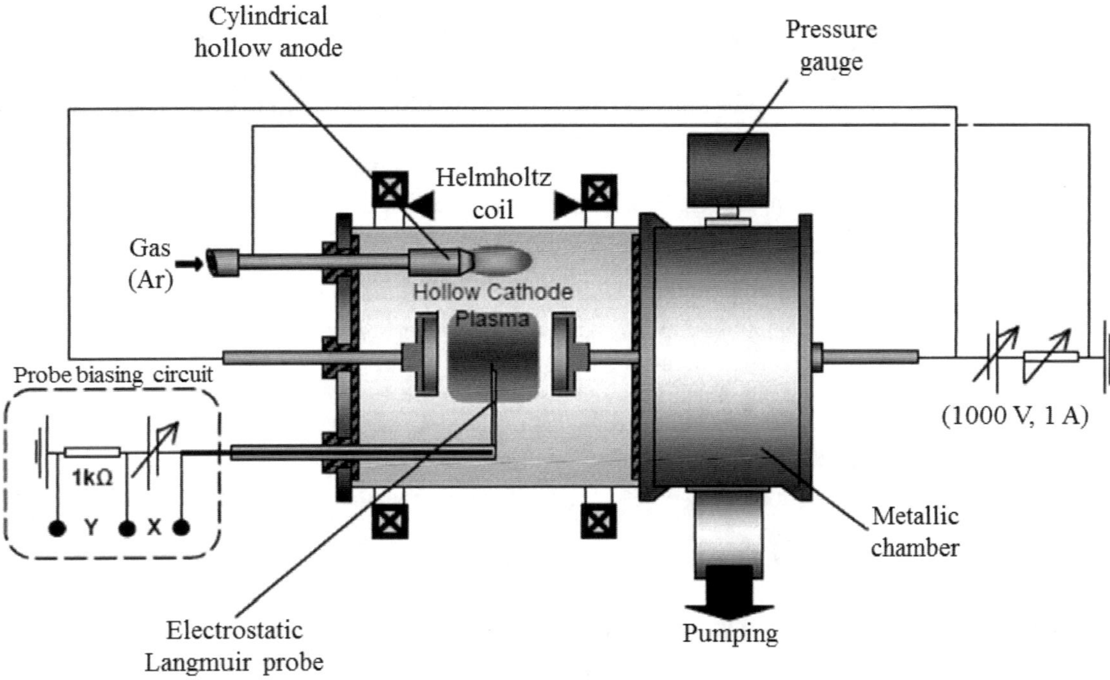


**Figure 1.** Schematic diagram of the plane hollow-cathode discharge apparatus.

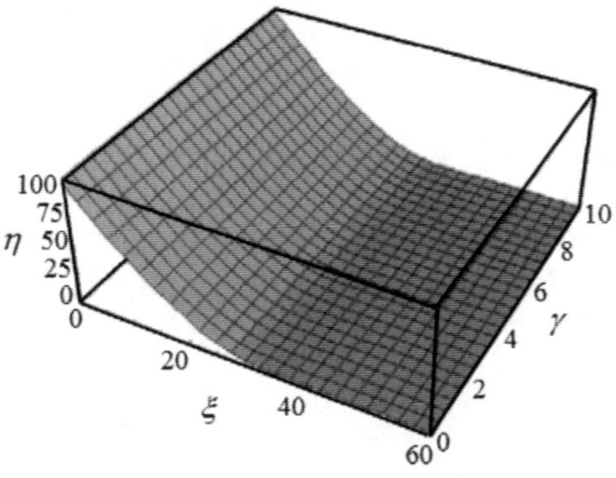


**Figure 2.** The normalized electric potential η as a function of distance normalized to the plasma Debye length and secondary electron coefficient, with $kT_e = 4\ eV$, $n_o = 1.0 \times 10^{16}\ m^{-3}$, $\eta_C = 104$.

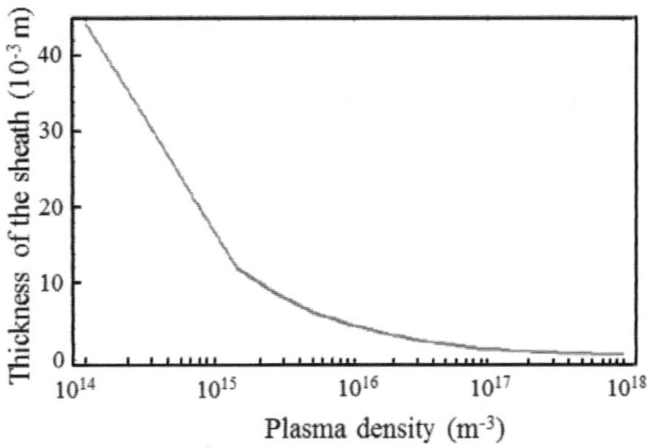


**Figure 3.** Thickness of the cathode sheath as function of the plasma density. The parameters used for making this simulation were: $kT_e = 4$ eV, $\eta_C = 104$ and $\gamma = 0.07$.

**Table 1.** Plasma parameters and the position of the sheath-negative glow interface for conditions I and II.

| Condition | $kT_e$ (eV) | $n_o(10^{16}m^{-3})$ | $\lambda_D$ ($10^{-3}$ m) | Position of interface ($10^{-3}$ m) |
|---|---|---|---|---|
| I | 7.7±0.9 | 1.9±0.3 | 0.15±0.08 | 3.3±0.5 |
| II | 5.9±0.6 | 2.3±0.2 | 0.12±0.06 | 2.8±0.5 |

Journal of Physics: Conference Series **370** (2012) 012041 doi:10.1088/1742-6596/370/1/012041

The sheath-negative glow interface region can be also evaluated by determining the floating potential profile along the axis of the discharge, as shown in figures 4 and 5 for conditions I and II, respectively, with emphasis on the transition range of brightness at the sheath-negative glow interface region. The floating potential corresponds to the probe polarization potential for which current the probe current vanishes. It is supposed here that the floating potential profile has approximately the same qualitative behavior as the space potential profile, meaning that at the point where the electric potential of the medium begins to change, it is expected the same to happen with the floating potential of the electrostatic probe. The sheath-negative glow interface region can be identified from the numerical solution of (15), see figure 6, for comparisons with the experimental results determined from the floating potential profile along the axis of the discharge.

The floating potential profile inferred from the probe characteristic may be qualitatively compared with the theoretical curve of electric potential of the medium ( without the interference of the probe), obtained from the numerical solution of (16), as shown in figure 6 simulated for the condition II. In this case, the electron secondary emission coefficient, which is already known not to influence significantly the results for the electrical potential was kept fixed at $\gamma$ =0.38 for argon and aluminum electrode [11]. For the argon ion, supposedly the only ionic species present in the discharge, $\mu \cong$ 72819.7. The cathode potential fall, $-V_C$, is theoretically estimated as the sum of the discharge voltage ($V_d$) with the anode potential fall, generally near the ionization potential of the gas [12]. In the case of argon, this potential is 15.76 V. The electric potential along the cathode sheath region increases monotonically with the distance from cathode and tends to zero in the negative glow region where we have supposed the charge neutrality. The theoretical sheath thickness ($s$) was evaluated to be about 3.13 mm and therefore, within the range of sheath edge position shown in figure 5.

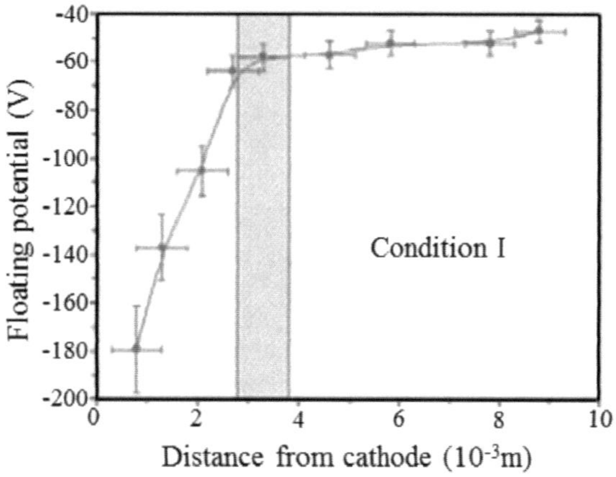


**Figure 4**. Floating potential as function of the distance from the cathode along the axis of the discharge; $P$=15 Pa, $V_d$= 431 V, $I_d$= 15 mA. The position of the sheath edge is within the range indicated by the red column.

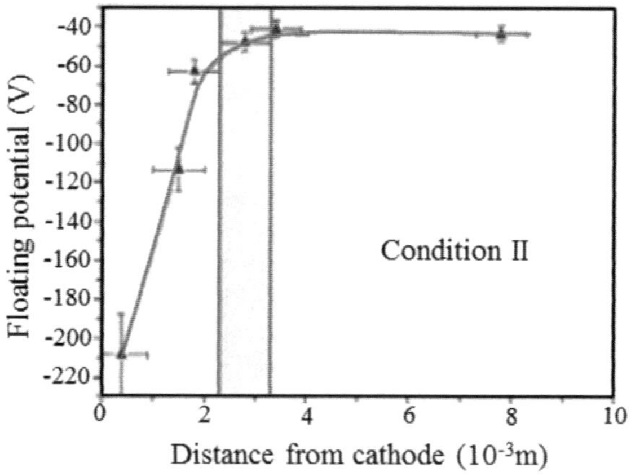


**Figure 5.** Floating potential as function of the distance from the cathode along the axis of the discharge; $P$ = 25 Pa, $V_d$ = 428 V, $I_d$ = 19 mA. The position of the sheath edge is within the range indicated by the blue column.

In figure 7, we can observe that the electric field is relatively small at distances larger than the sheath thickness, which is in agreement with the physical concept of the pre-sheath region, where a small electric field must exist to ensure the acceleration of ions coming from the plasma at a speed at least equal to the ion sound speed $c_s$. This electric field, yet of small intensity, is capable of producing significant modifications in charge densities of ions and electrons from plasma shortly before the entry in the cathode sheath region, as can be observed in figures 8 and 9.

Figure 9 also indicates that the electron number density in the sheath suffers a very drastic reduction governed by the ratio of Boltzmann equation (6) and practically vanishes at a distance of 2.5

mm from the cathode. The net electrical charge in figure 6 shows that for the parameters simulated, this density does not become negative at any point. Furthermore, we can explain the behavior of this profile: near the limiting plane between the sheath and negative glow, exists a maximum of net charge density. As the ions coming from the plasma have not yet been sufficiently accelerated, the continuity equation establishes that their density must be still high at this point, which is confirmed in figure 8.

On the other hand, the secondary electrons released from the cathode surface are accelerated through the cathode sheath towards the plasma region followed by an expressive reduction of their charge density, as shown in the figure 10. This observation indicates that the electrons from the plasma region, decelerated by the sheath potential, are the ones that play the dominant role for the strong reduction of the net charge density at the sheath-plasma boundary(see figure 11). On the left side of the maximum of the net charge density profile, the ions density is much higher than the total electron density. The secondary electron charge density decreases with the distance from cathode becoming three orders of magnitude lower than that of the ions in the plasma region. Overall, this means that the contribution of secondary electrons for the electrical potential profiles on the cathode sheath region is not significant, in agreement with the theoretical model.

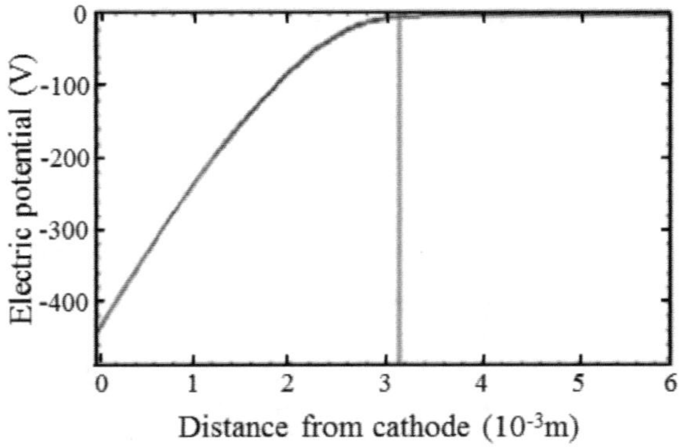


**Figure 6.** Electric potential profile along the $x$ axis; $kT_e = 5.9$ eV, $n_o = 2.3 \times 10^{16}$ m$^{-3}$, $V_C = -444$ V and $\gamma = 0.38$.

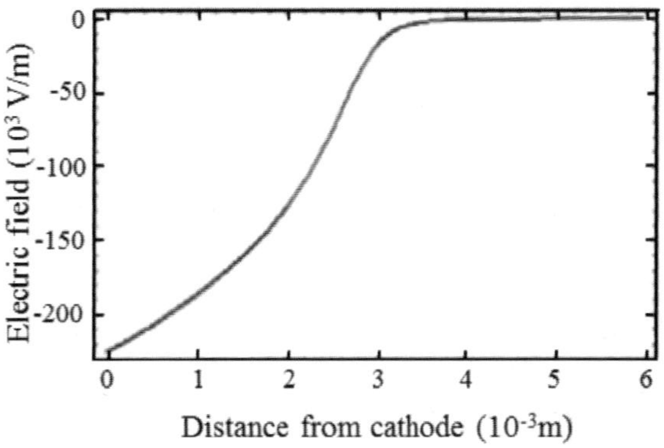


**Figure 7.** Electric field profile along the $x$ axis; $kT_e = 5.9$ eV, $n_o = 2.3 \times 10^{16}$ m$^{-3}$, $V_C = -444$ V and $\gamma = 0.38$.

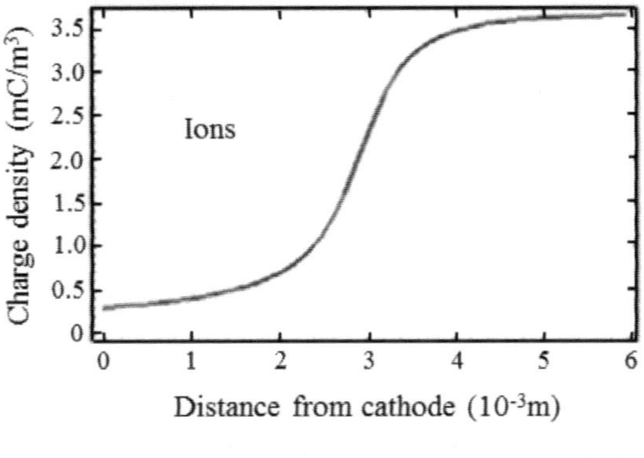


**Figure 8**. Variation of the ion charge density with the distance from the cathode along the $x$ axis; $kT_e = 5.9$ eV, $n_o = 2.3 \times 10^{16}$ m$^{-3}$, $V_C = -444$ V and $\gamma = 0.38$.

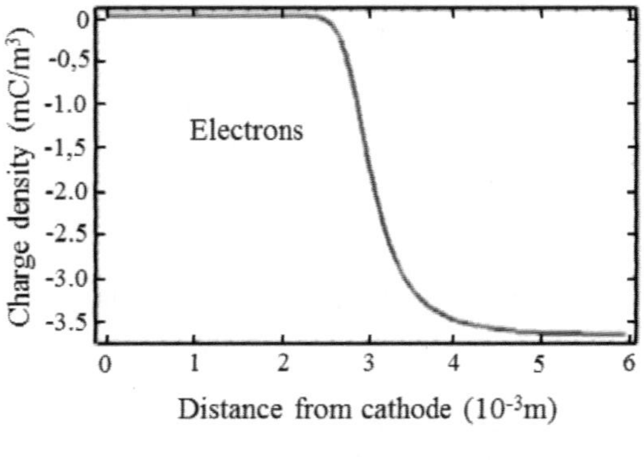


**Figure 9.** Variation of the electrons charge density with the distance from the cathode along the $x$ axis; $kT_e = 5.9$ eV, $n_o = 2.3 \times 10^{16}$ m$^{-3}$, $V_C = -444$ V and $\gamma = 0.38$.

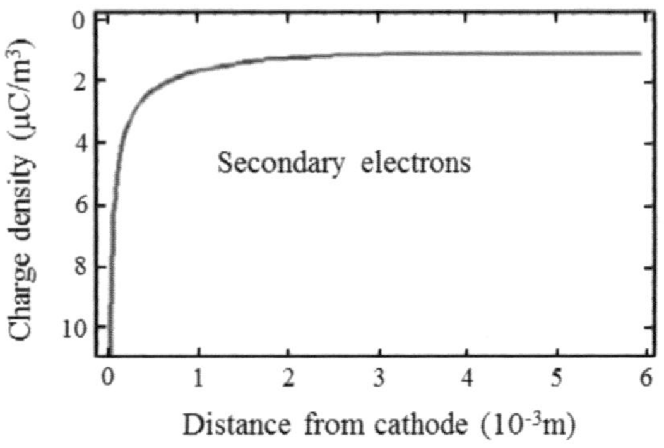


**Figure 10.** Variation of the secondary electrons charge density with the distance from the cathode along the $x$ axis; $kT_e = 5.9$ eV, $n_o = 2.3$ x $10^{16}$ $m^{-3}$, $V_C = -444$ V and $\gamma = 0.38$.

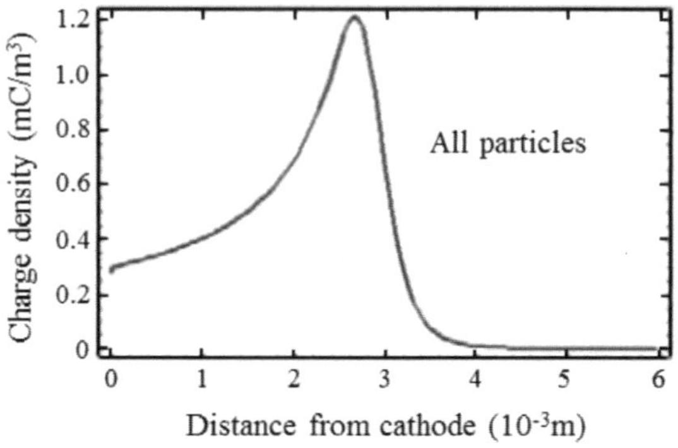


**Figure 11.** Variation of the total charge density with the distance from the cathode along the $x$ axis; $kT_e = 5.9$ eV, $n_o = 2.3$ x $10^{16}$ $m^{-3}$, $V_C = -444$ V and $\gamma = 0.38$.

## 5. Conclusion

The non-collisional model for the electrical potential near the cathode of a HCD in conjunction with experimental data obtained from a Langmuir probe indicates a weak relationship of the cathode sheath potential profile with the secondary electron coefficient (γ). We include atypical values of γ to show that significant changes can occur in the profiles of electric potential only in the case of a net negative charge near the cathode sheath - this means that in these hypothetical cases, the emission of secondary electrons from the cathode is so intense that, in the cathode region, the number density of these electrons exceeds that of ions from the plasma. The floating potential profiles inferred from the current-voltage characteristic of a single Langmuir probe are qualitatively similar to the theoretical electrical potential profiles. The position of the plasma-cathode sheath interface determined from the simulated model shows a good agreement with the experimental data. This transition region allows a good estimative of the thickness of the cathode sheath.

## References


[1] Hershkowitz N 2005 *Physics of Plasmas* **12** 55502–1

[2] Pessoa R S, Tezani L L, Maciel H S, Petraconi G and Massi M 2010 *Plasma Sources Science and Technology* **19** 025013 (7pp)

[3] Pessoa R S, Sismanoglu B N, Amorim J, Maciel H S and Petraconi G 2007 *Gas Discharges: Fundamentals and Applications* vol 1 ed J Amorim (Kerala, India: *Transworld Research Network*) pp 175–90

[4] Bárdős L 1996 *Surface and Coatings Technology* **86–87** 648–56

[5] Chapman B. 1980 *Glow discharge processes: Sputtering and plasma etching* (New York: J. Wiley & Sons)

[6] Donkó Z 2001 *Physical Review E* .**64** 26401–1

[7] Bogaerts A and Gijberls R 2002 *Plasma Sources Science and Technology* **11** 27–36

[8] Wang D, Liu D and Liu J 2000 *Journal of Applied Physics* **88** 1276

[9] Chen F 1984 *Introduction to Plasma Physics and Controlled Fusion* (New York: Plenum Press)

[10] Franklin R N 1976 *Plasma phenomena in gas discharges* (Oxford: Clarendon Press)

[11] Phelps A V and Petrovic Z L 1999 *Plasma Sources Science and Technology* **8** R21–R44

[12] Grill A 1994 *Cold Plasma in Materials Fabrication: From Fundamentals to Applications* (New York: Institute of Electrical and Electronics Engineers Press)

Journal of Physics: Conference Series **370** (2012) 012042 doi:10.1088/1742-6596/370/1/012042

# Bi-dust solitary waves

**J Puerta[1] and P Martin[2]**

[1]Departamento de Física, Universidad Simón Bolívar, Apdo: 8900, Caracas - Venezuela
[2]Departamento de Mecánica, Universidad Simón Bolívar, Apdo: 8900, Caracas - Venezuela

E-mail: jpuerta@usb.ve

**Abstract.** Propagations of non linear solitary waves in bi-dust plasma systems is analyzed. In the present treatment, one of the dust particles is assumed to be much smaller than other one, in such a way that there is enough time for this particle to reach quasi thermal equilibrium. Maxwell-Boltzmann factors are therefore applied for the density distribution of electrons, ions and light dust grains. The treatment of the problem can now be made by the method of the pseudo-potential taking into account temperature effects as function of the density of the heavy grain. In the limit where the light grain density tends to zero the effects found by other authors when only one kind of grains is re-obtained. Several numerical calculations for different values of the characteristic parameters will be shown.

## 1. Introduction

Soliton waves are very important nonlinear phenomena and play a principal role in plasma physics nowadays. This work is concerned with that kind of theories (solitary waves, vortices, voids, etc) but in the case of non-envelope solitons with two kind of dust grains. The interest in such systems is growing because they are present in different situations like in planetary systems, comets etc. The properties of this nonlinear structures of low amplitudes have been established, and they are governed by the KdV equation. Here it is assumed that one of the grains, the smaller ones, are in quasi-thermal equilibrium, and therefore can be treated with a Boltzmann factor. Such plasmas can be created in the laboratory [1] and they exist in planetary systems[2] as we have mentioned before.

## 2. Theoretical model

We are interested in the analysis of the properties of non-envelope solitary waves in non-magnetized bi-dust plasmas using a multi-fluid description together with the pseudo potential method. However, if two kind of particles exist in our system, new features concerned to this system are to be considered due to fact that one of the grain is assumed to be heavier and the smaller one will be considered to stay in thermal equilibrium as also are electrons and ions. Therefore, the whole dynamic resides on the heavier particle which are governed by the nonlinear dust continuity, momentum and energy relations considering $V_{Td} \ll V_p \ll V_{Te}, V_{Ti}$.[3]

$$\frac{\partial n_{d2}}{\partial t} + \frac{\partial}{\partial x}(n_{d2} v_{d2}) = 0 \tag{1}$$

$$\frac{\partial v_{d2}}{\partial t} + v_{d2}\frac{\partial v_{d2}}{\partial x} = \frac{Z_{d2} e}{m_{d2}}\frac{\partial \Phi}{\partial x} - \frac{1}{n_{d2} m_{d2}}\frac{\partial P_{d2}}{\partial x} \tag{2}$$

$$\frac{\partial P_2}{\partial t} + v_{d2}\frac{\partial P_2}{\partial x} + 3P_{d2}\frac{\partial v_{d2}}{\partial x} = 0 \tag{3}$$

$$4\pi e n_e - 4\pi e Z_i n_i + 4\pi e Z_{d1} n_{d1} + 4\pi e Z_{d2} = \frac{\partial^2 \phi}{\partial x^2}, \tag{4}$$

where $Z_i = 1$ , $n_{d1}$ is the density of the light impurity, $\chi$ is a dimensionless variable, and $n_2 = n_{d2/Z_{d2}n_{02}}$.

On the other hand

$$\begin{aligned} n_i &= n_{io}e^{-\frac{e\phi}{KT_i}} \quad n_{d1} = n_{o1}e^{\frac{e\phi}{KT_{d1}}} \qquad n_e = n_{eo}e^{\frac{e\phi}{KT_e}} \\ n_{od1} &= n_{o1} \qquad n_{od2} = n_{o2} \qquad \sigma_{d1} = \frac{Z_{d1}T_i}{T_{d1}} \quad \varphi = \frac{e\phi}{k_B T_i}. \end{aligned} \tag{5}$$

Now, in order to solve the equation (4) we first transform it into a dimensionless system getting

$$\frac{\partial^2 \chi}{\partial x^2} = \mu_e e^{\sigma_i \varphi} - \mu_i e^{-\varphi} + \mu_{d1} e^{\sigma_{d1}\varphi} + n_2 \tag{6}$$

where

$$\begin{aligned} \mu_e &= \frac{n_{oe}}{Z_{d2}n_{o2}}; \qquad \mu_i = \frac{noi}{Z_{d2}n_{o2}}; \quad \mu_{d1} = \frac{Z_{d1}no1}{Z_{d2}n_{o2}} \\ \lambda^2_{Dm2} &= \frac{K_b T_i}{4\pi Z_{d2} n_{o2} e^2}; \qquad \sigma_i = \frac{T_i}{T_e} \end{aligned} \tag{7}$$

In order to make a time-independent analysis we use the stretching transformation $\xi = \chi - \tau M$ where $\xi$ is normalized by $\lambda_{Dm2}$, and $M$ is defined by *Mach Number* = solitary wave speed$/C_{d2}$. If we use the steady state solution, boundary conditions $n_{d2} \to 1$, $v_{d2} \to 0$, $\varphi \to 0$ and $d\varphi/d\xi \to 0$ at $\xi \to \pm\infty$ can be used. Hence, (6) can be written as

$$\frac{1}{2}\left(\frac{\partial \varphi}{\partial \xi}\right)^2 + V(\varphi) = 0. \tag{8}$$

where $V(\varphi)$ is the Sagdeev potential. In our case, we are interested in the effect of the light particle $(\mu_{d1}, \sigma_{d1})$ on the Sagdeev potential, because this one is responsible of particle trapping. In order to describe these properties in an appropriate way, it is useful to observe that equation (1) can be written in terms of $\xi$ in the form

$$(M - v_2) = \frac{\partial n_2}{\partial \xi} = n_2\frac{\partial v_2}{\partial \xi}, \tag{9}$$

and, therefore,

$$n_2 = \frac{M}{M - v_2} \tag{10}$$

where $t = \frac{\tau}{\omega_{pd2}}$, $\chi = \frac{X}{\lambda_{Dm2}}$, and $v_{d2} = v_2 C_{d2}$.

Now from (2) and (3) written in terms of $\xi$, we get

$$(M^2 + 3\sigma_{d2} + 2\varphi)n_2^2 + 3\sigma_{d2}n_2^4 - M^2 = 0 \tag{11}$$

and

$$n_2 = \frac{1}{\sqrt{6}}\sqrt{3 + \frac{2\varphi + M^2}{\sigma_2} - 1/\sigma_2\sqrt{12M^2\sigma_2 + (M^2 + 3\sigma_2 + 2\varphi)^2}} \tag{12}$$

Now, (6) can be integrated to obtain the Sagdeev potential plus a constant that can be determined through the boundary conditions explained before

$$\begin{aligned} V(\varphi) &= \frac{1}{6}\left(-6e^{-\varphi}\mu_i - \frac{6e^{\sigma_{d1}\varphi}\mu_{d1}}{\sigma_{d1}} - \frac{6e^{\sigma_i\varphi}\mu_e}{\sigma_i} - \right. \\ &\quad \sqrt{\frac{2}{3}}\sqrt{\frac{M^2 + 3\sigma_2 + 2\varphi - \sqrt{-12M^2\sigma_2 + (M^2 + 3\sigma_2 + 2\varphi)^2}}{\sigma_2}} \\ &\quad \left. \times \left\{2M^2 + 6\sigma_2 + 4\varphi + \sqrt{-12M^2\sigma_2 + (M^2 + 3\sigma_2 + 2\varphi)^2}\right\}\right) \end{aligned} \tag{13}$$

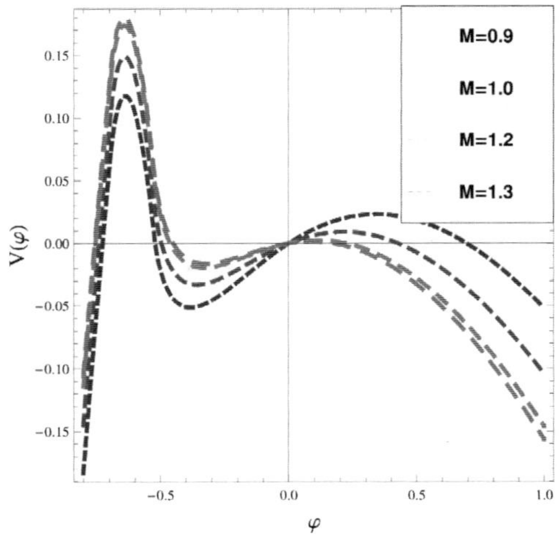


**Figure 1.** Behavior of the Sagdeev potential $V(\varphi)$ for $\sigma_i = 0.05$, $\sigma_{d1} = 0.09$, $\sigma_{d2} = 0.02$, $\mu_{d1} = 0.0$.

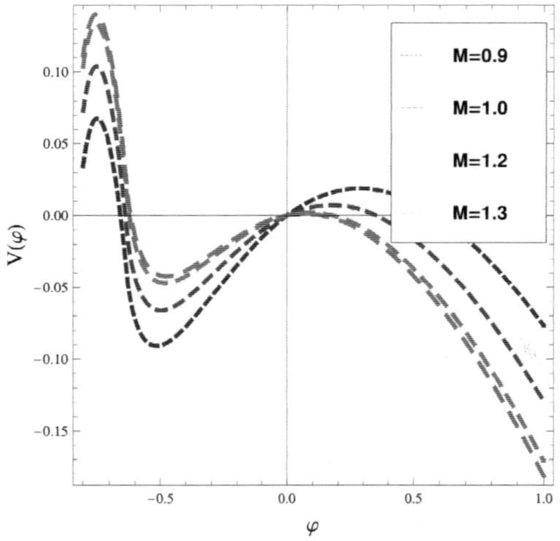


**Figure 2.** Behavior of the Sagdeev potential $V(\varphi)$ for $\sigma_i = 0.05$, $\mu_e = 0.018$, $\sigma_{d1} = 0.002$, $\mu_{d1} = 0.05$, $\sigma_{d2} = 0.002$.

## 3. Results and conclusions

In figure 1 is shown the case when small particles are present in the system. We recover properties of the Sagdeev potential found by other authors, where it is used for the cases $M = 1.2, 1.25$ and 1.3 [4] [5]. However, in figures 2-4 we present the behavior of the Sagdeev potential $V(\varphi)$, for $\sigma_i = 0.05$, $\sigma_{d1} = 0.002$ and M as before, but for $\mu_{d1} = 0.05$, 0.002 and 0. The effect was a singular reduction of the potential and also a significative decrease of the trapping phenomenon was accused. We assume that this effect is due to a reduction of the equilibrium as result of the presence of the smaller component and also a decrease of the trapping process.

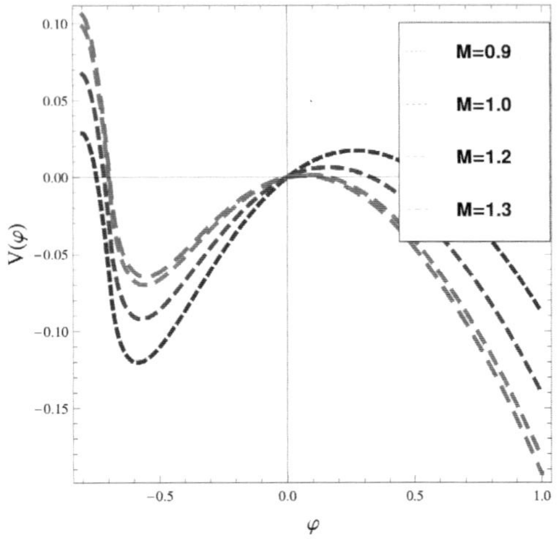


**Figure 3.** Behavior of the Sagdeev potential $V(\varphi)$ for $\sigma_i = 0.05$, $\mu_e = 0.018$, $\sigma_{d1} = 0.002$, $\mu_{d1} = 0.09$, $\sigma_{d2} = 0.002$.

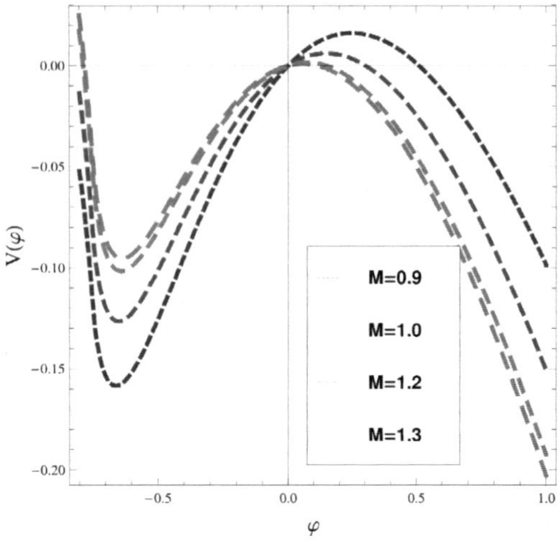


**Figure 4.** Behavior of the Sagdeev potential $V(\varphi)$ for $\sigma_i = 0.05$, $\mu_e = 0.018$, $\sigma_{d1} = 0.002$, $\mu_{d1} = 0.1$, $\sigma_{d2} = 0.002$.

## Acknowledgments
Work supported by grant DID-G22, Universidad Simón Bolívar, Caracas, Venezuela

## References
[1] Carstensen J, Greiner F, Hou L and Piel A 2010 *IEEE Trans. Plasma Sci.* **38** 788
[2] Shukla P K 2002 *Introduction to Dusty Plasma Physics* ed P K Shukla and A A Mamun (Bristol: IOP) pp 195-205
[3] Rao N N, Shukla P K and Yu M Y 1990 *Planet. Space Sci.* **38** , 543
[4] Singh S V and Rao N N 1997 *Phys. Lett. A* **235** 164
[5] Mamun A A 1999 *Astrophys. Space Sci.* **268** 443

Journal of Physics: Conference Series **370** (2012) 012043 doi:10.1088/1742-6596/370/1/012043

# The role of pressure anisotropy in the turbulent intracluster medium

**M S Nakwacki[1,2], E M de Gouveia Dal Pino[1], G Kowal[1] and R Santos-Lima[1]**

[1]Instituto de Astronomía, Geofísica e Ciências Atmosféricas, Universidade de São Paulo, Rua do Matão 1226, Cidade Universitária, 05508-090 São Paulo, Brazil
[2]Present address: Instituto de Astronomía y Física del Espacio, UBA-CONICET, 1428, CABA, Argentina

E-mail: sole@astro.iag.usp.br

**Abstract.** In low-density plasma environments, such as the intracluster medium (ICM), the Larmour frequency is much larger than the ion-ion collision frequency. In such a case, the thermal pressure becomes anisotropic with respect to the magnetic field orientation and the evolution of the turbulent gas is more correctly described by a kinetic approach. A possible description of these collisionless scenarios is given by the so-called kinetic magnetohydrodynamic (KMHD) formalism, in which particles freely stream along the field lines, while moving with the field lines in the perpendicular direction. In this way a fluid-like behavior in the perpendicular plane is restored. In this work, we study fast growing magnetic fluctuations in the smallest scales which operate in the collisionless plasma that fills the ICM. In particular, we focus on the impact of a particular evolution of the pressure anisotropy and its implications for the turbulent dynamics of observables under the conditions prevailing in the ICM. We present results from numerical simulations and compare the results which those obtained using an MHD formalism.

## 1. Introduction

Magnetic fields play an important role in the development of large-scale structure in the Universe, and in recent years their presence in galaxy clusters has been unambiguously proved. According to the standard scenario of structure formation, galaxy clusters are built-up by gravitational merger of smaller units, such as groups and sub-clusters. They are composed of hundreds of galaxies in a Mpc-size region, and the Intra Cluster Medium (ICM) is filled with hot and rarefied gas, emitting in the soft-X ray domain through optically thin bremsstrahlung, magnetic fields and relativistic particles. Magnetic fields in the ICM are investigated through synchrotron emission of cluster-wide radio sources and from the study of the Rotation Measure of radio galaxies revealing the turbulent nature of these fields [1].

Moreover, the ICM is weakly magnetized and nearly collisionless, i.e., the gyro-frequency of both the electrons and the ions is much greater than the collision frequency (under these conditions the mean free path of ions is $\approx$ 30 kpc in the hot ICM, [2]). Plasmas with such characteristics are known to present anisotropic pressures with respect to the magnetic field orientation (see e.g. [3–5]), whose imprints can survive for considerably long periods compared with the dynamical timescales of the system. The presence of mergers, accretion, active galactic nuclei (AGN), galactic winds and instabilities cause complex flows that generate shocks, discontinuities and shear. All these processes lead to the generation of turbulence, which transports and amplifies the magnetic fields present [6].

Understanding the role and evolution of magnetic fields in clusters of galaxies is of significant importance for many questions including the origin of cluster magnetic fields, the interaction of AGN with the ICM, and physical processes operating within the ICM plasma. Which (if any) physical mechanism dominates in the ICM depends sensitively upon the magnitude and distribution of turbulence, which is currently only poorly understood [7].

The fact that the ICM is a low density environment makes the typical MHD description not reliable. As a consequence, these astrophysical plasmas are known to present anisotropic pressures with respect to the magnetic field orientation, which can be generated by several different processes, such as kinetic pressure of cosmic rays, supernovae explosions, stellar winds or anisotropic turbulent motions (see [5, 8]). On small scales, this turbulence is often expected to consist of highly anisotropic fluctuations with frequencies small compared to the ion cyclotron frequency. For a number of applications, the small scales are also collisionless, so a kinetic treatment of the turbulence is necessary, thus making a Kinetic MHD (KMHD) description more appropriate.

The purpose of this contribution is to show and discuss some recent advances in the KMHD description of the ICM plasma. After a brief discussion of the properties and applicability of the Chew-Golberger-Low (CGL) closure, we will describe the KMHD framework, which does include a pressure anisotropy in the description of plasma systems. We will then discuss the impact of this pressure anisotropy on the turbulent evolution of MHD observables, using as an illustrative case a particular variation in time for the pressure anisotropy ratio that reaches the isotropic, thus MHD, state. We will show numerical results corresponding to both developments and briefly discuss implications for this scenario.

## 2. The KMHD model

In order to determine the influence of pressure anisotropy on the turbulent evolution of the plasma in the ICM, in this work we use an MHD formalism with a Chew-Golberger-Low double-isothermal closure, the so called CGL closure [9], as implemented in the numerical code [8, 10–12]. As already mentioned, AGNs and the random motion of galaxies are the main sources of kinetic energy of the ICM plasma, and may be considered the main reason for the eventual pressure anisotropy observed in this medium. Because the kinetic energy injection is continuous the timescale of anisotropic radiative loses is much larger than the timescale associated with plasma instabilities, thus rendering the CGL closure applicable. Here, we will only consider this regime since we are interested in the large scale properties of the ICM for which this condition is satisfied.

In order to study the magnetic field dynamics in the ICM we simulate turbulence in a periodic box of 1 Mpc of size with $128^3$ grid points solving the set of double-isothermal KMHD equations in a conservative form given by the equations

$$\frac{\partial \rho}{\partial t} + \nabla.(\rho \vec{V}) = 0 \tag{1}$$

$$\frac{\partial(\rho \vec{V})}{\partial t} + \nabla.[\rho \vec{V}\vec{V} + (P + \frac{B^2}{8\pi})I - \frac{1}{4\pi}\vec{B}\vec{B}] = f \tag{2}$$

$$\frac{\partial \vec{B}}{\partial t} - \nabla \times (\vec{V} \times \vec{B} = 0 \tag{3}$$

where the term $f$ is a random solenoidal large-scale driving force representing the turbulence driving, which is driven at wave scale $k = 2.5$, the scale injection in the model (2.5 times smaller than the size of the box), and $P = p_\perp \hat{I} + (p_\parallel - p_\perp)\hat{b}\hat{b}$ is the pressure tensor with components

$p_{\parallel}$ and $p_{\perp}$ parallel and perpendicular to the magnetic field direction ($\hat{b} = \vec{B}/|\vec{B}|$) (see [8]). To close the system a CGL closure [9] is used, which yields:

$$\frac{\partial(\rho\vec{V})}{\partial t} + \nabla.[(a_{\perp}^2\rho + \frac{B^2}{8\pi})I - (1-\alpha)\vec{B}\vec{B}] = f \tag{4}$$

with $p_{\perp,\parallel} = a_{\perp,\parallel}^2\rho$, where $a_{\perp,\parallel}$ are constants and represent speeds of sound along the perpendicular and parallel directions to the magnetic field, and $\alpha = (p_{\parallel} - p_{\perp})/(2P_{mag})$, where $P_{mag} = \frac{B^2}{8\pi}$ is the magnetic pressure.

## 3. Numerical simulations

In this work, we do not take into account viscosity and diffusion in the equations. The scale at which the dissipation starts to act is defined by the numerical diffusivity of the scheme. The numerical integration of the system evolution governed by the KMHD equations were performed by using the second-order shock-capturing Godunov-scheme code and the time integration was done with the second-order Runge-Kutta method [8, 11–13].

The spatial coordinates are given in units of a typical length $L_0$. The density $\rho$ is normalized by a reference density $\rho_0$, and the velocity field $V$ by a reference velocity $V_0$. The constant sound speed $c_s$ is also given in units of $V_0$ , and the magnetic field $\vec{B}$ is measured in units of $V_0\sqrt{4\pi\rho_0}$. Time $t$ is measured in units of $L_0/V_0$.

To go one step further in the comparison between KMHD and MHD, the simulation includes a variation in time for the pressure anisotropy. The values for the parallel and perpendicular sound speeds are set equal to 1.0 and 3.0, respectively, keeping them constant until $t = 2.0$, when the turbulence may be considered as fully developed. From $t = 2.0$, there is a change of pressure anisotropy (reducing the perpendicular sound speed and increasing the parallel one) according to the following law $a_{\perp}(t) = a_{\perp 0} + (a_{\perp 0} - a_{\perp\infty})/(1 - t_{half}/t)$, where $a_{\parallel 0} = 1.0$ and $a_{\perp 0} = 3.0$ are the initial values of the sound speed, $a_{\perp\infty} = 1.5$ is the perpendicular sound speed in the limit $t_{\infty}$, and $t_{half} = 6.3$ is the time after which the perpendicular speed decays to the value of $(a_{\perp 0} + a_{\perp\infty})/2$. As the parallel sound speed also changes, the total square sound speed, $a_{\parallel}^2(t) + 2a_{\perp}^2(t)$, is kept constant and equal to the initial value, i.e. the total pressure is kept constant and there is an interplay between the parallel and perpendicular pressures. In this way, we change the pressure anisotropy in time starting from a low value ($\sim 0.3$) and ending up close to the isotropic value, thus reaching the MHD regime. The chosen final value of the anisotropy parameter reaches the value $\sim 0.98$ corresponding to the value in the ICM that is expected from theoretical considerations [14–16]. Figure 1 shows the evolution of the anisotropy parameter.

We also consider the conditions prevailing in the ICM: superalfvenic and subsonic regime with $B_0\hat{x} = 0.1$ as the initial magnetic field.

### *3.1. Results*

The results obtained for the evolution of the physical observables, namely velocity, magnetic field and density show strong differences between the KMHD and standard MHD models. It is interesting to note that kinetic instabilities play a role in the evolution of turbulence under the particular conditions taken into account. In figures 2-7, we present the density, velocity intensity and magnetic field intensity in the central slices of the computational domain obtained for the KMHD model (first line), as as well as for the MHD model (second line) for comparison at t=2 when the turbulence can be considered as fully developed and the pressure anisotropy rate starts changing. It is seen that for the three quantities that we consider the structures obtained in KMHD are significantly smaller than the ones obtained in MHD. Moreover, density, magnetic field intensity and velocity fluctuations are slightly smaller in the isotropic case.

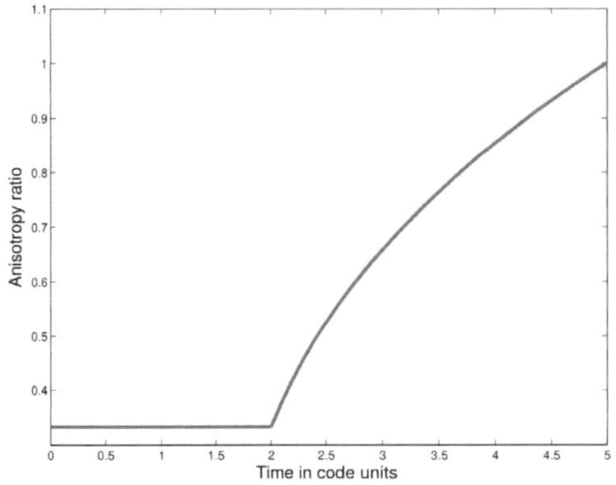


**Figure 1.** Evolution of the speed of sound ratio, $a_{\parallel}/a_{\perp}$. See the text for further details.

**Figure 2.** Central slices (projected on the walls of the computational domain) showing the plasma density at t=2 for KMHD.

**Figure 3.** Central slices (projected on the walls of the computational domain) showing magnetic field intensity at t=2 for KMHD.

**Figure 4.** Central slices (projected on the walls of the computational domain) showing the velocity intensity at t=2 for KMHD.

**Figure 5.** Idem figure 2 for MHD.

**Figure 6.** Idem figure 3 for MHD.

**Figure 7.** Idem figure 4 for MHD.

It is interesting to quantify the impact of the pressure anisotropy change on the evolution of the three observables mentioned above. To this end, we show in figures 8-13 density, velocity intensity and magnetic field intensity in the central slices of the computational domain obtained for the KMHD model (first line), as well as for the MHD model (second line) at t=5. Comparing the results obtained at both times it is seen that the size of the structures in KMHD evolved towards their MHD counterpart. This is reasonable because at t=5 the pressure becomes almost isotropic in the KMHD model. However some differences are still observable due to the action of mirror instabilities during the evolution of the plasma. The mirror instability is responsible for changes in the velocity distribution by slowing the gas and reducing the effective sonic Mach number (see also [8]). Another feature that can be observed from figures 8-13 is that the number of structures that develop in KMHD is considerably larger than that obtained in MHD. Our results show that even at late times important imprints of the pressure anisotropy are clearly observed, thus indicating that for a more accurate description of the evolution of this plasma the effects of pressure anisotropy should be included in the simulations.

**Figure 8.** Idem figure 2 at t=5.

**Figure 9.** Idem figure 3 at t=5.

**Figure 10.** Idem figure 4 at t=5.

**Figure 11.** Idem figure 5 at t=5.

**Figure 12.** Idem figure 6 at t=5.

**Figure 13.** Idem figure 7 at t=5.

The correlation between different scales can be characterized by the energy spectrum, whose evolution is shown in figure 14 for both formalisms. It is seen that the spectrum in KMHD evolves toward its MHD counterpart. At late times, the slope of the spectra of both formalisms is similar. Besides providing crucial information on scale correlations of the plasma, the analysis of the simulated energy spectra is also important because the slope of the spectra is one of the quantities that can be indirectly determined by observations using synthetic rotation maps with the observed polarization coming from radio sources in the ICM [17]. The results of previous numerical works based on MHD [18] seems to favor a power law spectrum with a $k \sim -5/3$ slope, providing support for the use of this value in the construction of synthetic rotation maps (see e.g. [1]). In figure 14 we have added the Kolmogorov slope of $-5/3$ (black line). We note that our results are preliminary and, therefore, further studies are still required to determine the spectrum slope more precisely.

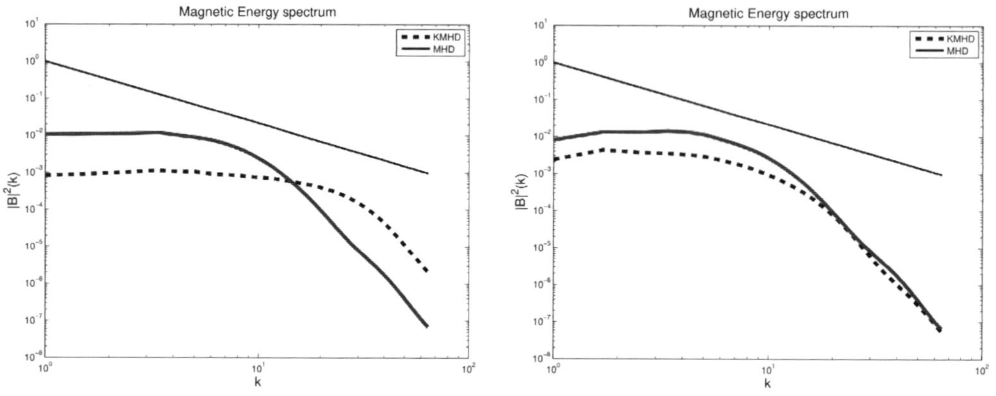


**Figure 14.** Magnetic energy spectrum corresponding to KMHD (dashed black line) and MHD (solid blue line) at $t$=2 (left) and $t$=5 (right). The black line corresponds a slope of $-5/3$.

## 4. Conclusions

As an example of the application of the KMHD with a variable $a_{\parallel}/a_{\perp}$, we compare this formalism with the MHD one performing turbulence simulations for astrophysical conditions, namely those prevailing in the ICM, under a Godunov scheme. We find that pressure anisotropy changes have significant impact on the turbulent properties of nearly collisionless magnetized plasmas, such as the ICM. Particularly, the three observables presented here (plasma density, magnetic field and velocity field) show a much more granular structure in the KMHD formalism. The differences can be understood from the action of the mirror instabilities which are fully developed for values of the pressure anisotropies studied here ($a_{\parallel}/a_{\perp} \leq 2$) (see figures 2-13).

The development of kinetic instabilities, such as these, which arise in the KMHD regime from pressure anisotropy, causes the accumulation of energy at the smaller scales. This may have important consequences on the formation of large scale structures in the intergalactic medium (IGM) and the ICM, and also in the dynamo amplification of cosmic magnetic fields, e.g. [6, 8, 19]. A more detailed study of the relation between pressure anisotropy and structure formation in the ICM and IGM is currently in progress [6, 20].

## Acknowledgments

This work is partially supported by the Brazilian agencies FAPESP and CNPq.

## References

[1] Bonafede A, Feretti L, Murgia M, Govoni F, Giovannini G and Vacca V 2010 *ArXiv e-prints* (*Preprint* 1009.1233)

[2] Schekochihin A A and Cowley S C 2006 *Physics of Plasmas* **13** 056501 (*Preprint* arXiv:astro-ph/0601246)

[3] Barakat A R and Schunk R W 1982 *Plasma Physics* **24** 389–418

[4] Krall N A and Trivelpiece A W 1973 USA: McGraw–Hill

[5] Quest K B and Shapiro V D 1996 *Journal of Geophys. Res.* **1012** 24457–24470

[6] Nakwacki M S, de Gouveia dal Pino E M, Kowal G and Santos-Lima R 2012 in preparation

[7] Bogdanović T, Reynolds C S and Massey R 2011 *The Astrophysical Journal* **731** 7 (*Preprint* 1005.2193)

[8] Kowal G, Falceta-Gonçalves D A and Lazarian A 2011 *New Journal of Physics* **13** 053001 (*Preprint* 1012.5125)

[9] Chew G F, Goldberger M L and Low F E 1956 *Royal Society of London Proceedings Series A* **236** 112–118

[10] Kowal G and Lazarian A 2010 *The Astrophysical Journal* **720** 742–756

[11] Kowal G, Lazarian A and Beresnyak A 2007 *The Astrophysical Journal* **658** 423–445 (*Preprint* arXiv:astro-ph/0608051)

[12] Kowal G, Lazarian A, Vishniac E T and Otmianowska-Mazur K 2009 *The Astrophysical Journal* **700** 63–85 (*Preprint* 0903.2052)

[13] Falceta-Gonçalves D, Lazarian A and Kowal G 2008 *The Astrophysical Journal* **679** 537–551 (*Preprint* 0801.0279)

[14] Rosin M S, Schekochihin A A, Rincon F and Cowley S C 2011 *MNRAS* **413** 7–38 (*Preprint* 1002.4017)

[15] Schekochihin A, Cowley S, Kulsrud R, Hammett G and Sharma P 2005 Magnetised plasma turbulence in clusters of galaxies *The Magnetized Plasma in Galaxy Evolution* ed K T Chyzy, K Otmianowska-Mazur, M Soida, & R-J Dettmar pp 86–92

[16] Howes G G, Cowley S C, Dorland W, Hammett G W, Quataert E and Schekochihin A A 2006 *The Astrophysical Journal* **651** 590–614 (*Preprint* arXiv:astro-ph/0511812)

[17] Murgia M, Govoni F, Feretti L, Giovannini G, Dallacasa D, Fanti R, Taylor G B and Dolag K 2004 *Astron. Astrophys.* **424** 429–446 (*Preprint* arXiv:astro-ph/0406225)

[18] Jones T W, Porter D H, Ryu D and Cho J 2011 *Memorie della Societa Astronomica Italiana* **82** 588 (*Preprint* 1101.4050)

[19] Santos-Lima R, de Gouveia Dal Pino E M, Lazarian A, Kowal G and Falceta-Gonçalves D 2011 Dynamo in the Intra-Cluster Medium: Simulation of CGL-MHD Turbulent Dynamo *IAU Symposium* (*IAU Symposium* vol 274) pp 482–484 (*Preprint* 1102.5139)

[20] Santos-Lima R, de Gouveia dal Pino E M, Kowal, Nakwacki M S, Falceta-Gonçalves D and Lazarian A 2012 in preparation

Journal of Physics: Conference Series **370** (2012) 012044 doi:10.1088/1742-6596/370/1/012044

# Nonlinear stationary structures in nonthermal plasmas

**L A Rios**[1] **and R M O Galvão**[1,2]

[1] Centro Brasileiro de Pesquisas Físicas and Instituto Nacional de Ciência e Tecnologia de Sistemas Complexos, Rua Xavier Sigaud 150, Rio de Janeiro 22290-180, Brazil
[2] Instituto de Física, Universidade de São Paulo, São Paulo 05508-900, Brazil

E-mail: `rios@cbpf.br`

**Abstract.** In plasmas with two distinct populations of hot and cold electrons waves with wave frequency larger than the ion plasma frequency can be generated. In the nonlinear regime, the existence of rarefaction waves and shocks in two-electron temperature isothermal plasmas has been analyzed in the past. In the present work we investigate the conditions for the existence of rarefaction waves and shocks in nonthermal two-electron plasmas. Here the cold and hot electrons are modeled by the Maxwellian and $\kappa$ distributions, respectively. Some preliminary results are presented, and the influence of electron nonthermality is discussed.

## 1. Introduction

It is well known that laboratory and space plasmas can contain distinct populations of hot and cold electrons [1]. In two-electron plasmas, electron-acoustic waves (EAWs) with wave frequency larger than the ion plasma frequency can be generated [2]. In their classical paper of 1978, Bezzerides, Forslund and Lindman [3] investigate the nonlinear regime and analyze the existence of rarefaction waves and shocks in a two-electron temperature isothermal plasma. The study of rarefaction waves (and shocks) is important for a variety of problems in plasma physics, including the so-called current-free double layers [4].

A double layer (DL) consists of a positive/negative Debye sheath, connecting two quasineutral regions of a plasma. These nonlinear structures can be found in a variety of plasmas, from discharge tubes to space plasmas. A DL may be regarded as a BGK equilibrium in some cases, for which certain conditions must be fulfilled. The best known of these structures is the strong Langmuir DL, which is characterized by two counterstreaming plasmas, carrying a large electric current across the DL. The current-free double layer (CFDL) constitutes a different group, for which there is no trapped ion population. Contrary to the Langmuir DL, the CFDL is weak, with $\varphi < k_B T_h/e$ ($\varphi$ is the potential drop across the layer and $T_h$ is the temperature of the hot electron population). It is worth to mention that in general the plasma distributions near a DL are strongly non-Maxwellian [5].

As a preparatory step for a deeper investigation of CFDLs, we follow the steps of reference [3] and analyze the conditions for the existence of rarefaction waves and shocks in nonthermal two-electron plasmas. The dynamics of the plasma is described by the fluid equations, with the cold and hot electrons modeled by the Maxwellian and $\kappa$ distributions, respectively. The family of $\kappa$ distributions has been employed to analyze and interpret data on different plasma environments, like the solar wind [6] and the Earth's ionosphere [7]. The reduced form of the standard $\kappa$ distribution is equivalent to the distribution function obtained from the maximization of the Tsallis entropy, the $q$ distribution [8], with the parameter $\kappa$ measuring the deviation from the Maxwellian equilibrium (in the limit $\kappa \to \infty$ the Maxwellian distribution is recovered). Here some preliminary results of our work are presented, and the influence of the superthermal electrons present in the long tails of the $\kappa$ distribution is discussed.

## 2. Model equations

First we consider the general rarefaction wave problem. The ions are assumed to obey the cold hydrodynamic equations, and the ion and electron densities are related through Poisson's equation

$$\frac{\partial^2 \varphi}{\partial z^2} = -4\pi e \left[N_i - N_e(\varphi)\right]. \tag{1}$$

The electron number density can be written as

$$N_e(\varphi) = N_c(\varphi) + N_h(\varphi), \tag{2}$$

where

$$N_c(\varphi) = N_{c0} e^{e\varphi/k_B T_c}, \tag{3}$$

$$N_h(\varphi) = N_{h0}\left[1 - \frac{e\varphi}{(\kappa - 3/2)\, k_B T_h}\right]^{-(\kappa-1/2)}, \tag{4}$$

with $\kappa > 3/2$. In the above equations $T_c$ is the temperature of the cold electrons and $N_{c0} + N_{h0} = N_0$. In the limit $\kappa \to \infty$ we obtain the Boltzmann distribution (3) also for the hot electrons .

Introducing the similarity parameter $\xi = (z/t)/c_h$, where $c_h = (k_B T_h/m_i)^{1/2}$, and following reference [3] we obtain the equation that governs the electrostatic potential $\phi$, i.e.

$$\frac{d\phi}{d\xi}\left(1 + \frac{1}{2}\frac{dc_s^2}{d\phi}\right) + c_s = 0, \tag{5}$$

where $\phi = e\varphi/k_B T_h$ and

$$c_s^2 = \frac{dP_e}{dn_e} \equiv \frac{n_e(\phi)}{\left\{n_c(\phi)\,\tau + n_h(\phi)\,\frac{(\kappa-1/2)}{(\kappa-3/2)}\left[1 - \frac{\phi}{(\kappa-3/2)}\right]^{-1}\right\}}, \tag{6}$$

with $n_e(\varphi) = N_e(\varphi)/N_0$ and $n_i = N_i/N_0$. The above expression is the square of the normalized speed of sound, where $\tau = T_h/T_c$ and $P_e \equiv P_e(\phi)$ is the normalized pressure, which obeys the relation $dP_e(\phi)/d\phi = n_e(\phi)$. A complete solution for (5) exists provided $\phi$ is a single valued function of $\xi$. However, multiple valued solutions occur when

$$\frac{dc_s^2}{d\phi} + 2 \le 0, \tag{7}$$

with the equality defining the onset of the singularity in the rarefaction wave.

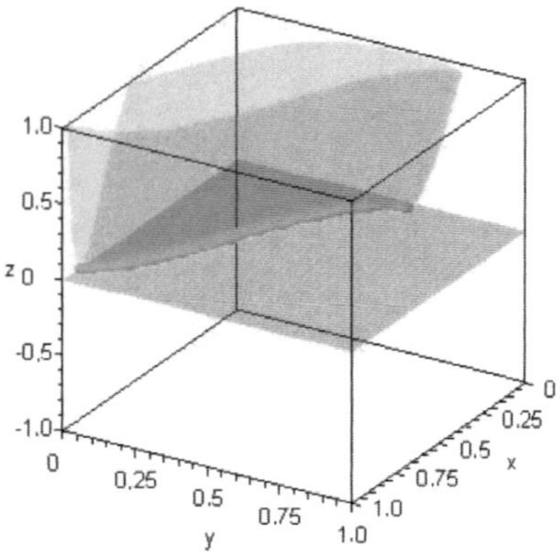


**Figure 1.** $\kappa = 500$ and $\tau = 10$.

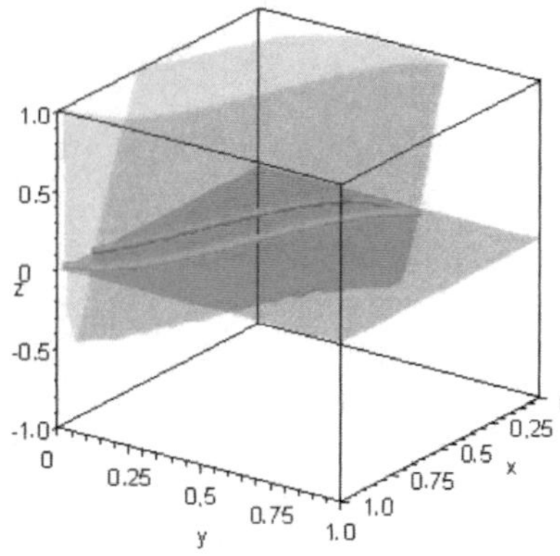


**Figure 2.** $\kappa = 500$ and $\tau = 12$.

## 3. Rarefaction waves and shocks

As discussed by Bezzerides, Forslund and Lindman, the threshold for rarefaction shocks and the onset of the singularity in the rarefaction wave are expressed by the same condition, $dc_s^2/d\phi + 2 = 0$. When both electron populations are modeled via Maxwellian distributions, condition (7) reduces to $\tau \geq 5 + \sqrt{24} \approx 9.9$ [3]. This can be seen in figure 1, where we plot $z = 2 + dc_s^2/d\phi$ as a function of $x = \alpha = N_{h0}/N_0$ and $y = -\phi$ for $\kappa = 500$ (Maxwellian limit). For $\tau = 10$ we observe that $z = 0$ (onset of the singularity) for a broad range of $\alpha$'s. For $\tau = 12$ (figure 2) we notice that $z \leq 0$ also for a broad range of $\alpha$'s, with the formation of the shock between the two extremes of $\phi$. However, as $\kappa$ decreases (figure 3, $\kappa = 5$), we notice the singularity starts to appear (for a short range of $\alpha$'s) only for $\tau \approx 11$. As $\tau$ increases (figure 4), we have the formation of the shock for all the values of $\alpha$. For longer tails (figure 5), the singularity

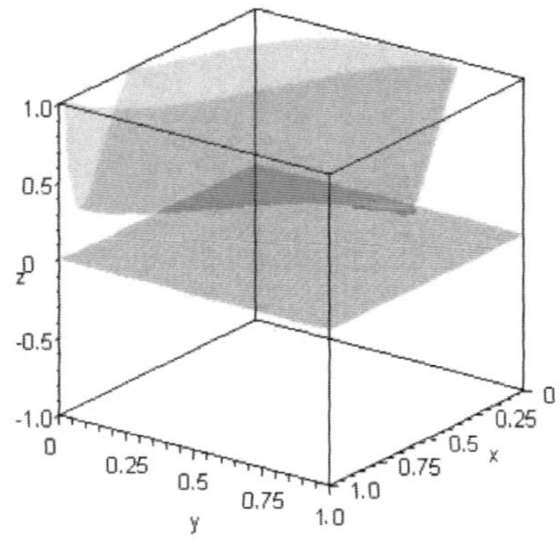


**Figure 3.** $\kappa = 5$ and $\tau = 11$.

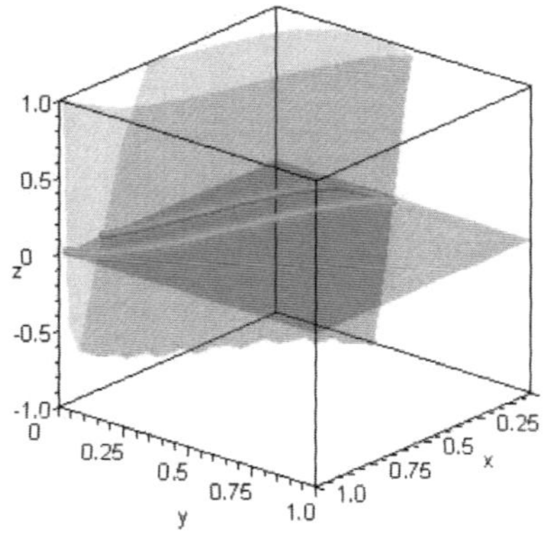


**Figure 4.** $\kappa = 5$ and $\tau = 16$.

appears only for a larger value of $\tau$ ($\approx 13$). For the same value of $\tau$, we observe that electron nonthermality seems to "disturb" the formation of the shock: as $\kappa$ decreases, the shock does not

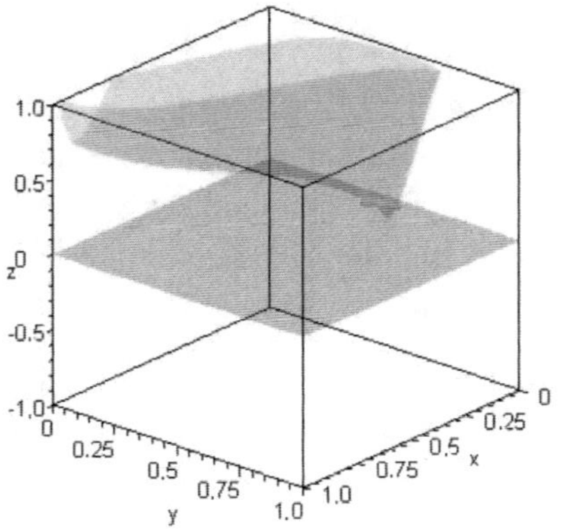


**Figure 5.** $\kappa = 2.5$ and $\tau = 13$.

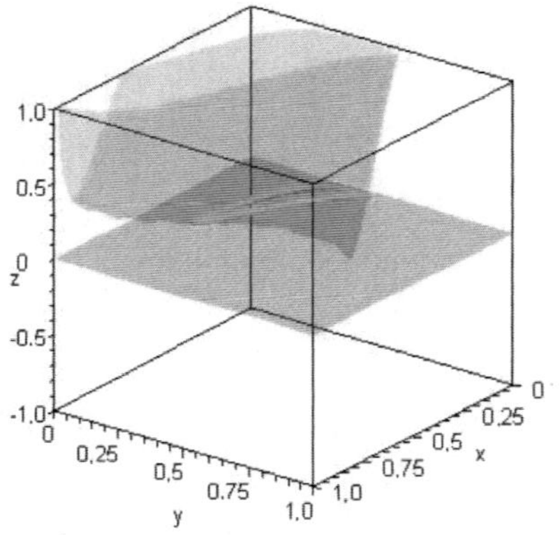


**Figure 6.** $\kappa = 2.5$ and $\tau = 16$.

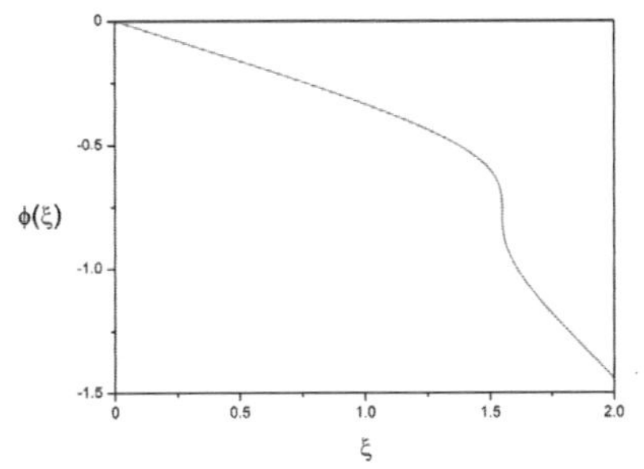


**Figure 7.** $\kappa = 500$, $\alpha = 0.01$ and $\tau = 9.9$.

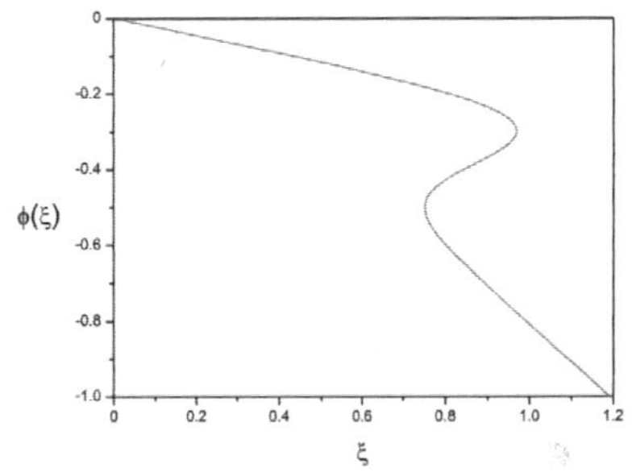


**Figure 8.** $\kappa = 500$, $\alpha = 0.01$ and $\tau = 20$.

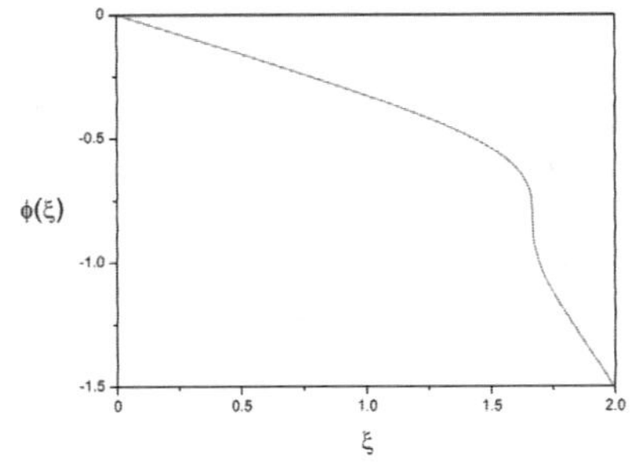


**Figure 9.** $\kappa = 2.5$, $\alpha = 0.01$ and $\tau = 10.2$.

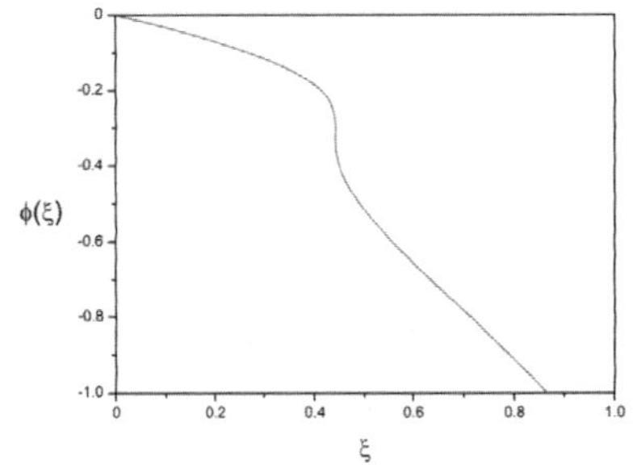


**Figure 10.** $\kappa = 2.5$, $\alpha = 0.2$ and $\tau = 14.2$.

appear for all the values of $\alpha$ (figures 4 and 6).

Solving (5) numerically, we can analyze the different profiles obtained for the electrostatic potential $\phi$. In figures 7 and 8 we present the results for $\kappa = 500$, $\alpha = 0.01$ and $\tau = 9.9$ and 20, respectively. In figure 7 we notice the eminent formation of the shock (singularity), while in figure 8 it is observed that $\phi$ is not a single valued function of $\xi$. This discontinuity in the profile of the electrostatic potential represents the shock formation: in a reference frame moving with velocity $\xi \approx 0.9$ we observe two different values for $\phi$. For such a small value of $\alpha$ $(= 0.01)$ and $\kappa = 2.5$, the onset of the singularity appears for $\tau = 10.2$ (figure 9). As $\alpha$ grows $(\alpha = 0.2$, figure 10), the shock becomes eminent only for $\tau = 14.2$. Is is also noticed that the shock starts to appear for smaller values of $\xi$ as $\alpha$ becomes larger.

## 4. Conclusions

The presented results indicate that electron nonthermality (represented by the parameter $\kappa$) influences the onset of the singularity in the rarefaction wave and the formation of the shock. For distributions with longer tails (small $\kappa$) the formation of the shock becomes eminent only for larger values of $\tau$ when compared to the Maxwellian case. It is also noticed that, for a fixed $\tau$, a decrease in $\kappa$ implies the disappearance of the shock for larger values of $\alpha$.

## Acknowledgments

This work was partially supported by CAPES, FAPERJ and CNPq/INCT-SC.

## References

[1] Montgomery D S, Focia R J, Rose H A, Russell D A, Cobble J A, Fernández J C and Johnson R P 2001 *Phys. Rev. Lett.* **87** 155001; Ferrante G, Zarcone M, Uryupina D S and Uryupin S A 2003 *Phys. Plasmas* **10** 3344; Pottelette R, Ergun R E, Treumann R A, Berthomier M, Carlson C W, McFadden J P and Roth I 1999 *Geophys. Res. Lett.* **26** 2629
[2] Gary S P and Tokar R L 1985 *Phys. Fluids* **28** 2439
[3] Bezzerides B, Forslund D W and Lindman E L 1978 *Phys. Fluids* **21** 2179
[4] Ahedo E and Sánchez M M 2009 *Phys. Rev. Lett.* **103** 135002
[5] Block L P 1978 *Astrophys. Space Sci.* **55** 59
[6] Collier M R, Hamilton D C, Gloeckler G, Bochsler P and Sheldon R B 1996 *Geophys. Res. Lett.* **23** 1191; Pierrard V, Maksimovic M and Lemaire J 1999 *J. Geophys. Res.* **104** 17021
[7] Bahcivan H, Cosgrove R B and Tsunoda R T 2006 *J. Geophys. Res.* **111** A07306
[8] Tsallis C 1988 *J. Stat. Phys.* **52** 479; Tsallis C 1995 *Physica* A **221** 277

Journal of Physics: Conference Series **370** (2012) 012045 doi:10.1088/1742-6596/370/1/012045

# Error analysis in the electron temperature measurements in TCABR

**J H F Severo[1], F O Borges[2], M P Alonso[3], R M O Galvão[1], V C Theodoro[1], L A Berni[4], L C Jeronimo[1], J I Elizondo[1], A C A Figueiredo[3], M Machida[5], I C Nascimento[1], Yu K Kuznetsov[1], E K Sanada[1], O C Usuriaga[1] and M Tendler[6]**

[1] Institute of Physics, University of São Paulo, Rua do Matão, s/n, 05508-900 SP, Brazil.
[2] Laboratório de Plasma e Espectroscopia Atômica, Instituto de Física, Universidade Federal Fluminense (UFF), 24210-340, Niterói, RJ, Brazil.
[3] Instituto de Plasmas e Fusão Nuclear, Instituto Superior Técnico, Associação Euratom/IST 1049-001, Lisboa, Portugal.
[4] Laboratório Associado de Sensores e Materiais, INPE, 12227-010, São José dos Campos, SP, Brazil.
[5] Universidade Estadual de Campinas, 13083-970 Campinas, SP, Brazil.
[6] Royal Inst Technol, Alfven Lab, Stockholm, Sweden.

E-mail: jhsevero@if.usp.br

**Abstract.** An analytical method is proposed to evaluate the experimental uncertainty in the electron temperature measurements in the TCABR tokamak. Solving the integral equation resulting from the convolution of two functions, one representing, the scattered light and the other the spectral apparatus function, i.e., the polychromator, an analytical expression for the electron temperature is obtained, from which the uncertainty in the measured value is readily evaluated. The results show that the major contribution to the error comes from the noise in the signal; the uncertainties in the filters parameters do not contribute significantly to the total error.

## 1. Introduction

The TCABR tokamak is a machine with a broad research program on the physics of magnetically confined plasmas; in particular the interaction of RF waves in the region of Alfvén frequencies [1], stochastic processes at the plasma edge, transport barriers [2-3], plasma rotation [4-6], MHD activity [7], and diagnostic development [8-9]. The Thomson scattering system of TCABR [10] has been developed under a scientific collaboration between the University of São Paulo (Brazil) and the Instituto Superior Técnico (Lisbon, Portugal).

Thomson scattering is the most straightforward and widely used diagnostic in controlled thermonuclear research to measure the electron temperature and density. It was first applied to measure the electron temperature in a tokamak in 1969 by a British-Soviet team [11-12] and thereafter it became a standard technique.

In the Thomson diagnostic, the electron temperature is obtained from the shape of the spectrum $s(\lambda)$ of the scattered light, where $\lambda$ is the wavelength. However, any apparatus used to measure the spectrum distribution introduces distortions in the spectrum, which have to be properly taken into account. Mathematically, this distortion can be computed from the convolution of the apparatus function $a(\lambda)$ with the spectral distribution of scattered light, i.e. [13],

$$f(\lambda) = \int_{-\infty}^{+\infty} s(\lambda')a(\lambda - \lambda')d\lambda'. \tag{1}$$

The Thomson Scattering system of TCABR is composed of a vertically oriented Nd:Glass laser beam ($\lambda_i = 1054nm$), a polychromator equipped with three interference filters coupled to silicon APDs (avalanche photodiodes), and a telescope that collimates the scattered light into an optical fiber that transmits this light to the polychromator [8].

Solving (1) for the different interference filters and taking the ratio of the corresponding solutions, an analytical expression for the electron temperature is obtained from which the uncertainty in the measurement can be calculated.

The remainder of this paper is organized as follows. The experimental setup is described in section 2. In section 3, a general expression for electron temperature as a function of signal ratio is derived. The results and discussions are presented in section 4. Section 5 contains the conclusions. Appendix A is dedicated to the evaluation of the uncertainties.

## 2. Experimental setup

A schematic representation of the Thomson scattering diagnostic of the TCABR tokamak ($R = 0.61m$, $a = 0.18m$) is shown in figure 1. The Nd:Glass laser is Q-switched by a Pockels cell triggered to deliver one pulse per tokamak shot. The plasma volume is probed by the laser beam ($\lambda_i = 1054nm$) in the vertical direction. The scattered light is collected from an $18mm$ long chord localized at the plasma center by an objective lens and transmitted to the polychromator by an optical fiber. The polychromator is composed of band-pass interference filters, to split the light into three spectral bands coupled to three APDs. The detectors integrate the scattered light in three different spectral intervals corresponding to pass-bands of the interference filters. The relative responses of the three polychromator channels are shown in the figure 2. These curves were obtained after calibration and include all system parameters, such as interference filters transmittance, detector quantum efficiency, amplifier gain, etc.. Figure 2 shows also the spectral distribution of the scattered light for an electron temperature of $500eV$ [14].

The transmittance of the filters is approximately 75% and the APDs (RCA type CA30956E) have a quantum efficiency of $85-40\%$ in the range $800-1060nm$. These specifications guarantee the high spectral sensitivity of the diagnostic.

The plasma temperature is evaluated from the signal ratio of polychromator channels, $R = P_{S,i}/P_{S,j}$, where $P_{S,i}$ is the scattered power detected by the $i-th$ channel of the polychromator. This method eliminates the dependence on the power of the incident laser, solid angle, plasma density, spectral response of detector etc. leaving only the dependence on the electron temperature.

## 3. Electron temperature as function of signal ratio

The theory of Thomson scattering is well described in the literature, and its application to fusion plasmas has been analyzed in detail [15]. According to this reference, as the laser beam propagates through the plasma the power $P_S$ scattered into the wavelength range $\lambda_1-\lambda_2$ by free electrons within the solid angle of observation $\Delta\Omega$ is given by

$$P_S = P_0 r_0^2 n_e \Delta L \Delta\Omega \int_{\lambda_1}^{\lambda_2} f(\lambda, T_e)\, d\lambda, \tag{2}$$

where $P_0$ is the power of the incident laser, with wavelength $\lambda_I$, $r_0$ is the classical electron radius, $n_e$ is the plasma density, $\Delta L$ is the length of the scattering region as observed by

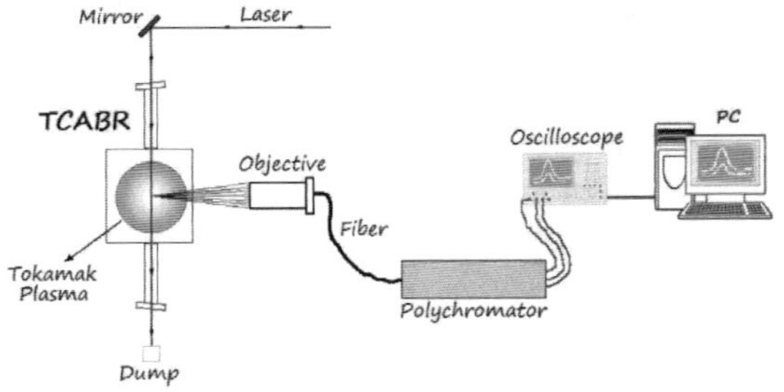


**Figure 1.** Schematic representation of the experimental arrangement of the Thomson scattering system in the TCABR tokamak.

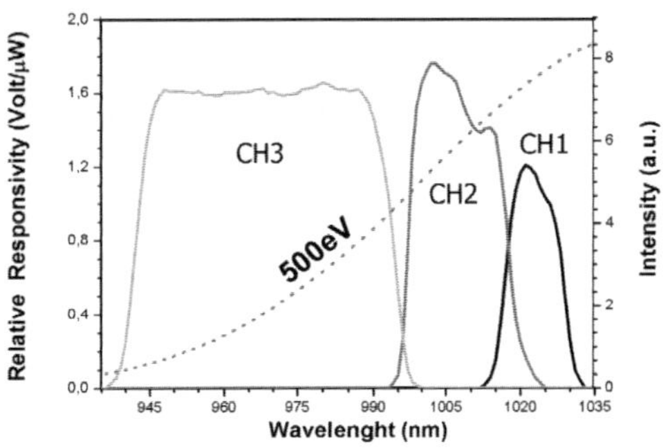


**Figure 2.** Electron temperature as function of signal ratio.

the collection optics (18 $mm$ in the TCABR). The function $f(\lambda, T_e)$ is given by (1). The function $a(\lambda)$ is sometimes called the transfer function and includes all possible changes in the amplitude and phase of the signal related to the detector characteristics, as quantum efficiency, circuit characteristics, amplifier gain, distortions of the input signals due to diffraction on the diaphragms of the optical system of the spectral apparatus, its aberrations, the finite aperture widths, the time lag of the recording instrument etc., as schematically represented in figure 3.

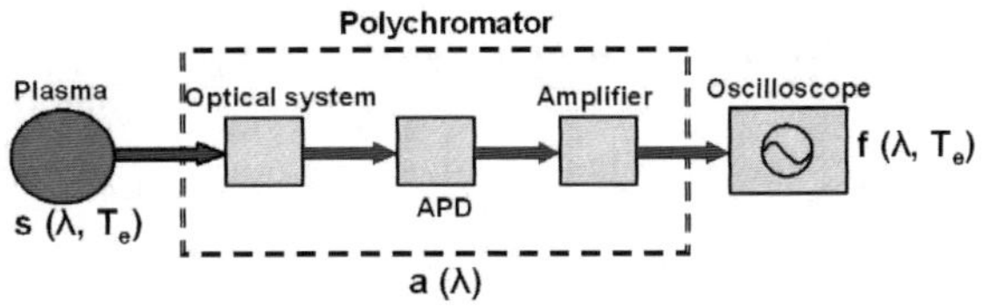


**Figure 3.** Schematic representation of signal transformation during the measurements process.

The function $s(\lambda, T_e)$ in equation (1) is well described in the literature and has the following form [15]:

$$s(\lambda, T_e) = \frac{1}{\Delta\lambda_e \lambda_I \sqrt{\pi}} \left( \lambda_I - 3.5(\lambda - \lambda_I) + \frac{(\lambda - \lambda_I)^3}{\Delta\lambda_e^2} \right) \exp\left[ -\frac{(\lambda - \lambda_I)^2}{\Delta\lambda_e^2} \right], \qquad (3)$$

where $\Delta\lambda_e = \frac{2V_T\lambda_I}{c}\sin\left(\frac{\theta}{2}\right)$, $c$ is the speed of light, $V_T = \sqrt{\frac{2K_BT_e}{m_e}}$ is the electron thermal velocity, $T_e$ the electron temperature, $k_B$ the Boltzmann constant, $m_e$ the electron mass, and $\theta$ is the observation angle between the electrical field of the incident radiation and the scattering direction.

To evaluate the uncertainty in electron temperature, it is necessary to first obtain an expression for the electron temperature as function of all system parameters e.g., the integral in (1) has to be carried out, inserted into (2), and the resulting integral inverted to explicitly obtain an expression for the electron temperature.

One difficulty with this procedure is that the apparatus function $a(\lambda)$, which is obtained through a calibration process, is known only numerically, as shown in figure 2 for the three interference filters of the TCABR Thomson scattering system. However, considering the global shape of the calibration curves, one realizes that they can be represented as sum of Gaussians, i.e.,

$$a(\lambda) = \sum_{i=1}^{n=3} \sum_{j=1}^{m} g_{i,j}(\lambda) \tag{4}$$

where $g_{i,j}(\lambda) = 1/(\sigma_{i,j}\sqrt{2\pi}) \exp\left[-0.5(\lambda - \lambda_{0,i,j})^2/\sigma_{i,j}^2\right]$, $n$ is the number of interference filters used in the diagnostic, and $m$ is the number of Gaussian functions used to adjust the spectral dependence of each interference filter curve.

A general method to carry out the integration in (1) is based upon using the Fourier transforms of the two functions in the integrand,

$$s(\lambda) = \frac{1}{\sqrt{2\pi}} \int_{-\infty}^{+\infty} S(\omega) e^{i\omega\lambda} d\omega; \qquad a(\lambda) = \frac{1}{\sqrt{2\pi}} \int_{-\infty}^{+\infty} A(\omega) e^{i\omega\lambda} d\omega. \tag{5}$$

Then, (1) can be rewritten as

$$f(\lambda) = \frac{1}{2\pi} \int_{-\infty}^{+\infty} S(\omega) A(\omega) e^{i\omega\lambda} d\omega, \tag{6}$$

where

$$A(\omega) = \frac{1}{\sqrt{2\pi}} \int_{-\infty}^{+\infty} a(\lambda) e^{-i\omega\lambda} d\lambda = \frac{1}{\sqrt{2\pi}} \int_{-\infty}^{+\infty} \sum_{i=1}^{n} \sum_{j=1}^{m} g_{i,j}(\lambda) e^{-i\omega\lambda} d\lambda, \tag{7}$$

is the Fourier transform of the apparatus function. Using the Gaussian expression given by (4), it follows that

$$A(\omega) = \sum_{i=1}^{n} \sum_{j=1}^{m} G_{i,j}(\omega) \qquad G_{i,j}(\omega) = \exp\left[-\left(i\lambda_{0,i,j}\omega + \frac{\omega^2 \sigma_{i,j}^2}{2}\right)\right]. \tag{8}$$

Equation (6) can be rewritten in the following way

$$f(\lambda) = \frac{1}{2\pi} \int_{-\infty}^{+\infty} \left[A(\omega) - 1 + 1\right] S(\omega) e^{i\omega\lambda} d\omega,$$

that is,

$$f(\lambda) = \frac{1}{2\pi} \int_{-\infty}^{+\infty} \left[A(\omega) - 1\right] S(\omega) e^{i\omega\lambda} d\omega + s(\lambda). \tag{9}$$

Substituting equation (8) into (9) yields

$$f(\lambda) = \frac{1}{2\pi} \sum_{i=1}^{n} \sum_{j=1}^{m} \int_{-\infty}^{+\infty} \left[G_{i,j}(\omega) - 1\right] S(\omega) e^{i\omega\lambda} d\omega + s(\lambda). \tag{10}$$

The term in brackets in this equation is very small and vanishes in the case of ideal spectral apparatus. In most practical cases, when the half-width of the scattered spectral distribution of light is much wider than the half-width from instrumental broadening, the distortion caused by the real apparatus is very small and this term can be neglect. So, expanding the Fourier transform of the Gaussian function in a Taylor series and substituting in (10) yields

$$f(\lambda) \simeq \frac{1}{2\pi}\sum_{i=1}^{n}\sum_{j=1}^{m}\int_{-\infty}^{+\infty}\left[-i\omega\lambda_{0,i,j} - \frac{1}{2}\omega^2\left(\sigma_{i,j}^2 + \lambda_{0,i,j}^2\right)\right] S(\omega)e^{i\omega\lambda}d\omega + s(\lambda). \tag{11}$$

Taking into account that

$$\frac{ds\,(\lambda)}{d\lambda} = \frac{1}{2\pi}\int_{-\infty}^{+\infty} i\omega S(\omega)e^{i\omega\lambda}d\omega \qquad \text{and} \qquad \frac{d^2s\,(\lambda)}{d\lambda^2} = -\frac{1}{2\pi}\int_{-\infty}^{+\infty}\omega^2 S(\omega)e^{i\omega\lambda}d\omega,$$

it follows that,

$$f(\lambda) \simeq s(\lambda) - \sum_{i=1}^{n}\sum_{j=1}^{m}\lambda_{0,i,j}\frac{ds\,(\lambda)}{d\lambda} + \frac{1}{2}\sum_{i=1}^{n}\sum_{j=1}^{m}\left(\sigma_{i,j}^2 + \lambda_{0,i,j}^2\right)\frac{d^2s\,(\lambda)}{d\lambda^2}. \tag{12}$$

The function $s(\lambda)$ has a Gaussian shape for temperatures below $500eV$. The second and third terms in parentheses in (3) are responsible for the relativistic corrections and can be neglected for temperatures below $500eV$. Therefore the function $s(\lambda)$ can be approximately expressed in the following form:

$$s\,(\lambda) \simeq \frac{1}{\Delta\lambda_e\sqrt{\pi}}\exp\left[-\frac{(\lambda-\lambda_I)^2}{\Delta\lambda_e^2}\right], \tag{13}$$

where $\Delta\lambda_e = 2\sqrt{2K_BT_e/m_e}\lambda_I\sin^2(\theta/2)/c$ and $\lambda_I$ is wavelength of the incident beam.

Using these results, it follows that the scattered power $P_{S,i}$ measured in each channel is given by

$$P_{S,i} = E_0\left[\int_{\lambda_{1,i}}^{\lambda_{2,i}} s(\lambda)d\lambda - \sum_{j=1}^{m}\lambda_{0,i,j}\int_{\lambda_{1,i}}^{\lambda_{2,i}}\frac{ds\,(\lambda)}{d\lambda}d\lambda + \frac{1}{2}\sum_{j=1}^{m}\left(\sigma_{i,j}^2 + \lambda_{0,i,j}^2\right)\int_{\lambda_{1,i}}^{\lambda_{2,i}}\frac{d^2s\,(\lambda)}{d\lambda^2}d\lambda\right], \tag{14}$$

where $E_0 = P_0r_0^2n_e\Delta L\Delta\Omega$.

Expanding the function $s(\lambda)$ in a Taylor series around $\lambda = \lambda_{C,i}$ where $\lambda_{C,i}$ represents the central wavelength of $i-th$ interference filter, the power $P_{S,i}$ for each channel can be written as

$$P_{S,i} \simeq \frac{P_0r_0^2n_e\Delta L\Delta\Omega W_ib_i}{\Delta\lambda_e^3\sqrt{2\pi}}\exp\left[-\frac{(\lambda_{C,i}-\lambda_I)^2}{2\Delta\lambda_e^2}\right]\left(\frac{a_i}{W_ib_i}\Delta\lambda_e^2 + 1\right) \tag{15}$$

where:

$$a_i = W_i - \sum_{j=1}^{m}\lambda_{0,i,j}$$

and

$$b_i = \left[\frac{W_i^2}{24} + (\lambda_{C,i}-\lambda_I)\sum_{j=1}^{m}\lambda_{0,i,j} - \frac{1}{2}\sum_{j=1}^{m}\left(\sigma_{i,j}^2 + \lambda_{i,j}^2\right)\right]$$

Let $R = P_{S,2}/P_{S,1}$ express the signal ratio between channels 2 and 1, which is a function of the electron temperature and the filters parameters, such as the central wavelength $\lambda_{C,i}$ and the waveband $W_i$ , etc.. Solving this equation with respect to $T_e$ yields,

$$T_e = \frac{-K_1 \pm \sqrt{K_1^2 - 4K_2K_0}}{2K_2}, \tag{16}$$

where:

$$K_2 = A\left(\frac{\lambda_I^2 2k_B}{c^2 m_e}\right)^2,$$
$$K_1 = B\left(\frac{\lambda_I^2 2k_B}{c^2 m_e}\right),$$
$$K_0 = C,$$

and

$$A = 2\left(a_1 W_2 b_2 - a_2 W_1 b_1\right),$$
$$B = 2W_1 W_2 b_1 b_2 \ln\left(\frac{RW_1 b_1}{W_2 b_2}\right),$$
$$C = W_1 W_2 b_1 b_2 \left[\left(\lambda_{C,2} - \lambda_I\right)^2 - \left(\lambda_{C,1} - \lambda_I\right)^2\right].$$

Let now $R = \left(P_{S,3} + P_{S,2}\right)/P_{S,1}$, in this case the electron temperature can be expressed as:

$$T_e = \frac{-K_1 \pm \sqrt{K_1^2 - 4K_2 K_0}}{2K_2}, \tag{17}$$

where:

$$K_2 = A\left(\frac{\lambda_I^2 2k_B}{c^2 m_e}\right)^2,$$
$$K_1 = B\left(\frac{\lambda_I^2 2k_B}{c^2 m_e}\right),$$
$$K_0 = C,$$

$$A = 2\left[a_1\left(W_2 b_2 + W_3 b_3\right) - W_1 b_1\left(a_2 + a_3\right)\right],$$
$$B = 2W_1 b_1\left(W_2 b_2 + W_3 b_3\right)\ln\left(\frac{RW_1 b_1}{W_2 b_2 + W_3 b_3}\right),$$
$$C = W_1 b_1\left(W_2 b_2 + W_3 b_3\right)\left[\left(\lambda_M - \lambda_I\right)^2 - \left(\lambda_{C,1} - \lambda_I\right)^2\right],$$

and

$$\lambda_M = \frac{\lambda_{C,2} W_2 + \lambda_{C,3} W_3}{W_2 + W_3}.$$

## 4. Results and discussions

The plasma temperature in the TCABR tokamak is evaluated from the signal ratios of the different polychromator channels. Equation (16) gives the electron temperature as a function of the signal ratio between channels 2 and 1, while equation (17) gives the electron temperature as a function of the signal ratio of channel 2 plus channel 3 and channel 1. For the ratio between channels 3 and 1 all indexes 2 in equation (16) should be changed to index 3. These equations show the dependence of electron temperature on the signal ratio and the filters parameters, such as $\lambda_{C,i}$, $W_i$, $\lambda_{0,i,j}$ and $\sigma_{i,j}$ that are obtained after fitting the curves in figure 2 with a least squares method.

Table 1 gives the central wavelength $\lambda_{0,i,j}$ and bandwidth $\sigma_{i,j}$ of each Gaussian used in equation (4) to express the apparatus function.

Once all parameters in equations (16) and (17) have been obtained, after fitting the curves in figure 2, the dependence of electron temperature on the signal ratio can be calculated. The ratios $R = (P_{S,3} + P_{S,2})/P_{S,1}$ and $R = P_{S,3}/P_{S,1}$ as function of $T_e$ $[eV]$ are shown in figures 4 and 5, respectively. In these figures is also shown the dependence of electron temperature on the signal ratio obtained by computation. The solid line shows the computed results while the dashed line shows the calculated results from equation (16) and (17).

**Table 1.** Table of parameters obtained by least squares fitting of the curves 1, 2 and 3 in figure 2.

| Curve 1 in figure 2 | |
|---|---|
| $\lambda_{0,1,j}$ $[nm]$ | $\sigma_{1,j}$ $[nm]$ |
| $1019.15 \pm 0.03$ | $3.52 \pm 0.09$ |
| $1023.02 \pm 0.08$ | $5.93 \pm 0.13$ |
| $1028.17 \pm 0.03$ | $3.87 \pm 0.04$ |
| Curve 2 in figure 2 | |
| $\lambda_{0,2,j}$ $[nm]$ | $\sigma_{2,j}$ $[nm]$ |
| $999.21 \pm 0.09$ | $2.70 \pm 0.20$ |
| $1001.8 \pm 0.8$ | $4.4 \pm 0.6$ |
| $1006.2 \pm 0.3$ | $6 \pm 3$ |
| $1012.7 \pm 0.6$ | $(90 \pm 13) \cdot 10^{-1}$ |
| $1016.43 \pm 0.14$ | $3.5 \pm 0.6$ |
| Curve 3 in figure 2 | |
| $\lambda_{0,3,j}$ $[nm]$ | $\sigma_{3,j}$ $[nm]$ |
| $945.2 \pm 0.3$ | $5.0 \pm 0.8$ |
| $(9491 \pm 17) \cdot 10^{-1}$ | $7 \pm 4$ |
| $(9556 \pm 22) \cdot 10^{-1}$ | $11 \pm 9$ |
| $(9689 \pm 21) \cdot 10^{-1}$ | $17 \pm 12$ |
| $983.1 \pm 0.9$ | $11 \pm 5$ |
| $990.1 \pm 0.3$ | $(64 \pm 13) \cdot 10^{-1}$ |
| $994.06 \pm 0.15$ | $(23 \pm 12) \cdot 10^{-1}$ |

The disagreement that occurs for $T_e > 600eV$ in both figures is related to the spectral distribution of the scattered light $s(\lambda)$, which we take as pure Gaussian (equation (13)), and relativistic corrections are not being taken in to account, since the electron temperature in the TCABR tokamak is $< 600eV$.

From the analytical expression for the electron temperature given by equations (16) and (17), the uncertainty in the electron temperature can be calculated by applying the standard method

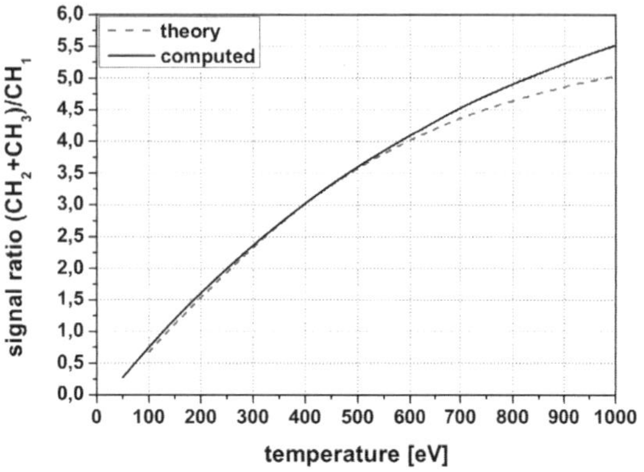


**Figure 4.** The solid line shows the computed results of $R = (P_{S,3} + P_{S,2})/P_{S,1}$, and the dashed line shows the calculated results from equation (16) and (17).

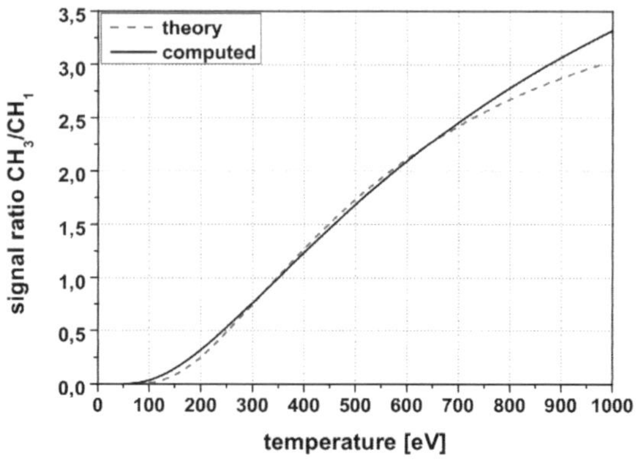


**Figure 5.** The solid line shows the computed results of $R = P_{S,3}/P_{S,1}$, and the dashed line shows the calculated results from equation (16) and (17).

of error propagation. The standard deviation $\sigma^2_{T_e}$ is given by,

$$\sigma^2_{T_e} = \left(\frac{\partial T_e}{\partial R}\right)^2 \sigma^2_R + \left(\frac{\partial T_e}{\partial \lambda_{C,1}}\right)^2 \sigma^2_{\lambda_{C,1}} + \left(\frac{\partial T_e}{\partial \lambda_{C,2}}\right)^2 \sigma^2_{\lambda_{C,2}} + \left(\frac{\partial T_e}{\partial \lambda_{C,3}}\right)^2 \sigma^2_{\lambda_{C,3}} + \left(\frac{\partial T_e}{\partial W_1}\right)^2 \sigma^2_{W_1} + \left(\frac{\partial T_e}{\partial W_2}\right)^2 \sigma^2_{W_2} + \left(\frac{\partial T_e}{\partial W_3}\right)^2 \sigma^2_{W_3} + \sum_{j=1}^{m}\left(\frac{\partial T_e}{\partial \lambda_{0,1,j}}\right)^2 \sigma^2_{\lambda_{0,1,j}} + \sum_{j=1}^{m}\left(\frac{\partial T_e}{\partial \lambda_{0,2,j}}\right)^2 \sigma^2_{\lambda_{0,2,j}} + \sum_{j=1}^{m}\left(\frac{\partial T_e}{\partial \lambda_{0,3,j}}\right)^2 \sigma^2_{\lambda_{0,3,j}} + \sum_{j=1}^{m}\left(\frac{\partial T_e}{\partial \sigma_{1,j}}\right)^2 \sigma^2_{\sigma_{1,j}} + \sum_{j=1}^{m}\left(\frac{\partial T_e}{\partial \sigma_{2,j}}\right)^2 \sigma^2_{\sigma_{2,j}} + \sum_{j=1}^{m}\left(\frac{\partial T_e}{\partial \sigma_{3,j}}\right)^2 \sigma^2_{\sigma_{3,j}} \quad (18)$$

Applying the uncertainties given in table 1 in equation (18), where each term is given in the appendix A, and taking into account that the uncertainty in the signal ratio $\sigma_R$ is about 10%, the evaluated uncertainty in the electron temperature is given in table 2.

**Table 2.** Uncertainties in the electron temperature.

| $P_{S,2}/P_{S,1}$ | $\frac{\sigma_{T_e}}{T_e}$ | $P_{S,3}/P_{S,1}$ | $\frac{\sigma_{T_e}}{T_e}$ | $(P_{S,3}+P_{S,2})/P_{S,1}$ | $\frac{\sigma_{T_e}}{T_e}$ |
|---|---|---|---|---|---|
| $(60 \pm 27) \cdot 10^1 eV$ | 45% | $600 \pm 99 eV$ | 16.5% | $(60 \pm 12) \cdot 10^1 eV$ | 20% |
| $400 \pm 87 eV$ | 22% | $400 \pm 41 eV$ | 10% | $400 \pm 41 eV$ | 10% |
| $200 \pm 21 eV$ | 11% | $200 \pm 11 eV$ | 11% | $200 \pm 10 eV$ | 5% |
| $100 \pm 6 eV$ | 6% | $100 \pm 4 eV$ | 4% | $100 \pm 4 eV$ | 4% |

It is important to note that the uncertainty in the signal ratio $\sigma_R$ that is related to the noise level in the signal depends on plasma density. For low density plasmas $n_e \simeq 1.10^{13} cm^{-3}$, the uncertainty $\sigma_R$ can be of the order of 15%.

From table 2 it can be concluded that for high temperatures the best option to measure electron temperature from the signal ratio is $P_{S,3}/P_{S,1}$, which gives the smallest uncertainty in the electron temperature.

From table 2 it can be also concluded that for low plasma temperatures the uncertainty is lower. However, in TCABR the electron temperature is nowadays measured just in one spatial point at the plasma center, therefore for the plasma edge, where temperature maybe of order of $100eV$, plasma density is lower and consequently the uncertainty in electron temperature will be higher.

## 5. Conclusions

In this work a procedure to estimate the uncertainty in the electron temperature measured by the Thomson scattering diagnostic in TCABR has been presented. The results obtained with an analytic expression for the electron temperature as function of the signal ratio and other optical parameters of the polychromator show good agreement with results obtained directly from the computation of (1) and (2).

From figures 4 and 5 it can be concluded that the best option to evaluate the electron temperature is from the ratio of channel 2 plus channel 3 and channel 1 ($P_{S,3}/P_{S,1}$). However, error analysis shows that the uncertainty for this case is 5% higher.

Analysis of the uncertainties in electron temperature is dominated by uncertainties in parameter $R$, the signal ratio, which depends on noise in each channel. So the uncertainties in the filters and detector parameters have small influence on the electron temperature.

## Acknowledgments

This work was partially supported by the Foundation for Supporting Research of the State of São Paulo (FAPESP projects 2010/50496-4 and 2011/50773-0), and FINEP (project 01.11.0092.00), Brazil.

## Appendix A.

In this appendix each derivative in equation (18) is calculated. Firstly the case $R = P_{S,2}/P_{S,1}$ is calculated, and then the case $R = (P_{S,3} + P_{S,2})/P_{S,1}$ is calculated. For the case $R = P_{S,3}/P_{S,1}$ all indexes 2 must be changed to 3.

**1)** $R = P_{S,2}/P_{S,1}$:

Let

$$\begin{aligned}
\gamma &= \frac{\lambda_I^2 2k_B}{c^2 m_e} \\
\beta &= \left(1 + K_1/\sqrt{K_1^2 - 4K_2K_0}\right) \\
\delta &= \left(\frac{K_0}{K_2\sqrt{K_1^2 - 4K_2K_0}} + \frac{K_1 + \sqrt{K_1^2 - 4K_2K_0}}{2K_2^2}\right) \\
L_{C,n} &= \lambda_{C,n} - \lambda_I
\end{aligned}$$

Then

$$\frac{\partial T_e}{\partial R} = -\gamma\beta\frac{W_1 b_1 W_2 b_2}{RK_2} \tag{1A}$$

$$\frac{\partial T_e}{\partial \lambda_{C,1}} = A_{\lambda_{C,1}} + B_{\lambda_{C,1}} + C_{\lambda_{C,1}} \tag{2A}$$

where:

$$A_{\lambda_{C,1}} = -\gamma\beta\frac{W_1 W_2 b_2}{K_2}\left[\ln\left(\frac{RW_1 b_1}{W_2 b_2}\right)+1\right]\sum_{j=1}^{3}\lambda_{0,1,j}$$

$$B_{\lambda_{C,1}} = -2\gamma^2\delta W_1 a_2\sum_{j=1}^{3}\lambda_{0,1,j}$$

$$C_{\lambda_{C,1}} = \frac{W_1 W_2 b_2}{\sqrt{K_1^2-4K_2K_0}}\left[(L_{C,2}^2-L_{C,1}^2)\sum_{j=1}^{3}\lambda_{0,1,j}-2b_1L_{C,1}\right]$$

$$\frac{\partial T_e}{\partial \lambda_{C,2}} = A_{\lambda_{C,2}}+B_{\lambda_{C,2}}+C_{\lambda_{C,2}} \tag{3A}$$

where:

$$A_{\lambda_{C,2}} = -\gamma\beta\frac{W_1 W_2 b_1}{K_2}\left[\ln\left(\frac{RW_1 b_1}{W_2 b_2}\right)-1\right]\sum_{j=1}^{5}\lambda_{0,2,j}$$

$$B_{\lambda_{C,2}} = 2\gamma^2\delta W_2 a_1\sum_{j=1}^{5}\lambda_{0,2,j}$$

$$C_{\lambda_{C,2}} = \frac{W_1 W_2 b_1}{\sqrt{K_1^2-4K_2K_0}}\left[(L_{C,2}^2-L_{C,1}^2)\sum_{j=1}^{5}\lambda_{0,2,j}+2b_2L_{C,2}\right]$$

$$\frac{\partial T_e}{\partial W_1} = A_{W_1}+B_{W_1}+C_{W_1} \tag{4A}$$

where:

$$A_{W_1} = -\gamma\beta\frac{W_2 b_2}{K_2}\left[\ln\left(\frac{RW_1 b_1}{W_2 b_2}\right)+1\right]\left(b_1+\frac{W_1^2}{12}\right)$$

$$B_{W_1} = 2\gamma^2\delta a_2\left[-b_1-\frac{W_1^2}{12}+\frac{W_2 b_2}{a_2}\right]$$

$$C_{W_1} = \frac{W_2 b_2\left(L_{C,2}^2-L_{C,1}^2\right)}{\sqrt{K_1^2-4K_2K_0}}\left(b_1+\frac{W_1^2}{12}\right)$$

$$\frac{\partial T_e}{\partial W_2} = A_{W_2}+B_{W_2}+C_{W_2} \tag{5A}$$

where:

$$A_{W_2} = -\gamma\beta\frac{W_1 b_1}{K_2}\left[\ln\left(\frac{RW_1 b_1}{W_2 b_2}\right)-1\right]\left(b_2+\frac{W_2^2}{12}\right)$$

$$B_{W_2} = 2\gamma^2\delta a_1\left[b_2+\frac{W_2^2}{12}-\frac{W_1 b_1}{a_1}\right]$$

$$C_{W_2} = \frac{W_1 b_1}{\sqrt{K_1^2-4K_2K_0}}\left[b_2+\frac{W_2^2}{12}\right]$$

$$\frac{\partial T_e}{\partial \lambda_{0,1,j}} = A_{\lambda_{0,1,j}} + B_{\lambda_{0,1,j}} + C_{\lambda_{0,1,j}} \tag{6A}$$

where:

$$\begin{aligned}
A_{\lambda_{0,1,j}} &= -\gamma\beta\frac{W_1 W_2 b_2}{K_2}\left[\ln\left(\frac{RW_1 b_1}{W_2 b_2}\right)+1\right](L_{C,1}-\lambda_{0,1,j})\\
B_{\lambda_{0,1,j}} &= -2\gamma^2\delta\left[b_2 W_2 + a_2 W_1\left(L_{C,1}-\lambda_{0,1,j}\right)\right]\\
C_{\lambda_{0,1,j}} &= \frac{W_1 W_2 b_2\left(L_{C,2}^2-L_{C,1}^2\right)(L_{C,1}-\lambda_{0,1,j})}{\sqrt{K_1^2-4K_2K_0}}
\end{aligned}$$

$$\frac{\partial T_e}{\partial \lambda_{0,2,j}} = A_{\lambda_{0,2,j}} + B_{\lambda_{0,2,j}} + C_{\lambda_{0,2,j}} \tag{7A}$$

where:

$$\begin{aligned}
A_{\lambda_{0,2,j}} &= -\gamma\beta\frac{W_1 W_2 b_1}{K_2}\left[\ln\left(\frac{RW_1 b_1}{W_2 b_2}\right)-1\right](L_{C,2}-\lambda_{0,2,j})\\
B_{\lambda_{0,2,j}} &= 2\gamma^2\delta\left[a_1 W_2\left(L_{C,2}-\lambda_{0,2,j}\right)-W_1 b_1\right]\\
C_{\lambda_{0,2,j}} &= \frac{W_1 W_2 b_1\left(L_{C,2}^2-L_{C,1}^2\right)(L_{C,2}-\lambda_{0,2,j})}{\sqrt{K_1^2-4K_2K_0}}
\end{aligned}$$

$$\frac{\partial T_e}{\partial \sigma_{1,j}} = A_{\sigma_{1,j}} + B_{\sigma_{1,j}} + C_{\sigma_{1,j}} \tag{8A}$$

where:

$$\begin{aligned}
A_{\sigma_{1,j}} &= \gamma\beta\sigma_{1,j}\frac{W_1 W_2 b_2}{K_2}\left[\ln\left(\frac{RW_1 b_1}{W_2 b_2}\right)+1\right]\\
B_{\sigma_{1,j}} &= 2\gamma^2\delta\sigma_{1,j}W_1 a_2\\
C_{\sigma_{1,j}} &= -\frac{\sigma_{1,j}W_1 W_2 b_2\left(L_{C,2}^2-L_{C,1}^2\right)}{\sqrt{K_1^2-4K_2K_0}}
\end{aligned}$$

$$\frac{\partial T_e}{\partial \sigma_{2,j}} = A_{\sigma_{2,j}} + B_{\sigma_{2,j}} + C_{\sigma_{2,j}} \tag{9A}$$

where:

$$A_{\sigma_{2,j}} = \gamma\beta\sigma_{2,j}\frac{W_1 W_2 b_1}{K_2}\left[\ln\left(\frac{RW_1 b_1}{W_2 b_2}\right)-1\right]$$

$$\begin{aligned}
B_{\sigma_{2,j}} &= -2\gamma^2\delta\sigma_{2,j}W_2 a_1\\
C_{\sigma_{2,j}} &= -\frac{\sigma_{2,j}W_1 W_2 b_1\left(L_{C,2}^2-L_{C,1}^2\right)}{\sqrt{K_1^2-4K_2K_0}}
\end{aligned}$$

**2)** $R = (P_{S,3} + P_{S,2})/P_{S,1}$:

$$\frac{\partial T_e}{\partial R} = -\gamma\beta \frac{W_1 b_1 (W_2 b_2 + W_3 b_3)}{R K_2} \tag{10A}$$

$$\frac{\partial T_e}{\partial \lambda_{C,1}} = A_{\lambda_{C,1}} + B_{\lambda_{C,1}} + C_{\lambda_{C,1}} \tag{11A}$$

where:

$$A_{\lambda_{C,1}} = -\gamma\beta \frac{W_1 (W_2 b_2 + W_3 b_3)}{K_2} \left[ \ln\left( \frac{R W_1 b_1}{W_2 b_2 + W_3 b_3} \right) + 1 \right] \sum_{j=1}^{3} \lambda_{0,1,j}$$

$$B_{\lambda_{C,1}} = -2\gamma^2 \delta W_1 (a_2 + a_3) \sum_{j=1}^{3} \lambda_{0,1,j}$$

$$C_{\lambda_{C,1}} = \frac{W_1 (W_2 b_2 + W_3 b_3)}{\sqrt{K_1^2 - 4K_2K_0}} \left[ \left( L_{CM}^2 - L_{C,1}^2 \right) \sum_{j=1}^{3} \lambda_{0,1,j} + 2 b_1 L_{C,1} \right]$$

$$\frac{\partial T_e}{\partial \lambda_{C,2}} = A_{\lambda_{C,2}} + B_{\lambda_{C,2}} + C_{\lambda_{C,2}} \tag{12A}$$

where:

$$A_{\lambda_{C,2}} = -\gamma\beta \frac{W_1 W_2 b_1}{K_2} \left[ \ln\left( \frac{R W_1 b_1}{W_2 b_2 + W_3 b_3} \right) - 1 \right] \sum_{j=1}^{5} \lambda_{0,2,j}$$

$$B_{\lambda_{C,2}} = 2\gamma^2 \delta W_2 a_1 \sum_{j=1}^{5} \lambda_{0,2,j}$$

$$C_{\lambda_{C,2}} = \frac{W_1 W_2 b_1}{\sqrt{K_1^2 - 4K_2K_0}} \left[ \left( L_{CM}^2 - L_{C,1}^2 \right) \sum_{j=1}^{5} \lambda_{0,2,j} + 2 \frac{(W_2 b_2 + W_3 b_3)}{W_2 + W_3} L_{CM} \right]$$

$$\frac{\partial T_e}{\partial \lambda_{C,3}} = A_{\lambda_{C,3}} + B_{\lambda_{C,3}} + C_{\lambda_{C,3}} \tag{13A}$$

where:

$$A_{\lambda_{C,3}} = -\gamma\beta \frac{W_1 W_3 b_1}{K_2} \left[ \ln\left( \frac{R W_1 b_1}{W_2 b_2 + W_3 b_3} \right) - 1 \right] \sum_{j=1}^{7} \lambda_{0,3,j}$$

$$B_{\lambda_{C,3}} = 2\gamma^2 \delta W_3 a_1 \sum_{j=1}^{7} \lambda_{0,3,j}$$

$$C_{\lambda_{C,3}} = \frac{W_1 W_3 a_1}{\sqrt{K_1^2 - 4K_2K_0}} \left[ \left( L_{CM}^2 - L_{C,1}^2 \right) \sum_{j=1}^{7} \lambda_{0,3,j} + 2 \frac{(W_2 b_2 + W_3 b_3)}{W_2 + W_3} L_{CM} \right]$$

$$\frac{\partial T_e}{\partial W_1} = A_{W_1} + B_{W_1} + C_{W_1} \tag{14A}$$

where:

$$\begin{aligned}
A_{W_1} &= -\gamma\beta\frac{(W_2b_2 + W_3b_3)}{K_2}\left[\ln\left(\frac{RW_1b_1}{W_2b_2 + W_3b_3}\right) + 1\right]\left(b_1 + \frac{W_1^2}{12}\right)\\
B_{W_1} &= 2\gamma^2\delta\,(a_2 + a_3)\left[-b_1 - \frac{W_1^2}{12} + \frac{(W_2b_2 + W_3b_3)}{a_2 + a_3}\right]\\
C_{W_1} &= \frac{(W_2b_2 + W_3b_3)\left(L_{CM}^2 - L_{C,1}^2\right)}{\sqrt{K_1^2 - 4K_2K_0}}\left(b_1 + \frac{W_1^2}{12}\right)
\end{aligned}$$

$$\frac{\partial T_e}{\partial W_2} = A_{W_2} + B_{W_2} + C_{W_2} \tag{15A}$$

where:

$$\begin{aligned}
A_{W_2} &= -\gamma\beta\frac{W_1b_1}{K_2}\left[\ln\left(\frac{RW_1b_1}{W_2b_2 + W_3b_3}\right) - 1\right]\left(b_2 + \frac{W_2^2}{12}\right)\\
B_{W_2} &= 2\gamma^2\delta a_1\left[b_2 + \frac{W_2^2}{12} - \frac{W_1b_1}{a_1}\right]\\
C_{W_2} &= \frac{W_1b_1\left(L_{CM}^2 - L_{C,1}^2\right)}{\sqrt{K_1^2 - 4K_2K_0}}\left[b_2 + \frac{W_2^2}{12} + 2W_3L_{CM}\frac{(W_2b_2 + W_3b_3)\left(\lambda_{C,2} - \lambda_{C,3}\right)}{\left(L_{CM}^2 - L_{C,1}^2\right)(W_2 + W_3)^2}\right]
\end{aligned}$$

$$\frac{\partial T_e}{\partial W_3} = A_{W_3} + B_{W_3} + C_{W_3} \tag{16A}$$

where:

$$A_{W_3} = -\gamma\beta\frac{W_1b_1}{K_2}\left[\ln\left(\frac{RW_1b_1}{W_2b_2 + W_3b_3}\right) - 1\right]\left(b_3 + \frac{W_3^2}{12}\right)$$

$$\begin{aligned}
B_{W_3} &= 2\gamma^2\delta a_1\left[b_3 + \frac{W_3^2}{12} - \frac{W_1b_1}{a_1}\right]\\
C_{W_3} &= \frac{W_1b_1\left(L_{CM}^2 - L_{C,1}^2\right)}{\sqrt{K_1^2 - 4K_2K_0}}\left[b_3 + \frac{W_3^2}{12} + 2W_2L_{CM}\frac{(W_2b_2 + W_3b_3)\left(\lambda_{C,3} - \lambda_{C,2}\right)}{\left(L_{CM}^2 - L_{C,1}^2\right)(W_2 + W_3)^2}\right]
\end{aligned}$$

$$\frac{\partial T_e}{\partial \lambda_{0,1,j}} = A_{\lambda_{0,1,j}} + B_{\lambda_{0,1,j}} + C_{\lambda_{0,1,j}} \tag{17A}$$

where:

$$\begin{aligned}
A_{\lambda_{0,1,j}} &= -\gamma\beta\frac{W_1\,(W_2b_2 + W_3b_3)}{K_2}\left[\ln\left(\frac{RW_1b_1}{W_2b_2 + W_3b_3}\right) + 1\right]\left(L_{C,1} - \lambda_{0,1,j}\right)\\
B_{\lambda_{0,1,j}} &= -2\gamma^2\delta\left[W_1\,(a_2 + a_3)\left(L_{C,1} - \lambda_{0,1,j}\right) + (W_2b_2 + W_3b_3)\right]\\
C_{\lambda_{0,1,j}} &= \frac{W_1\,(W_2b_2 + W_3b_3)\left(L_{CM}^2 - L_{C,1}^2\right)\left(L_{C,1} - \lambda_{0,1,j}\right)}{\sqrt{K_1^2 - 4K_2K_0}}
\end{aligned}$$

$$\frac{\partial T_e}{\partial \lambda_{0,2,j}} = A_{\lambda_{0,2,j}} + B_{\lambda_{0,2,j}} + C_{\lambda_{0,2,j}} \tag{18A}$$

where:

$$\begin{aligned}
A_{\lambda_{0,2,j}} &= -\gamma\beta\frac{W_1W_2b_1}{K_2}\left[\ln\left(\frac{RW_1b_1}{W_2b_2+W_3b_3}\right)-1\right](L_{C,2}-\lambda_{0,2,j})\\
B_{\lambda_{0,2,j}} &= 2\gamma^2\delta\left[a_1W_2\left(L_{C,2}-\lambda_{0,2,j}\right)+W_1b_1\right]\\
C_{\lambda_{0,2,j}} &= \frac{W_1W_2b_1\left(L_{CM}^2-L_{C,1}^2\right)\left(L_{C,2}-\lambda_{0,2,j}\right)}{\sqrt{K_1^2-4K_2K_0}}
\end{aligned}$$

$$\frac{\partial T_e}{\partial \lambda_{0,3,j}} = A_{\lambda_{0,3,j}} + B_{\lambda_{0,3,j}} + C_{\lambda_{0,3,j}} \tag{19A}$$

where:

$$\begin{aligned}
A_{\lambda_{0,3,j}} &= -\gamma\beta\frac{W_1W_3b_1}{K_2}\left[\ln\left(\frac{RW_1b_1}{W_2b_2+W_3b_3}\right)-1\right](L_{C,3}-\lambda_{0,3,j})\\
B_{\lambda_{0,3,j}} &= 2\gamma^2\delta\left[a_1W_3\left(L_{C,3}-\lambda_{0,3,j}\right)+W_1b_1\right]\\
C_{\lambda_{0,3,j}} &= \frac{W_1W_3b_1\left(L_{CM}^2-L_{C,1}^2\right)\left(L_{C,3}-\lambda_{0,3,j}\right)}{\sqrt{K_1^2-4K_2K_0}}
\end{aligned}$$

$$\frac{\partial T_e}{\partial \sigma_{1,j}} = A_{\sigma_{1,j}} + B_{\sigma_{1,j}} + C_{\sigma_{1,j}} \tag{20A}$$

where:

$$\begin{aligned}
A_{\sigma_{1,j}} &= \gamma\beta\sigma_{1,j}W_1\frac{(W_2b_2+W_3b_3)}{K_2}\left[\ln\left(\frac{RW_1b_1}{W_2b_2+W_3b_3}\right)+1\right]\\
B_{\sigma_{1,j}} &= 2\gamma^2\delta\sigma_{1,j}\left(a_2+a_3\right)\\
C_{\sigma_{1,j}} &= -\frac{\sigma_{1,j}W_1\left(W_2b_2+W_3b_3\right)\left(L_{CM}^2-L_{C,1}^2\right)}{\sqrt{K_1^2-4K_2K_0}}
\end{aligned}$$

$$\frac{\partial T_e}{\partial \sigma_{2,j}} = A_{\sigma_{2,j}} + B_{\sigma_{2,j}} + C_{\sigma_{2,j}} \tag{21A}$$

where:

$$\begin{aligned}
A_{\sigma_{2,j}} &= \gamma\beta\sigma_{2,j}\frac{W_1W_2b_1}{K_2}\left[\ln\left(\frac{RW_1b_1}{W_2b_2+W_3b_3}\right)-1\right]\\
B_{\sigma_{2,j}} &= -2\gamma^2\delta\sigma_{2,j}W_2a_1\\
C_{\sigma_{2,j}} &= -\frac{\sigma_{2,j}W_1W_2b_1\left(L_{CM}^2-L_{C,1}^2\right)}{\sqrt{K_1^2-4K_2K_0}}
\end{aligned}$$

$$\frac{\partial T_e}{\partial \sigma_{3,j}} = A_{\sigma_{3,j}} + B_{\sigma_{3,j}} + C_{\sigma_{3,j}} \tag{22A}$$

where:

$$A_{\sigma 3,j} = \gamma\beta\sigma_{3,j}\frac{W_1W_3b_1}{K_2}\left[\ln\left(\frac{RW_1b_1}{W_2b_2+W_3b_3}\right)-1\right]$$
$$B_{\sigma 3,j} = -2\gamma^2\delta\sigma_{3,j}W_3a_1$$
$$C_{\sigma 3,j} = -\frac{\sigma_{3,j}W_1W_3b_1\left(L_{CM}^2-L_{C,1}^2\right)}{\sqrt{K_1^2-4K_2K_0}}$$

## References

[1] Ruchko L F et al 2002 *Braz. J. Phys.* **32(1)** 57-64
[2] Nascimento I C et al 2005 *Nucl. Fusion* **45** 796
[3] Nascimento I C et al 2007 *Nucl. Fusion***47** 1570
[4] Severo J H F Tsypin V S, Galvão R M O, Nascimento I C, Tendler M and Fagundes A N 2002 *Braz. J. Phys* **32(1)** 13-9
[5] Severo J H F, Nascimento I C, Tsypin V S and Galvão R M O 2003 *Nucl. Fusion* **43** 1047-1056
[6] Severo J H F et al 2009 *Nucl. Fusion* **49** 115026
[7] Severo J H F, Nascimento I C, Tsypin V S, Kuznetov Yu K, Saettone E A, Vannucci A, Galvão R M O, Tendler M and Mikhailovskii A B 2004 *Phys. Plasma* **11(2)** 846-848
[8] Alonso M P, Figueiredo A C A, Borges F O, Elizondo J I, Galvão R M O, Severo J H F, Usuriaga O C, Berni L A and Machida M 2010 *Rev. Sci. Inst.* **81** 10D529
[9] Severo J H F, Nascimento I C, Kuznetov Yu K, Tsypin V S, Galvão R M O and Tendler M 2007 *Rev. Sci Inst.* **78** 043509
[10] Alonso M P et al 2010 *J. Phys.: Conf. Series* **227** 012027
[11] Forrest, M J, Peacock N J, Robinson D C, Sannikov V V and Wilcock P D 1970 *Culham Plasma Physics Laboratory Report CLM-R 107* (Abingdon)
[12] Anashin A M, Gorbunov E P, Ivanov D P, Lysenko S E, Peacock N J, Robinson D C, Sannikov V V and Strelkov V S 1971 *Sov. Phys. - JETP* **33** 1127
[13] Rautian S G 1958 *Sov. Phys. Uspekhi* **66(1)** 245
[14] Alonso M P, Figueiredo A C A, Berni L A, Varandas C A F 2008 *IEEE Trans. Plasma Sci.* **36(4)** 1094
[15] Sheffield J 1975 *Plasma Scattering of Electromagnetic Radiation* (New York: Academic press)

Journal of Physics: Conference Series **370** (2012) 012046 doi:10.1088/1742-6596/370/1/012046

# Maximization of hydrogen incorporation on a-Si:H using optical emission spectroscopy

**T D O Moura[1,2], D O Miranda[2,3], R J Santana[1,2], R T Proença[1] and J R T Branco[1]**

[1]CETEC-Technological Center of Minas Gerais Foundation,
José Candido da Silveira Avenue, 2000, Horto, Belo Horizonte-MG, Brazil
[2]REDEMAT. Tiradentes square, 20, Downtown, Ouro Preto-MG, Brazil
[3]IFMG – Federal Institute of Education, science and technology of Minas Gerais, Michael Pereira de Souza Avenue, 3007 - Campinho - Congonhas – MG, Brazil

E-mail: thiagodaniel_moura@hotmail.com

**Abstract**. The use of hydrogenated amorphous silicon, a-Si:H, for improving heterojunction solar cell efficiency and decreasing cost has found increasing interest in recent years. Hydrogen incorporation in thin films of amorphous silicon is crucial for obtaining semiconductor quality because it passivates Si dangling bonds in the structure. In the present work the plasma optical emission during film growth by DC magnetron sputtering in Ar-H plasma was used to maximize hydrogen incorporation the a-Si and the effects of hydrogenation were evaluated, as part of a systematic investigation of different vapor a-Si:H deposition processes, physical and chemical. The experiments were performed using the Magnetron Sputtering technique varying the DC power and Ar-H gas rates. The grown films were characterized by Raman scattering, FT-IR spectroscopy and AFM. The results show that maximizing the plasma emission from Si-H species enhances hydrogen incorporation in amorphous films. The results also show the possibility of automatically controlling plasma composition for optimum hydrogen incorporation through optical emission spectroscopy (OES).

## 1. Introduction

In recent decades energy generation through the use of solar power has been becoming a significant and competitive supplier to the world's electricity market. The generation of energy from solar radiation is becoming a mature technology, but its high cost limits its widespread use. Efforts have been made to make the technology more accessible, especially in developing countries like Brazil, and also to get more efficient devices. One example is a heterostructure based on the junction of an amorphous silicon layer grown on a crystalline silicon with an intrinsic a-Si:H layer between the p-n layers [1]. This kind of solar cells has achieved efficiency above 20% [2,3].

The use of amorphous silicon (a-Si) as an emitter layer in photovoltaic heterojunctions solar cells reduces production costs because the p-n junction can be produced at lower temperatures compared to crystalline silicon (c-Si) [3].

In pure a-Si, there is a large concentration of defects, in the order of $10^{21}$ per $cm^3$, called dangling bonds. One way to reduce these defects is to passivate them with hydrogen which decrease the defect density from about $10^{21}cm^{-3}$ to $10^{15}$ - $10^{16}cm^{-3}$. In a-Si:H this represents less than one dangling bond per million of silicon atoms [4-6]. The understanding of the incorporation and stability of hydrogen in a-Si:H has been the subject of extensive research [3-6].

In this work optical emission spectroscopy has been used to study the behavior of active plasma species during Si sputtering in a Ar + $H_2$ atmosphere, to better understand the corresponding hydrogen incorporation in the a-Si:H thin films, that happens in such a process.

## 2. Experiment

The experiments were performed in a Balzers Magnetron Sputtering reactor BAS 450M (figure 1) with a 150W DC power supply, under different levels of gas Ar/$H_2$ flow, ranging from 100% Ar and 20% Ar/$H_2$. The substrate temperature was constant, at 100°C, while the base pressure of the chamber and the working pressure were set at $9.0\times10^{-7}$ mbar and approximately $10^{-4}$mbar, respectively. The films were deposited over glass and over electronic grade p-type Silicon substrates (100).

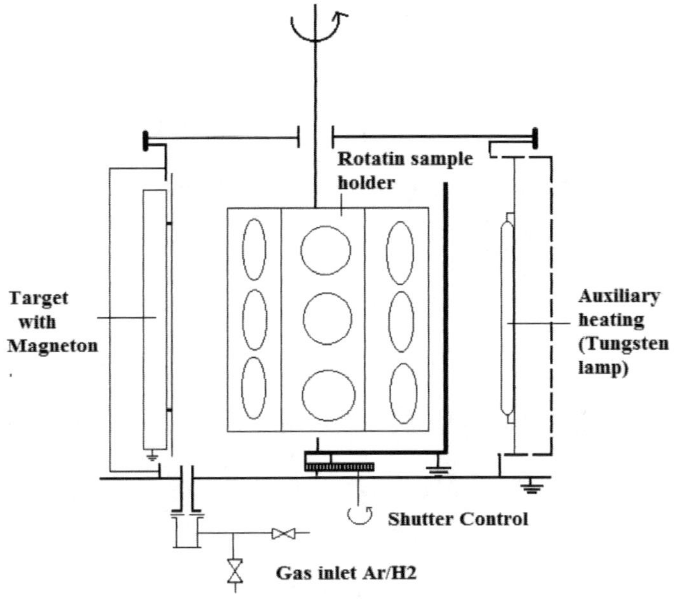


**Figure 1.** Scheme of the magnetron BAS chamber.

### *2.1. Optical emission analysis*

The plasma optical emission analysis was carried out using a EP200Mmd from Verity Instruments, with a diffraction grating for spectral range between 200 and 900nm. The plasma emission under sputtering power of 150, 300 and 450W was monitored.

The observation of optical emission was concentrated on getting relevant information on the process of hydrogenation of amorphous silicon films, by means of the following transitions: ArI 4p – 4s, H$\alpha$ (n=3), H$\beta$ (n=4) and SiH $A^2\Delta - X^2\Pi$.

### *2.2. Thin film analysis*

The surface morphology of the sample was investigated by Atomic Force Microscopy (AFM) using a MultiMode microscope with a Nanoscope II controller. It was equipped with a SiN tip and operated in tapping mode.

Raman scattering experiments were performed in the back-scattering geometry configuration, using the 632.8 nm line of an He–Ne laser at room temperature.

In order to calculate the hydrogen percentage in the thin film we used the infrared spectroscopy technique, through which it is possible to determine the percentage of hydrogen bonded in the a-Si:H structure. The process consists of calculating the area between the stretching modes for SiHx (2000 – 2090 $cm^{-1}$) [7]:

$$S = \int \frac{\alpha(\omega)}{\omega} d\omega \tag{1}$$

Where the absorption coefficient $\alpha$ is obtained using the intensity ratio for a given thickness $d$:

$$\alpha = \frac{1}{d} \ln\left(\frac{I_0}{I}\right) \tag{2}$$

Then the sample density in the substrate is:

$$N_x = A_w \cdot S \tag{3}$$

Where: $N_x$ is the film density, and $A_\omega$ is a proportionality constant. Using these expressions we can calculate the hydrogen percentage formula by means of [7]:

$$C_x = \frac{N_x}{N_S + N_x} \tag{4}$$

Where $c_x$ is the percentage of the material in the film and $N_s$ is the substrate density.

The microstructure factor defined as the ratio between the intensity related to the stretching mode of monohydrits ($I_{2090}$) with respect to the sum of monohydrits and polyhydrits ($I_{2000}$ + $I_{2090}$) is given by the equation 5 [8].

$$R_{2090} = \frac{I_{2090}}{I_{2000} + I_{2090}} \tag{5}$$

The hydrogen incorporation at the film and the microstructure factor were performed using FTIR spectroscopy

## 3. Experimental results

### *3.1. The light emission from the plasma*

We carried out monitoring of the main peaks in order to infer about the process of hydrogenation of amorphous silicon films, namely the peaks ArI 4p – 4s, Hα (n=3), Hβ (n=4) and SiH $A^2\Delta - X^2\Pi$. Figure 2 shows the evolution of these peaks with the variation of both, the Ar percentage and the power.

The effect of the power source in the behavior of the optical emission lines was observed for the case in which both gases vary. Figure 3 shows the changes in intensity for different gas percentages and power source used in the system. The relevant effect of power source on the intensity can be also observed.

For each power source the behavior of the curves are similar. Intensity variation is greater for the greater powers source as might be expected from the analysis of the graphs in figure 3. Only the behavior of the curve for SiH has a small difference from the others. It is noted that below 50% of Ar the variation of the Power source of 300W becomes greater than for a power of 450W and the variation the 150W curve is much lower compared to the others. These are important indicators for the production of films of a-Si:H in which the points of greatest emission intensity of SiH line lead to higher rates of H bonded to the structure of amorphous Si, which consequently passivate more dangling bonds [9].

Figure 3 shows the rate of variation of the optical emission intensity for the data presented in figure 2. The idea of these graphs is to analyze the "speed" with which the emission intensity varies for each percentage of gas and energy. It will be important to support the analysis of the power effect on the growth of the films.

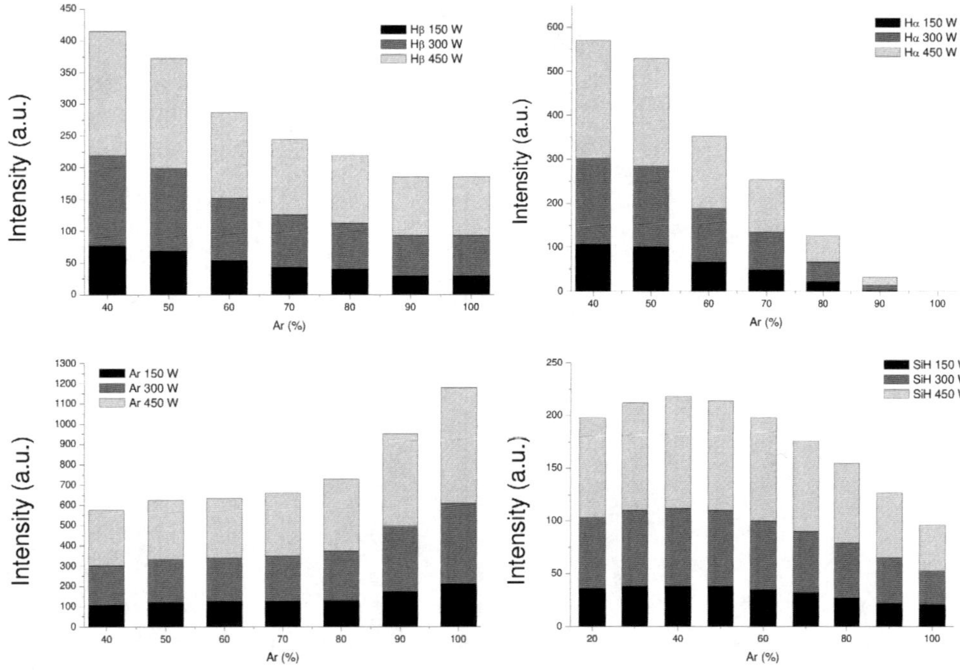


**Figure 2.** Effect of power on the intensity of the Hα, Hβ, Ar e SiH peaks.

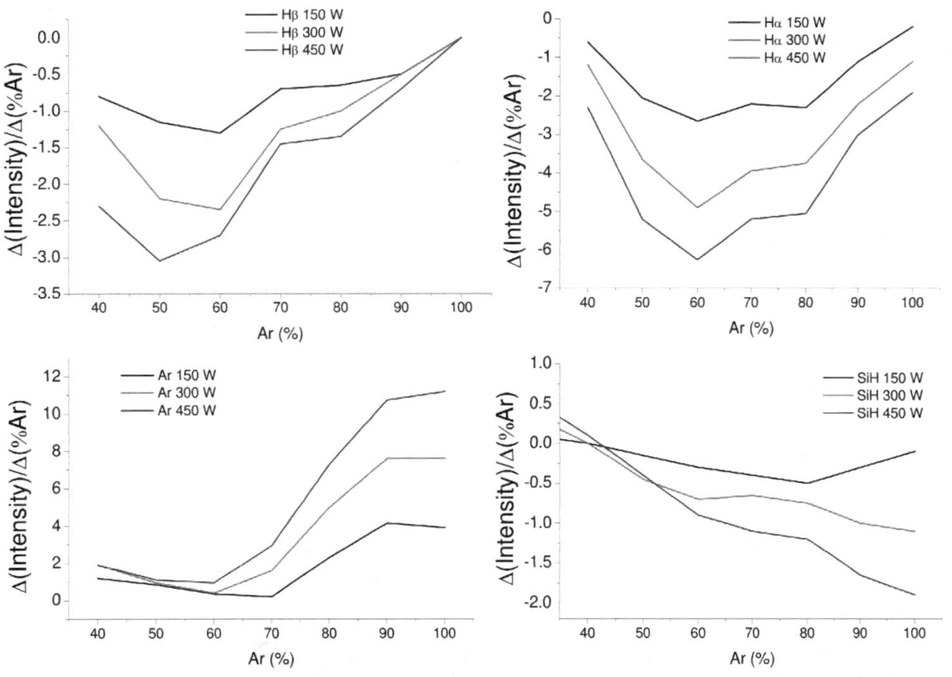


**Figure 3.** Rate of variation of the optical emission intensity with the %Ar.

Journal of Physics: Conference Series **370** (2012) 012046 doi:10.1088/1742-6596/370/1/012046

The pattern of the SiH line emission, increasing the intensity reaching a maximum and decreasing changing the Ar-H ratio determined the range of gases to grow the films as shown in figure 4 for the power of 150W.

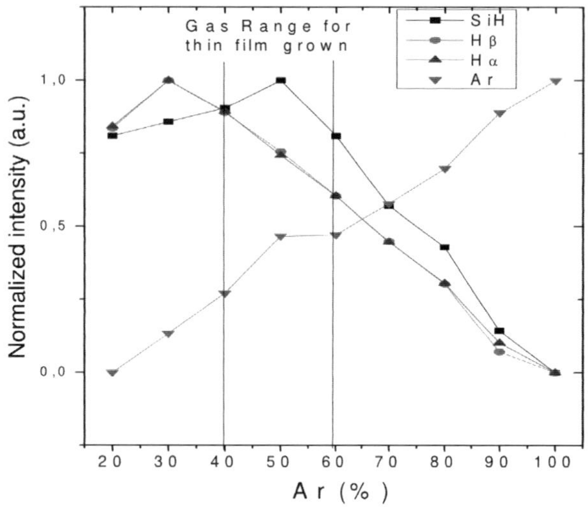


**Figure 4.** Evolution of the normalized optical emission lines with the Ar-$H_2$ concentration in the process for 150W.

*3.2. Thin film properties*

The following analysis are related to properties of the films grown at the highlighted region on figure 4. The surface morphology and the root-mean-square roughness (RMS) were obtained from AFM images. The AFM images showed a rough (film) and compact thin films. Figure 5 shows the topography images of the samples.

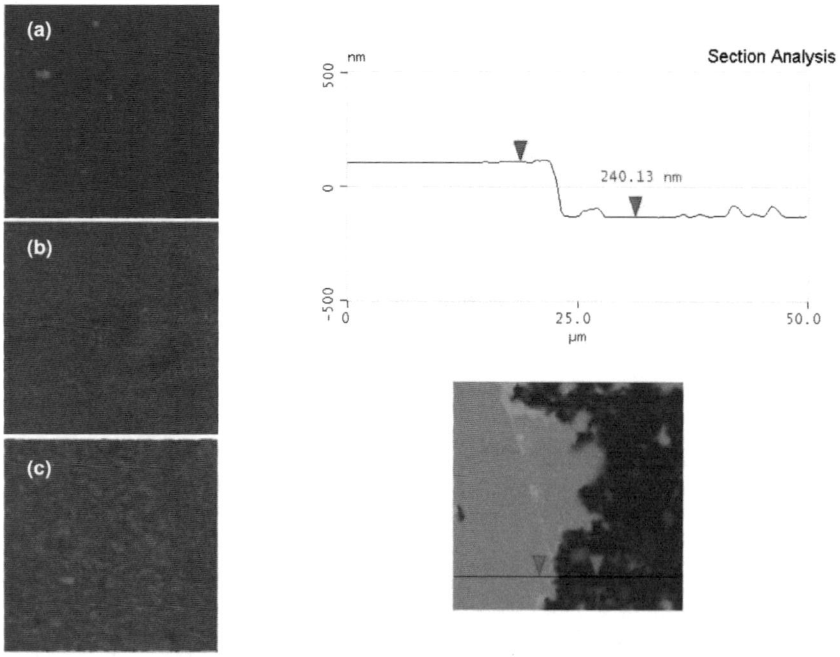


**Figure 5.** Topography image for: (a) 40%, (b) 50%, (c) 60% of $H_2$ content and the section analysis of a step to get the thickness.

The RMS of the sample with higher flow (60%) was approximately 0.5 nm and the thickness of the film was 220 nm, for a sample with 50% the RMS was 0.3 nm and the step was 73.5 nm, and for a sample with 40% 0.4 nm to RMS and 72 nm thickness. The increasing of the rugosity corresponds to the high flux of the $H_2$. A possible meaning for it should be that when the hydrogen partial pressure increases, it induces crystallization as previously showed [10].

Figure 7 shows the Raman shift of the samples (a) and the de-convolution of the vibrational modes (b). The pattern showed is due the TA, LA, LO and TO1 vibrations mode related to a-Si phase.

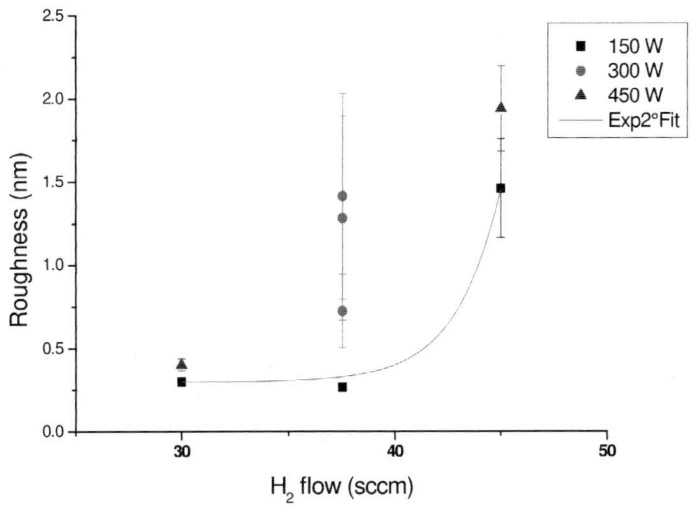


**Figure 6.** Roughness as a function of the flow for different powers.

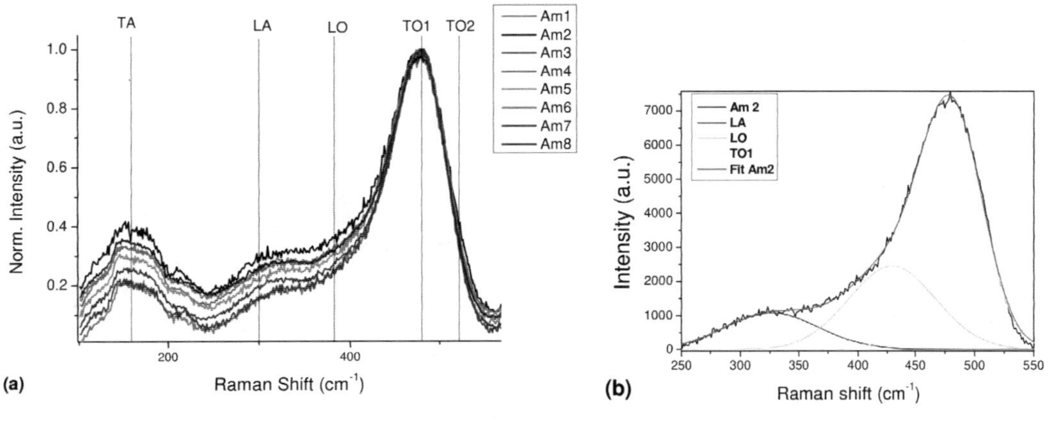


**Figure 7.** Raman shift for all samples (a) and the de-convolution for the sample 2.

The FTIR analysis is presented on figure 8 with the complete spectrum (a), the range related to the Si-H stretching modes (b) and the de-convolution of the samples grown at 150W (c). From the figure 8 (b) one can see the centre of the peaks varying from 2000 to 2100 cm$^{-1}$, and (c) the amplitude de-convoluted for each sample is changing. This gives rise to the microstructural factor ($R_{2090}$) and is related to the polyhydrides (Si – $H_x$) and monohydrides (Si – H) presented on the structure of the a-Si.

Figure 9 presents the hydrogen incorporation at the films for the different parameters. As we can see the pattern of the hydrogen at the film e similar to the pattern of the optical emission line related to the SiH emitting at the plasma. One can see a high level of incorporation of hydrogen 20 e 30%.

Figure 10 shows the microstructure factor at the studied range of atmosphere plasma composition. This graph means that highest $R_{2090}$, more polyhydrides (Si – $H_x$) is present at the structure of the film. Which is worst for good electrical properties in the film.

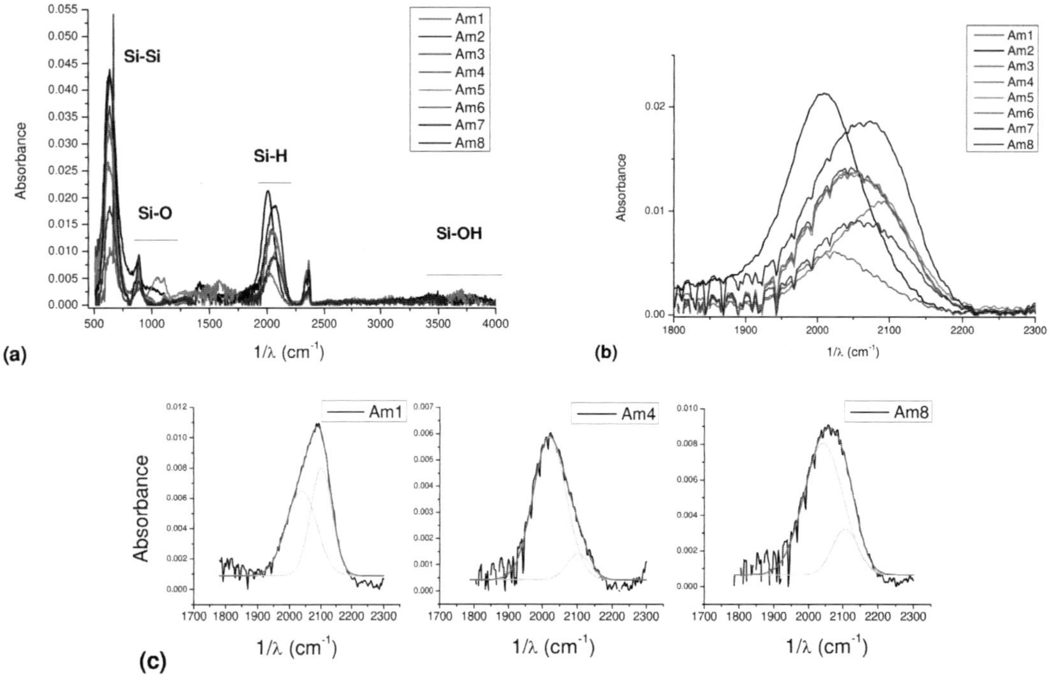


**Figure 8.** (a) Absorbance spectrum as a function of the wavenumber (b) Si-H / Si – $H_x$ stretching modes (c) de-convolution of the samples grown at 150W.

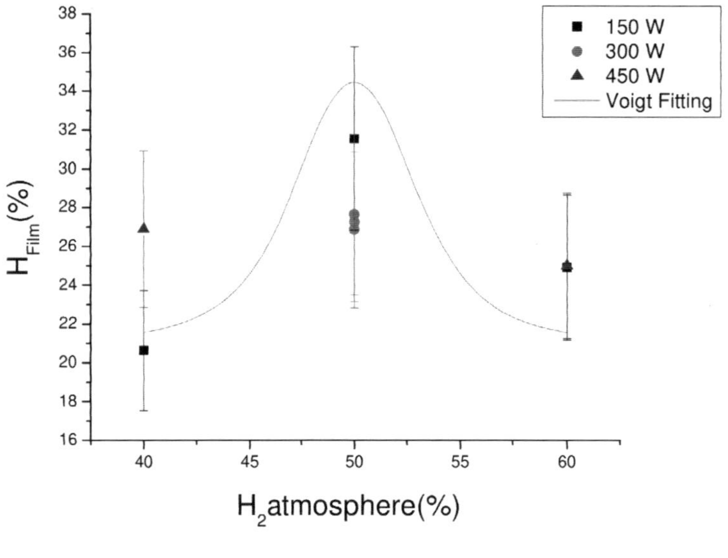


**Figure 9.** H% incorporated on the a-Si:H thin films as a function of the $H_2$ at the deposition atmosphere and the power.

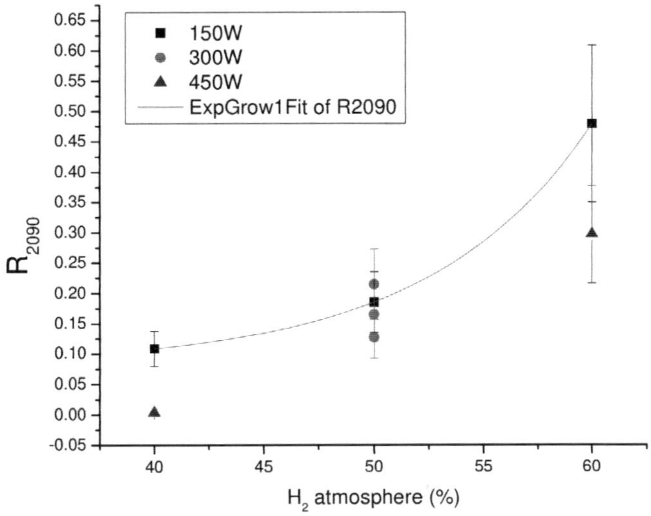


**Figure 10.** Microstructure factor ($R_{2090}$) of the films.

### *3.3. Discussions*

The pattern of the SiH line emission at the plasma was repeated in the pattern of incorporation of hydrogen in the film grown in the same range of the partial pressure of the gases. Comparing the figures 6, 9 and 10 one can see that at higher partial pressure of hydrogen more polyhydrites incorporate to the film, but the total hydrogen at the structure of the films gets a maximum for 50%. In other words, the excess of hydrogen that increase the deposition rate, damage the topology of the film and increase the microstructural factor. If the partial pressure of hydrogen is increased one can induce the crystallization of the film, as previously showed [10]. This can indicate some aspect related to the growth mechanism of the a-Si:H thin film. As showed by Abdesselem et al. [11], at low temperature the growth mechanism is governed by the incorporation of SiHx (x=0-1) coming from the plasma, but the ratio between mono-poly hydrides is very important to high performance. In this way the role of the excess of Hydrogen at the plasma is very important and the optical emission coming from the plasma helped to found it.

On the graph of incorporation of hydrogen in the film (figure 9), note that at the highest partial pressure of Hydrogen (60%), there is no difference at the amount incorporated for the extreme powers of 450 and 150. The power of 300W at 50% of the gases showed slight difference. Comparing with Figure 3, where one can see a higher decay rate for the powers of 300 and 450W, we can associate the more abrupt decrease in the light emission of species of interest (Si-H) with the reduction of incorporation of the same.

## 4. Conclusions

A sputtering process that allows high H content in the a-Si:H has been done. The study showed that monitoring of plasma species is effective to predict the pattern of hydrogenation of a-Si. It is also shown that maximum hydrogenation is mainly a function of the H content in the atmosphere and there is an optimum level of H dilution in the H-Ar plasma that corresponds to a maximum in the SiH species plasma emission. It is also shown that the H content corresponding to this optimum dilution, as well as the level of silicon hydrogenation, increases with sputtering power for low partial pressure of H but it does not change significantly for high partial pressure of Hydrogen. It is shown that the excess of hydrogen significantly affects the topography of the film and increases the microstructural factor. Finally it is shown that the effects of changing process variables on the hydrogenation can be monitored by the optical emission from plasma, which makes this technique a potential tool to control the production of a-Si:H.

## References

[1] Page M R, Iwaniczko E, Xu Q Y, Roybal L, Hasoon F, Wang Q and Crandall R S 2011 *Thin Solid Films* **519** 4527-30

[2] Mishima T, Taguchi M, Sakata H and Maruyama E 2011 *Solar Energy Materials & Solar Cells* **95** 18–21

[3] Tanaka M, Okamoto S, Tsuge S and Kiyama S 2003 *Proc. Photovoltaic Energy Conversion* (Osaka-Japan) Vol.1 955-8

[4] Stradins P 2010 *Proc. 35th IEEE Photovoltaic Specialists Conference* (Honolulu, Hawaii), (35th IEEE PVSC) p 142-5

[5] Burrows M Z, Das U K, Opila R L, De Wolf S and Birkmire R W 2008 *J. Vac. Sci. Technol.* A **26** 683

[6] Zeman M 2006 *Thin Film Solar Cells Fabrication, Characterization and Applications* vol 1, ed Poortmans J and Arkhipov V (Chichester: John Wiley & Sons Ltd)

[7] Saleh R, Munisa L and Beyer W 2002 *Phys. J. IPS* **A5** 0210

[8] Ben Abdelmoumen A, Cherfi R, Kechoune M and Aoucher M 2008 *Thin Solid Films* **517** 369–71

[9] Matsuda A, Nakagawa K, Tanaka K and Matsumura M 1980 *J. of Non-Crystal. Solids.* **35-36** 183-8

[10] Miranda D O, Moura T D O, Santana R J, Guimarães G R, Karger E R S, Diniz A S A C and Branco J R T 2010, *Proc. 35th IEEE Photovoltaic Specialists Conference* (Honolulu, Hawaii), (35th IEEE PVSC) p 003706–8

[11] Abdesselem S, Aida M S, Attaf N and Ouahab A 2006 *Physica B: Physics of Condensed Matter* **373** 33-41

# Diagnostics of parameters by optical emission spectroscopy and Langmuir probe in a discharge of Ar/$N_2$/$CH_4$ ternary mixture

**A Guerrero[1,2], L Salazar-Flores[1], C Torres-Segundo[2], H Martínez[1], P G Reyes[2] and F Castillo[3]**

[1]Instituto de Ciencias Físicas, Universidad Nacional Autónoma de México, Apartado postal 48-3, 62251, Cuernavaca, Morelos, México
[2]Facultad de Ciencias, Universidad Autónoma del Estado de México Instituto Literario 100, Centro, 50000, Toluca, Estado de México, México
[3]Instituto de Ciencias Nucleares, Universidad Nacional Autónoma de México Ciudad Universitaria 1000, Coyoacan, D.F., México

E-mail: hm@fis.unam.mx

**Abstract**. A low-pressure DC plasma discharge sustained in a 85.0%$N_2$-12.0%Ar- 3.0%$CH_4$, 82.3%$N_2$-13.7%Ar- 4.0%$CH_4$ and 79.6%$N_2$-15.4%Ar- 5%$CH_4$ ternary mixtures is studied. These plasmas were generated in the total pressure range of 1.0 to 10.0 Torr, power of 7 W and 50 l/min flow rate gas. The diagnostic has been made by optical emission spectroscopy (OES) using a spectrometer. The species observed were: $N_2$; $N_2^+$; $CH^+$; CN; $C_2$; H and Ar. We present the behavior of the bands and lines intensities as a function of the pressure and concentration of Ar, $N_2$ and $CH_4$ gases. Also, we display more quantitatively the ratios of intensities of $N^+_2$ (391.44 nm), CN (392.08 nm), and H (486.13 nm) to that of the $N_2$ (337.13 nm) as function of pressure. The ratios show an almost constant behavior as a function of the pressure. Being the CH/$N_2$ ratio more highest and H/$N_2$ ratio the lowest one.

## 1. Introduction

The $CH_4$/$N_2$/Ar mixtures represent a model atmosphere of Titan, the largest satellite of Saturn. Unlike other moons in the solar system, Titan has a substantial atmosphere and offers an interesting comparison with Earth and other planets. Experimental studies of electrical discharges [1] has been using methane and nitrogen (the two major constituents of the Titan atmosphere) for decades since such discharges appear to be a good methodology for simulating the chemistry of Titan's atmosphere. Moreover, Titan is a test of our understanding of the organic chemistry of planetary atmospheres, since its atmospheric bulk composition is ($N_2$:$CH_4$:Ar). Recently, we have been able to see the emergence of papers devoted to the formation of particles in various gas discharges such as hydrocarbon plasmas [2], also to the study of methane decomposition in a nitrogen plasma due to a large number of applications in various domains such as the simulation of Titan's atmosphere [3], in deposition of diamond-like carbon layers [4], in optical and electrical characterization of an atmospheric pressure microplasma jet [5], in the conversion of natural gas to $C_2$ hydrocarbons [6], in the properties and reactions of the CN radical [7] and in metal nitriding [8]. One of the most interesting questions about Titan's atmosphere is the influence of electrical activity on the organic chemistry of its nitrogen and methane rich atmosphere. On the other hand, cold plasma offer a milder environment because it is only the reactive species, ions and electrons located in the ionization zone, which are subjected to high temperatures while the neutral molecules remain far from the ionization zone, but ready to interact with those species anyway. So, the goal of this work, it is to explore the optical emission spectroscopy (OES) in a low-pressure plasma discharge of $Ar/N_2/CH_4$ ternary mixture. Also, the dependence of the principal species observed on gas pressure are presented and discussed.

## 2. Experimental setup

The experimental apparatus and technique to measure the DC plasma emission spectrum is identical to that described in previous work [9]. A brief description is reported here. The system consists of two stainless steel circular plane electrodes, 3 mm thick and 30 mm in diameter. The electrodes are positioned at the center of the reaction chamber with 5 mm gap spacing. The gas was injected into the reaction chamber through the bottom flange. The same gas connection was used for the pressure sensor (MKS, Type 947 convection enhanced Pirani). The DC plasma was produced in a concentration of 82.3%$N_2$-13.7%Ar- 4.0%$CH_4$ ternary mixture at pressures between 1.0 and 10.0 Torr. Ultra pure $CH_4$, $N_2$ and Ar gases (Praxair 99.99%) were used in the course of the measurements. A continuous dynamic flow of $N_2$-Ar-$CH_4$ mixtures were let in the system through Matheson flowmeter model FM1000. The flow rate of gases was 50 l/min. The discharge power supply was maintained at an output of 280 Volts and a current of 25 mA (7 W). The base pressure of the plasma chamber of $3.39\times10^{-3}$ $m^3$ volume was $2.0\times10^{-2}$ Torr, achieved using a mechanical pump. The vacuum chamber was purged with the working gas several times at a pressure of 1.0 Torr in order to remove the background gas. In order to monitor the plasma discharge, in front of the lateral plane of the electrodes was located an optical fiber (Solarization-resistant UV and fiber diameter size of 400 μm) that was connected to the entrance aperture of a high-resolution Ocean Optics Inc. Spectrometer Model HR2000CG-UV-NIR (101.6 mm focal length, 5 μm inlet and outlet slits, and 300 lines-$mm^{-1}$) in the spectral range 200-1100 nm at a resolution of ~0.35 nm (an OFLV-200-1100 order-sorting filter was installed to eliminate second- and third-order effects) with a Sony ILX511B linear silicon CCD array (2048 individual pixels with 14 μm × 200 μm of pixel size and sensitivity of 75 photons/count at 400 nm). The wavelength accuracy of the spectrometer was calibrated using an Ar Source (Ocean Optics Inc.). The spectra data were obtained with a 9 s integration time. Spectral response of the CCD spectrometer has been measured using a reference light source (deuterium lamp for UV, and Ocean Optics LS-1-CAL for visible and IR). All spectra as well as particular line intensities are corrected to the spectral response of the CCD spectrometer.

## 3. Results and discussion

To investigate the plasma properties, the plasma emission was measured with a CCD-array spectrometer (Ocean Optics) with an optical fiber though the Pyrex glass chamber. The plasma-ions flux reaching the cathode can be evaluated from $J = I_D/S$, where $I_D$ is the current discharge (25 mA) and $S$ is the electrode area (7.07 $cm^2$). So, the plasma- ions flux is $\sim1.32\times10^{18}$ $cm^{-2}$ $min^{-1}$. A typical OES measurement of mixture glow discharge plasma at a pressure of 5.0 Torr is displayed in figure 1, at 7 W of power. This allowed analysis of the most luminous area, which corresponds to the negative glow near the cathode dark space. Only the most intense spectral lines and bands of the plasma within the 200-1000 range are quoted [10]. The principal species observed were: at 315.93, 337.13, 375.54, 399.84 and 405.94 nm for $N_2$ ($C^3\ \Pi_u$-$B^3\Pi_g$); at 354.89, 358.21, 376.16, 391.44 and 427.81 nm for $N_2^+$($B^2\Sigma_u^+$-$X^2\Sigma_g^+$); at 381.06 nm for $CH^+$($A^1\Pi$-$X^1\Sigma$); at 392.08 nm for CN($B^2\Sigma$-$X^2\Sigma$); at 428.15 nm and CN($B^2\Sigma$-$A^2\Pi$); at 436.52, 592.34, 653.40, 667.73 and 708.32 nm for $C_2$($A^3\Pi_g$-$X'^3\Pi_u$); at 392.59 nm for $C_2$($A^3\Pi_g$-$X'^3\Pi_u$); at 486.14 nm for $H_\beta$; at 486.13 and 656.27 nm for $H_\alpha$; at 739.23, 751.47, 763.51, 772.42, 794.82 and 811.53 nm for Ar.

Also, several species were observed with low intensity, those are: $CN^+$, $CN_2$, $C_3H_3$, CH, $H_2$, C, $CH_2$, $NH_2$, $NH_3$, $C_2H_2^+$ and NH, which are not displayed in figure 2, in order to have a better picture of the principal species.

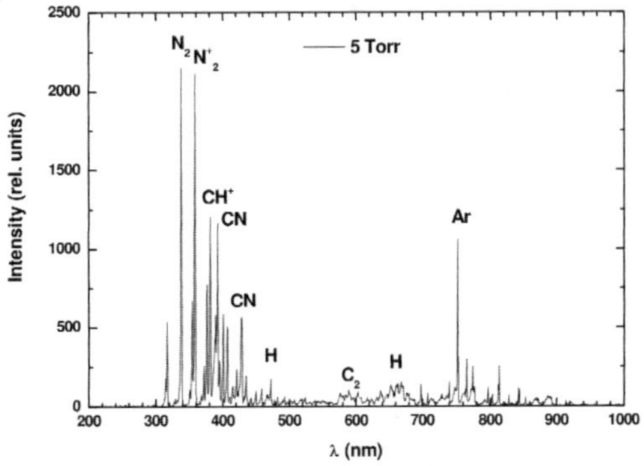


**Figure 1.** Emission spectra of $N_2$-$CH_4$-Ar plasma at 5.0 Torr.

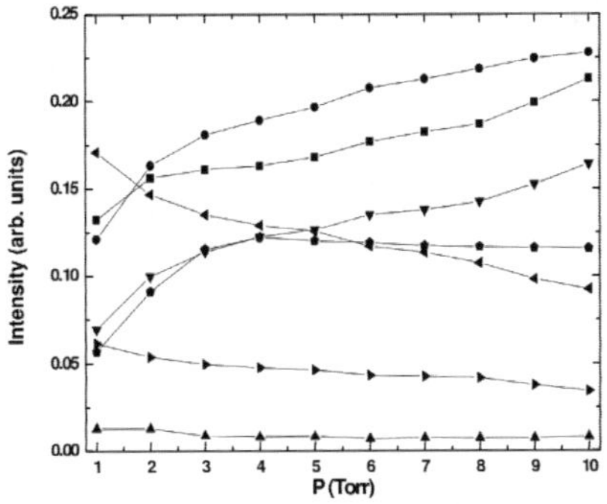


**Figure 2.** Intensity of the some emission bands and lines as a function of pressure. ●, $N_2^+$; ■, $N_2$; ◆, $C_3$; ▼, $CH^+$; ⬟ , Ar; ◀, $C_2$; ▶, H; ✶, $C^+$.

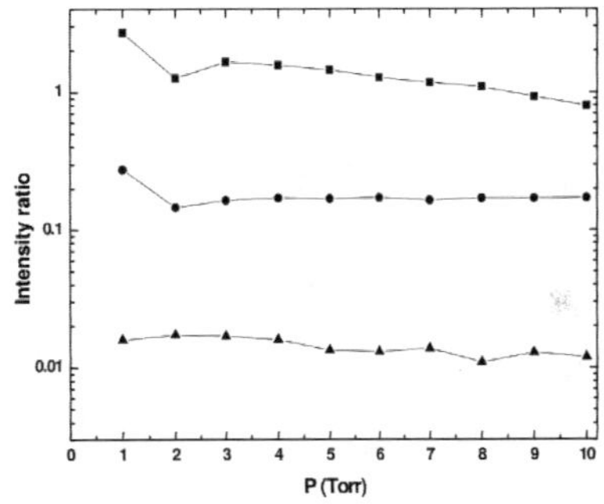


**Figure 3.** Ratio of intensities as a function of pressure. ■, $CN/N_2$; ●, $N_2^+/N_2$; ▲, $H/N_2$.

At the present plasma condition (V = 280 Volts, d = 5 mm, P = 1-10 Torr) and Tg = 300 K, the gas density (N) is between $0.32\times10^{17}$ and $3.21\times10^{17}$ $cm^{-3}$, which with ne ~ $10^{10}$ $cm^{-3}$ [9], we obtained an ionization degree of $10^{-7}$. As the ionization degree is $< 10^{-4}$, it may neglect electron-electron interaction [11]. Also, in this case the reduced electric field *E/N* (*E* is the field strength) is 174.4 Td (at P = 10.0 Torr) and 1744.5 Td (at P = 1.0 Torr). The measurement conditions, such as the integration time of the spectrometer and the position of the optical fiber, were fixed for every measurement. The limitation factor on long-time measurements was the deposition of yellow-orange layer on the electrodes causing sparks in the discharge and that avoid measuring the electron temperature and ion density. The influence and elimination of this kind of deposition requires further investigations. Although the intensities are not calibrated, the relative intensity is still valid for discussions below. In order to observe more carefully the effect of the pressure on the OES measurements, the pressure dependence of several emission bands and lines are displayed in figure 2 All bands and lines have an increasing behavior within the present pressure range. This can be explained by the fact that at higher pressures ion-neutral reactions become more important [11].

Journal of Physics: Conference Series **370** (2012) 012047 doi:10.1088/1742-6596/370/1/012047

Figure 3 shows more quantitatively the ratios of intensities of $N^+_2$ (391.44 nm), CN (392.08 nm), and H (486.13 nm) to that of the $N_2$ (337.13 nm) as function of pressure. As it can be seen, the $CN/N_2$, $N_2^+/N_2$ and $H/N_2$ ratios show an almost constant behavior as a function of the pressure. Being the $CN/N_2$ ratio more highest and $H/N_2$ ratio the lowest one. The presence of the $N_2^+$ ion in the pressure plasma suggests that the ion–molecule and molecule-molecule reactions in the gas phase are likely to play a dominant role at the present pressures. The constant ratios of excited species at different pressures, may suggest that there is no change of the chemical reactions in the gas phase significantly. That means that the rate coefficients of the chemical reactions increase as a function of the pressure in the same proportion to keep the relative ratios of the excited chemical species almost constant.

## 4. Conclusions

An experimental study of optical emission spectroscopy of a low-pressure DC plasma discharge sustained in a 85.0%$N_2$-12.0%Ar- 3.0%$CH_4$, 82.3%$N_2$-13.7%Ar- 4.0%$CH_4$ and 79.6%$N_2$-15.4%Ar-5.0%$CH_4$ ternary mixtures were studied. These plasmas were generated in the total pressure range of 0.5 to 10.0 Torr, power of 7 W and 50 l/min flow rate of gases. The principal species observed were from the molecular bands of $N_2$, $N_2^+$, $CH^+$, CN, $C_2$ and $C_3$, and atomic lines of $H_\beta$, $H_\alpha$ and Ar. All bands and lines have an increasing behavior within the present pressure range, which can be explained by the fact that at higher pressures ion-neutral reactions become more important. The ratios show an almost constant behavior as a function of the pressure. Being the $CH/N_2$ ratio more highest and $H/N_2$ ratio the lowest one. The variations of excited species at different pressures may change the subsequent chemical reactions in the gas phase significantly. The present results suggest that the ion–molecule and molecule-molecule reactions in the gas phase are likely to play a dominant role in the present pressures.

## Acknowledgments

The authors are grateful to A. González, O. Flores, and José Rangel for technical assistance. This research was partially sponsored by DGAPA IN-105010, CONACyT 128714.

## References

[1] Nacimento L F C, Mota R P, Valença G P, Teixeira J C, Algatti M A, Honda R Y and Bortoleto J R R 1998 *Planet. Space Sci.* **46** 969
[2] Chen G Y, Stolojan V, Silva S R P, Herman H and Haq S 2005 *Carbon* **43** 704
[3] Lara L M, Banaskiewicz M, Rodrigo R and Lopez-Moreno J J 2002 *Icarus* **158** 191
[4] Herrebout D, Bogaerts A, Yan M and Gijbels R 2001 *J. of Appl. Phys.* **90** 570
[5] Yanguas-Gil A, Focke K, Benedikt J and von Keudell A 2007 *J. of Appl. Phys.* **101** 103307
[6] Oumghar A, Legrand J C, Diamy A M and Turillon N 1995 *Plasma Chem. Plasma Process.* **15** 87
[7] Janča J, Tálský A, Krčma F, Hochard L and Ricard A 1994 *Plasma Chem. Plasma Process.* **14** 197
[8] Ricard A, Oseguera-Pena J E, Falk L, Michel H and Gantois M 1990 *IEEE Trans. Plasma Sci.* **18** 940
[9] Garcia-Cosio G, Martinez H, Calixto-Rodriguez M and Gomez A 2011 *Journal of Quantitative Spectroscopy and Radiative Transfer* **112** 2787
[10] Pearse R W B and Gaydon A G 1976 *The identification of molecular spectra* (Great Britain: University Printing House Cambridge)
[11] Janda M, Martišovitš V, Morová M, Machala Z and Hensel K 2007 *Eur. Phys. J. D* **45** 309

# On the dynamic behavior of the anode–arc–root at the nozzle surface in a non-transferred plasma torch

**L Prevosto[1,*], M Risso[1], D Infante[1], E Cejas[1], H Kelly[1,2,*] and B Mancinelli[1]**

[1]Grupo de Descargas Eléctricas, Departamento Ing. Electromecánica, Universidad Tecnológica Nacional, Regional Venado Tuerto, Laprida 651, 2600 Venado Tuerto, Santa Fe, Argentina
[2]Instituto de Física del Plasma (CONICET), Departamento de Física, Facultad de Ciencias Exactas y Naturales (UBA) Ciudad Universitaria Pab. I, 1428 Buenos Aires, Argentina
[*]Member of CONICET

E-mail: prevosto@waycom.com.ar

**Abstract** The dynamic behavior of the anode–arc–root at the nozzle surface of a plasma torch was experimentally investigated in this work. A gas ($N_2$) vortex–stabilized non–transferred arc torch with a thoriated tungsten rod (2wt %) cathode (3.2 mm diameter) and a coaxial anode (5 mm diameter, 30 mm length) was used in the experiment. By using a sweeping Langmuir probe in floating condition, the voltage of the plasma jet outside the nozzle was inferred. Arc voltage waveforms were also obtained. Data have been obtained for an arc current of 100 A and a gas flow rate of 30 Nl min$^{-1}$. A typical sawtooth shape (i.e., restrike mode) (with a fluctuating level of $\approx \pm 25$ %) and a dominant frequency of $\approx 6.5$ kHz was observed in the arc voltage waveforms, which is attributed to anode–arc–root movements along the anode surface followed by a restrike at a certain point close to the cathode. By performing a time correlation between the probe and arc voltage oscillograms together with simple estimations, the amplitude of the movement of the arc–root along the anode surface as well its velocity were inferred.

## 1. Introduction

Atmospheric pressure thermal plasma jets generated in direct current (dc) non–transferred arc plasma torches are used in a number of applications like plasma surface modifications, spray coatings, material synthesis, and waste treatment [1].

The electric arc inside these torches can be divided into three parts: the cathode region (located right in front of the cathode), the arc column, and the anode region (located right in front of the anode). The cathode arc root is quite stable for hot (thermionic) cathodes, but the anode–arc–root usually moves along the anode surface. This movement is in general desirable, as it allows a more uniform heating of the anode surface (increasing its lifetime); but in plasma spray applications it also causes a non–uniform heating of the particles injected in the plasma jet, which is undesirable.

The dynamics of the arc inside dc non–transferred arc plasma torches are mostly the result of the unbalance between the magnetohydrodynamic force, produced by the local curvature of the current path and the arc self-induced magnetic field, and the flow drag, caused by the interaction of the

Published under licence by IOP Publishing Ltd

incoming dense, cold gas and the hot, low density, plasma arc [2]. If the magneto–hydrodynamic force equals the drag force, then a steady attachment is obtained, characterized by a negligible movement of the arc. If the drag force is larger than the electromagnetic force, then the flow will drive the arc downstream. At the same time as the arc moves downstream, the arc length increases and so do the total voltage drop and the magnitude of the electrical field around the arc. If the voltage drop between a certain point of the stretched arc and the anode exceeds the local breakdown voltage value, a new breakdown and, hence, a new arc attachment forms somewhere upstream the original attachment. The corresponding arc voltage exhibits a typical transient restrike mode (sawtooth shape), which is attributed to the above described arc dynamics.

The dynamic behavior of the anode–arc–root at the nozzle surface of a non–transferred plasma torch was experimentally investigated in this work. The amplitude of the movement of the arc–root along the anode surface as well as its time–averaged velocity was inferred by using a floating Langmuir probe and time resolved arc voltage measurements.

## 2. Experimental set–up

A gas ($N_2$) vortex–stabilized non–transferred arc torch with a thoriated tungsten rod (2wt %) cathode (3.2 mm diameter) and a coaxial anode (5 mm diameter, 30 mm length) was used in the experiment. The operating torch conditions were: arc current of 100 A and a gas flow rate of 30 Nl min$^{-1}$.

A schematic of the employed experimental set–up is shown in figure 1.

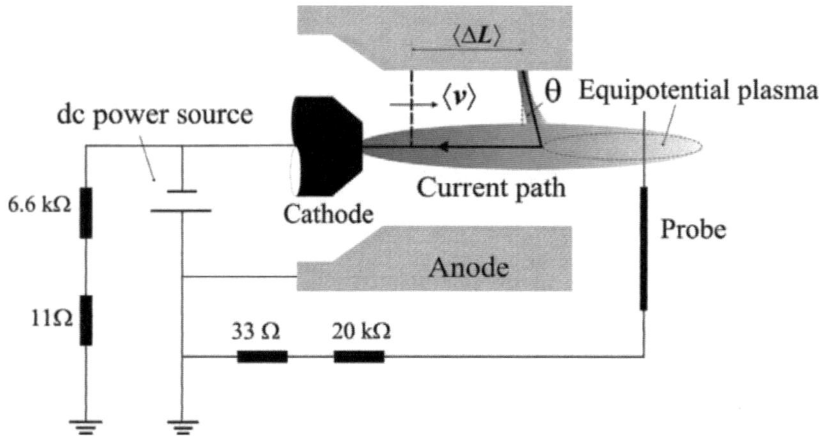


**Figure 1**. Schematic of the employed experimental set–up.

The employed Langmuir probe system consisted in a rotating aluminum disk carrying one single probe mounted in the radial outward direction of the disk. The cylindrical probe was made of tungsten wire with a radius of 250 μm. The probe was swept trough the arc at constant velocity (17 m s$^{-1}$) at 3.5 mm from the nozzle exit in unbiased (floating) condition. This condition was obtained by connecting the probe to ground through a large resistance (20 kΩ) that limited the collected current to a very low value, and using and additional low value resistance (33 Ω) to register a signal proportional to the probe voltage ($V_p$). The arc voltage waveforms ($V_c$) were also obtained by connecting the cathode to ground through a large resistance (6.6 kΩ), and using and additional low value resistance (11 Ω) to register a signal proportional to the arc voltage. Both voltage signals were registered using a two-channel Tektronix TDS 1002 B oscilloscope with a sampling rate of 500 MS/s and an analogical bandwidth of 60 MHz. A detailed description of the employed probe system can be found elsewhere [3].

## 3. Experimental results and discussion

Typical $V_c$ and $V_p$ waveforms corresponding to the above torch operating conditions are given in figure 2. The arc voltage exhibits a characteristic sawtooth shape (i.e., restrike mode) with a fluctuating level of ≈ ± 25 %, with respect to its time–averaged value, and a dominant frequency ($\langle f \rangle$) of ≈ 6.5 kHz. This phenomenon is attributed to a motion of the arc root driven by the flow, followed by a sudden restrike near the cathode (indicated in the figure 1 with a dotted vertical line as a cold gas layer breakdown), between some point of the arc column and the anode. This results in sawtooth shaped voltage traces because there is an almost linear relationship between the arc voltage and the arc length [4].

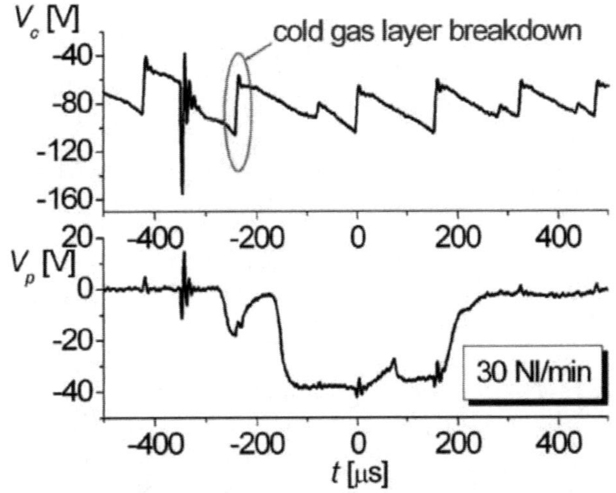


**Figure** 2. Correlated arc and plasma floating voltages.

The time–correlated $V_p$ signal is also given in figure 2. The relevant part of this signal extends from ≈ - 200 μs to ≈ 200 μs, which corresponds to the passage of the probe through the arc. Taking into account that in this kind of torches the difference between the plasma potential and the floating potential is quite small [5], a plasma potential value ($V_s$) of ≈ -35 V can be inferred for the whole plasma jet emanating from the torch. This voltage value represents the voltage drop between the arc column and the anode–arc–root (see the straight path tilted an angle $\theta$ in figure 1). Neglecting the small anode voltage drop, the average electric field strength along such path is given by

$$\langle E \rangle \approx \frac{V_S}{R}\cos(\theta), \tag{1}$$

where $R$ is the nozzle radius. Furthermore, by setting $\langle E \rangle$ equal to the uniform electric field strength along the arc, the time–averaged restrike length ($\Delta L$) can be estimated as

$$\langle \Delta L \rangle \approx \frac{\langle \Delta V_C \rangle}{\langle E \rangle}, \tag{2}$$

where $\langle \Delta V_c \rangle$ is the time–averaged peak to peak value of the sawtooth shape of the arc voltage.

The corresponding time–averaged drift velocity ($\langle v \rangle$) is given by

$$\langle v \rangle = \langle \Delta L \rangle \langle f \rangle. \tag{3}$$

The spatial–averaged arc electric field strength, together with the time–averaged restrike length as well the drift velocity, are shown in figure 3 as functions of the model parameter ($\theta$). As it can be seen, $\Delta L$ varies in the range 3–7 mm ($\langle v \rangle$ is around 25–50 m s$^{-1}$) while $\langle E \rangle$ remains relatively high, in the range of 7–14 V mm$^{-1}$.

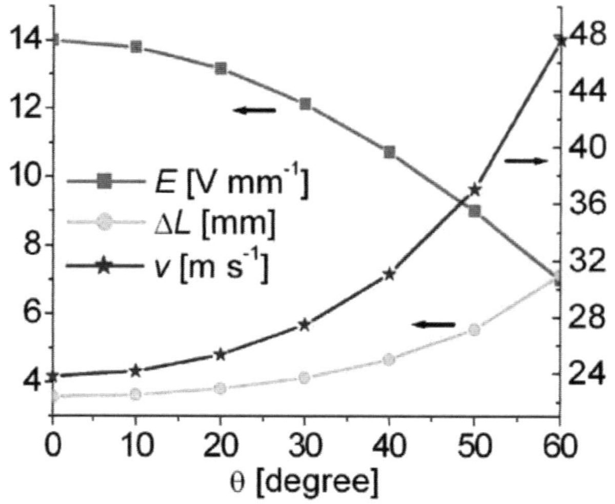


**Figure 4**. Calculated arc electric field strength, restrike length and restrike velocity.

## 4. Conclusions

The dynamic behavior of the anode–arc–root at the nozzle surface of a plasma torch was experimentally investigated in this work by using a floating sweeping Langmuir probe and time resolved arc voltage measurements. By performing a time correlation between the probe and arc voltage oscillograms together with simple estimations, a range of values for the arc electric field, the amplitude of the restrike length and the drift velocity of the arc were inferred. The estimated values of the restrike length are in accordance with numerical simulations [6].

**Acknowledgements**

This work was supported by grants from the Universidad de Buenos Aires (PID X108), CONICET (PIP 5378) and Universidad Tecnológica Nacional (PID 1389).

## References

[1] Boulos M, Fauchais P, Pfender E 1994 *Thermal Plasmas, Fundamentals and Applications Vol. 1* (New York: Plenum Press)
[2] Fauchais P 2004 *J. Phys. D: Appl. Phys.* **37** R86
[3] Prevosto L, Kelly H and Mancinelli B 2008 *IEEE Trans. Plasma Sci.* **36** 263
[4] Fauchais P and Vardelle A 1997 *IEEE Trans. Plasma Sci.* **25** 1258
[5] Raizer Y P 1991 *Gas Discharge Physics* (Berlin: Springer)
[6] Trelles J P, Chazelas C, Vardelle A and Heberlein J V R 2009 *J. Therm. Spray Technol.* **15** 524

Journal of Physics: Conference Series **370** (2012) 012049 doi:10.1088/1742-6596/370/1/012049

# XUV radiation from gaseous nitrogen and argon target laser plasmas

**P Vrba[1], M Vrbová[2], P Brůža[2], D Pánek[2], F Krejčí[2,3], M Kroupa[3] and J Jakůbek[3]**

[1]Institute of Plasma Physics, Academy of Sciences, Za Slovankou 3, 182 00 Prague 8, Czech Republic
[2]Faculty of Biomedical Engineering, Czech Technical University in Prague, Sítná 3105, Kladno 2, Czech Republic
[3]Institute of Experimental and Applied Physics, Czech Technical University in Prague, Horská 3a/22, 128 00 Prague 2, Czech Republic

E-mail: vrba@ipp.cas.cz

**Abstract**. Laser plasma created in gaseous target is studied as a source of radiation in the "water window" wavelength range. Plasma is created by focusing an 800 mJ/7 ns Nd:YAG laser pulse into the gas-puff target. Using nitrogen gas results in emission of an intense quasi-monochromatic radiation with the wavelength 2.88 nm, corresponding to the quantum transition 1s2p –> $1s^2$ of helium –like nitrogen ion. The emission spectrum with argon target covers all the water window range. Laboratory and computer experiments have been performed for both target gases. The spatial distributions of emitted energy in the water window spectral range were compared. The total emitted energy with argon was one order higher than with nitrogen.

## 1. Introduction

We intend to develop a new laboratory source of XUV radiation appropriate for university biological research laboratory. Namely, a source of intensive monochromatic radiation in the "water window" spectral region (2.33 – 4.37 nm) may be useful for imaging of small biological objects in water environment and as a source emitting continuous spectra in the vicinity of carbon K absorption edge may be used to study carbon molecular bonds via absorption spectroscopy. Such sources are not readily available today outside synchrotron facilities. We have decided to extend our previous research activities and to develop a source of XUV radiation based on laboratory plasmas properly heated either by capillary pinching discharge or by focused infrared laser radiation [1, 2].

Results of our study accomplished with nitrogen and argon plasma laboratory radiation source made by Laser-Laboratorium Göttingen (LLG) [3] are presented here. Recently, we have measured XUV emission spectra and proved the quasi-monochromatic emission at wavelength 2.88 nm with nitrogen and quasi-continuous emission spectrum with argon target. Here we pay attention to time profiles of emitted XUV radiation intensity and spatial distribution of emitted XUV energy. Results of computer simulations and experiments are summarized and compared.

**Figure 1.** Experimental setup: Nd:YAG laser (1), interaction chamber (2), gas valve assembly (3), CCD camera for plasma monitoring (4), IR laser light absorber (5), large chamber compartment for imaging and time-resolved measurements (6).

## 2. Experiment

The experimental setup is shown in figure 1. The target material - nitrogen and argon gas, respectively - is injected in short bursts through a conical tungsten nozzle into a vacuum chamber using a piezoelectric needle valve. The valve assembly is located at the top of the vacuum chamber. The plasma is generated by the impact of high energy pulses from an Nd:YAG laser focused onto the gaseous target by a 100 mm focal length plan-convex lens. The laser beam enters the chamber from a side, perpendicular to the direction of gas injection. The diameter of the focal spot is approx. 60 $\mu$m. The wavelength of the laser output is 1064 nm. The energy of the pulses is 0.8 J and the pulse width 7 ns. The source was operated at 2 Hz repetition rate. For both temporal and imaging measurements, a 0.8 $\mu$m thick titanium foil filter was employed to block radiation outside the spectral region of interest.

### *2.1. Measured time dependences of emitted power*

The waveforms were recorded using InfiniiVision DSO7104A digital storage oscilloscope (Agilent, USA) with an analogue bandwidth of 1 GHz and a sampling rate of 4 GSamples/s. The temporal profile of the infrared excitation pulses was measured using a biased InGaAs photodiode DET01CFC (Thorlabs GmbH, Germany) with a 1.2 GHz bandwidth. The photodiode was placed to detect the excitation radiation scattered from the focusing lens. The temporal profile of the infrared pulse presented in figure 2 is an average of 100 recorded waveforms.

Time dependencies of emitted power in the XUV spectral region were recorded using a fast XUV-sensitive photodiode AXUV20HS1 (IRD Inc., USA) placed at the 250 mm distance from the source. A reverse bias voltage of 90 V was applied. The inbound radiation power was reduced by a pinhole aperture with 20 $\mu$m diameter placed 3 mm from the source. An additional titanium foil filter with 0.4 $\mu$m thickness was used to ensure the spectral purity of the detected radiation. The measured time profiles of infrared pulses from the excitation laser and XUV pulses generated from the plasma are shown in Figure 2.

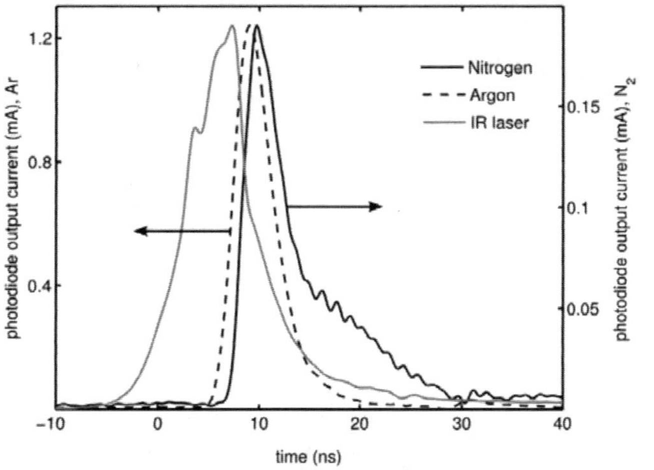


**Figure 2.** Temporal profiles of measured emitted XUV radiation power from nitrogen (solid line), argon (dashed), and IR power from the excitation laser (solid grey).

*2.2. Spatial profiles of emitted energy*

Spatial distribution of energy emitted in the XUV region by the laser generated plasma was recorded in a pinhole imaging setup [5]. A 30 μm pinhole aperture was placed 34 mm from the object (plasma), followed by the titanium filter. The object-screen distance was 996 mm, thus the resulting magnification of the imaging setup was approx. 30 times. A hybrid semiconductor pixel device Timepix was used as the screen. The resolution of the Timepix chip is 256×256 pixels with pixel size 55 $\mu$m. The details and advantages of using Timepix device for detection of soft X-rays are given in [6].

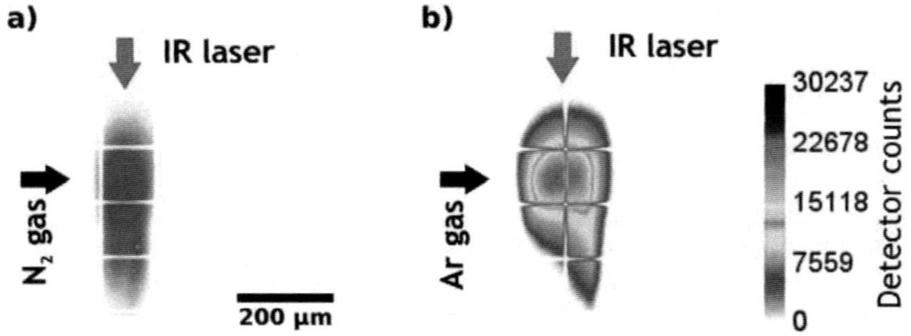


**Figure 3.** The spatial distributions of XUV radiation energy recorded using Timepix detector in a pinhole imaging setup for nitrogen (left) and argon (right) gas target.

The spatial profiles of XUV radiation emission recorded for nitrogen and argon gas targets, respectively, are shown in Figure 3. Both profiles were acquired by integration of 100 consecutive single-shot images recorded at 2 Hz repetition rate. The energy emitted from the nitrogen plasma is clearly smaller than that from argon plasma. Metal mesh supporting the titanium foil filter can be recognized as it is superimposed on the plasma images.

Journal of Physics: Conference Series **370** (2012) 012049 doi:10.1088/1742-6596/370/1/012049

## 3. Computer modeling

The spatial development of plasma created by the laser beam was simulated using 2D RMHD (Radiation- Magneto-Hydro-Dynamic) code Z* [4]. The non-stationary, non-equilibrium radiation physics are introduced in this code for modeling the transient laser plasma properties in 2D geometry. The axis of symmetry of the simulations coincides with the laser beam axis.

For the purpose of calculation, the gas stream cone is approximated by a plasma disk target [5] (2D object) with axial symmetry along the laser beam (see Figure 4). The *z*-coordinate is oriented in the opposite direction to the laser beam propagation. Note that the laser beam is propagating horizontally in our experimental setup.

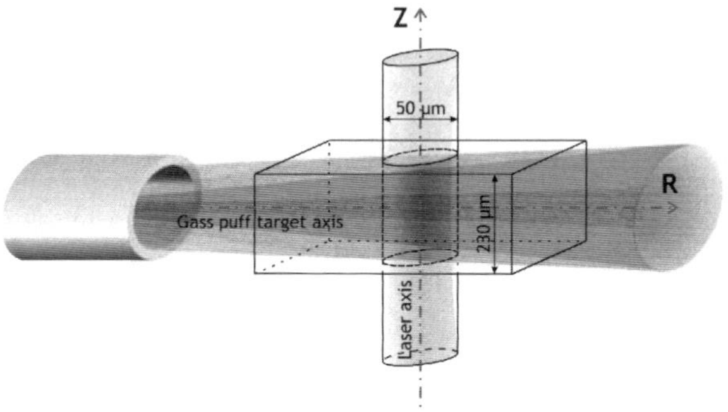


**Figure 4.** Gas puff laser plasma source geometry.

### 3.1. *Input data for the Z*code*

The input laser pulse time profile is presumed to be Gaussian with the peak position at 10 ns and the calculation begins at -15 ns. The main input parameters are listed in table 1.

**Table 1.** Laser and target parameters.

| **Nd:YAG laser** | | | | **Target** | |
|---|---|---|---|---|---|
| Wavelength | 1064 nm | Beam diameter | 6 mm | Thickness | 0.2 nm |
| Laser energy | 800 mJ | Focus position | 2.5 mm | Radius | 2.5 mm |
| Pulse FWHM | 8 ns | Focal spot diameter | 0.05 mm | Pressure $N_2$ | 1 – 100 bar |
| Focal length | 100 mm | Pulse time delay | 0 ns | Pressure Ar | 10 bar |

### 3.2. *Spatial distributions of plasma quantities*

The Z* code allows to determine all plasma field quantities at given temporal and spatial coordinates; plasma mass density $\rho$, longitudinal and radial velocities $v_z$ and $v_r$, electron density $N_e$, electron and ion temperatures $T_e$ and $T_i$ during the laser pulse. The spatial distributions of selected field quantities with nitrogen initial pressure 10 bar at three different times (at 4 ns – corresponding to the laser pulse leading edge, 8 ns – peak of laser power and 18 ns- after the laser pulse) are shown in Figure 5.

At the very beginning of the laser pulse (~ 1 ns) a deep crater is formed and the gas target is "drilled trough" at ~ 4 ns. The mass density compression wave moves in the radial direction in the gas layer. After the end of the laser pulse the compression wave vanishes (see the 1$^{st}$ row of figure 5). The created plasma moves in the opposite direction to the laser beam reaching the highest value of longitudinal velocity $v_{z,max}$ = -180 km.s$^{-1}$ at the time ~ 4 ns. When the crater is formed, the plasma starts moving in both directions. The peak value of electron plasma density is achieved at ~8 ns i.e. in the region of the compression wave (see the 2$^{nd}$ row of Figure 5).

Changes of electron temperature $T_e$ are seen from the 5$^{th}$ row of Figure 5. During a short period before the deep crater is formed the increase of temperature is remarkable in front of the gas target. The peak value of electron temperature is $T_e$ ~ 130 eV is achieved at the peak of laser pulse. Spatial distribution of electron temperature is almost the same on both sides of the target sheet and fades away with increasing time. The plasma is non-isothermal in the central part of the plasma cloud.

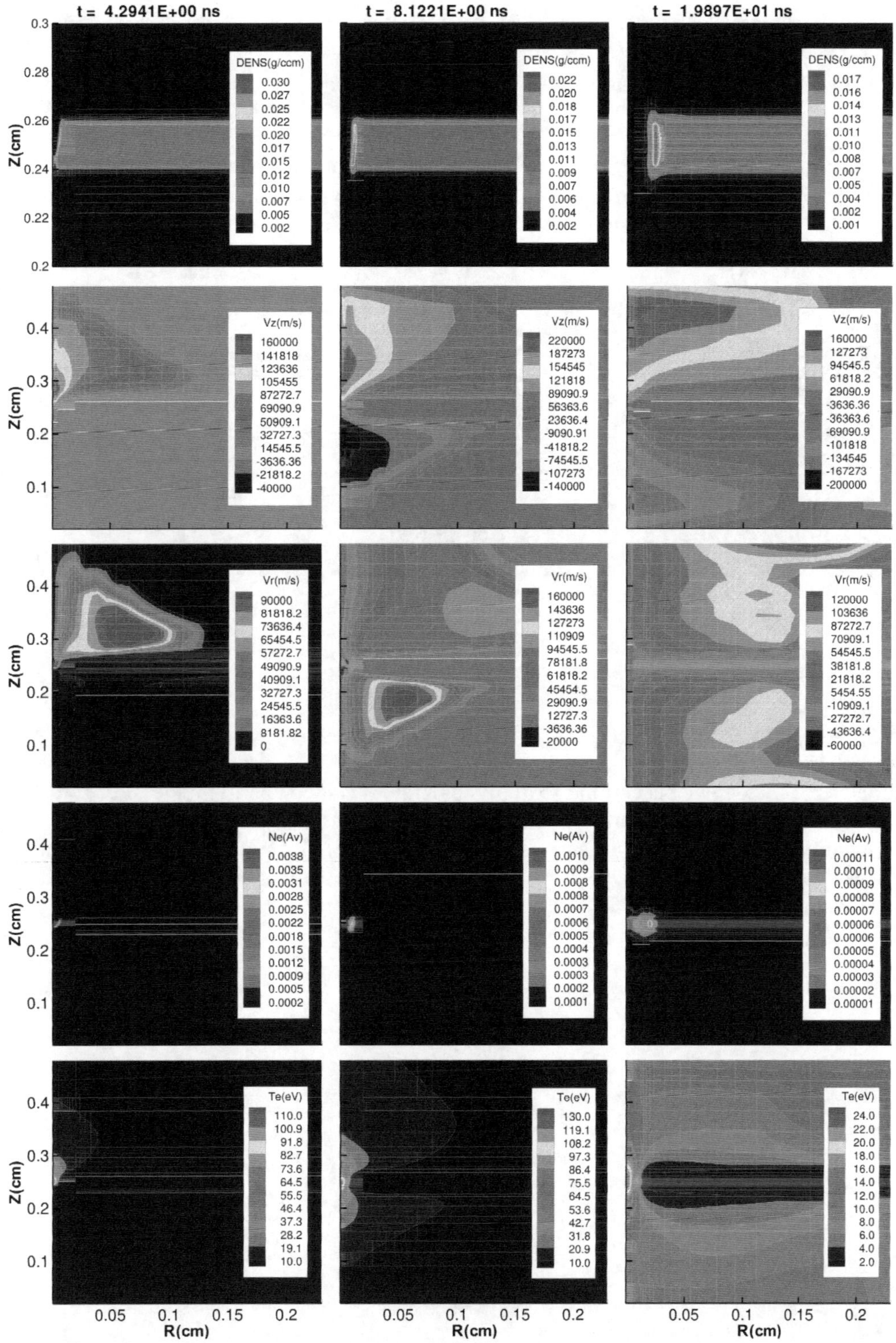


**Figure 5.** Spatial distribution of selected plasma quantities at various times for initial nitrogen pressure 10 bar: 1 - Mass plasma density $\rho$, 2 - Longitudinal velocity $v_z$, 3 - Radial velocity $v_r$, 4 - Electron density $N_e$ (in Av units), 5 - Electron temperature $T_e$.

Journal of Physics: Conference Series **370** (2012) 012049 doi:10.1088/1742-6596/370/1/012049

### *3.3. Spatial distributions of emitted energy*

The code Z* enables us also to determine the spatial development of plasma radiation, together with all its spectral characteristics. The integrated power emitted in the 2.88 nm helium – like nitrogen spectral line at ~ 4, ~ 8 and ~ 20 ns, evaluated for the above used parameters, is seen from Figure 6. At the very beginning of the pulse $t < 1$ ns the volume of radiation is very small. In the radial direction its radius is practically the same as the laser beam radius 25 $\mu$m and in the longitudinal direction it is ~ 45 $\mu$m. The volume of the luminous region and the local amplitude of the emitted radiation grow up to the time of the peak value of the laser power and are stable later on.

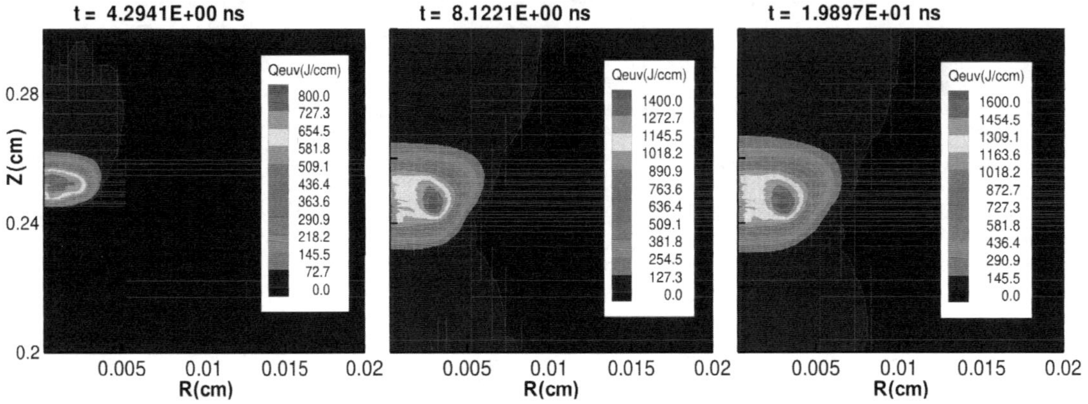


**Figure 6.** Spatial distributions of integrated power emitted in wavelength band 2.876 - 2.886 nm at various times; for nitrogen target initial pressure 10 bar.

### *3.4. Pressure dependence of emitted energy*

We have performed repetitive simulations for various initial nitrogen gas pressures and the above used other parameters. The dependence of the spatially integrated emitted energy in the three selected bands is seen from Figure 7. In the pressure region from 1 bar to 50 bar the emitted energy is increasing with increasing initial pressure. If the pressure is about 50 bar the highest expected energy in the quasi-monochromatic spectral line is about 1.5 mJ whereas for the pressure about 5 bars the estimated output is one order lower, corresponding to $10^{11}$-$10^{12}$ photons/steradian.

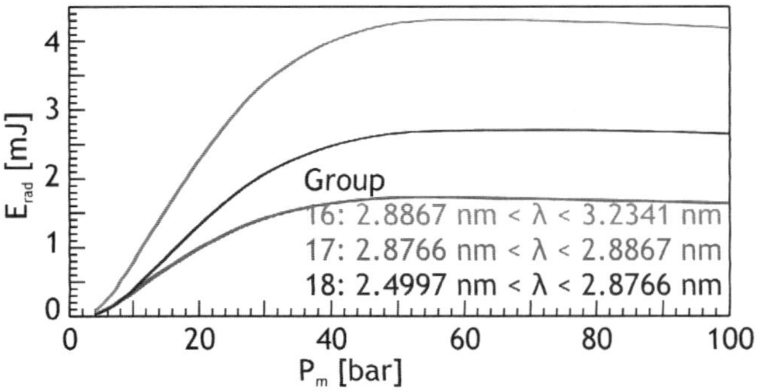


**Figure 7.** Pressure dependences of radiation energy emitted in selected spectral bands.

## 4. Comparison of simulated results for argon and nitrogen

We have compared the evaluated results for nitrogen and argon performed under identical initial conditions. The initial pressure of 10 bars has been presumed. The output emission in the spectral wavelength region limited by the transmission of the titanium foil was taken into account. For nitrogen target we have taken into account only one spectral line or wavelength range 2.8766 – 2.8867 nm and for argon the wavelength range 2.0664 – 3.0996 nm.

*4.1. Evaluated time dependences of absorbed laser power and emitted radiation power*

Spatial integration over the laser plasma cloud provides the information about total absorbed laser power and total emitted power in selected spectral bands, see Figure 8. The peak value of absorbed power is almost two times higher with the argon gas than with the nitrogen one. The peak value of the emitted radiation in the whole radiation spectrum is more than 4 times higher with argon gas than with nitrogen. The peak value of emitted power in the single 2.88 nm line of helium -like nitrogen is about $8\times10^4$ watts, whereas the power emitted by the argon plasma, when filtered by the titanium foil, is estimated to be $3\times10^6$ watts.

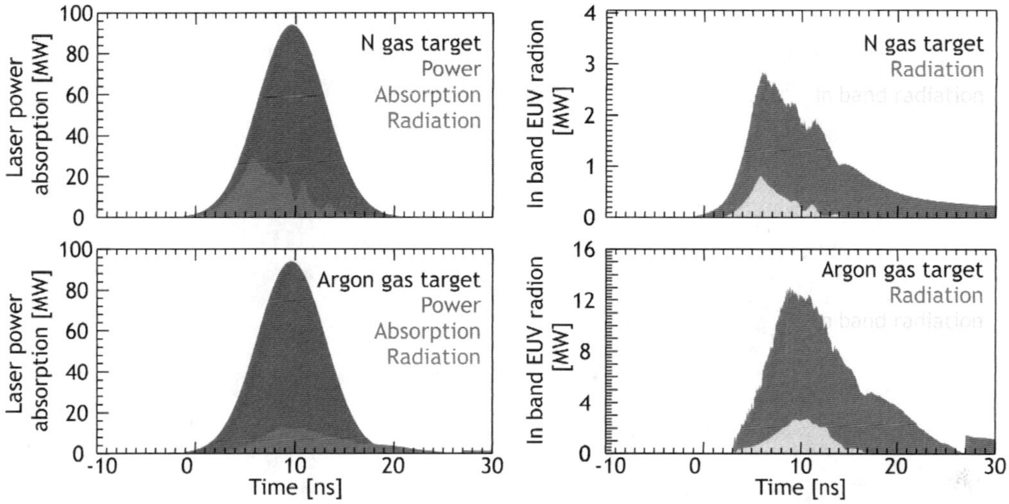


**Figure 8.** Time dependences of laser power (blue filling), absorbed power (green filling), emitted radiation power (orange filling) and XUV radiation power (yellow filling) from nitrogen and argon targets at pressures 10 bar (XUV output power for nitrogen is 10 times magnified).

*4.2. Evaluated spatial profiles of emitted energy in selected spectral bands*

The evaluated spatial distributions of energy emitted in selected spectral bands by the plasma cloud are shown as Figure 9. The plasma source may be roughly approximated by a cylinder 450 μm long with diameter of 150 μm if nitrogen is considered and by a cylinder 850 μm long with diameter of 400 μm with argon. The estimated peak value of the emission is roughly one order higher with argon than with nitrogen.

## 5. Conclusions

We have a studied laser plasma as a source of XUV radiation and compared the results of computer and laboratory experiments for nitrogen and argon. The calculated time profiles of XUV radiation power and spatial distributions of luminous areas are in a good agreement with the experimentally recorded data. The initial density of the gas target during the experiment was not known. Therefore, an estimated value had to be used for the simulations. Performed computer simulations show strong increase of emitted XUV energy with pressure in the range from 1 to 50 bars. Thus, to achieve higher output energy of XUV radiation, higher gas density in the jet is needed.

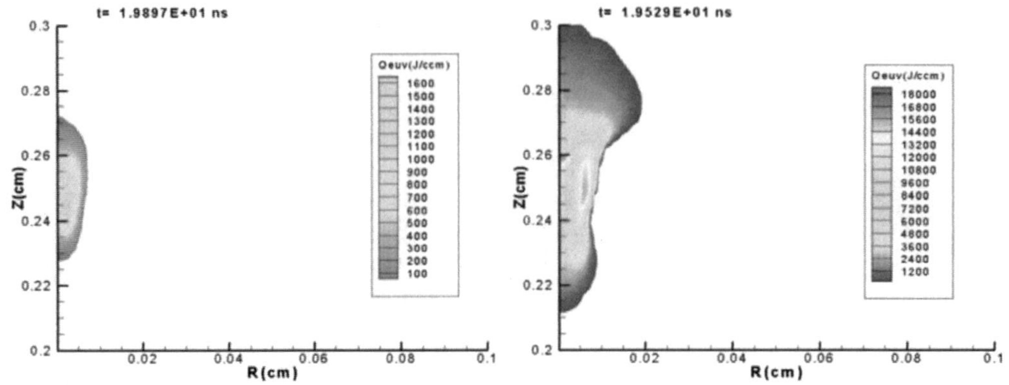

**Figure 9.** Estimated XUV radiation energy spatial distributions emitted by nitrogen and argon targets when observed behind the titanium foil; initial pressures are the same for both gases.

**Acknowledgements**
This research has been supported by the grants of Ministry of Education Youth and Sports CR under the projects LC528 and LA 08024. The spatial diagnostics presented here were carried out under the Medipix Collaboration and were supported by the Project 6840770040 of the Ministry of Education, Youth and Sports of the Czech Republic.

**References**

[1] Vrba P, Zakharov S V, Jancarek A, Vrbova M, Nevrkla M and Kolar P 2011 *J. Electr. Spectr. and Rel. Phenomena* **184** 335
[2] Kolar P et al. 2011 Preprint **0511** CTU FBMI
[3] Peth C, Barkusky S and Mann K 2008 *J. Appl. Phys.* **41** 105202
[4] Zakharov S V, Novikov V G and Choi P 2005 in *EUV Source for Lithography* (SPIE Press) 223
[5] Vrba P, Vrbova M, Bruza P and Panek D 2011 *BOX Workshop 1* Prague
[6] Krejci F, Kroupa M, Jakubek J and Bruza P 2011 *IEEE Nucl. Sci. Symp. MIConference*

Journal of Physics: Conference Series **370** (2012) 012050 doi:10.1088/1742-6596/370/1/012050

# CN violet system spectrum used as a molecular pyrometer and the influence noise to signal ratio on the temperature values

**H Nassar**

Lebanese University, FSP, section IV, Karak Zahlé, Lebanon

E-mail: nassar_hossein@hotmail.com

**Abstract.** The CN violet system ($B^2\Sigma^+$- $X^2\Sigma^+$) molecular emission spectrum is frequently observed in plasma sources containing hydrocarbons and nitrogen mixture. We have simulated the spectrum of (0,0) and (1,1) bands of this system for different rotational and vibrational temperatures. The influence of the noise to signal ratio has been studied, if the noise to signal ratio is about 10% we found an error of 6% at temperature 3000K and 10% at 6000K.

## 1. Introduction

The violet system of CN molecule is a good tool for diagnosing plasmas containing this molecule, the molecular constants and the lines structure are well-known, and it is present in most of the plasma sources containing hydrocarbon and nitrogen in mixture for a temperature between 2000K and 6000K.

Czernykowski (1987) has simulated CN violet system to measure temperature in Electrobrûleurs plasma [1]. Bontemps-Faure in 1997 uses this system for temperature measurements in arc plasma [2]. In 2004, Lino Da Silva used the violet system of CN to study the radiative properties of the plasma surrounding a vehicle through a planetary atmosphere at hypersonic speed [3].

In the previous publications, we have proposed a numerical method based on the simulation of molecular spectrum for studying the distribution temperature in modified arc Maeker [4] and [5], and in transferred arc [6]. To determine $N_2^+/N_2$ ratio and the temperature distribution in Glidarc [7], we have made the simulation of both systems first negative and second positive one simultaneously.

In this paper, we have used this method based on the simulation of the (0,0) and (1,1) bands of CN molecular violet system. The method proposed permit to evaluate, by comparing point to point a real spectrum with the simulated one, temperature and apparatus function, approximated by the gauss function (the half-width at *1/e* height). Moreover, it is shown, by noised spectra simulation, the influence of noise to signal ratio at the calculated temperature values.

## 2. Simulation of the ($B^2\Sigma^+$- $X^2\Sigma^+$) spectrum of CN

### *2.1. The violet system of CN molecule*

A complete description of the spectral terms of the diatomic molecules may be found, e.g. in [9] and [10], or in our previous publications [5] and [7].

The violet system of molecule diatomic CN corresponds to radiative transitions between the $B^2\Sigma^+$ and $X^2\Sigma^+$ electronic state ( $\Lambda = 0$) at 3.32 eV and 0 eV respectively. We are interested in the (3858Å,

3888Å) spectral range covering the (0,0) and (1,1) bands. The coupling is type (b) Hund case [9]. The absolute rotational emission line intensity of the molecule is given by

$$\varepsilon_{ul} = C\sigma^4 S_{J'J''} \exp\left[ -\frac{hcF(J')}{KT} \right] \qquad (1)$$

where $C$ is a normalization constant, $\sigma$ is the wave number, $S_{J'J''}$ is the Hönl-London factors, $h$ and $c$ are the conventional physical constants, $F(J')$ is the spectral terms and $T$ is the temperature (marks ' concern the lower spectral term and the marks " are for the upper one). Formulae for $S_{J'J''}$ factors for the (b) Hund case are taken from Kovacs [9] and the spectral terms are those of Budo [10].

*2.2. Numerical simulation*

Line intensities are calculated when a state of thermodynamic equilibrium is fully reached and the corresponding temperature of the gas, considered from a macroscopic and thermodynamic point of view, are the same. In reality, within laboratory plasmas, we usually have to deal with a local thermal equilibrium state instead of thermodynamic equilibrium state.

It is possible, when the population of discrete energy levels follow Boltzmann's Law, to attribute a corresponding temperature for each corresponding energy distribution; electronic, $T_e$, vibrational $T_v$ and rotational $T_r$. In a non-equilibrium plasma $T_r$, $T_v$ and $T_e$ may differ from the kinetic temperature $T_g$ of heavy species. Taking into account the strong coupling between translational and rotational energy states, the temperature derived from experiments is, in a general case, close to the gas kinetic temperature [11] and [12].Thus, $T_g$ can be deduced from $T_r$, measured by using the rotational structure of molecular spectra.

All the spectral constants are taken from Herzberg [8]. Computed spectrum of individual lines for a given temperature is then convoluted with our apparatus function approximated by the gauss function whose DX (the half-width at 1/e height):

$$I_0 = \int_{\nu_0-\varepsilon}^{\nu_0+\varepsilon} I_\nu \exp\left[ -\frac{(\nu-\nu_0)^2}{DX^2} \right] d\nu \qquad (2)$$

where $\tau$ =2.5 DX and $\nu_0$ is the wave number of the line.

An integrated cell (pixel) of simulated spectrum is chosen according to the cell of experimental spectrum under consideration. We can, following the situation, group several cells in one.

Practically, for each given set of values (DX, *Tr,*), the line intensities of this band are normalized to its maximum. Examples of simulated spectra, normalized to 100, are shown in figures 1, 2 and 3 for different sets of (*Tv*, *Tr,* DX) values.

A quick analysis of these simulated spectra shows that and vibrational temperature has a great influence on the height of the band head (1.1) figure 1.

The apparatus function has essentially influenced the resolution of the rotational line structure. Particularly, if DX is too high, only the shape of the band can be used to determine the rotational temperature. The rotational temperature has a significant influence on the rotational line structure.

## 3. Numerical method for rotational temperature evaluation

To evaluate the *Tr* from a given real spectrum, we have chosen to compare this spectrum point-by-point with spectra simulated at different values of *DX Tr.* In our experiments the following procedure was applied. (i) The measured spectrum was corrected for the continuous background, which was assumed to be linear in the studied spectral range. (ii) The experimental spectral intensity was normalized to the maximum value, which was set to 100. (iii) The experimental data were compared with that of the simulated spectrum. The minimization procedure was used to minimize $E(T_r,DX)$ as a function of *Tr*, *and DX* defined as:

$$E(T_r, DX) = \frac{1}{N}\left[\sum_{i=1}^{N}\left(I_i^r - I_i^s\right)^2\right]^{1/2} \tag{3}$$

$I_i^r$ et $I_i^s$ are respectively the intensities of the real and simulated i-th pixel of the spectrum containing $N$ points. If the choice of the spectrum reference origin (background) is a bad one, the result will be affected. To avoid this error, we fix several origins, and our computer program searches the best origin to find the *E(T)* at minimum.

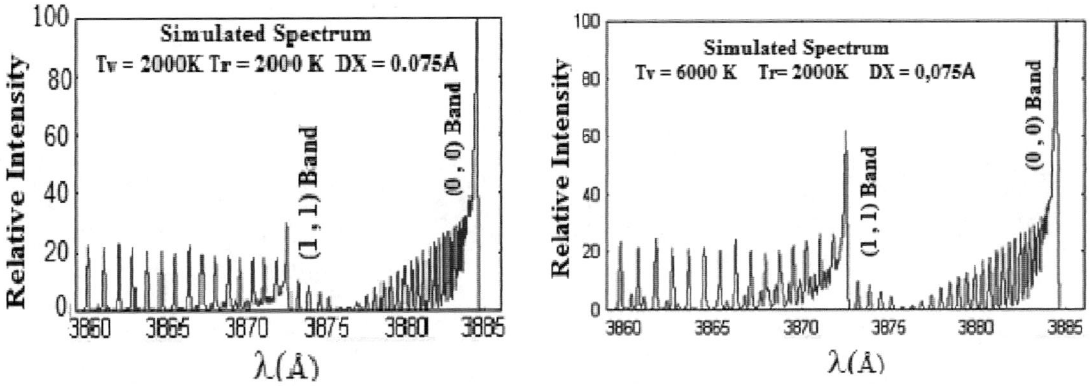


**Figure 1.** Effect of the vibration temperature *Tv* (*Tr*=2000K DX = 0.075Å).

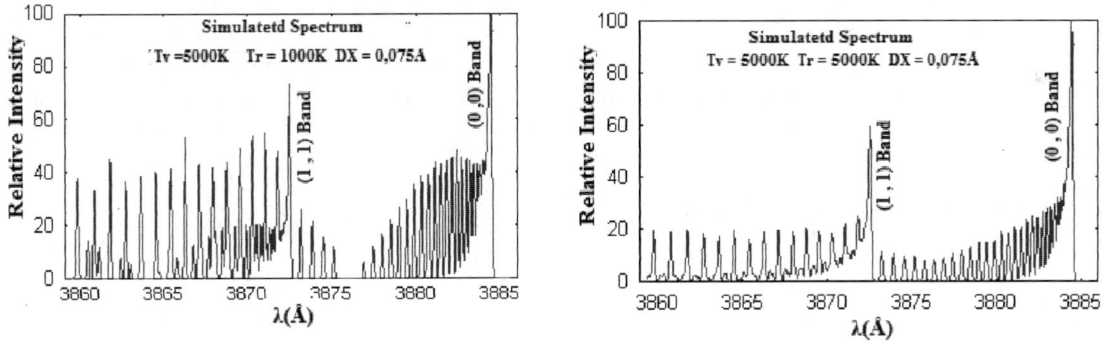


**Figure 2.** Effect of rotational temperature *Tr* (*Tv*=5000K DX = 0.075Å).

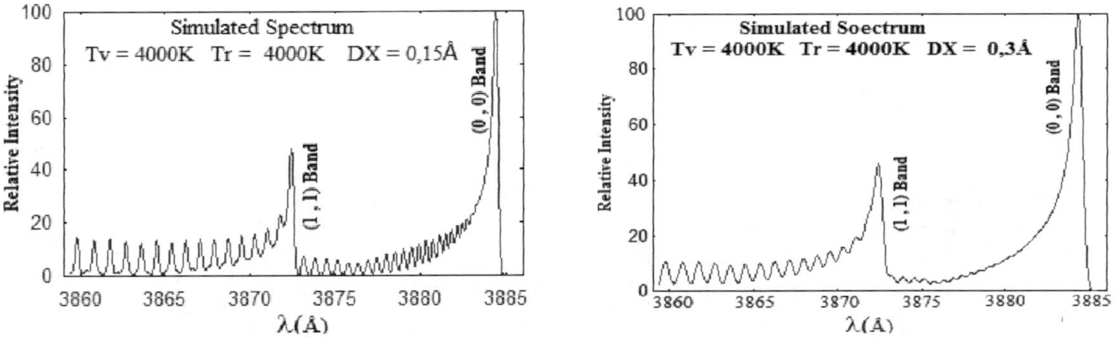


**Figure 3.** Effect of the apparatus function *DX* (*Tr* = *Tv* = 4000K).

## 4. Application to simulated noisy spectra

In order to test the sensitivity of the temperature estimator, we applied this method when a noisy spectrum was used instead of the experimental one. The synthetic spectra were calculated for known apparatus functions, temperatures, and noise was introduced using a random number generator. Examples of simulated spectra noised, normalized to 100, are shown in figure 4 for different sets of N/S ratio. We can see that the rotational spectra are very deformed for the high N/S values.

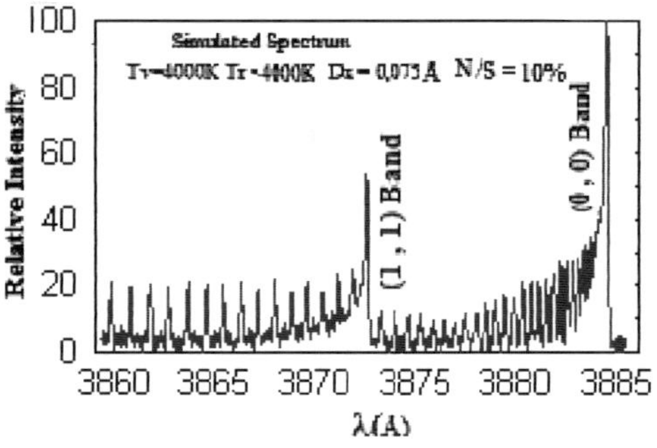


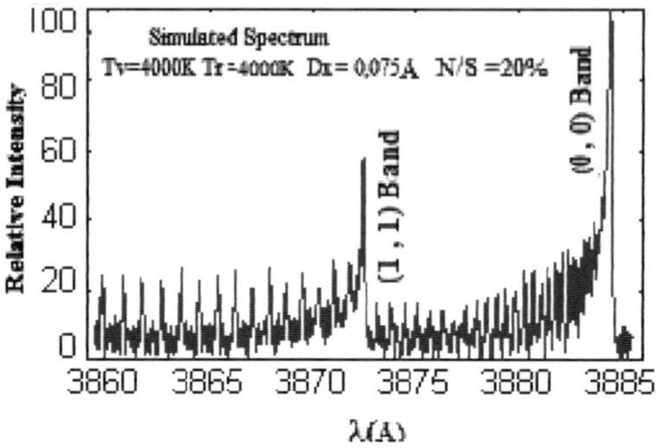


**Figure 4.** Effect of noise to signal ratio (N/S) ($Tv = Tr$ =4000K and $DX$ =0.075Å).

## 5. Results and discussion

Calculations were performed for noise-to-signal ratios N/S equal to 1%, 5%, 10% and 20% and for temperatures of 3000 and 6000 K. The results obtained are given in table 1.

The results are satisfactory since the relative error in the temperatures is of the same order of magnitude as the noise-to-signal ratio, even if the last one was relatively high. Given that when the concentration of emitting molecule is relatively weak, the spectrum of molecular emission seems immersed in the continuous background noise. In other words, the ratio noise / signal depends strongly on the molecular concentration, we can therefore, using this method to determine at what temperature the emission is maximum and hence its concentration.

**Table 1**. Evolution of the relative error on the temperature with noise/signal ratios.

| | $\Delta T/T$% | |
|---|---|---|
| Noise/signal % | 3000k | 6000k |
| 1% | 1.5% | 3% |
| 5% | 4% | 7% |
| 10% | 6% | 10% |
| 20% | 11.5% | 18.5% |

## References

[1] Czernichowski A and Chapelle J 1986 *Journée d'Etudes "Les Electrobrûleurs"* Société Française des Thermiciens Les Renardières France 30.10 8 pp
[2] Bontemps-Faure G 1997 *Thèse de l'Université de Blaise Pascal*
[3] Da Silva L 2004 *Thèse Université d'Orléans*
[4] Nassar H and Czernichowski A 1990 *J. de Phys.* **51** C5-289
[5] Nassar H 1991 *Thèse Université d'Orléans*
[6] Nassar H, Meguernes K, Czernichowski A and Chapelle J 1995 *Proc ISPC-12* **2** 651
[7] Nassar H, Pellerin S, Musiol K, Martinie O, Pellerin N and Cormier J-M 2004 *J. Phys. D: Appl. Phys.* **37** 1904–16
[8] Herzberg G 1962 *Spectra of diatomic molecules Molecular Spectra and Molecular Structure* vol 1 (New York:Van Nostrand Reinhold)

[9] Kovacs I 1969 *Rotational structure in the spectra of Diatomic Molecules* (London: Adam Hilger)
[10] Budo A 1963 *Z. Phys.* **105** 582
[11] Motret O, Hibert C, Pellerin S and Pouvesle J-M 2000 *J.Phys. D: Appl. Phys.* **33** 1493–8
[12] Neumann W 1975 *Progress in Plasmas and Gas Electronics* vol 1, ed Rompe R and Steenbeck M (Berlin: Akademie-Verlag) pp 54–65

Journal of Physics: Conference Series **370** (2012) 012051 doi:10.1088/1742-6596/370/1/012051

# Laser induced aluminiun plasma analysis by optical emission spectroscopy in a nitrogen background gas

**J C Chamorro, J Uzuriaga and H Riascos**

Grupo Plasma, Láser y Aplicaciones, Universidad Tecnológica de Pereira, A.A. 097, Pereira, Colombia

E-mail: hriascos@utp.edu.co

**Abstract.** We studied an Al plasma generated by a Nd:YAG laser with a laser fluence of 4 $J/cm^2$, a wavelength of 1064 nm, energy pulse of 500 mJ and 10 Hz repetition rate. We studied their spectral characteristics at various ambient nitrogen pressures by optical emission spectroscopy (OES). The $N_2$ gas pressure was varied from 20 mTorr to 150 mTorr. In Al plume, both atomic and ionic spectra were observed. The electron temperature and electron number density of the plume as of the function ambient gas pressure were determined. The electron temperature was calculated by using the Boltzmann-plot method and the number density was calculated considering the stark effect as dominating on the emission lines.

## 1. Introduction

Pulsed laser deposition (PLD) has emerged as a potential technique for the fabrication of thin films and nanostructures of various materials because of the ability to control the size and shape of nano-deposits by varying the laser parameters, the gas atmosphere, target-substrate distance and substrate temperature [1-5]. This technique is highly suitable for depositing thin films at a relative high deposition rate and low cost. However, the interaction of laser light with the solid targets is a complicated process and is not yet completely understood. It consists of different stages: the laser ablation of the target; plasma generation; laser interaction with the plasma; plasma expansion and collision with a substrate. Laser-produced plasma is transient in nature with characteristic parameters that evolve quickly and are highly dependent on irradiation conditions such as incident laser intensity and pulse duration, laser wavelength, irradiation spot size, ambient gas composition, and ambient pressure. The stages of the plasma generation, laser interaction with plasma and plasma expansion play a very important role in the thin film growth process. In fact, during temporal evolution of the laser induced plasma (LIP), excitation and ionization of the evaporated material occurs, therefore, it is important to define its thermodynamic parameters, such as electron number density *Ne* and temperatures of electrons *Te*. In order to optimize and control the thin film growth process it is necessary to study the dependence of these parameters of the LIP on the deposition conditions and to develop suitable diagnostics. For measuring and controlling the parameters of the LIP the method of OES is used extensively [6]. It is based on the study of the spectral distribution of line intensity and broadening in emission spectra. In this paper we report on optical emission spectroscopic studies of AlN plasma plume accompanying pulsed laser ablation.

Atomic species are identified through the characteristic emissions in recorded spectra. The electron temperature *Te* and number density *Ne* are determined from relative atomic and ionic line intensity

Journal of Physics: Conference Series **370** (2012) 012051 doi:10.1088/1742-6596/370/1/012051

measurements using the Boltzmann plot method. The properties of the created plume were investigated by acquiring spectra at different ambient pressures.

## 2. Experimental setup

The plasma was produced by ablating a Al target in a $N_2$ background gas. We used a short-pulse, Q-switched Nd:YAG laser which provided 7 ns, 500 mJ laser pulses at a wavelength of 1064 nm with a 10 Hz repetition rate. The experiment took place in a stainless steel vacuum chamber configured as a six-way cross of 10 cm inner diameter tubes forming a central 10 cm diameter target chamber evacuated to $10^{-5}$ mTorr, and flushed with pure nitrogen (99.99%) at different flow rates during experiments. The laser beam was focused with a 20 cm focal length lens on a Aluminum target (99.99%) with a 45° angle of incidence, giving an energy density of approximately 4 J/cm$^2$. Nitrogen gas (99.99%) was injected to the chamber as a reactive gas, and the flow of such was accurately controlled by a needle valve. During laser ablation, nitrogen pressure was varied from 20 to 150 mTorr as measured by a vacuum gauge. The substrate was placed at a distance of 60 mm from the target. The films were deposited on silicon substrates that had previously been ultrasonically cleaned in acetone. Emission spectra were collected, imaged onto spectrograph TRIAX 550™ with 1200 grooves/mm grating and a resolution of 0.025 nm using two 10 and 13 cm focal length lenses positioned outside the reactor. For time-integrated spectra measurements, the spectrograph was equipped with a CCD linear sensor (1024 × 256). Most of the spectra were collected with integration times of 0.5 s.

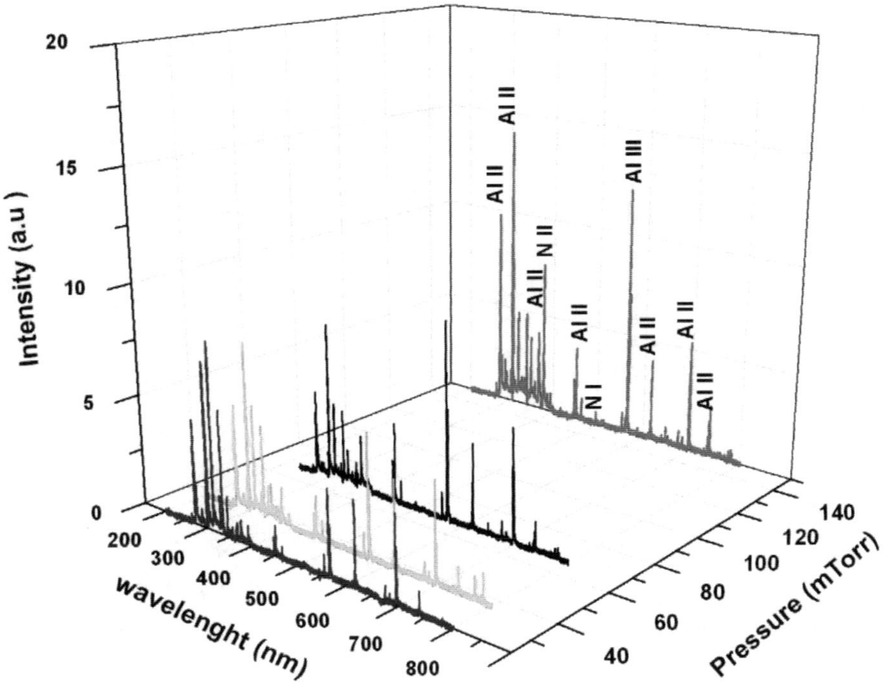


**Figure 1.** Optical emission spectra of laser induced aluminum plasma at different nitrogen pressures.

## 3. Results and discussion

### *3.1. Analysis of plasma*

Figure 1 shows optical emission spectra of the plume, recorded at various pressures of nitrogen gas. The optical emission spectra of plasma plume were recorded in the wavelength range from 200 to 800

Journal of Physics: Conference Series **370** (2012) 012051 doi:10.1088/1742-6596/370/1/012051

nm at different pressures; the general structure does not vary much with the $N_2$ pressure. In the spectra shown in figure 1, the most intense lines are related to the emission from the aluminum species. Apparently, the main emitting species in the laser- produced plasma are excited single (Al II) and double ionized (Al III) ones. The lines with the highest intensity in the plasma spectrum are Al III at 563.15 nm and Al II at 273.4 nm, 302.66 nm, 376.25 nm, 452.64 nm, 616.15 nm and 695.02nm. In addition, the most intense lines of nitrogen gas observed are neutral and multiple-ionized atoms of nitrogen; the lines are at 376.260 nm (NIII), 282.364 nm (NII) and 496.398 nm (NI).

### *3.2. Determination of electron temperature*

Determination of electron temperature was made assuming LTE resulting from collisions in the plasma, so that the populations of the bound states follow the Boltzmann distribution. The relative line intensities from a particular state can then be used to calculate the electron temperature of the plasma [7-8].

$$\ln\left(\frac{I_{mn}\lambda_{mn}}{A_{mn}g_{mn}}\right) = \ln\left(\frac{N}{z}\right) - \left(\frac{E_m}{kT_e}\right) \tag{1}$$

where $\lambda_{mn}$ is the transition wavelength, $I_{\text{mn}}$ the intensity of the observed transition line, $A_{mn}$ the transition probability, $g_{mn}$ the degeneracy of the upper level, $E_{mn}$ the energy of the upper level, $k$ the Boltzmann constant, $N$ the total number of states, $Z$ the partition function, and $Te$ the electron temperature. For the transition, the upper state is labeled as $m$ and the lower state by $n$. The slope $(-1/\kappa Te)$ of the plot of $ln(I_{mn}\lambda_{mn}/A_{mn}g_{mn})$ versus $E_m$ yields the temperature.

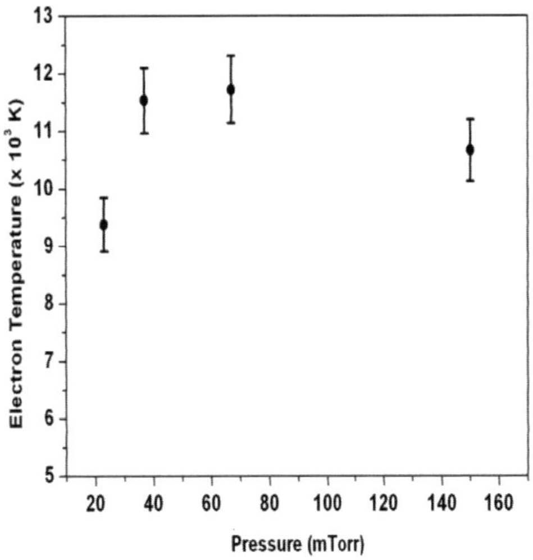


**Figure 2.** Calculation of electron temperature using the Boltzmann-plot method at different pressures.

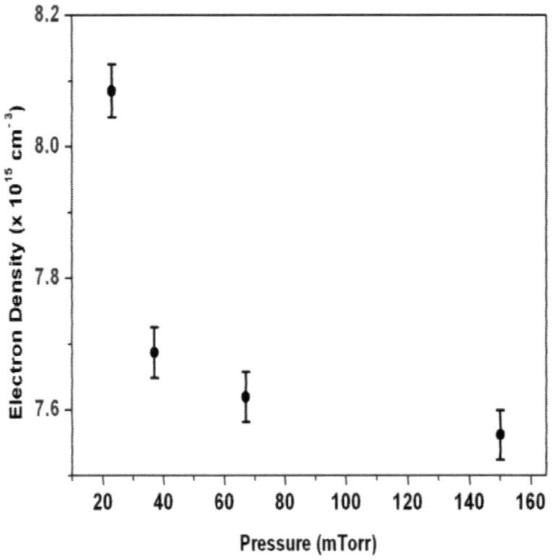


**Figure 3.** Variation of plasma electron density at different nitrogen pressures, calculated from Stark broadening.

### *3.3. Estimation of electron density*

There are three possible broadening mechanisms that can occur in the laser-ablated plume. The Doppler broadening, resulting from motion of the atoms, the Stark-broadening from collisions with charged species, and the resonance broadening arising from collisions between the neutral species exciting strong resonance lines [7]. Estimation of electron density depends principally upon the measurement of the Stark and Doppler broadening of some optical lines [8]. Doppler broadening is estimated by using the formula $(\Delta\lambda)_{1/2} = 7.16\times10^{-7}(Te/M)^{1/2}$, where Te is the electron temperature in Kelvin and M the atomic mass [9]. In our case, Doppler broadening can be ignored. Stark broadening of an isolated line in the plasma was, thus, used to estimate electron density in the plasma. The $S^5d^1D$ $-S^{16}f^1F$ transition of atomic Aluminum (Al II) at 617.2949 nm was used for electron density measurements. The Stark-broadened profiles were recorded at various pressures and fitted to the Lorentzian profile. FWHM $(\Delta\lambda)_{1/2}$ of the Stark broadened of a line is related to the electron density through the expression [7, 9].

$$\Delta\lambda_{1/2} = 2W\left(\frac{n_e}{10^{16}}\right) + 3.5A\left(\frac{n_e}{10^{16}}\right)^{1/4}\left(1-\frac{3}{4}N_D^{-1/3}\right)W\left(\frac{n_e}{10^{16}}\right) \tag{2}$$

this is the sum of two (electron impact and ion-impact) terms. $A$ is the ion broadening parameter and $W$ is the electron impact-width parameter. Both $W$ and $A$ are nearly temperature independent [8]. $N_D$ is the number of particles in a Debye sphere is given by

$$N_D = \frac{1.72\times10^9\left[T_e(eV)\right]^{3/2}}{\left[n_e\left(cm^{-3}\right)\right]^{1/2}} \tag{3}$$

the contribution is almost entirely due to electron impact, and the half-width of the Stark broadened transition can be estimated by the first term in equation (2) to a good approximation [10]. Expression (2) reduces to:

$$\Delta\lambda_{1/2} = 2W\left(\frac{n_e}{10^{16}}\right) \tag{4}$$

where $n_e$ is the electron density in $cm^{-3}$; the $W$ and $A$ parameters can be found in [7, 11].

**Table 1.** Spectroscopic parameters of Al II transition lines used to calculate electron temperature Te.

| | λ (nm) | $A_{ki}(s^{-1})$ | $E_i$ (eV) | $E_k$ (eV) | Configuration | $J_i$ - $J_k$ | $g_i$ - $g_k$ |
|---|---|---|---|---|---|---|---|
| **Al II** | 270.6435 | 3.326e+06 | 13.071346 | 17.651076 | 3*s*4*p* - 3*s*7*d* | 0 - 1 | 1 - 3 |
| **Al II** | 373.4715 | 2.954e+06 | 15.062085 | 18.383310 | 3*s*4*d* - 3*s*11*f* | 2 - 3 | 5 - 7 |
| **Al II** | 617.2949 | 5.103e+06 | 16.604633 | 18.612722 | $S^5d^1D - S^{16}f^1F$ | 2 - 3 | 5 - 7 |
| **Al II** | 697.3449 | 7.445e+05 | 16.604633 | 18.382210 | $S^5d^1D - S^{11}f^3F$ | 2 - 3 | 5 - 7 |

## 4. Conclusions

We used the optical emission spectroscopy method to analyze the spectra plasma generated by Nd:YAG pulsed-laser of nanosecond pulsed inciding on Al target in a nitrogen atmosphere. Emission spectra were mainly due at single ionized and double ionized Al species, also we observed neutral and single ionized Nitrogen species. The electron temperature $T_e$ has been determinated from the Boltzmann plot method and electron number density $N_e$ is estimated from the stark broadened profile of the spectra lines. The electron density found to be decrease with increase in Nitrogen pressure; besides, it is observed that the electron temperature achieves a maximum value at 67 mTorr of Nitrogen gas pressure.

**References**

[1] Chrisey B and Hubler G, editors, 1994 *Pulsed Laser Deposition of Thin Films* (New York: John Wiley and Sons)
[2] Shukla G and Khare A 2009 *Applied Surface Science* **255** 8730–8737
[3] Haglund R F 1998 *Laser Ablation and Desorption* vol 30 (New York: Academic)
[4] Miller J C 1998 *Laser Ablation and Desorption* (New York: Academic)
[5] Bauerle D 2003 *Laser Processing and Chemistry* (Berlin: Springer)
[6] Aragón C and Aguilera J A 2008 *Spectrochimica Acta Part B* **63** 893–916
[7] Griem H R 1964 *Plasma spectroscopy* (New York: McGraw-Hill)
[8] Bekefi G 1976 *Principles of laser plasmas* (New York: Wiley-Interscience)
[9] Wiese W L 1965 In: Huddlestone RH, Leonard SL (eds) *Plasma diagnostics techniques* (New York: Academic)
[10] Khare A and Thareja R K 1988 *IEEE J Quant Electr* **24** 2525
[11] http://physics.nist.gov, NIST Atomic Spectra Database, Kurucz output Atomic Spectral Line Database

Journal of Physics: Conference Series **370** (2012) 012052 doi:10.1088/1742-6596/370/1/012052

# Plasma properties of 355 nm and 532 nm laser-ablated molybdenum target at atmospheric pressure

**E M A Sternberg[1], N A S Rodrigues[2], J Amorim[3] and M E Sbampato[2]**

[1]Technological Institute of Aeronautics, PO Box 6050, 12228-900, Brazil
[2]Institute for Advanced Studies, PO Box 6044, 12231-970, Brazil
[3]National Laboratory of Science and Technology of the Bioethanol, PO Box 6170, 13083-970, Brazil

E-mail: manu.sternberg@gmail.com

**Abstract.** This paper presents the results obtained in Mo laser ablation experiments performed in air. Two lasers were used: one Nd:YAG AVIA 355-X emitting pulses with 0.205 mJ energy and 25 ns duration in 355 nm, and one Quantronix Hawk-532-40-M emitting pulses with 2.05 mJ energy and 110 ns duration in 532 nm. The Mo targets were fixed to a XY table, controlled by a computer program which moves permanently the table in order to avoid cratering. The plume was imaged by a 4x telescope and the light emission was collected at the image position by a fiber-optic and coupled to a Triax 550 Jobin-Yvon spectrometer. The spectra analysis was performed assuming the Local Thermal Equilibrium approximation and calculating the plasma temperature by using the Boltzmann plot method using neutral Mo spectral lines. The plasma temperature was obtained for different positions along the expansion axis, which allowed obtaining the electron population distribution as a function of the distance from the target. The plasma temperature along the expansion axis allowed evaluating the evolution of the excited states population when the plume expands.

## 1. Introduction

Currently, laser ablation has been used in applications in which accuracy and fine tuning of parameters are needed, such as in Medicine (as instrument on gastroenterological and ophthalmic surgery through ablation of tissues and organs [1-2]) or in the industry (in material processing [3-8], thin film deposition and isotopic separation [9]) for example.

On the other hand, nuclear medicine applies radioactive materials for diagnostic and treatment of patients. Radioisotopes are used like tracers in image medical diagnostics due to some propitious physical properties such as radiation emission and time decay. When radioisotopes are combined with chemical compounds of specific metabolic, biochemical and biological characteristics, they become radiopharmaceuticals. SPECT (Single Photon Emission Computed Tomography) is one of the techniques which use the radiation emitted by radiopharmaceuticals, mainly $^{99m}$Tc, to produce images for medical diagnostics [2, 10-11]. Molybdenum-99 is the only reliable source for the production of technetium-99m, accounting for 80% of nuclear medicine diagnostics. Table 1 shows the production mechanisms of molybdenum-99, however these processes present low efficiency due to the losses, the variety of reaction sub-products and the high cost. So, it is necessary to develop ways to optimize the molybdenum-99 production, to meet the market demand for diagnostics imaging.

**Table 1.** Reactions of production of $^{99}$Mo.

| Production mechanisms | Reaction | Physical Principle |
|---|---|---|
| Cyclotron | $^{96}Zr(\alpha,n)^{99}Mo$ | Neutron zirconium activation |
| | $^{100}Mo(p,pn)^{99}Mo$ | Proton bombardment of 100 molybdenum |
| | $^{100}Mo(d,dn)^{99}Mo$ | Deuteron bombardment of 100 molybdenum |
| Reactor | $^{98}Mo(n,\gamma)^{99}Mo$ | Neutron activation of 98 molybdenum |
| | $^{235}U(n,f)^{99}Mo$ | Uranium fission |

According to the research line developed at IEAv (Institute for Advanced Studies) on isotopic separation, it is intended to separate the molybdenum-99 through laser ablation of irradiated molybdenum metal target followed by selective photo ionization. To this end, a nanosecond pulsed Nd:YAG laser focuses on a molybdenum target and generates a plasma plume that is composed of neutral and ionized atoms. The part of the plume related to neutral atoms must be separated from the ionized part and, after that, an arrangement of laser beams is applied to separate the isotopes according their ionization energies (selective photo ionization) [12-15].

In this paper, it is discussed the molybdenum plasma plume production and characterization. An important goal in this process was to determine the proportion of neutral and ionized atoms on the material resulting of ablation and the population of neutral atoms excited states. The characterization of plasma plume was made by Optical Emission Spectroscopy [16-20] and the spectra were acquired for different positions along the expansion axis. Doing so, it was possible to quantify parameters such as plasma temperature, through Boltzmann plot method, to compare theoretical and experimental spectra considering the relevance of some physical processes (blackbody radiation, recombination, bremsstrahlung radiation), and to calculate the population of excited states of the plasma.

## 2. Laser ablation

Laser ablation consists in material removal from a target by incidence of a laser, commonly a pulsed laser. This process results from focusing a laser beam onto a surface inducing heating, ejection and ionization of the material and forms a plasma jet. For metallic samples, most of the energy of laser beam, nearly 40%, is transferred to superficial layers of the sample, which corresponds to $1/\varphi$, known as optical absorption length, in which φ is optical absorption coefficient. For molybdenum, optical absorption length is 14 nm. The heat flows into the material, thus, it is important to define the thermal diffusion length $(L_D)$, equation (1).

$$L_D = \sqrt{4t_P\alpha} \tag{1}$$

where, $\alpha$ is thermal diffusivity and $t_p$ is laser pulse duration ($t_p$ =25 ns and 110 ns). For molybdenum, thermal diffusion length is nearly 2.0 μm.

Figure 1 shows the interaction between laser and the material. The sample is reached by a sequence of laser pulses and the surface temperature increases due energy absorption. The process following the surface heating is phase transition of part of the solid forming a liquid region. After that, laser beam continues focused at same portion of the target until it absorbs sufficient energy to reach latent heat of evaporation. Considering that laser pulse duration is too short, the sample volume that was molted and evaporated does not suffer displacements and heat does not conduct into the target. The vapor reaches very high temperature, increasing the collision rate between vapor particles leading to ionization. The interaction between laser and the sample, considering the ionized atoms, free electrons and neutral particles in vapor, sustain the plasma and increase its density [21-24].

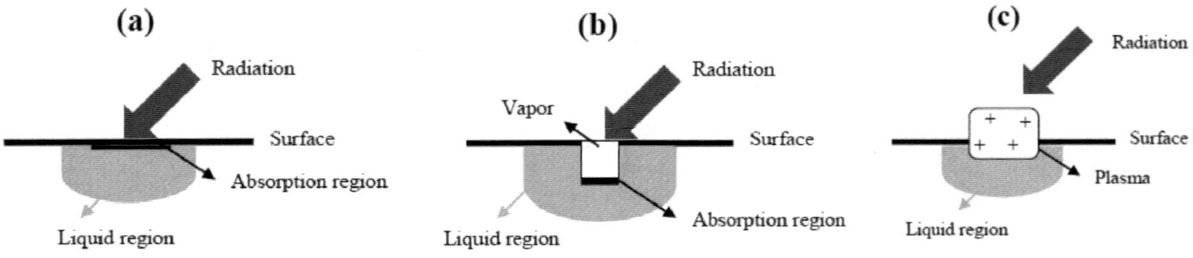


**Figure 1.** Interaction between laser and solid. At (a) fusion process is represented; at (b) evaporation and at (c) thermo ionization.

After the end of the laser pulse, the density of ejected material is approximately $10^{21}$ atoms/cm³, comparable to solid density, and is strongly ionized. At this point, the energy exchange between the plasma particles is due to collisions. The plasma plume behaves like a fluid and its expansion is ruled out by a hydrodynamic model. At a certain distance of the target, the plasma achieves local thermal equilibrium.

## 3. Plasma spectroscopy

Spectroscopy allows non intrusive methods for the measurement of species population and plasma temperature [21, 24-25]. However, it is required that the plasma is in Local Thermal Equilibrium (LTE), such that the intensity of emitted light for a given transition is proportional to transition upper state population.

### *3.1. LTE requirement*

The radiation emitted by the plasma escapes from the material and thus there is a deviation from Planck's equilibrium. However, when the energy exchange is dominated by collisions the resulting stationary state can be described for the same laws of complete thermal equilibrium, with exception for the emitted radiation. This plasma is in LTE [26] which is characterized by:

- The electron population density in each state is determined for collision processes.
- The velocity distribution of free electron is maxwellian.
- The electron population distribution of the states is determined for Boltzmann distribution while, to ionization cases, it is determined for Saha distribution.

The criterion for the electron density $(N_e)$ required for a local thermal equilibrium between two states separated in energy by $\Delta E$ [eV] is a function of electronic temperature $T_e$ [K], given by equation (2):

$$N_e\left[cm^{-3}\right] \geq 1.6\times 10^{12}\left(T_e[K]\right)^{\frac{1}{2}}\left(\Delta E[eV]\right)^3 \quad (2)$$

### *3.2. Temperature measurement*

Electron temperature can be determined using Boltzmann distribution, equation (3), through a method developed by Griem [25-26], that consider the ratio of intensities between two emission lines of spectrum. Equation (3) shows the emitted spectral line intensity $(I_{nm})$ that is a measure of the population of the corresponding energy in the plasma, under LTE approximation.

$$I_{nm} = \frac{hc}{4\pi\lambda_{nm}} A_{nm} g_n \frac{1}{U(T)} e^{\left(-\frac{E_n}{k_B T}\right)} \quad (3)$$

where $\lambda_{nm}$, $A_{nm}$ and $g_n$ are wavelength, transition probability and statistical weight of the upper level, $E_n$ is the energy of the level n and $U(T)$ is the partition function. Using the ratio of two emission lines (1 and 2), the electron temperature can be calculated, equation (4).

$$\frac{I_1}{I_2} = \frac{g_1 A_1 \lambda_2}{g_2 A_2 \lambda_1} e^{\left(-\frac{E_1 - E_2}{k_B T}\right)} \quad (4)$$

## 4. Experimental setup

This work presents molybdenum optical emission spectra for experiments performed in air, using two Nd:YAG lasers: Avia 355-X and Hawk-532-40-M, the first one, emitting pulses with 0.205 mJ energy and 25 ns duration in 355 nm, and the second, emitting pulses with 2.05 mJ energy and 110 ns duration in 532 nm.

The Mo targets were fixed to a XY table, controlled by a computer program which moves permanently the table in order to avoid cratering. The plume was imaged by a 4x telescope and the light emission was collected at the image position by a fiber-optic and coupled to a Triax 550 Jobin-Yvon spectrometer. Figure 2 presents a scheme of experimental setup [27-31].

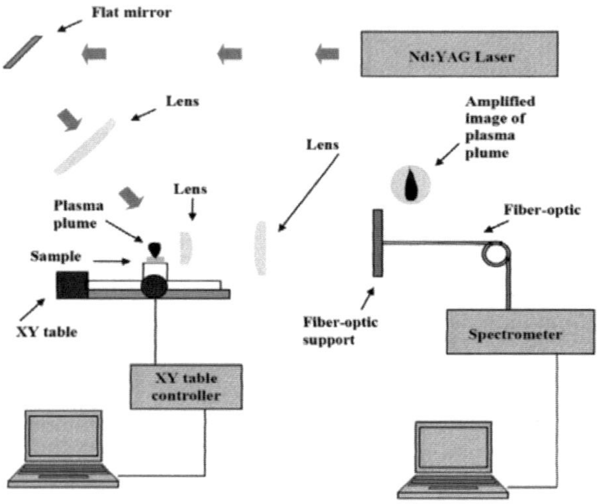


**Figure 2.** Experimental setup for spectra acquisition, performed in air.

## 5. Results and discussion

The theoretical spectra construction, figure 3, begins with data files generated through fundamental data obtained at NIST (National Institute of Standards and Technology) website [32]. A table that contains wavelengths of emission lines, λ, related with transition probabilities (which are given by Einstein coefficient, A), degeneracy of energy levels, (2J+1), energy values of the levels involved on transitions and electronic configurations is used to assign an intensity to the emission lines. So, using data acquired at NIST on equation (3), theoretical spectra are constructed.

In figure 3, it can be seen that emission at 400-450 nm is due de-excitation of neutral atoms while emission at ultraviolet region is due de-excitation of ionized atoms. For selective photo ionization, the plume must be neutral and monatomic and this explains the choice of interest region in 400-450 nm.

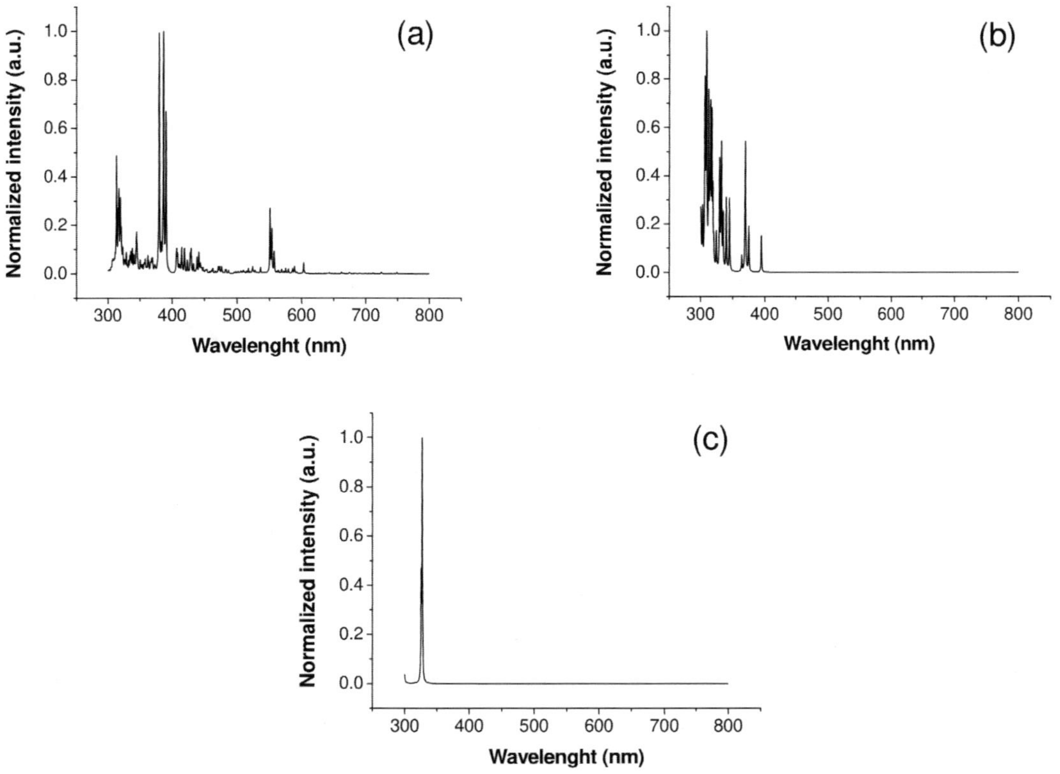


**Figure 3.** Theoretical molybdenum spectra. (a) Emission lines for neutral atoms; (b) emission lines for singly ionized atoms and (c) emission lines for doubly ionized atoms.

Quantitative data were acquired assuming Local Thermal Equilibrium approximation and excitation temperature was calculated through intensity ratio between two emission lines, equation (4). Two neutral molybdenum emission lines were considered for each 5 nm spectral range and the measurement was made for five different positions along the expansion axis.

Figure 4 and figure 5 represent the excitation temperature along the plasma plume. Despite the large error bars, it can be seen that the behavior of the values of temperature does not change for two lasers: the value of temperature has a maximum at the center and is minimum at the ends of the plume, that agree with literature [33]. The average temperature is 15000 K.

Another point that should receive attention is the electron population distribution. Figure 6 and figure 7 show the evolution of electron population for $4d^4$ 5s 6s state. In the first portion of the plot, Boltzmann distribution was considered. After it, the plasma reaches a non-collision regime, in which the intensity of emission light decreases and it can be attributed to spontaneous emission.

During the experiments, other important results were also obtained:

- The height of plume was measured and is nearly 2mm.
- Plasma "survives" just a short period after the laser pulse ends. The plasma lifetime is, approximately, 250 ns.
- The target must move relative to the laser avoiding cratering. This movement produces instability in spectra amplitude, but does not affect the intensity ratio of emission lines.

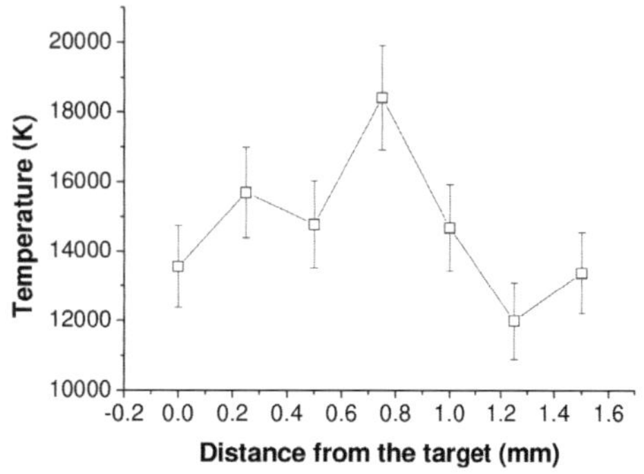


**Figure 4.** Excitation temperature along the plasma plume for 355 nm laser.

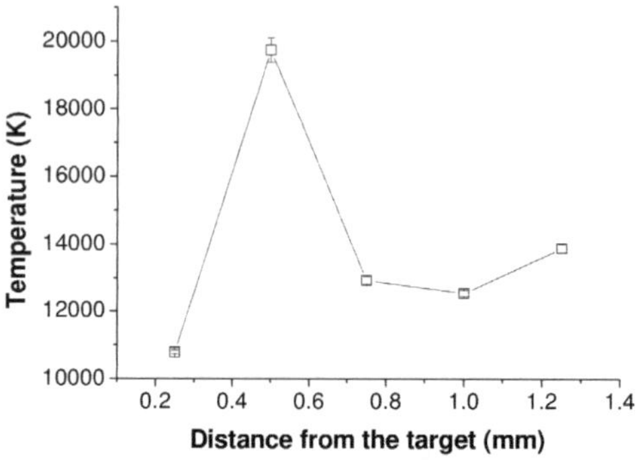


**Figure 5.** Excitation temperature along the plasma plume for 532 nm laser.

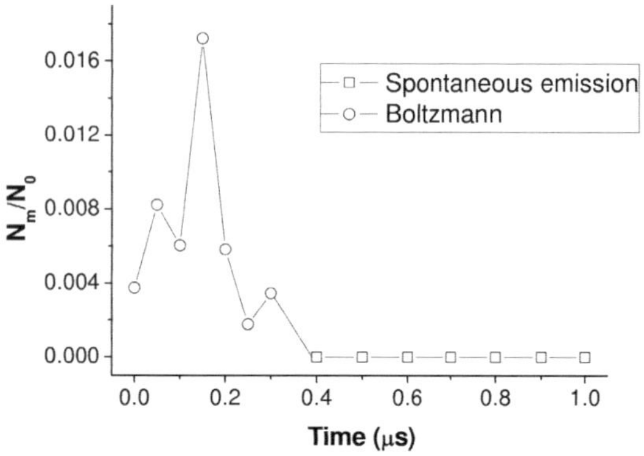


**Figure 6.** Electron population distribution for $4d^4 5s6s$ state. For 355 nm laser.

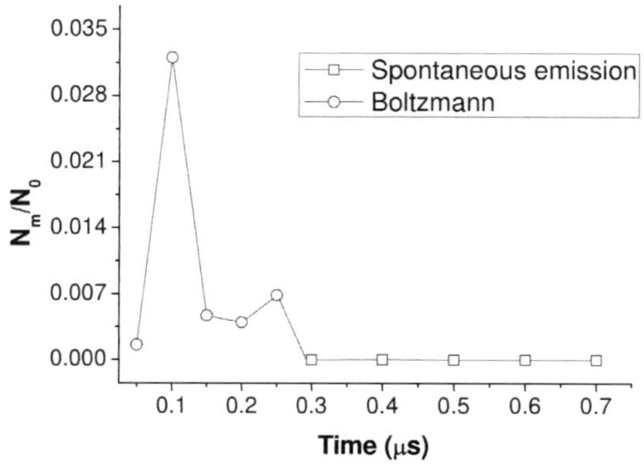


**Figure 7.** Electron population distribution for $4d^4 5s6s$ state. For 532 nm laser.

## 6. Conclusions

In this work, optical emission spectroscopy of laser-ablated molybdenum was performed in air. The generated plasma was considered at local thermal equilibrium, so that plasma temperature and electron population distribution were determined by Boltzmann distribution. The average temperature is 15000 K, and it has a maximum at the center of the plasma plume and a minimum at the ends. Through electron population distribution, it was observed that, near the target surface, both atoms and ions are in excited states and tends to decay to the ground or metastable state as the plume expands.

Future work will be devoted to improve the accuracy of the measurements and to demonstrate, more precisely, the molybdenum plasma properties in air. Moreover, it is necessary to establish the optimal experimental conditions for evaporation and data acquisition for subsequent measurement in vacuum.

## References

[1] Vogel A and Venugopalan V 2003 *Chemical Review* **103** 577-644

[2] Soonawala D, Amin T, Ebmeier K P, Steele J D, Dougall N J, Best J, Migneco O, Nobili F and Scheidhauer K 2002 *NeuroImage* **17** 1193-202

[3] Cheng J, Perrie W, Sharp M, Edwardson S P, Semaltianos N G, Dearden G and Watkins K G 2009 *Applied Physics A* **95** 739-46

Journal of Physics: Conference Series **370** (2012) 012052 doi:10.1088/1742-6596/370/1/012052


[4] Gurlui S, Agop M, Nica P, Ziskind M and Focsa C 2008 *Physical Review E* **78** 026405
[5] Virt I S, Shkumbatyuk T P, Kurilo I V, Rudyi I O, Lopatinskyi T Ye, Linnik L F, Tetyorkin V V and Phedorov A G 2010 *Semiconductors* **44** 544-9
[6] Scott C D, Arepalli S, Nikolaev P and Smalley R E 2001 *Applied Physics A: Materials Science & Processing* **72** 573-80
[7] Srinivasan R and Braren B 1989 *Chemical Review* **89** 1303-16
[8] Morales A M and Lieber C M 1998 *Science* **279** 208
[9] Pronko P P, VanRompay P A, Zhang Z and Nees J A 1999 *Physical Review Letters* **83** 13
[10] Boyd R E 1982 *Int. J. Appl. Radiat. Isot.* **33** 801-9
[11] Hatazawa J, Shimosegawa E, Toyoshima H, Ardekani B A, Suzuki A, Okudera T and Miura Y 1999 *Stroke* **30** 800-6
[12] Schwab C, Damiao A J, Silveira C A B, Neri J W, Destro M G, Rodrigues, N A S and Riva R 1998 *Progr. Nucl. Energy* **33** 217-64
[13] Lucas D M, Ramos A, Home J P, McDonnell M J, Nakayama S, Stacey J P, Webster S C, Stacey D N and Steane A M 2004 *Physical Review A* **69** 012711
[14] Kaschejev N A, Polyakov L A and Tunin V V 1993 *Nuclear Instruments and Methods in Physics Research A* **334** 27-32
[15] Chuvilin D Yu, Meister J D, Abalin S S, Ball R M, Grigoriev G Yu, Khvostionov V E, Markovskij D V, Nordyke H W and Pavshook V A 2003 *Journal of Radioanalytical and Nuclear Chemistry* **257** 1 59–63
[16] Amoruso S, Bruzzese R, Spinelli N, Velotta R, Vitiello M and Wang X 2003 *Physical Review B* **67** 224503
[17] Sherril M E, Mancini R C, Bailey J, Filuk A, Clark B, Lake P and Abdallah J Jr 2007 *Physical Review E* **76** 05640
[18] Narayanan V and Thareja R K 2004 *Applied Surface Science* **222** 382–93
[19] Sakka T, Iwanaga S, Ogata Y H, Matsunawa A and Takemoto T 2000 *Journal of Chemical Physics* **112** 19 8645-53
[20] Bye C A and Scheeline A 1993 *Applied Spectroscopy* **47** 12 2022-30
[21] Amoruso S, Bruzzese R, Spinelli N and Velotta R 1999 *J. Phys. B: At. Mol. Opt. Phys.* **32** R131–72
[22] Lorazo P, Lewis L J and Meunier M 2006 *Physical Review B* **73** 134108
[23] Gamaly E G, Rode AV and Luther-Davies B 1999 *Journal of Applied Physics* **85** 8 4213-21
[24] Hafez M, Khedr M, Elaksher F and Gamal Y 2003 *Plasma Source Science Technology* **12** 185-98
[25] Lacroix D, Jeandel G and Boudot C 1997 J. *Appl. Phys.* **81** 10
[26] Griem H 1963 *Physical Review* **131** 3
[27] *User Manual V2.3*. The EMC Team. November 8, 2009
[28] http://www.oceanoptics.com/technical/hr4000.pdf accessed 02/02/2012
[29] http://www.jyhoriba.com accessed 03/02/2012
[30] http://sales.hamamatsu.com/index.php?id=13189618&language=1&undefined accessed 02/02/2012
[31] http://www.pco.de/pco/php/products accessed 02/02/2012
[32] http://www.nist.gov/physlab/data/asd.cfm accessed 03/02/2012
[33] Capitelli M, Casavola A, Colonna G and De Giacomo A 2003 *Spectrochimica Acta Part B* **59** 271–89

Journal of Physics: Conference Series **370** (2012) 012053 doi:10.1088/1742-6596/370/1/012053

# Measurements of plasma edge electron temperature and density using visible spectroscopy in NOVA-UNICAMP tokamak

**F do Nascimento and M Machida**

Instituto de Física "Gleb Wataghin", UNICAMP, Campinas, S. P., Brazil

E-mail: fellype@gmail.com

**Abstract.** The electron temperature ($T_e$) and density ($n_e$) at the edge of NOVA-UNICAMP tokamak plasma were determined along the discharge using the concept of particle confinement time ($\tau_P$) uniqueness and spectroscopic measurements of hydrogen Balmer series emissions. We have used three absolutely intensity calibrated spectrometers with photomultipliers for simultaneous measurements of hydrogen alpha, beta and gamma emissions throughout the discharges. With the use of data from Johnson and Hinnov's table, we have performed an interactive method to find electron temperatures and densities that satisfy the $\tau_P$ uniqueness to obtain the temporal evolution of $T_e$ and $n_e$ parameters. The results achieved are in agreement with the expected values for these parameters at the edge of the NOVA-UNICAMP tokamak plasma.

## 1. Introduction

In plasma physics the development of diagnostic tools and methods to obtain plasma parameters, like temperatures, densities and confinement times, is an important matter. In this work we present a diagnostic method used to determine the local electron temperature ($T_e$) and density ($n_e$) in the NOVA-UNICAMP tokamak (NUt) plasma using visible spectroscopy, which has the advantage to be a passive method. The NOVA-UNICAMP [1] is a small iron core tokamak with major radius $R$ = 0.30 m, plasma radius $a$ = 0.06 m, toroidal field $B_T$ = 0.8 T and characteristic plasma current of $I_P$ = 10 kA.

The use of hydrogen spectral emissions for estimation of particle confinement times ($\tau_P$) in tokamaks is a well known technique. The $\tau_P$ calculation depends on $T_e$ and $n_e$ values and it is our starting point to make the reverse path, that is, the use of $\tau_P$ values to determine electron densities and temperatures.

Each tokamak has a characteristic value for the particle confinement time in a given place of the plasma and its value should be independent of the spectral emission observed. It means that the $\tau_P$ value must be unique in a certain plasma position.

Combining the absolute brightness measurements of the hydrogen alpha, beta and gamma emissions and data from Johnson and Hinnov's tables [2] we have developed an interactive method to find $T_e$ and $n_e$ values that satisfy the $\tau_P$ uniqueness. Using this method it was possible to determine electron densities and temperatures for various time instants along the tokamak discharges.

The main concepts of this diagnostic were developed firstly by Daltrini [3], where the measurements of hydrogen emissions were taken in different tokamak discharges, whereas, in this work the hydrogen emissions, $H_\alpha$, $H_\beta$ and $H_\gamma$, were measured simultaneously using three absolutely intensity calibrated spectrometers equipped with photomultipliers, avoiding therefore the problem of discharge reproducibility. Furthermore, the light emitted by the plasma was collected using three optical fibers, instead of the lens collection system used before, and a collimation system was adopted to have some degree of spatial definition on measurements.

## 2. Theoretical model

In pure hydrogen plasma and in the equilibrium condition ($dN_e/dt = 0$) we can use the particle confinement time theory [2,4,5] and combine it with the hydrogen spectral emissions to find the following expression for $\tau_P$ (considering only the Balmer series):

$$\tau_P[H_{j2}] = \frac{N_e}{4\pi^2 l r_0 \overline{R}[H_{j2}]B_{j2}} \tag{1}$$

where $N_e$ is the total number of electrons in the plasma, $l$ is the toroidal length, $r_0$ the peak position of hydrogen emissions ($r_0 \approx 0.8a$ in NUt) and $B_{j2}$ is the brightness of the $H_{j2}$ emission ($j > 2$). The $R[H_{j2}]$ factor is the average ionization rate per $H_{j2}$ photon emission rate (for instance, $H_{32}$, $H_{42}$ and $H_{52}$ stands for $H_\alpha$, $H_\beta$ and $H_\gamma$ emissions respectively). The $R[H_{j2}]$ factor can be written as [2,3,4]:

$$R[H_{j2}] = \frac{1}{j^2}\frac{n_e S}{A_{j2} r_1(j)} exp[(I_1 - I_j)/T_e] \tag{2}$$

where $I_j$ is the ionization potential for the $j$ level, $A_{j2}$ is the transition probability from level $j$ to level 2 and $T_e$ is in eV. The $S$ and $r_1(j)$ coefficients, that depends on electron temperature and density, are given in the work of Johnson and Hinnov [2].

In practice, the calculation of $\tau_P$ values itself are not necessary, being sufficient to calculate the ratio between confinement times for different hydrogen emissions, and to find electron temperatures and densities values that satisfy the relationship:

$$\frac{\tau_P[H_\alpha]}{\tau_P[H_\beta]} = \frac{\tau_P[H_\beta]}{\tau_P[H_\gamma]} = 1 \tag{3}$$

This reduces to calculate only:

$$\frac{\tau_P[H_\alpha]}{\tau_P[H_\beta]} = \frac{\overline{R}[H_\beta]B[H_\beta]}{\overline{R}[H_\alpha]B[H_\alpha]} \tag{4}$$

and

$$\frac{\tau_P[H_\beta]}{\tau_P[H_\gamma]} = \frac{\overline{R}[H_\gamma]B[H_\gamma]}{\overline{R}[H_\beta]B[H_\beta]} \tag{5}$$

Now we have two independent equations which allow us to find the $T_e$ and $n_e$ parameters.

The equations (4) and (5) show that the ratios between particle confinement times depends only on the measured brightness and its respective $R$ factors, avoiding difficulties inherent to the determination of $N_e$, $l$ and $r_0$ parameters of equation (1).

## 3. Analysis method

After calculate the $R$ coefficients, we made a fine interpolation of this values in function of $T_e$ and $n_e$ in order to improve the precision of our measurements.

With the brightness of $H_\alpha$, $H_\beta$ and $H_\gamma$ emissions and the $R$ coefficients, we choose, firstly, an electron temperature to start the interactions and then we make calculations of $\tau_P$ ratios (using equations 4 and 5) for a wide range of electron densities, changing it numerically and trying to find $n_e$

values that satisfy the $\tau_P$ uniqueness. Sample curves of this procedure are shown in figure 1. As we can see in this figure, we have two different values, $n_e' \approx 8.0\times10^{12}\text{cm}^{-3}$ and $n_e'' \approx 6.5\times10^{12}\text{cm}^{-3}$, one for each ratio curve, of electron density satisfying particle confinement time ratios equals to one. The meaning of this fact is that we have used an incorrect value of electron temperature and another attempt with a different value of $T_e$ is required in order to satisfy the equation (3).

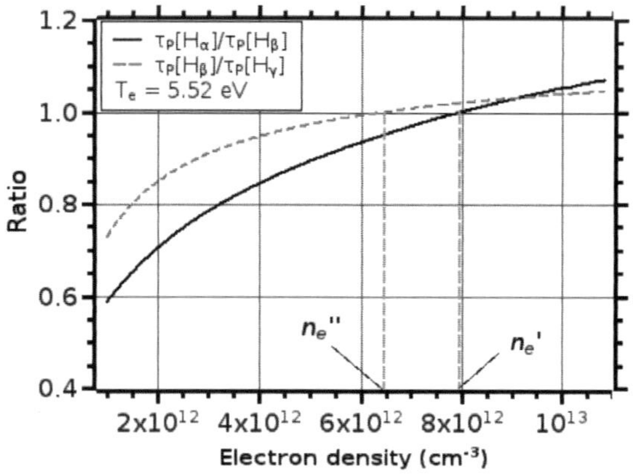


**Figure 1.** Curves of ratios between particle confinement times in function of $n_e$, for an electron temperature of 5.52 eV. The $n_e'$ value satisfy the condition $\tau_P[H_\alpha]/\tau_P[H_\beta]=1$while $n_e''$ satisfy $\tau_P[H_\beta]/\tau_P[H_\gamma]=1$.

Repeating this procedure for various temperature values, we can observe the behavior of $n_e'$ and $n_e''$ in function of $T_e$. Figure 2 shows these two curves.

The cross point between the two curves above determines the local $T_e$ and $n_e$ values in the plasma that satisfy the $\tau_P$ uniqueness for a given time instant. The values of $T_e$ and $n_e$ in figure 2 are 6.32 eV and $8.0\times10^{12}\text{cm}^{-3}$, respectively.

This procedure must be repeated for other time instants of the discharge in order to obtain the temporal evolution of $T_e$ and $n_e$.

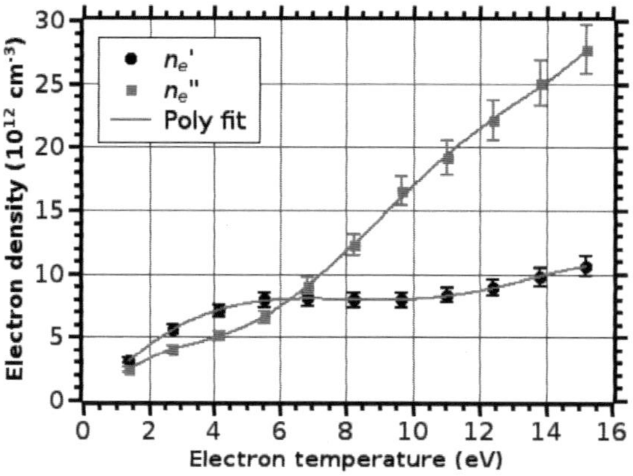


**Figure 2.** Curves of densities values that satisfy the $\tau_P$ uniqueness for each $T_e$ value.

## 4. Results and conclusions

Using the method previously described, we have determined the temporal evolution of the electron temperature and density at the edge of NOVA-UNICAMP tokamak plasma. The $T_e$ and $n_e$ results, shown in figure 3, were obtained in a 7.4 ms tokamak discharge plateau with 12 kA plasma current.

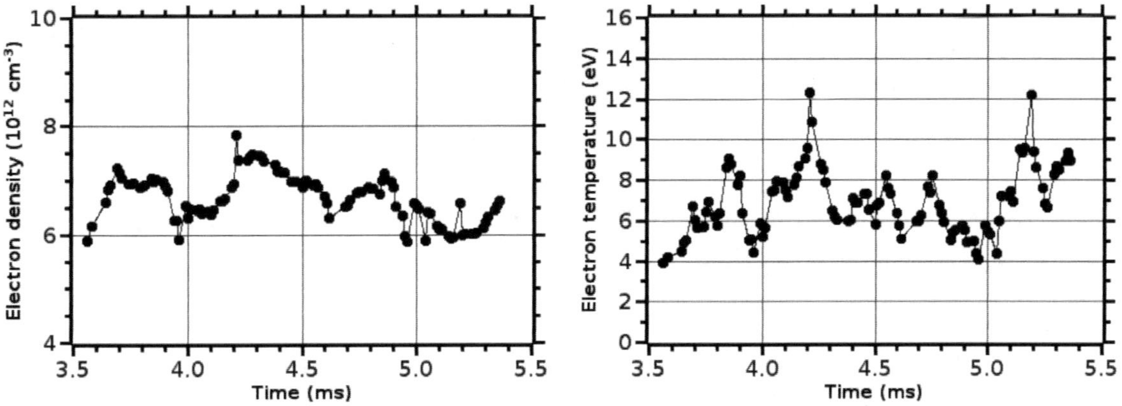


**Figure 3.** Time evolution of electron density (left) and temperature (right) for a tokamak shot with a work pressure of approximately $1.08\times10^{-4}$ Torr.

The average electron temperature obtained in this discharge was $6.9 \pm 0.8$ eV while the average electron density was $(6.7 \pm 0.5)\times10^{12}$cm$^{-3}$. The results of $T_e$ and $n_e$ temporal evolution are in agreement with the expected values for these parameters at the edge of the NOVA-UNICAMP tokamak plasma [6]. This indicates that the proposed diagnostic method can be used to monitor the electron densities and temperatures at the edge of tokamak plasmas.

The next step is to apply this technique, using simultaneous measurements of hydrogen emissions, in tokamaks larger than the NOVA-UNICAMP and to compare the results with other diagnostic tools like Langmuir probes and microwave interferometer in the same discharge and radial position.

**Acknowledgments**
This work was supported by CAPES.

**References**
[1] Daltrini A M, Machida M et al. 2002 *Brazilian J. of Physics* **32** 26-9
[2] Johnson L C and Hinnov E 1973 *J. Quant. Spectrosc. Radiat. Transfer* **13** 333-58
[3] Daltrini A M and Machida M 2005 *Review of Scientific Instruments* **76** 053508
[4] Machida M et al 1995 *Brazilian Journal of Physics* **25** 7-13
[5] Griem H R 1964 *Plasma Spectroscopy* (USA: McGraw-Hill Book Company)
[6] Monteiro M J R, Machida M, Daltrini A M et al 2002 *Brazilian J. of Physics* **32** 54-6
[7] Wiese W L, Smith M W and Glennon B M 1966 *Atomic Transition Probabilities* Vol. I – Hydrogen Through Neon – NSRDS-NBS 4

Journal of Physics: Conference Series **370** (2012) 012054 doi:10.1088/1742-6596/370/1/012054

# Plasma temperature measurement in a hybrid discharge by using optical diagnostics

**F Parada-Becerra[1], R Cabanzo[1], V Dugar-Zhabon[1], P Tsygankov[2], E Mejía-Ospino[1] and E D V Niño[1,3]**

[1]Grupo de Investigación en Física y tecnología del Plasma y Corrosión – FITEK, Universidad Industrial de Santander, A.A. 678, Bucaramanga, Santander, Colombia
[2]Bauman Moscow State Technical University, Moscow, Rusia
[3]Grupo de Investigación en tecnología del plasma – GINTEP, Universidad Pontificia Bolivariana, A.A. 2932, Bucaramanga, Santander, Colombia

E-mail: vdougar@uis.edu.co

**Abstract.** Of the plasma methods used to improve the tribological properties of the solid surface [1-3], the treatment of metal pieces in the hybrid discharge, which is a combination of high voltage and electric arc discharges, seems very promising [4,5]. This method is developed in the Plasma Physics and Technology Laboratory at the Universidad Industrial de Santander (Colombia). In our work, the hybrid discharge is ignited in tungsten vapor. The plasma temperature and atomic composition of the discharge are measured through its optical spectrum.

## 1. Introduction

The hybrid technology of surface treatment includes two steps: the vaporization and ionization of metal atoms in an arc high current discharge and putting the obtained metal vapor under high voltage pulses. The samples to be treated serve as the cathode, thus being bombarded by a flow of high energy ions. So these ions turn to be responsible for modifying the surface characteristics. For the surface processing technologies, the atomic composition and temperature are the essential characteristics of the processing plasma. Some limitations of the electric probe methods in their use in the metal vapor discharges and their impossibility to determine the atomic composition prompt to resort to the optical diagnostics.

## 2. Experiment

The hybrid discharges are initiated in the MOSMET reactor [6-10] which is shown in figure 1. The reactor chamber of $70\times70\times80\ cm^3$ is evacuated up to 0.1 mPa by a 1500 l/s turbo molecular pump. The cathode of the arc vaporizer disposed on the upper wall of the chamber (see figure 1) is made of 99.9% pure titanium. Titanium as a treatment material is chosen because of its anticorrosive properties and compatibility with organic matter. The arc discharge current is maintained at a level of 150 A with a voltage of 30 V. The high voltage pulses are applied within the area between the cathode situated on the lower wall of the chamber and the chamber stainless walls. The 10 kV pulses are of 0.25 ms duration at 30Hz repetition frequency.

The spectral diagnostic system comprises an optical fiber cable which receives the discharge light and transports it to a DK480 spectrometer with a diffraction grid of 1200 $mm^{-1}$ and a CCD Princeton Instruments camera of 1024×128 pixels. The optical spectra of the hybrid discharge are observed in two positions of the fiber optic receiver: in and outside the chamber. In the latter case the light emission of the discharge passes through the sapphire window in a lateral chamber wall on the fiber optic receiver.

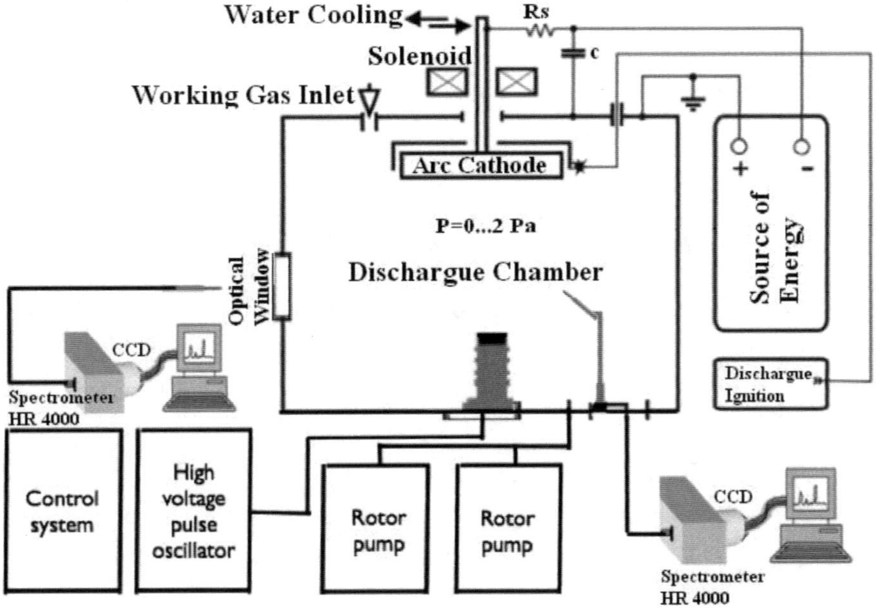


**Figure 1.** Scheme of the MOSMET reactor with the spectroscopic diagnostic line.

## 3. Results and discussion

Figures 2 and 3 show the spectra emitted by the hybrid plasma in the above mentioned positions of the optical fiber receiver. The spectra, which are practically identical, are comprised of TI and TiII lines, which evidences that the arc consists predominantly of ions and neutral Ti atoms. In these spectra, TiIII lines or ions of higher charges are not observed although there should be triple charge ions and even ions of a higher charge states in accordance with the Saha thermodynamic equilibrium formula in the arc discharge. The problem of detection of the multicharged ion spectral lines is related to the fact that these lines fall in the ultraviolet diapason where the sensitivity of the used spectrometric scheme falls down. For the ion implantation processes, it is important to determine which multicharged species are in the incident flux. In order to solve this problem, we are planning to expand the detectable spectral range to 200 nm.

The plasma temperature is calculated taking into consideration TiI lines and the parameters used for these calculations are presented in tables 1 and 2.

The plasma temperature is calculated by using the spectroscopic lines of titanium atoms TiI whose characteristics are presented in table 1 and table 2.

Here:

$$I_{\lambda}^{ki} = Fn_k A_{ki} = FC_s g_k \frac{e^{-\frac{E_k}{K_B T}}}{U_s(T)} A_{ki} \tag{1}$$

$I_{\lambda}^{ki}$ is the integrated intensity of a spectral line at a wavelength, emitted at the transition between the states $k$ and $i$, $C_s$ is the concentration of the radiative atoms, $F$ is the experimental parameter which determines the optic system efficiency to collect the radiation, $g_k$ is the statistic weight of the excited electronic state up to energy of $E_k$, $k_{\beta}$ is the Boltzman constant, $T$ is the plasma temperature, and $U_s$ is the partition function for the radiative species.

Figures 4 and 5 show the behavior of the hybrid plasma temperature calculated in accordance with the data presented in tables 1 and 2.

As it is shown in figures 4 and 5, the temperatures calculated from the data obtained by the receiver placed in the chamber volume can be estimated as 7127 ± 200K and by the receiver which is outside of chamber is 7574 ± 200K.

## 4. Conclusions and recommendations

The plasma temperature of a hybrid titanium discharge is measured with a spectroscopic method. The data obtained by two mentioned optical fiber receivers practically coincide and can be estimated as 7300 K.

The present optical study was done through the spectral window of 480 – 540nm and in order to identify the spectral lines of multicharged titanium ions (TiIII and TiIV), in case they exist, we are planning to increase the spectral range to 200-600nm.

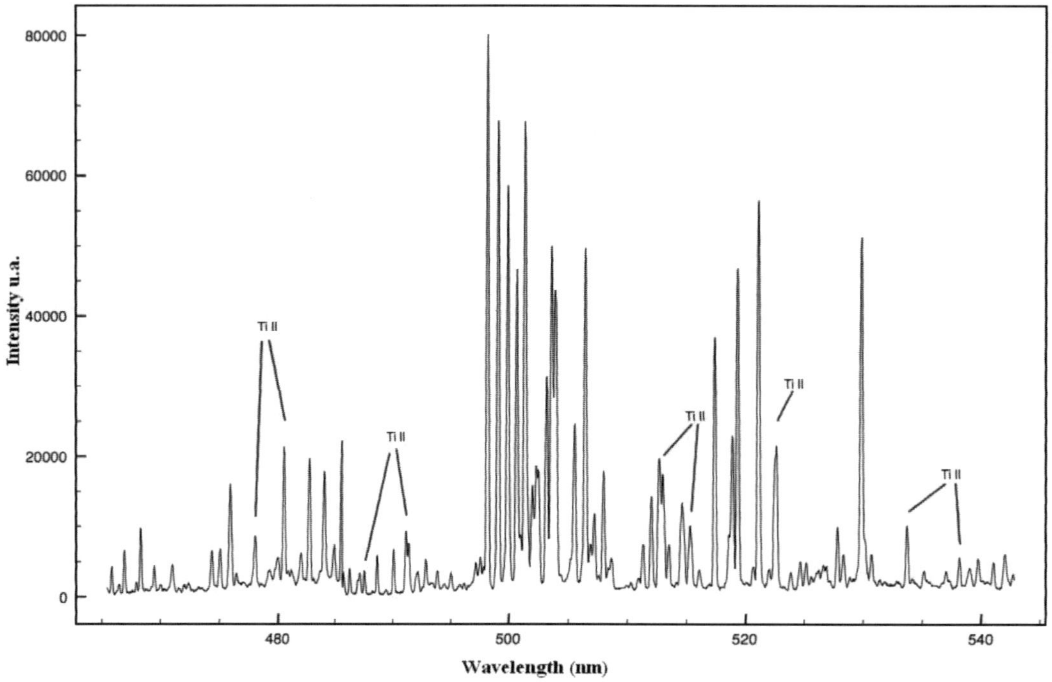


**Figure 2.** Spectrum obtained by the receiver situated in the interior of the treatment chamber.

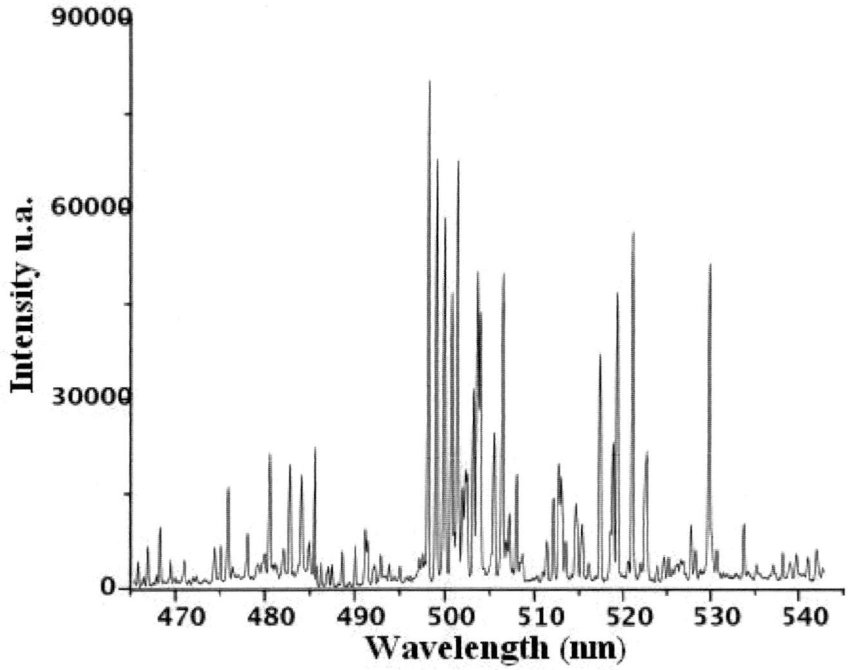


**Figure 3.** Spectrum obtained by the receiver situated at the exterior of the treatment chamber.

**Table 1.** TiI line characteristics obtained with help of the optical fiber receiver situated interior the discharge chamber.

| Wavelength [nm] | $I_\lambda^{ki}$ [ua] | $E_{k\,\mathrm{k}}$ [eV] | $g_{k\,\mathrm{k}}$ | $A_{ki}$ | $Ln\left(\frac{I_\lambda^{ki}}{g_k A_{ki}}\right)$ |
|---|---|---|---|---|---|
| 482.03396 | 3154.80143 | 4.0738 | 7 | 1.49E+07 | -10.40610 |
| 484.06795 | 6974.59070 | 3.4600 | 5 | 1.76E+07 | -9.44282 |
| 485.58196 | 6092.05817 | 4.8081 | 15 | 5.20E+07 | -11.76006 |
| 500.73246 | 13459.96140 | 3.2936 | 7 | 4.92E+07 | -10.14984 |
| 517.40183 | 9847.26358 | 2.3957 | 5 | 3.80E+06 | -7.56500 |

**Table 2.** TiI line characteristics obtained with help of the optical fiber receiver situated at the exterior the discharge chamber.

| Wavelength [nm] | $I_\lambda^{ki}$ [ua] | $E_{k\,\mathrm{k}}$ [eV] | $g_{k\,\mathrm{k}}$ | $A_{ki}$ | $Ln\left(\frac{I_\lambda^{ki}}{g_k A_{ki}}\right)$ |
|---|---|---|---|---|---|
| 482.03396 | 2350.30026 | 4.0738 | 7 | 1.49E+07 | -10.70048 |
| 484.06795 | 7044.73286 | 3.4600 | 5 | 1.76E+07 | -9.43281 |
| 485.58196 | 6092.05817 | 4.8081 | 15 | 5.20E+07 | -11.76006 |
| 500.73246 | 13373.87940 | 3.2936 | 7 | 4.92E+07 | -10.15626 |
| 517.40183 | 7364.04823 | 2.3957 | 5 | 3.80E+06 | -7.85558 |

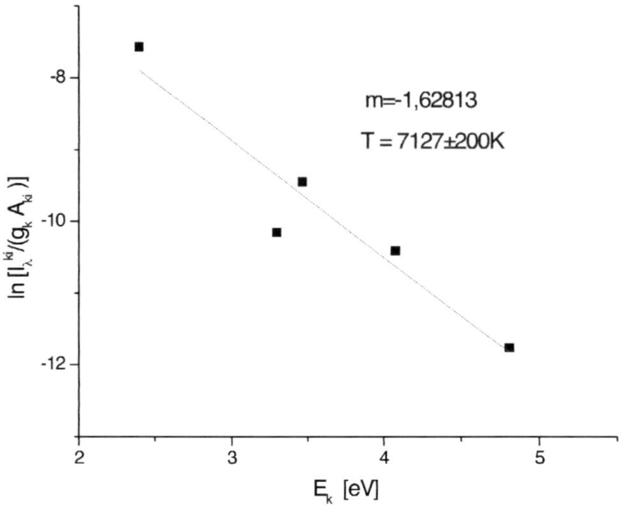


**Figure 4.** Graph for the temperature calculations: the fiber optical receiver is interior of the discharge chamber.

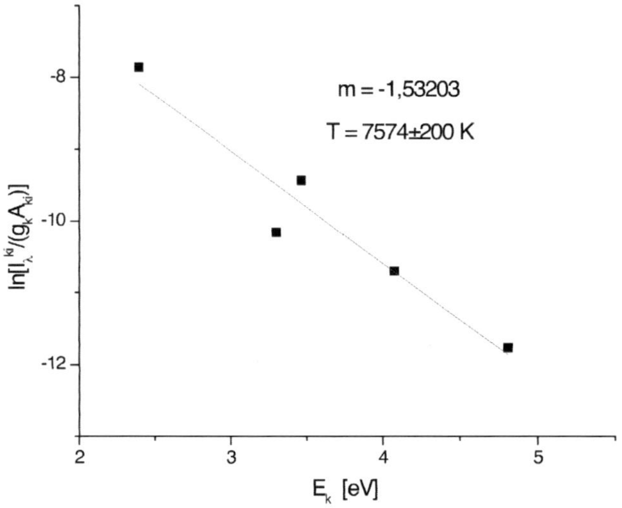


**Figure 5.** Graph for the temperature calculations: the fiber optical receiver is exterior of the discharge chamber.

**Acknowledgments**
This work was supported by Colciencias under project No 1102-06-17623 and the mobility program 2011 of the Universidad Industrial de Santander.

**References**

[1] Anders A A 2007 *Brief Modern History of Cathodic Arc Coating in 50 Years of Vacuum Coating Technology and the Growth of the Society of Vacuum Coaters* ed Donald M. Mattox and Vivienne Harwood Mattox (Society of Vacuum Coaters) chapter 16
[2] Hantzsche E 2003 *IEEE Trans. Plasma Sci.* **31** 799

[3] Ohring M 1992 *The Materials Science of Thin Films* (London: Academic Press Ltd.) chapter 3 p 5
[4] Current M I, Yamada I and Cheung N W 1992 *New Technologies for Ion Implantation* (Ion Implantation Technology) ed J F Ziegler (North-Holland)
[5] Khvesyuk V I and Tsygankov P A 1997 *Surface and Coatings Technology* **96** 68
[6] Tsygankov P, Plata A, V Niño Ely D, Ochoa C, Parada F, Chacón C and Dugar-Zhabon V D 2011 *Revista Colombiana de Física* **43** 458-62
[7] Dougar-Jabon V D, Dulce Moreno J and Tsygankov P A 2002 *Review of Scientific Instruments* **73** 828
[8] Valbuena Niño E D, Peña D, Salinas D V and Chinchilla L F 2011 *Revista Iteckne* **8** 31-6
[9] Valbuena Niño E D, Dulcé H J and Dugar-Zhabon V 2010 *Revista Colombiana de Física* **42** 387-92
[10] Dulce Moreno J H, Tsygankov P A, Dugar-Zhabon V D, V Niño E D and Parada F F 2011 *Puente Revista Científica* **5** 21-6

Journal of Physics: Conference Series **370** (2012) 012055 doi:10.1088/1742-6596/370/1/012055

# RF plasma parameter determination by a Langmuir multipoint double probe array

**I A Rojas-Olmedo[1], R López-Callejas[1,2], A de la Piedad-Beneitez[1], R Valencia-Alvarado[2], R Peña-Eguiluz[2], A Mercado-Cabrera[2], S R Barocio[2], A E Muñoz-Castro[2] and B G Rodríguez-Méndez[2]**

[1]Instituto Tecnológico de Toluca, AP 890, Toluca, Estado de México, Mexico
[2]Instituto Nacional de Investigaciones Nucleares, Plasma Physics Laboratory, AP 18-1027, 11801 México DF, Mexico

E-mail: regulo.lopez@inin.gob.mx

**Abstract:** A multipoint double Langmuir (MDL) probe system, which is exempt from interference, has been designed and assembled to be applied to an RF plasma. The system provides the measurement of fundamental plasma parameters such as density, temperature, plasma potential, etc. on the basis of the Bohm Approximation Theory and the Orbital Movement Limit. Thus, one pair of the MDL system is selected so as to consider the right plasma parameters within the prevailing pressure-power intervals. Both the hardware and software of the system have been applied to the modification of material properties by means of the PIII process.

## 1. Introduction

Radio-frequency (RF) plasmas are widely applied to material processing, such as plasma immersion ion implantation (PIII or PSII). These plasmas are generally characterized by means of Langmuir probes in some of their kinds, simple (SP), double (DP) or triple (TP), whose *I-V* curve contains information about the electron temperature ($T_e$), particle density ($n$), etc. In contrast to SP, DP is constituted by two electrodes whose voltages are mutually referenced (electrically floated electrodes) which is particularly convenient in the extremely noisy RF plasma environment with constant ample plasma potential ($V_p$) fluctuations [1]. Probes have been developed in different geometries (planar, spherical, cylindrical, etc.) [2,3]. Several studies have shown that $n$ can be overestimated by a SP [4] whereby a DP validation would apply. The present report describes the measurement of particle density at one single point by means of a multipoint double Langmuir (MDL) probe array in an environment similar to an RF plasma discharge reactor.

## 2. Setup

The probe radius to Debye length ratio $\xi_p=R_p/\lambda_D$ value is relevant, as shown by Laframboise [5]. This relation is important at the moment to choose the probe radius ($R_p$), when selecting the collision theory approximation to describe mathematically the attraction and recollection of species towards the probe, such as the Bohm approximation and the orbital movement limit (OML) [6]:

$$n_i = \frac{I_{isat}}{0.61 A_S e}\sqrt{\frac{m_i}{kT_{eV}}} \tag{1}$$

$$n_{OML} = \sqrt{\frac{[-\Delta(I_i^2)/\Delta V_p]m_i}{0.2e^3 A_p^2}} \tag{2}$$

The first one is useful when $\xi_p \leq 3$ [6], whereas, with $\xi_p \geq 10$, the OML is dependable [7]. It follows a considerable $3 < \xi_p < 10$ gap. Chen [8] has developed a plot relating $\xi_p$ and $n$ in this region for different $R_p$ values. According to this, $n < 1\text{x}10^{17}$ m$^{-3}$ can be measured at $\xi_p \leq 3$ provided that $R_p < 0.1$ mm. Likewise, with $R_p > 0.4$ mm, one can measure $n \geq 1\text{x}10^{17}$ m$^{-3}$ values provided that $\xi_p \geq 10$. Thus, the $\xi_p$ gap could be eluded at specific $R_p$ values.

The discharge reactor used in the present study is constituted by a 50 cm long glass cylinder, with a 10 cm radius and endowed with two flat lids, where plasma is produced a low pressure with argon working gas, by an RF generator operating at 13.56MHz. The multipoint probe contains three sets of Tungsten (S1, S2, S3) of one same length ($L_1$=$L_2$=$L_3$= 5.56 mm) and three different radii ($r_1$=0.090 mm, $r_2$= 0.435 mm and $r_3$= 0.640 mm). Each point is insulated and supported by an alumina tube thanks to an especially designed coupling (figure 1).

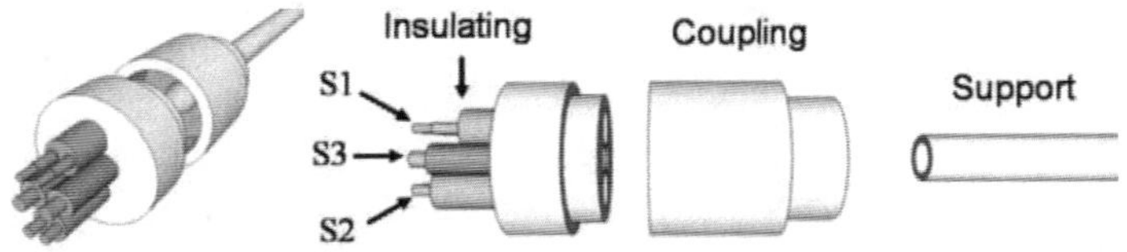


**Figure 1.** Main parts of the multipoint probe with one same length and three different radii.

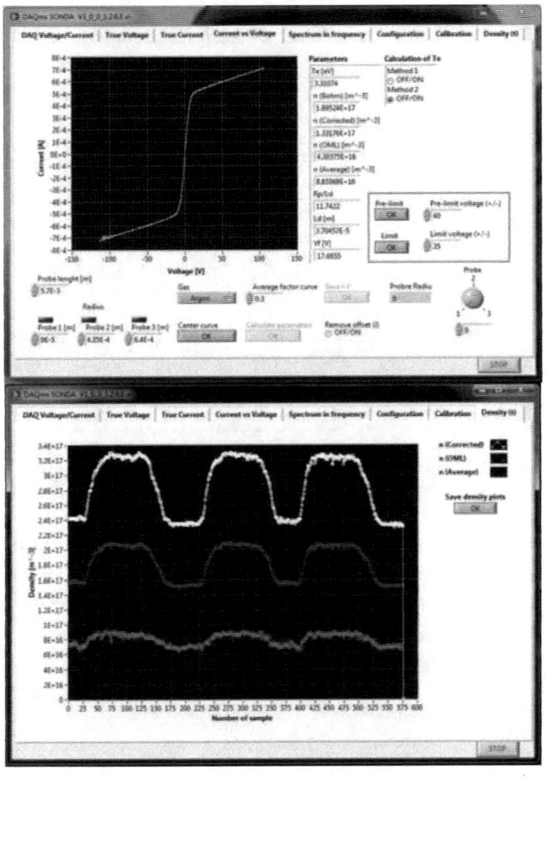

(a)

(b)

**Figure 2.** Plasma characterization program. (a) Calculated $n$ during the voltage scanning (b) Response of the RF plasma density to a sudden pressure change.

The characteristic *I-V* curve is obtained due to a data acquisition system [9] in order to specify the $n$ value at one point and then to get the information in similar conditions. A PC program was developed so as to calculate the plasma parameters *in situ*, visualizing the $n$ results. Figure 2(a) shows the program main window with the input parameters (voltage limit, probe radius $R_p$, etc.). Some output parameters are also listed, such as the $T_e$ value resulting from both the saturation current slope technique [10] and the mathematical approximation to the *I-V* curve [7], the correction to the $I$ signal shift created in practice by potential grads [10-12]. The program is capable to calculate three different values of the density: the first one was calculate from the Bohm approximation though a convergence iterative algorithm ($n_{\text{corrected}}$) [7], the second one by OML theory ($n_{\text{OML}}$) and finally the average between two mentioned above ($n_{\text{average}}$). Sudden pressure changes (up and down) were deliberately produced to test the PC program, the outcome is presented in figure 2(b).

## 3. Experimental results

The MDL was laterally introduced to the centre of the reactor varying the argon plasma pressure from 1 Pa to 30 Pa as well as the power (100-500 W). Figure 3 presents the ensuing plasma characterization results from the PC program, at constant 400 W, the pressure varying between 10 Pa and 30 Pa.

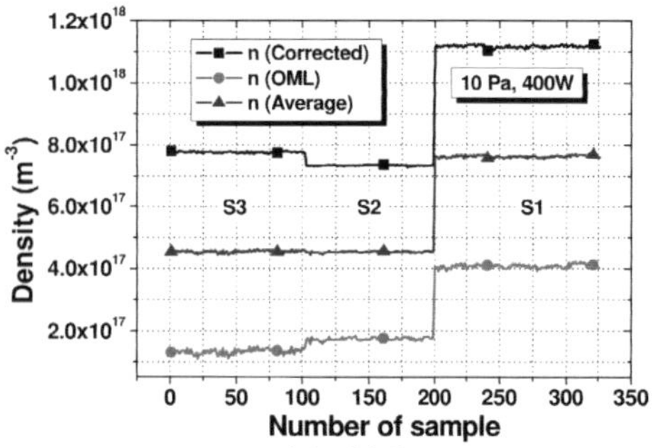


(a) 10 Pa

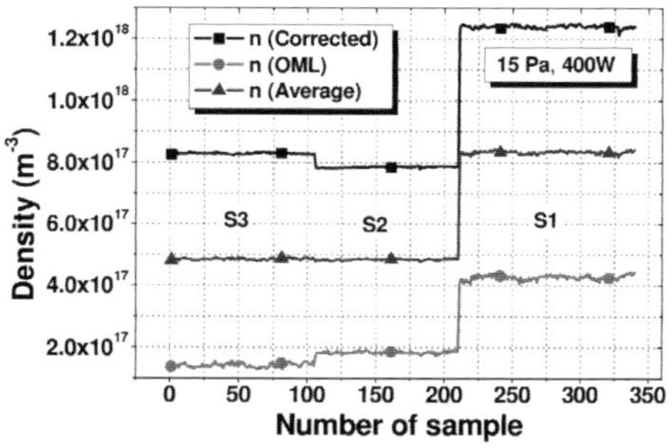


(b) 15 Pa

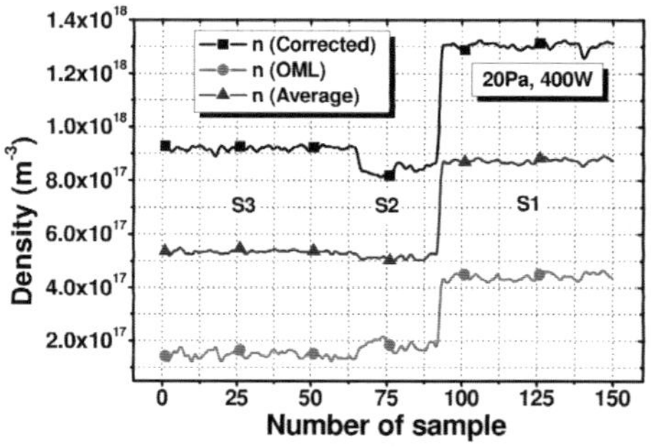


(c) 20 Pa

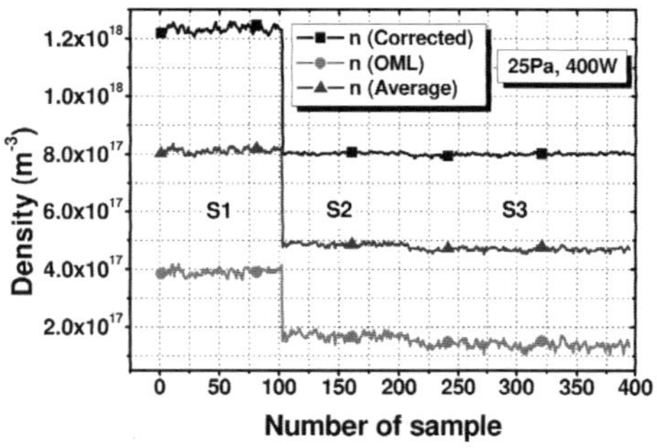


(d) 25 Pa

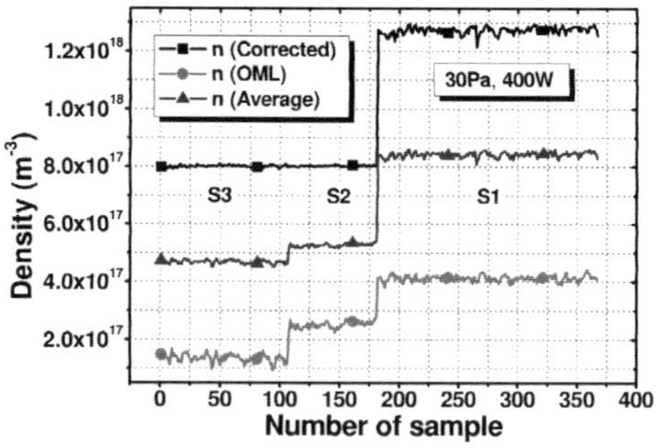


(e) 30Pa

**Figure 3.** Density using multipoint Langmuir probe with constant 400 W powerat different pessures.

At 1 Pa, $T_e \sim 3.5$ eV which decays to 2.5 eV as the pressure increases up to 10 Pa, in agreement with Herman *et al* [7]. The density, according to the Bohm approximation ($n_{\text{corrected}}$) using the probe radii $R_p = r_2$, $r_3$, are very alike whereas, with $R_p = r_1$, $n$ turns out to be overestimated given the gap ($3 < \xi_p < 10$) and the $R_p < 0.1$ mm region. However, the average of the $n_{\text{corrected}}$ and $n_{\text{OML}}$ values coincides with the results from $R_p = r_2$, $r_3$, as seen in figure 3.

Figure 4 displays the Bohm approximation results with $\xi_p \leq 3$, the OML theory for $\xi_p \geq 10$ and their average over $3 < \xi_p < 10$, with a minimal discrepancy between curves. The $R_p = r_2$ results have been selected to plot the pressure versus density plasma response (figure 5) indicating that the maximal density takes place within 17.5 Pa and 25 Pa, according to the power input.

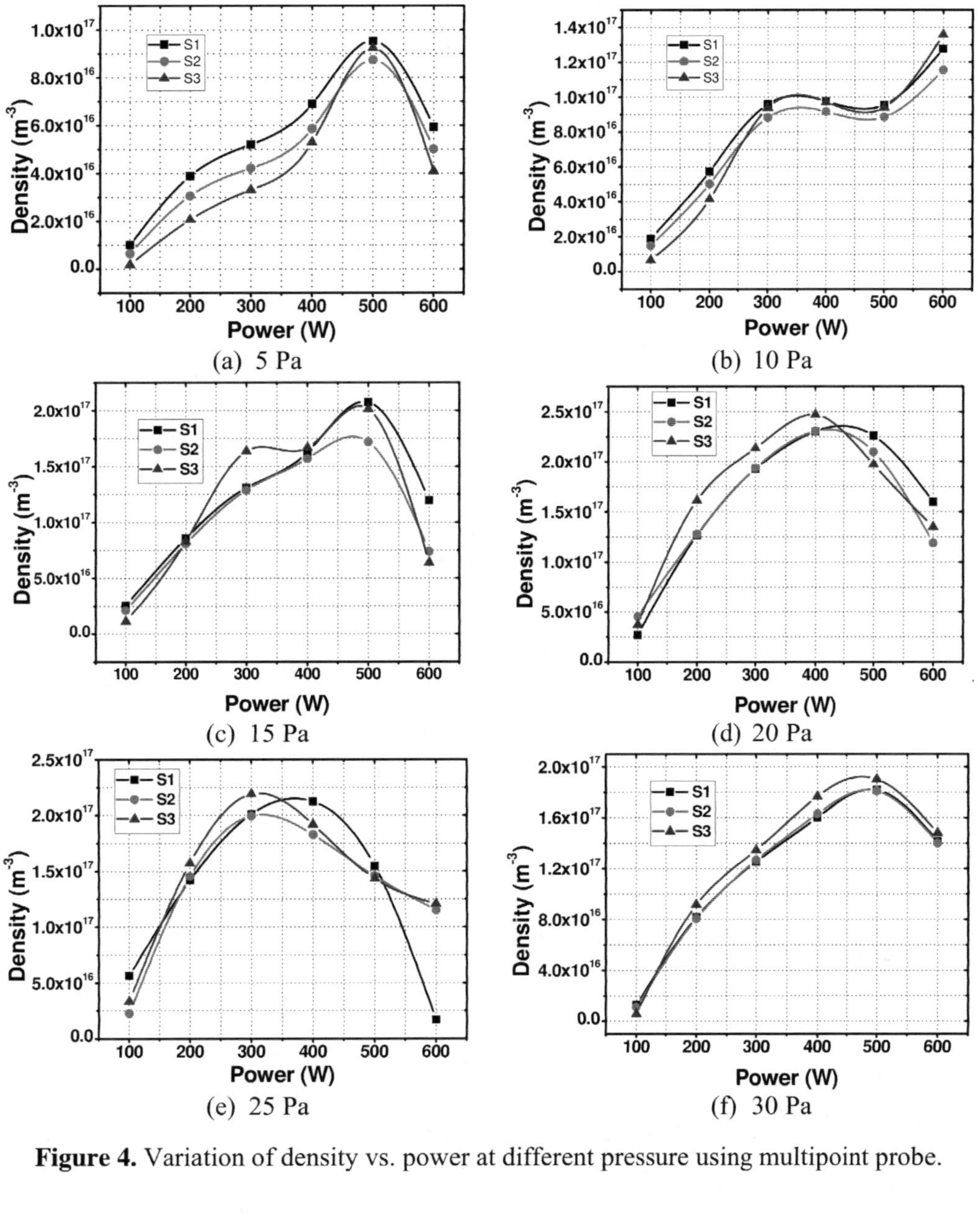


(a) 5 Pa (b) 10 Pa

(c) 15 Pa (d) 20 Pa

(e) 25 Pa (f) 30 Pa

**Figure 4.** Variation of density vs. power at different pressure using multipoint probe.

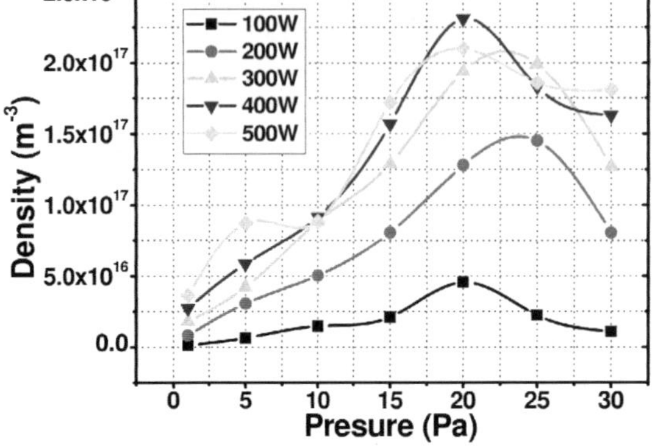


**Figure 5.** Density response vs. pressure.

## 4. Conclusions

A multipoint double Langmuir (MDL) probe array has enabled to deduce the *I-V* characteristic curves at one single point using three probe points of diver dimensions operating in the same environment. The $R_p$ = $r_2$ probe provides a good estimation of the density following the Bohm approximation ($n_{corrected}$). In the case of $R_p = r_1$ the OML theory has to be applied instead, according to the $\xi_p$ value in order to avoid the overestimation of the density. The specifically developed PC program not only calculated accurately the plasma parameters *in situ* but also to follow up the density behavior when the value of some of the remaining variables is modified. By plotting the pressure versus density plasma response when $R_p = r_2$, the maximal density is achieved between 17.5 Pa and 25 Pa.

## Acknowledgements

This work has received financial support from CONACYT, Mexico. The authors are grateful to the technical collaboration received from P. Angeles Espinoza, M. T. Torres Martínez, and I. Contreras Villa.

## References

[1] Jauberteau J L and Jauberteau I 2008 *Rev. Sci. Instrum.* **79** 033505
[2] Johnson J D and Holmes A J T 1990 *Rev. Sci. Instrum.* **61** 2628
[3] Hiroharu Fujita and Masanori Akazaki 1974 *Jpn. J. Appl. Phys.* Vol. **13**
[4] Castro R M, Cirino G A, Verdonck P, Maciel H S, Massi M; Pisani M B and Mansano R D 1999 *Contrib. Plasma Phys.* **39** 235-46
[5] Ruzic D N 1994 *Electric Probes for Low Temperature Plasmas* The American Vacuum Society Education Committee, New York, NY
[6] Tuszewski M and Tobin J A 1996 *Plasma Sources Sci. Technol.* **5** 640
[7] Herman D A and Gallimore A D 2008 *Rev. Sci. Instrum.* **79** 013302
[8] Chen F F 2001 *Physics of Plasmas* **8** 3029
[9] Rojas I A, López R, de la Piedad A, Valencia-Alvarado R, Peña-Eguiluz R, Mercado-Cabrera A, Barocio S R, Muñoz-Castro A E and Rodríguez-Méndez B G 2011 *Surface & Coatings Technology* **205** S397–S401
[10] Silla H, Dougherty T J 1972 *Combustion and Flame* **18** 65-74
[11] Ling Yi-Ming 2006 *J. Phys D: Appl. Phys* **39** 3305-9
[12] Welzel Th, Dunger Th, Kupfer H and Richter F 2004 *J. Appl. Phys.* **96** 6994

Journal of Physics: Conference Series **370** (2012) 012056 doi:10.1088/1742-6596/370/1/012056

# Optical emission spectra of ZnMnO plasma produced by a pulsed laser

**J Uzuriaga, J C Chamorro, R A Marín and H Riascos**

Grupo Plasma, Láser y Aplicaciones, Universidad Tecnológica de Pereira A.A. 097, Pereira, Colombia.

E-mail: hriascos@utp.edu.co

**Abstract.** Optical emission lines from the plasma generated by a laser ablation process have been investigated to gather information on the nature of the chemical species present. In particular, the experiments were carried out during the laser ablation of a ZnMnO target, in presence of a controlled oxygen atmosphere. The resolved emission spectra are dominated by the atomic neutral or singly ionized emission lines from Zn species, while the Mn I line was detected in emission spectrum at 40 mTorr only. The background continuum, intensities and widths of all observed lines increased with increasing gas pressure. The electron density and electron temperature were calculated for various gas pressures. The electron density was found to increase with the increment of the argon gas pressure, whereas electron temperature decreased. The electron temperature and density are found to be similar to those obtained in the well-known pulsed laser deposition (PLD) process.

## 1. Introduction

Pulsed Laser Deposition (PLD) of material processing is a technique which has given good results for various applications of thin films of wide range materials [1–6]. In PLD, a pulsed laser is focused onto a target of the material to be deposited. For sufficiently high laser energy density, each laser pulse vaporizes or ablates a small amount of the material creating a plasma plume. The ablated material is ejected from the target in a highly forward-directed plume. In PLD, the properties of the film growth are mostly affected by laser-ablation plasma; the composition of the plume is nearly identical to that of the target although there could be some mismatch due to incongruent melting of the target and/or intentional supply of atoms from the ambient gas. When the deposition is performed in a reactive ambient, in addition, the chemical reactions of the ablated species with the species in the reactive ambient play a crucial role in film deposition. Hence, the characteristics of the plume produced by pulsed laser ablation and the understanding of the processes involved in film deposition are of great interest. In this respect, it would be important to elucidate the correlation between plasma parameters and deposited film properties [7], one of the ultimate goals of the research on PLD mechanisms. The characterization of laser-ablation plasmas through the determination of their main parameters, as electron temperature and number densities of the different species present in the plasma, has gained high interest in recent years, allowing to improve the process control and providing better understanding of these complex and versatile spectroscopic sources. The most widely used method to determine the temperature of laser-ablation plasmas has been the Boltzmann plot method, which provides the excitation temperature from the measurement of line emissions

Journal of Physics: Conference Series **370** (2012) 012056 doi:10.1088/1742-6596/370/1/012056

from a single species. In its simplest form, only two emission lines, which differ on their upper level energies, are necessary to obtain the excitation temperature [9]. In some works, Boltzmann plots have been obtained from a higher number of emission lines [8–11]; in this case, a good correlation of the experimental data to a linear fitting indicates the validity of the Boltzmann equation. In some cases, the measured line intensities corresponded to the emission from the whole plasma [12–14], so the Boltzmann plot method provided a single value of the excitation temperature of the plasma. Also, time resolution is used in light detection, as the plasma temperature is known to experience a rapid decrease due to the plasma expansion. In other cases, the emission from the plasma is spatially resolved in order to obtain the temperature at a given distance from the sample [9]. In this work, we present experimental results on the laser ablation of a ZnMnO target in Oxygen atmosphere, calculating the temperature and density of plume by using the optical emission spectroscopy method. Optical emissions from the excited molecules, as well as atomic species—both ionic and neutral—were observed as a function of the gas pressure and temperature of vacuum chamber.

## 2. Experiment

The oxygen gas was introduced at the vacuum chamber to obtain a pressure of several mTorr. The chamber was first pumped at a pressure of $10^{-6}$ Torr with a turbo- molecular pump.

An Nd:YAG pulsed laser operating at 1064 nm and a pulse width of 9 ns was used as a power source for ablation of ZnMnO target. The power of the laser pulse was about 1.2 W and the elliptical spot was below 0.05 $cm^2$. The plasma plume was created by illuminating the focused laser pulse onto the target at an angle of 45°C with respect to the normal direction of the target by a plane-convex lens of a 22 cm focal length. The target had a set rotating speed of 2.4 rpm to avoid the non-uniform pitting of the target and the target-to-substrate distance was 60 mm. The light emitted by the plume was measured along the direction perpendicular to the ejection through a spectrometer (Jobin Yvon Triax550) with a spectral resolution of 2 nm. The spectrometer was equipped with a 1200 lines per mm diffraction grating. The Optical Emission Spectroscopy (OES) measurements were carried out at the different temperature and oxygen pressures.

## 3. Plume characteristics

Figures 1 and 2 show optical emission spectra of the plume recorded at various pressures of oxygen gas and at different chamber temperature. The optical emission spectra of plasma plume were recorded in the wavelength range from 400 to 800 nm at various gas pressures and chamber temperature. The optical spectroscopy showed that the plasma emission is mainly due to atomic and singly ionized Zn for both gas pressures and chamber temperature variation. The observed line intensities are higher in ambient gas as compared to temperature. Since the plasma expands in ambient gas highly energetic particles produced in the early stage of plasma undergo collisions with the ambient atmosphere resulting in enhanced cooling. The strongest emissions were observed around 481.05 nm and 636.23 nm from the neutral Zn atoms (Zn I) and 773.25 nm from the singly ionized Zn atoms (ZnII) in the all gas pressure, others neutral spectral lines such as 468.01 nm, 472.1 nm, and the ionic line 492.4 nm appear in all pressures except at 13 mTorr. Practically at 13 mTorr there are a few presences of Zn emission lines.

We detected also following many weaker lines: three belonging to Mn I at 602.23 nm, 754.15 nm, and 768.02 nm, and of O I 610.62 nm. It is interesting to note that for the spectra at room temperature and 13 mTorr and at 300°C and 46 mTorr appear 481.05 nm, 636.23 nm and 773. 25 nm lines only.

Significant variations in the intensities of emission lines of various atomic and ionic species of the plasma were observed with changes ambient gas pressures, it is observed that both electronically excited Zn atoms and the singly ionized Zn atoms peaks increase linearly with

**Table 1.** The strongest observed emission lines of the plume.

| Wavelength | Transitions | $g_k$ | $g_i$ | $A(s^{-1})$ | $E_k(eV)$ |
|---|---|---|---|---|---|
| 468.01 | $4s5s^3S_1 \rightarrow 4s4p^3P_0$ | 3 | 1 | $1.55x10^7$ | 6.654 |
| 472.21 | $4s5s^3S_1 \rightarrow 4s4p^3P_1$ | 3 | 3 | $4.58x10^7$ | 6.654 |
| 481.05 | $4s5s^3S_1 \rightarrow 4s4p^3P_2$ | 3 | 5 | $7.00x10^7$ | 6.654 |
| 636.23 | $4s4d^1D_2 \rightarrow 4s4p^1P_1$ | 5 | 3 | $4.65x10^7$ | 7.744 |

gas pressure as shown in figure 2. The strongest observed emission lines are listed in table 1 with corresponding spectroscopic data [15].

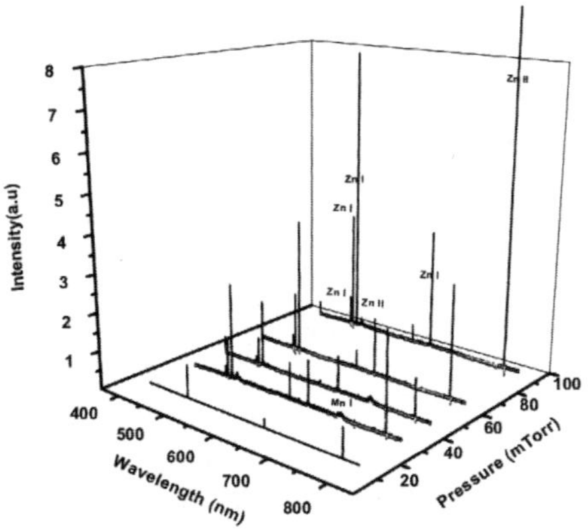


**Figure 1.** Spectra at various pressures.

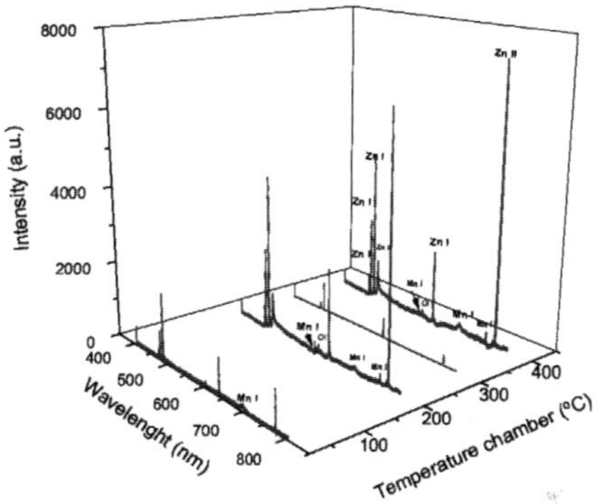


**Figure 2.** Spectra at different chamber temperature.

## 4. Electron temperature

The characterization of the laser-induced plasma through the determination of their main parameters, i.e., electron temperature (Te) and electron number density (Ne), is important to understand the dynamics of the process. Assuming plasma in local thermodynamic equilibrium (LTE), the electron temperature can be estimated using ratio of intensity of transitions belonging to the same atomic species and same ionization state for the optically thin plasma given by

$$\frac{I_1}{I_2} = \frac{A_1 g_1 \lambda_2}{A_2 g_2 \lambda} exp\left(-\frac{E_1 - E_2}{k_B T_e}\right) \tag{1}$$

where I, $\lambda$, and E are the intensity, wavelength (nm), and excitation energy of upper state (eV) of the corresponding OES peaks; g and A are the statistical weight of upper state and Einstein transition probability ($s^{-1}$); k is the Boltzmann constant; and T is the excitation temperature of the species under investigation. Subscripts 1 and 2 refer to the transition from respective states. The slope of the plot of

$$ln\frac{I_{mn}\lambda_{mn}}{g_m}A_{mn} \tag{2}$$

against $E_m$ gives the electron temperature (eV).

Journal of Physics: Conference Series **370** (2012) 012056 doi:10.1088/1742-6596/370/1/012056

To calculate the electron temperature of the plasma, we have used four (Zn I) lines at 468.01, 472.25, 481.05 and 636.3 nm using the Boltzmann plot method and is shown in figures 3 and 4. Errors are bound to be present in the determination of the plasma temperature by this method; therefore the temperature is determined with approximately 15% uncertainty, mainly coming from the transition probabilities and the measurement of the integrated intensities of the spectral lines.

The electron temperature was calculated as a function of the gas pressure as well as the chamber temperature. In the former, the temperature varies from 11500 K to 38500 K, presenting the maximum electron temperature at 46 mTorr, whereas, in the latter it varies from 18200 K to 39300 K as gas pressure varied from 13 to 95 mTorr and the chamber temperature varied from room temperature to 400 °C respectively as shown figures 3 and 4. The electron temperature increases up to 46 mTorr and for higher pressures decreases. As shown on figure 4, in contrast with the case chamber temperature, the electron temperature behave very different, there were no a clear trend of increasing electron temperature with increasing chamber temperature.

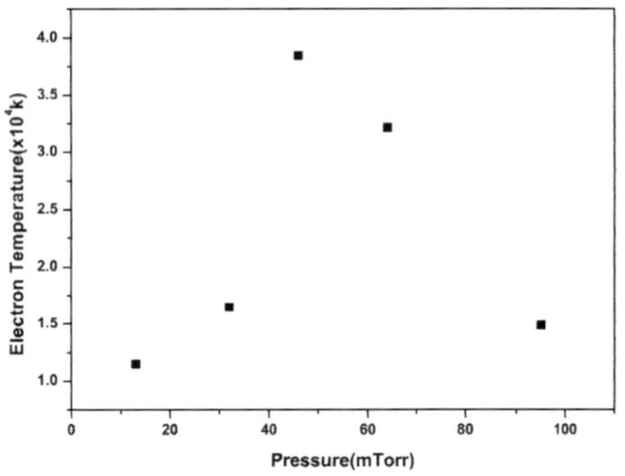


**Figure 3.** Electron temperatures as a function of gas pressure.

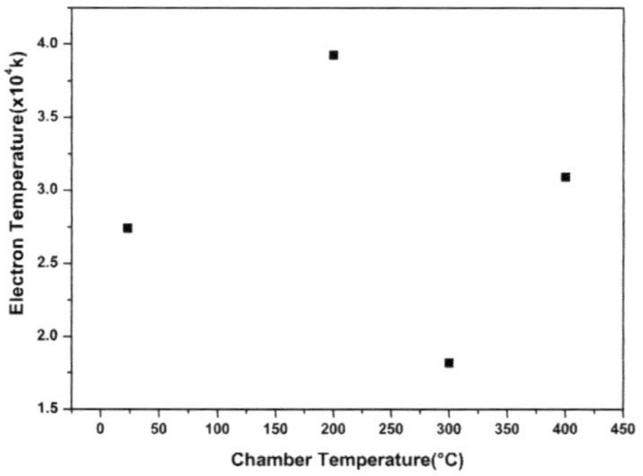


**Figure 4.** Electron temperatures as a function of chamber temperature and constant gas pressure.

## 5. Electron number density

Spectral width of lines is composed of three contributions, namely, Doppler broadening, resonance pressure broadening and Stark broadening. It is assumed that the Stark contribution largely dominates the broadening of spectral lines in LTE.

The broadening of well-isolated lines from non-hydrogen neutral atoms and singly ionized ions is predominantly caused by electron impact. Since the perturbations caused by ions is negligible compared to electrons, the ion correction factor can safely be neglected. In this work, the ZnMnO plasma was assumed to be optically thin. With these simplifications the full width at half-maximum (FWHM) of the Zn line is related to plasma density (Ne) by the relationship [16, 17].

$$\Delta\lambda_{\frac{1}{2}} = 2\omega\left(\frac{N_e}{10^{16}}\right) \tag{3}$$

where W is electron impact parameter [18]. From the above equation, it is clear that the electron density Ne is proportional to $\Delta\lambda_{1/2}$. Strong Zn II spectral lines were also observed in the case of ablation in oxygen, which confirms that the zinc plasma is actually formed. The zinc plasma formed above the target surface may absorb part of the laser energy and decrease the efficiency of the laser energy available for ablation of the target material.

Figure 5 shows the electron number density determined using Stark broadening of isolated lines of Zn at different gas pressures. At the pressure of 46 mTorr a maximum density was observed equal to 1.4 x $10^{16}$ $cm^{-3}$, for higher pressures the density decreases. Figure 6 shows that the electron density values are not much affected by the presence of the temperature chamber, the electron density seems to be constant and equal to 1.44 $10^{16}$ $cm^{-3}$ at 46 mTorr.

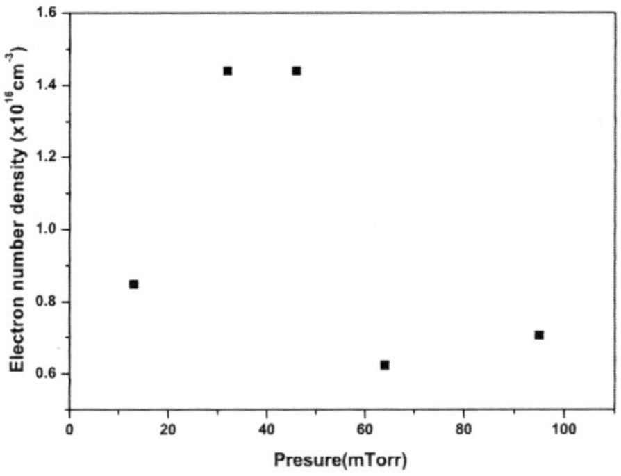


**Figure 5.** Plasma density determined by Stark broadening for diferent gas pressure.

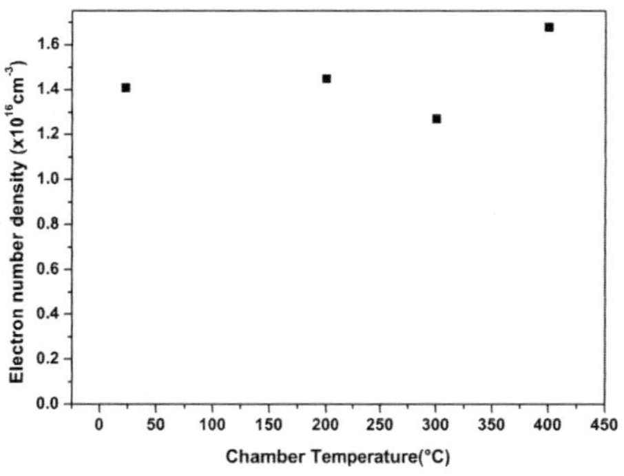


**Figure 6.** Plasma density determined by Stark broadening for diferent chamber temperature.

## 6. Conclusions

The electron temperature was calculated as a function of gas pressure as well as chamber temperature, ratio of intensity of transitions belonging to the same atomic species and same ionization state for the optically thin plasma, Its value as function of gas pressure and temperature chamber varied slightly. The electron density seems to be constant and equal to 1.44 $10^{16}$ $cm^{-3}$ at 46 mTorr for temperature chamber and gas pressure.

## References

[1] Harilal S 2007 *J. Appl. Phys.* **102** 1
[2] Geohegan D et al 2006 *Photon-based Nanoscience and Nanobiotechnology* (Springer)
[3] Wang X et al 2002 *Chem. Phys. Lett* **353**.
[4] Ferrando V et al 2003 *Supercond. Sci. Technol.* **16** 241
[5] Dauscher A, Thomy A and Scherrer H 1996 *Thin Solid Films* **280** 61-66
[6] Miller J and Richard F 1998 *Laser Ablation and Desorption* (UK: Academic Press)
[7] Saji J, Joshy N and Jayaraj M 2006 *J. Appl. Phys.* **104** 053307
[8] Aragón C and Aguilera J A 2008 *Spectrochim. Acta* Part B **63** 893-916
[9] Shukla G and Khare A 2009 *Appl. Surf. Sci.* **255** 8730-8737
[10] Shaikh N M, Hafeez S, Baig M 2007 *Spectrochim. Acta* Part B **62** 1311-1320
[11] Griem H R 1964 *Plasma Spectroscopy* (McGraw-Hill Book Company).
[12] Harilal S 2001 *Appl. Surf. Sci.* **172** 103
[13] Voevodin A, Jones J and Zabinski J 2000 *J. Appl. Phys.*. **88** 1088
[14] Amoruso S, Bruzzese R, Spinelli N and Velotta R 1999 *J. Phys. B* **32** R131
[15] NIST Atomic Spectra Database, http://physics.nist.gov, Kurucz output Atomic Spectral Line database
[16] I. Gornushkin B, King L A, Smith B W, Omenetto N and Winefordner J D 1999 *Spectrochim. Acta* Part B **54** 1207
[17] Marr G V 1969 *Plasma Spectroscopy* (Elsevier Publishing Company)
[18] Dimitrijević M S and Sahal-Bréchot S 1999 *Serb. Astron. J.* **160** 21-33

Journal of Physics: Conference Series **370** (2012) 012057 doi:10.1088/1742-6596/370/1/012057

# Soft x-ray measurements in the FN-II dense plasma focus device for different anode configurations

**C Rojo-Blanco[1], F Castillo-Mejía[2], J Rangel-Gutiérrez[2] and J J E Herrera-Velázquez[2]**

[1] Facultad de Ciencias, Universidad Nacional Autónoma de México
Circuito Exterior s/n, Ciudad Universitaria, 04510, México, D.F., México
[2] Instituto de Ciencias Nucleares, Universidad Nacional Autónoma de México
A P 70-543, Ciudad Universitaria, 04511 México, D F, México

E-mail: herrera@nucleares.unam.mx

**Abstract.** A study of the soft x-ray emission is presented, for a low energy (4.8 kJ) dense plasma focus device. Three Quantrad Si PIN-diodes with differential filter combinations of Be, Al, Ti, Ni, and Mo are employed as time-resolved x-ray detectors. The x-ray flux in different energy windows is measured as function of the deuterium filling pressure. A comparison is made for three anode configurations: (a) hollow, (b) flat, and (c) 2mm diameter W needle.

## 1. Introduction

A common feature in dense plasma focus devices is the production of a short-lived but high-temperature and high-density plasma, and its capability of emitting high-intensity short-duration pulses of x-rays, fast neutrons and energetic charged particles [1-3]. An important part of the experimental studies on x-ray and particle emission from plasma foci is oriented to interesting applications such as contact microscopy, x-ray and electron beam lithography, generation of soft x-ray spectral lines of highly charged heavy metal ions [4], and material research. Additionally, plasma foci are particularly attractive for x-ray generation with enhanced efficiency due to the following virtues: operational simplicity, compact size, portability, cost-effectiveness, and high versatility in doing both basic and applied research. The intense x-ray pulses produced by bremsstrahlung radiation from localized electron beams and from hot spots are excellent candidates for radiography of dynamic biological specimens [5] and metallic objects. X-rays in a plasma focus may have different origins. There are soft x-rays, which arise from thermal bremsstrahlung in the plasma and can be observed within the chamber by PIN diodes, registered on film with vacuum pinhole cameras, covering the pinholes with light-tight filters, or combining both techniques using the diodes instead of the film. Since there are usually impurities in the plasma, line spectra can also be observed. Seeding impurities, such as argon or xenon, is a common practice. Zakaullah et al., for example, have studied the efficiency of argon K series line radiation in an argon operated device. However, the most energetic radiation comes from an electron beam well in the region of tens of keV, generated within the plasma column, when it collides with the inner electrode. Such radiation can be observed outside the chamber and used for radiographic purposes.

## 2. Experimental set-up

This work was performed using the FN-II small plasma focus device, operated at the Instituto de Ciencias Nucleares, UNAM, whose details have been described in Ref. 6. For this work, it was operated at 38 kV with 4.8 kJ stored energy. The anode is made of oxygen-free copper, 40 mm long, with a 50 mm diameter. The coaxial cathode is formed by 12 copper rods, 8 mm diameter each, arranged in a squirrel cage configuration around a circle of 100 mm diameter. The insulator is an annular Pyrex® tube, 12 mm long, located at the base of the anode. As reported in Ref. 6, the neutron yield and its uniformity increased considerably when the diameters of the insulator and the inner electrode matched, probably due to an improvement of the current sheath breakdown, aided by the edge effect of the electrode close to the insulator. The energy storage is provided by four 1.863 μF capacitors in parallel, and the discharge is triggered by a simple mushroom electrode spark gap. Throughout the present work, the plasma focus was operated in its neutron optimized regime corresponding to a 2.7 Torr pressure of pure deuterium gas. In this regime a peak 350 kA focus current and an average yield of $1.6\times10^{8}$ neutrons/shot have been routinely obtained. The evolution of the current derivative is monitored by a Rogowski coil. Two silver foil calibrated activation counters, one at 20° from the axis, and the other at 90°, were used to measure the neutron yield.

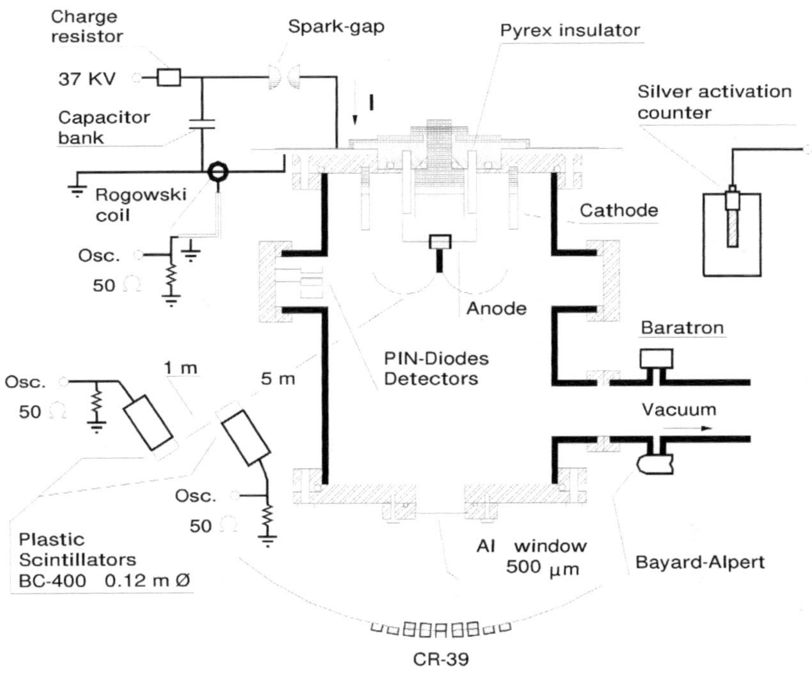


**Figure 1.** Schematic of the FN-II dense plasma focus and its diagnostics.

A schematic diagram of the plasma focus electrode configuration and diagnostics is shown in figure 1. The three anode shapes employed in this experiment were a hollow anode, a flat electrode with a 2 mm diameter W needle on the tip, and flat anode as shown in figure 2. The main diagnostics for x-ray investigation is a combination of time-resolved and time-integrated detectors. In order to obtain information regarding the x-ray emission in different shots, Quantrad PIN-diodes of 250 μm active layer thickness along with suitable absorption filters were used. The PIN-diodes were cross-calibrated using the same filter. These detectors were placed in the side-on direction at 16 ± 0.1 cm from the anode axis and 1.5 ± 0.1 cm below the anode tip. For time integrated analysis, a pinhole camera equipped with five 200 μm diameter pinhole apertures was used. The multipinhole camera was also placed in the radial position, viewing the plasma region as well as the tip of the anode. Simple spectral analysis was conducted using a *Ross filter* set Ti (25 μm), Al (25 μm), Be (25 μm), Ni (25 μm) and Mo (25 μm) along with the PIN-diode detectors. To estimate the contribution of soft x-rays in the 2.0 – 2.5 keV range, the selected *Ross filter* set is Al (25 μm) and Ni (25 μm), and for the x-ray emission in the energy range of 2.5 – 5 keV, the selected *Ross filter* pair is Ti (25 μm) and Mo (25 μm).

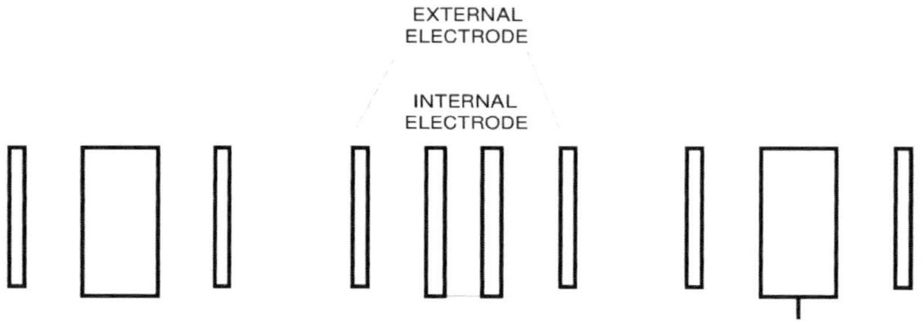


**Figure 2.** Illustration of the three electrode geometries used: flat electrode, hollow electrode and needle on tip electrode.

## 3. Experimental results and discussion

Figure 3 shows the x-ray images of two different shots, obtained with the multi-pin-hole camera, for the flat and the needle on tip electrode.

Typical signals from the PIN detectors are shown in figure 4, with two different sets of 25 micron filters: (a) molybdenum, titanium and beryllium, and (b) beryllium, titanium and molybdenum.

The variations of signal intensity (area under the curves) recorded by the detectors masked with Al (25μm) and Ni (25μm) filters and Ti (25 μm) and Mo (25 μm) filters, respectively, is described in figures 5, 6, and 7.

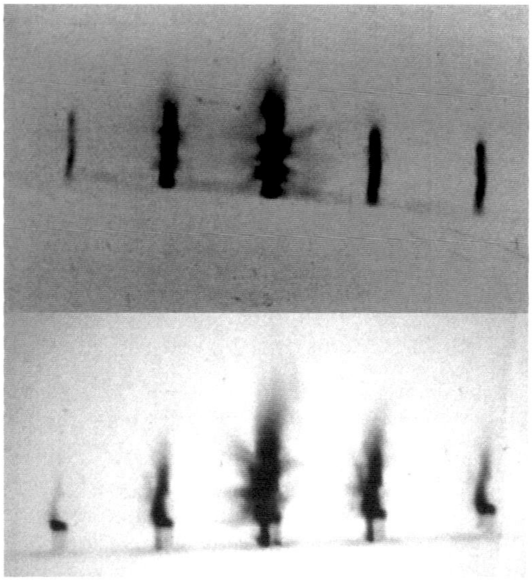

**Figure 3.** Images of the x-ray emissions from the plasma column, as observed with the multi-pin-hole camera. Each hole is filtered by aluminium (from left to right) 25, 10, 5, 15 and 20 micron thick. The picture above was taken with a flat electrode, and the one below with a 2mm diameter tungsten needle.

Journal of Physics: Conference Series **370** (2012) 012057 doi:10.1088/1742-6596/370/1/012057

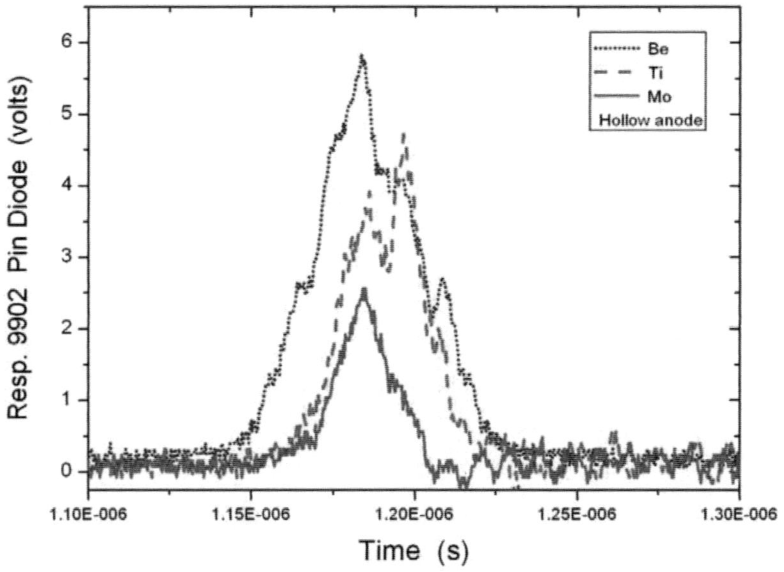


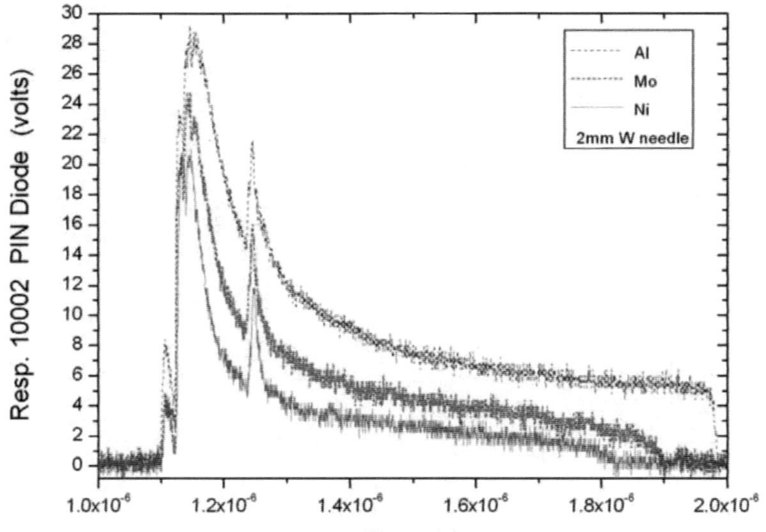


**Figure 4.** PIN signals with molybdenum, titanium and beryllium filters on the top figure, and beryllium, titanium and molybdenum. The thickness is 25 microns in all cases.

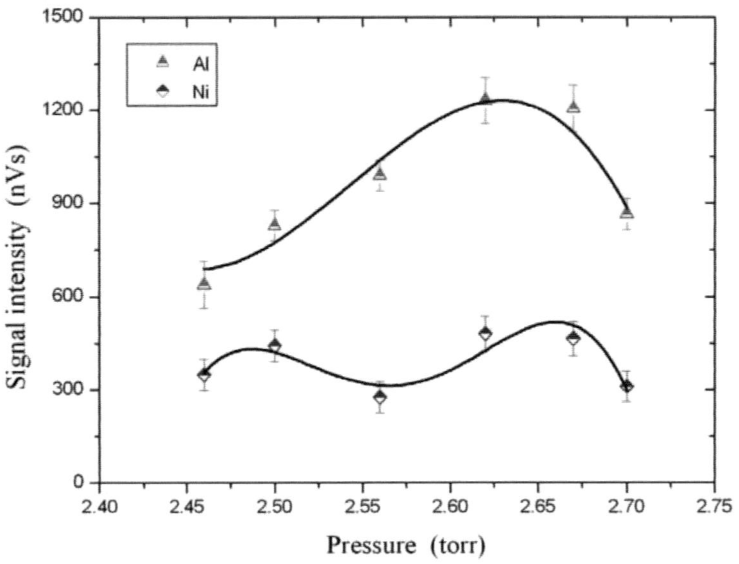


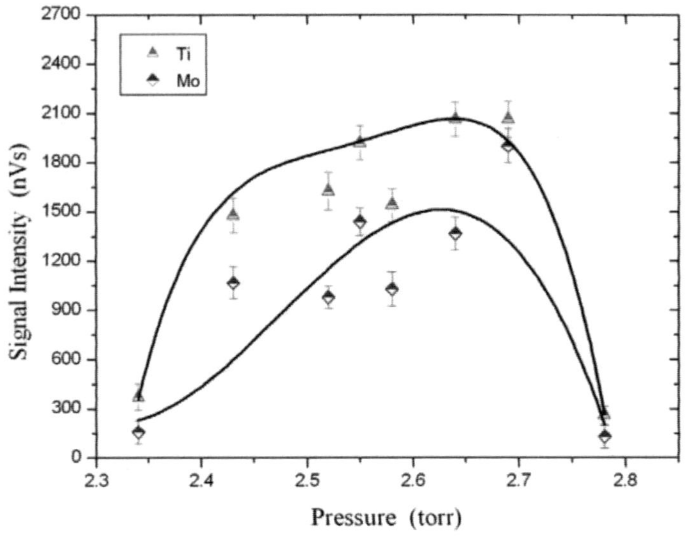


**Figure 5.** Averaged signal emissivity for the case of the hollow electrode, with Al and Ni filters (above), and Ti and Mo filters (below).

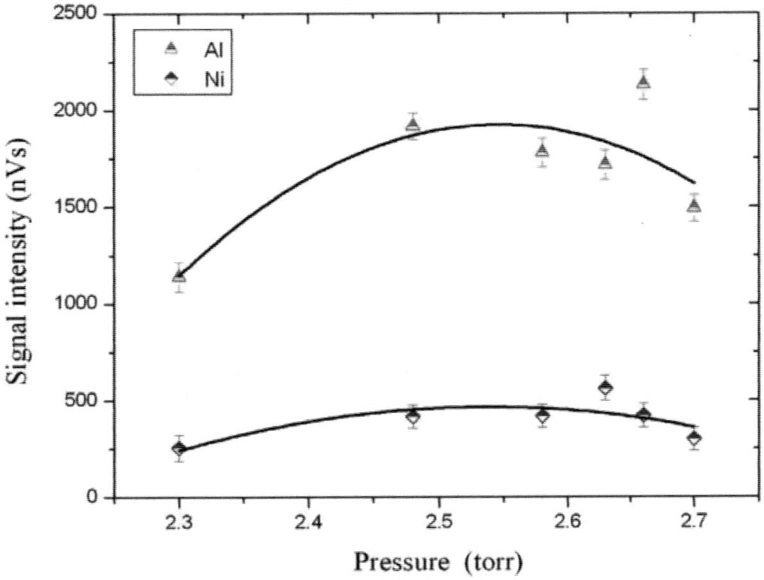


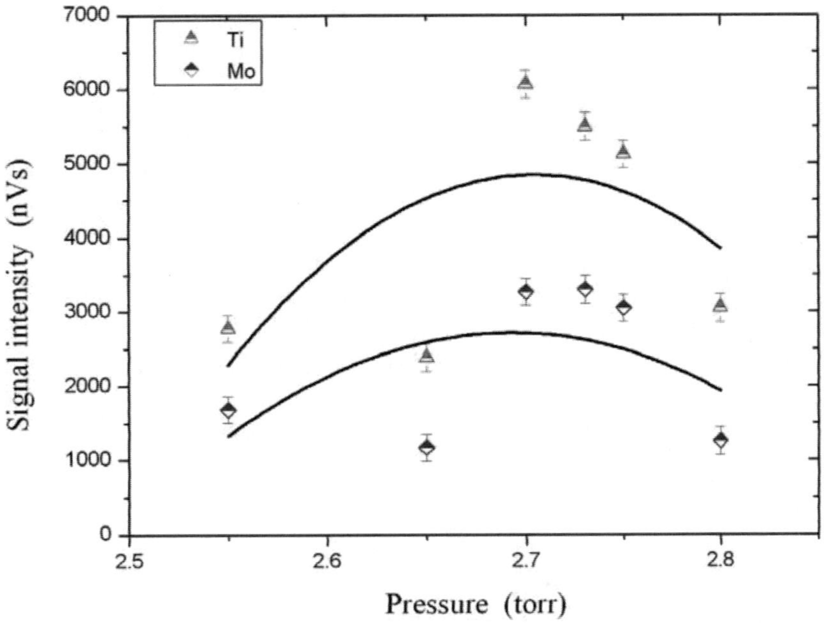


**Figure 6.** Averaged signal emissivity for the case of the flat electrode, with Al and Ni filters (above), and Ti and Mo filters (below).

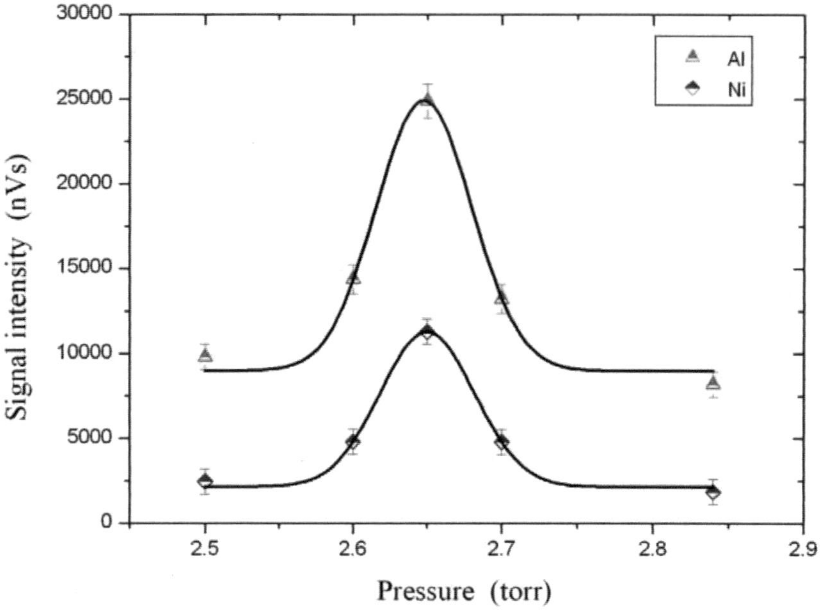


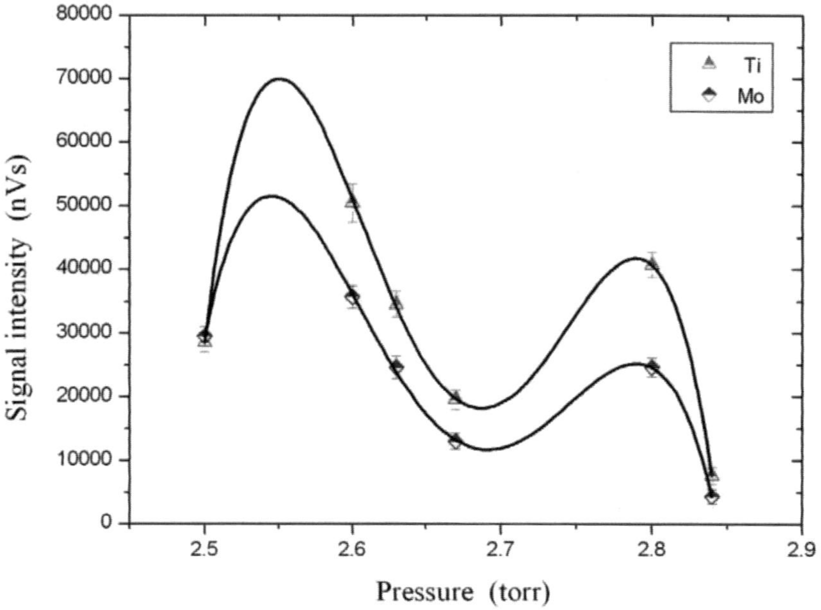


**Figure 7.** Averaged signal emissivity for the case of the "needle on electrode", with Al and Ni filters (above), and Ti and Mo filters (below).

From these results, it can be concluded that x-rays in the 2.0 – 5 keV energy range are emitted from our low energy (4.85 kJ) dense plasma focus, operated with deuterium as the filling gas.

Since the pressure of the filling gas increases from shot to shot in the range of tenths of Torrs, it was also found interesting to see if there was any systematic trend of the soft x-ray emissivity which would differ for each electrode geometry. Here we define emissivity as the area under the signal, so the units we use are nanoseconds × Volts. Figures 5, 6 and 7 show this parameter, averaged over 100 shots, for each filter and for each electrode geometry. The fit is made using Origin® software, choosing the best fit polynomial. Although these data need further validation, a clear difference in the qualitative behavior is observed for each geometry. Further analysis will be needed in order to understand the significance of this, and determine the convenience of using each kind of electrode, according to the intended application.

**Acknowledgments**

This work was partially supported by PAPIIT-DGAPA-UNAM IN120409 and ANPCYT of Argentina.

**References**

[1] Mather J W 1965 *Phys Fluids* **8** 366
[2] Sadowski M, Herold H and Shakhater M 1984 *Phys. Lett. A* **105** 117
[3] Hussain S, Shafq M, Ahmad R, Waheed A and Zakaullah 2005 *Plasma Sources Sci. and Technol.* **14** 61
[4] Hirano K *et al.* 1988 *Japan J. Appl. Phys.* **27** 89
[5] Castillo-Mejía F, Milanese M M, Moroso R L, Pouzo J O and Santiago M A 2001 *IEEE Trans. Plasma Sci.* **29** 921
[6] Castillo F, Herrera J J E, Rangel J, Alfaro A, Maza M A, Sakaguchi V, Espinosa G and Golzarri J I 2002 *Brazilian J. Phys.* **37** 3
[7] Zakaullah M *et al.* 2000 *Plasma Sources Sci. Technol.* **9** 592

Journal of Physics: Conference Series **370** (2012) 012058 doi:10.1088/1742-6596/370/1/012058

# Multiple pinching in the miniature plasma focus nanoPLADEMA

**M Barbaglia[1], L Soto[2] and A Clausse[1]**

[1]CNEA-CONICET and Universidad Nacional del Centro,7000 Tandil, Argentina
[2]Comisión Chilena de Energía Nuclear, Santiago, Chile

E-mail: barbagli@exa.unicen.edu.ar

**Abstract**. This work presents, evidence of multiple pinching deduced from electrical diagnoses taken during discharges on the Plasma-Focus nanoPLADEMA, of extremely low energy. The dimensions of the anode are 2mm OD and 0.5mm length, the insulator is an alumina tube 4 mm OD and 2.3 mm length. The cathode is open, so the current sheet expands radially free.

## 1. Introduction

Low energy Plasma Foci (PF) are attracting increasing interest for their potential applications and portability as neutron and x-ray sources. Comparing with isotopic neutron sources, PF have unique qualities as a source of x-ray/neutron pulses, emitting only on operator demand, with low risk of radiation contamination and low cost of operation and maintenance. PF devices not only are interesting for basic plasma research but also for industrial applications [1] ranging from tailored soft x-ray sources [2,3] and soft x-ray microlithography [4] to hard x-ray introspective imaging of metallic pieces [4-9], neutron production and applications [9,10] and plasma thrusters [11] among others.

The exploration of the lowest energy possible in the plasma focus devices was, among others, a main goal to achieve in a cooperation research project between the National Atomic Energy Commissions from Argentina and Chile. Two similar devices were designed and built in each of these countries [12]. The Chilean device was named “Nanofocus” (not to be confused with the 100J PF reported by Milanese *et al.* [13]), and the Argentine device was called “*NanoPLADEMA*”. The present work was conducted in the latter.

## 2. Experimental arrangement

The schematic of the apparatus is shown in figure 1. A pair of 200 mm diameter brass electrodes acts as a final capacitor to drive the discharge. A 1.6 mm Cooper tube covered with alumina is attached to the center of the anode plate and passes through a small hole in the cathode center. Four dielectric PVDF films, 80 μm thick, were placed between both plates. The measured capacity is 4.9 nF. The total dimensions of the discharge chamber are 20cm×20cm×5cm. A transmission line attached to a spark-gap connects the discharge camera with the primary energy source that is accumulated in the capacitor 3 in figure 1. Figure 2 shows a photograph of the complete system. The high voltage current source to energize capacitor 3 (figure 1) and the vacuum system are not displayed in the picture.

## 3. Results

The device was operated in 5 mbar of Hydrogen, charging at 5kV. The current derivative passing through the anode, dIa/dt, was measured with an independent Rogowski coil placed inside the anode plate. Meanwhile, the current derivative history that flows to and from the camera, was measured with another Rogowski coil, dIe/dt. It should be noted that the current inside the chamber and the current in the external circuit are not equal during transients due to charge accumulation in the plate capacitor. Both Rogowski coils have similar temporal response.

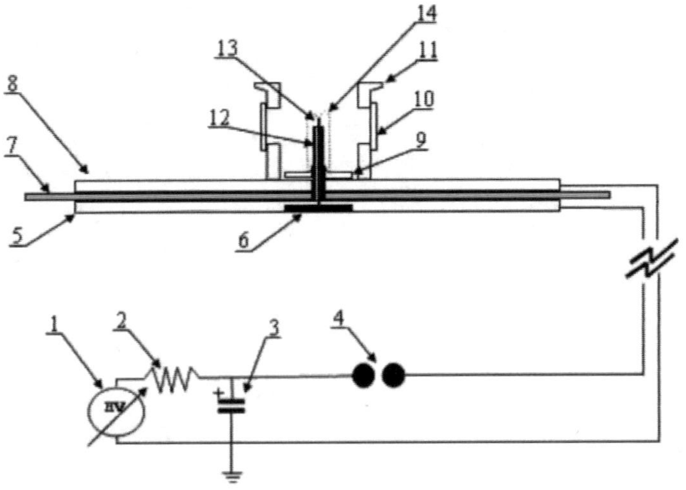


**Figure 1**: Schematic diagram of the Plasma Focus nano-Pladema. 1: power supply, 2: charging resistance, 3: capacitor 4.9 nF, 4: spark-gap, 5: anode, 8 cathode, 5 and 8: pair of 200 mm diameter brass electrodes forming the capacitor to drive the discharge (5 nF), 6: anode, 7: dielectric, four PVDF films of 80 μm thickness, 9: cathode, 10: optical windows, 11: discharge chamber, 12: alumina tube, 13: anode, 14: plasma sheath between anode and cathode.

Figure 3 shows the signals of the anode voltage and current derivative in the anode and the external circuit, recorded during a discharge at 5 mbar of hydrogen. Three spikes of the anode current derivative can be identified at 0.6 μs, 0.85 μs and 1.45 μs. These events correlate with smaller spikes of the anode voltage. The absence of spikes in the external current derivative precludes the interpretation of the spikes as line reflections. The usual interpretation of these correlated spikes in PF devices is the occurrence of sudden inductance changes. Since by construction the device cannot produce changes in the inductances, it is reasonable to conclude that the spikes displayed in figure 3 are brought about by plasma inductance changes inside the discharge chamber, probably associated with pinching phenomena.

## 4. Conclusions

The spikes observed in the anode Rogowski signal indicate the occurrence of multiple pinching in different periods of the discharge oscillation. The absence of spikes in the external current derivative precludes the interpretation of the spikes as line reflections, thus supporting the present claim. Similar multi-pinch phenomena were already reported in previous experiments in small Plasma-Focus devices [14] also in the same work, it was reported the hard x-ray production in multi-pinch phenomena. To the present, the physical explanation of this sort of phenomena in small PF devices is a pending task.

Currently the neutron and x-ray emissions are being studied in discharges in deuterium. However, the expected yield is under the detection threshold of typical detectors used in ordinary plasma focus research. New techniques are under study to guarantee the accuracy of the measurement.

**Figure 2**: Photography of the plasma focus device *nanoPLADEMA*.

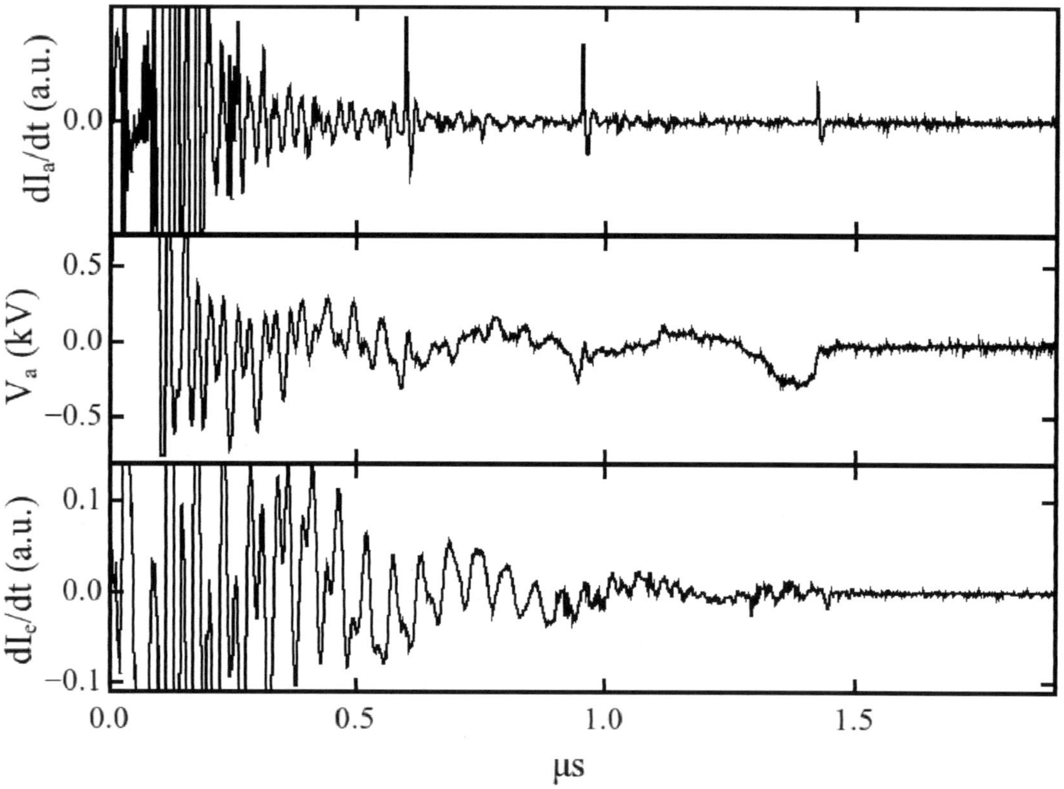


**Figure 3**: Signals of a discharge from *nanoPLADEMA* at 5 mbar of hydrogen, 5kV, showing the current derivative through the anode (dIa/dt), the anode voltage (Va) and the current derivative through the external circuit (dIe/dt).

## References

[1] Moreno C, Venere M, Barbuzza R, Del Fresno G, Ramos C, Bruzzone H, Florido P, González J and Clausse A 2002 *Braz. J. Phys.* **32** 20–5

[2] Zakaullah M, Alamgir K, Shafiq M, Sharif M, Waheed A and Murtaza G 2000 *J. Fusion Energy* **19** 143–57

[3] Hussain S, Ahmad S, Khan M Z, Zakaullah M, and Waheed A 2004 *J. Fusion Energy* **22** 195–200

[4] Lee S, Lee P, Zhang G, Feng X, Gribkov V, Liu M, Serban A and Wong T K S 1998 *IEEE Trans. Plasma Sci.* **26** 1119–26

[5] Moreno C, Clausse A, Martinez J F, Llovera R and Tartaglione A 2001 *Nukleonika* **46** S33–4

[6] Venere M, Moreno C, Clausse A, Barbuzza R and Del Fresno G 2001 *Nukleonica* **46 (Supplement 1)** S93

[7] Raspa V, Sigaut L, Llovera R, Cobelli P, Knoblauch P, Vieytes R, Clausse A and Moreno C 2001 *Braz. J. Phys.* **34** 696–9

[8] Silva P, Moreno J, Soto L, Birstein L, Mayer R and Kies W 2003 *Appl. Phys. Lett.* **83** 3269

[9] Raspa V, Moreno C, Sigaut L and Clausse A 2007 *J. App. Phys.* **102** 123303

[10] Tartaglione C, Ramos C, Clausse A and Moreno C 2004 *Braz. J. Phys.* **34** 1756–8

[11] Scheuer J T, Schoenberg K F, Gerwin R A, Hoyt R P, Henins I, Moses R W Jr, Black D C and Mayo R M 2001 *IEEE Trans. Plasma Sci.* **22** 1015–33

[12] Soto L, Pavez C, Moreno J, Barbaglia M O and Clausse A 2009 *Plasma Sources Sci. Technol.* **18** 015007

[13] Milanese M, Moroso R and Pouzo J 2003 *European Physical Journal D* **27** 77-81

[14] Barbaglia M, Bruzzone H, Ríos I, Acuña H, González J and Clausse A 2010 *Plasma Phys. & Controll. Fusion* **52** 032001

Journal of Physics: Conference Series **370** (2012) 012059 doi:10.1088/1742-6596/370/1/012059

# Dynamics of the expansion discharge originated by a dense plasma focus

**M Martínez-Fuentes[1], H Ortiz-Uribe[2], F Castillo-Mejía[1], M Nieto-Pérez[2], J J E Herrera-Velázquez[1] and J Rangel-Gutiérrez[1]**

[1]Instituto de Ciencias Nucleares, Universidad Nacional Autónoma de México, Apartado Postal 70-543, 04510, México D. F., México
[2]Centro de Investigación en Ciencia Aplicada y Tecnología Avanzada, Instituto Politécnico Nacional, Unidad Querétaro, Cerro Blanco 141, Colinas del Cimatario, Querétaro, QRO 76090, México

E-mail: marco.martinez@nucleares.unam.mx

**Abstract**. The expansion phase of the discharge in Dense Plasma Focus devices might be used in order to simulate the interaction between edge plasmas and wall facing materials in fusion devices. Preliminary results of the study of this phase with the use of a triple a Langmuir probe are presented. A review of the use of triple Langmuir probes for the study of flowing plasmas is made.

## 1. Introduction

Dense plasma focus devices have been and will continue to be extensively studied, in their first three stages, namely, the breakdown, rundown, and focusing of the plasma column. However, one of its features has been neglected; namely, the expansion of its discharge along the vacuum chamber, after the compression of the plasma column, which is akin to those in plasma guns. This work is motivated by the possibility of using this stage as a source of energetic plasma, which might simulate the effect of high density plasmas colliding on walls in magnetic confinement devices. This may be important from the materials study standpoint. For this purpose, it is necessary to be able to determine the dynamics of the discharge, for which measurements of density, temperature and speed are needed. We show preliminary results obtained using a triple Langmuir probe (TLP). As part of the work, a review of such probes is made. A study of the convenience of using the probe in voltage [1] and current modes [2, 3], which is reported to be best suited for a high electromagnetic noise environment, is made. The present work is done along the axis of the FN-II dense plasma focus, which is a 4.8 kJ device.

## 2. Langmuir probes

The Langmuir probe is the most familiar of the electrical diagnostic techniques, consisting on one or more conductors immersed in the plasma. The main variants are single, double and triple. The single probe consists on a conductor (usually cylindrical or spherical) immersed in the plasma. When there is a plasma flow, it is usually placed parallel to its direction. In the case of the single probe, there is an external voltage applied between the probe and one electrode. From sweeps of an applied voltage, the floating potential can be obtained, and consequently the electron temperature ($T_e$) and electron density

($n_e$) [1,4,5]. However, there are cases when the use of a single probe is difficult. For example, when there is no convenient ground or reference plane to use as a counter-electrode. In these instances the use of a double probe is indicated. In that case, the return current source comes from a second probe immersed in the plasma, and the external voltage is applied between the probes. Both single and double probes are suitable for flowing and steady plasmas, in the collisionless regime.

In the triple probe a third electrode is added in close proximity to a double probe, and is used in order to measure the floating potential independently. Triple probes offer an advantage over single and double Langmuir probes, allowing simultaneous measurement of plasma parameters without the need for voltage sweeps that can be limiting in pulsed plasma environments, specially if the discharges are not reproducible from shot to shot [1,2,4,6].

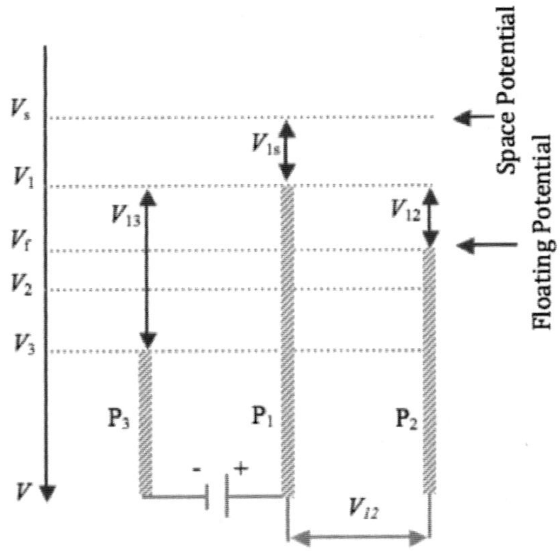


**Figure 1(a).** Diagram of voltage-mode triple Langmuir probe.

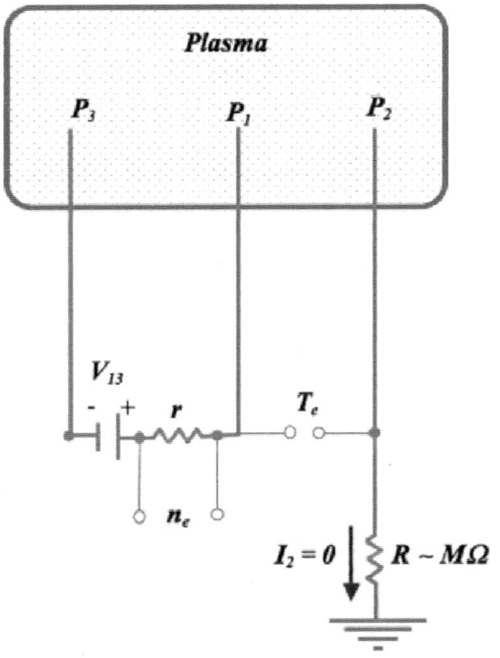


**Figure 1(b).** Electric circuit for a triple Langmuir probe operating in voltage-mode.

Triple probes can be operated in either voltage or current modes. In voltage mode operation there is an external potential applied only between two probes while the third one is floating (figure 1).Temperature $T_e$ and density $n_e$ are obtained from

$$\frac{1}{2} = \frac{1 - \exp\left[-\frac{eV_{13}}{kT_e}\right]}{1 - \exp\left[-\frac{eV_{12}}{kT_e}\right]} \tag{1}$$

$$n_e = \frac{J_i}{\exp\left[-\frac{1}{2}\right] e \sqrt{\frac{kT_e}{m_i}}} \tag{2}$$

where $J_i$ is the ion current density, given by

$$J_i = \frac{1}{A_p} \frac{I_{13}}{\exp\left[-\frac{e(V_{12} - V_{13})}{kT_e}\right] - 1} \tag{3}$$

where $A_P$ is the parallel area of the cylindrical probe, and $I_{13}$ is the current between probes 1 and 3, measured over the resistance *r*.

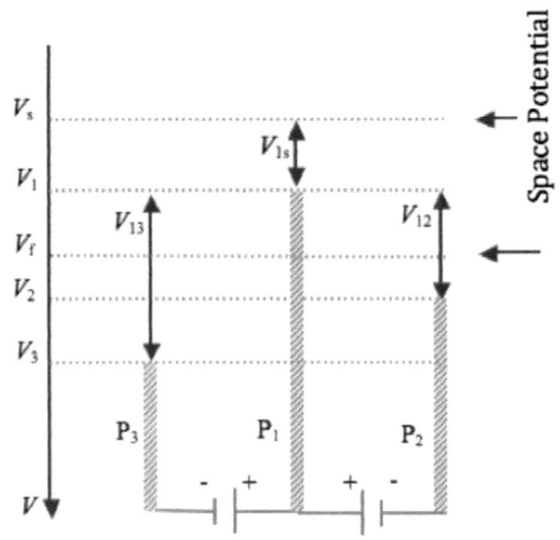


**Figure 2(a).** Diagram of a current-mode triple Langmuir probe.

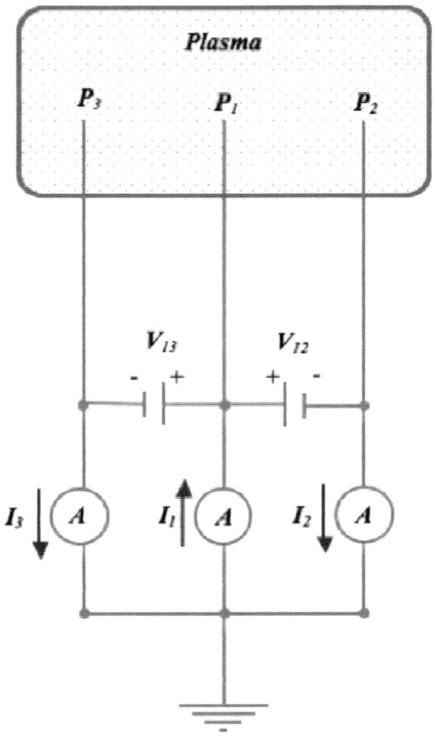


**Figure 2(b).** Electric circuit for a triple Langmuir probe operating in current-mode.

In current mode there are external potentials applied between the three probes, as shown in figure 2. The potential $V_{1s}$, and $T_e$ and $n_e$ are obtained by. $J_{e0}$ is the random electron current equation (7). The triple probe, referred operated in current mode, is suitable for flowing, steady, and unsteady plasmas in the collisionless regime. The implementation of current mode originates from the fact that the voltage needed in the voltage mode operation (voltage that comes from the probe which is floating) is a high impedance measurement that can be susceptible to electromagnetic noise in pulsed plasmas [1-6].

$$I_1 = A_p J_{e0} \exp\left[-\frac{eV_{s1}}{kT_e}\right] - A_p e n_e \sqrt{\frac{kT_e}{m_i}} \exp\left[-\frac{1}{2}\right] \tag{4}$$

$$I_2 = A_p J_{e0} \exp\left[-\frac{e(V_{s1} + V_{12})}{kT_e}\right] - A_p e n_e \sqrt{\frac{kT_e}{m_i}} \exp\left[-\frac{1}{2}\right] \tag{5}$$

$$I_3 = A_p J_{e0} \exp\left[-\frac{e(V_{s1}+V_{13})}{kT_e}\right] - A_p e n_e \sqrt{\frac{kT_e}{m_i}} \exp\left[-\frac{1}{2}\right] \quad (6)$$

where

$$J_{e0} = e n_e \sqrt{\frac{kT_e}{2\pi m_e}} \quad (7)$$

In the TLP theory for plasma, consisting of electrons and a single species of ions, it is assumed that the electrodes are immersed in a plasma with different temperatures for each species, $s = e, i$, with distribution function for species is Maxwellian distribution functions:

$$f_s(r,v,t) = n_s(r,t)\left(\frac{m_s}{2\pi kT_s}\right)^{\frac{3}{2}} \exp\left[-\frac{m_s(v_s - U_s)}{2kT_s}\right]$$

where $m_s$ is the mass, $v_s$ is the velocity, $n_s(r, t)$ is the number density, $U_s(r, t) = \langle v_s \rangle$ is the average velocity (brackets represent an average over the distribution function) and $T_s$ is the temperature. Furthermore, we assume that the angle between the plasma flow velocity and the axis of the probe (a cylindrical probe is considered) is given by an angle $\theta$ (figure 3).

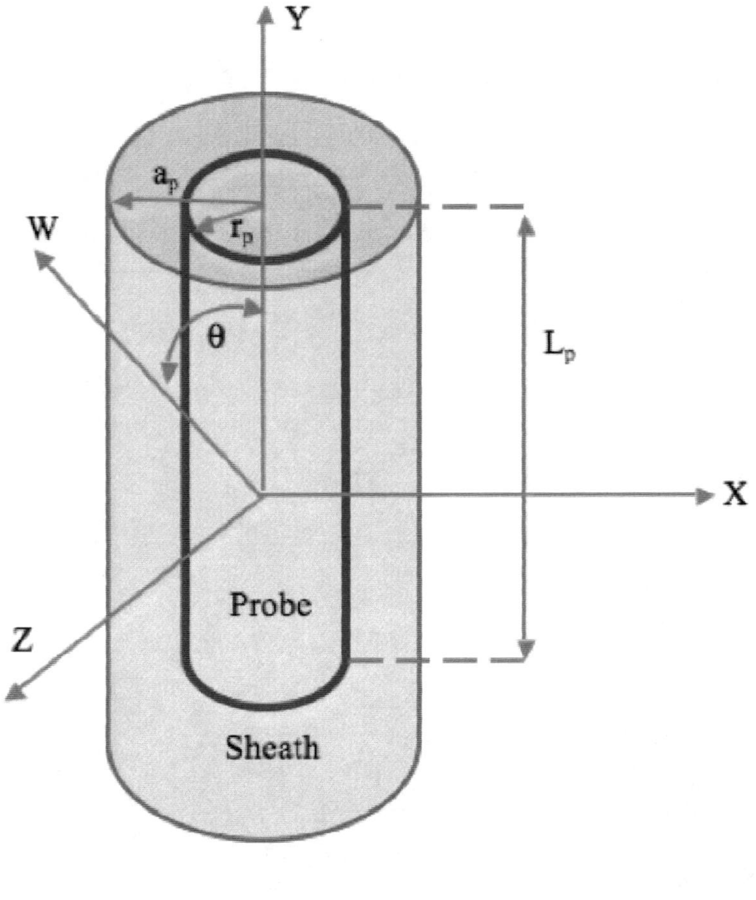


**Figure 3.** Probe and its sheath. Where $W$ is the velocity of the plasma.

Theory [1,2,4-6] requires that the cylindrical probe is working in collisionless plasma, so the Knudsen number for a probe of radius ($r_p$) must satisfy

$$Kn_{sl} = \frac{\lambda_{sl}}{r_p} >> 1$$

where $\lambda_{sl} = \lambda_{ei}, \lambda_{ee}, \lambda_{en}, \lambda_{ii}, \lambda_{in}$, are the mean free paths for collisions of species s and l. Therefore, it must meet:

$$r_p << \lambda_{ei}, \lambda_{ee}, \lambda_{en}, \lambda_{in}, \lambda_{ii}$$

It also requires that the sheath of the plasma has to be collisionless (where $a_p$ is the width of the sheath of the plasma):

$$a_p << \lambda_{ei}, \lambda_{ee}, \lambda_{en}, \lambda_{in}, \lambda_{ii}$$

The Debye length $\lambda_D$ (assuming a quasi-neutral plasma $n_e = n_i$ ) given by:

$$\lambda_D = \sqrt{\frac{\varepsilon_0 k T_e}{e^2 n_i}}$$

should also be smaller than both $r_p$ and $a_p$. It is also required that the width of the sheath is smaller than the space ($d_p$) between probes, i.e. there is no interaction between the sheaths of the probes.

## 3. Plasma Focus FN-II

The device on which this work was done is the Fuego Nuevo-II dense plasma focus, which operates at the Instituto de Ciencias Nucleares, UNAM, and has been described in [7]. Both inner and outer electrodes are made of oxygen-free copper, and the latter is a squirrel cage made out of 12 bars. Their length is 40 mm, and the diameters of the inner and outer electrodes are 50 and 100 mm respectively. The insulator is a 19 mm *Pyrex®* tube, whose outer diameter matches the 50 mm of the inner electrode. The usual operating voltage is 38 kV, with a stored energy in the capacitor bank of 4.8 kJ. The current has been measured to be 350 kA. The pressure range in which these experiments were made is 2.5-2.6 Torr.

## 4. Experimental set-up and results

A voltage-mode TLP was placed inside the vacuum chamber, along the axis of the plasma focus, at three different distances from the inner electrode ($\Delta x$ = 16 cm, 21 cm, 26 cm) (figure 4). The device was operated with pure deuterium at ~ 2.6 Torr, and the data were collected with Tektronix TDS3034 oscilloscopes.

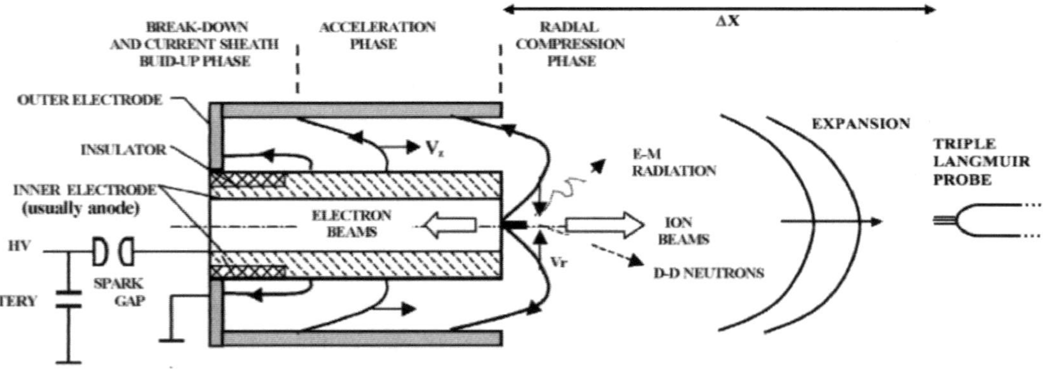


**Figure 4.** Sketch of a plasma focus. The expansion of the discharge, and the location of the TPL.

Figures 6 and 7 show the behavior of the signals of $V_{12}$ (as defined in figure 1) and the potential measured in the resistance $r$, respectively, when $r$ was varied with values of 1MΩ, 100 kΩ, 10 kΩ, 1kΩ, 100 Ω, 10 Ω and 1 Ω (always with $V_{13} = 100\ V$).

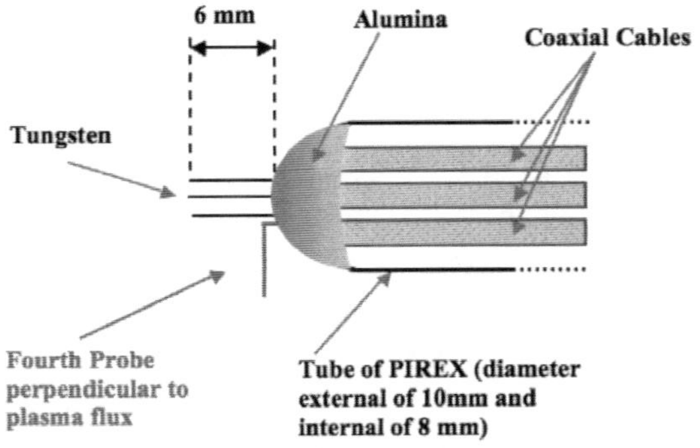


**Figure 5(a).** Sketch of triple Langmuir probe.

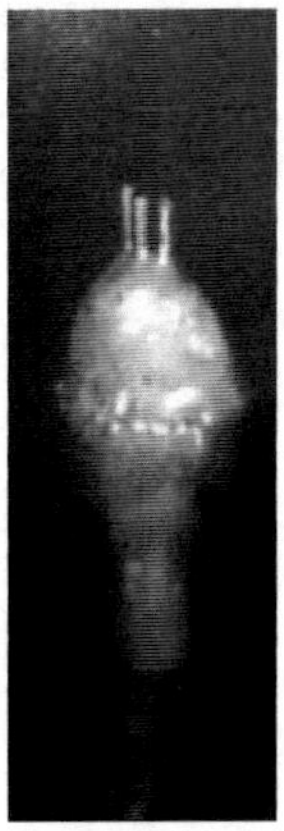

**Figure 5(b).** Photograph of TLP inside the vacuum chamber.

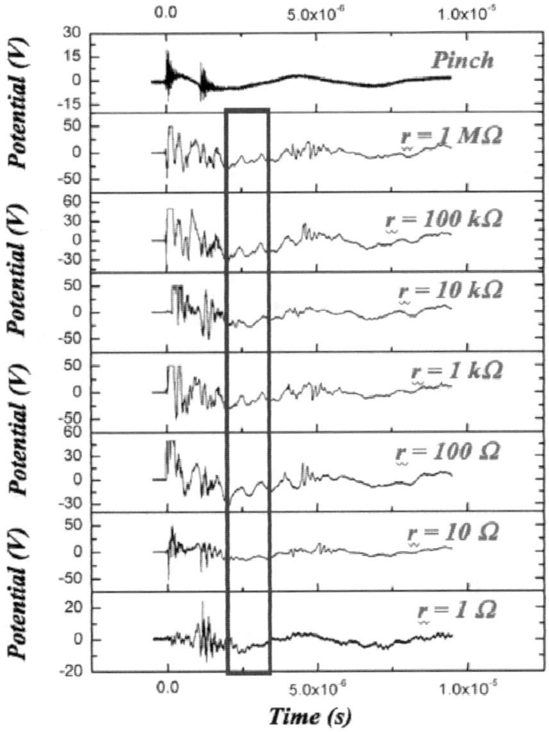


**Figure 6.** Signal of $V_{12}$ for different values of $r$. The first curve is the signal of the Rogowski coil which measures the derivative of the current. The two peaks mentioned in the text are encased in the blue box. The peaks inside the blue square show the moment when the discharge passes across the probe.

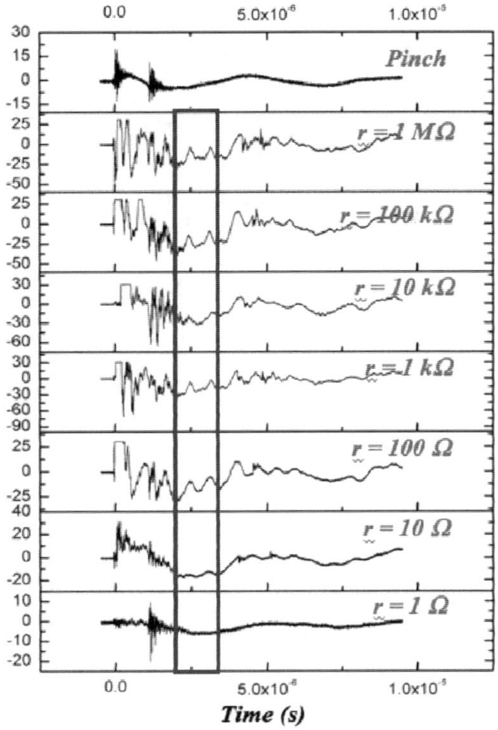


**Figure 7.** Signal of the potential measured in the resistance $r$, when $r$ has different values. The first curve is the signal of the Rogowski coil. The two peaks mentioned in the text are encased in the blue box. The peaks inside the blue square show the moment when the discharge passes across the probe.

The purpose of varying the value of $r$ is to find out how electromagnetic noise may be reduced. Although it decreases, it is possible to see in the curves that there are two peaks after the pinch, that do not vanish with the different values of $r$. These are encased in figures 6 and 7 in blue boxes. We interpret them as a sign of the discharge front passing across the TLP.

Figure 8 shows that there is a phase difference ($\Delta t$) between the three curves (at the start of the first peak of the discharge that passes across the TLP), this allows the determination of the value of the average velocity (between curves) of the expansion:

$$\Delta v_{16\text{ cm}-21\text{ cm}} = (5.7\pm3.6)\times10^5\text{ m/s}$$

$$\Delta v_{21\text{ cm}-26\text{ cm}} = (5.3\pm3.5)\times10^5\text{ m/s}$$

Using the Voltage-Mode [3-9] allows measuring the electron temperature and the density of electrons in the zone when the discharge passes across the TLP (figure 9).

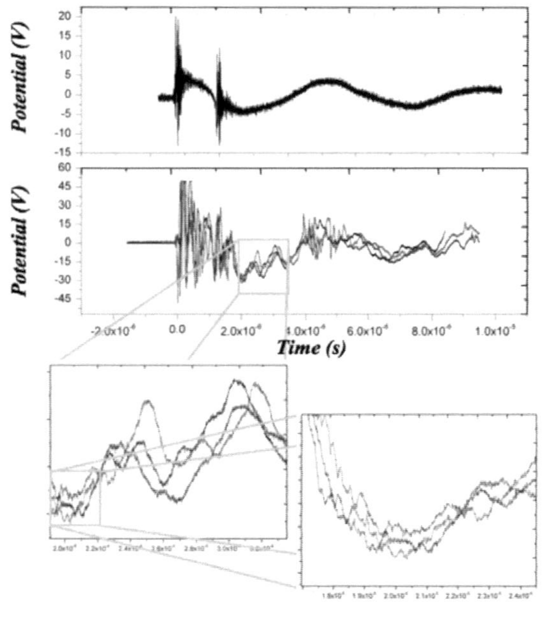


**Figure 8.** The upper signal comes from the Rogowski coil. Signals below show probe signals at different distances ($\Delta x$ = 26 cm, 21 cm, 16 cm). The red square is the zone where the discharge passes across the TLP. The blue line is at $\Delta x$ = 16 cm, the red line is at $\Delta x$ = 21 cm, and the black line is at $\Delta x$ = 26 cm. (Right) Zoom in the zone where discharge passes across the TLP.

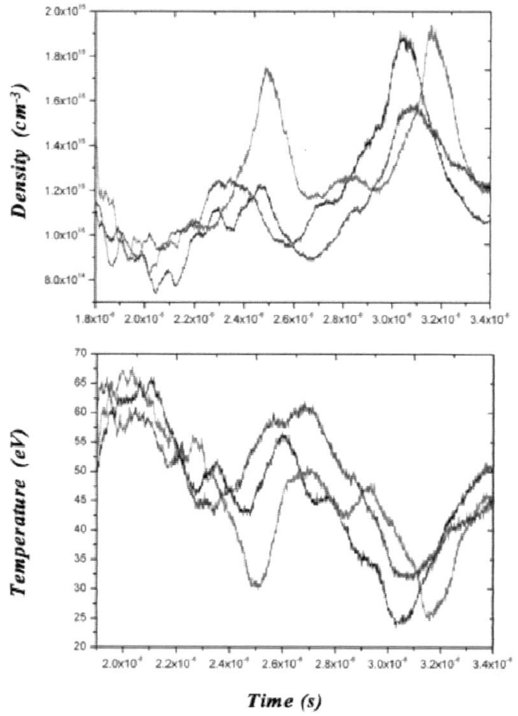


**Figure 9.** (Left) Curves of electron density and (right) electron temperature. Both curves were calculated in the zone when the discharge passes across the TLP (red square, see figure 8).

## 5. Conclusions and future work

The dense plasma focus device has been extensively studied in the breakdown, rundown and pinch phases. The first two are of paramount importance for a good pinch performance, when most of the interesting phenomena occur; soft and hard x-ray emission, collimated ion and electron beams, and fusion reactions, when deuterium is used as a filling gas. However, the there is a fourth phase beyond the pinch formation, which is usually neglected, and which we have called here the expansion discharge phase. While it may not be of relevance for the usual research of plasma foci, this is a very energetic plasma which may be of interest in order to simulate the interaction between the plasma and wall-facing materials.

Although it has not been reported in the literature to our knowledge, there are at least two groups (The PACO team at Tandil, Argentina, and the CCHEN group at Santiago, Chile) that have recently followed the evolution of the discharge at this phase with Schlieren photography, and have found the formation of a plasma dome, which evolves like a shock wave, out of which a bubble emerges on the axial direction. This was also reported in pictures in the early days of plasma focus research by Bernard et al. [10], although no attention was paid to the bubble, since the interest was focused in the neutron production. It is probably produced by a plasma jet, which emerges from the focus region, although it is preceded by an ion beam, as has been established in previous studies by Lepone et al. [11]. Since the speed of this bubble is larger than that of the dome, which roughly has the same speed as the plasma sheath before the pinch, this may help to explain the high speed reported in this paper. On the other hand, it is important to notice that the speeds obtained using two different distances are consistent.

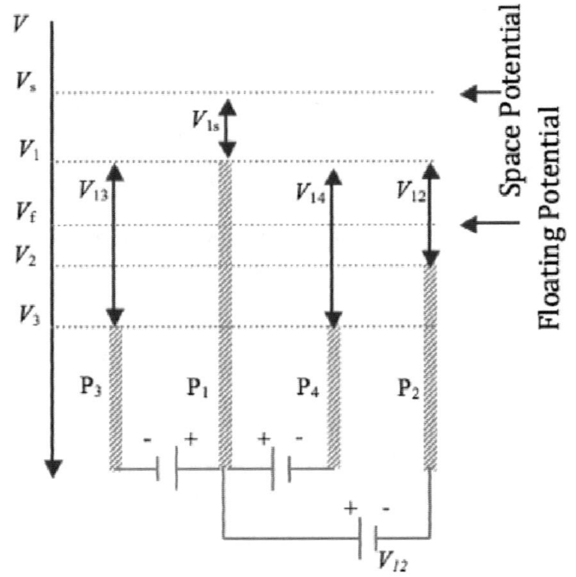


**Figure 10(a).** Diagram of Current-Mode QLP.

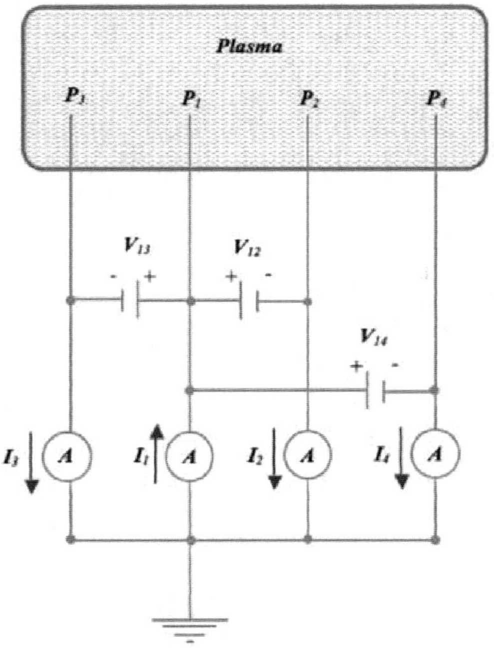


**Figure 10(b).** Electric circuit.

Several improvements need to be performed in our experiment. In order to get good signals, reliable enough to make temperature and density measurements, it will be necessary to operate the probe in current mode, rather than in voltage mode, which implies the use of two, rather than one power supply. We have also found that it will be necessary to use higher current power sources. A further improvement would be to use a fourth crossed probe in the flowing plasma. This adaptation of a quadruple Langmuir probe (QLP) should allow the determination of the ratio ($S_s$) between ion speed ($U_s$) and the most probable ion velocity $C_s$, [12-16] (figure 10). This would be accomplished by solving

$$U_s(r,t)=\langle v_s \rangle \tag{8}$$

$$c_s = \sqrt{\frac{2kT_s}{m_s}} \tag{9}$$

$$S_s = \left(\frac{U_s}{c_s}\right) \tag{10}$$

$$I_4 = A_p J_{e0} \exp\left[-\frac{e(V_{s1}+V_{14})}{kT_e}\right] - A_\perp e n_e \sqrt{\frac{kT_e}{2\pi m_e}} \frac{2}{\sqrt{\pi}} \exp\left[-S_i^2\right] \sum_{n=0}^{\infty} \left[\frac{S_i}{n!}\right]^2 \Gamma\left(n+\frac{3}{2}\right) \tag{11}$$

**Acknowledgments**
This work was partially supported by PAPIIT-DGAPA-UNAM IN120409

**References**

[1] Chen S and Sekiguchi T 1965 *J. Appl. Phys.* **36** 2363
[2] Gatsonis N A, Byrne L T, Zwahlen J C, Pencil E J and Kamhawi H 2004 *IEEE Transactions on Plasma Science* **32** 2118
[3] Manos D M 1985 *J. Vac. Sci. Technol.* **A3** 1059
[4] Lohete-Holtgreven W 1995 *Plasma Diagnostics* (American Vacuum Society Classics, American Institute of Physics)
[5] Auciello O and Flamm D L 1989 *Plasma Diagnostics, Discharge Parameters and Chemistry* Volume 1 (Academic Press, Inc.)
[6] Gatsonis N A, Zwahlen J C, Wheelock A, Pencil E J and Kamhawi H 2004 *J. Propulsion Power* **20** 243
[7] Castillo F, Herrera J J E, Rangel J, Alfaro A, Maza M A, Sakaguchi V, Espinosa G and Golzarri J I 2002 *Brazilian J. Phys.* **37** 3
[8] Theiler C 2011 *Review of Scientific Instruments* **82** 013504
[9] Riccardi C, Longoni G, Chiodini G and Fontanesi M 2001 *Rev. Sci. Instrum.* **72** 461
[10] Bernard A, Coudeville A, Jolas A, Launspach J and de Mascureau J 1975 *Phys Fluids* **18** 180
[11] Lepone A, Kelly H, Lamas D and Márquez A 1999 *Applied Surface* **143** 124
[12] Eckman R F, Byrne L, Gatsonis N A and Pencil E J 2001 *J. Propulsion Power* **17** 762
[13] Peterson E W and Talbot L, 1970 *Amer. Inst. Aeronaut. Astronaut. J* **8** 2215
[14] Laframboise J 1966 *Theory of Cylindrical and Spherical Langmuir Probes in a Collisionless Plasma at Rest* (Toronto: Univ. Toronto Inst. Aerospace Studies)
[15] Kanal M. 1964 *Journal of Applied Physics* **35** 1697
[16] Johnson B H and Murphree D L 1969 *Amer. Inst. Aeronaut. Astronaut. J* **7** 2028

# Understanding the insulator conditioning process in fast plasma focus discharges

**A Tarifeño-Saldivia[1,3], E Ramos-Moore[2], P Ferrari[2] and L Soto[1,3]**

[1]Comision Chilena de Energía Nuclear, Casilla 188-D, Santiago, Chile
[2]Facultad de Física, Pontificia Universidad Católica de Chile, Santiago 7820436, Chile
[3]Center for Research and Applications in Plasma Physics and Pulsed Power, P[4], Chile

E-mail: atarifeno@cchen.cl

**Abstract.** The presence of metallic deposition and oxides at the surface of insulators used in plasma focus discharges is expected to have a significant effect in the conditioning process that leads to neutrons and hard x-rays emissions. In this work, the atomic concentrations of the main elements present in these insulators are studied by energy dispersive x-ray (EDX) analysis. Preliminary results are presented. Surface oxidation was observed after the conditioning process. A significant concentration of metallic deposition from the anode is observed after a working cycle of a thousand shots.

## 1. Introduction

In the present, there is a renewed interest in the plasma community for the study of plasma focus discharges due to its potential as high intensity and compact source of fusion neutrons and x-rays. The plasma focus (PF) is a transient electrical discharge produced in an arrangement of coaxial electrodes separated by an insulator. The discharge is driven by a capacitive pulsed power generator through a spark-gap switch. During breakdown, a plasma sheet is formed over the insulator surface by a gliding discharge. It is well known that the efficiency for production of radiation in PF depends on a previous conditioning of the insulator. This process includes usually a series of high-pressure discharges. In spite of the intense research done on PF discharges, the breakdown phase has not been studied in full detail and, to the best of our knowledge; the physical mechanism behind the conditioning process has not been studied in detail. In this work, energy dispersive x-ray (EDX) analysis is used to study the atomic concentration of the deposited material on the insulator surface. The discharges were produced in the PF-400J device and the EDX analysis was carried out on alumina insulators just after the conditioning process and after a long-term working cycle.

## 2. Breakdown and role of the insulator

In PF discharges, the breakdown is triggered by fast transient discharges ($dV/dt > 1$ kV/ns). Thus, the breakdown voltage can be two orders of magnitude higher than the breakdown voltage calculated from the Pashen's law [1]. The efficiency of the PF discharge for production of radiation depends, among other things, on the formation of a homogeneous plasma sheet during the breakdown stage [1]. The proper formation of the sheet is promoted by the existence of charge carriers in the space close to the insulator surface. Several techniques have been proposed to improve sheet formation, including use of a knife edge on the cathode plate, continuous pre-discharge and ionizing radiation [2, 3, 4]. A metallic deposition along the insulator is also beneficial to the formation of the sheet. In effect, this fact has been proved by studies on Pyrex insulators, where the formation of metallic patches of diameter less than 10μm has been observed after long-term working cycles [3, 5]. The geometry and distribution of the patches on the Pyrex surface also influence the proper working of the discharge. Therefore, the insulator surface should be conductive enough to provide electrons for ionization, but without losing the insulating properties that allow the formation of high electric field gradients which are responsible of electron acceleration during breakdown. On the other hand, conditioning has been proposed to be cleaning of the insulator surface [1], but it is not clear if either removal or chemical modification is the main mechanism involved.

## 3. Experimental setup and study

The experiments were carried out using the PF-400J device [6]. This device is a capacitor bank with the following characteristics parameters: 880 nF, 27 kV, 38 nH short circuit inductance, 370 J, 300 ns quarter of period, 140 kA maximal current. The electrode configuration consists of an inner electrode of 1.2 cm diameter, and an outer electrode 2.6 cm diameter. Anode and cathode are made from stainless steel AISI 304. The alumina insulator has an effective length of 2.3 cm while the effective length of the anode is 0.7 cm. The discharges were operated in repetitive mode (~0.1 Hz) in hydrogen at pressures in the 3-25 mbar range. Emission of hard x-ray was monitored using a plastic scintillator (BC408) and a photomultiplier tube at 1.5m from the pinch.

Three alumina insulators were studied (A1, A2 and A8). Before the discharges, A1 and A2 were cleaned with methanol and tissue paper. A8 was subjected to an annealing process (2 h at 210 °C) and then cleaned with methanol in an ultrasonic bath. Insulators A1 and A2 were studied under a discharge working cycle starting at 24mbar, and A8 starting at 25 mbar. After five shots, the pressure was decreased by around 2mbar consecutively. Conditioning of the insulator was assumed to be complete when the first hard x-rays signal was observed from the scintillator. The total number of shots for A1, A2 and A8 were 267, 36 and ~1000 respectively. Scanning electron microscopy (SEM) and EDX were performed in the system LEO 1420VP. The electron beam was operated at different acceleration voltages in order to study the atomic concentration of the insulator as a function of the information depth.

## 4. Results and discussion

Insulator A1 was used to compare conditioning and working cycle of the insulator. After conditioning, x-ray detection frequency ($Rp_X$) was observed to be dependent on the operational pressure. The operation pressure regime of the device is 9 to 11 mbar, with $Rp_X$ ranging from 40% to 80%. These values are consistent with previous characterizations of the device. Comparison of the conditioning process on insulators A1 and A2 showed a reasonable reproducible behavior. The first indication of a rapid compression of the plasma sheet (dip) appeared after 13 and 16 shots in A1 and A2, respectively. On the other hand, x-ray emission was detected on A1 and A2 after 32 and 40 shots, respectively. In figure 1, the EDX results for the insulator just after conditioning are shown. The insulator area under analysis was between 1 to 6 $\mu m^2$ with a maximal deep of around 1.1μm.

The effect of the PF discharges in the surface of the insulator during the conditioning process is clearly seen in figure 1(a), which shows a high concentration of oxygen and carbon compared with the virgin insulator. Despite the variations of carbon and aluminum concentrations, the concentration of oxygen is found to be almost uniform along the insulator, suggesting the increase of the oxidation state of the surface due to the discharges. This is also seen as the darker zone visible in the picture of the insulator after conditioning in figure 1(d). As shown in figure 1(c), the atomic profile of the anode corresponds to the main constituents of the steel: Fe, Cr and Ni. It is interesting to note that the metallic deposition is almost negligible after conditioning. Only small amounts (< 0.4%) of metallic deposition near the top-edge (figure 1(b)) of the insulator are observed using 30 kV in A2.

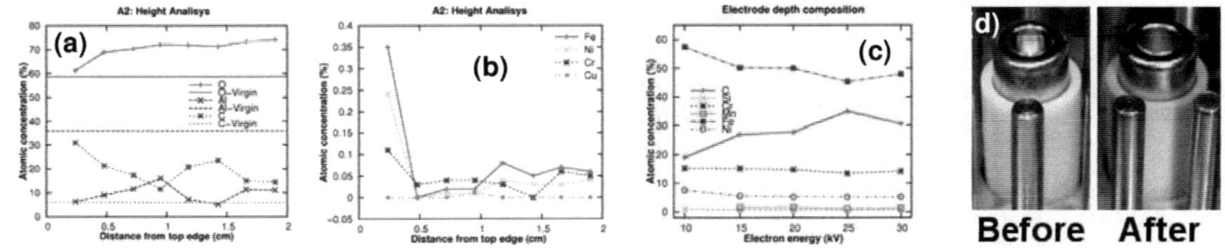

**Figure 1.** EDX analysis of insulator A2 after 36 PF-shots (a) and (b) and its respective anode (c). The atomic concentration of (a) oxygen (OKα), aluminum (AlKα) and carbon (CKα), and (b) iron (FeKα), nickel (NiKα), chromium (CrKα) and copper (CuKα) are shown as a function of the distance from the top-edge of the insulator. (c) Atomic concentration of the anode as a function of the electron energy. The lower concentration of carbon at low information depth (E < 20kV) is due to the methanol used for cleaning before the analysis. (d) Optical picture of the insulator before and after conditioning.

After a long-term working cycle, the effect of multiple discharges is the increase of the concentration of iron by an almost uniform metallic deposition from vaporization of the anode. As shown in figure 2(c), for a high information depth (30 kV) in A8, the iron concentration is similar to the highest values observed in insulator A2. Moreover, when decreasing the information depth (10 kV), Fe concentrations as high as 4% are observed along the insulator surface.

For both high (30 kV) and low (10 kV) information depths, the analysis of A8 show a similar behavior for the main constituents of the insulator (figures 2(a) and 2(b)). The correlation between the increase of the aluminum and oxygen signals and the decrease of the carbon signal is due to an effect of the x-rays bulk absorption, and indicates that the oxygen and aluminum atoms are bonded within the insulator bulk. Nevertheless, this is not observed in the area close to the top anode edge. This is an oxygen-rich and iron-free region, which is modified due to the proximity to the anode. The processes behind this phenomenon and their effects on the rupture and efficiency of the formation of the plasma sheet remain to be understood.

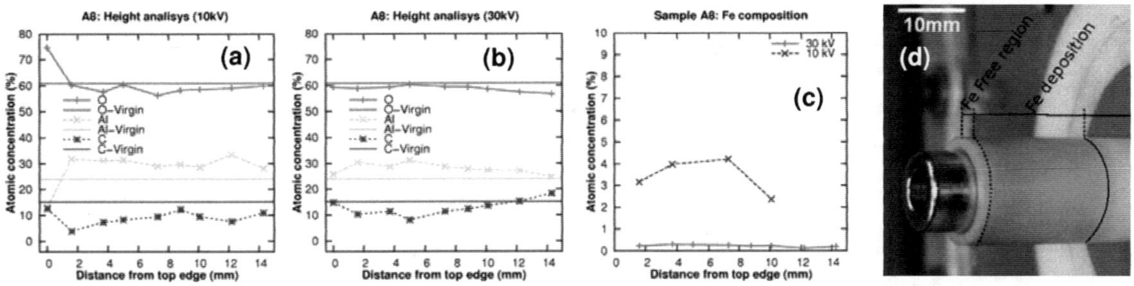

**Figure 2.** EDX analysis of insulator A8 (~1000 shots). Figure (a) and (b) show the atomic concentration of oxygen (OKα), aluminum (AlKα) and carbon (CKα) at 10 and 30 kV of beam acceleration voltage, respectively. (c) Comparison of the iron concentration at different electron beam voltage. (d) Image of insulator A8 after working cycle.

## 5. Final remarks and future work

Several questions arise from the present results. The role of carbon and oxygen during the conditioning process and its role on the efficiency for production of radiation should be understood. In addition, the effect of very small concentrations of metallic deposition on the breakdown should be studied. A systematic study, including verification of the present results, correlation of efficiency for radiation production with morphology and composition of deposition along the insulator, and characterization of the breakdown process is the next step in this research.

As preliminary conclusions, we can state that the chemical effect of PF-discharges during the insulator conditioning was successfully measured by EDX analysis. Low metallic deposition from the anode and surface oxidation were found along the insulators after the conditioning process. The spatial distribution of the related elements depends on the number of shots and the proximity to the top anode edge. Moreover, several questions arise from the present results. The role of carbon and oxygen during the conditioning process and their role on the efficiency for production of radiation should be understood. In addition, the effect of very small concentrations of metallic deposition on the breakdown will be systematically studied. Morphology and concentration analyses will be correlated with the efficiency for radiation production and the breakdown process in the next steps of this research.

**Acknowledgements**

We would like to thank the financial support from Facultad de Física, Pontificia Universidad Católica de Chile and FONDECYT (grant 1110940).

**References**

[1] Bernard A, Bruzzone H, Choi P, Chuaqui H and Gribkov V *et al.* 1998 *J. Moscow Phys. Soc.* **8** 93
[2] Lu M, Tsinchi Yang M, Luo C and Miyamoto T 2001 *Trans. Plasma Sci.* **29** 973
[3] Kies W 1986 *Plasma Phys. Control. Fusion* **28** 1645
[4] Ahmad S, Hussain S, Sadiq M, Shafiq M, Waheed A and Zakaullah M 2006 *Plasma Phys. Control. Fusion* **48** 745
[5] Feugeas J N 1989 *Journal of Applied Physics* **66** 3467
[6] Silva P, Moreno J, Soto L, Birstein L, Mayer R E and Kies W 2003 *Applied Phys. Lett.* **83** 3269

# Preliminary results of eigenvalue and eigenvector treatment for kink instabilities in tokamaks

**F Konefka, P Martin and J Puerta**

Departamento de Física, Universidad Simón Bolívar, Apdo: 8900, Caracas - Venezuela

E-mail: konefka@gmail.com

**Abstract.** A mathematical method is presented for the treatment of MHD instabilities in a simplified model, but illustrative of the plasmas in a tokamak. First order perturbation treatment of MHD equilibrium equations has been performed, leading to a set of differential equations and associated dispersion relations characteristic of kink modes. Using appropriate boundary conditions these equations are solved and the frequency eigenvalues are found in the stable region. The unstable growth rates have been also found with more accuracy than in the usual functional treatment using the energy integral.

## 1. Introduction

The ideal magnetohydrodynamic (MHD) internal kink modes are the most dangerous instabilities leading to the sawtooth instability and the breaking of the plasma confinement [1]. Analytic results usually give more general results than those obtained by transport codes, but these are difficult to obtain and some approximations have to be performed, usually well adapted to actual experiments in tokamaks. Most of the analytic results are limited to circular cross sections and large aspect ratio approximation [2]. However, recent works have been performed for non-circular section using numerical methods [3]. Mikhailovskii [4] directly solves the differential equation for the case $m >> 1$ (Suydam modes) in the marginal stability limit. The simplest treatment is the classic one where the toroidal plasma is approximated by a periodic circular cylindrical one in the z-coordinate. The most complete information on the MHD kink instability is performed with these approximations using the energy principle.

## 2. Theoretical model

In the classical transport the toroidal geometry of the tokamak is replaced by a cylindrical one with a periodic conditions on the z-coordinate. Now the toroidal magnetic lines become helical lines and the equilibrium toroidal and poloidal magnetic fields become the axial and circular magnetic fields $B_z\hat{z}$ and $B_\theta\hat{\theta}$, respectively. Axis symmetry assumptions, now means that those fields are not functions of $z$, and the equilibrium uniform axial current is given by

$$\mathbf{J_0} = (0, 0, \frac{2w}{\mu_0}) \tag{1}$$

Now, because of Ampére law and circular cross section, the magnetic field $\mathbf{B_0}$ becomes

$$\mathbf{B_0} = (0, B_\theta, B_z) \tag{2}$$

where

$$\mathbf{B}_\theta = wr \tag{3}$$

The radial component of the momentum equation is given now by

$$\mathbf{J} \times \mathbf{B} = \nabla p \tag{4}$$

where

$$p = \frac{w^2(a^2 - r^2)}{\mu_0} \tag{5}$$

When a perturbation displacement field $\overrightarrow{\xi}(\overrightarrow{r}, t)$ is considered the linearized MHD equations lead to a differential equation,(8). This equation can be treated by considering the eigenvalues and eigenvectors, as well as the usual time dependence $e^{i\omega t}$

$$-\varrho_0\, \omega^2 \overrightarrow{\xi} = F(\overrightarrow{\xi}) = \lambda \overrightarrow{\xi} \tag{6}$$

Here $\overrightarrow{\xi}$ is only a function of the position and $F(\overrightarrow{\xi})$ is an operator given by

$$F(\overrightarrow{\xi}) = \nabla(\overrightarrow{\xi} \cdot \nabla p_0 + \gamma p_0 \nabla \cdot \overrightarrow{\xi}) + \frac{1}{\mu_0}(\nabla \times \mathbf{B_0}) \times (\nabla \times (\overrightarrow{\xi} \times \mathbf{B_0})) + \frac{1}{\mu_0}(\nabla \times (\nabla \times (\overrightarrow{\xi} \times \mathbf{B_0}))) \times \mathbf{B_0} \tag{7}$$

Since $\overrightarrow{\xi}$ is univaluated around $\theta$ and periodic along $z$, then the dependence in $\theta$ and $\xi_z$, is given by the usual $e^{i(m\theta)+kz}$ function, where the poloidal number $m$ is an integer and $k$ can be written as

$$\begin{aligned} k &= \frac{2\pi n}{R} \\ \vec{\xi} - \xi(\vec{r})e^{i(m\theta+kz)} &= \left[\xi_r(r)\hat{r} + \xi_\theta(r)\hat{\theta} + \xi_z(r)\hat{z}\right] e^{m\theta+kz)} \\ \nabla \cdot \vec{\xi} &= \left[\frac{1}{r}\frac{\partial(r\xi_r)}{\partial r} + im\xi_\theta(r) + ik\xi_z(r)\right] e^{m\theta+kz)} \end{aligned} \tag{8}$$

where $R$ is the major radius. Furthermore, if the radial component of the perturbation $\xi_r$ is real, then the amplitude $\nabla \cdot \overrightarrow{\xi}$ must be also real, and therefore the remaining two components $\xi_\theta$ and $\xi_z$ must be pure imaginary. In this way the vectorial equation Eq(4) can be reduced to three differential equations in the variables $\xi_\theta(\overrightarrow{r})$, $\xi_z(\overrightarrow{r})$ and $\xi_r(\overrightarrow{r})$. The first two equations are:

$$\sum_{i=1}^{i=3} P_i \sin(kz + m\theta) = 0$$

$$P_1 = -\frac{(B_z^2 m + B_z k r^2 w + m(a^2 - r^2)w^2\gamma)\xi_r(r)}{r^2\mu_0} + \frac{B_z r(m^2 + k^2 r^2)w\xi_z(r)}{r^2\mu_0}$$

$$etc. \tag{9}$$

$$\sum_{i=1}^{i=3} A_i \sin(kz + m\theta) = 0$$

$$A_1 = \frac{(wB_z mkw(-a^2\gamma + r^2(1+\gamma)))\xi_r(r) + r(m^2w^2 + \lambda\mu_0)\xi_z(r)}{r\mu_0}$$

$$etc \quad (10)$$

Before writing down the radial equations it is important to note that both poloidal and toroidal equations depend on the functions $\xi_\theta(r)$ and $\xi_z(r)$, but not on their derivatives . Then, it is straightforward to solve the $2 \times 2$ system defining $\xi_\theta(r)$ and $\xi_z(r)$. This is possible thanks to our choice of phase, since all terms are proportional to $\sin(kz + m\theta)$ and not to a mix of this one and $\cos(kz + m\theta)$ terms. The solutions are given as follows

$$\xi_\theta(r) \sum_{i=1}^{i=2} D_i = \sum_{i=1}^{i=4} N_i$$

$$D_1 = (a^2 - r^2)(m^2 + k^2r^2)w^2(B_z kmw)^2$$

$$D_2 = \lambda\mu_0((m^2 + k^2r^2)(B_z^2 + w^2(r^2 + (a^2 - r^2)\gamma)) + r^2\lambda\mu_0)$$

$$N_1 = B_z kw^2(-mr^2(B_z k + 2mw) + a^2(B_z km + 2w(m^2 + k^2r^2))\gamma\xi_r(r)$$

$$etc. \quad (11)$$

The z component is also given by:

$$\xi_z(r) \sum_{i=1}^{i=3} D_i = \sum_{i=1}^{i=3} K_i$$

$$K_1 = rw(kw\gamma(a^2 - r^2)(B_z^2k^2 - m^2w^2) - \lambda\mu_0(B_z m + kr^2w))\xi_r(r)$$

$$etc \quad (12)$$

Following this, it is possible to replace the poloidal and toroidal components by the radial one, deriving a second order differential equation for the radial component of the vector field only. This can be written as

$$\sum_{i=1}^{i=5} R_i = 0$$

$$R_1 = -\frac{(B_z^2(1 + k^2r^2) + 2B_z kmwr^2 + (-1 + m^2) + \lambda\mu_0 r^2)}{r^2\mu_0}$$

$$etc \quad (13)$$

This final radial differential equation is

$$M_1(\omega^2, r)\xi_r''(r) + M_2(\omega^2, r)\xi_r'(r) + M_3(\omega^2, r)\xi_r(r) = 0, \quad (14)$$

where the expressions of $M_i(\omega^2, r)$ are not given here, because they are to long.

Journal of Physics: Conference Series **370** (2012) 012061 doi:10.1088/1742-6596/370/1/012061

## 3. Results and conclusions

The eigenvalue method has been introduced as a more accurate alternative to the energy integral method. This mathematical method was applied to a very simple plasma cylindrical discharge of constant safety factor for the sake of simplicity as a proof of concept. The long calculations involved even in this scenario shed doubts on its applicability to more interesting toroidal discharges. Here it is shown that the way to obtain the eigenvalues and eigenvectors from the differential equation in the radial component of the perturbation field is by using the property that the eigenfunctions are just the solutions, which are finite at the main axis. Eigenfunctions and eigenvalues are simultaneously determined. The eigenvalues are determined digit by digit, in this way the accuracy increases step by step, but you never get the exact result. For the simple case here considered, the onset of instabilities appears at the same boundary value of the axial magnetic field as those calculated with the energy method. Figures 1 and 2 show some examples of the obtained $\xi_r(r)$. Growth rates (figure 3) were calculated by requiring the solution to be finite at the axis.

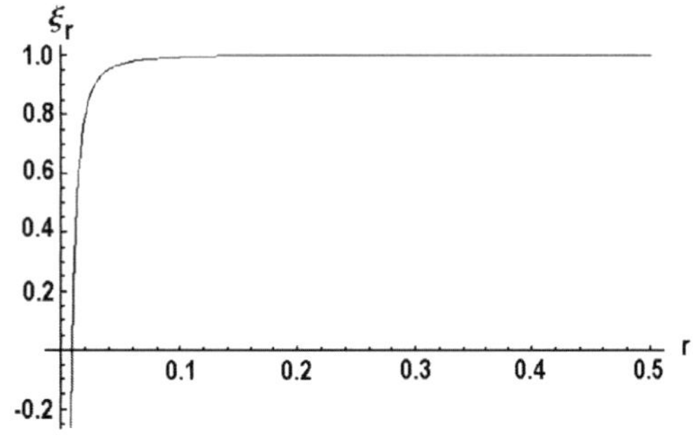


**Figure 1.** Numerical solution of $\xi_r(r)$ against minor radius with $\lambda = 3.110^{10} Kg\ m^{-3}s^{-2}$.

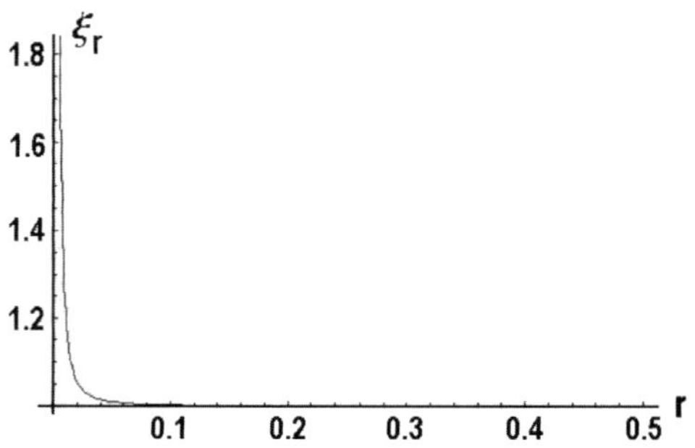


**Figure 2.** Numerical solution of $\xi_r(r)$ against minor radius with $\lambda = 3.110^{10} Kg\ m^{-3}s^{-2}$. The divergence at $r = 0$ is due to an inaccurate selection of the eigenvalue. It does not reflect the real eigenfunction.

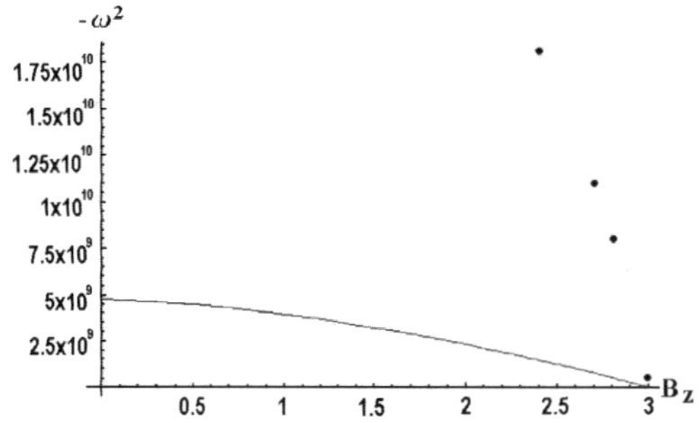


**Figure 3.** Instability growth rates versus toroidal magnetic fields for the energy method (solid line) and eigenvalues method (dots).

Journal of Physics: Conference Series **370** (2012) 012061 doi:10.1088/1742-6596/370/1/012061

The apparent divergence shown in figures 1 and 2 does not reflect the real eigenfunction. The numerical solution of the model at the axis always goes to $+\infty$ or $-\infty$ at the axis. However when there is a change in sign is when you get an additional digit for the real eigenvalue.The important result is that in the line of marginal stability($\lambda$ equal to zero) the eigenvalues obtained for both methods coincide. However, when growth rates are calculated in the unstable region, the calculation with the energy principle gives rates that are always smaller than those calculated with the actual method. Furthermore, the function $\nabla \cdot \overrightarrow{\xi}$ is zero at the marginal stability line in both methods, but not in the unstable region, where the energy method considers also the divergence as zero, but it is not. Our calculation of $\nabla \cdot \overrightarrow{\xi}$ is performed by using the eigenvalues and they are obtained from the differential equation. The eigenfunctions are independent of the plasma model: compressible or incompressible. Only when the eigenvalues are zero, is when $\nabla \cdot \overrightarrow{\xi}$ is also zero, and in the other cases $\nabla \cdot \overrightarrow{\xi}$ is different than zero. Though no a general proof is presented, these results seem more general and probably are also right in toroidal geometries.

The eigenvectors in the energy principle $\xi_r(r)$ are taken as a step function, this assumption is not corroborated by the present calculations. The actual function calculated here with several digits of accuracy, looks like a step function, but there is a slowly and continuous decreasing from the axis to the plasma border or scrape off layer, that is, it is in someway near to a step function, but its pattern is more physical. Additionally, the similarity is limited to the shape of radial function $\xi_r(r)$, while the poloidal and axial functions, $\xi_\theta(r)$ and $\xi_z(r)$, are fundamentally different.

## Acknowledgments
Work supported by grant DID-G22, Universidad Simón Bolívar, Caracas, Venezuela

## References
[1] Shafranov V D 1958 *Sov. Physics JETP* **6** 545
[2] Shafranov V D 1970 *Sov. Physics Tech. Phys.* **15** 175
[3] Lutgens H and Bondeson A 1992 *Nuclear Fusion* **32** 1625
[4] Mikhailovskii A B 1998 *Instabilities in a Confined Plasma* (Bristol: Institute of Physics Publishing)

# Design and construction status of the TPM-1U tokamak

**M Nieto-Perez[1], D Hernandez-Arriaga[1] and F Ceballos-Soto[2]**

[1]CICATA Queretaro – IPN. Cerro Blanco 141, Col. del Cimatario, Queretaro, QRO, 76090 Mexico
[2]Instituto Tecnológico Superior de Misantla, Misantla, VER, 93821 Mexico

E-mail: mnietop@ipn.mx

**Abstract.** TPM-1U is a small size tokamak, with a vacuum vessel of 0.15 m minor radius and 0.4 m major radius, with an expected magnetic field in the torus centre of 0.5 T and a plasma current above 150 kA. One of the goals of the TPM-1U is to establish the scientific and technological base for the study of magnetically confined high temperature plasmas in Mexico, and help bring cohesion to the efforts in this field that have been scattered to date. Another goal is to contribute to the generation of high level human resources that can help the community grow in size. In addition to these two strategic goals, the machine will also have technical and scientific goals such as plasma-wall interaction studies, development of new diagnostics and exploration of new magnetic topologies. In this work, the design for the pulsed power system that will be used to feed the set of toroidal coils and design considerations for the central solenoid will be presented. Calculations for the dimensioning of the central solenoid used for ohmic heating are also presented, as well as preliminary considerations for the implementation of microwave power injection into the plasma.

## 1. Introduction

During the last 50 years, there has been steady progress on the road to generate thermonuclear fusion on earth, and the main approach that has been pursued to achieve this goal has been magnetically confined plasmas. Today, very large machines with long pulses have already demonstrated that it is possible, and the next step is to build a device that can demonstrate net energy generation, and this prototype is already being constructed [1]. Despite this tremendous advance, not all issues have been resolved and there are still many scientific and technological challenges to make controlled nuclear fusion a reality. The study of some of these open questions, in particular those where the risk of experiment is too high to be run in a large facility, is better suited for small experimental facilities, where the necessary reconfiguration might be easier to achieve and implement, the number of people affected by a malfunction is low and the cost of recovery is modest. Hence, small machines still play an important role in the path towards energy from nuclear fusion [2], as testbed of new ideas, development and improvement of diagnostic techniques and as training instruments for the development of human resources.

In the particular case of Mexico, previous efforts have been done to build and operate a small tokamak [3-6]. However, this machine had many operational issues that made it unreliable and it ceased to operate around the year 2000. Parallel to that effort, a larger machine was also designed and built, but due to a shortage of both funds and personnel that machine was only partially constructed and was left unfinished. Recently, the infrastructure of this machine was transferred to CICATA-IPN, where there is an active group in the field of magnetically confined plasmas and where there are better opportunities for obtaining funding.

## 2. The TPM-1U tokamak

Figure 1 shows a rendering of the TPM-1U vacuum vessel, which has a 0.15 m minor radius and a 0.5 m major radius, formed by four 90° elbows, two of which are electrically isolated from each other by use of a Viton O-ring to make the seal and plastic washers on the bolts joining these two sections. The vessel has 10 ports of 3.75 cm in diameter, oriented in different positions, and one port for pumping which mates into an 8" CF flange. The vacuum is achieved using a Pfeiffer Balzers TPU-520 turbomolecular pump, with a nominal pumping speed of 520 liters per second for nitrogen; this pump is backed by a Balzers DUO 016 mechanical pump. Monitoring of the vacuum environment is done with an Accu-Quad quadrupole mass spectrometer, with a range of 1 - 100 amu. Preliminary measurements of the achievable vacuum level show that a vacuum of 5 x $10^{-7}$ Torr can be achieved without any conditioning of the chamber, and the main impurities are water and air.

For supplying the necessary power to the tokamak, there are 60 Maxwell capacitors, with a maximum charging voltage of 7.5 kV and a capacitance of 20 μF each, so a maximum total energy of 33 kJ can be stored in this capacitor bank. To charge it, a 20 kV DC power supply with up to 500 mA current capacity is available. In the following sections it will be shown that this power is sufficient to drive the toroidal filed of the machine for pulses of up to 10 ms, as well as providing the ohmic transformer power used to drive the plasma current.

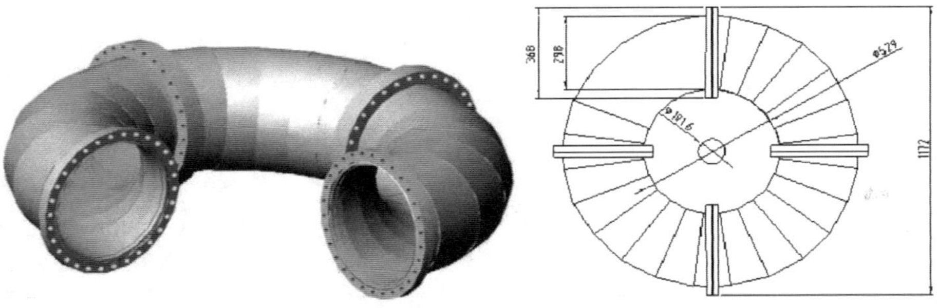


**Figure 1.** Rendering of the TPM-1U vacuum vessel with one of the sectors removed, and a schematic with dimensions (access ports not shown).

## 3. TPM-1U Subsystems

### *3.1. Toroidal field*

In order to confine high temperature plasmas, it is necessary to create a magnetic field in the machine along the largest curvature, known as the toroidal direction. To generate this field in TPM-1U, a set of 16 coils 40 cm in diameter is being considered. Each coil has 6 layers with 6 turns each of 5 AWG magnet wire with double insulation. Each coil requires 8 m of wire, which translates into approximately 8 kg of wire per coil. The field produced by this coils is shown in figure 2, both in the equator plane and in a vertical plane. The magnetic field generated by the addition of a plasma column with 12 cm radius and a parabolic current density profile with 5 MA/m$^2$ in the centre and vanishing right at the vessel wall. For these conditions, the current in the defined plasma column is on the order of 150 kA, and the calculated radial safety factor in the midplane and across the plasma column is shown in figure 3. It can clearly be seen that for the chosen plasma current profile, the stable portion of the plasma is bounded by the peak in safety factor at a plasma radius of 4 cm, which is 1/3 of the set value. In order to achieve a good safety factor value in the edge, the current would have to drop by a factor of 10, to 15 kA, reducing the ohmic heating significantly.

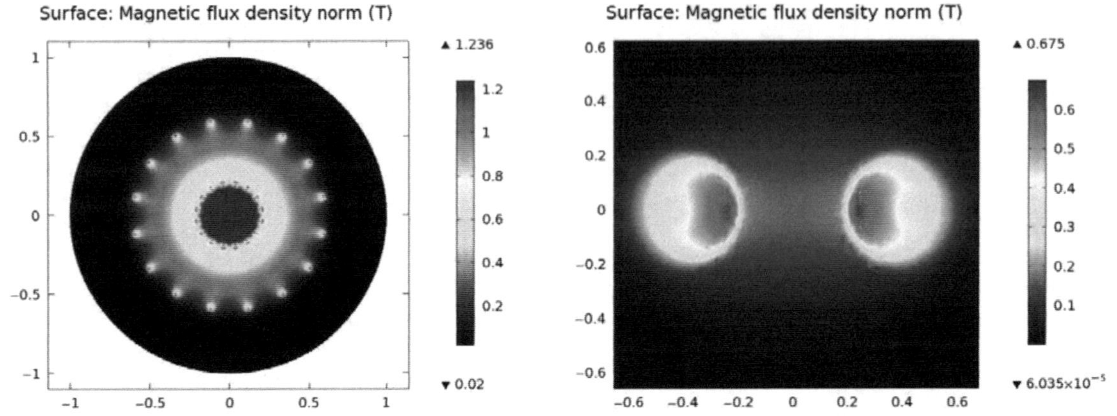


**Figure 2.** Vacuum magnetic field of the TPM-1U tokamak, along the equatorial plane (left) and on a vertical plane (right).

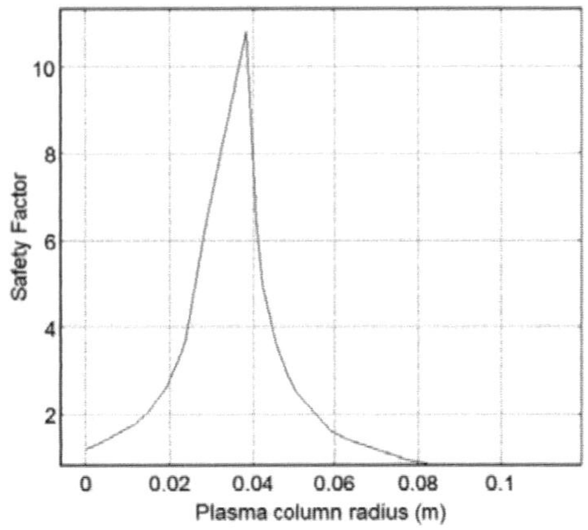


**Figure 3**. Safety factor across the plasma column along a radial line in the equatorial plane.

The main requirement for the toroidal field is a maximum intensity of 0.5 T in the centre of the torus for at least 2 ms. hence, a suitable pulsed forming network (PFN) needs to be constructed. The PFN has two main advantages: 1) it can store the energy required, and 2) this energy can be delivered as a rectangular pulse [7]. The PFN consists of capacitors and inductors connected in a ladder configuration, and the values of the different elements in it have an effect on the pulse shape and duration.

For the generation of a rectangular pulse, the network consists of inductors and capacitors with the same values connected in cascade [7, 8]. The capacitor bank used to store the energy is charged up to 7.5 kV, and the inductance of the set of 16 toroidal coils is estimated to be 250 μH. Simulations were performed using the ATP software, and a particular case of these simulations is shown in figure 4, where the left panel shows the circuit used and the right one shows the current pulse obtained. As it can be seen, the network can generate a pulse with a rise time of only a few μs with duration of a few ms.

The load resistor affects the value of the peak voltage output. To achieve the power output of the pulse forming network, a Marx generator is attractive because it multiplies the voltage, so that the PFN will produce the waveform and the Marx generator can multiply the output voltage. For the PFN-Marx configuration, the output pulse parameters as a function of network parameters can be expressed by the equations in table 1.

Figure 5 shows different pulses of current obtained for different load inductance values. It can be seen that larger load inductances damp the initial oscillations of the pulse, at the cost of increasing the initial rise time. A plateau starting at t=1 ms can be seen, and it has a duration of 1 ms; the addition of more stages to the PFN may allow the extension of the pulse to the desired 2 ms.

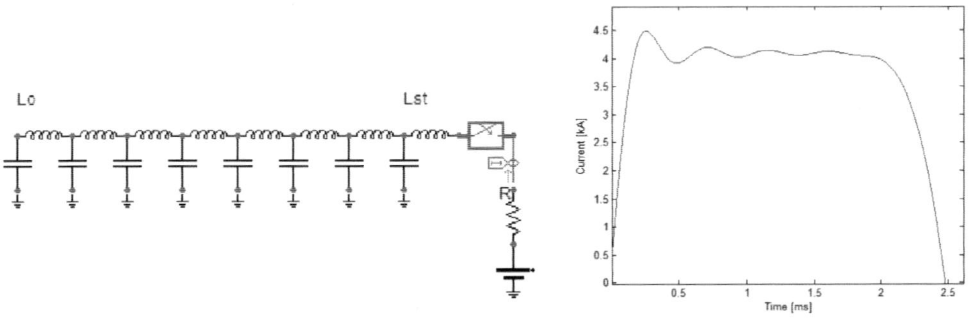


**Figure 4.** The pulse forming network (left) and the resulting current pulse (right).

**Table 1**. Parameters for pulse in terms of PFN parameters.

| Parameter | Expression |
|---|---|
| Pulse length | $\tau = 2k\sqrt{LC}$ |
| Characteristic impedance | $Z_M = n\sqrt{L/C}$ |
| Pulse voltage | $V = \frac{nV_0}{2}$ |
| Delivered power | $P_L = \frac{nV_0^2}{4\sqrt{L/C}}$ |

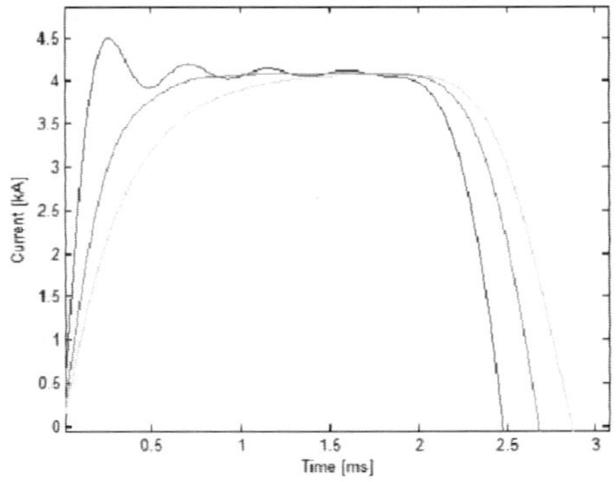


**Figure 5**. Effect of load inductance on pulse shape for the PFN shown in figure 4.

*3.2. Ohmic heating.*

The solenoid is a part of the TMP-1U prototype, and it is the component responsible for inducing high currents (> 100 kA) in the plasma confined within the torus, in order to achieve heating to several tens of millions of degrees for a short time (microseconds). As a starting point, it is assumed that the maximum available space is a circular area of 101.6 mm in diameter; the reason for this limitation is the internal space available on the toroidal vacuum chamber already available. Another consideration is that the signal for the solenoid power will have the minimum rise time to maximize flux swing. Even though a trapezoidal signal is used as a first approximation, the more realistic and easily achievable sine wave is used, considering only half the cycle. It is assumed that this waveform is obtained by discharging an 800μF capacitor bank. In figure 6 a comparison between the trapezoidal and sinusoidal signals shows the similarity between the two signals. A better approximation of the trapezoidal signal can be obtained using Fourier series, which gives the infinite series that approximates the desired waveform; this is valid considering that only half cycle of the sine signal is used, and that the signal is generated by a single-shot capacitor bank discharge.

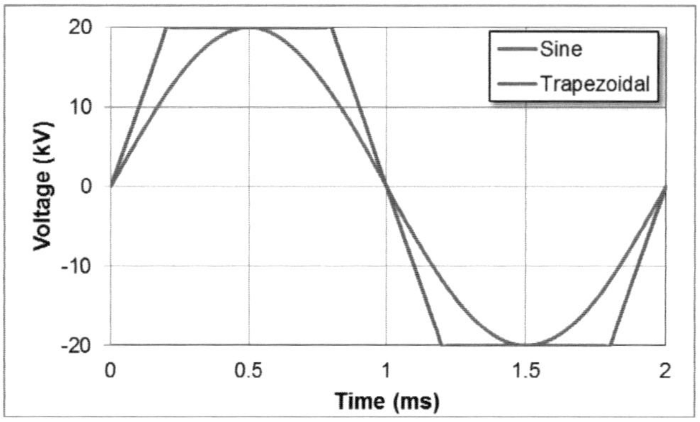


**Figure 6**. Comparison between a trapezoidal signal and an equivalent sine wave.

Some parameters for solenoid construction are still free, such as the number of turns that must carry the density at which the coil works the type and size of conductors and their insulation, and the material of the core. According to Faraday's law, the voltage induced in a coil wound on set of a magnetic core is given by:

$$e_{ind} = -N\frac{d\phi}{dt} \tag{1}$$

where $N$ is the number of turns carrying the winding and $\phi$ is the magnetic flux flowing through the core. This equation is valid for an induction coil which experiences a magnetic field effect; the same equation allows us to represent the case when a voltage $V(t)$ is applied to the coil, the only difference being the sign, since in the case of an induced voltage it is negative according to Lenz's Law; the flux, however, remains the same in both cases. For the case of a solenoid to which a potential $V$ is applied, the expression is then:

$$V(t) = N\frac{d\phi}{dt} \tag{2}$$

However, in the case of a time-changing voltage, the result will be a circulating magnetic flux in accordance with the implementation of the Right Hand Rule and this in turn will vary with time in the same way as the applied voltage. If the voltage applied is a sine function with frequency $\omega$, the magnetic flux is:

$$\phi(t) = \phi_{\max}\sin(\omega t) \tag{3}$$

Substituting the magnetic flux in equation (2) yields:

$$V(t) = N\phi_{\max}\,\omega\cos(\omega t) = N\phi_{\max}(2\pi f)\cos(2\pi f t) \tag{4}$$

In the above equation, the maximum magnetic flux is given in Wb. Considering that the application of the maximum potential generates the maximum flux, that condition determines the maximum solenoid area required to drive the peak magnetic flux. Instead of using instantaneous values, effective values are used, which allow the comparison between a sinusoidal signal and its equivalent direct

Journal of Physics: Conference Series **370** (2012) 012062 doi:10.1088/1742-6596/370/1/012062

current. This is indicative of how much energy the signal carries. Writing the maximum values in terms of effective values and removing the cosine variation with time, the above equation becomes:

$$V = \sqrt{2}\pi f N\phi \tag{5}$$

If this last equation is written in terms of the flux density $B$ instead of the magnetic flux, then it takes the following form:

$$V_t = \sqrt{2}\pi f B A \tag{6}$$

where $A$ is the area where the magnetic flux will flow, and $V_t$ is the volt-per-turn value, defined as the potential induced by each of the N turns ($V/N$). Rearrangement of the equation finally yields an expression for the solenoid cross sectional area:

$$A = \frac{V_t}{\sqrt{2}\pi f B} \tag{7}$$

If we apply this equation to our case where the input is the half cycle of a periodic signal with a 200 μs period, or a frequency of 5 kHz, the flux density and the volts per turn value are still free to be set, with the restriction that the maximum cross sectional area available for the solenoid is roughly a 100 mm diameter circle. This means that the number of layers needs to be low, so a large number of turns would imply a long solenoid; this is advantageous in this case, since a long solenoid would have low impedance and hence be more efficient in energy transfer.

The power of a coil is given by the product of voltage between its terminals and the current flowing through it, i.e.:

$$S = VI\cos(\theta) \tag{8}$$

This formula allows us to calculate the apparent power, in this formula the power factor depends on the load that will be powered by the coil, with three distinct cases: it is unity in the case of resistive load, less than one for highly inductive loads with a delay on the current waveform with respect to the voltage, and also less than 1 for capacitive loads with the current waveform ahead in time with respect to the voltage. The power factor really is nothing more than a way to quantify the effect of the phase shift of current with the voltage; the coil power can be estimated using any of the two windings, either the primary (solenoid) or secondary (plasma); as the plasma is a very complicated impedance when subjected to high currents, the only thing assumed initially is that the current to be circulating in the secondary is 150 kA.

### *3.3. Microwave heating considerations*

The proposed mechanism to inject power into TPM-1U is electron cyclotron resonant heating (ECRH). Heating using X mode is complicated due to both the requirement of inboard launching to avoid deposition near the edge and the frequency required to achieve the resonance (14 GHz) [9]. Since the plasma density is expected to be low, injection from the outboard is simpler and sources at a lower frequency are less expensive, the natural decision is to perform ECRH using the O mode initially. The heating efficiency will be then dictated only by the plasma density, and interesting experiments regarding wave mode conversion to achieve Bernstein wave heating [10] may be possible. Microwave generators on the S band (2.45 GHz) are relatively cheap and readily available due to their extensive commercial application; even though this frequency range may not be very well suited for plasma heating, its price and hardware availability enables a relatively quick implementation of relevant research activities regarding interaction of microwaves with magnetically confined plasmas. Such activities may include microwave diagnostic development, non-inductive current drive studies or fundamental studies on mode conversion and wave propagation at lower field and no plasma current [11], as well as power deposition (lower hybrid [9]) at this frequency. For the X-mode heating and higher field operation, which for our central field value falls on the $K_u$ band, the VTX5783G3

TWT from CPI [12] seems like a good prospect, with a 100 μs pulse length and an operating frequency of 10 GHz (resonant at 0.4 T); but being many times more expensive than the S band sources its implementation will be a long term project. Overall, it is deemed that microwave heating will be more likely implemented once good operation with just ohmic heating is achieved, and only once the planned diagnostics for the machine (and in particular the Thompson scattering setup) come online in order to have the necessary tools to measure power deposition via microwaves.

## 4. Conclusions

The TPM-1U tokamak has begun construction at CICATA Queretaro. It is a small machine with high aspect ratio similar to ISTTOK and TORPEX. It is expected to provide up to 150 kA of plasma current and 0.5 T of magnetic field in the centre. Vacuum and power infrastructure is already available to drive the toroidal field and the central solenoid, and a PFN to obtain a long toroidal magnetic field pulse has already been designed. Basic parameters for a prototype toroidal coil prototype are currently being evaluated.

## Acknowledgments

The authors wishing to acknowledge assistance from COFAA-IPN for providing partial funding that allowed the attendance of authors to the meeting where this work was presented.

## References

[1] http://www.iter.org
[2] Gryaznevich M P, del Bosco E, Malaquias A, Mank G, van Oost G, He Y, Hegazy H, Hirose A, Hron M, Kuteev B, Ludwig G O, Nascimento I C, Silva C and Vorobyev G M 2005 *Nucl. Fusion* **45** S245
[3] Ramos J, de Urquijo J, Meléndez L et al 1983 *Rev. Mex. Fis.* **29** 551
[4] Valencia R, de la Rosa J, Camp E, Contreras G and Muhl S 2002 *J. Nuc. Mat.* **305** 232
[5] López-Callejas R, Benítez-Read J S, Longoria-Gándara L C, Pacheco-Sotelo J O, Valencia-Alvarado R, Tamayo F J, Valdés A, Fernández M C and Serrano F 2001 *AIP Conf. Proc.* **563** 173
[6] Melendez L, Chavez E, Barocio S R, Flores A, Cruz G J, Lopez R and Olayo M G 2000 *IEEE Trans. Plasma Sci.* **28** 1676
[7] Bluhm H *Pulsed Power systems: Principles and Applications* (Berlin: Springer)
[8] Li H, Ryoo H, Kim J, Rim G, Kim Y and Deng J 2009 *IEEE Trans. Plasma Sci.* **37** 190
[9] England A C, Eldrige O C, Knowlton S F, Porkolab M and Wilson J R 1989 *Nuc. Fusion* **29** 1527
[10] Podoba Y Y, Laqua H P, Warr G B, Schubert M, Otte M, Marsen S, Wagner F and Holzhauer E 2007 *Phys. Rev Lett.* **98** 255003
[11] Fasoli A, Chavan R, Fasel D, Joye B, Llobet X, Marmillod P, McGrath M et al. 2002 *44th APS Annual Meeting of the Division of Plasma Physics* Orlando FL
[12] http://www.cpii.com/product.cfm/1/17/120

Journal of Physics: Conference Series **370** (2012) 012063 doi:10.1088/1742-6596/370/1/012063

# Rayleigh-Taylor analysis in a laser-induced plasma

**R A Marin, C A Gonzales and H Riascos**

Grupo de Plasma, Láser y Aplicaciones, Universidad Tecnológica de Pereira, A.A 097, Pereira, Colombia

E-mail: adr_1407@hotmail.com

**Abstract.** We report the conditions (plasma parameters) under which the Rayleigh-Taylor Instability (RTI) develops in an Al plasma produced by a Nd:Yag pulsed laser with a fluence range of 1 to $4\,J/cm^2$, wavelength of $1064nm$ and $10Hz$ repetition rate. The used data correspond to different pressure values of the ambient N atmosphere. From previous works, we took the RTI growth rate form as $\eta^2 = -K\,a\left(\frac{\rho_p-\rho_g}{\rho_p+\rho_g}\right)$. From the perturbation theory the instability amplitude is proportional to $e^{-\eta t}$. Using the drag model, we calculated the plume dynamics equations integrating the instability term and plotted the instability growth profile with the delay time values to get critical numbers for it, in order to show under which conditions the RTI appears.

## 1. Introduction

When a high power laser impinges on a target and the laser fluence exceeds a certain threshold, the irradiated zone is vaporized sending material to the environment in the form of plasma [1]. Generally this procedure is carried out in a closed chamber. This plasma is called plasma plume, which also can be confined by an specific gas pressure within the chamber [2], resulting in an interaction between the plasma plume particles and the gas atoms. Because of such interaction a few phenomena happen that are studied in various physics fields such as in the new material science, in the deposition of thin films, and theoretical plasma physics. In a relatively dense plasma, collisions can happen between the plasma species and the background gas, making it possible the occurrence of phenomena like plasma plume bifurcation, self-generated magnetic field, shock waves, and certain types of instabilities. A common instability, the Rayleigh-Taylor instability (RTI), could be one of the causes of the self-generated magnetic field in the plume and, of course, another cause for the premature plasma collapse. The RTI dynamics can be explained considering an interface between two incompressible and immiscible fluids with different densities, the denser fluid lying on the lighter one [3]. We present some calculus about the growth rate of the RTI, using the drag model and plot the trend curve of this growth rate.

## 2. Model

The RTI for laser produced plasmas at high pressures occurs as a plasma of increasing density in a very narrow region, spreads in the background gas [4]. The physics of RTI can be described through the Drag Model (DM), considering a constant related to the deceleration phenomena, called Stopping Coefficient. The equation of motion describing this dynamics has the form,

$$\frac{dv}{dt} = -\beta v \tag{1}$$

Where $v$ is the plume expansion velocity and $\beta$ is the stopping coefficient. Its solution is [4, 5],

$$R(t) = R_0(1 - e^{-\beta t}) \tag{2}$$

With $R_0$ the stopping distance (distance at which the plume comes to rest).

## 3. Results and discussions

The DM makes it possible to calculate the acceleration term of the instability growth rate. From Sharma and Thareja works we have estimated the maximum velocity and the $R_0$ values, for a range of pressures between 1 to 70 $Torr$ [4, 6]. The maximum velocity value is $6,5 \times 10^6\, cm/s$ and the $R_0$ values are shown in table 1.

**Table 1.** $R_0$ values for a time of 100 to 200 $ns$.

| Pressure ($Torr$) | $R_0$ ($m$) |
|---|---|
| 1 | $8,7\times10^{-4}$ |
| 10 | $3,2\times10^{-3}$ |
| 20 | $3,2\times10^{-3}$ |
| 30 | $3,1\times10^{-3}$ |
| 40 | $3,0\times10^{-3}$ |
| 50 | $2,9\times10^{-3}$ |
| 60 | $2,8\times10^{-3}$ |
| 70 | $2,7\times10^{-3}$ |

To determine the growth rate of the RTI from perturbation theory, we took the instability amplitude as $e^{-\eta t}$ [7], where the growth rate is defined as [4],

$$\eta^2 = -K\, a \left[\frac{\rho_p - \rho_g}{\rho_p + \rho_g}\right] \tag{3}$$

with $K$ the wave number at the plasma front acceleration, and $\rho_p$ and $\rho_g$ the plasma and gas density, respectively. According to the $R_0$ estimated values, and using the momentum equation,

$$\frac{d}{dt}\left[\left(m_0 + \frac{4}{3}\pi R^3 \rho_g\right) u\right] = 0 \tag{4}$$

with $m_0 = 2,1 \times 10^{-11}\, kg$ the estimated ablated mass, we calculated the gas density as shown in table 2.

Journal of Physics: Conference Series **370** (2012) 012063 doi:10.1088/1742-6596/370/1/012063

**Table 2.** Gas density $\rho_g$ calculated at different pressure values.

| **Pressures** $(Torr)$ | **Gas Density** $\rho_g$ $(kg/m^3)$ |
|---|---|
| 1 | $2{,}175\times10^{-3}$ |
| 10 | $4{,}371\times10^{-5}$ |
| 20 | $4{,}372\times10^{-5}$ |
| 30 | $4{,}808\times10^{-5}$ |
| 40 | $5{,}305\times10^{-5}$ |
| 50 | $5{,}873\times10^{-5}$ |
| 60 | $6{,}525\times10^{-5}$ |
| 70 | $7{,}277\times10^{-5}$ |

For $\eta^2 > 0$, it can be shown that the plasma front acceleration is given by,

$$a = \frac{d^2R(t)}{dt^2} = \left|R_0\beta^2e^{-\beta t}\right| \tag{5}$$

with different parameters calculated above, we can plot the instability growth rate as a function of pressure, as shown in figure 1.

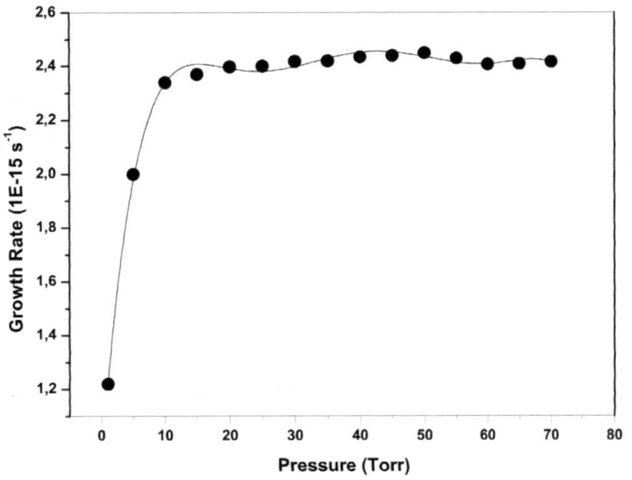


**Figure 1.** Growth rate behavior at different pressure values.

In figure 1 we can see an interesting fast-growing phenomena occurring in a pressure range of 1 to 10 $Torr$, however at higher pressures the growth rate becomes more stable. Like the growth rate, the instability amplitude has a similar behavior with the increasing pressure, due to the instability amplitude form, the plasma-gas interface collapses faster with increasing pressure up to certain pressure value.

## 4. Conclusions

As we expected, the instability starts to grow as a function of pressure, the interesting part occurs approximately between 0 and $20\,Torr$. Thanks to the drag model we calculated the acceleration term as a function of time from the growth rate equation and saw how it behaves for an increasing background gas pressure. Considering the instability behavior at different pressure values, it would be hard to estimate the Rayleigh-Taylor instability at pressures lower than $1\,Torr$, even so, the instability has a violent growth between $0,5$ and $10\,Torr$ and this should be targeted for future studies. Finally we have been able to analyze the instability behavior as a function of pressure.

## References

[1] Rifai K, Vidal F and Johnston T W 2007 *Phys. Plasmas* **14** 082311
[2] Abdeli-Messaci S, Kerdja T, Lafane S and Malek S 2009 *Spectrochimica Acta Part B* **64** 968-73
[3] Freidberg J P 2007 *Plasma Physics and Fusion Energy* (UK: Cambridge University Press)
[4] Sharma A K and Thareja R K 2005 *Appl. Surface Sci.* **243** 68-75
[5] Sharma A K and Thareja R K 2007 *Appl. Surface Sci.* **253** 3113-21
[6] Sharma A K and Thareja R K 2004 *Appl. Phys. Lett.* **84** 4490-3
[7] Chen F F 1984 *Introduction to Plasma Physics and Controled Fusion: Plasma Physics* (New York: Plenum Press)

# Power supply instrumentation for pulsed dielectric barrier discharges

**V E Quiroz Velázquez[1], R López Callejas[1, 2], B G Rodríguez Méndez[2], R Peña Eguiluz[2], A de la Piedad Beneitez[1], A E Muñoz Castro[2], S R Barocio[2], A Mercado Cabrera[2] and R Valencia Alvarado[2]**

[1] Instituto Tecnológico de Toluca, AP 890, Toluca, Estado de México, México
[2] Instituto Nacional de Investigaciones Nucleares, AP 18-1027, CP 11801 México DF

E-mail: regulo.lopez@inin.gob.mx

**Abstract**. The design and implementation of a pulsed high voltage supply intended to the production and control of pulsed dielectric barrier discharges are reported. The instrumentation includes three independently built DC sources coupled to Flyback-like converters using three 1:50 high voltage transformers. The system is capable of supplying voltages up to 70 kV at a 100-2000 Hz repetition rate, delivering 1-500 μs wide pulses. The system has been applied to the development of pulsed dielectric barrier discharges in a stainless steel coaxial reactor 30 cm long and with a 2.54 cm diameter. The inner nickel electrode diameter is 0.005 cm and is embedded in alumina. The discharges have been carried out in room pressure air. Discharges have been implemented. The discharge is made is a water environment for purposes of bacterial elimination.

## 1. Introduction

Advanced oxidation technology (AOT) is based on the use of oxidize radicals such as ozone ($O_3$), hydrogen peroxide ($H_2O_2$) and hydroxide ($OH^-$) in order to remove both organic and inorganic compounds from the environment. One shortcoming of this technique is that the radicals are not produced *in situ* whereby their often depleted population diminishes the effectiveness of the process [1]. However, a cold (or out of equilibrium) plasma produced at room pressure could perform the oxidation process required by the production of oxidize radicals. One case in point is the dielectric barrier discharge (DBD) technique which has often been applied to the control and removal of volatile organic compounds [2], elimination of bacteria [3] and decontamination of gases [4] among other uses. The present report describes the electronic and mechanical instrumentation required to carry out DBDs aimed at the elimination of bacteria in water, which implies voltages of 30-50 kV magnitudes, operation frequencies in the 500-2000 Hz range and pulse widths ($t_w$) in the order of micro seconds.

## 2. Experimental set-up

The array shown in figure 1 has been conceived in order to conduct PDBD discharges. It is constituted by three sections: a pulsed power source (PPS), a data acquisition system capable of monitoring the behavior of the discharge electrical parameters, and a coaxial discharge reactor (figure 2) consisting in a stainless steel cathode, a Ni/Cu anode and an alumina ($Al_2O_3$) dielectric barrier.

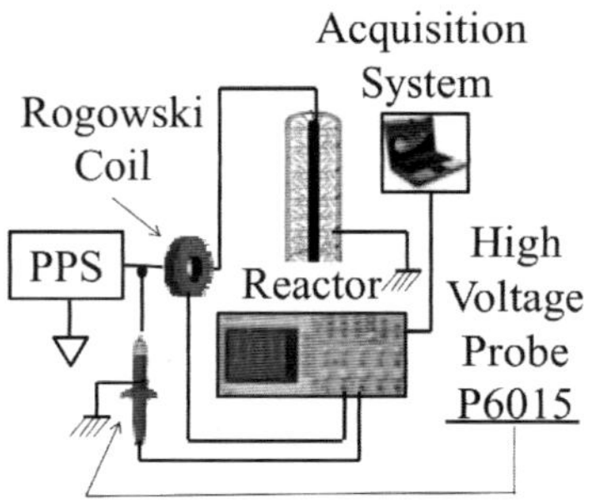


**Figure 1.** Pulsed Power System for DBD.

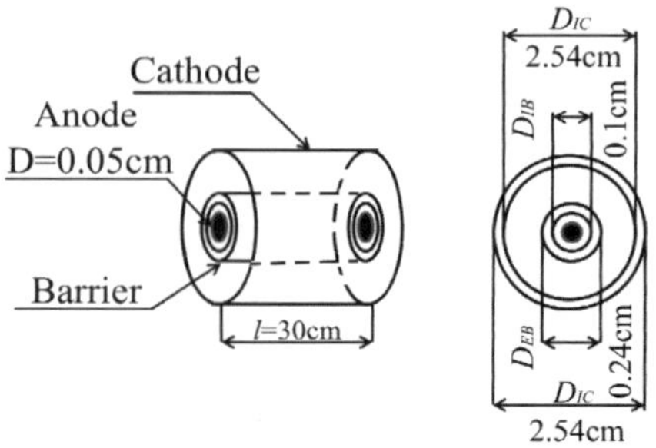


**Figure 2.** Reactor for Pulsed Discharge Barrier Dielectric (PDBD).

An electric circuit that models the electrical evolution of the discharges has been studied elsewhere [5]. The circuit takes into consideration: (1) the physical dimensions and its contents, $R_m$ is a function of the water conductivity and $C_m$ refers to capacitance reactor, (2) parameters previous to breakdown, $R_S$ involves the resistance between electrodes and $C_S$ refers to a capacitive which depends on the gap and the voltage applied, and (3) during breakdown, $R_b$ and $L_b$ represent the inductance and the discharge resistance respectively [6] in the case of a pulsed corona discharge (PCD). The circuit in figure 3 incorporates the parameters $C_B$ and $R_B$ that estimate the $Al_2O_3$ dielectric capacitance and resistance during the discharge. Thus, $R_B$ can be expressed by [7]:

$$R_B = (\sigma_A e)^{-1} \tag{1}$$

where $\sigma_A$ is the electric conductivity of the barrier (~$1\times10^{-11}$ S/m) and $e$ denoted its thickness (~0.0014m) whereby $R_B$ ~140 MΩ. The dielectric layer entails a capacitance given by [7]:

$$C_B = \frac{\varepsilon_r \varepsilon_0 A_r}{\delta} \tag{2}$$

where $\varepsilon_r\varepsilon_0$ is the absolute permittivity of the material (~$8.8541\times10^{-11}$ F/m), $A_r$ its area (~1.885 $m^2$) and $\delta$ represents the ratio between the diameters of the barrier that covers the anode (~0.876), so that $C_B$ ~190 pF. On its part, $R_m$ [Ω] can be put in the form [8]:

$$R_m = \frac{\ln\left(\frac{D_{IC}}{D}\right)}{2\pi\sigma_m l} \tag{3}$$

where $l$ is the reactor length (0.3 m), $D_{IC}$ is the internal cathode diameter (0.0254 m) and $D$ is the diameter of the anode (0.0005 m), $\sigma_m$ is the prevailing conductivity, 0.0005 S/m in the case of sweet water and $1\times10^{-13}$ S/m in the air. Then, it follows from equation (3) that $R_w$~4.17 kΩ (water) and $R_A$=20.81 TΩ (air). As to $C_m$, according to [8]:

$$C_m = \frac{2\pi\varepsilon l}{\ln\left(\frac{D_{EB}}{D_{IB}}\right)} \tag{4}$$

where $D_{EB}$ is the external diameter (0.00254 m) and $D_{IB}$ is the inside diameter (0.001 m) of the dielectric barrier respectively, $\varepsilon$ is the permittivity of the medium where the discharge takes place, for water $\varepsilon_w$ ~$7.08334\times10^{-10}$ F/m and for air $\varepsilon_A$ ~$8.8541\times10^{-12}$ F/m. Then, from equation (4), $C_w$ ~340 pF (water) and $C_A$ ~4.25 pF (air). $C_S$ and $R_S$ describe, as mentioned, the state previous to breakdown,

Journal of Physics: Conference Series **370** (2012) 012064 doi:10.1088/1742-6596/370/1/012064

characterized by a capacitive transient and an inductive one created by the electric current that precedes the discharge [9]. $C_S$ and $R_S$ depend on the gap between electrodes ($d_g$) and the voltage between them ($U$). The value of $R_S$ [Ω] becomes [10]:

$$R_S = \frac{d_g}{n q \mu A_S} \tag{5}$$

where $q$ is the elementary charge (~$1.60256\times10^{-19}$ C), $n$ is the electron density namely ~$1\times10^{26}$ m$^{-3}$ [10], and $\mu$ is the mobility (~$1\times10^{-5}$ m$^2$/V·s) [11]. Thus, if $d_g$ ~0.01245 m and the streamer area $A_S$ ~$3.1416\times10^{-8}$ m$^2$ [13] then, given equation (5), $R_S$ ~2473 Ω. Now, $C_S$ can be expressed as [9]:

$$C_S = \frac{\mu_d}{U}\left(d_g\right) \tag{6}$$

where $U$ ~70 kV, $\rho_d$ is the streamer charge ~$1\times10^{-6}$ C/m [9], and (6) $C_S$ ~177 fF. The breakdown onset is estimated by means of $R_b$ and $L_b$, which depend in turn on the current conveyed by the streamer [12]. $R_b$ [Ω] is determined by [13]:

$$R_b = \frac{d_g}{\sigma_c \pi r_s^2} \tag{7}$$

where $r_s$ is the streamer effective radius (~$1\times10^{-4}$ m [14]) and $\sigma_c$ is the conductivity of the ionized medium given by [9]:

$$\sigma_c = 1.5\times10^3\, T^{3/2} \tag{8}$$

here $T$ is the temperature (~2 eV [9]), and, therefore, from (8) $\sigma_c$ ~4242.7 S/m whereas, from (7), $R_b$~94 Ω. $L_b$ can be described by [9]:

$$L_b = \frac{\mu_0 d_g}{2\pi}\left[\frac{1}{4} + \ln\left(\frac{d_f}{r_S}\right)\right] \tag{9}$$

where $\mu_0$ is the vacuum permeability ~ $4\pi\times10^{-7}$ N/A$^2$ and $d_f$ is the gap between streamer and the null electric field point. Then, $d_f$~#, and, from (9), $L_b$~20.56 nH.

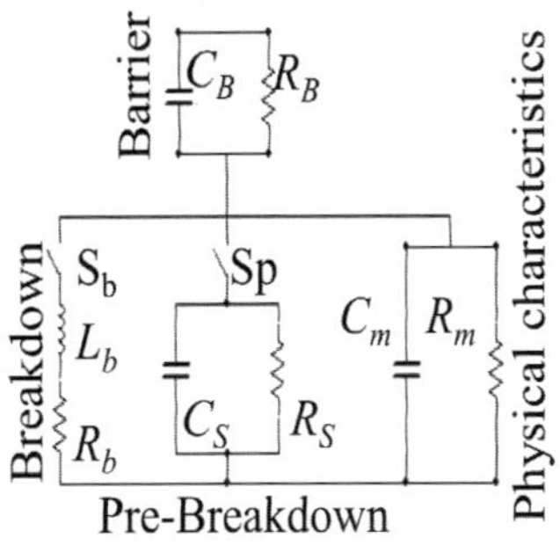


**Figure 3.** Equivalent PDBD circuit.

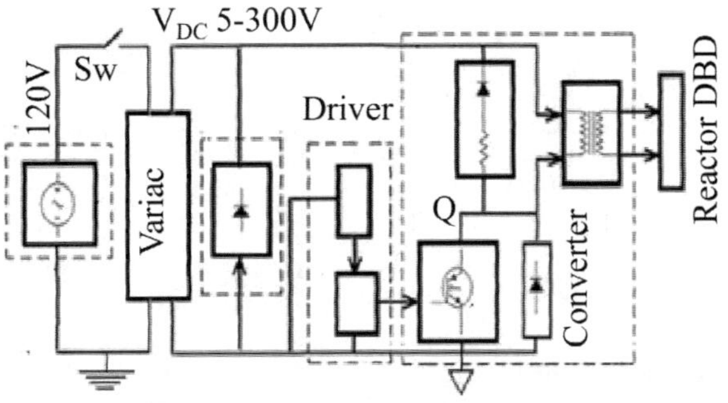


**Figure 4.** PPS sub-system.

The pulsed power source (PPS) circuit developed to conduct PDBDs is displayed in Figure 4. It is structured as three identical sub-systems separated by relays labeled as Sw. Each sub-system c006Fntains a DC voltage supply ($V_{DC}$) operating within 5-300V in order to feed a flyback converter constituted by a semiconductor (HGTG27N120N transistor) operating as a Q switch, and by a 1:50

pulse transformer (PT) characterized by the parameters: $L_1 = 20.1$ mH, $L_2 = 40$ mH, $L_m = 20.1$ mH, $L_f = 55$ µH, $R_1 = 70.6$ mH, $R_2 = 196$ Ω, $C_{D2} = 90.48$ pF, monitored by a 4263B *Agilent* LRC bridge operated at 1 and 10 kHz.

One driver has been implemented in each sub-system, formed by a pulse width ($t_w$) modulator circuit type PWM (SG3524) specifically designed to generate $t_w$ values within 1-500 µs, at frequencies in the order of 100-2000 Hz, which provides for the recovery time of water subjected to electric discharges [15]. The PWM supplies the commutation signal for the controller (M57962L), and through it, maintains the electric parameters of the Q switch.

## 3. Results and discussion

The simulation of the PPS with the three subsystems connected as shown in figure 5 led to a ratio value n:1:150. Figures 6(a), 6(c), 6(b) and 6(d) exhibit the waveforms simulated and experimental where the output generated voltage corresponds to ~28 kV at a ~60 mA current. Figures 7(a), 7(b), 7(c) and 7(d) describe the experimental voltage and current behavior at several frequencies and pulse widths in order to identify the regime of maximal energy transference during the discharge. Thus, the effectiveness of the system is given by $\eta = Vout_{rms}Iout_{rms}/Vin_{rms}Iin_{rms}$ [16], where, according to the measurements taken during the discharge represented in figures 6.a and 6.c, $Vout_{rms}$ ~3.62 kV and $Iout_{rms}$ ~11.54 mA. As $Vin_{rms}$ ~68.737 V and $Iin_{rms}$ ~0.6 A, then the efficiency is ~82%.

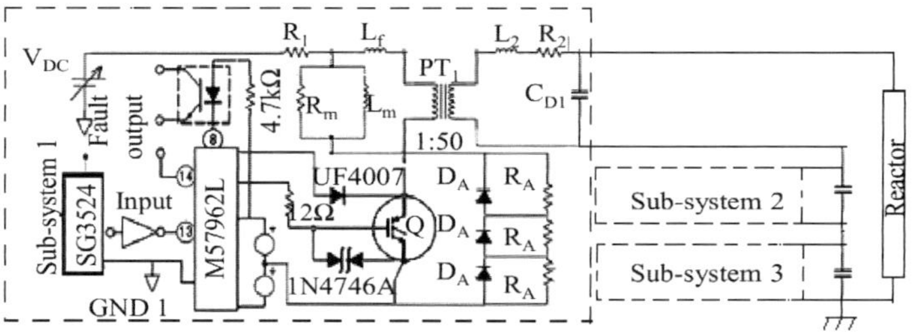


**Figure 5.** PPS model.

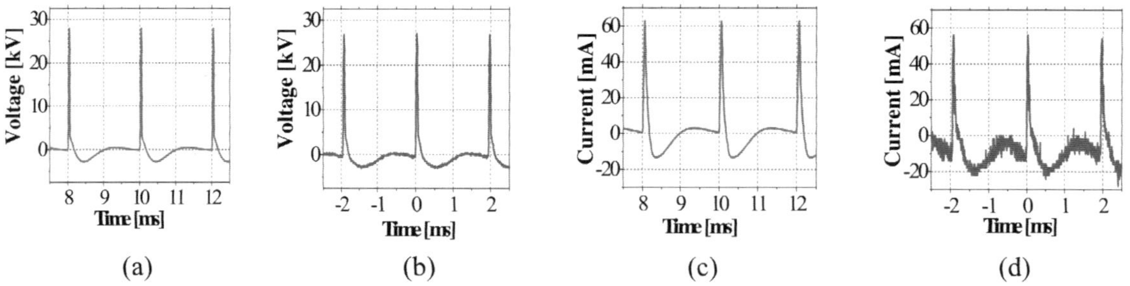


**Figure 6.** Simulation waveforms for: (a) voltage, (c) current. Experimental waveforms: (b) voltage and (d) current. In both, simulation and experimental $t_w$=50 µs and f=1 kHz.

## 4. Conclusions

An electrical model capable of predicting the operational characteristics of a voltage/current controlled PDBD plasma source has been developed. The flexibility of the proposed model enables one to adjust several electrical parameters like voltage, width pulsed and frequency. Also, it allows to modify discharge parameters such as pre-breakdown and breakdown time. In addition, it can change electrical characteristics that involve the physical reactor model. The electric and electronic instrumentation has been carried out and can operate one, two or three subsystems connected simultaneously. This arrangement provides an output voltage in the order of 70 kV. Experimental tests have been carried out in a cylindrical reactor with room pressure air obtaining an efficiency of around 82%. The system has been applied to a PDBD in order to generate non-thermal plasma aimed at a treatment for the removal of microorganisms in water.

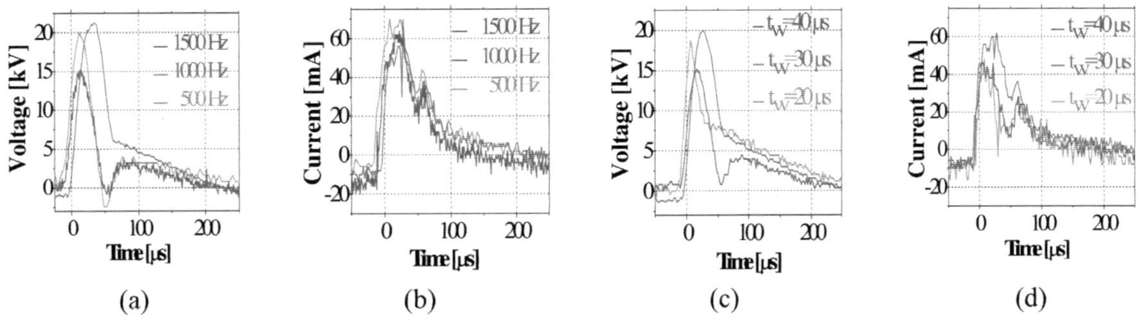

**Figure 7.** Waveforms experimental for several frequencies with at pulse width ~40μs: (a) voltage, (b) current; and several pulse widths with a frequency ~1 kHz: (c) voltage and (d) current.

**Acknowledgements**
This project was financially supported by CONACYT Mexico. The technical assistance of P. Angeles Espinoza, I. Contreras Villa and M. T. Torres Martinez.

**References**

[1] Fridman A 2008 *Plasma Chemistry* (United States of America: Cambridge University) pp 450-750
[2] Dorai Rajesh and Kushner M J 2002 *J. Phys. D: Appl. Phys* **35** 2954–68
[3] Gweon Bomi, Kim D B, Moon S Y and Choe W 2009 *Current Applied Physics* **9** 625–8
[4] Dong-Joo Kim and Kyo-Seon Kim 2003 *IEEE Transactions on Plasmas Science* **31** 227–35
[5] Rodriguez-Mendez B G, López-Callejas R, Peña-Eguiluz R, Mercado-Cabrera A, Valencia-Alvarado R, Barocio S R., de la Piedad-Beneitez A, Benitez-Red J S and Pacheco-Sotelo J O 2008 *IEEE Transactions on Plasma Science* **36** 185–91
[6] Gao Lan, Larsson A, Cooray V and Scuka V 1999 *IEEE Transactions on Dielectrics and Electrical Insulation* **6** 35–42
[7] Han-Jun Oh, Yongsoo Jeong, Su-Jung Suh, Young-Jik Kim and Choong-Soo Chi 2003 *Journal of Physics and Chemestry of Solids* **64** 2219–25
[8] Grenier J, Jayaram S H, El-Hag A H and Kazerani M 2005 *A study on effect of medium conductivity on its electric strength under different source conditions in nanosecond regimes* (Ontario: Waterloo Univ.) pp 261–64
[9] Grearson W D 1999 *IEEE Transactions on Industry Applications* **35–2** 359-65
[10] Jones H M and Kunhardt E E 1996 *Nanosecond pre-breakdown and breakdown phenomena in water: influence of pressure, conductivity, and ionic sheath formation* (Roma, Italy) pp 15–19
[11] Kupershtokh A L and Karpov D I 2006 *Technical Physics Letters* **32** 406–9
[12] Fofana I and Béroual A 1996 *IEEE Transactions on Dielectrics and Electrical Insulation* **3** 273–82
[13] Kefu Lui, Qiong Hu, Jian Qui and Houxiu Xiao 2005 *IEEE Transactions on Plasma Science* **33** 1182–85
[14] Lisitsyn I V, Nomiyama H, Katsuki S and Akiyama H 1999 *IEEE Transactions on Dielectrics and Electrical Insulation* **6** 351–6
[15] Xiao S, Kolb J, Kono S, Katsuki S, Joshi R P, Laroussi M and Schoenbach K H 2004 *IEEE Trans. Dielectrics Electrical Insulation* **1** 604–12
[16] Rashid M H 1988 *Power Electronics* 2th edition (USA: Prentice Hall) p 52–55

# Low cost sonoluminescence experiment in pressurized water

**L Bernal[1], M Insabella[1] and L Bilbao[2,*]**

[1]LADOP, University of Mar del Plata, Argentina
[2]INFIP, University of Buenos Aires and CONICET, Argentina
[*]Member of CONICET

E-mail: bernal@mdp.edu.ar

**Abstract:** We present a low cost design for demostration and mesurements of light emmision from a sonoluminescence experiment. Using presurized water introduced in an acrylic cylinder and one piezoelectric from an ultrasonic cleaner, we are able to generate cavitacion zones with emission of light. The use of argon to pressurize the water improves the emission an the light can be seen at naked eye in a softlit ambient.

## 1. Introduction

Sonoluminescence is a phenomenon which is caused by acoustic energy concentration in a liquid medium. The acoustic energy is concentrated by the geometry of the system up to about 12 orders of magnitude [1-4], resulting in high-energy concentration zones in the fluid. The high energy density zones give rise to the formation of bubbles, by increasing the intensity of the acoustic wave, which begin to grow in amplitude. For a given threshold energy the bubble collapses in the compression cycle, resulting in the heating and ionizing of the gas, thus producing emission of light. On some geometric configurations a stationary oscillating bubble can remain synchronous with the stationary sound field [5], making it easy to study it by optical and spectroscopic methods. There are reports of measured temperatures up to 50,000°K [1] and even higher, which is of great interest for applications in high-speed/high-temperature chemical reactions [2]. The method we use to generate the acoustic stationary wave in the fluid employs a piezoelectric transducer. As a comment, there exists another method, purely mechanical, called liquid hammer [4].

## 2. Experimental setup

The experiment was built in a student laboratory using standard materials. We first experiment with a glass balloon, but it was difficult to obtain a stable configuration using one piezoelectric. Instead, the use of a cylindrical geometry reduces the mechanical and coupling problems and gives us more parameters to control. The scheme of the experiment is shown in figure 1, the cylinder oriented in the $z$-axis has an outer diameter of 100 mm, a wall of 5 mm thickness, and a length of 415 mm. This configuration can resist 2.5bar without problems. With the cylinder filled and pressurized we found it too hard to make measurements at different water levels, so we decided to use a 2 liter buffer vessel to improve the design. The power source for the piezoelectric was built ad-hoc in order to be able to control and fix the frequency output in a range of 20 to 60 kHz. A conventional autoresonance scheme did not match these requirements. The power output was typically 150 W at 600 V. The piezoelectric was coupled to the water through an aluminium disc that was cemented with epoxy resin. The water facing side of this disc was worn away to an ellipsoid to focus the ultrasonic energy. Both the aluminium disc at the bottom and a polyamide disc at the top were sealed with silicon O-rings. To register light emitted we use a Hamamatsu photomultiplier and a digital oscilloscope Agilent with 2 channels and 100 MHz bandwidth.

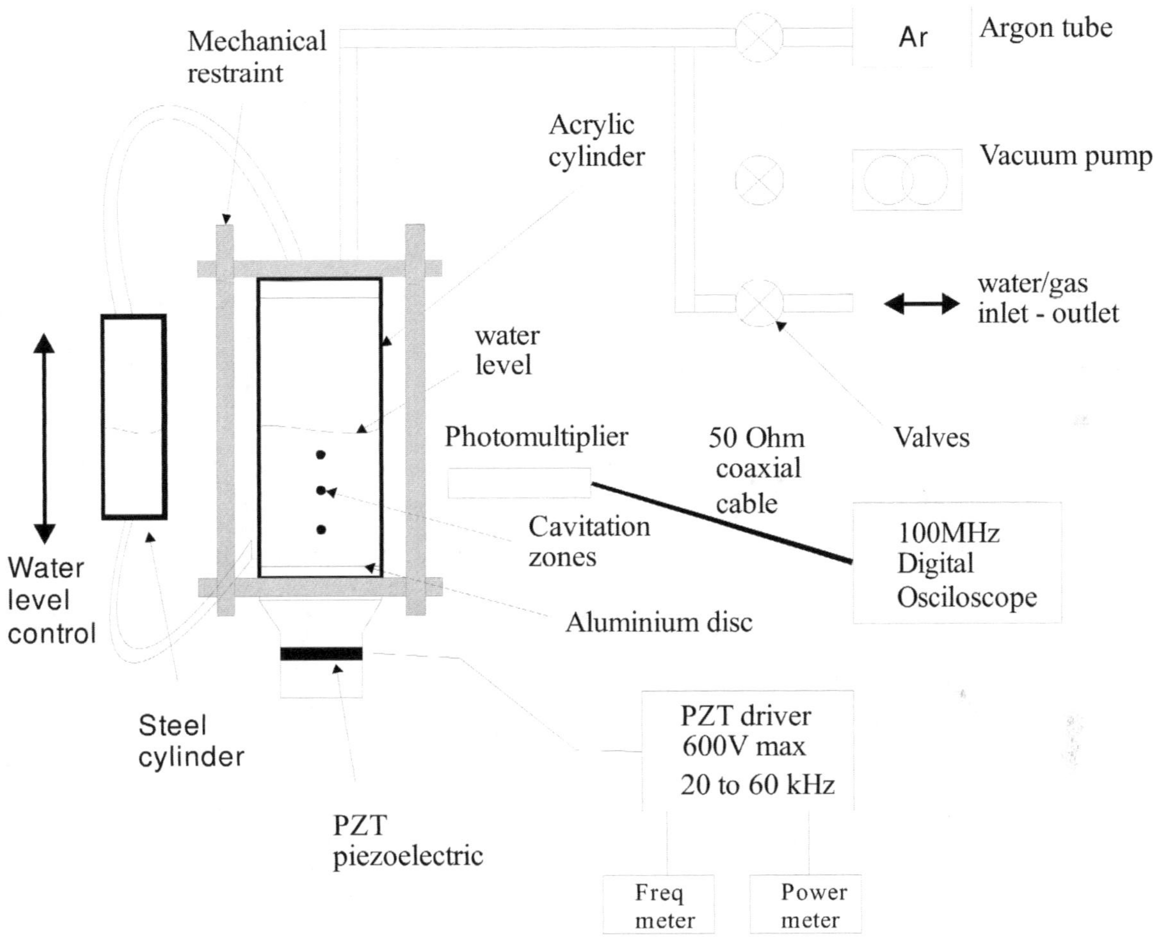


**Figure 1.** Experimental setup.

## 3. Measurements

Before introducing the distilled water into the acrylic cylinder we made vacuum with a mechanical pump until stable conditions are reached. At this point we close the pump valve, and then we let the water run into the cylinder through the in-out valve. The amount of water filled into the cylinder and the steel vessel is approximately five litters. With the water filled we made vacuum again to degas it and during this process we applied low-power to the piezoelectric to improve the gas liberation. Finally, we close the vacuum valve and fill the system with the argon desired pressure. The water level is adjusted moving up or down the steel vessel.

The pressure was varied from 1 to 2.5 bar and the optimum emission of light was found at maximum pressure. Probably, at this pressure the argon concentration increases with time and the cavitation regime begins to change, until the emission of light ends. Lower pressures gives less cavitation power (less emission of light) but more stable operation. In the cases when the concentration of Ar is too high, an increase of the temperature can improve the behavior. We believe that this change in temperature changes the concentrations of Ar nearer its optimum range. As also the vapor pressure changes with temperature, we are not sure which process dominates.

We register different signals to show both the long time spectrum and the shape of an individual pulse. In figure 2 can be seen the resulting spectrum obtained from the oscilloscope and converted to a linear scale (Agilent show dB scale). Unexpectedly, there exists a very intense peak at half the piezoelectric frequency, but we need more diagnostics in order to clarify this fact.

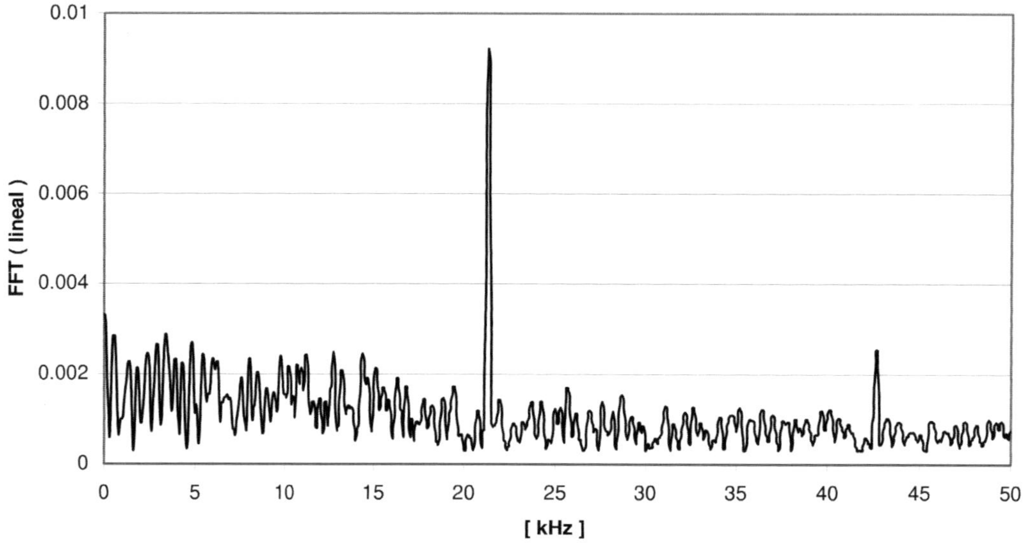


**Figure 2.** FFT spectrum of light pulses. Frequency operation 42.7 kHz.

Using the lowest time base scale of 5 ns/div we registered an individual light pulse in figure 3. Clearly the only conclusion that we can make is that the light pulse is well below the resolution of the oscilloscope and the phototube, probably below the nanosecond scale. The noise (dark) pulses of the phototube are below 0.1 V.

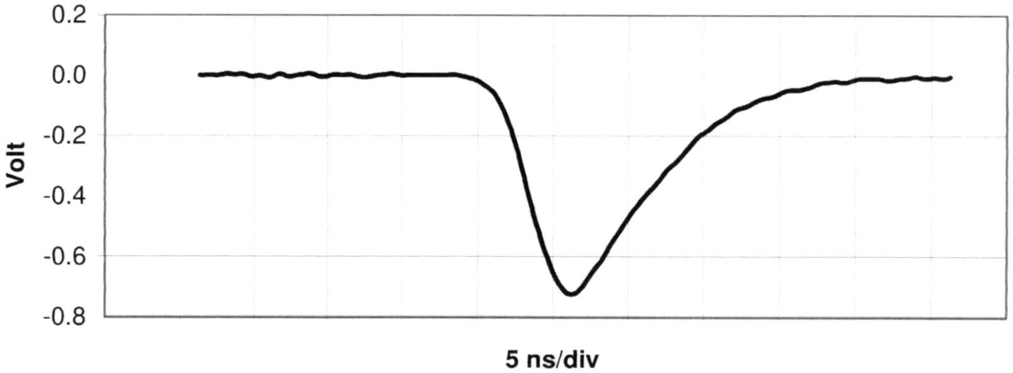


**Figure 3.** Single emission pulse 100MHz bandwidth oscilloscope.

We made several photographs of the emitted light with a Nikkon DS100 camera and the best results were obtained taking 30 s exposure time. In order to show both the spot light and the experiment in the same exposition we illuminate the ambient with a soft light. The results are shown in figure 4. The piezoelectric is at the bottom of the picture.

**Figure 4.** Light pulse from a bubble, taken with 30 s exposition time. A soft backlight was used to illuminate the experiment.

## 4. Resonant model

We observe that there are two main resonance modes in the cylinder, the axial and the radial one. The addition of these modes fixes a discrete set of heights and frequencies to work with. A simple calculation can be made to obtain those conditions. If $n$ and $m$ are integers

$$n = \frac{2rf}{c_s},$$

$$h = \frac{\left(m - \frac{1}{2}\right)r}{n},$$

where $r$ is the radio of the cylinder, $h$ the water level and $c_s$ the speed of sound in water, and as the piezoelectric works in the range 40-44 Khz, the only possible value for the conditions of the experiment is $n = 3$. In this way $h = (m\text{-}1/2)\times 15$ mm, and so we can use different heights to experiment the resonant cavitation.

When the pressure of the system is increased we need to slightly increase the frequency to maintain the conditions of resonance. That can be explained assuming a little change in the sound velocity of water.

The shape of the cavitation zone changes with the argon dilution and we need almost one day with the argon pressurized to achieved emission conditions, but as we said before, if the Ar concentration in water is too high the cavitation regime changes and light emission stops.

## 5. 1D pressure model

We compute a one dimensional model in the $z$-axis taking into account the attenuation of sound with distance by a parameter $\delta$. The constant pressure is represented by $P_0$ and the ongoing wave pressure from the piezoelectric by $\Delta P$.

$$P = P_0 + \Delta P_1 \cos(kz - \Omega t) + \Delta P_2 \cos(kz + \Omega t + \Phi) \tag{1}$$

$$\Delta P_1 = \Delta P \exp\left(-\frac{z}{\delta}\right) \tag{2}$$

$$\Delta P_2 = \Delta P \exp\left(-\frac{2h-z}{\delta}\right) \tag{3}$$

and for the dynamics we use

$$\frac{dv_z}{dt} = \frac{1}{\rho}\left(\frac{\partial}{\partial z} p(z,t) - \alpha v_z\right) \tag{4}$$

where $\alpha$ is a friction (viscosity) coefficient, $\rho$ an artificial density, $h$ the water level, $\Omega$ the angular frequency, $k$ the wave number and $\Phi = \pi\text{-}2kh$ the shift phase of the reflexted wave. This equation describes an hypothetical particle which moves due to the pressure gradients.

We solve (4) coupled with (1) using different initial conditions. As it can be seen in figure 5, the bubbles are trapped at the pressure antinodes ($dp/dz = 0$).

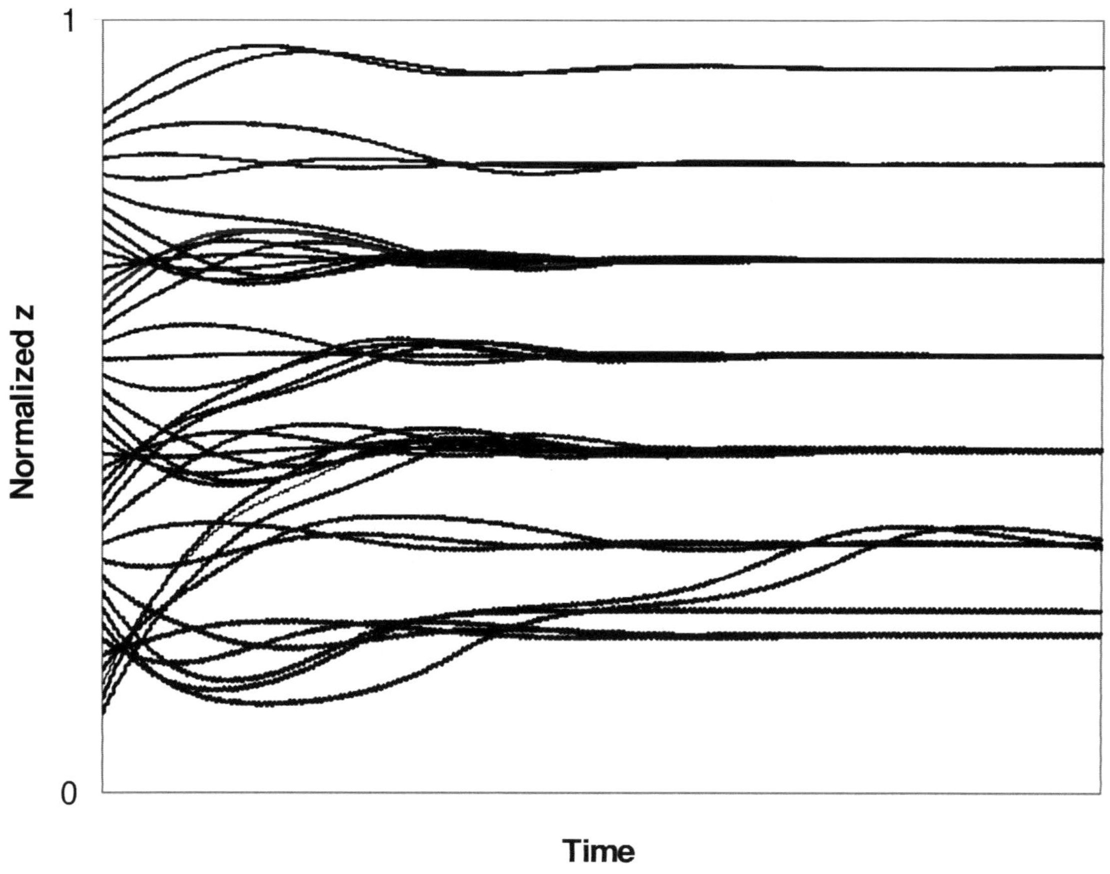


**Figure 5.** Simulated trapped bubbles. Evolution of $z$ coordinate for different initial positions.

We can observe in this simulation that the behavior of trapped antinodes changes with $z$, making the regions near the gas-water interface more effective than the others. This can be explained by the effect of $\delta$ that makes it more difficult for a pure antinode to appear near the piezoelectric, as the condition $dp/dz = 0$ is less well satisfied. With this in mind, we can partially understand why the antinodes that are near the surfaces are the brightest ones.

## 6. Conclusions

We have presented a low cost experiment for the demonstration and study of sonoluminescence phenomena that allows to change the frequency, power, liquid level and gas type and pressure. It is the first experiment of this type built at the Mar del Plata University, and is very useful as a training experiment for advanced students. (Although is a good practice to show theoretical and numerical resolutions). In the future we expect to make an spectral analysis of the emmited light, and quantitative measurements of gas diluted in the water. There are few references to the use of a pressurized ambient, so it seems to be a good line to explore.

## References

[1] Urteaga R 2008 *Concentración de energía en sonoluminiscencia* PhD Thesis, Inst. Balseiro, Universidad Nacional de Cuyo. Argentina
[2] Brenner M P, Harvard U, Hilgenfeldt S and Lohse D 2002 *Rev. of Mod. Phys.* **74** 425
[3] Barber B P and Putterman S J 1992 *Phys. Rev. Lett.* **69** 3839
[4] Chakravarty A, Georghiou T, T. Phillipson E and Walton A J 2004 *Phys. Rev. E* **69** 066317
[5] Barber B and Putterman S J 1991 *Nature* **352** 318-20

Journal of Physics: Conference Series **370** (2012) 012066 doi:10.1088/1742-6596/370/1/012066

# Major issues in the design and construction of the stellarator of Costa Rica: SCR-1

**J Mora, V I Vargas, L F Villegas, L Barillas, J I Monge and L Rivas**

Instituto Tecnológico de Costa Rica

E-mail: jaime.plasma.tec@gmail.com

**Abstract**. This paper aims at briefly describing the design and construction issues of the stellarator of Costa Rica 1 (SCR-1). The SCR-1 is a small modular Stellarator for magnetic confinement of plasma developed by the Plasma Physics Group of the Instituto Tecnológico de Costa Rica (ITCR). The SCR-1 is based on the small Spanish Stellarator UST_1 (Ultra Small Torus 1), created by Eng. Vicente Queral. These mains issues consist of the size of the Stellarator, closeness between coils, coupling of ECH to the vacuum chamber and the device for support. The size has become a problem because the vacuum chamber does not allow a lot of space to attach diagnosis devices, the heating system, the vacuum system and the very same support of the chamber. As a result of this lack of space, the Stellarator's coils are placed very close to each other; this means that two of the coils around of the vacuum chamber clash and cannot be placed as designed. The issue regarding the coupling of the ECH (electron cyclotron radio-frequency) to the vacuum chamber comprises the fact that the wave guide with rectangular shape does not match the CF port with circular shape on the vacuum chamber. In addition, the device for supporting the Stellarator has presented a challenge because of its size and the placement of the coils; in other words, there is not enough space between the ports and coils in the Stellarator to place appropriately the device for support.

## 1. Introduction

This paper deals with the major issues of the construction of the stellarator of Costa Rica (SCR-1). This device is being built by the Plasmas Physics Group of the Instituto Tecnológico de Costa Rica, based on the Spanish Stellerator UST-1 (Ultra Small Torus 1).

Some of the characteristics of the SCR-1 are the following. It will be a 2-field period modular stellarator with an aspect ratio of 6. Low shear configuration with core and edge rotational transform equal to 0.32 and 0.28. This plasma will be confined by a magnetic field ($B \approx 90$ mT) given by 12 modular coils with 12 turns each, carrying a current of 725 A per turn providing a total toroidal field (TF) current of 8.7 kA-turn per coil. Typical length of the plasma pulse will be between 4 s to 10 s.

## 2. Major issues

The main construction issues consist of the size of the Stellarator, closeness between coils, coupling of ECH to the vacuum chamber and the device for support. The size has become a problem because the vacuum chamber does not allow a lot of space to attach diagnosis devices, the heating system, the vacuum system and the very same support of the chamber.

As a result of this lack of space, the Stellarator's coils are placed very close to each other; this means that two of the coils around of the vacuum chamber clash and cannot be placed as designed.

Journal of Physics: Conference Series **370** (2012) 012066 doi:10.1088/1742-6596/370/1/012066

The issue regarding the coupling of the ECH (electron cyclotron radio-frequency) to the vacuum chamber comprises the fact that the wave guide with rectangular shape does not match the CF port with circular shape on the vacuum chamber. In addition, the device for supporting the Stellarator has presented a challenge because of its size and the placement of the coils.

*2.1. Coils clash*

The clash of coils is a result of previous simulations of heating of coil cord [1] which prove that for the SCR-1 coils of 6 wires with 1088 A (see figure 1) cannot be used to sustain a discharge of 10s duration. This forced the reduction of the current and the increase of the amount of turns per coil; a total of 12 turns per coil with 725 A. However, the size of the coils was not considered for it would produce a clash between them avoiding the construction of the device.

To solve this it was proposed to use wire to coil motors to accommodate better the coils and by doing so reducing their size. In addition, the coils will be rotated one degree, to reduce the clash (see figure 2). Then, the associated simulations are done and confirmation that the field lines remain closed is obtained [2].

**Figure 1.** Coils with multiple clashes between them.

**Figure 2.** Coils not clashing between them; coil's height = 43.2 mm; coil width = 14.4 mm; GAGE = AWG#4.

*2.2. Connection of the ECH system to the vacuum chamber*

The heating system ECH (2.45 GHz) for the SCR-1 forces to reduce the rectangular area of the WR340 waveguide to connect it to the circular viewport in the CF port, trying to reduce loses and allowing the wave to travel through the waveguide to the toroidal chamber. There are three possibilities under analysis: (1) Locate in the market a transition of WR340 waveguide to circular (see figure 3); (2) Use a system with Adapter WR340 Waveguide to Coax and antenna; or (3) locate a port with the WR340 shape in the vacuum chamber and vacuum windows for microwave with WR340 waveguide shape.

For option 1, it was identified a transition of WR340 waveguide to circular of 85 mm; to that end it was located a CF port of 6" and a viewport of 89 mm of visible area. The quartz viewport was selected because this material allows the passing of microwaves and supports the power of the ECH (5 KW). Besides this material, the possibility of using sapphire was under study because it also supports the work conditions; nevertheless, sapphire greatly increases costs and, therefore, is discarded. The other two options were discarded due to the fact that the first option solved the problem.

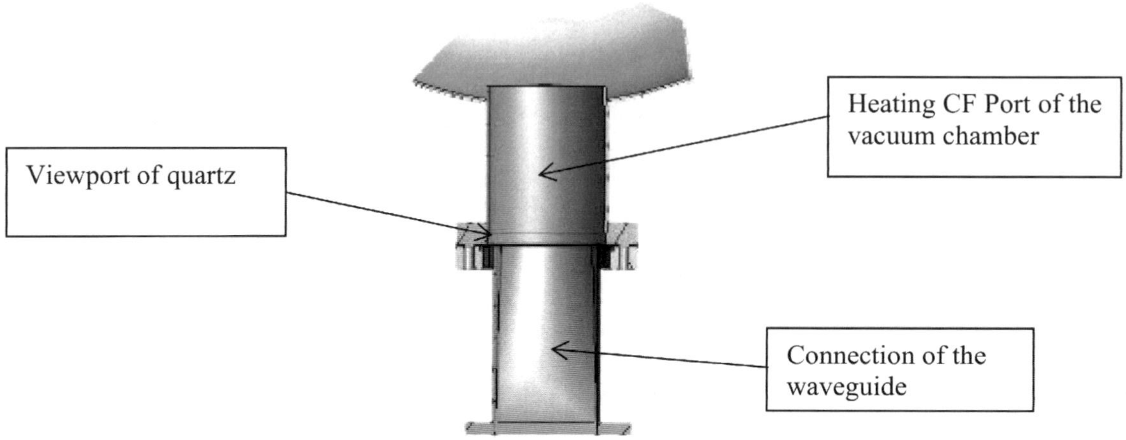


**Figure 3.** Connection from the ECH to the Vacuum chamber

*2.3. Space to place the vacuum ports*

The space between coils limits the size of the ports that can be placed. The maximum CF allowed size for the SCR-1 is 6 inches. To reduce the distance between the vacuum pump and the toroidal chamber, a vacuum port was placed on the bottom. Initially, a CF port of 6 inches was considered to be placed but due to the reduced space to perform the lines for coils it was changed for a CF port of 4 ½ inches. This only affects the molecular conductance of the system.

The CF port of 6 inches makes difficult to do the lines for the coils with a milling machine (see figure 4). This presents a challenge because the coils will be made with a CNC milling machine. As observed in figure 5, there is more room for the milling machine to move and do the canals for the coils.

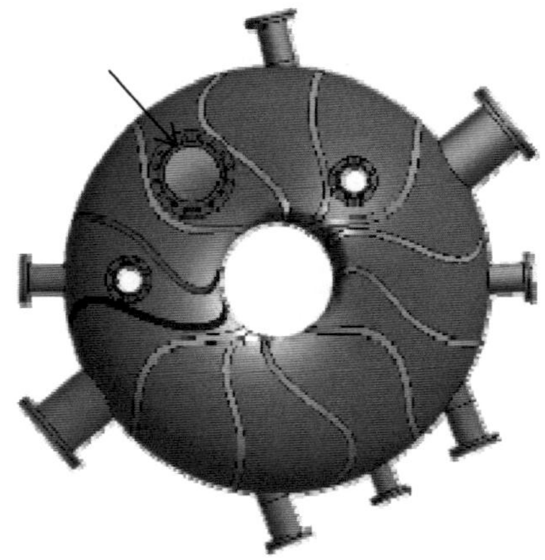

**Figure 4.** Vacuum chamber with a CF port of 6 inches.

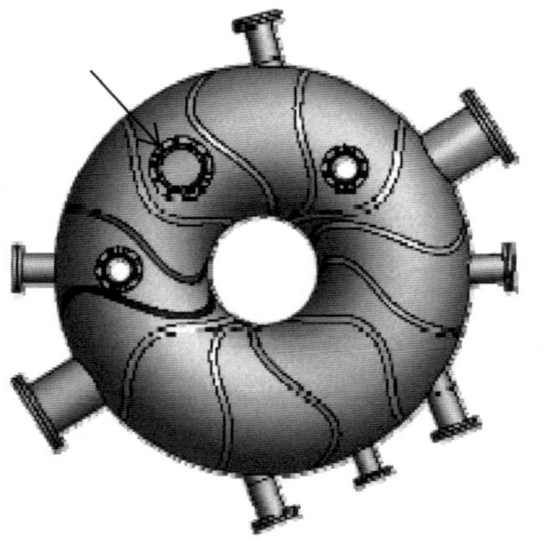

**Figure 5.** Vacuum chamber with a CF port of 4 ½ inches.

*2.4. Space between coils to place ports*

The space between coils limits the placement of ports and it is important to guarantee that any port is placed on top of the coils. Ports must be away from the coils to place correctly the coils. The placement of the viewports is a challenge. As shown in figure 6, viewport 1 barely allows seeing the interior of the stellarator. Also viewports 2 and 3 allow seeing more details within the vacuum chamber but without the desired vision range.

The solution is to look for the appropriate and easy placing of these ports to allow a useful vision range, the current placement of the visual ports allows seeing the ports in which the diagnostics will be placed. These diagnostics include the Langmuir probe enabling to see when the probe enters the plasma. Also, the placement of those ports is to enable seeing the 6" ports in which the experiment of mapping surfaces will be used.

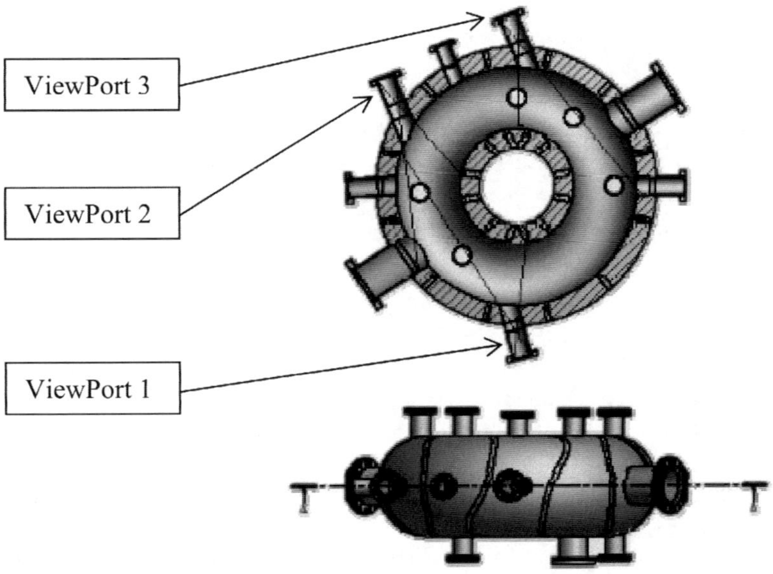


**Figure 6.** Cross-section view of the vacuum chamber that shows the vision range of the ports.

*2.5. Vacuum chamber support*

The reduced size of the SCR-1, with an outer diameter of the toroidal chamber of 75.06 cm, and the little space between coils make it difficult to locate support points and equipment as the vacuum pump. To solve this, a special support was designed.

This design allows having a removable work area around it in case that equipment needs to be placed in this space.

Besides that, the support will be screwed to the table with protruding edges at the bottom of the vacuum chamber. These edges will be screwed to the support on the points signaled on figure 7. In addition, the support points are open to leave space for the vacuum chamber and provide more stability to the table. Also, the support for the Langmuir probe was designed so that it does not interfere with other equipment as shown in figure 7.

## 3. Conclusions

The construction challenge of a small stellarator constitutes that the space on the vacuum chamber is limited and makes it hard to manipulate all the equipment around the stellarator. The big difference between designing a normal stellarator and a small stellarator is that with the big one there are many spaces between coils to place ports. The SCR-1 has 16 ports which limited the amount of diagnostics that can be placed on it.

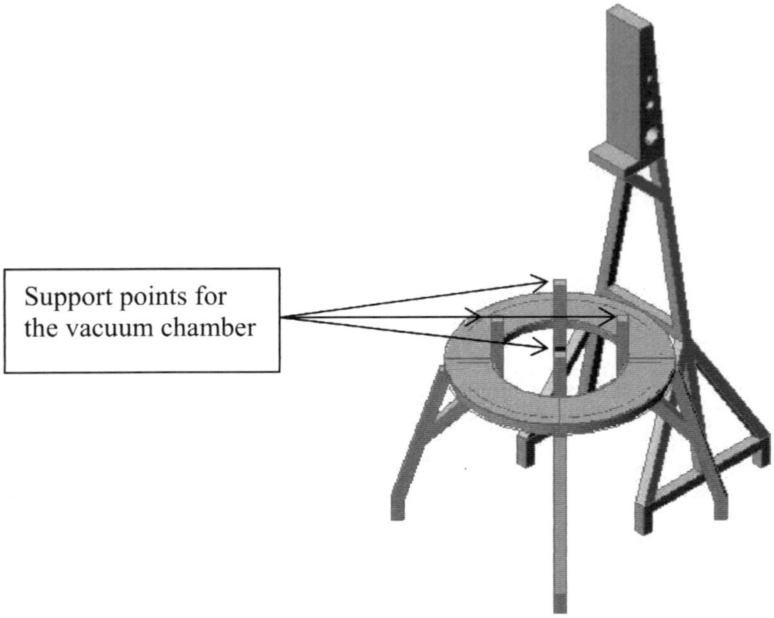


**Figure 7.** Group of supports for the SCR-1 and support of the Langmuir probe in its final position.

**References**

[1] Barillas L, Vargas V I, Alpizar A, Asenjo J, Carranza, J M, Cerdas F, Chávez A, Gutiérrez R, Monge J I, Mora J, Morera J, Peraza H, Queral V, Rojas C, Rozen D, Saenz F, Sánchez G, Sandoval M, Trimiño H, Umaña J, Villegas L F 2010 *SCR-1: Design and Construction of a Small Modular Stellarator for Magnetic Confinament of Plasma.* 15th International Congress on Plasma Physics (ICPP 2010) in combination with the 13th Latin American Workshop on Plasma Physics (LAWPP 2010), Santiago, Chile

[2] Vargas V I, Asenjo J, Sura A , García S, Cerdas F, Chacón D, Chacón J, Fatjó S, Jiménez J, Rivas L and Stradi B 2011 *Simulations for designing of a Small Modular Stellarator for Magnetic Confinement of Plasma.* 14th Latin American Workshop on Plasma Physics (LAWPP 2011), Mar del Plata, Argentina

Journal of Physics: Conference Series **370** (2012) 012067 doi:10.1088/1742-6596/370/1/012067

# Paschen law for argon glow discharge

**C Torres[1], P G Reyes[1], F Castillo[2] and H Martínez[3]**

[1] Facultad de Ciencias, Universidad Autónoma del Estado de México, Estado de México, Mexico
[2] Instituto de Ciencias Nucleares, Universidad Nacional Autónoma de México, México D.F., Mexico
[3] Instituto de Ciencias Físicas, Universidad Nacional Autónoma de México, Apartado Postal 48-3, 62210, Cuernavaca, Morelos, Mexico

E-mail: ctorress@uaemex.mx

**Abstract.** Electric discharge between two electrically charged surfaces occurs at a well-defined, gas-dependent combination of atmospheric pressure and the distance between those surfaces, as described by Paschen's law. The understanding of when the discharge will occur in an Ar discharge is essential basic knowledge. A glow discharge apparatus was used in this experiment of Ar discharge at a pressure range between 2.0 Torr and 12 Torr, a power of 20 W and 40 l/min flow rate of gases. The optical emission spectroscopy was carried out in the wavelength range of 200 to 1100 nm. Here, we present experimentally measured plasma Paschen curves for Ar gas and compare our results of breakdown voltages with the literature. The minimum voltage measured for a discharge in Ar atmosphere was 215 ± 2.2 at 0.7 Torr-cm, which agree with previous measurements.

## 1. Introduction

Friedrich Paschen developed a law in 1889, which is known as the Paschen's curve describing the breakdown voltage as a function of the electrode spacing or gap ($d$), operating pressure ($p$), and gas composition [1]. The breakdown voltage is a function of the product of the pressure $p$ and the inter-electrode distance $d$: $V_{rup} = f(p \times d)$. The mathematical formulation of Paschen's curve is derived from Townsend's description of the basic charge generation processes including electron impact ionization ($\alpha$ process) and secondary electron emission from the cathode due primarily to ion bombardment (the $\gamma$ process), though other bombardment processes may play a role [2]. Historically, Paschen's curve has proved to be accurate for large gaps and at low pressures [3], but it is often acknowledged that it fails to describe behavior at extremely low or high $pd$ values [4].

## 2. Experimental details

The experiments were carried out with a discharge chamber of the Laboratorio de Física Avanzada, Facultad de Ciencias, UAEM. The apparatus to measure the plasma emission spectrum is similar to that described elsewhere [5-7]. The discharge cell consists of two movable parallel electrodes enclosed in a stainless steel vacuum chamber. The two electrodes were made of copper disc, with 2.5 cm in diameter. The electrodes are positioned at the centre of the reaction chamber with 20 mm of gap spacing. The cell was pumped down by a vacuum system (Turbomolecular pump Alcatel and ATP80 Varian D5302) to a base pressure of $10^{-6}$ Torr. A continuous dynamic flow of Ar gas (ultra-pure,

Journal of Physics: Conference Series **370** (2012) 012067 doi:10.1088/1742-6596/370/1/012067

Praxair 99.99%) was let in the system through a needle valve at the desired pressures. A DC glow discharge was activated between the two electrodes at a pressure of 2.0 Torr. The discharge power supply (Spellman SA4) was maintained at an output of 0.5kV and a current of 40mA (20 W), which was measured using a digital Fluke multimeter model 8846A.The other lateral flange was a quartz window, used to monitor the active species generated in the glow discharge by plasma emission spectroscopy; the spectrum of the emission (200-1100 nm) was obtained using a spectrometer Ocean Optics HR4000CG-UV-NIR and an optical fiber P600-2-SR. The wavelengths interval was 0.75 nm and the well time of 500ms. The data were obtained in a single accumulation with a 5.0 s integration time. The DC source was connected to the electrodes and an oscilloscope, in order to measure the breakdown voltage. To measure the breakdown voltage the chamber was first filled to a certain pressure; then the voltage on upper electrode was slowly, manually increased until breakdown was indicated by an increase in current on the ammeter and a spike in the voltage displayed on the scope. This breakdown voltage was recorded, after which the voltage on the electrode was increased further to ensure that a discharge was occurring. Then, the electrode voltage was reduced to 0 V, the pressure was increased (or decreased) and the procedure was repeated. Measurements were taken from 2.0 to 12.0 Torr, which was sufficient to obtain a curve, to determine the lowest breakdown point.

## 3. Results

A typical emission spectroscopy measurement of Ar glow discharge plasma at a pressure of 2.0 Torr is displayed in figure 1.

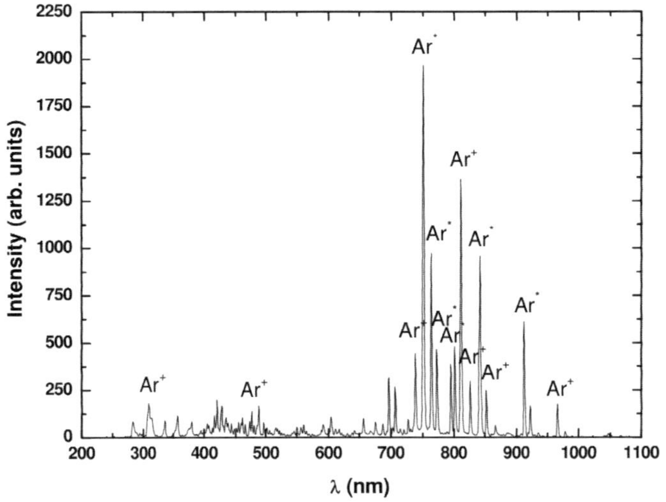


**Figure 1.** Emission spectra measurements of Ar at pressure of 2.0 Torr.

All lines appearing in the spectrum have been identified as Ar* and $Ar^+$ species.

For discharges containing argon, models have been built and are considered to be valid over a wide range of conditions [8]. More accurately measured cross sections have been used in the models in recent years [8-9]. According to those models, the most important processes involved with high-lying (3p, 4p, and 5p) excited levels are the electron impact excitation and ionization from the ground state,

$$e + Ar_g \rightarrow e + Ar_i, \qquad (1)$$

$$e + Ar \rightarrow 2e + Ar^+,$$

the electron impact transition between excited levels,

$$e + Ar_i \rightarrow e + Ar_j,$$
$$e + Ar_j \rightarrow e + Ar_i, \quad (2)$$

and spontaneous radiation,

$$Ar_j \rightarrow Ar_i + h\nu \quad (3)$$

Here $Ar_g$ are atoms in the ground state. $Ar_i$ and $Ar_j$ are atoms in metastable states or excited levels.

Figure 2 shows the breakdown voltage in a d-c discharge at a pressure of 2.0 Torr with an electrode separation of 2.0 cm. It can be observed the step in voltage caused by the generation of the glow discharge.

The Paschen curve was built by varying the distance between electrodes for a given pressure. In this case the curve shows a decreasing behavior of the voltage as function of the product between the distance and the pressure up to the point *pd*= 0.8 Torr cm, after which the voltage has a nearly linear increase (figure 3).

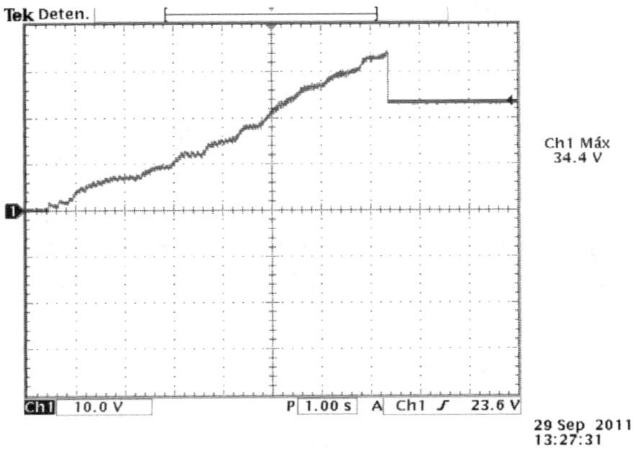


**Figure 2.** Breakdown voltage.

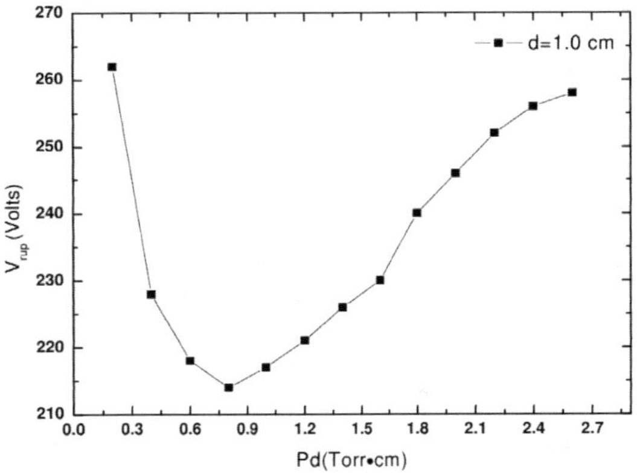


**Figure 3.** Paschen´s curve obtained varying the distance between cooper electrodes.

Figure 4 illustrates Paschen curves for four different inter-electrode distances (0.5 cm to 2.0 cm) and reflects the Townsend breakdown mechanism in gases, that is, it related to the non-spatial uniformity of the electric field.

We measured the minimum voltage for discharge in Ar atmosphere of (215 ± 2.2) V at 0.7 Torr-cm, which agree with previous measurements. However, at the same pd value, the breakdown voltage differs depending on the inter-electrode distance d.

From figure 4, it observed that the four curves have similar behavior between 0.1 Torr-cm and 0.3 Torr-cm, and at pd values much bigger than 0.3 Torr-cm the curves are separated in two groups, at the inter-electrode distances of 0.5 Torr-cm and 2.0 Torr-cm has almost the same behavior and their absolute values of the voltage are much bigger than the curves at inter-electrode distances of 1.0 Torr-cm and 1.5 Torr-cm, which have the same behavior.

## 4. Conclusions

We obtained experimentally Paschen's curves for an Ar glow discharge. Also, it found that the breakdown voltage depends on a function between the product of the pressure and electrode distance, like Paschen's law established.

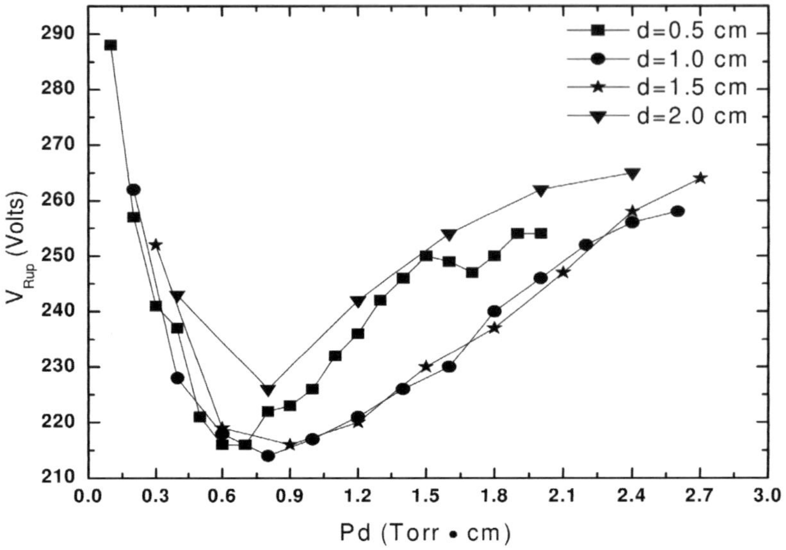


**Figure 4.** Paschen´s curve obtained varying the Ar gas pressure and distance between cooper electrodes.

**Acknowledgments**
We are grateful to Nayeli Rodriguez Osorio (FC-UAEM) and Anselmo González Trujillo (ICF-UNAM) for her technical assistance. This research was supported by DGAPA IN105010, CONACyT 128714 and student grant Enlace-UAEM.

**References**

[1] Paschen F 1889 *Analytical Physics* **37** 69
[2] Townsend J 1915 *Electricity in Gases* (Oxford: Oxford University Press)
[3] Dakin T W, Luxa G, Oppermann G, Vigreux J, Wind G and Winkelnkemper H 1974 *Electra.* **32** 61
[4] Nasser E 1971 *Fundamentals of Gaseous Ionization and Plasma Electronics* (New York: Wiley-Interscience)
[5] Martínez H and Yousif F B 2008 *Eur. Phys. J. D***46** 493
[6] Reyes P G, Mendez E F, Osorio-Gonzalez D, Castillo F and Martínez H 2008 *Physica Status* **5** 907
[7] Méndez-Martínez E F, Reyes P G, Osorio-González D, Castillo F and Martínez H 2010 *Plasma Science and Technology* **12** 314
[8] Yanguas-Gil A, Cotrino J and González-Elipe A R 2006 *J. Appl. Phys.* **99** 33104
[9] Bultel A, van Ootegem B, Bourdon A and Vervisch P 2002 *Phys. Rev. E***65** 46406

Journal of Physics: Conference Series **370** (2012) 012068 doi:10.1088/1742-6596/370/1/012068

# The silver counter neutron detector revisited

**A Llanquihuen-Martinez[2], A Tarifeño-Saldivia[1,3], J Moreno[1,3] and L Soto[1,3]**

[1]Comision Chilena de Energía Nuclear, Casilla 188-D, Santiago, Chile
[2]Universidad Andrés Bello, República 220 tercer piso, Santiago, Chile
[3]Center for Research and Applications in Plasma Physics and Pulsed Power, P[4], Chile

E-mail: atarifeno@cchen.cl

**Abstract**. The silver counter neutron detector is revisited. The detector characterization methodology based on the use of an isotopic source is revised. The effect of the moderator geometry on the counting efficiency of the detector is experimentally studied. Preliminary results are presented.

## 1. Introduction

The so-called 'silver activation counter' is one of the most common detection system used for measurement of the neutron strength in pulsed fast neutron sources. In this detector, the energy of fast neutrons is decreased to epithermal and thermal energies by a moderator. The moderated neutrons activate a silver foil and the subsequent beta decay is monitored by a Geiger Müller counter tube (GM). This system was first proposed by Lanter and Bannerman in the late sixties [1]. In plasma focus research, these systems have been used since forty years. The state of the art calibration technique is usually referred to the methodologies reported by Gentilini et al [2]. By using these methods, the calibration constant is usually obtained with experimental errors of the order of 20 – 30%. For small laboratories, the most suitable methodologies reported in the work by Gentilini presents several systematic errors, which are related mostly with gamma counting. In a recent work by Moreno et al [3], a new calibration method is outlined. The methodology aims to deal with systematic errors and is based only on the use of a continuous neutron source as reference. In this case, the experimental error in the calibration constant was reduced to less than 10%.

In this work, the characterization of the detector is revised and the response of the detector is studied as a function of the moderator geometry. Preliminary results of this study are presented.

## 2. Counting model an efficiency

The accumulated counts as a function of time when irradiating the detector by a pulsed source are given by

$$G(T) = Z_c Y\left[a_1\left(1-e^{-\lambda_1 t}\right)+\left(1-a_1\right)\left(1-e^{-\lambda_2 t}\right)\right] \tag{1}$$

where $Y$ is the number of neutrons impinging on the detector, $\lambda_1$ and $\lambda_2$ are the decay constant of both stable isotopes present in pure silver ($^{108}$Ag and $^{110}$Ag). Constants $a_1$ and $Z_c$ are obtained by

Published under licence by IOP Publishing Ltd

Journal of Physics: Conference Series **370** (2012) 012068 doi:10.1088/1742-6596/370/1/012068

characterization of the detector using a continuous source [3]. $Z_c$ is the counting efficiency of the detector

$$Z_c = \frac{4\pi \, R_{sat}}{I_o \Omega F(\alpha)} \tag{2}$$

where $I_0$ is the intensity of the reference continuous source and $F(\alpha)$ is the source anisotropy correction [4], $\Omega$ is the solid angle subtended by the detector and $R_{sat}$ is the measured neutron counting rate at saturation. The counting efficiency is dependent on the efficiency for moderating neutron energy from fast to epithermal and thermal, the geometry of the silver foil and the beta counting efficiency of the Geiger tube. Therefore, while keeping the same silver foil geometry and the Geiger tube, the detector counting efficiency will depend ultimately on the moderator material and its geometry.

Analogously to equation (1), when irradiating the detector until saturation, the accumulated counts after removal of the source are by

$$G(t) = M\left[A_1\left(1-e^{-\lambda_1 t}\right)+\left(1-A_1\right)\left(1-e^{-\lambda_2 t}\right)\right] \tag{3}$$

By using non-linear least squares, the constants $M$ and $A_1$ are obtained from the experimental decays. Thus,

$$a_1 = \frac{A_1\lambda_1}{\left(A_1\lambda_1 + (1-A_1)\lambda_2\right)} \tag{4}$$

and

$$R_{sat} = M\left(A_1\lambda_1 + (1-A_1)\lambda_2\right) \tag{5}$$

The term $4\pi/I_0F(\alpha)$ in equation (2) can be considered a constant while keeping the same relative angle between source and detector. Thus, in order to seek improvements for the counting efficiency, the quotient $R_{sat}/\Omega$ should be studied as a function of the moderator geometry for a given moderator material.

## 3. Experimental setup and results

We build a polyethylene moderator with variable geometry as shown in figure 1. The main piece is a $4\times4\times13$ cm$^3$ parallelepiped which contains the silver foil and the Geiger tube inside. The axis of the Geiger tube is along the parallelepiped axis. The moderator geometry is modified by adding pieces at the outer surface. Two configurations of the detector were studied (C1, C2). They are shown in figures 2 and 3. In C1, the counter axis intersects the source, while in C2 the counter axis is perpendicular to the line which intersects the source. A $^{252}$Cf neutron source was used for the study. The source is placed in a holder at $D = 20$ cm from the counter face and aligned with its center as shown in figure 2 and 3. The GM is fed by a 900 V voltage source. The acquisition system consists in a Canberra 2015A amplifier, connected to a Canberra multiport II MP2-1E in multi scalimeter mode. The irradiation time was set to 800 s, while the data acquisition during decay, after removing the source, was set to 1000 s. This period was found be enough to allow obtaining the background counting rate after acquisition of decays. The analysis of data was done using the non-linear least squares routine provided by GNUPLOT [5]. From the present data, constant $a_1$ was found to have a mean value of $0.86\pm0.009$, which is consistent with previous studies [3]. In figures 4 and 5 the experimental results from configurations C1 and C2, respectively, are shown. Configuration C1 shows a narrow optimum around a side length of 12 cm. In the case of configuration C2, it shows a flat optimum from 8-12 cm. Configuration C1 presents a slightly higher counting efficiency than C2 (12 cm side length). However, high counting efficiencies can be achieved with less than a half of the moderator material in configuration C2 (8 cm side height).

**Figure 1.** Moderator with variable geometry.

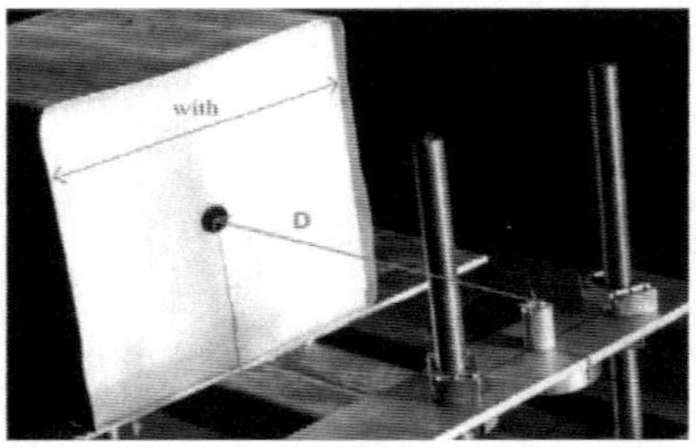


**Figure 2.** C1: Axis of the GM counter intersects the source.

**Figure 3.** C2: Axis of the GM counter is perpendicular to the line which intersects the source.

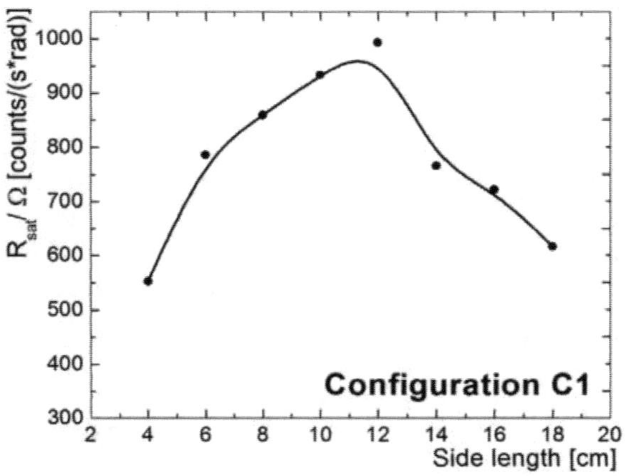


**Figure 4.** Experimental results for configuration C1 (see figure 2), $R_{sat}/\Omega$ as function of side length.

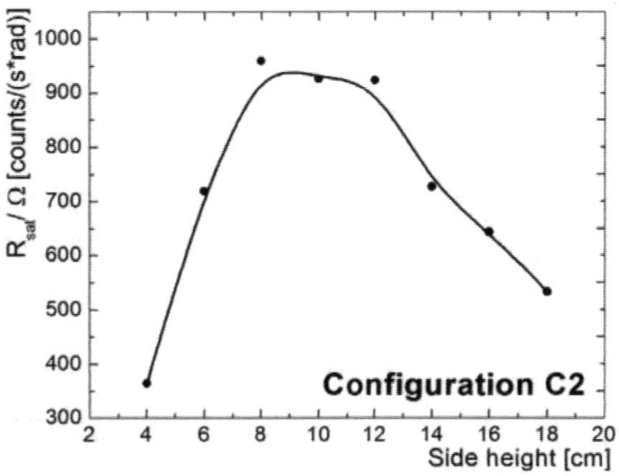


**Figure 5.** Experimental results for configuration C2 (see figure 3), $R_{sat}/\Omega$ as function of side height.

## 4. Remarks and future work

This work is a part of a systematic effort to improve the calibration methodologies and efficiency of this kind of detection system. Future work is related with the validation of the found optimum geometries for experiments when irradiation is done at larger distances ($D > 1$m). The effect on the counting efficiency by adding a frontal moderator in configuration C1 and the linearity of the counting efficiency when adding more than one counter at different positions inside the moderator will be also studied.

**Acknowledgments**

We would like to thank the financial support from FONDECYT grant 1110940.

**References**

[1] Lander R J and Bannerman D E 1968 *Rev. Sci. Instruments* **39** 1588

[2] Gentilini A, Steinmetz K, Tacchia M, Antonini D, Arcipiani B, Moioli P, Pedretti E and Scafé R 1980 *Nucl. Instr. And Meth.* **172** 541
[3] Moreno J, Tarifeño-Saldivia A, Pavez C and Soto L 2010 *AIP Conf. Proc.* **1265** 505
[4] Eisenhauer C M, Hunt J B and Schwartz R B 1985 *Radiation Protection Dosimetry* **10** 43
[5] http://www.gnuplot.info

# AUTHOR INDEX

# AUTHOR INDEX

# AUTHOR INDEX